SOLVED PROBLEMS IN PHYSICS

Volume 2

S.L. SRIVASTAVA

ATLANTIC
PUBLISHERS & DISTRIBUTORS (P) LTD

Published by

ATLANTIC

PUBLISHERS & DISTRIBUTORS (P) LTD

7/22, Ansari Road, Darya Ganj, New Delhi-110002
Phones : +91-11-40775252, 23273880, 23275880, 23280451
Fax : +91-11-23285873
Web : www.atlanticbooks.com
E-mail : orders@atlanticbooks.com

Branch Office
5, Nallathambi Street, Wallajah Road, Chennai-600002
Phones : +91-44-64611085
E-mail : chennai@atlanticbooks.com

Printed in India at Nice Printing Press, A-33/3A, Site-IV,
Industrial Area, Sahibabad, Ghaziabad, U.P.

PREFACE

While teaching various courses in physics to the B.Sc. and M.Sc. students over the last thirty-seven years, it has been my experience that a large number of students do not have a clear understanding of the basic concepts of the subject. A slight deviation from what had been taught to them is generally beyond the grasp of average students. It is my firm belief that solving numerical problems can go a long way in aiding comprehension of the underlying basic principles of physics provided that these are suitably designed and properly arranged. It has been my endeavour to fulfil these requirements.

In our country, Intermediate Science (+2 level) is the minimum qualification for appearing in the entrance examinations to different courses in engineering and medical institutions. The minimum qualification for those appearing in civil services examination is graduation and for those appearing in the national eligibility test is a master degree from a recognised University. Students appearing in these examinations will find this book particularly useful. It offers more than two thousand solved problems with complete solutions in a very lucid style and has been written keeping in mind the special needs of aspiring students. It is hoped that the book will also be very useful for those students who seek to make career in the fields of medicine, engineering, biology and earth science. I will consider my efforts amply rewarded if the book is able to instil in the students a sense of confidence and inspiration for these examinations.

S.L. SRIVASTAVA

Contents

10

HEAT

10.1. A clock with a brass pendulum keeps a correct period of 2 s at 0°C. How much time will the clock lose per day when the temperature is 30°C? Coefficient of linear expansion of brass = 0.000019/°C.

Solution: Let T and T' be the time periods of the brass pendulum at 0°C and 30°C respectively. Then,

$$\frac{T'}{T} = \sqrt{\frac{l_{30}}{l_0}}$$

Now $l_{30} = l_0(1 + \alpha \cdot 30)$

On substituting the value of l_{30} in the above equation, we have

$$\frac{T'}{T} = (1 + \alpha \cdot 30)^{1/2} = 1 + 15\,\alpha = 1 + 15 \times 0.000019 = 1.000285$$

$T' = 2.000570$ since $T = 2$ s

Loss of time = 2.00057 – 2 = 0.00057 s

$$\text{Total loss in time for a period of 24 hours} = \frac{24 \times 60 \times 60 \times 0.00057}{2.00057}$$

$$= 24.6 \text{ s}$$

10.2. A clock with a brass pendulum keeps a correct period of 2 s at 25°C. How much time will the clock gain per day when the temperature falls to 0°C. Coefficient of linear expansion of brass = 0.000019/°C.

Solution: Let T and T' be the time periods of the brass pendulum at 0° and 25°C respectively. Then

$$\frac{T'}{T} = \sqrt{\frac{l_{25}}{l_0}} = \sqrt{\frac{l_0(1+25\alpha)}{l_0}} = 1 + 12.5\,\alpha = 1 + 12.5 \times 0.000019 = 1.0002375$$

$$T = \frac{2}{1.0002375} \text{ since } T' = 2 \text{ s}$$

$$\text{Gain in time} = 2 - \frac{2}{1.0002375} = \frac{2 \times 0.0002375}{1.0002375} = \frac{0.000475}{1.0002375}$$

Total gain in time for a period of 24 hours

$$= \frac{24 \times 60 \times 60}{2} \times 1.0002375 \times \frac{0.000475}{1.0002375}$$

$$= 12 \times 3{,}600 \times 0.000475$$

$$= 20.52 \text{ s}$$

10.3. How much should the temperature of an iron ring of diameter 100 cm be raised in order to fit is on a wooden disc of diameter 100.5 cm? The coefficient of linear expansion of iron = 12×10^{-6}/°C.

Solution: In linear expansion the length l_t at any temperature t is given by

$$l_t = l_1 (1 + \alpha\, \Delta t)$$

Here $l_t = 2\pi \times \dfrac{100.5}{2}$ and $l_1 = 2\pi \times \dfrac{100}{2}$

On substituting these values in the above equation, we have

$$2\,\pi \times \frac{100.5}{2} = \frac{2\pi\, 100}{2} (1 + \alpha\, \Delta t)$$

or $\;0.5 = 100\, \alpha\, \Delta t$

$$\therefore \Delta t = \frac{0.5}{100 \times 12 \times 10^{-6}} = 416.6°\text{C}$$

10.4. An aluminium pendulum is 5 seconds fast every day at a temperature of 20°C and 15 seconds slow at 40°C. Calculate the coefficient of linear expansion of aluminium.

Solution: In a day there are 86,400 seconds. At 20°C the pendulum takes 86,400 – 5 seconds to complete n oscillations, whereas at 40° takes 86,400 + 15 seconds for the same number of oscillations. The two time periods are given by

$$T_1 = 2\pi \sqrt{\frac{l(1 + \alpha \cdot 20)}{g}} = \frac{86{,}395}{n}$$

and $$T_2 = 2\pi \sqrt{\frac{l(1 + \alpha \cdot 40)}{g}} = \frac{86{,}415}{n}$$

$$\therefore \frac{86{,}415}{86{,}395} = \sqrt{\frac{(1 + \alpha \cdot 40)}{(1 + \alpha \cdot 20)}} \cong 1 + 10\,\alpha$$

$$\therefore \alpha = \frac{86{,}415 - 86{,}395}{10 \times 86{,}395} = \frac{20}{10 \times 86{,}395} = 2.315 \times 10^{-5}/°\text{C}$$

10.5. In a system an iron and a brass rod are placed side by side. What should be their lengths at 0°C for the two rods to show a constant difference in length equal to 8 cm at any temperature?

$\alpha_{Fe} = 11 \times 10^{-6}$/°C and $\alpha_{\text{brass}} = 19 \times 10^{-6}$/°C

Solution: Let l_{Fe} and l_{brass} be the lengths of the iron and brass rods at 0°C.

$l_{Fe} - l_{brass} = 8$ cm

At any temperature t, the lengths of the two rods are given by

$$l'_{Fe} = l_{Fe}(1 + \alpha_{Fe}\, t)$$

and $$l'_{brass} = l_{brass}(1 + \alpha_{brass}\, t)$$

On subtracting the two equations, we have

$$l'_{Fe} - l'_{brass} = l_{Fe} - l_{brass} + (l_{Fe} \cdot \alpha_{Fe} - l_{brass} \cdot \alpha_{brass})\, t$$

but $$l'_{Fe} - l'_{brass} = l_{Fe} - l_{brass}$$

Therefore $$l_{Fe} \cdot \alpha_{Fe} = l_{brass} \cdot \alpha_{brass}$$

or $$l_{Fe} = \frac{l_{brass}\, \alpha_{brass}}{\alpha_{Fe}} = l_{brass}\, \frac{19}{11}\text{ cm}$$

Now $$l_{Fe} - l_{brass} = l_{brass}\frac{19}{11} - l_{brass} = 8$$

$$\therefore 8\, l_{brass} = 11 \times 8$$

$$\therefore l_{brass} = 11\text{ cm}$$

$$l_{Fe} = 19\text{ cm}$$

10.6. A glass bowl of volume 300 cm^3 is completely filled with mercury at 10°C. Calculate the mass of mercury which will overflow when the temperature is raised to 60°C. Coefficient of cubical expansion of glass = 1.2×10^{-5}/°C, coefficient of cubical expansion of mercury = 18×10^{-5}/°C and density of mercury = 13.6 g/cm^3.

Solution:

Change in the volume of glass bowl $= 300 \times 1.2 \times 10^{-5} \times (60 - 10)$

$= 15{,}000 \times 1.2 \times 10^{-5}$

$= 18 \times 10^{-2}$ cm^3

Change in the volume of mercury $= 300 \times 18 \times 10^{-5} \times (60 - 10)$

$= 15{,}000 \times 18 \times 10^{-5}$

$= 270 \times 10^{-2}$ cm^3

Volume of mercury which will overflow $= (270 - 18) \times 10^{-2}$ cm^3

$= 2.52$ cm^3

Mass of mercury which will overflow $= 13.6 \times 2.52$ g $= 34.27$ g

10.7. A glass hydrometer reads specific gravity 0.92 in a liquid at 45°C. Find out its reading at 15°C. Coefficient of cubical expansion of the liquid = 0.000525/°C and that of glass 0.000024/°C.

Solution: Let V_{45} and V_{15} be the volumes of liquid at 45°C and 15°C respectively and d_{45} and d_{15} its densities at the same respective temperatures.

Now $d_{15} = d_{45}\,[1 + \gamma_l\,(45 - 15)] = d_{45}\,(1 + 0.000525 \times 30) = 1.01575\,d_{45}$

and $V_{15} = V_{45}\,[1 - \gamma_g\,(45 - 15)] = V_{45}\,(1 - 0.000024 \times 30) = 0.99928\,V_{45}$

Mass of liquid at 15°C $= d_{15} \times V_{15}$

Now $d_{15} \times V_{15} = 1.01575\,d_{45} \times 0.9928\,V_{45}$

$$\therefore\ d_{15} = \frac{1.01575 \times 0.99928 \times V_{45}}{V_{15}} \times 0.92 = 0.9345$$

10.8. A cubical block of length $l = 5$ cm and mass 187.5 g floats in a liquid. It begins to sink at a temperature of 35°C. Assuming the density of liquid at 0°C to be $\rho_0 = 1.605$ g/cc, find the coefficient of volume expansion of the liquid.

Solution: As the block begins to sink, the upward thrust acting on it is equal to its weight, i.e., $V\rho_{35}\,g = mg$

or $$\rho_{35} = \frac{m}{V} = \frac{187.5}{125} = 1.5\ \text{g/cc}$$

Now $\rho_0 = \rho_{35}\,(1 + \gamma \cdot 35)$

$$\therefore\ \gamma = \frac{\rho_0 - \rho_{35}}{35 \cdot \rho_{35}} = \frac{1.605 - 1.5}{35 \times 1.5} = \frac{0.105}{35 \times 1.5} = 2 \times 10^{-3}/°\text{C}$$

10.9. A solid sphere floating in a liquid has 35% of its volume above the liquid surface at 10°C. When the temperature of the liquid is raised to 90°C, the floating sphere has only 30% of its volume above the surface of the liquid. Calculate the volume coefficient of expansion of the liquid.

Solution: Let V be the volume of the sphere. The upward thrust acting on the body is given by

$$d_{10} \times V_{10} = d_{90} \times V_{90}$$

where d_{10} and d_{90} represents the densities of the liquid at 10° and 90°C respectively.

Now $d_{90}\,[1 + \gamma\,(90 - 10)] \times 0.65\,V = d_{90} \times 0.7V$

or $$1 + \gamma\,80 = \frac{70}{65}$$

or $$\gamma = \frac{5}{80 \times 65} = \frac{1}{13 \times 80} = 9.615 \times 10^{-4}/°\text{C}$$

10.10. The height of a barometer is 762 mm according to a brass scale which is correctly graduated at 0°C. The temperature at the time of reading is 20°C. Find the true pressure. Coefficient of linear expansion of brass = 19 × 10^{-6}/°C, coefficient of cubical expansion of mercury = 182 × 10^{-6}/°C.

Solution:

Length of the scale at 20°C = 762 $(1 + 20 \times 19 \times 10^{-6})$ mm

$$\rho_{20} = \text{density of mercury at } 20°\text{C} = \frac{\rho_0}{1+20\times 182\times 10^{-6}}$$

$$\text{True pressure at } 20°\text{C} = \frac{762\left(1+20\times 19\times 10^{-6}\right)\times \rho_0 \times g}{\left(1+20\times 182\times 10^{-6}\right)}$$

This must be equal to $h_0 \rho_0 g$,

or $h_0 = 762\ [1 + 20 \times 19 \times 10^{-6}]\ [1 - 20 \times 182 \times 10^{-6}]$

$= 762\ [1 - 20\ (182 - 19) \times 10^{-6}]$

$= 762\ (1 - 0.00326)$

$= 762 \times 0.99674 = 759.5$ mm

10.11. A mercury thermometer is constructed in order to measure any required temperature between 0°C and 100°C, using a glass tube of internal radius 0.2 mm. The length of the tube between the marks for 0°C and 100°C, $l = 25$ cm. Determine the internal volume of the bulb and the stem below the mark 0°C. Coefficient of cubical expansion of mercury = 0.00018/°C and the coefficient of linear expansion of glass = 0.000009/°C.

Solution: Let V_0 be the volume of mercury at 0°C.

The volume of mercury at 100°C – volume of mercury at 0°C

$= \pi r^2 l = 3.14 \times (0.02)^2 \times 25$ cc

$= 3.14 \times 0.01$

$= 0.0314$ cc

$$\gamma_a = \text{coefficient of apparent expansion} = \frac{0.0314}{V_0 \times 100} = \frac{0.000314}{V_0}/°\text{C}$$

$\gamma_a = 3\alpha = 3 \times 0.000009 = 0.000027/°\text{C}$

Coefficient of cubical expansion of mercury = coefficient of apparent expansion of mercury + coefficient of volume expansion of glass

$$\therefore 0.00018 = 0.000027 + \frac{0.000314}{V_0}$$

or $$180 = 27 + \frac{314}{V_0}$$

or $$V_0 = \frac{314}{153} = 2.052 \text{ cc}$$

10.12. A metallic bob weighs 50 g in air. If it is immersed in a liquid at a temperature of 25°C, it weighs 45 g. When the temperature of the liquid is raised to 100°C, it weighs 45.1 g. Calculate the coefficient of cubical expansion of the liquid, given the coefficient of linear expansion of metal is 12×10^{-6}/°C. (IIT 1973)

Solution: Let V_{25} and V_{100} be the volumes of the bob at 25 and 100°C respectively. Then

$$V_{100} = V_{25}[1+3\alpha(100-25)] = V_{25}(1+36\times75\times10^{-6}) = 1.0027\,V_{25}$$

Now $V_{25}\cdot d_{25}$ = upward thrust at 25°C = 50 – 45 = 5

or $$d_{25} = \frac{5}{V_{25}}$$

Further, $V_{100}\times d_{100}$ = upward thrust at 100°C = 50 – 45.1 = 4.9

$$\therefore d_{100} = \frac{4.9}{V_{100}}$$

We know that $d_{25} = d_{100}[1+\gamma(100-25)]$

or $$\frac{d_{25}}{d_{100}} = 1+75\gamma = \frac{V_{100}\times5}{V_{25}\times4.9} = \frac{1.0027\times5}{4.9} = 1.023$$

$$\therefore \gamma = \frac{0.023}{75} = 3.066\times10^{-4}/°\text{C}$$

10.13. A one litre glass flask contains some mercury. It is found that at different temperatures the volume of air inside the flask remains the same. What is the volume of mercury in this flask if the coefficient of linear expansion of glass is 9×10^{-6}/°C and coefficient of volume expansion of mercury $= 1.8\times10^{-4}$/°C.

Solution: Let V_1 be the volume of mercury V be the volume of the flask and V_2 be the volume of air, then

$$V = V_1 + V_2$$

For constancy we must have

$$\Delta V = \Delta V_1$$

$$V\gamma_{glass}\times\Delta\theta = V_1\gamma_{mercury}\cdot\Delta\theta;\ \gamma_{glass} = 3\alpha_{glass} = 3\times9\times10^{-6}$$

$$1000\times27\times10^{-6} = V_1\times1.8\times10^{-4}$$

$$\therefore V_1 = \frac{27\times10^{-3}}{18\times10^{-5}} = 150\text{ cc}$$

10.14. A sinker of weight W_0 has an apparent weight W_1 when weighed in a liquid at a temperature t_1 and W_2 when weighed in the same liquid at a temperature t_2. The coefficient of cubical expansion of the material of the sinker is β. What is the coefficient of volume expansion of the liquid.

(IIT 1978)

Solution: $W_0 - W_1$ = weight of liquid displaced

$= V_1 d_1 g$; V_1 = volume of sinker at t_1°C

d_1 = density of liquid at t_1°C

Similarly

$$W_0 - W_2 = V_2 d_2 g$$

$$\frac{W_0 - W_1}{W_0 - W_2} = \frac{V_1 d_1}{V_2 d_2}$$

$$V_2 = V_1 [1 + \beta (t_2 - t_1)],\ d_1 = d_2 [1 + \gamma (t_2 - t_1)]$$

$$\therefore \quad \frac{W_0 - W_1}{W_0 - W_2} = \frac{V_1 d_2 \left[1+\gamma(t_2 - t_1)\right]}{V_1 d_2 \left[1+\beta(t_2 - t_1)\right]}$$

$$(W_0 - W_1) [1 + \beta (t_2 - t_1)] = (W_0 - W_2) [1 + \gamma (t_2 - t_1)]$$

$$\gamma = \frac{W_2 - W_1}{(t_2 - t_1)(W_0 - W_2)} + \beta \frac{W_0 - W_1}{W_0 - W_2}$$

10.15. A piece of metal weighs 46 cm in air. When it is immersed in a liquid of sp. gravity 1.24 at 37°C it weighs 30 gm. When the temperature of the liquid is raised to 42°C, the metal piece weighs 30.5 gm. The sp. gravity of the liquid at 42°C is 1.20. Calculate the coefficient of linear expansion of the metal. (IIT 1974)

Solution:

Loss in weight of metal = 46 – 30 = 16 gm = weight of liquid displaced

Volume of the liquid displaced = $\frac{16}{1.24}$ cc

This is equal to the volume of metal at 27°C

$$V_{27} = \frac{16}{1.24}$$

Similarly the volume of liquid displaced at 42°C = $\frac{46 - 30.5}{1.20}$ cc

This is equal to the volume of metal at 42°C.

$$V_{42} = \frac{15.5}{1.20} \text{ cc}$$

$$V_{42} = V_{27} [1 + \gamma (t_2 - t_1)]$$

$$\frac{15.5}{1.2} = \frac{16}{1.24} (1 + \gamma \cdot 15)$$

$$\gamma = \frac{1}{15} \times \left[\frac{15.5}{1.2} \times \frac{1.24}{16} - 1\right]$$

$$= \frac{1}{15} \times \frac{1}{960}$$

$$\alpha = \frac{1}{3 \times 15 \times 960} = 2.315 \times 10^{-5}/°\text{C}$$

10.16. A sphere of diameter 7 cm and mass 266.5 gm floats in a bath of liquid. As the temperature is raised, the sphere begins to sink at temperature of 35°C. If the density of liquid is 1.527 gm/cc at 0°C, find the coefficient of cubical expansion of the liquid. Neglect the expansion of sphere.

Solution:

$$\text{Density of sphere} = \frac{266.5}{\frac{4}{3}\pi(3.5)^3}$$

$$= 1.483$$

This will also be equal to density of liquid at 35°C

Using relation

$$d_0 = d_{35}(1 + \gamma \cdot 35)$$

$$1 + \gamma \cdot 35 = \frac{1.527}{1.483}$$

$$\gamma = \frac{1.527 - 1.483}{35 \times 1.483}$$

$$= 0.00083/°\text{C}$$

10.17. An ideal gas of density ρ is enclosed in a vessel of volume V at 0°C. As a result of deflective sealing some amount of gas has leaked out and due to this the gas pressure inside the vessel decreases by an amount Δp. Find the mass of the gas leaked out of the vessel.

Solution: Let p_0 be the atmospheric pressure then

$$\frac{\Delta p}{p_0} = \frac{\Delta \rho}{\rho} = \frac{\Delta m}{V\rho}; \ \Delta p = \frac{\Delta m}{V}$$

$$\therefore \Delta m = \frac{V\rho \cdot \Delta p}{p_0}$$

10.18. A mixture of two different gases of molecular weight M_1 & M_2 is filled in a vessel of volume V at a pressure p and temperature t°C. The mass of the mixture is m. Find the ratio of the mass of the two gases in the mixture.

Solution: Let m_1 & m_2 be the mass of the two gases having molecular weight M_1 & M_2 and p_1 & p_2 be their pressures. Then

$$m_1 + m_2 = m \text{ and } p_1 + p_2 = p$$

$$\text{Now} \quad p_1V = \frac{m_1}{M_1} \cdot RT$$

$$\text{and} \quad p_2V = \frac{m_2}{M_2} \cdot RT$$

$$\text{Now} \quad p_1V + p_2V = pV = RT\left[\frac{m_1}{M_1} + \frac{m_2}{M_2}\right]$$

$$\frac{M_1 pV}{RT} = m_1 + \frac{M_1}{M_2}\cdot m_2$$

$$m = m_1 + m_2$$

On substracting the two we have

$$\left(\frac{M_1 pV}{RT} - m\right) = \left(\frac{M_1}{M_2} - 1\right)m_2$$

$$\therefore m_2 = \left(\frac{M_1 pV}{RT} - m\right)\Big/\left(\frac{M_1}{M_2} - 1\right)$$

$$m_1 = m - m_2 = m - \frac{\left(\frac{M_1 pV}{RT} - m\right)}{\left(\frac{M_1}{M_2} - 1\right)}$$

$$= \frac{m\frac{M_1}{M_2} - m - \frac{M_1 pV}{RT} + m}{\left(\frac{M_1}{M_2} - 1\right)}$$

$$= \frac{M_1 pV}{RT}\left[\frac{m}{M_2}\frac{RT}{pV} - 1\right]\Big/\left(\frac{M_1}{M_2} - 1\right)$$

$$\frac{m_1}{m_2} = \frac{\frac{M_1 pV}{RT}\left(\frac{m}{M_2}\frac{RT}{pV} - 1\right)}{\frac{M_1 pV}{RT}\left(1 - \frac{m}{M_1}\frac{RT}{pV}\right)} = \frac{\left(1 - \frac{m}{M_2}\frac{RT}{pV}\right)}{\left(\frac{mRT}{M_1 pV} - 1\right)}$$

10.19. A vessel of volume $2 \times 10^{-2}\text{m}^3$ contains a mixture of hydrogen and helium at 47°C temperature and 4.15×10^5 N/m^2 pressure. The mass of the mixture is 10^{-2} kg. Calculate the masses of hydrogen and helium in the given mixture. (Roorkee 1994)

Solution: In view of Q. No. 10.18 we have

$$\frac{m_1}{m_2} = \frac{\left(1 - \frac{m}{M_2}\frac{RT}{pV}\right)}{\left(\frac{m}{M_1}\frac{RT}{pV} - 1\right)}$$

here m_1 = mass of hydrogen

and m_2 = mass of helium

M_1 = mol wt of hydrogen

M_2 = mol wt of helium

On putting the values

$$\frac{1-\dfrac{10^{-2}}{4\times 10^{-3}}\times\dfrac{8.3\times(273+47)}{2\times 10^{-2}\times 4.15\times 10^{5}}}{\left(\dfrac{10^{-2}}{2\times 10^{-3}}\dfrac{8.3\times(273+47)}{2\times 10^{-2}\times 4.15\times 10^{5}}-1\right)}=\frac{1-\dfrac{8.3\times 3.2}{8\times 4.15}}{\left(\dfrac{8.3\times 3.2}{4\times 4.15}-1\right)}$$

$$\frac{1-0.8}{1.6-1}=\frac{0.2}{0.6}=\frac{1}{3}$$

$$m_1+m_2 = 10^{-2}\text{ kg}$$

$$m_1+3m_1 = 10^{-2}\text{ kg}$$

$$m_1 = 2.5\times 10^{-3}\text{ kg}$$

$$m_2 = 7.5\times 10^{-3}\text{ kg}$$

10.20. A vertical hollow cylinder of height 1.52 m is filled with a movable piston of negligble mass and thickness. The lower half portion of the cylinder contains an ideal gas and the upper half is filled with mercury. The cylinder is initially at 300 K. When the temperature is raised, half of the mercury comes out of the cylinder. Find this temperature, assuming the thermal expansion of mercury to be negligible. (Roorkee 1993)

Solution: The initial and final positions of the piston in the cylinder is shown in the Figs. 10.1a & 10.1b.

$$P_i=P+\frac{h}{2}\rho g;\quad P=\text{atmospheric pressure}$$

$$\rho=\text{density of mercury}$$

$$V_i=\frac{h}{2}\cdot A;\quad A=\text{area of cross-section of the piston}$$

$$P_f=P+\frac{h}{4}\rho g$$

$$V_f=\frac{3h}{4}\cdot A$$

According to gas Law we have

$$\frac{P_i\cdot V_i}{T_i}=\frac{P_f\cdot V_f}{T_f}$$

$$T_f=\frac{P_f\cdot V_f\cdot T_i}{P_i\cdot V_i}$$

On putting the values we have

$$T_f=\frac{\left(P+\dfrac{h}{4}\rho\cdot g\right)\cdot\dfrac{3h}{4}\cdot A\cdot 300}{\left(P+\dfrac{h}{2}\rho g\right)\times\dfrac{h}{2}\cdot A}$$

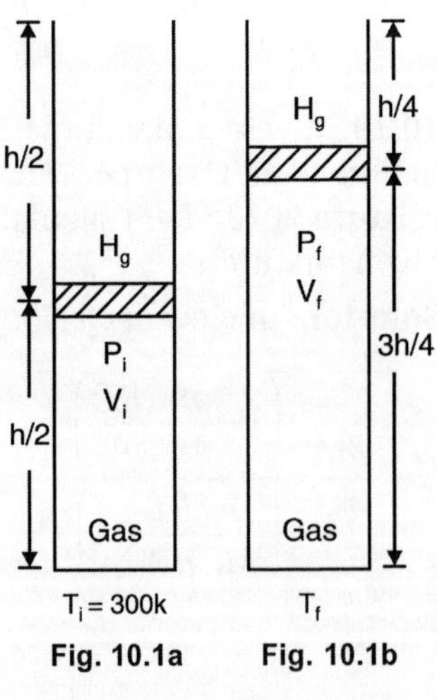

Fig. 10.1a Fig. 10.1b

$$= \frac{3}{2} \times 300 \frac{\left[0.76\,\rho g + \frac{1.52}{4}\rho g\right]}{\left[0.76\rho g + \frac{1.52}{2}\rho g\right]}$$

$$= \frac{450\,[0.76 + 0.38]}{0.76 + 0.76}$$

$$= \frac{450 \times 1.14}{1.52}$$

$$= 337.5\text{K}$$

10.21. A vessel contains a mixture of two ideal gases of respective masses m_1 and m_2 and molecular weight M_1 and M_2. Assuming the temperature of the mixture as T and pressure as p_0 find the density of the mixture.

Solution: Let p_1 and p_2 be the pressures of the two gases and V be the volume of the vessel, then

$$p_1 V = \frac{m_1}{M_1} RT$$

and $$p_2 V = \frac{m_2}{M_2} RT$$

$$(p_1 + p_2)\, V = p_0 V = \left(\frac{m_1}{M_1} + \frac{m}{M_2}\right) RT$$

$$\rho = \text{density of the mixture} = \frac{m_1 + m_2}{V} = \frac{(m_1 + m_2)\, p_0}{\left(\frac{m_1}{M_1} + \frac{m_2}{M_2}\right) RT}$$

10.22. A mixture of three ideal gases with respective number of mols n_1, n_2 & n_3 and molecular weight M_1, M_2 & M_3 is prepared and is placed in a vessel of volume V at temperature T. The mass of the mixture is m. Find the pressure and molecular weight of the mixture.

Solution: For the three gases we have the following equations

$$p_1 V = \frac{m_1}{M_1} \cdot RT\,,\ p_2 V = \frac{m_2}{M_2} \cdot RT \text{ and } p_3 V = \frac{m_3}{M_3} \cdot RT$$

where m_1, m_2 & m_3 are the masses of the three gases

$$p = p_1 + p_2 + p_3 = \frac{1}{V}\ [n_1 RT + n_2 RT + n_3 RT]$$

$$= \frac{RT}{V}\ [n_1 + n_2 + n_3]$$

Further $n_1 + n_2 + n_3 = \frac{m}{M}$; M = molecular weight of the mixture

$$\therefore M = \frac{m}{n_1 + n_2 + n_3}$$

or $$M = \frac{n_1 M_1 + n_2 M_2 + n_3 M_3}{n_1 + n_2 + n_3}$$

10.23. A long glass tube open at both the ends is immersed in a tank containing mercury so that 10 cm of its lengths is above the mercury level (Fig. 10.2a). After the upper end is sealed, the tube raised further by 50 cm. What length of the tube will now be occupied by the air column? Atmospheric pressure = 76 cm of mercury.

Solution: Let a be area of cross-section of the tube.

Initial pressure of air = 76 cm of mercury

Initial volume of air = $10a$

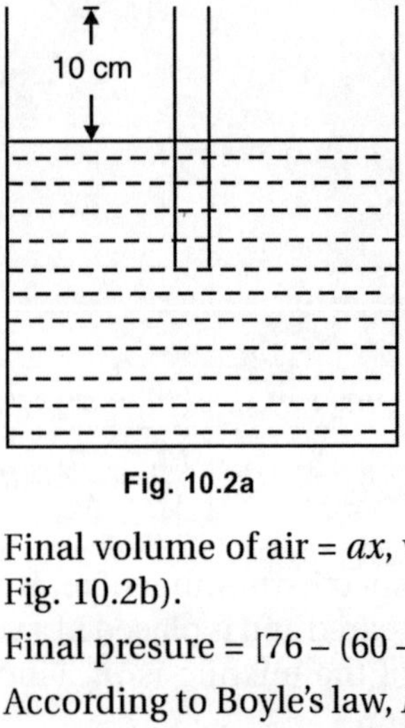

Fig. 10.2a

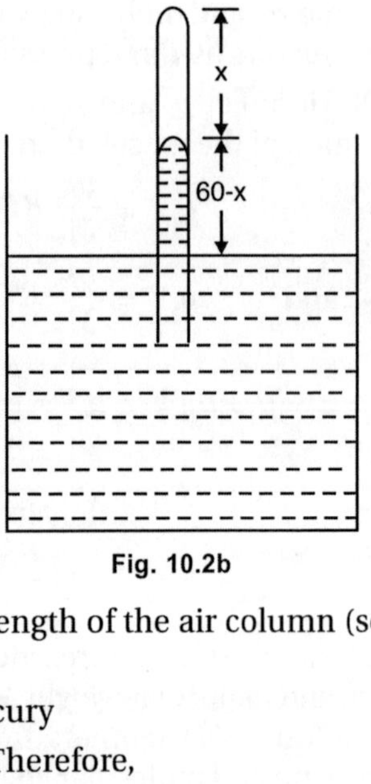

Fig. 10.2b

Final volume of air = ax, where x is the length of the air column (see Fig. 10.2b).

Final presure = $[76 - (60 - x)]$ cm of mercury

According to Boyle's law, PV = constant. Therefore,

$$76 \times 10a = ax \times [76 - (60 - x)]$$

or $$760 = x^2 + 16x$$

or $$x^2 + 16x - 760 = 0$$

$$x = \frac{-16 \pm \sqrt{256 + 3040}}{2}$$

$$= \frac{-16 \pm \sqrt{3296}}{2}$$

$$= \frac{57.41 - 16.00}{2} = 20.7 \text{ cm}$$

10.24. A glass capillary tube, sealed at both ends is 100 cm long. It lies horizontally with the middle 10 cm containing mercury. The two ends of the tube (which are equal in length) contain air at 27°C and at a pressure of 76 cm of mercury. The tube is kept in a horizontal position such that the air column at one end is at 0°C and the other end is maintained at 127°C. Calculate the length of air which is at 0°C and its pressure. Neglect the change in volume of mercury and glass. (IIT 1975)

Solution: Let a be area of cross-section of the tube.

Initial volume of air on each side of the mercury pellet = 45 a (see Fig. 10.3a). In Fig. 10.3b, the volume of air on the left-hand side of the mercury pellet = xa. Since the mercury pellet is at rest, the pressure P_1 and P_2 are equal. Therefore,

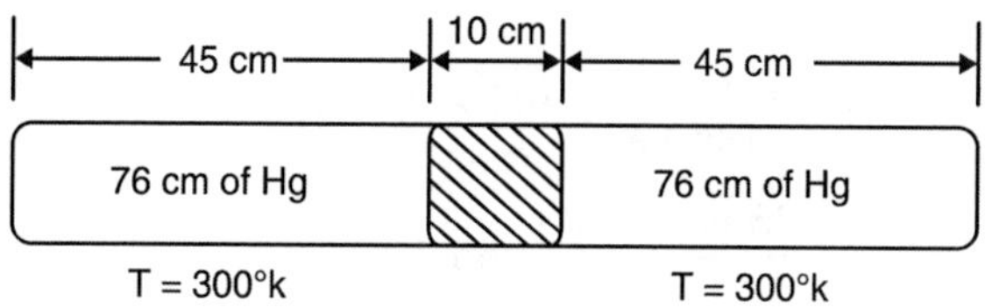

Fig. 10.3a

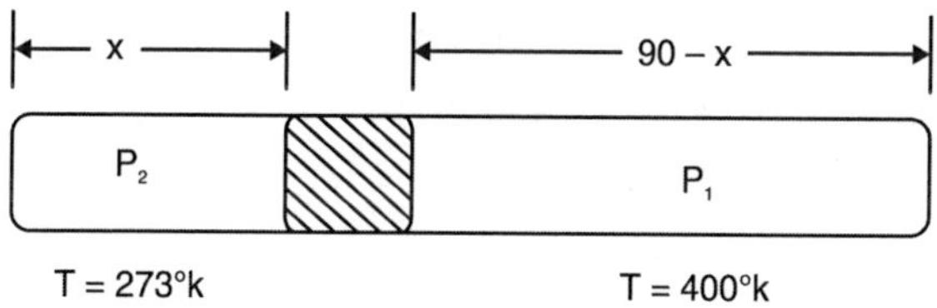

Fig. 10.3b

$$\frac{ax \cdot P_2}{273} = \frac{a(90-x)P_1}{400}$$

or $400x = 273 \times 90 - 273\,x$

or $673\,x = 24{,}570$

or $x = 24{,}570/673 = 36.51$ cm

Now $$\frac{45a \times 76}{300} = \frac{36.5 \times a \times P_2}{273}$$

or $$P_2 = \frac{45 \times 76 \times 273}{300 \times 36.5} = 85.27 \text{ cm of mercury}$$

10.25. In the middle of a narrow horizontal tube sealed at both ends is a column of mercury 10 cm long. Both halves of the tube contains air under a pressure of 76 cm of mercury. By what distance will the mercury column be displaced if the tube is held vertical? The length of the tube is 100 cm. (IIT 1978)

Solution: Let a be the area of cross-section of the tube. When the tube is horizontal, the both halves will have the same volume = 45 a and pressure = 76 cm of mercury. Now when the tube is held vertical the pressure of the air column above the mercury pellet = P cm of mercury. Volume of air above the mercury column = $(45 + x)\ a$ where x is the distance through which the mercury column is displaced. The pressure of air below the mercury column = $(P + 10)$ cm of mercury. Volume of air below the mercury columns = $(45 - x)\ a$.

Now, according to Boyle's law

$$45a \times 76 = P \times (45 + x)\ a = (P + 10) \times (45 - x)\ a$$

$$P = \frac{45 \times 76}{(45 + x)}$$

Further $$45 \times 76 = \left[\frac{45 \times 76}{(45 + x)} + 10\right] \times (45 - x)$$

or $$45 \times 76 = \frac{(45 \times 86 + 10x)}{(45 + x)}(45 - x)$$

or $$45 \times 76 \times 45 + 45 \times 76 \times x$$

$$= 45 \times 45 \times 86 + 45 \times 10x - 45 \times 86x - 10x^2$$

or $$10x^2 + 45\ (76 + 86)\ x - 450\ x = 45 \times 45 \times 10$$

or $$x^2 + 684 - 2025 = 0$$

or $$x = \frac{-684 \pm \sqrt{684^2 + 8100}}{2}$$

$$= \frac{-684 \pm 690}{2} = 3 \text{ cm}$$

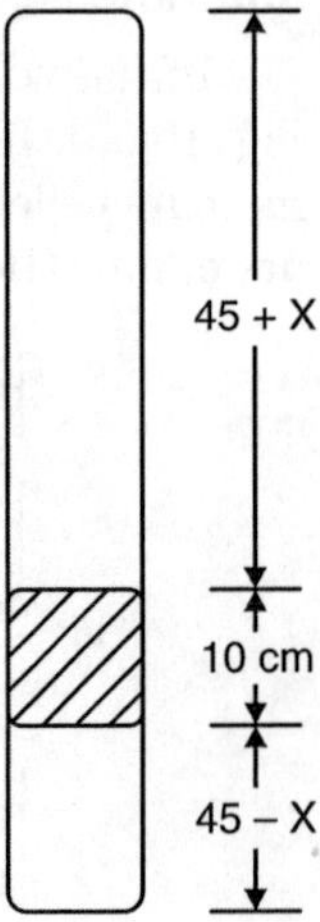

Fig. 10.4

10.26. A thin tube of uniform cross-section is sealed at both ends. When it lies horizontally, the middle 5 cm length contains mercury and the two equal ends contain air at the same pressure P. When the tube is held at an angle of 60°C with the vertical, then the lengths of the air columns above and below the mercury column are 46 cm and 44.5 cm respectively. Calculate the pressure P in cm of mercury. The temperature of the system is kept at 30°C. (IIT 1986)

Solution: When the tube is kept horizontal then the total length of air column = 46 + 44.5 = 90.5 cm

The length of each air column = 45.25 cm

Let a be the area of cross-section of the tube. According to Boyle's law,

$P \times 45.25 \times a = P_1 \times 44.5 \times a = P_2 \times 46 \times a$ (see Fig. 10.5)

Further $P_1 = P_2 + 5 \cos 60 = P_2 + 2.5$

Now $P_2 = P_1 - 2.5 = P_1 \times \dfrac{44.5}{46}$

or $46\,P_1 - 46 \times 2.5 = 44.5\,P_1$

or $1.5\,P_1 = 46 \times 2.5$

or $P_1 = \dfrac{46 \times 2.5}{1.5}$

Now $P = \dfrac{44.5 \times P_1}{45.25}$

$= \dfrac{44.5 \times 46 \times 2.5}{45.25 \times 1.5}$

$= 75.39 = 75.4$ cm

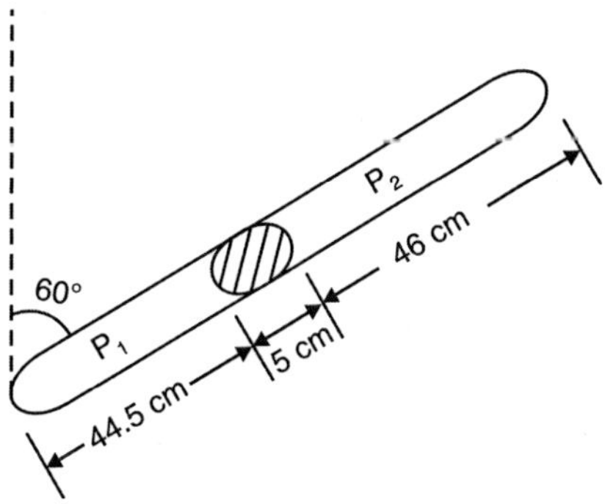

Fig. 10.5

10.27. A barometer contains some air above the mercury. It provides readings for the measurement of two pressures as 67 and 69 cm of mercury. The corresponding true barometric readings are 71 and 74 cm of mercury. Find: (i) the length of the faulty barometer tube, and (ii) the true barometric reading corresponding to the reading of 68 cm as recorded by the faulty barometer.

Solution: Let l be the length and a be the area of cross-section of the faulty barometer tube.

Volume V_1 of the air above 67 cm of mercury column = $(l - 67)\,a$. Pressure P_1 of the air above the mercury column of the faulty barometer = $71 - 67 = 4$ cm of Hg.

Volume V_2 of the air above 69 cm of mercury column $(l - 69)\,a$

Pressure P_2 of the air above the mercury column of the faulty barometer $= 74 - 69 = 5$ cm of mercury

According to Boyle's law,

$P_1V_1 = P_2V_2$

or $4 \times (l - 67)\,a = 5 \times (l - 69)\,a$

or $4l - 268 = 5l - 345$

or $l = 345 - 268 = 77$ cm

(ii) Let the true reading be h. Then

$(77 - 67)\,a \times 4 = (77 - 68)\,(h - 68) \times a$

or $\dfrac{40}{9} + 68 = h$

$68 + 4.4 = h$

or $h = 72.4$ cm of mercury.

10.28. A horizontally placed cylinder of length 80.5 cm contains a 0.5 cm thick piston which divides it into two equal sections. The cylinder is closed at both ends. Each section of the cylinder maintained at the freezing point of ice contains same mass of air at a pressure of 1 atmosphere. On heating the air confined in the left section up to 27°C the piston moves towards the right through some distance. Calculate the final pressure in each section of the cylinder.

Solution: Let P be pressure in the two parts of the cylinder and a be area of cross-section of the piston. Let x be the distance through which the piston moves towards the right when the air is heated. The volume of air in the two sections is $(40 + x)\,a$ and $(40 - x)\,a$ cm^3 respectively. According to the gas law,

$$\frac{P(40+x)a}{300} = \frac{P(40-x)a}{273}$$

or $\quad 40 \times 273 + 273x = 40 \times 300 - 300x$

or $\quad 573x = 40 \times (300 - 273) = 40 \times 27$

$$\therefore x = \frac{40 \times 27}{573} = 1.885 = 1.9 \text{ cm}$$

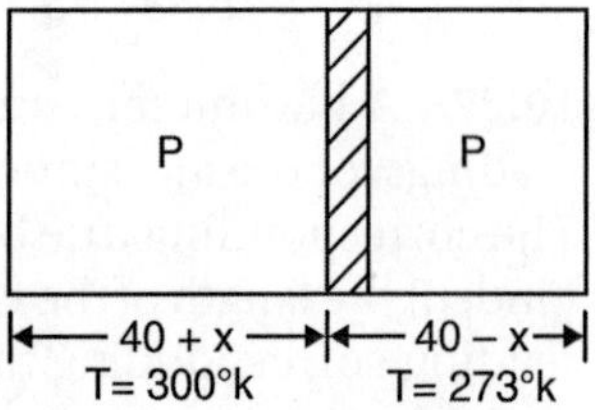

Fig. 10.6

Now $\quad \dfrac{40a \times 1}{273} = \dfrac{P \times (40 + 1.9) \times a}{300}$

$$\therefore P = \frac{300 \times 40}{273 \times 41.9} = 1.049 = 1.05 \text{ atmosphere}$$

10.29. A glass tube of length 20 cm sealed at one end is lowered to a depth of 5 cm in a tank containing mercury. At 0°C the height of the mercury column inside the tube is 5 cm above the surface of mercury in the tanks (see Fig. 10.7). To what temperature should the tube at the sealed end be heated in order to remove all the mercury from the tube? The atmospheric pressure is 75 cm of mercury.

Solution: Let a be the area of cross-section of the tube. At 0°C the volume of air enclosed = $10a$ and the pressure of air is 70 cm of mercury. At a temperature θ°C, the volume of air = $20a$ and the pressure of air = 75 + 5 = 80 cm of mercury.

Now according to the gas law,

$$\frac{70 \times 10a}{273} = \frac{80 \times 20a}{273 + \theta}$$

or $\quad 7 \times 273 + 7\theta = 16 \times 273$

$7\theta = 9 \times 273$

$\theta = 351$°C

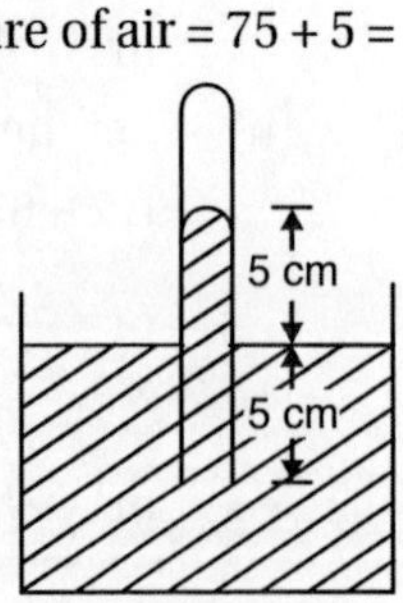

Fig. 10.7

10.30. An air bubble located at the bottom of a 30 m deep pond rises upto its surface covered by a thin layer of ice. The temperature at the bottom of the pond is 7°C. Calculate the percentage increase in the volume of the bubble. Density of the water of the lake = 1.01×10^3 kg/m³. Atmospheric pressure = 1.01×10^5 N/m². Acceleration due to gravity = 10 m/s².

Solution: Let V and V' be the volumes of air bubble at the bottom and top of the pond here the temperatures are 7°C and 0°C respectively.

Total pressure on bubble when it is at the bottom of the pond

$$= (P + 30 \times 1.01 \times 10^3 \times 10)\ \text{N/m}^2$$

Where P is the atmospheric pressure.

Pressure at the top of the pond = P

Now according to the gas law, we have

$$\frac{(P + 30 \times 1.01 \times 10^3 \times 10)V}{280} = \frac{PV'}{273}$$

On dividing the above equation by P, we have

$$\left(1 + \frac{30 \times 1.01 \times 10^3 \times 10}{1.01 \times 10^5}\right)V = \frac{280}{273}V'$$

$$\frac{4 \times 273}{280} = \frac{V'}{V}$$

or
$$\frac{4 \times 273 - 280}{280} = \frac{V' - V}{V}$$

or
$$\frac{812}{280} \times 100 = \text{percentage increase in volume of the air bubble}$$

$$= 290\%$$

10.31. An air bubble located at the bottom of a lake rises up to its surface where its volume becomes 11 times its initial value. Calculate the depth of the lake. Density of the water of the lake = 1.02×10^3 kg/m³. Atmospheric pressure = 75 cm of mercury. Acceleration due to gravity = 9.8 m/s².

Solution: Let V be the volume of the air bubble lying at the bottom of the lake at a depth of H m.

Total pressure on the bubble when it is at the bottom of lake

$$= (0.75 \times 13.6 \times 10^3 \times 9.8 + H \times 1.02 \times 10^3 \times 9.8)\ \text{N/m}^2$$

Pressure on the bubble when it is at the surface of the lake

$$= 0.75 \times 13.6 \times 10^3 \times 9.8$$

Volume of the bubble at the surface = $11\ V$

According to Boyle's law

$$(0.75 \times 13.6 \times 10^3 \times 9.8)\ 11\ V$$
$$= (0.75 \times 13.6 \times 10^3 \times 9.8 + H \times 1.02 \times 10^3 \times 9.8) V$$

or $\quad 0.75 \times 13.6 \times 11 = 0.75 \times 13.6 + H \times 1.02$

or $\quad 1.02\, H = 7.5 \times 13.6$

or $$H = \frac{7.5 \times 13.6}{1.02} = 100 \text{ m}$$

10.32. A 40 cm long test tube of area of cross-section a = 0.4 cm^2 and weight W = 9 gf is submerged in water after its upper end is sealed (see Fig. 10.8). What force should be applied at the upper sealed end of the tube in order to hold it at a depth of h = 13.6 cm below the surface of the water? Atmospheric pressure = 75 cm of mercury. Density of water = 1 g/cm^3.

Solution: In this system three forces are acting: (i) force F (to be applied), (ii) weight W of the tube, and (iii) the force F' which expels the tube. Under the conditions of equilibrium the sum of the three forces will be equal to zero, i.e.,

$$F' - W - F = 0$$

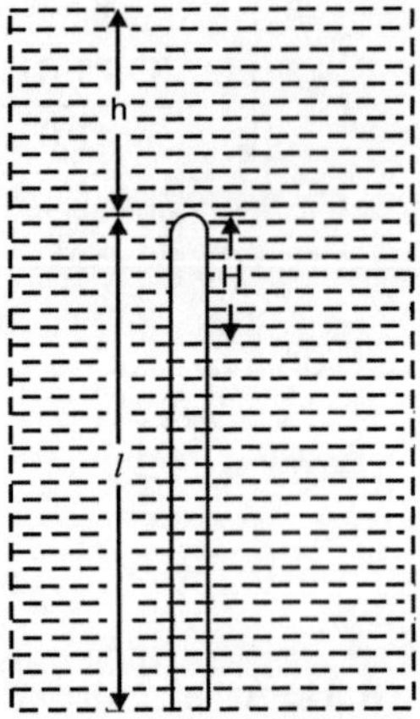

Fig. 10.8

Let H be the length of air column when the tube is inside water at a depth of h. The force $F' = Hdg \cdot a$ where a is the area of cross-section of the test tube and d is the density of water. The same force F' can be evaluated as the difference of force at the two ends of the tubes, i.e.,

$$F' = p_1 a - (p_0 + dgh)\, a$$

where p_1 is the pressure of the air column when the tube is at a depth h and p_0 is the atmospheric pressure.

Further $p_0 la = p_1\, Ha$

Thus $$\frac{p_0\, ldg}{p_1} = p_1 - (p_0 + dgh)$$

or $$p_1^2 - (p_0 + dgh)\, p_1 - p_0 ldg = 0$$

$$p_1 = \frac{p_0 + dgh \pm \sqrt{(p_0 + dgh)^2 + 4 p_0 ldg}}{2}$$

$$\therefore F' = \frac{a}{2}(p_0 + dgh) + \frac{a}{2}\sqrt{(p_0 + dgh)^2 + 4 p_0\, ldg} - (p_0 + dgh)\, a$$

$$F = F' - W = \frac{a}{2}\left[\sqrt{(p_0 + dgh)^2 + 4 p_0\, ldg} - (p_0 + dgh)\right] - W$$

On substituting the values of various terms we have

$$F = 0.2\left[\sqrt{(75\times 13.6+13.6)^2+4\times 75\times 13.6\times 40}-(75\times 13.6+13.6)\right]-9$$

$$= 0.2\left(\sqrt{1.068\times 10^6+0.163\times 10^6}-1033.6\right)-9 = 15.18-9 = 6.18\text{ gf}$$

10.33. A vertical cylinder closed at both the ends contains a piston which can easily move in the vertical direction. The piston divides the volume into two parts and each part contains one mol of a gas. In equilibrium at 47°C the ratio of the volume of the upper and lower part is $\eta = 5$. Calculate the temperature at which the ratio of the two volumes reduces to $\eta' = 3$.

Solution: Figs. 10.9a & 10.9b show the initial and final pressure and volume of gas.

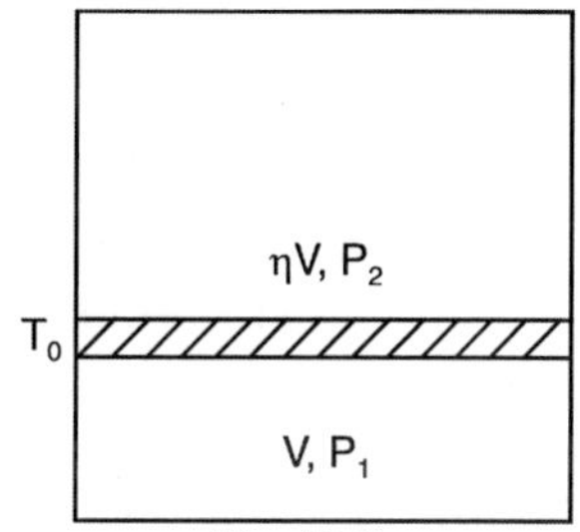

Fig. 10.9a

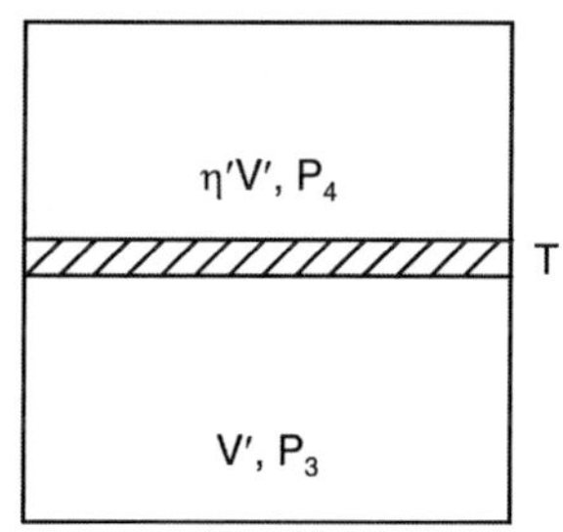

Fig. 10.9b

According to gas law, we have

$$\frac{\eta V\cdot P_2}{T_0} = \frac{\eta' V'\cdot P_4}{T}$$

or $$T = T_o\frac{\eta'\cdot V'\cdot P_4}{\eta\cdot V\cdot P_2}$$

Further $(\eta+1)V = (\eta'+1)V'$

In view of this

$$T = \frac{T_0\cdot\eta'(\eta+1)P_4}{\eta(\eta'+1)P_2}$$

Now $P_1 = \eta P_2$

$\therefore P_1 - P_2 = (\eta-1)P_2$

Similarly $P_3 = \eta' P_4$

or $P_3 - P_4 = (\eta'-1)P_4$

In equilibrium we have

$$P_1 = \frac{mg}{A}+P_2;$$ m = mass of the piston and A = area of cross-section of the piston.

and $$P_3 = \frac{mg}{A}+P_4$$

In view of the above $P_1 - P_2 = P_3 - P_4$

or $(\eta-1)P_2 = (\eta'-1)P_4$

or $$\frac{(\eta-1)}{\eta'-1} = \frac{P_4}{P_2}$$

or $$T = T_0 \frac{\eta'(\eta+1)\cdot(\eta-1)}{\eta(\eta'+1)(\eta'-1)}$$

p $$= T_0 \frac{\eta'(\eta^2-1)}{\eta(\eta'^2-1)}$$

On putting the values of various terms we have

$$T = \frac{320\times 3(5^2-1)}{5(3^2-1)}$$

$$= \frac{320\times 3\times 24}{5\times 8}$$

$$= 576 \text{ K}$$

$$= 303°\text{C}$$

10.34. An ideal gas is enclosed in a volume V. The vessel is evaluated under isothermal conditions by using a common air pump. At each piston stroke a volume ΔV of the gas is swept out. Calculate the number of strokes required to reduce the pressure in the vessel η times.

Solution: Let p_0 be the initial pressure then at the first stroke the pressure p_1 is given by

$$p_1 (V+\Delta V) = p_0 V$$

or $$p_1 = \frac{p_0 V}{V+\Delta V}$$

At second stroke we have pressure p_2 given by

$$p_2 (V+\Delta V) = p_1 V$$

or $$p_2 = p_0 \left(\frac{V}{V+\Delta V}\right)^2$$

At n strokes we have the pressure p_n as

$$p_n = p_0 \left(\frac{V}{V+\Delta V}\right)^n$$

$$\left(\frac{V+\Delta V}{V}\right)^n = \frac{p_0}{p_n} = \eta$$

$$n \log_e \left[1+\frac{\Delta V}{V}\right] = \log_e \eta$$

or $$n = \frac{\log_e \eta}{\log_e \left[1+\frac{\Delta V}{V}\right]}$$

10.35. Air is enclosed in a vessel of volume V at a pressure P_0. The vessel is evacuated at a rate Q m³/sec by using a common air pump under isothermal conditions. Find the pressure inside the vessel at an instant t.

Solution:

$$pV = \text{constant}$$

$$pdV + Vdp = 0$$

$$\therefore \frac{dV}{V} = -\frac{dp}{P}$$

or
$$\frac{dV}{dt}\cdot\frac{dt}{V} = -\frac{dp}{P}$$

$$-Q\cdot\frac{dt}{V} = +\frac{dp}{P}$$

$$-\frac{Q}{V}\int_0^{\tau} dt = \int_{P_0}^{P}\frac{dp}{P}$$

$$-\frac{Q}{V}\cdot t = \log\frac{P}{P_0}$$

or
$$P = P_0 e^{-\frac{Qt}{V}}$$

10.36. A gas is enclosed in a vessel of volume V at a pressure P_0. The vessel is evacuated isothermally at a rate Q m³/sec using a vacuum pump. Calculate the time in which a pressure $\frac{P_0}{\eta}$ is attained inside the vessel.

Solution:

$$pV = \text{constant}$$

$$pdV + Vdp = 0$$

$$\frac{dV}{V} = -\frac{dp}{P}$$

$$\frac{dV}{dt}\cdot\frac{dt}{V} = -\frac{dp}{P}$$

$$\int_0^{\tau} dt = \frac{V}{Q}\int_{P_0}^{P_0/\eta}\frac{dp}{p}$$

$$t = \frac{V}{Q}\log_e^{\eta}$$

10.37. Two uniform bore tubes *abcd* & *efgh* are cemented to construct a vertical tube as shown in the Fig. 10.10. The two pistons A & B attached to an extensive thread are capable of moving in their sections. One mol of an ideal gas is enclosed in the space between the two pistons. The combined mass of the two pistons is m = 10 kg. The difference in the cross-sectional area of the two piston ΔA = 15 cm^2. The outside air pressure p_0 = 1 atmosphere. Through what temperature range must the gas between the pistons be heated in order to shift the piston through a distance L = 10 cm?

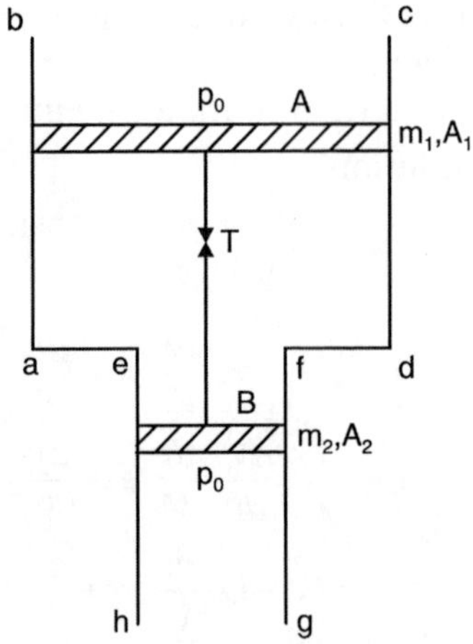

Fig. 10.10

Solution: Let us consider the lower piston.

$$p \cdot A_2 + m_2 g = p_0 A_2 + T;$$

m_2 = mass of lower piston.

A_2 = area of cross-section of the lower piston.

T = tension in the thread.

p = pressure of the gas.

or $\quad T = (p - p_0) A_2 + m_2 g$

Consider the upper piston

$p_0 A_1 + T + m_1 g = pA$; $\quad m_1$ = mass of upperpiston

A_1 = area of cross-section of the upper piston.

$$T = (p - p_0) A_1 - m_1 g$$

or $\quad (p - p_0) A_2 + m_2 g = (p - p_0) A_1 - m_1 g$

$$(m_1 + m_2) g = mg = (p - p_0)(A_1 - A_2) = (p - p_0)\Delta A$$

or $\quad p = p_0 + \dfrac{mg}{\Delta A}$

Increase in volume = $L\,[A_1 - A_2]$

Work done = $pdV = pL\,(A_1 - A_2] = RdT$; dT rise in temp.

$$\Delta T = \frac{pL\Delta A}{R} = \left(p_0 + \frac{mg}{\Delta A}\right) \cdot \frac{L}{R} \cdot \Delta A$$

$$\Delta T = (p_0 \cdot \Delta A + mg)\,\frac{L}{R}$$

On putting the values of various terms we have

$$\left[10^5 \times 15 \times 10^{-4} + 10 \times 9.8\right] \times \frac{10^{-1}}{8.3} = \frac{150 + 98}{83} = \frac{248}{83}$$

$$\cong 3°\text{C}$$

10.38. How much ice at –15°C should be added to 0.6 kg of water at 20°C in order to bring down its temperature to 0°C. Specific heat of ice = 0.5, latent heat of ice = 80 cal/g.

Solution: Let the amount of ice required at –15°C = m kg

Heat given by water = 0.6 × 1 × (20 – 0) = 12 k cal

Heat taken by m kg of ice to increase its temperature to 0°C

$= m \times 0.5 \times [0 - (-15)] = 7.5\, m$ k cal

Heat taken by m kg of ice to convert into m kg of water at 0°C

$= 80\, m$ k cal

Now heat given = heat taken

12 k cal = (80 + 7.5) m k cal

$$\therefore\ m = \frac{12}{87.5} = 0.1372 \text{ kg}$$

10.39. A piece of 0.3 kg of ice at –15°C is dropped into 0.2 kg of water at 20°C. Find the amount of water and ice in the mixture in thermal equilibrium. Specific heat of ice = 0.5, latent heat of ice = 80 k cal/kg.

Solution: In thermal equilibrium, the system will acquire a temperature of 0°C.

Heat given by 0.2 kg of water = 0.2 × 1 × 20 = 4 k cal

Heat taken by 0.3 kg of ice $= 0.3 \times 0.5 \times [0 - (-15)] + 0.3 \times 80$

$= 2.25 + 24 + 26.25$ k cal.

However 26.25 k cal of heat is not available. This indicates that only a part of ice will be converted into water.

Let us assume that m kg of ice is converted into water.

Heat taken by 0.3 kg of ice at – 15°C to transform into ice at 0°C

$= 0.3 \times 0.5\ [0 - (-15)] = 2.25$ kcal.

Heat taken by m kg out of 0.3 kg of ice at 0°C to transform into water at 0°C = $m \times 80$ k cal

Thus 4 k cal = 2.25 k cal + 80 m k cal

$$\text{or} \qquad m = \frac{1.75}{80} = 0.022 \text{ kg}$$

Therefore, at thermal equilibrium, the mixture will contain 0.278 kg of ice and 0.222 kg of water.

10.40. A 20 g copper calorimeter contains 30 g of ice at –20°C. Steam at 100°C is passed into the calorimeter in order to change ice into water, and then heat it to 40°C. Calculate the amount of steam condensed in the calorimeter. Latent heat of steam = 540 cal/g, latent heat of ice = 80 cal/g, and specific heat of copper = 0.09 cal/g/°C.

Solution: Let the amount of steam condensed by m g.

Heat given by m g of steam = 540 m

Heat given by m g of water to cool up to 40°C

$= m \times 1 \times (100 - 40) = 60\,m$

Heat taken by the calorimeter $= 20 \times 0.09 \times [40 - (-20)]$

$= 1.8 \times 60 = 108$ cal

Heat taken by 30 g of ice at – 20°C to transform into ice at 0°C

$= 30 \times 1 \times [0 - (-20)] = 600$ cal

Heat taken by 30 g of ice at 0°C to have change of state

$= 30 \times 80 = 2{,}400$ cal

Heat taken by 30 g of water at 0°C to raise its temperature to 40°C

$= 30 \times 1 \times (40 - 0) = 1{,}200$ cal

Thus $(540 + 60)\,m = 108 + 600 + 2{,}400 + 1{,}200$

$600\,m = 4{,}308$ cal

$\therefore\ m = 7.18$ g

10.41. A 60 W bulb in an incandescent state is immersed into a glass vessel containing 600 cm^3 of water and a thermometer. In a duration of 3 min a rise of temperature of 4°C is recorded by the thermometer. What is the percentage of the energy of the bulb released from the container $J = 4.2$ J/cal.

Solution: Energy consumed by the bulb $= 60 \times 3 \times 60 = 10{,}800$ J

Heat taken by water $= 600 \times 1 \times 4 \times 4.2$

$= 10{,}080$ J

Energy released from the container $= 10{,}800 - 10{,}080 = 720$ J

$$\text{Percentage of this energy} = \frac{720}{10{,}800} \times 100 = 6.67\%$$

10.42. An aluminium container of mass 100 g contains 200 g of ice at –20°C. Heat is added to the system at a rate of 100 calories per second. What is the temperature of the system after 4 min? Specific heat of ice = 0.5 cal/g/°C, specific heat of aluminium = 0.2 cal/g/°C. Latent heat of fusion of ice = 80 cal/g.

Solution: Let θ be the final temperature of the system.

Heat taken by 200 g of ice at –20°C to transform into ice at 0°C

$= 200 \times 0.5 \times [0 - (-20)] = 2{,}000$ cal

Heat taken by 200 g of ice at 0°C to have a change of state

$= 200 \times 80 = 16{,}000$ cal

Heat taken by 200 g of water to raise its temperature up to θ°C

$= 200 \times 1 \times (\theta - 0) = 200\,\theta$

Heat taken by aluminium container to raise its temperature to θ

$$= 100 \times 0.2 \times [\theta + 20] = 20\,\theta + 400$$

Total heat given $= 100 \times 4 \times 60 = 24{,}000$ cal.

Thus $2{,}000 + 16{,}000 + 200\,\theta + 20\,\theta + 400 = 24{,}000$

or $\quad 220\,\theta = 24{,}000 - 18{,}400$

$$= 5{,}600$$

$$\therefore \theta = \frac{560}{22} = \frac{280}{11} = 25.45°C$$

10.43. An earthen pitcher loses one gram of water per minute due to evaporation. If the water equivalent of pitcher is 0.5 kg and the pitcher contains 9.5 kg of water, calculate the time required for the water in the pitcher to cool to 28°C from its original temperature of 30°C. Neglect radiation effect. Average latent heat of vaporization of water in this range is 580 cal/g.

Solution: Let the required time be t min.

Total heat going out in t min $= 580 \times t$ cal

The same is obtained from the water and pitcher and is equal to

$$= (0.5 \times 1{,}000 + 9.5 \times 1{,}000)\,(30 - 28)$$

$$= 20{,}000$$

$$\therefore t = \frac{20{,}000}{580} = \frac{1{,}000}{29} = 34.48 \text{ min}$$

10.44. A lump of 400 g of wet snow is dropped into 1,500 cm^3 of water of 20°C. Due to exchange of heat all snow melts and the final temperature of water becomes 5°C. Calculate the amount of water contained in the lump of snow. Specific heat of water is 1 and latent heat of ice = 80 cal/g.

Solution: Of the 400 g of lump of snow let there be x g of water at 0°C.

Heat given by water $1{,}500 \times 1 \times 15$ cal

Heat taken by ice to have the change of state $= (400 - x)\,80$ cal

Heat taken by 400 g of water at 0°C to increase its temperature to 5°C

$$= 400 \times 1 \times 5 \text{ cal}$$

According to conservation of energy

$$1{,}500 \times 15 = (400 - x)\,80 + 400 \times 5$$

or $\quad 22{,}500 = 32{,}000 - 80\,x + 2{,}000$

04 $\quad 80\,x = 34{,}000 - 22{,}500$

$$80\,x = 11{,}500$$

or $\quad x = \dfrac{1{,}150}{8} = 143.75$ g

10.45. In an industrial process, 10 kg of water per hour is to be heated from 20°C to 80°C. To do this, steam at 150°C is passed from a boiler into a copper coil immersed in water. The steam condenses in the coil and is returned to the boiler as water at 90°C. How many kg of steam are required per hour. Latent heat of steam = 540 cal/g.

Solution: Let the amount of steam required be m g.

Heat given by m g of steam to cool from 150 to 100°C = $m \times 1 \times 50$ cal

Heat given by m g of steam at 100°C to have a change of state

= 540 m cal

Heat given by m g of water at 100°C to cool to 90°C

= $m \times 1 \times (100 - 90) = 10\,m$ cal

Total heat given = 600 m cal

Heat taken by 10 kg of water to raise its temperature from 20°C to 80°C = $10 \times 1{,}000 \times 1 \times 60$ cal

Thus,

$$600\,m = 600 \times 1{,}000$$

$$\therefore m = 1{,}000 \text{ g} = 1 \text{ kg}$$

10.46. Calculate the gas constant R for 1 g of oxygen. Density of hydrogen at NTP = 0.09 g/lit, density of mercury = 13.6 g/cc and $g = 10$ m/s².

Solution:

Density of oxygen at $NTP = 16 \times 0.09 \times 10^{-3}$ g/cc

Volume of 1 g of oxygen at $NTP = \dfrac{1}{16 \times 0.09 \times 10^{-3}}$ cc

At NTP, $T = 273$ K

and atmospheric pressure $P = 0.76 \times 13.6 \times 10^3 \times 10$

$$R = \frac{PV}{T} = \frac{0.76 \times 13.6 \times 10^4}{273} \times \frac{1000}{16 \times 0.09} \times 10^{-6}$$

$$= \frac{76 \times 136}{16 \times 273 \times 9} = 0.2629 \text{ J/g/K}$$

10.47. A gas leaks from a 5 litre cylinder because of a faulty valve. As a result, the gas pressure drops by 2.9×10^3 Pa. The temperature in the cylinder remains unchanged at 17°C. How many molecules escape from the cylinder? Boltzmann constant $K = 1.38 \times 10^{-23}$ J/K.

Solution: Let the initial pressure of the gas be p_1. At this pressure let there be n_1 molecules. When the pressure is changed from p_1 to p_2, let there be n_2 molecules. The number of molecules thus escaped = $n_1 - n_2$.

From the gas law we have the equations.

$$p_1 v = n_1 kT$$

and $p_2 v - n_2 kT$

Now $(p_1 - p_2)v = (n_1 - n_2)\ kT$

or $$\Delta n = n_1 - n_2 = \frac{(p_1 - p_2)v}{kT} = \frac{\Delta pv}{kT}$$

On substituting the values of the various terms, we have

$$\Delta n = \frac{2.9\times10^3\times5\times1000\times10^{-6}}{1.38\times10^{-23}\times290}$$

$$= \frac{5}{1.38}\times10^{21} = 3.6\times10^{21}$$

10.48. What will be the increase in the mass of air in a room if the atmospheric pressure changes from 9.84×10^4 to 10.1×10^4 Pa and the air temperature remains unchanged and equal to 273 K? The size of the room $= 4 \times 5 \times 2.5$ m^3. Density of air at *NTP* = 1.29 kg/m^3. Boltzmann constant $k = 1.38 \times 10^{-23}$ J/K, and Avogadro's number $= 6.02 \times 10^{23}$/mol.

Solution: Let p_1 and p_2 be the initial and final pressures of the air in the room and let there be n_1 and n_2 molecules in the room corresponding to these pressures.

From the gas law we have two equations

$$p_1 V = n_1 kT$$

and $p_2 V = n_2 kT$

Now $(p_2 - p_1)\ V = (n_2 - n_1)\ kT$. Here $n_2 - n_1$ represents the additional number of air molecules due to increase of pressure. Therefore,

$$n_2 - n_1 = \frac{\Delta pV}{kT} = \frac{0.26\times10^4\times50}{1.38\times10^{-23}\times273}$$

Mass of air at *NTP* = 1.29×22.4 kg

Mass of 6.02×10^{26} molecules = 1.29×22.4 kg

Mass of $\dfrac{0.26\times10^4\times50}{1.38\times10^{-23}\times273}$ molecules

$$= \frac{0.26\times10^4\times50\times1.29\times22.4}{1.38\times10^{-23}\times273\times6.02\times10^{26}}$$

$$= \frac{130\times1.29\times22.4}{1.38\times273\times6.02} = 1.656 \text{ kg}$$

10.49. Calculate the root mean square velocity of air molecules at 100°C. Density of air at *NTP* = 1.29 kg/m^3. Atmospheric pressure = 76 cm of mercury.

Solution: The root means square velocity of air molecules at *NTP* is given by

$$c_0 = \sqrt{\frac{3P}{d}} = \sqrt{\frac{3 \times 0.76 \times 13.6 \times 10^3 \times 9.8}{1.29}}$$

The root mean square velocity of air at 100°C is given by

$$C_{100} = C_0 \sqrt{\frac{273+100}{273}}$$

$$= \sqrt{\frac{3 \times 0.76 \times 13.6 \times 10^3 \times 9.8 \times 373}{1.29 \times 273}}$$

$$= 567.2 \text{ m/s}$$

10.50. The velocity of sound in air at 0°C is $v = 330$ m/s. The specific heat of air at constant pressure = 0.2382 k cal/kg/K. Find the value of the mechanical equivalent of heat.

Solution: The velocity of sound v is given by

$$v = \sqrt{\frac{\gamma p}{d}}$$

Assuming air to behave like a perfect gas, we have

$$Pv = RT$$

or $\frac{PM}{d} = RT$ where M is the molecular weight and d is the density of air

$$\therefore R = \frac{PM}{Td} \quad \text{or} \quad \frac{R}{M} = \frac{1}{T}\frac{v^2}{\gamma}$$

In view of the law of equipartition of energy, $C_v = (5/2)\,(R/M)$ (assuming air to be composed of diatomic molecules)

$$\therefore C_v = \frac{5}{2}\frac{v^2}{T\gamma} = \frac{2.5 \times 330^2}{273\gamma} \text{ J/kg/K}$$

$$= \frac{2.5 \times 330^2}{273\gamma J} \text{ cal/kg/K}$$

Now $$\frac{C_p}{C_v} = \gamma = \frac{0.2382 \times 10^3}{2.5 \times 330^2} \times 273\,\gamma J$$

or $$J = \frac{2.5 \times 330^2}{0.2382 \times 10^3 \times 273} = 4.185 \text{ J/cal}$$

10.51. The gases constituting air are mainly diatomic. Find the values of C_p and C_v for air assuming it to be a diatomic gas. Density of air at

$NTP = 1.293$ kg/m^3, $J = 4.186$ J/cal, and atmospheric pressure $p = 76$ cm of mercury.

Solution: Assuming air to behave like a perfect gas, we have

$$PV = RT$$

or $$P\frac{M}{d} = RT$$

or $$\frac{P}{dT} = \frac{R}{M}$$

According to the law of equipartition of energy, a diatomic molecule has five effective degrees of freedom

$$\therefore C_v = \frac{5}{2}\frac{R}{M}$$

Now $$C_v = \frac{5}{2}\frac{R}{M} = \frac{5}{2}\frac{P}{dT} = \frac{5}{2}\times\frac{0.76\times 13.6\times 10^3\times 9.8}{1.293\times 273\times 4.186}\text{ cal/kg/K}$$

$$= 0.1713\text{ k cal/kg/K}$$

$$C_p = \frac{5}{2}\frac{R}{M} + \frac{R}{M} = \frac{7}{2}\frac{R}{M} = \frac{7}{5}C_v = \frac{7\times 0.1713}{5}$$

$$= 0.2398\text{ k cal/kg/K}$$

10.52. One g mol of nitrogen at 27°C and at a pressure of 76 cm of mercury is enclosed in a cubical vessel of volume 1 m^3. At this temperature it is assumed that the molecules of the gas are moving with their root mean square velocity. Calculate the rate of collision on a wall of the vessel. $K = 1.38\times 10^{-23}$ J/K, $N = 6.02\times 10^{23}$ molecules/mol.

Solution: We know that the kinetic energy (1/2) mC^2 of a single molecule is equal to (3/2) KT

i.e., $$\frac{1}{2}mC^2 = \frac{3}{2}KT$$

or $MC_{rms}^2 = 3\,KNT$ where, M is molecular weight

or $$C_{rms} = \sqrt{\frac{3KNT}{M}} = \sqrt{\frac{3\times 1.38\times 10^{-23}\times 6.02\times 10^{23}\times 300}{28\times 10^{-3}}}$$

$$= 516.75\text{ m/s.}$$

Rate of change of momentum $= 2m\,C_{rms}n$

$$= \frac{2\times 28\times 10^{-3}}{6.02\times 10^{23}}\times 516.75\times n$$

$$= \text{Force,}$$

where n is number of collisions per second

or $$\frac{2\times 28\times 10^{-3}\times 516.75\times n}{6.02\times 10^{23}} = 0.76\times 13.6\times 10^3\times 9.8$$

or $$n = \frac{0.76\times 13.6\times 10^3\times 9.8\times 6.02\times 10^{23}}{2\times 28\times 10^{-3}\times 516.75} = 2.107\times 10^{27} \text{ per second}$$

10.53. A tyre is pumped to a pressure of 2 atmospheres at 27°C. When it suddenly bursts, calculate the resulting change in temperature, γ for air = 1.4.

Solution: Sudden burst of a tyre is an adiabatic process. In adiabatic changes, the temperature and pressure are related by the equation

$$T^{\gamma} p^{1-\gamma} = \text{Constant}$$

Let T_1 be the initial temperature and $2P$ be the initial pressure. As the tyre bursts, let T_2 be the final temperature and P be the final pressure. Here P will be the atmospheric pressure.

we have, $T_1^{\gamma}\,(2P)^{1-\gamma} = T_2^{\gamma}\,P^{1-\gamma}$

or $$T_2 = T_1 2^{\frac{1-\gamma}{\gamma}}$$

$$= 300\times 2^{-\frac{0.4}{1.4}}$$

$$= \frac{300}{2^{2/7}}$$

$$= 246.1 \text{ K}$$

Change in temperature = $T_1 - T_2$ = 300 – 246.1 K = 53.9°C

10.54. A volume of gas at 27°C expands adiabatically until its volume becomes twice that of the original value. Calculate the final temperature of the gas. Given that $\gamma = 1.4$.

Solution: In an adiabatic process, the temperature and volume are related by the equation

$$TV^{\gamma-1} = \text{constant}$$

Let T_1, V_1 be the initial temperature and volume of the gas and T_2, V_2 be the final temperature and volume of the gas.

Now $$T_1V_1^{\gamma-1} = T_2V_2^{\gamma-1}$$

or $$T_2 = T_1\left(\frac{V_1}{V_2}\right)^{\gamma-1} = \frac{300\times 1}{2^{0.4}}$$

$$= 227.4 \text{ K}$$

10.55. One mole of an ideal gas is subjected to a process in which the pressure varies with volume via relation.

(a) $p = p_0 - aV^2$ (b) $p_0 e^{-bV}$

here p_0, a and b are positive constants.

Calculate the maximum attainable temperature for each of the two processes.

Solution:

$$p = p_0 - aV^2$$

For a perfect gas, we have

$$pV = RT$$

$$\frac{RT}{V} = p_0 - aV^2$$

or $$RT = p_0 V - aV^3$$

$$R\frac{dT}{dV} = p_0 - 3aV^2 = 0$$

$$\therefore V = \sqrt{\frac{p_0}{3a}}$$

$R\frac{d^2T}{dV^2} = -6aV$ hence with $V = \sqrt{\frac{p_0}{3a}}$ we have the maxima conditions.

$$p = p_0 - a \cdot \frac{p_0}{3a} = \frac{2p_0}{3}$$

$$\therefore T = \frac{pV}{R} = \frac{2}{3}\frac{p_0}{R} \cdot \sqrt{\frac{p_0}{3a}}$$

(b) $$p = p_0 e^{-bV}$$

$$\frac{RT}{V} = p_0 e^{-bV}$$

$$R\frac{dT}{dV} = p_0 \left[e^{-bV} + Ve^{-bV} \times -b\right] = 0$$

$$\therefore V = \frac{1}{b}$$

$$\therefore p = p_0 e^{-b \times \frac{1}{b}} = \frac{p_0}{e}$$

or $$T = \frac{pV}{R} = \frac{p_0}{ebR} \quad \frac{\pi\eta\left(r_2^4 - r_1^4\right)\theta^2}{4l}$$

10.56. One mole of an ideal gas is subjected to a process in which the temperature varies with volume via relation.

$T = T_0 + aV^2$; where T_0 and a are positive constants.

Find the maximum pressure that can be attained in this process.

Solution:

$$PV = RT = RT_0 + aRV^2$$

$$\therefore P = \frac{RT_0}{V} + aRV$$

$$\frac{dP}{dV} = -\frac{RT_0}{V^2} + aR = 0$$

$$\therefore V = \sqrt{\frac{T_0}{a}}$$

$$T = T_0 + a \cdot \frac{T_0}{a} = 2T_0$$

$$P = \frac{RT}{V} = R \cdot 2T_0 \sqrt{\frac{a}{T_0}} = 2R\sqrt{aT_0}$$

10.57. A tall cylindrical vessel filled with air is located in the earths gravitational field. The temperature of air varies with height in order to keep its density constant throughout the volume. Calculate the drop in temperature at a height of 1 km above the bottom of the vessel. Molecular mass of air = 0.029 kg.

Solution: Consider a layer of thickness dh at a height h. Let P and $P + dP$ be the air pressure on the two sides of the layer under consideration (see Fig. 10.11).

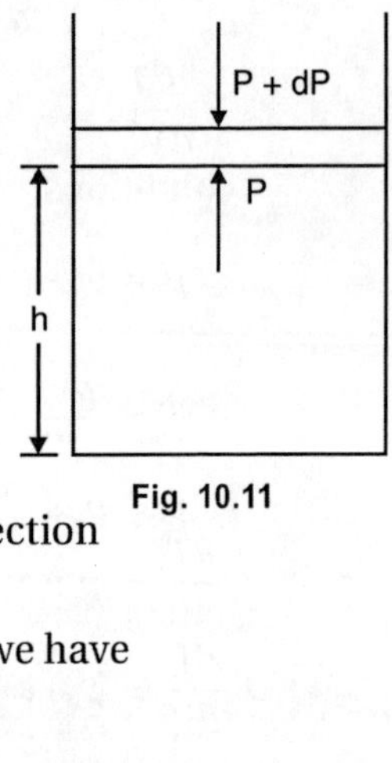

Fig. 10.11

Mass of the layer = $A\, dh\, \rho$; A = area of cross-section

ρ = density

On equating the force in the vertical direction we have

$$(P + dP)A + A dh\, \rho \cdot g = PA$$

or $\quad -dP = dh\, \rho \cdot g$

Treating air to be an ideal gas we have

$$P = \rho \frac{RT}{M}; \; M = \text{molecular mass}$$

$$\frac{dP}{dh} = \frac{\rho R}{M} \cdot \frac{dT}{dh} = -\rho g$$

or $$\frac{dT}{dh} = -\frac{Mg}{R}$$

or $$dT = -\frac{Mg}{R} \times dh$$

On substituting the values we have

$$= -\frac{0.029 \times 9.8 \times 10^3}{8.3}$$

$$= -34.2°C$$

10.58. Calculate the temperature gradient for air assuming that irrespective of height the pressure p and density ρ of air are related as $\frac{p}{\rho^n}$ = constant.

Solution: Let $-dp$ be the decrease in pressure corresponding to an increase in height dh. Then if ρ be the density of air at that point considered we have

$$-dp = dh\,\rho \cdot g$$

Now $$p = C\rho^n$$

$$\frac{dp}{dh} = Cn \cdot \rho^{n-1} \cdot \frac{d\rho}{dh}$$

or $$-Cn\rho^{n-1} \cdot \frac{d\rho}{dh} = g\rho \qquad (1)$$

Assuming air as an ideal gas we have

$$p = \rho \frac{RT}{M}$$

$$C\rho^n = \rho \frac{RT}{M}$$

or $$C\rho^{n-1} = \frac{RT}{M}$$

or $$C(n-1)\rho^{n-2} \cdot \frac{d\rho}{dh} = \frac{R}{M} \cdot \frac{dT}{dh} \qquad (2)$$

On dividing the two equations we have

$$-\frac{n}{n-1} = \frac{Mg}{RdT/dh}$$

or $$\frac{dT}{dh} = -\frac{Mg(n-1)}{nR}$$

10.59. Assuming that the air close to the surface of earth is under normal conditions and its temperature is independent of height. Calculate the air pressure at a height of 10 km above the surface. Atmospheric pressure = 10^5 N/m^2. Density of air = 1.29 kg/m^3.

Solution: Let dp be the decrease in pressure due to an increase of height dh then we have

$-dp = dh \cdot \rho \cdot g$

Assuming air as perfect gas we have

$$p = \rho \frac{RT}{M}$$

$$-dp = dh \cdot \frac{Mg}{RT} \cdot p$$

or $$\frac{dp}{p} = -\frac{Mg}{RT} \cdot dh$$

On integration we have

$$\int_{p_0}^{p} \frac{dp}{p} = -\frac{Mg}{RT} \cdot \int_0^h dh$$

$$\log_e \frac{p}{p_0} = -\frac{Mgh}{RT}$$

$$\log_{10} p - \log_{10} p_0 = -\frac{Mgh}{2.303\, RT} = -\frac{\rho gh}{p_0 \times 2.303}$$

On putting the values we have

$$\log_{10} p - \log_{10} 10^5 = -\frac{1.29 \times 9.8 \times 10^4}{10^5 \times 2.303}$$

$$\log_{10} p - 5 = -0.5489$$

$$\log_{10} p = 4.4511$$

$$p = 2.826 \times 10^4 \text{ N/m}^2$$

$$= 0.2826 \text{ atmosphere}$$

10.60. Calculate the difference in heights at which the air densities at the temperature 0°C differ *e* times by assuming the temperature of air to be independent of the height.

Solution: Let $-dp$ be the change in pressure for an increase of height dh.

$$-dp = \rho\, g\, dh$$

Assuming air to be a perfect gas we have $p = \rho \frac{RT}{M}$

$$\therefore -dp = -\,d\rho \frac{RT}{M} = \rho g dh$$

$$\int \frac{d\rho}{\rho} = \int -\frac{Mg}{RT} \cdot dh$$

$$\log \rho = -\frac{Mg}{RT} \cdot h + c$$

At $h = h_1$ $\rho = e\,\rho_0$, and at $h = h_2$; $\rho = \rho_0$; ρ_0 = density at 0°C

$$\log_e e\,\rho_0 = -\frac{Mgh_1}{RT} + c$$

$$\log_e \rho_0 = -\frac{Mgh_2}{RT} + c$$

$$\log_e e = 1 = \frac{Mg}{RT}(h_2 - h_1)$$

or $$h_2 - h_1 = h = \frac{RT}{Mg}$$

10.61. A tall vertical cylindrical tube of uniform cross-sectional area A and height h is filled with an ideal gas at temperature T. Assuming the gas pressure at the bottom of the base as p_0 and the temperature of the gas to be independent of height find the mass of the gas confined into the vessel.

Solution: Let dp be the decrease in pressure due to an increase in height dh then

$$-dp = \rho g \cdot dh$$

Further $$p = \rho\frac{RT}{M}; M = \text{molecular weight}$$

$$-dp = dh \cdot \frac{Mg}{RT} \cdot p$$

or $$\int_{p_0}^{p} \frac{dp}{p} = -\int_0^h \frac{Mg}{RT} \cdot dh$$

$$p = p_0\, e^{-\frac{Mgh}{RT}}$$

$$\text{Change of pressure} = \Delta p = p_0 - p = p_0\left[1 - e^{-\frac{Mgh}{RT}}\right]$$

$$\text{Force} = \Delta p \cdot A = mg = p_0 A\left[1 - e^{-\frac{Mgh}{RT}}\right]; m = \text{mass of gas enclosed}$$

$$\therefore m = \frac{p_0 A}{g}\left[1 - e^{-\frac{Mgh}{RT}}\right]$$

10.62. An ideal gas filled in a very tall cylindrical tube is placed in the earth's gravitational field. Assuming the temperature of the gas to be independent of height. Calculate the height at which the centre of gravity of the gas is located.

Solution: Consider a layer of thickness dh at a height h from the base.

$$dm = \text{mass of the layer} = \frac{A \cdot dp}{g} = \rho g \cdot dh \cdot \frac{A}{g}; A = \text{surface area}$$

$$\therefore\ dm = A\rho \cdot dh$$

$$Y = \frac{\int h \cdot dm}{\int dm} = \frac{\int A\rho \cdot hdh}{\int A\rho \cdot dh}$$

$$= \frac{\int_0^\infty h\rho dh}{\int_0^\infty \rho dh}$$

Pressure varies as

$$p = p_0 e^{-\frac{Mgh}{RT}} \text{ and } p = \rho \frac{RT}{M}$$

$$Y = \frac{\int_{p_0}^{0} -\frac{RT}{Mg} \cdot \log\left(\frac{p}{p_0}\right) \cdot \frac{Mp}{RT} \cdot -\frac{RT}{Mg} \cdot \frac{dp}{p}}{\int_{p_0}^{0} \frac{Mp}{RT} \cdot -\frac{RT}{Mg} \cdot \frac{dp}{p}}$$

$$= \frac{RT}{Mg} \frac{\int_{p_0}^{0} \log\left(\frac{p}{p_0}\right) dp}{\int_{p_0}^{0} -dp}$$

$$= \frac{RT}{Mg}$$

10.63. An ideal gas filled in a tall vessel is located in the earth's gravitational field. The pressure at the base of the vessel is p_0. Find the pressure of the gas at a height h if its temperature varies with height according to the relation

(i) $T = T_0\,(1 - ah)$, (ii) $T = T_0\,(1 + ah)$

where a is a positive constant.

Solution:

(i) $-dp = \rho g dh$

$$p = \frac{\rho RT}{M} = \frac{\rho RT_0}{M}[1 - ah]$$

$$-dp = \frac{p \cdot g \cdot M \cdot dh}{RT_0(1 - ah)}$$

or $$\int_{p_0}^{p} \frac{dp}{p} = -\frac{Mg}{RT_0} \int_0^h \frac{dh}{1 - ah}$$

$$\log \frac{p}{p_0} = \frac{Mg}{aRT_0} \log(1 - ah)$$

$$= \log(1 - ah)^n;\ n = \frac{Mg}{aRT_0}$$

$$\therefore p = p_0 (1 - ah)^n$$

(ii) $$-dp = \rho g \cdot dh$$

$$p = \frac{\rho RT_0}{M}(1+ah)$$

$$-dp = \frac{p \cdot Mg}{RT_0}\frac{dh}{(1+ah)}$$

$$\int_{p_0}^{p}\frac{dp}{p} = -\frac{Mg}{RT_0}\int_0^h \frac{dh}{1+ah}$$

$$\log\frac{p}{p_0} = -\frac{Mg}{aRT_0}\log(1+ah)$$

$$= \log(1+ah)^{-n};\ n = \frac{Mg}{aRT_0}$$

$$p = \frac{p_0}{(1+ah)^n}$$

10.64. A horizontal cylinder closed at one end is rotated about a vertical axis passing through open end with a constant angular velocity ω. The atmospheric pressure is p_0 and the temperature is T. Find the pressure inside the cylinder at a distance x from the axis of rotation.

Solution: Consider a layer of thickness dx at a distance x from the axis of rotation (see Fig. 10.12).

Centripetal force acting on the layer $= Adx\, \rho \cdot \omega^2 x$;

A = area of cross-section of the tube.

ρ = density of air.

In view of the figure we have

$$A \cdot dp = \rho\, \omega^2\, xdx$$

Assuming air to be an ideal gas $\rho = \dfrac{M \cdot P}{RT}$

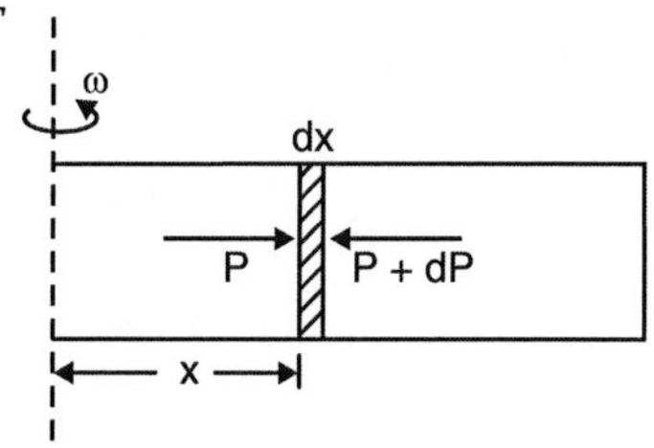

Fig. 10.12

$$dP = \frac{M \cdot P}{RT}\cdot \omega^2 \cdot xdx$$

$$\int_{P_0}^{P}\frac{dP}{P} = \frac{M\omega^2}{RT}\cdot\int_0^x xdx$$

$$\log_e\frac{P}{P_0} = \frac{M\omega^2}{2RT}\cdot x^2$$

or $$P = P_0 e^{-\frac{M\omega^2 x^2}{2RT}}$$

10.65. Using a vacuum pump a low pressure of $4.15 \times 10^{-10}\ N/m^2$ is attained inside a vessel at 27°C. Calculate the number of molecules per c.c. at this pressure, $R = 8.3$ J/mol/K, Avogadro number $N = 6 \times 10^{23}$.

Solution:

$$PV = nKT; K = \frac{R}{N}$$

$$\therefore n = \frac{PV}{KT}$$

Put $V = 1 \text{ m}^3$

$$n = \frac{PN}{R \cdot T}$$

On putting the values we have

$$n = \frac{4.15 \times 10^{-10} \times 6 \times 10^{23}}{8.3 \times 300}$$

$$n' = \text{no. of molecules per c.c.} = \frac{4.15 \times 2}{8.3} \times \frac{10^{11}}{10^6}$$

$$= 10^5$$

10.66. At a temperature of $T = 1600$ K, $m = 2$ gm of nitrogen is filled in a vessel of volume $V = 5 \times 10^{-3}$ m^3. At this temperature $\eta = 25\%$ of molecules are dissociated into atom. Calculate the gas pressure.

Solution: M gm of gas contains N mols of nitrogen

$\therefore$ number of molecules of nitrogen in m gm $= \dfrac{mN}{M}$

molecules dissociated into atoms $= \dfrac{mN\eta}{M}$

No. of nitrogen molecules left $= \dfrac{mN}{M} - \dfrac{mN\eta}{M}$

Now $p_1 V = \dfrac{mN}{M}(1-\eta) RT$

Where p_1 is the pressure due to nitrogen molecule and

$$p_2 V = \frac{2mN}{M} \eta RT$$

where p_2 is the pressure due to nitrogen atom.

$$\text{Net pressure} = p = p_1 + p_2 = \frac{mNRT}{MV}[1+\eta]$$

$$= \frac{mRT}{M'V}[1+\eta]$$

Where $\dfrac{N}{M} = \dfrac{1}{M'}$

On substituting the values we have

$$p = \frac{2\times10^{-3}\times8.3\times1600(1+0.25)}{28\times10^{-3}\times5\times10^{-3}}$$
$$= \frac{2\times8.3\times16\times1.25\times10^{5}}{28\times5}$$
$$= \frac{2\times8.3\times16\times1.25}{28\times5} \text{ atmosphere}$$
$$= 2.37 \text{ atmosphere}$$

10.67. A parallel beam of n molecules/volume of a gas moving with a velocity v strikes elastically on a wall at an angle θ to its normal. What pressure will be exerted by the beam on the wall.

Solution: Let the surface area of the wall = S

Component of area along $AC = S\cos\theta$ (see Fig. 10.13)

Momentum of a molecule = mv

Consider a cylinder of length v and area $S\cos\theta$

Volume of the cylinder = $Sv\cos\theta$

Number of molecules contained into it

$= nSv\cos\theta$

Incident momentum of the molecule contained in this cylinder per sec = $mnSv^2\cos\theta$

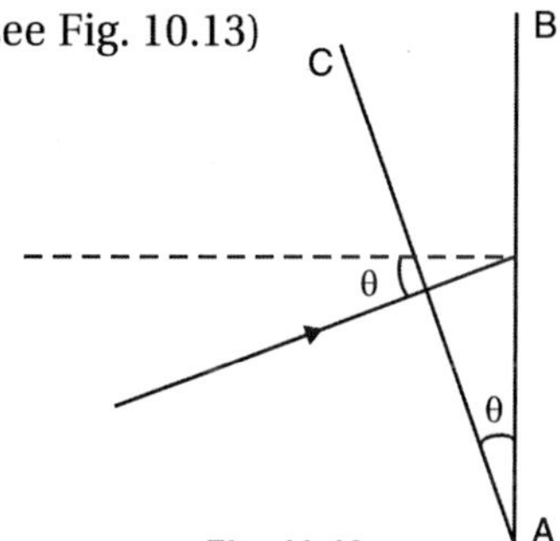

Fig. 10.13

Component of incident momentum/sec perpendicular to AB

$= nmSv^2\cos\theta\cdot\cos\theta$

Momentum after reflection/sec = $mnSv^2\cos^2\theta$

Force = $2\,mnSv^2\cos^2\theta$

Pressure = $2\,mnv^2\cos^2\theta$

10.68. Air is a mixture of diatomic gases. The velocity of sound in air is 330 m/sec. Assuming the density of air at N.T.P. equal to 1.29×10^{-3} gm/cc. Calculate the degree of freedom. Atmospheric pressure = 10^5 N/m^2.

Solution:

Velocity of sound $v = \sqrt{\frac{\gamma P}{d}}$

or $$\frac{d\cdot v^2}{P} = \gamma = \frac{C_P}{C_V} = \frac{\frac{x}{2}R+R}{\frac{x}{2}R} = 1+\frac{1}{x/2}$$; x is the degree of freedom.

$$x = \frac{2}{\left(\frac{dv^2}{P} - 1\right)}$$

On substituting the values we have

$$x = \frac{2}{\left(\frac{1.29 \times 10^{-3} \times 10^{-3} \times 330^2}{10^5 \times 10^{-6}} - 1\right)}$$

$$= \frac{2}{\left(\frac{1.29 \times 330^2}{10^5} - 1\right)}$$

$$= \frac{2}{1.4 - 1} = 5$$

10.69. Determine ratio of velocity of sound v in a mono & diatomic gas to the root mean square velocity of the molecules in the two gases.

Solution:

$$\text{Velocity of sound } v = \sqrt{\frac{\gamma P}{d}}$$

$$v_{rms} = \sqrt{\frac{3P}{d}}$$

$$\therefore \frac{v}{v_{rms}} = \sqrt{\frac{\gamma}{3}} = \sqrt{\frac{5}{9}} \text{ for mono atomic gas}$$

$$\text{and } \sqrt{\frac{7}{15}} \text{ for diatomic gas}$$

10.70. Nitrogen gas is filled in a thermally insulated vessel at 27°C. The vessel moving with a velocity of 150 m/sec is suddenly stopped. Calculate the percentage change in the gas pressure. Mol wt of Nitrogen = 28 gm.

Solution:

$$PV = RT$$

$$(P + dP)\,V = R\,(T + dT)$$

$$\frac{P + dP}{P} = \frac{T + dT}{T}$$

$$\text{or} \quad \frac{dP}{P} = \frac{dT}{T} = \frac{MC_V dT}{MC_V T} = \frac{\frac{1}{2}Mv^2}{\frac{5}{2}RT} = \frac{Mv^2}{5RT}$$

On substituting the values we have

$$\% \text{ change } = \frac{dP}{P} \times 100 = \frac{28 \times 10^{-3} \times 150^2}{5 \times 8.3 \times 300} \times 100$$

$$= 5.06\%$$

10.71. 14 gm of nitrogen is enclosed in a vessel at a temperature of 27°C. Calculate the amount of heat to be transferred to the gas in order to double its root mean square velocity. Mol wt of Nitrogen = 28 gm.

Solution:

$$C_{300} = C_0 \sqrt{T}$$

$$2C_{300} = C_0 \sqrt{T_1}$$

$$4 = \frac{T_1}{T}$$

$$3 = \frac{T_1 - T}{T} = \frac{\Delta T}{T}$$

Now $Q = \frac{5}{2} R \cdot \frac{m}{M} \cdot \Delta T$

$$= \frac{5}{2} R \cdot \frac{m}{M} \times 3T$$

On substituting the values we have

$$Q = \frac{5 \times 8.3 \times 14 \times 10^{-3} \times 3 \times 300}{2 \times 28 \times 10^{-3}}$$

$$= 9337.5 \text{ J}$$

10.72. Two different adiabatic paths for the same gas intersect two isothermals at T_1 and T_2 as shown in the $P - V$ diagram in Fig. 10.14. How does the ratio $\frac{V_a}{V_d}$ compare with the ratio $\frac{V_b}{V_c}$? (Roorkee 1983)

Solution: Let V_a, V_b, V_c and V_d be the volume of the gas which is under operation at point *a*, *b*, *c* and *d*, respectively.

Points *b* and *c* lies on the adiabatic curve.

Therefore, $T_1 V_b^{\gamma-1} = T_2 V_c^{\gamma-1}$

Similarly, points *a* and *d* lie on another adiabatic curve for which we have a similar relation given by

$$T_1 V_a^{\gamma-1} = T_2 V_d^{\gamma-1}$$

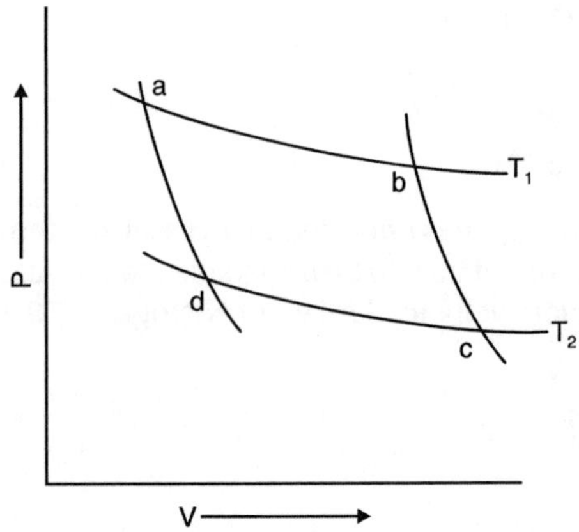

Fig. 10.14

On dividing the two relations, we have

$$\left(\frac{V_c}{V_d}\right)^{\gamma-1} = \left(\frac{V_b}{V_a}\right)^{\gamma-1}$$

or $$\frac{V_c}{V_d} = \frac{V_b}{V_a}$$

or $$\frac{V_a}{V_d} = \frac{V_b}{V_c}$$

10.73. One mol of a diatomic gas is at N.T.P. The gas is compressed adiabatically to $\frac{1}{\eta}$ of its original volume. Find the mean rotational kinetic energy of the molecule in the final state.

Solution: In an adiabatic process we have

$$T_0 V_1^{\gamma-1} = T_2 V_2^{\gamma-1} = T_2 \left(\frac{V_1}{\eta}\right)^{\gamma-1}$$

$$\therefore T_2 = T_0 \eta^{\gamma-1} = T_0 \eta^{2/5} \text{ since } \gamma = \frac{7}{5}$$

Mean rotational K.E. $= 2 \times \frac{1}{2} KT_2$; K = Bottzman constant

$$= K \cdot T_0 \eta^{2/5}$$

10.74. How will the rate of collisions of a diatomic molecule against the vessel's wall will change if the gas is expanded adiabatically η times.

Solution: If n be rate of collision, T be the temperature and P be the pressure then

$$n\sqrt{T} = P$$

Let n_1 be the initial rate of collision at temperature T_1 and pressure p_1, and n_2 be final rate of collision at temperature T_2 and pressure p_2.

We have $$\frac{n_1\sqrt{T_1}}{n_2\sqrt{T_2}} = \frac{p_1}{p_2}$$

Further $$\frac{p_1V_1}{T_1} = \frac{p_2V_2}{T_2} = \frac{p_2\eta V_1}{T_2}$$

$$\therefore \frac{p_1}{p_2} = \eta \cdot \frac{T_1}{T_2}$$

For an adiabatic process we have

$$T_1V_1^{\gamma-1} = T_2V_1^{\gamma-1}\cdot\eta^{\gamma-1}$$

$$\therefore \frac{T_1}{T_2} = \eta^{\gamma-1}$$

On putting these values we have

$$\frac{n_1}{n_2} = \frac{p_1}{p_2}\cdot\sqrt{\frac{T_2}{T_1}}$$

$$= \eta\cdot\left(\frac{T_1}{T_2}\right)^{1/2}$$

$$= \eta\cdot\eta^{\frac{\gamma-1}{2}}$$

$$= \eta^{\frac{\gamma+1}{2}}$$

$$= \eta^{\frac{7/5+1}{2}} = \eta^{6/5}$$

or $$n_1 = n_2\eta^{\frac{1+i}{i}}; \; i = \text{no. of degrees of freedoms.}$$

10.75. Following a polytropic process the volume of a diatomic gas is increased η times. The molar heat capacity is equal to R. Calculate the final rate of collision of the molecules against the walls of the vessel.

Solution: For a polytropic process we have

$$pV^n = \text{Constt.}$$

$$\frac{RT}{V}\cdot V^n = \text{Constt.}$$

or $$TV^{n-1} = \text{Constt.}$$

On differentiation we have

$$dT\cdot V^{n-1} + (n-1)\,V^{n-2}\cdot dV\cdot T = 0$$

or $$dT = -(n-1)T\cdot\frac{dV}{V}$$

Further $$\delta Q = C_V dT + PdV$$

$$= \frac{R}{\gamma-1}dT + \frac{RT}{V}\cdot dV$$

$$= \frac{R}{\gamma-1}dT - \frac{R}{n-1}\cdot dT$$

Molar heat capacity $C = R = \dfrac{R}{\gamma-1} - \dfrac{R}{n-1}$

$$\therefore \frac{1}{n-1} = \frac{1}{\gamma-1} - 1 = \frac{3}{2};\ \gamma = \frac{7}{5}$$

$$2 = 3n - 3$$

$$n = \frac{5}{3}$$

Hence the process is represented by the relation

$$pV^{5/3} = \text{Constt.}$$

Further $$p_1V_1^{5/3} = p_2\eta^{5/3}\cdot V_1^{5/3}$$

or $$\frac{p_1}{p_2} = \eta^{5/3}$$

$$\frac{p_1V_1}{T_1} = \frac{p_2\eta V_1}{T_2}$$

$$\therefore \frac{p_1}{p_2} = \eta\cdot\frac{T_1}{T_2}$$

Let n_1 and n_2 be the initial and final rate of collision then

$$\frac{n_1\sqrt{T_1}}{n_2\sqrt{T_2}} = \frac{p_1}{p_2}$$

or $$\frac{n_1}{n_2} = \frac{p_1}{p_2}\cdot\sqrt{\frac{T_2}{T_1}} = \eta\cdot\left(\frac{T_1}{T_2}\right)^{1/2}$$

$$= \frac{\eta}{\sqrt{\eta}}\cdot\eta^{5/6} = \eta^{4/3}$$

$$= \eta^{\frac{i-1}{i-2}}; \; i = \text{no. of degree of freedom}$$

$$\therefore n_2 = \frac{n_1}{\eta^{\frac{i-1}{i-2}}}$$

10.76. A diatomic molecules expands in a polytropic process so that the initial and final rate of collision assume the same value. Find the molar heat capacity of the gas.

Solution:

Molar heat capacity $C = \dfrac{R}{\gamma - 1} - \dfrac{R}{n-1}$ (see Question 10.75)

For a polytropic process

$pV^n = \text{Constt.}$

Now $$\frac{n_1\sqrt{T_1}}{n_2\sqrt{T_2}} = \frac{p_1}{p_2}$$

$$n_1 = n_2 \text{ (given)}$$

$$\therefore \frac{p_1}{p_2} = \sqrt{\frac{T_1}{T_2}}$$

$$\frac{p_1 V_1}{T_1} = \frac{p_2 \eta V_1}{T_2} \quad \text{final volume} = \eta V_1$$

$$\frac{p_1}{p_2} = \eta \cdot \frac{T_1}{T_2}$$

$$\sqrt{\frac{T_1}{T_2}} = \eta\left(\frac{T_1}{T_2}\right)$$

$$\therefore \eta = \sqrt{\frac{T_2}{T_1}}$$

$$pV^n = \frac{RT}{V} \cdot V^n = \text{Constt.}$$

or $$TV^{n-1} = \text{Constt.}$$

$$T_1 V_1^{n-1} = T_2 \eta^{n-1} \cdot V_1^{n-1}$$

or $$\frac{T_1}{T_2} = \eta^{n-1}$$

or $$1 = \frac{T_2}{T_1} \cdot \eta^{n-1} = \eta^2 \cdot \eta^{n-1} = \eta^{n+1}$$

This is possible when $n = -1$

$$C = \frac{R}{\frac{7}{5}-1} + \frac{R}{2} = \frac{5}{2}R + \frac{R}{2}$$

$$\frac{6R}{2} = \frac{R}{2}(i+1)$$

10.77. The ratio of the molecular concentration of hydrogen and nitrogen at the surface of earth and at a height h is η_0 and η respectively. Assuming the temperature to be independent of height calculate the ratio of $\frac{\eta}{\eta_0}$ at the temperature T°K.

Solution: Let $-dp$ be the decrease in pressure due to increase in height dh then

$$-dp = \rho \cdot g \cdot dh$$

For a perfect gas

$$p = \rho \frac{RT}{M}$$

$$\frac{dp}{P} = -\frac{Mg}{RT} \cdot dh = \frac{dn}{n}; \text{ } n \text{ is molecular concentration}$$

$$\int_{n_1}^{n_2} \frac{dn}{n} = -\frac{M_1 g}{RT} \int_0^h dh; \text{ } M_1 = \text{molecular weight of hydrogen}$$

or $$\log_e \frac{n_2}{n_1} = -\frac{M_1 g}{RT} \cdot h$$

Similarly for nitrogen we have similar equation

$$\log_e \frac{n_2'}{n_1'} = -\frac{M_2 gh}{RT}; \text{ } M_2 = \text{molecular weight of nitrogen}$$

$$\log_e \frac{n_2}{n_1} - \log_e \frac{n_2'}{n_1'} = \frac{(M_2 - M_1)gh}{RT}$$

$$\log_e \frac{n_2}{n_1} \cdot \frac{n_1'}{n_2'} = \frac{(M_2 - M_1)gh}{RT}; \quad \frac{n_1}{n_1'} = \eta_0, \ \frac{n_2}{n_2'} = \eta$$

$$\log_e \frac{\eta}{\eta_0} = \frac{(M_2 - M_1)gh}{RT}$$

$$\therefore \frac{\eta}{\eta_0} = e^{\frac{(M_2 - M_1)gh}{RT}}$$

10.78. Two different gases of molecular masses m_1 & m_2 with $m_2 > m_1$ is filled in a tall cylindrical vessel. The concentrations of the gas at the bottom of the vessel are n_1 & n_2 respectively. The temperature of gas does not vary with height. Calculate the height at which the two gases will have the same molecular concentration.

Solution: Let $-dp$ be the decrease in pressure with an increase in height by dh. Then

$$-dp = \rho g \, dh$$

According to Boyle's Law we have

$$\frac{p}{p_0} = \frac{\rho}{\rho_0}$$

or $$dp = -p\frac{\rho_0}{p_0} \cdot g \cdot dh$$

or $$\frac{dp}{p} = -\frac{\rho_0}{p_0} \cdot g \cdot dh$$

Now $$\frac{dp}{p} = \frac{dn}{n}; \; n \text{ is molecular concentration}$$

$$\therefore \frac{dn}{n} = -\frac{\rho_0}{p_0} \cdot g \cdot dh$$

For one gas we have

$$\int_{n_2}^{x} \frac{dn}{n} = -\frac{m_2 n_2}{p_0 V} g \int_0^h dh; \quad \rho_0 = \frac{m_2 n_2}{V}$$

$$\log_e x - \log_e n_2 = -\frac{m_2 n_2 gh}{p_0 V} = -\frac{m_2 gh}{KT}; \; p_0 V = n_2 KT$$

Similarly for other gas

$$\int_{n_1}^{x} \frac{dn}{n} = -\frac{m_1 n_1}{p_0 V} g \cdot \int_0^h dh$$

$$\log_e x - \log_e n_1 = -\frac{m_1 gh}{KT}; \; p_0 V = n_1 KT$$

$$\log_e \frac{n_2}{n_1} = \frac{m_2 - m_1}{KT} \cdot gh$$

or $$h = \frac{KT \log_e \frac{n_2}{n_1}}{(m_2 - m_1) g}$$

10.79. A horizontal tube of length l closed at both ends contains a heavy gas of molecular weight M at temperature T. The tube rotates about a vertical axis passing through one of its ends. At what value of the angular velocity the ratio of the molecular concentration at the two ends is equal to η?

Solution: Consider an element of thickness dx at a distance x (see Fig. 10.15).

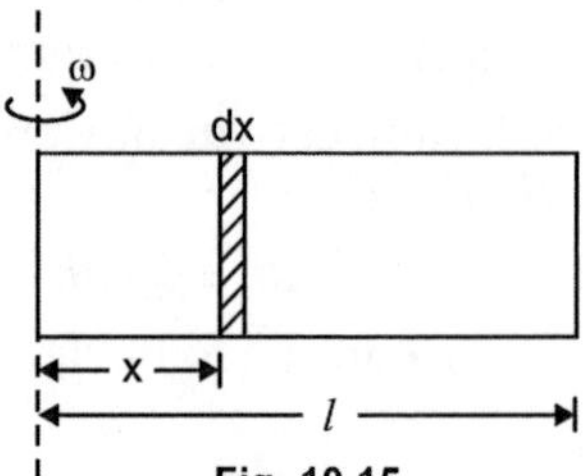
Fig. 10.15

Centripetal force $= x \cdot m\,\omega^2 \dfrac{n}{V} A \cdot dx$;

m = mass of a single molecule

A = area of cross-section of the tube

V = volume of the gas

n = number of molecules

$$\text{Pressure} = \frac{M \cdot \omega^2}{V} x \cdot dx$$

The molecular concentration corresponding to the above pressure is given by

$$-dn = \frac{M\omega^2\, x\, dx \cdot n}{P \cdot V}; P = \text{ pressure of the gas at which the concentration is } n.$$

$$-\int_{c\eta}^{c} \frac{dn}{n} = \frac{M\omega^2}{RT}\int_0^l x dx; PV = RT$$

$$\log_e n = \frac{M\omega^2 l^2}{2RT}$$

$$\therefore \omega = \sqrt{\frac{2RT \log_e n}{Ml^2}}$$

10.80. A thermally insulated vessel containing a gas of molecular weight M moves with a velocity v. Calculate the rise in temperature due to the sudden stoppage of the vessel.

Solution:

$$\frac{1}{2}Mv^2 = C_v \cdot dT$$

$$= \frac{R}{\gamma - 1} \cdot dT; \gamma = \frac{C_P}{C_V}$$

$$\therefore dT = \frac{1}{2}\frac{Mv^2}{R}(\gamma - 1)$$

10.81. Two vessels a & b filled with air are connected to each other through a valve. The thermodynamical variables of air in the two vessels are P_1, V_1, T_1 and P_2, V_2, T_2 respectively. The vessels are thermally insulated. Calculate the air pressure and temperature after the opening of the valve.

Solution: The number of molecules in the two vessels before the opening of the valve is given by

$$n_1 = \frac{P_1V_1}{RT_1} \text{ and } n_2 = \frac{P_2V_2}{RT_2}$$

The number of molecules in the two vessels after the valve is open is given by

$$n' = \frac{P_f V_1}{RT_f} \text{ and } n_2' = \frac{P_f V_2}{RT_f}; P_f = \text{final pressure}$$

$$T_f = \text{final temperature}$$

$$n_1 + n_2 = n_1' + n_2' = \frac{P_f (V_1 + V_2)}{RT_f}$$

$$\frac{P_1 V_1 T_2 + P_2 V_2 T_1}{T_1 \cdot T_2} = \frac{P_f (V_1 + V_2)}{RT_f}$$

Initial internal energy = Final internal energy

$$C_V \cdot \frac{P_1 V_1}{R} + \frac{C_V \cdot P_2 V_2}{R} = \frac{C_V \cdot P_f (V_1 + V_2)}{R}$$

$$\therefore P_f = \frac{P_1 V_1 + P_2 V_2}{V_1 + V_2}$$

$$T_f = \frac{(P_1 V_1 + P_2 V_2) T_1 T_2}{(P_1 V_1 T_2 + P_2 V_2 T_1)}$$

10.82. A gaseous mixture consists of n_1 mols of a monoatomic gas and n_2 mols of a diatomic gas. Find the value of $\gamma = C_P/C_V$ for the mixture.

Solution: Let C_P' and C_V' be the specific heats for monoatomic gas and C_P'' and C_V'' be the specific heats for diatomic gas.

C_P and C_V for the mixture.

$$C_P = \frac{n_1 C_P' + n_2 C_P''}{n_1 + n_2}; \; C_V = \frac{n_1 C_V' + n_2 C_V''}{n_1 + n_2}$$

$$\frac{C_P}{C_V} = \frac{n_1 C_P' + n_2 C_P''}{n_1 C_V' + n_2 C_V''}$$

$$= \frac{\dfrac{n_1 \gamma_1}{C_V''} + \dfrac{n_2 \gamma_2}{C_V'}}{\dfrac{n_1}{C_V''} + \dfrac{n_2}{C_V'}}; \; \gamma_1 = \frac{C_P'}{C_V'} \text{ and } \gamma_2 = \frac{C_P''}{C_V''}$$

$$(\gamma_1 - 1) = \frac{R}{C_V'} \text{ and } \gamma_2 - 1 = \frac{R}{C_V''}$$

$$\therefore \frac{C_P}{C_V} = \frac{n_1 \gamma_1 (\gamma_2 - 1) + n_2 \gamma_2 (\gamma_1 - 1)}{n_1 (\gamma_2 - 1) + n_2 (\gamma_1 - 1)}$$

10.83. One mol of a diatomic gas at a temperature T_0 is adiabatically compressed to increase its pressure η times. Find the final temperature of the gas and the amount of work done on the gas.

Solution: Let initial temperature and pressure be T_0 and P_0 and the final temperature and pressure be T and ηP_0, then

$$T^{\gamma} P_0^{\gamma-1} \cdot \eta^{1-\gamma} = T_0^{\gamma} \cdot P_0^{1-\gamma}$$

$$\therefore T = T_0 \eta^{\frac{\gamma-1}{\gamma}}$$

$$\text{Work done} = \frac{R}{\gamma-1}\left[T_o \eta^{\frac{\gamma-1}{\gamma}} - T_0\right]$$

$$= \frac{RT_0}{\gamma-1}\left[\eta^{\frac{\gamma-1}{\gamma}} - 1\right]$$

10.84. One mol of a diatomic gas is enclosed in a cylinder piston system. The initial pressure, volume and temperature of the gas is P_1, V_1 & T_1. The gas is compressed first adiabatically and then isothermally to acquire a volume V_1/η. Find the ratio of the works in the two compressions.

Solution: In adiabatic process we have $TV^{\gamma-1}$ = Constt.

The initial and the final temperature are related by the equation

$$T_1 V_1^{\gamma-1} = T_2\left(\frac{V_1}{\eta}\right)^{\gamma-1}$$

or $$T_2 = \eta^{\gamma-1} \cdot T_1$$

Work done on the gas during an adiabatic process $= \dfrac{-R}{\gamma-1}(T_1 - T_2)$

$$= \frac{RT_1}{\gamma-1}\left[\eta^{\gamma-1} - 1\right]$$

Work done on the gas during isothermal process $= -RT_1 \log_e \dfrac{V_1}{\eta V_1}$

$$= RT_1 \log_e \eta$$

$$\text{Ratio} = \frac{\eta^{\gamma-1} - 1}{(\gamma-1)\log_e \eta}$$

10.85. One mole of a gas is sealed in a vessel of volume V. The pressure of the gas is P and temperature is T. The gas is cooled and its temperature falls by ΔT. Find (i) the change in the internal energy (ii) the amount of heat lost by the gas.

Solution:

$$\frac{PV}{T} = R$$

According to first law of thermodynamics we have

$$\delta Q = dU + \delta W$$

Since volume is constant $\delta W = 0$

$$\therefore \delta Q = dU = C_V \cdot \Delta T = \frac{R}{\gamma - 1} \Delta T$$

$$dU = -\frac{PV}{T} \cdot \frac{\Delta T}{\gamma - 1}$$

$$\delta Q = -dU$$

10.86. One mol of a gas is heated at a constant pressure. Calculate the amount of heat added to the gas in order to obtain a work A done by the gas.

Solution:

$$\begin{aligned}
\delta Q &= dU + \delta W \\
&= C_V dT + RdT \\
&= (C_V + R)\, dT \\
&= RdT\left(1 + \frac{C_V}{C_P - C_V}\right) \\
&= RdT\left(1 + \frac{1}{\gamma - 1}\right) \\
&= \frac{\gamma RdT}{\gamma - 1} = \frac{\gamma A}{\gamma - 1}
\end{aligned}$$

10.87. To one mol of an ideal gas a quantity of heat Q is added at a constant pressure. As a consequence of this its temperature is raised by ΔT. Find (i) the increase in internal energy (ii) the work done by the gas (iii) the value of γ.

Solution:

$$PV = RT$$

$$\text{Work done} = PdV = RdT$$

$$dU = Q - RdT$$

$$\gamma = \frac{C_P dT}{C_V dT} = \frac{Q}{Q - RdT}$$

10.88. Two mols of an ideal gas having pressure P_0, volume V_0 and temperature T_0 were cooled at constant volume. The pressure reduced to P_0/η. The gas is now heated at a constant pressure. In this process its temperature got back to its initial value. Find the amount of heat absorbed by the gas.

Solution:

$$\frac{P_0V_0}{T_0}=\frac{P_0}{\eta}\cdot\frac{V_0}{T}$$

$$\therefore T=\frac{T_0}{\eta}$$

Since the system attains the same temperature $dU=0$

$$\therefore Q=\delta W=2RdT=2R\left[T_0-\frac{T_0}{\eta}\right]$$

$$=2RT_0\left[1-\frac{1}{\eta}\right]$$

10.89. A horizontally placed cylinder closed at both the ends contains a piston which can slide freely along its length. Initially the piston lies in the midway and thus separates the inside space into two equal parts each of volume V_0. In each part one mol of an ideal gas is enclosed at a pressure P_0 and temperature T_0. Now the piston moves towards right through an isothermal process. Calculate the amount of work done in order to increase the volume of section 1 to η times the volume of section 2.

Solution: Let the volume of section 2 be V' then

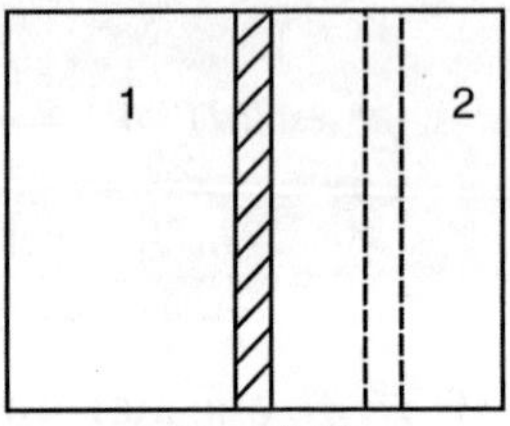

Fig. 10.16

$$\eta V'+V' = 2V_0 \text{ (see Fig. 10.16)}$$

or $$V' = \frac{2V_0}{(\eta+1)}$$

$$P_1V_1 = P_2V_2=RT_0=P_0V_0$$

$$\text{Work done} = P_1dV_1-P_2dV_2$$

$$= P_0V_0\int_{V_0}^{\eta 2V_0/\eta+1}\frac{dV_1}{V_1}-P_0V_0\int_{2V_0/\eta+1}^{V_0}\frac{dV_2}{V_2}$$

$$= P_0V_0\left[\log\frac{2\eta}{\eta+1}-\log\frac{(\eta+1)}{2}\right]$$

$$= -P_0V_0\log\frac{(\eta+1)^2}{4\eta}$$

10.90. A gas confined in a vessel of volume V flows out into a vacuum through an opening whose diameter is much smaller than the free path of the molecules. The area of the opening is S. The process is isothermal and is at a temperature T. Taking the molecular mass of the gas as M find the time τ in which the pressure of the gas in the vessel diminishes η times.

Solution: Let n be the molecular density of the gas. The number of molecules having velocity lying between c and $c+dc$ and passing through the hole in a direction lying within a solid $d\omega$ and making an angle θ with the normal to the hole in time dt is given by

$$S \cdot 4\pi n a^3 e^{-bc^2} c^2 dc \cdot \frac{d\omega}{4\pi} \cos\theta\, c \cdot dt$$

where $a = \sqrt{\frac{b}{\pi}}$ and $b = \frac{m}{2KT}$; m = mass of a single molecule.

For the entire beam of gas lying between θ and $\theta + d\theta$,

$d\omega = 2\pi \sin\theta \cdot d\theta$

Total number of molecules escaping in time dt is given by

$$S \cdot 4\pi n a^3 \int_0^\infty c^3 e^{-bc^2} dc \int_0^{\pi/2} \frac{1}{2} \sin\theta \cdot \cos\theta\, d\theta \cdot dt = S \cdot \frac{n}{4} \bar{c} \cdot dt;$$

$\bar{c}$ = average speed

Mass of the gas escaping out in time dt

$$\Delta m = \frac{S}{4} nm\bar{c}\, dt = \frac{S}{4} \cdot \rho\bar{c}\, dt$$

Now $$\int_{P_0}^{P_0/\eta} \frac{dp}{p} = \frac{d\rho}{\rho} = \frac{\Delta m}{V\rho} = \frac{S}{4V} \bar{c} \cdot dt = \frac{-S}{4V} \left(\frac{8RT}{\pi M}\right)^{1/2} \int_0^\tau dt$$

$$\log \eta = \frac{S}{4V} \sqrt{\frac{8RT}{\pi M}} \cdot \tau$$

$$\tau = \frac{4V}{S} \log \eta \cdot \sqrt{\frac{\pi M}{8RT}}$$

10.91. A rarefied gas is contained in a vessel at a pressure p. Determine the velocity of outflow v of the gas into a vacuum through a small aperture S_0 if the molecules of the gas are governed by Maxwell's law of distribution of velocities.

Solution: In the light of Question 10.90, the number of molecules going out in time dt is given by

$$-dN = S_0 4\pi n a^3 \int_0^\infty c^3 e^{-bc^2} dc \int_0^{\pi/2} \frac{1}{2} \sin\theta \cos\theta \cdot d\theta \cdot dt$$

$$v = -\frac{dN}{dt} = \frac{S_0 n\bar{c}}{4}; \quad \bar{c} = \text{mean molecular velocity}$$

$$= \frac{S_0}{4} \frac{N'}{V} \cdot \bar{c}; \quad N' = \text{total number of molecules}$$

$$= \frac{pS_0 N'\bar{c}}{4N'KT}$$

$$= \frac{pS_0 \bar{c}}{4KT}$$

10.92. A small round aperture with a cross-section S has been made in a vessel containing perfect gas. Find the number of particles that fall on a round disc of radius R situated at a distance h from the aperture. The plane of the disc is parallel to the plane of the aperture. The centers of S and the disc lie on a straight line perpendicular to the plane of the aperture. The molecules of gas are governed by the Maxwell's law of distribution of velocities.

Solution: Let n be the molecular density of the gas. Number of molecules having velocity lying between c and $c + dc$ and coming out of the hole and striking the disc in a direction lying within the solid angle $d\omega$ and making an angle θ with the normal to disc per sec is given by

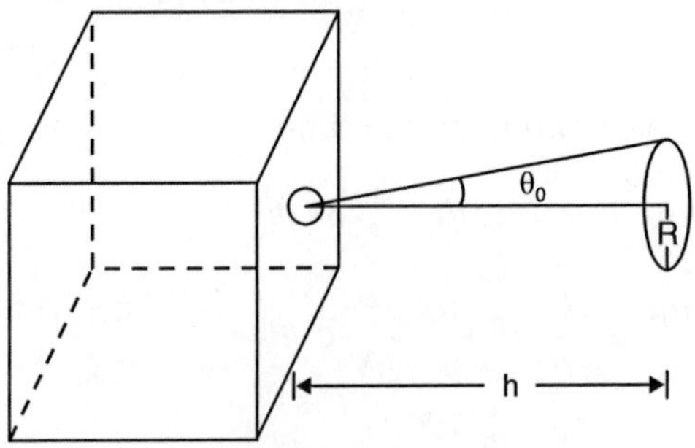

Fig. 10.17

$$S \cdot 4\pi n a^3 e^{-bc^2} c^2 dc \cdot \frac{d\omega}{4\pi} \cdot c \cos\theta$$

For the beam lying between θ and $\theta + d\theta$; $d\omega = 2\pi \sin\theta \cdot d\theta$

Total no. of particles coming out

$$= \int_0^\infty S \cdot 4\pi n a^3 e^{-bc^2} c^3 dc \cdot \int_0^{\theta_0} \frac{1}{2} \sin\theta \cdot \cos\theta \cdot d\theta$$

$$= \frac{Sn}{2} \cdot \sqrt{\frac{2KT}{\pi m}} \cdot \sin^2\theta_0$$

$$= \frac{S \cdot N}{2V} \cdot \sqrt{\frac{2KT}{\pi m}} \cdot \frac{R^2}{R^2 + h^2};$$ $V =$ volume of the gas and

$N =$ total number of particles contained.

10.93. Consider a close vessel containing some gas at a certain pressure and temperature. The gas is to be communicated to another vessel through a long and narrow pipe. The second vessel contains the same gas at the same temperature but at a lower pressure. Calculate the mass of the gas flowing per sec.

Solution: Let p_1 and p_2 be the pressures at the two ends of the tube. Let R and l respectively denote the radius and length of the tube. If the pressure in the second vessel is not too low then the mean free path will be small in comparison to the dimensions of the orifice. The flow of the gas will be governed by viscosity.

$$\text{Mass of the gas flowing per sec} = m = \frac{\pi R^2 \rho}{8\eta l}(p_1 - p_2)$$

where ρ = mean density of the gas and η = coefficient of viscosity of the gas

10.94. Five hundred calories of heat is given to a one mol of a diatomic gas at constant pressure. Determine the amount of workdone.

Solution:

$$\delta Q = dU + \delta W$$

$$C_P dT = C_V dT + RdT$$

For a diatomic gas $C_P = \frac{7}{2}R$

$$\frac{7}{2}RdT = 500 \times 4.2$$

$$\therefore RdT = \text{amount of work done} = \frac{100 \times 4.2 \times 2}{7}$$

$$= 600 \text{ Joule.}$$

10.95. Work equal to 16 kgf.m was obtained during the expansion of a diatomic gas at constant pressure. Calculate the amount of heat added to the gas.

Solution:

Work done = $RdT = 16 \times 9.8$ Joule

$$\text{Amount of heat added} = \frac{7}{2}RdT = \frac{7}{2} \times 16 \times 9.8$$

$$= 7 \times 8 \times 9.8$$

$$= 548.8 \text{ Joule}$$

10.96. One litre of an ideal gas in standard conditions expands isothermally to a volume of 2 litres by absorbing heat from a source. Calculate the amount of work obtained during expansion.

Solution: Amount of heat added is equal to work. Since the change in internal energy is zero.

$$\therefore \delta Q = \delta W = \frac{1000}{22400} \times 2.303\, RT \log_{10} \frac{V_2}{V_1}$$

$$= \frac{1}{22.4} \times 2.303 \times 8.3 \times 273 \log_{10} 2$$

$$= \frac{2.303 \times 8.3 \times 273 \times 0.3010}{22.4} = 70.1 \text{ Joule.}$$

10.97. Three mols of a diatomic gas at 0°C were expanded isothermally to $n = 4$ times of its initial volume. In order to acquire the same initial pressure the gas is heated at constant pressure. The total heat supplied to the gas is $Q = 60.4$ kJ. Find the value of γ.

Solution: During isothermal process amount of work is given by

$$3RT_0 \log_e n = 3R \cdot 273 \cdot \log_e 4$$

$$P_1V_1 = P_2V_2 = nV_1P_2$$

$$\therefore P_2 = \frac{P_1}{n}$$

$$\frac{P_2}{T_0} = \frac{P_1}{T}$$

$$\therefore T = \frac{P_1}{P_2} \cdot T_0 = nT_0$$

$$dU = 3C_V(n\,T_0 - T_0)$$

$$= 3\,(n-1)\,C_V \cdot T_0$$

$$Q = 3RT_0 \log_e n + 3\,(n-1)\,C_V \cdot T_0$$

$$\frac{Q}{3RT_0} - \log_e n = \frac{(n-1)C_V}{R}$$

$$\frac{R}{C_V} = \frac{(n-1)}{\dfrac{Q}{3RT_0} - \log_e n}$$

$$\frac{C_P - C_V}{C_V} = \frac{(n-1)}{\dfrac{Q}{3RT_0} - \log_e n}$$

$$\gamma - 1 = \frac{(n-1)}{\dfrac{Q}{3RT_0} - \log_e n}$$

$$\gamma = 1 + \frac{(n-1)}{\dfrac{Q}{3RT_0} - \log_e n}$$

On substituting the values we have

$$\gamma = 1 + \frac{(4-1)}{\dfrac{60.4 \times 10^3}{3 \times 8.3 \times 273} - 2.303 \log_{10} 4}$$

$$= 1 + \frac{3}{8.8853 - 1.3866} = 1 + \frac{3}{7.4987} = 1.4$$

10.98. Find the amount of heat required to raise the temperature of one mole of an ideal gas by ΔT in view of the fact that the volume of the gas varies with temperature via relation $V = \alpha / T$.

Solution:

$$Q = C_V \Delta T + PdV$$

$$V = \frac{\alpha}{T}$$

$$dV = -\frac{\alpha}{T^2} \cdot \Delta T$$

$$Q = C_V \Delta T - \frac{P\alpha}{T^2} \cdot \Delta T$$

For an ideal gas $PV = RT$

$$\frac{P\alpha}{T} = RT$$

or $$\frac{P\alpha}{T^2} = R$$

$$\therefore Q = C_V \Delta T - R \Delta T$$

$$= \Delta T (C_V - C_P + C_V)$$

$$= \Delta T (2C_V - C_P)$$

$$= \Delta T \left(\frac{2R}{\gamma - 1} - \frac{\gamma R}{\gamma - 1} \right); C_V = \frac{R}{\gamma - 1} \text{ and } C_P = \frac{\gamma R}{\gamma - 1}$$

$$= R \Delta T \frac{(2 - \gamma)}{\gamma - 1}$$

10.99. One mole of an ideal gas is taken through a process in which pressure and volume varies according to the relation $P = KV$. Find the molar specific heat of the gas.

Solution:

$$PV = RT$$

$$KV^2 = RT$$

Work done $= PdV = K \cdot V \cdot dV$

Net work done $= \int_0^V K V dV = \frac{KV^2}{2}$

$$= \frac{RT}{2}$$

$$Q = C_V T + \frac{RT}{2}$$

$$\frac{dQ}{dt} = C = \text{molar specific heat} = C_V + \frac{R}{2}$$

10.100. Calculate the molar heat capacity of one mole of an ideal gas in a polytropic process PV^n = constt. For what value of n the molar heat capacity is negative.

Solution:

$$PV^n = \text{Constt.}$$

$$PV = RT$$

$$\therefore P = \frac{RT}{V}$$

or $$\frac{RT}{V}\cdot V^n = \text{Constt.}$$

or $$TV^{n-1} = \text{Constt.}$$

On differentiation we have

$$dT\cdot V^{n-1} + T(n-1)V^{n-2}\,dV = 0$$

$$VdT = -T(n-1)\,dV$$

$$C\cdot dT = C_V dT + PdV$$

$$= C_V dT + \frac{RT}{V}dV$$

$$= C_V dT + \frac{RT}{V}\times -\frac{VdT}{T(n-1)}$$

$$= dT\left[\frac{R}{\gamma-1} - \frac{R}{n-1}\right] = \frac{R(n-\gamma)\,dT}{(\gamma-1)(n-1)}$$

$$\therefore C = \frac{R(n-\gamma)}{(\gamma-1)(n-1)}$$

C is negative when $1 < n < \gamma$

10.101. Find the molar heat capacity of an ideal gas in a polytropic process PV^n = Constt. in which the volume is increased by α times and simultaneously the pressure is decreased by β times.

Solution:

$$PV^n = \frac{P}{\beta}(\alpha V)^n = \text{Constt.}$$

or $$\frac{\alpha^n}{\beta} = 1$$

or $$n\log\alpha = \log\beta$$

or $$n = \frac{\log\beta}{\log\alpha}$$

Molar heat capacity $C = \dfrac{R(n-\gamma)}{(\gamma-1)(n-1)}$

On putting the values of n we have

$$C = \frac{R\left[\log\beta/\log\alpha - \gamma\right]}{(\gamma-1)\left(\log\beta/\log\alpha - 1\right)}$$

10.102. One mole of an ideal gas expands in a polytropic process. The temperature of the gas falls by $-\Delta T$. Find (i) the amount of heat given to the gas (ii) the amount of work performed by the gas.

Solution:

(i) $PV^n = \text{Constt.}$

$$\frac{RT}{V}\cdot V^n = \text{Constt.}$$

or $TV^{n-1} = \text{Constt.}$

$$\Delta T\cdot V^{n-1} + (n-1)\,V^{n-2}\cdot\Delta V\cdot T = 0$$

or $$\Delta V = -\frac{V\Delta T}{T(n-1)}$$

$$\delta Q = C_V\cdot\Delta T - \frac{PV\Delta T}{T(n-1)} = \left(\frac{R}{\gamma-1} - \frac{R}{n-1}\right)\Delta T$$

$$= \frac{R(n-\gamma)\Delta T}{(n-1)(\gamma-1)}$$

(ii) Work done $= PdV$

$$= -\frac{PV\Delta T}{T(n-1)} = -\frac{R\Delta T}{(n-1)}$$

10.103. One mole of an ideal gas having initial volume V_0 expands upto volume ηV_0 through a process in which pressure varies with volume via relation $P = \alpha V$. Find (i) the increase in internal energy (ii) the work done by the gas.

Solution:

(i) Increase in internal energy is given by $C_V\cdot dT$

$PV = RT$

$\alpha V^2 = RT$

$2\alpha V\cdot dV = RdT$

Increase in internal energy $= C_V\cdot dT$

$$= \frac{2\alpha\cdot V\cdot dV}{R}\cdot C_V$$

$$\text{Total increase in internal energy} = \frac{2\alpha C_V}{R}\cdot\int_{V_0}^{\eta V_0} V dV$$

$$= \frac{\alpha C_V}{C_P - C_V}\cdot V_0^2\cdot\left(\eta^2 - 1\right)$$

$$= \frac{\alpha}{\gamma - 1}\cdot V_0^2\left(\eta^2 - 1\right)$$

(ii) Work done $= PdV$

$$= \int_{V_0}^{\eta V_0} \alpha V\cdot dV = \frac{1}{2}\alpha V_0^2\left[\eta^2 - 1\right]$$

10.104. One mole of an ideal gas having initial volume V_0 and temperature T_0 expands upto volume ηV_0 in such a way in which the amount of heat given is equal to decrease in the internal energy. Find (i) the molar heat capacity of the gas (ii) the equation through which the process is carried out (iii) the work done by the gas.

Solution:

$$\delta Q = C_V dT + PdV$$

$$\delta Q = -C_V dT = C_V dT + PdV$$

$$0 = 2C_V dT + PdV$$

$$= \frac{2R}{\gamma - 1}dT + PdV$$

$$= \frac{2RdT}{\gamma - 1} + \frac{RTdV}{V}$$

or $$\frac{dT}{T} + \frac{\gamma - 1}{2}\frac{dV}{V} = 0$$

On integration we have

$$\log T + \log V^{\frac{\gamma-1}{2}} = \text{Constt.}$$

or $$TV^{\frac{\gamma-1}{2}} = \text{Constt.}$$

Further $2\delta Q = PdV$

$$= TV^{\frac{\gamma-1}{2}} = \text{Constt.}$$

$$V^{\frac{\gamma-1}{2}}\cdot dT + T\cdot\frac{\gamma - 1}{2}V^{\frac{\gamma-3}{2}}dV = 0$$

or $$dV = -\frac{2dT\cdot V}{T(\gamma - 1)}$$

$$\therefore 2\delta Q = -\frac{2PV\,dT}{T(\gamma - 1)}$$

$$\delta Q = -\frac{RdT}{(\gamma - 1)}$$

$$\therefore \text{ molar heat capacity } C = -\frac{R}{\gamma - 1}$$

$$\text{Work done } W = \int PdV = \int -\frac{2R}{\gamma - 1} \cdot dT$$

$$= -\frac{2R}{\gamma - 1} \int_{T_0}^{T} dT = \frac{2R}{\gamma - 1} [T_0 - T]$$

$$\text{Further } T_0 V^{\frac{\gamma - 1}{2}} = T \eta^{\frac{\gamma - 1}{2}} \cdot V^{\frac{\gamma - 1}{2}}$$

$$\therefore T = T_0 \, \eta^{\frac{1-\gamma}{2}}$$

$$\therefore \text{ Work done} = \frac{2RT_0}{\gamma - 1} \left[1 - \eta^{\frac{1-\gamma}{2}} \right]$$

10.105. One mole of an ideal gas follows a process for the transformation in which pressure of the gas varies with temperature via relation $P = aT^{\beta}$. Where a & β are positive constants. The temperature of the gas increases by ΔT. Find (i) the work performed by the gas (ii) the molar heat capacity of the gas.

Solution:

(i) $$P = aT^{\beta}$$

$$\frac{RT}{V} = aT^{\beta}$$

or $$T^{\beta - 1} V = \text{Constt.}$$

$$\Delta V \cdot T^{\beta - 1} + (\beta - 1)\, T^{\beta - 2} \cdot \Delta T \cdot V = 0$$

or $$\Delta V = (1 - \beta)\, V \cdot \frac{\Delta T}{T}$$

$$\text{Work done} = P\Delta V = \frac{PV}{T} \Delta T\,(1 - \beta)$$

$$= R\,(1 - \beta) \cdot \Delta T$$

(ii) $$\delta Q = C_V \cdot \Delta T + R\,(1 - \beta)\,\Delta T$$

$$C = \frac{R}{\gamma - 1} + R\,(1 - \beta)$$

$$= \frac{\gamma R}{\gamma - 1} - R\beta$$

The molar capacity will be negative when $\beta > \dfrac{\gamma}{\gamma - 1}$

10.106. One mole of an ideal gas undergoes a process in which the internal energy varies with volume through relation $U = aV^{\beta}$. Calculate (i) the work performed by the gas (ii) the amount of heat required to increase the internal energy by an amount ΔU (iii) the molar heat capacity of the gas.

Solution:

(i) $$C_V \cdot T = U = aV^{\beta}$$

$$\frac{R}{\gamma-1} \cdot T = aV^{\beta}$$

or $$TV^{-\beta} = \text{Constt.}$$

$$\Delta T \cdot V^{-\beta} - \beta V^{-\beta-1} \cdot \Delta V \cdot T = 0$$

or $$\Delta T = \beta \cdot \frac{\Delta V}{V} \cdot T$$

$$\text{Work done} = PdV = \frac{P \cdot V \cdot \Delta T}{\beta \cdot T} = \frac{R}{\beta} \cdot \Delta T$$

$$\frac{R}{\gamma-1} \cdot \Delta T = \Delta U$$

$$\therefore \text{Work done} = \frac{\Delta U \cdot (\gamma-1)}{\beta}$$

(ii) $$\delta Q = \Delta U + \frac{\Delta U \cdot (\gamma-1)}{\beta} = \left[1 + \frac{(\gamma-1)}{\beta}\right] \Delta U$$

(iii) $$\delta Q = C_V \cdot \Delta T + P\Delta V = \frac{R}{\gamma-1} \cdot \Delta T + \frac{R}{\beta} \cdot \Delta T$$

$$\therefore C = \text{molar heat capacity} = \frac{R}{\gamma-1} + \frac{R}{\beta}$$

10.107. The molar heat capacity at a constant volume of an ideal gas is C_V. The gas undergoes a process in which the pressure varies with volume by relation $P = P_0 e^{\alpha V}$ and the temperature varies with volume by relation $T = T_0 e^{\alpha V}$. Find the molar heat capacity C of the gas in the two cases.

Solution:

(i) $$P = \frac{RT}{V} = P_0 e^{\alpha V}$$

$$\log\left(\frac{RT}{P_0}\right) = \log V + \alpha V$$

$$\frac{P_0}{RT} \cdot \frac{R}{P_0} \cdot dT = \frac{1}{V} \cdot dV + \alpha \cdot dV$$

or $$\frac{dT}{T} = \left(\frac{1+\alpha V}{V}\right) dV$$

$$\delta Q = C_V dT + PdV$$

$$= C_V dT + \frac{PV \cdot dT}{T(1 + \alpha V)}$$

$$\delta Q = \left(C_V + \frac{R}{1 + \alpha V}\right) dT$$

$$\therefore C = \frac{\delta Q}{dT} = \left(C_V + \frac{R}{1 + \alpha V}\right)$$

(ii) $$T = T_0 \, e^{\alpha V}$$

$$\log\left(\frac{T}{T_0}\right) = \alpha V$$

$$V = \frac{1}{\alpha} \cdot \log\left(\frac{T}{T_0}\right)$$

$$\delta Q = C_V dT + PdV$$

$$= C_V dT + \frac{P}{\alpha} \cdot \frac{T_0}{T} \cdot \frac{dT}{T_0} ; PV = RT$$

$$= C_V dT + \frac{PdT}{\alpha T}$$

$$\delta Q = C_V dT + \frac{R}{\alpha V} \cdot dT$$

or $$C = \frac{\delta Q}{dT} = C_V + \frac{R}{\alpha V}$$

10.108. One mole of an ideal gas follows a process in which pressure varies by relation $P = P_0 + \alpha / V$. The volume of the gas increases from its initial value V_1 to V_2. Find (i) the increase in internal energy (ii) the amount of work done (iii) the amount of heat added to the gas.

Solution:

(i) $$P = P_0 + \frac{\alpha}{V}$$

$$\frac{RT}{V} = P_0 + \frac{\alpha}{V}$$

$$RT = P_0 V + \alpha$$

$$RdT = P_0 dV$$

$$dU = C_V dT = \frac{R}{\gamma - 1} dT = \frac{P_0 dV}{\gamma - 1}$$

$$U = \int dU = \int_{V_1}^{V_2} \frac{P_0 dV}{\gamma - 1} = \frac{P_0}{\gamma - 1}(V_2 - V_1)$$

(ii) $$W = \int P dV = \int_{V_1}^{V_2} P_0 dV + \int_{V_1}^{V_2} \frac{\alpha}{V} dV$$

$$W = P_0 (V_2 - V_1) + \alpha \log_e \frac{V_2}{V_1}$$

(iii) $$\delta Q = C_v dT + P dV$$

$$= C_V dT + \frac{P \cdot R dT}{P_0}$$

$$= C_V dT + \left(P_0 + \frac{\alpha}{V}\right) \frac{R dT}{P_0}$$

$$= (C_V + R)\, dT + \frac{\alpha R}{P_0 V} \cdot dT$$

$$= \frac{\gamma R dT}{\gamma - 1} + \frac{\alpha R dT}{P_0 V}$$

$$\delta Q = \frac{\gamma}{\gamma - 1} P_0 dV + \frac{\alpha}{P_0 V} \cdot P_0 dV$$

$$Q = \int \delta Q = \frac{\gamma}{\gamma - 1} \cdot P_0 \int_{V_1}^{V_2} dV + \alpha \int_{V_1}^{V_2} \frac{dV}{V}$$

$$= \frac{\gamma}{\gamma - 1} \cdot P_0 (V_2 - V_1) + \alpha \log_e \frac{V_2}{V_1}$$

10.109. The temperature of one mole of an ideal gas having specific heat at constant pressure as C_p varies with volume via relation $T = T_0 + \alpha V$. In the process of transformation the volume of the gas increases from its initial value V_1 to V_2. Calculate the amount of heat added to the gas.

Solution:

$$T = T_0 + \alpha V$$

$$dT = \alpha \cdot dV$$

$$\delta Q = C_V dT + P dV$$

$$= C_V dT + \frac{R(T_0 + \alpha V)}{V} dV$$

$$= C_V dT + \frac{R(T_0 + \alpha V)}{V} \cdot \frac{dT}{\alpha}$$

$$= (C_V + R)\, dT + \frac{R T_0}{V \alpha} \cdot dT$$

$$= C_P dT + \frac{R T_0}{V \alpha} \cdot dT$$

$$Q = \int \delta Q = C_P \cdot \alpha \cdot \int_{V_1}^{V_2} dV + RT_0 \int_{V_1}^{V_2} \frac{dV}{V}$$

$$= \alpha\, C_P\, (V_2 - V_1) + RT_0 \log_e \frac{V_2}{V_1}$$

10.110. The molar heat capacity of an ideal gas varies via relations. (i) $C = C_V + \alpha T$, (ii) $C = C_V + \beta V$, (iii) $C = C_V + aP$. Find the equation of the process in terms of temperature and volume of the gas.

Solution:

(i)
$$C = C_V + \alpha T$$
$$\delta Q = dU + PdV$$
$$\therefore PdV = \alpha T dT$$
$$\frac{RT \cdot dV}{V} = \alpha\, T \cdot dT$$

On integration $\log V = \frac{\alpha}{R} T + \text{Constt.}$

$$= \log e^{\frac{\alpha T}{R}} + \text{Constt.}$$

$$\log V e^{-\frac{\alpha T}{R}} = \text{Constt.}$$

$$V e^{-\frac{\alpha T}{R}} = \text{Constt.}$$

(ii) $C = C_V + \beta V$

$$\beta V \cdot dT = PdV = \frac{RT}{V} \cdot dV$$

$$\frac{dT}{T} = \frac{R \cdot dV}{\beta V^2}$$

On integration

$$\log T = -\frac{R}{\beta V} + \text{Constt.}$$

$$\log T = \log e^{-\frac{R}{\beta V}} + \text{Constt.}$$

$$T e^{\frac{R}{\beta V}} = \text{Constt.}$$

(iii)
$$C = C_V + aP$$
$$a\, P \cdot dT = PdV$$
$$dV = adT$$
$$V = aT + \text{Constt.}$$

or $\quad (V - aT) = \text{Constt.}$

10.111. A gas ($\gamma = 1.4$) of 2 m^3 volume and at a pressure of 4×10^5 N/m^2 is compressed adiabatically to a volume of 0.5 m^3. Find its new pressure. Compare it with the pressure obtained if the compression were isothermal. Calculate the work done in each process.

Given, $4^{1.4} = 6.96$ and $\log_{10} \frac{1}{4} = -0.6021$. (Roorkee 1987)

Solution: In a adiabatic process, the pressure and volume are related by the equation

$$P_i V_i^\gamma = P_f V_f^\gamma$$

$$\therefore P_f = P_i \left(\frac{V_i}{V_f}\right)^\gamma = 4 \times 10^5 \times \left(\frac{2}{0.5}\right)^{1.4} = 4 \times 10^5 \times 4^{1.4}$$

$$= 4 \times 6.96 \times 10^5 \text{ N/m}^2 = 27.84 \times 10^5 \text{ N/m}^2$$

In an isothermal process, final pressure is given by

$$P_f = P_i \left(\frac{V_i}{V_f}\right) = 4 \times 10^5 \times \frac{2}{0.5} = 16 \times 10^5 \text{ N/m}^2$$

$$\text{Work done in an adiabatic process} = \frac{1}{\gamma - 1}\left(P_i V_i - P_f V_f\right)$$

$$= \frac{1}{0.4}\left(4 \times 10^5 \times 2 - 27.84 \times 10^5 \times 0.5\right)$$

$$= -14.8 \times 10^5 \text{ Joule}$$

$$\text{Work done in isothermal process} = 2.303 \times P_i V_i \log_{10} \frac{V_f}{V_i}$$

$$= 2.303 \times 4 \times 10^5 \times \log\left(\frac{0.5}{2}\right)$$

$$= -2.303 \times 8 \times 10^5 \times 0.6021$$

$$= -1.109 \times 10^6 \text{ Joule}$$

10.112. The insulated box shown in Fig. 10.18 has an insulated partition which can slide without friction along the length of the box. Initially each of the two chambers of the box has one mol of a monoatomic ideal gas ($\gamma = 5/3$) at a presure P_0, volume V_0 and temperature T_0. The chamber on the left is slowly heated by an electric heater so that its gas, pushing the partition, expands until the final pressure in both the chambers becomes $243\, P_0/32$. Determine (i) the final temperature of the gas in each chamber and (ii) the work done on the gas in the right chamber. (IIT 1984)

Solution:

(i) On heating the gas in the left chamber, the partition will move towards right and the gas in the right chamber will be

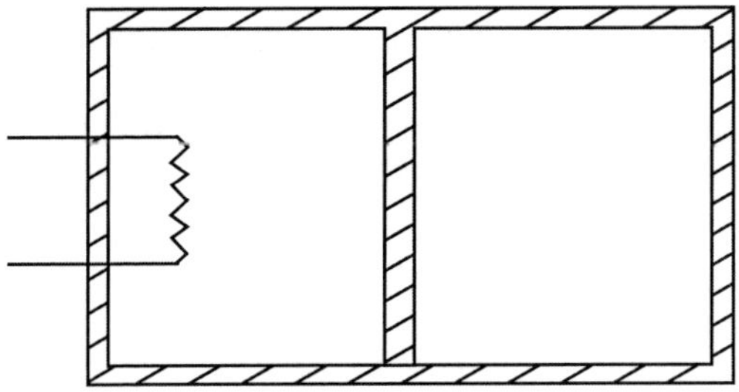

Fig. 10.18

adiabatically compressed. The final temperature T_2 and pressure P_2 in the second chamber are related by

$$T_0 P_0^{1-\gamma} = T_2^{\gamma} P_2^{1-\gamma}$$

or $$T_2 = T_0 \left(\frac{P_0}{P_2}\right)^{\frac{1-\gamma}{\gamma}} = T_0 \times \frac{1}{\left(\frac{32}{243}\right)^{-2/5}} = \frac{9}{4} T_0$$

Here P_0 and T_0 represent the initial pressure and temperature, respectively. The volume of the gas in the second chamber can be obtained by the relation

$$T_0 V_0^{\gamma-1} = T_2 V_2^{\gamma-1}$$

or $$V_2 = \left(\frac{T_0}{T_2}\right)^{\frac{1}{\gamma-1}} V_0 = \left(\frac{4}{9}\right)^{3/2} V_0 = \frac{8}{27} V_0$$

Volume of air in the first chamber $V_1 = 2V_0 - \frac{8}{27} V_0 = \frac{46}{27} V_0$

The temperature of gas in the first chamber can be obtained by the relation $\frac{P_0 V_0}{T_0} = \frac{P_1 V_1}{T_1}$

$$\therefore T_1 = \frac{P_1 V_1}{P_0 V_0} T_0 = \frac{243}{32} \times \frac{46}{27} T_0 = \frac{207}{16} T_0$$

(ii) Work done on the gas in the second chamber

$$= \frac{1}{\gamma-1}[P_0 V_0 - P_2 V_2]$$

$$= \frac{1}{\gamma-1}\left[P_0 V_0 - \frac{243}{32} P_0 \frac{8}{27} V_0\right]$$

$$= -\frac{15}{8} P_0 V_0; \gamma = 5/3$$

10.113. A lead bullet, moving with a velocity of 200 m/s strikes a target and is completely stopped by it. Due to this phenomenon, the kinetic energy of the bullet is converted into heat. If it is assumed that the heat produced is equally absorbed by the bullet and the target, find the rise in temperature of the bullet. Specific heat of lead = 0.032 cal/°C, $J = 4.2$ J/cal.

Solution: Let m kg be the mass of the bullet. The kinetic energy associated with the bullet $= \frac{1}{2} mv^2 = \frac{1}{2} \times m \times 4 \times 10^4$ Joule

According to the law of conservation of energy

$$\frac{0.5 \times 2m \times 10^4}{4.2} = m \times 10^3 \times 0.032 \times \theta; \ \theta \text{ is the rise in temperature}$$

$$\therefore \theta = \frac{10^4}{4.2 \times 10^3 \times 0.032} = 74.4°\text{C}$$

10.114. A hammer of mass 5 kg strikes a piece of lead of mass 500 g at 27°C with a velocity of 2 m/s. How many strokes of the hammer are required to melt the piece of lead. Specific heat of lead = 0.03 cal/°C. Latent heat of lead = 6 cal/g, $J = 4.18$ J/cal, melting point of lead 327°C.

Solution: Let us assume that n strokes are required for melting the lead. The total energy of the hammer in n strokes $= \frac{1}{2} \times 5 \times 4 \times n$ Joule $= \frac{10n}{4.18}$ cal.

The part of energy used in increasing the temperature of the piece of lead up to its melting point = 500 × 0.03 × (327 – 27) = 15 × 300 = 4500 cal

The remaining part 500 × 6 = 3000 cal is used in changing its state.

According to the law of conservation of energy

$$\frac{10n}{4.18} = 4500 + 3000 = 7500$$

$$n = 7.5 \times 418 = 3135$$

10.115. A piston-cylinder assembly contains 32 kg of oxygen at a pressure of 4 atmospheres and a temperature of 27°C. It undergoes a cycle in which it is first heated at constant volume until the pressure becomes 3 times, then it expands isothermally until the pressure drops to the original value and finally it is cooled isobarically back to its original temperature. Calculate the change in internal energy and the work done in each process. Atomic weight of oxygen = 16 and $R = 8.3$ J/mol/K.

Solution:

(i) The first operation is done under constant volume, hence no work will be done. Therefore, all heat added will be used in increasing the internal energy, i.e., $\delta Q = dU$, since $\delta W = 0$.

Now $\frac{p_i}{T_i} = \frac{p_f}{T_f}$ (since volume remains the same)

$\therefore \quad T_f = \frac{p_f T_i}{p_i} = 3 \times T_i = 3 \times (273 + 27) = 900$ K

Change in temperature = $\Delta T = T_f - T_i = 600$ K

For diatomic gas, $C_V = \frac{5}{2} R = \frac{5}{2} \times 8.3$ J/mol/K

The quantity of heat required to raise the temperature of the gas though 600 K at constant volume is given by

$$\delta Q = dU = C_V \Delta T = \frac{5}{2} R \Delta T = \frac{5}{2} \times 8.3 \times 600 = 1.245 \times 10^4 \text{ J}$$

(ii) The operation is an isothermal expansion, i.e., the temperature remains the same.

Now $p_i V_i = p_f v_f$

$$\frac{V_f}{V_i} = \frac{p_i}{p_f} = 3$$

Work done in an isothermal process = $2.303\, RT \log_{10}\left(\frac{V_f}{V_i}\right)$

$$= 2.303 \times 8.3 \times 900 \times \log_{10}^{3}$$

$$= 2.303 \times 8.3 \times 900 \times 0.4771$$

$$= 8.206 \times 10^3 \text{ J}$$

Change in internal energy = $dU = 0$

(iii) The third operation is done under constant pressure, i.e., $dp = 0$. Change of temperature in this operation $\Delta T = -600$ K

Change in internal energy

$$dU = C_v \Delta T = \frac{5}{2} R \Delta T = -\frac{5}{2} \times 8.3 \times 600 = -1.245 \times 10^4 \text{ J}$$

$$\delta Q = C_P \Delta T,\ C_P = 7/2\, R$$

$$\therefore \delta Q = 7/2 \times 8.3 \times -600$$

$$= -1.743 \times 10^4 \text{ J}$$

Work done $\delta W = \delta Q - \delta U = -1.743 \times 10^4 + 1.245 \times 10^4$

$$= -4.98 \times 10^3 \text{ J}$$

10.116. 1671 cm^3 of water vapour is formed from 1 cm^3 of water at atmospheric pressure (1.01×10^5 N/m^2) and 100°C. Latent heat of vaporisation is 540 cal/g. How much increase in the internal energy will take place when one gram of water is converted into vapour at atmospheric pressure ? $J = 4.2$ J/cal.

Solution: Change of phase is always associated with change in volume, so that work is always done on or by the system. Amount of work $= p\,(v_2 - v_1)$ where p is the atmospheric pressure and v_1 is the volume of 1 g of vapour and v_2 is specific volume.

If u_1 and u_2 be the internal energies of 1 g of water and 1 g of vapour, respectively then according to the first law of the thermodynamics, we have

$$Q = L = u_2 - u_1 + p\,(v_2 - v_1)$$

or $$u_2 - u_1 = L - p\,(v_2 - v_1)$$

$$= 540 - \frac{1.01 \times 10^5\,(1671 - 1) \times 10^{-6}}{4.2}$$

$$= 540 - \frac{1.01 \times 167}{4.2}$$

$$= 540 - 40.15$$

$$= 499.85 \text{ cal.}$$

10.117. A vertical cylinder containing a piston of weight 10 kgf and having base area $A = 7$ cm^2 is filled with gas at 7°C. The initial volume of the gas is 14 litres. How much heat is required to raise the temperature of the gas up to 27°C. The thermal capacity of the gas is 5 cal/degree.

Solution: Initial volume of the gas $V_i = 14 \times 1000 \times 10^{-6}$ m^3 $= 14 \times 10^{-3}$ m^3

On heating, the gas will expand against constant pressure. The final volume of the gas v_f is given by the relation

$$v_f = \frac{T_f V_i}{T_i} = \frac{(273 + 27) \times 14 \times 10^{-3}}{(273 + 7)} = \frac{300 \times 14 \times 10^{-3}}{280} = 15 \times 10^{-3} \text{ m}^3$$

Work done against constant force $= \dfrac{10 \times 9.8}{7 \times 10^{-4}} \times (15 - 14) \times 10^{-3} = 140$ J

$$= \frac{140}{4.2} \text{ cal} = 33.3 \text{ cal}$$

Quantity of heat required to raise the temperature of the gas by 20°C $= 20 \times 5 = 100$ cal

Therefore, heat added $= 100 + 33.3 = 133.3$ cal

10.118. A cylinder-piston assembly contains 10 g of helium. The gas is subjected to a very slow compression from an initial volume $V_1 = 30$ litres and pressure $P_1 = 3.2$ atmospheres to final volume $V_2 = 8$ litres and pressure $P_2 = 14.2$ atmospheres along a straight path CD, as shown in the Fig. 10.19.

Calculate the maximum temperature attained by the gas during the compression. $R = 8.3$ J/mol/K and 1 atmosphere = 10^5 N/m².

Solution: Considering the straight variation of pressure and volume, the pressure can be represented by the relation

$$P = AV + B$$

where A and B are the constants to be determined to know the exact values.

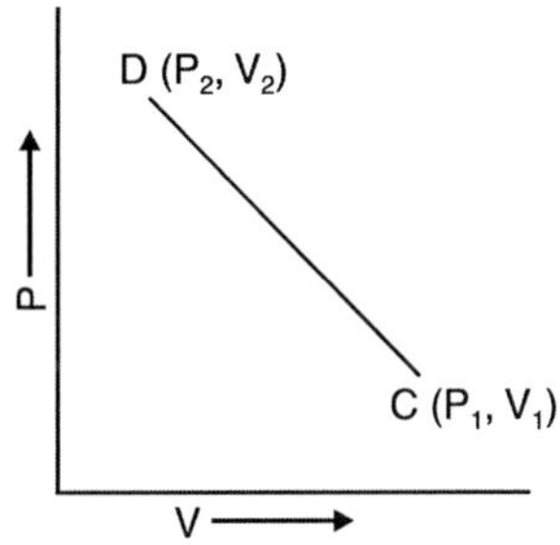

Fig. 10.19

With two pressures P_1 and P_2 and volumes V_1 and V_2 we can have two relations

$$P_1 = AV_1 + B$$

and $$P_2 = AV_2 + B$$

On solving the two equations, we have

$$A = \frac{P_1 - P_2}{V_1 - V_2} = \frac{3.2 - 14.2 \text{ atmosphere}}{(30 - 8) \times 10^{-3} \text{m}^3}$$

$$= -0.5 \times 10^8 \text{ N/m}^5$$

$$B = \frac{P_2V_1 - P_1V_2}{V_1 - V_2} = \frac{14.2 \times 30 - 3.2 \times 8}{30 - 8} = \frac{400.4}{22}$$

$$= 18.2 \text{ atmosphere} = 18.2 \times 10^5 \text{ N/m}^2$$

For an ideal gas $PV =$ constant $\times T$

Now $PV = AV^2 + BV$ or $AV^2 + BV -$ constant $\times T = 0$

The above equations suggests the V varies with temperature T.

Let α and β be the roots of the above equation, then

$$\alpha + \beta = -\frac{B}{A}$$

Now, when $\alpha = \beta$, V will have the maximum values as V_{max}

$$V_{max} = \frac{-B}{2A} = \frac{18.2 \times 10^5}{2 \times 0.5 \times 10^8} \text{ m}^3 = 18.2 \times 10^{-3} \text{ m}^3$$

Further $P_{max} = AV_{max} + B = (-0.5 \times 10^8 \times 18.2 \times 10^{-3} + 18.2 \times 10^5)$ N/m²

$$= 9.1 \times 10^5 \text{ N/m}^2$$

$$T_{max} = \frac{P_{max} V_{max} \times M}{mR}$$; m = mass of the gas, M = molecular weight

$$= \frac{9.1 \times 10^5 \times 18.2 \times 10^{-3} \times 4 \times 10^{-3}}{10 \times 10^{-3} \times 8.3} = 798.16 \text{ K}$$

10.119. A piston divides a closed gas cylinders into two parts. Initially the piston is kept pressed such that one part has a pressure P and volume $5V$ and the other part has pressure $8P$ and volume V. The piston is now left free. Find the new pressures and volumes for the adiabatic and isothermal processes. For this gas $\gamma = 1.5$ (Roorkee 1990)

Solution:

(i) In isothermal change

Let new pressure be P' and change in volume be ΔV, then

$$P \times 5V = P'(5V - \Delta V)$$

and $$8P \times V = P'(V + \Delta V)$$

$$\frac{5}{8} = \frac{5V - \Delta V}{V + \Delta V}$$

$$5V + 5\Delta V = 40V - 8\Delta V$$

$$13\Delta V = 35V$$

$$\therefore \Delta V = \frac{35}{13}V$$

$$\therefore \text{Volume } V_1 = 5V - \Delta V = 5V - \frac{35V}{13} = \frac{30V}{13}$$

$$\text{Volume } V_2 = V + \Delta V = V + \frac{35V}{13} = \frac{48V}{13}$$

(ii) In adiabatic change

Let P' be the new pressure

$$P(5V)^\gamma = P'(5V - \Delta V)^\gamma$$

and $$8PV^\gamma = P'(V + \Delta V)^\gamma$$

$$\frac{5^\gamma}{8} = \left(\frac{5V - \Delta V}{V + \Delta V}\right)^\gamma$$

$$\frac{5}{4} = \left(\frac{5V - \Delta V}{V + \Delta V}\right)$$

$$5V + 5\Delta V = 20V - 4\Delta V$$

$$9\Delta V = 15V$$

$$\therefore \Delta V = \frac{5V}{3}$$

$$\text{New volume } V_1 = 5V - \frac{5V}{3} = \frac{10V}{3}$$

$$V_2 = V + \frac{5V}{3} = \frac{8V}{3}$$

$$P' = P\left(\frac{5V}{10V/3}\right)^{3/2} = P\left(\frac{27}{8}\right)^{1/2} = 1.84P$$

10.120. Three moles of an ideal gas of 300 K are isothermally expanded to five times its volume and heated at this constant volume so that the pressure is raised to its initial value before expansion. In the whole process 83.14 KJ heat is required. Calculate the ratio C_P/C_V of gas. $\log_e 5 = 1.61$.

(Roorkee 1990)

Solution: In isothermal process we have

$$P_1V_1 = P_2V_2$$

$$\therefore \frac{P_1}{P_2} = \frac{V_2}{V_1} = \frac{5V_1}{V_1} = 5$$

Further $$\frac{P_2}{T_1} = \frac{P_3}{T_3} = \frac{P_1}{T_3}$$

$$\frac{P_2}{P_1} = \frac{T_1}{T_3} = \frac{1}{5}$$

$$\therefore T_3 = 5T_1 = 5 \times 300 = 1500 \text{ K}$$

According to first law of thermodynamics

$$\delta Q = dU + PdV$$

$$\delta Q = 3\, C_V\, dT + 3RT_1 \log_e \frac{V_2}{V_1}$$

$$= \frac{3R}{\gamma - 1} dT + 3RT_1 \log_e \frac{V_2}{V_1}$$

$$83.14 \times 10^3 = \frac{3 \times 8.3}{\gamma - 1} + 3 \times 8.3 \times 300 \times \log_e 5$$

On solving $\gamma = 1.42$

10.121. 10^{-4} kg oxygen is contained in a non-conducting cylindrical vessel of cross-sectional area 29.15×10^{-5} m^2 and volume 10^{-3} m^3. It is closed by a frictionless non-conducting piston which is free to move in a vertical direction. If the piston is slightly depressed and then released, show that the piston will execute simple harmonic motion. Find its frequency ignore atmospheric pressure and take $C_P/C_V = 1.4$ for oxygen. (Roorkee 1991)

Solution:

$$PV^{\gamma} = \text{Constt.}$$

$$P\gamma \cdot V^{\gamma-1} \cdot \Delta V + V^{\gamma} \cdot \Delta P = 0$$

$$\therefore \Delta P = \gamma P \frac{\Delta V}{V}$$

Restoring force $F = -\Delta P \cdot A = -\gamma PA \cdot \dfrac{\Delta V}{V}$

$$= \frac{-\gamma P A^2 x}{V}$$; $\Delta V = A \cdot x$ where x is the displacement of the piston.

M. acceleration $= \dfrac{-\gamma P A^2}{V}$ displacement

It is a case of simple harmonic motion.

$$\omega = \sqrt{\frac{\gamma \cdot PA^2}{MV}} = \sqrt{\frac{\gamma \cdot Mg \cdot A^2}{MV \cdot A}}$$

$$n = \frac{1}{2\pi}\sqrt{\frac{\gamma g A}{V}}$$

On putting the values we have

$$n = \frac{1}{2\pi}\sqrt{\frac{1.4 \times 9.8 \times 29.15 \times 10^{-5}}{10^{-3}}}$$

$$= \frac{1}{\pi}\ /\text{sec}$$

10.122. An ideal gas having initial pressure P, volume V and temperature T is allowed to expand adiabatically until its volume becomes 5.66 V while its temperature fall to $T/2$.

(i) How many degrees of freedom do the gas molecules have?

(ii) Obtain the work done by the gas during the expansion as a function of the initial pressure P and volume V. (IIT 1990)

Solution: For an adiabatic process we have

$$TV^{\gamma-1} = T_1V_1^{\gamma-1}$$

$$TV^{\gamma-1} = \frac{T}{2}(5.66V)^{\gamma-1}$$

$$(5.66)^{\gamma-1} = 2$$

$$(\gamma - 1)\log 5.66 = \log 2$$

$$\gamma = 1 + \frac{\log 2}{\log 5.66} = 1 + \frac{0.3010}{0.7528} = 1.4$$

The gas is therefore diatomic and the degrees of freedom = 5.

(ii) Work done during and adiabatic process is given by

$$W = \frac{1}{1-\gamma}(P_1V_1 - PV)$$

$$\frac{P_1V_1}{T_1} = \frac{PV}{T}$$

$$P_1 = \frac{P \cdot V}{5.66V}\frac{T/2}{T} = \frac{P}{11.32}$$

$$W = \frac{1}{1-1.4}\left[\frac{P \times 5.66V}{11.32} - PV\right]$$

$$= \frac{PV}{0.4}\left[1 - \frac{5.66}{11.32}\right] = \frac{PV}{0.8} = 1.25\,PV$$

10.123. A gas with an initial volume $V_1 = 3\ m^3$ expands upto $V_2 = 4\ m^3$. In this process its pressure p is no longer constant but varies with volume via equation. $p = p_0 e^{-\alpha(V-V_0)}$ where $p_0 = 6 \times 10^5$ Pa, $\alpha = 0.2\ m^{-3}$ and $V_0 = 2\ m^3$. Calculate the amount of work done by the gas.

Solution: Amount of work done $dw = pdv = p_0 e^{-\alpha(V-V_0)} \cdot dv$

$$\text{Net work done} = W = \int_{V_1}^{V_2} p_0 e^{-\alpha(V-V_0)} dv$$

$$= -\frac{p_0}{\alpha}\left| e^{-\alpha(V-V_0)} \right|_{V_1}^{V_2}$$

$$= \frac{6 \times 10^5}{0.2}\left[e^{-\alpha} - e^{-2\alpha}\right]$$

$$= 3 \times 10^6 \left[\frac{1}{e^{0.2}} - \frac{1}{e^{0.4}}\right]$$

$$= 444 \text{ KJ}$$

10.124. The internal energy of an imaginary gas is determined by the equation $U = aln\left(\frac{T}{T_0}\right) + bln\left(\frac{p}{p_0}\right)$ where $a = 3$ KJ, $b = 7$ KJ, $T_0 = 200$ K and $p_0 = 10$ KPa. On imparting 500 J of heat at constant pressure of $p = 10^5$ Pa the volume of the gas increases by $0.5 \times 10^{-3}\ m^3$. How will the temperature of the gas will change.

Solution:

$$\delta Q = dU + \delta W$$

$$\delta W = pdv = 10^5 \times 0.5 \times 10^{-3} = 50 \text{ J}$$

$$dU = 500 - 50 = 450$$

$$450 = a\frac{dT}{T} = a\frac{(T - T_0)}{T}$$

$$\frac{450}{a} = 1 - \frac{T_0}{T}$$

or $$\frac{T_0}{T} = 1 - \frac{450}{3000} = \frac{2550}{3000}$$

$\therefore$ $$T = \frac{3000}{2550} \cdot T_0 = 1.17 T_0$$

The final temperature of the gas will be 1.17 times the initial temperature.

10.125. The internal energy of an imaginary gas is governed by the relation $U = aln\left(\frac{T}{T_0}\right) + bln\left(\frac{V}{V_0}\right)$ where $a = 4$ KJ, $b = 5$ KJ, $T_0 = 200$ K, $V_0 = 10$ lit. A quantity of heat of 4000 J is given to the gas and as a result it expands against constant pressure $p = 10^5$ Pa from volume $V_1 = 20$ lit to $V_2 = 30$ lit. Determine the final temperature of the gas by assuming the initial temperature $T_1 = 300$ K.

Solution:

Work done = $pdv = 10^5 \times (30 - 20) \times 10^{-3} = 1000$ J

$$dU = 4000 - 1000 = 3000 = a ln\left(\frac{T_2}{T_0}\right) - a ln\left(\frac{T_1}{T_0}\right) + b ln\left(\frac{V_2}{V_0}\right) - b ln\left(\frac{V_1}{V_0}\right)$$

or $$3000 = a \times 2.303 \log_{10} \frac{T_2}{T_1} + b \times 2.303 \log_{10} \frac{V_2}{V_1}$$

$$3000 = a \times 2.303 \log_{10} \frac{T_2}{T_1} + 5000 \times 2.303 \times \log_{10} 1.5$$

or $$3000 - 2027.3 = 972.7 = a \times 2.303 \log_{10} \frac{T_2}{T_1}$$

or $$\log_{10} \frac{T_2}{T_1} = \frac{972.7}{4000 \times 2.303} = 0.1056 = \log_{10} 1.275$$

$$\therefore T_2 = T_1 \times 1.275 = 300 \times 1.275 = 382.5 \text{ K}$$

10.126. A cylinder piston system shown in the Fig. 10.20 contains one mole of an ideal diatomic gas. The area of cross-section of the cylinder is A. The mass of the piston is M. An electric heater attached to the wall of the cylinder supplies Q calories of heat per sec when connected to mains. Taking the atmospheric pressure as p_0 determine the velocity of the piston. It is assumed that the pressure remains unchanged and the system is thermally insulated.

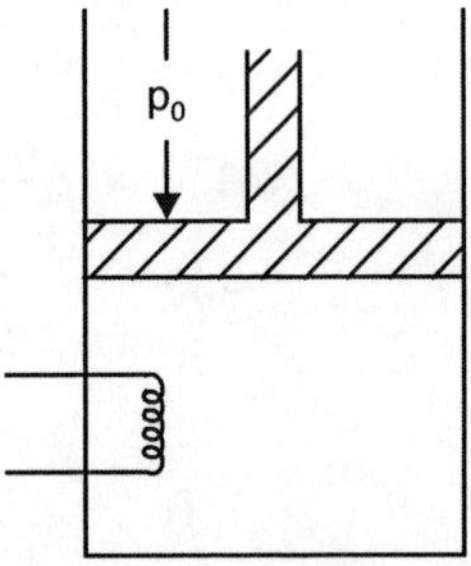

Fig. 10.20

Solution: For a diatomic gas we have

$$\frac{7}{2} RdT = \frac{5}{2} RdT + RdT$$

Now $$\frac{7}{2} (Mg + p_0 A)dx = Q \times 4.2\, dt$$

Where dx is the displacement of the piston in time dt.

$$\therefore v = \frac{dx}{dt} = \frac{8.4Q}{7(Mg + p_0 A)}$$

10.127. Three moles of an ideal gas performs a cyclic process as shown in the Fig. 10.21. The gas temperatures in different states are $T_1 = 400$K, $T_2 = 800$K, $T_3 = 2400$K and $T_4 = 1200$K. Find the amount of work done by the gas during the cycle.

Solution: Work along path 1 to 2 = 0

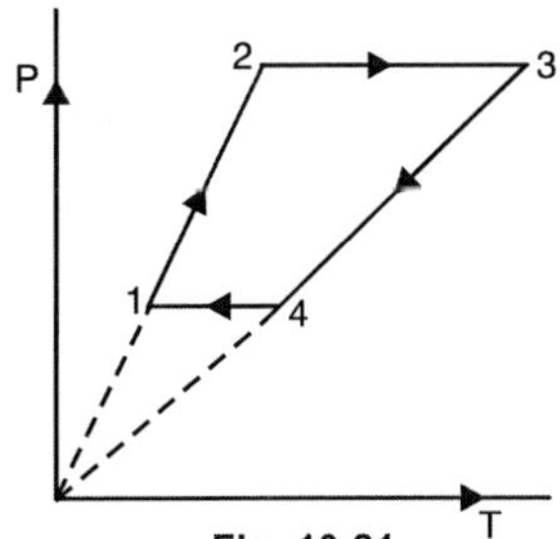

Fig. 10.21

Heat absorbed $= 3\,C_V\,(T_2 - T_1)$;

volume is constant

Work along path 2 to 3 $= 3R\,(T_3 \quad T_2)$

Heat absorbed $= 3\,C_V\,(T_3 - T_2) + 3R\,(T_3 - T_2)$

Work along path 4 to 3 $= 0$

Heat rejected $= -\,3C_V\,(T_3 - T_4)$

Work along path 4 to 1 $= -\,3R\,(T_4 - T_1)$

Heat rejected $= -\,3\,C_V\,(T_4 - T_1) - 3R\,(T_4 - T_1)$

$$\begin{aligned}\text{Net work } &= 3R\,(T_3 - T_2) + 3C_V\,(T_3 - T_2) + 3R\,(T_3 - T_2) - 3C_V\,(T_3 - T_4) \\ &\qquad - 3C_V\,(T_4 - T_1) - 3R\,(T_4 - T_1) \\ &= 3R\,(T_1 + T_3 - T_2 - T_4) \\ &= 3R\,(400 + 2400 - 800 - 1200) \\ &= 2400\,R \text{ Joule}\end{aligned}$$

10.128. A cyclic process is in the form of an ellipse as shown in Fig. 10.22. Determine the amount of heat obtained by the working body during one complete cycle.

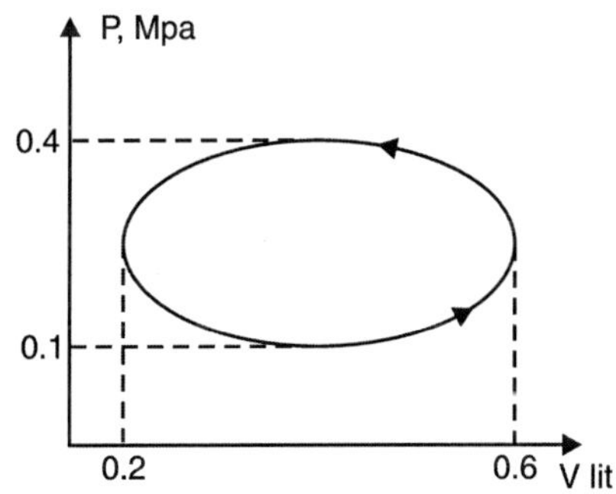

Fig. 10.22

Solution: The semi major axis of the ellipse $= a = 0.2 \times 10^{-3}$ m^3

The semi minor axis of the ellipse $= b = \dfrac{0.3}{2} \times 10^6$

$$\begin{aligned} Q = -\,\pi\, ab &= -\,3.14 \times 0.2 \times 10^{-3} \times 0.15 \times 10^6 \\ &= -\,3.14 \times 30 \\ &= -\,94.2 \text{ J}\end{aligned}$$

10.129. A reversible cycle completed by a thermodynamical system has the form as shown in Fig. 10.23 in a *T-S* diagram. Find the efficiency of the cycle.

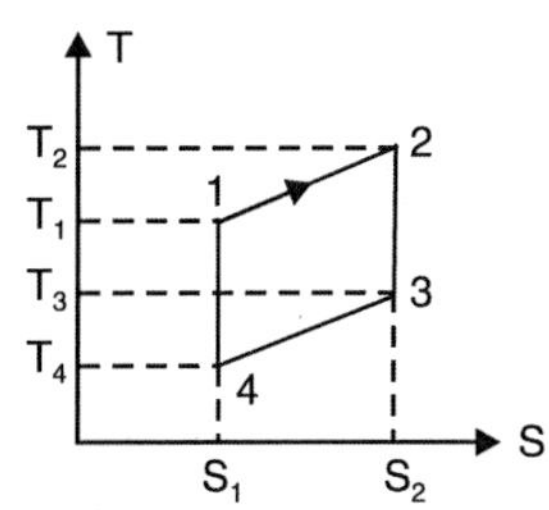

Fig. 10.23

Solution: Heat absorbed along the path 1 to 2 is given by

$$Q_1 = (S_2 - S_1)\,T_2 - (S_2 - S_1)\,\frac{T_2 - T_1}{2} = (S_2 - S_1)\frac{(T_1 + T_2)}{2}$$

Heat rejected along the path 3 to 4 = Q_2

$$= (S_2 - S_1)\frac{(T_3 - T_4)}{2} + (S_2 - S_1)\,T_4$$

$$= (S_2 - S_1)\frac{(T_3 + T_4)}{2}$$

$$\eta = 1 - \frac{Q_2}{Q_1} = 1 - \frac{T_3 + T_4}{T_1 + T_2}$$

10.130. Three moles of an ideal gas $\left(C_P = \frac{7R}{2}\right)$ at pressure P_A and temperature T_A is isothermally expanded to twice its initial volume. It is then compressed at constant pressure to its original volume. Finally the gas is compressed at constant volume to its original pressure P_A.

(a) Sketch $P-V$ and $P-T$ diagrams for the complete process.

(b) Calculate the net work done by the gas and net heat supplied to the gas during the complete process. (IIT 1991)

Solution: The $P-V$ and $P-T$ diagram is shown in Figs. 10.24a and 10.24b.

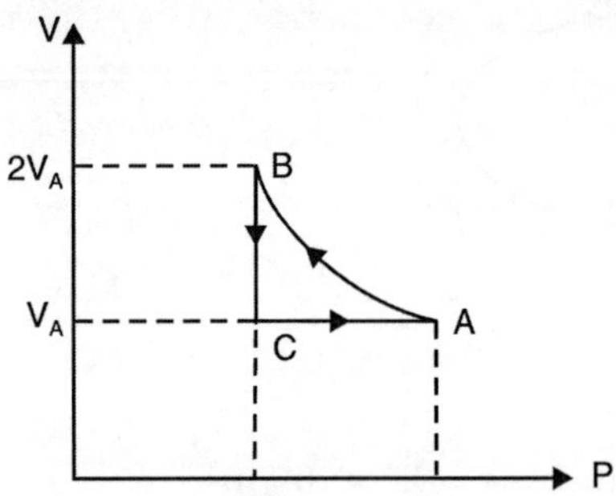

Fig. 10.24a

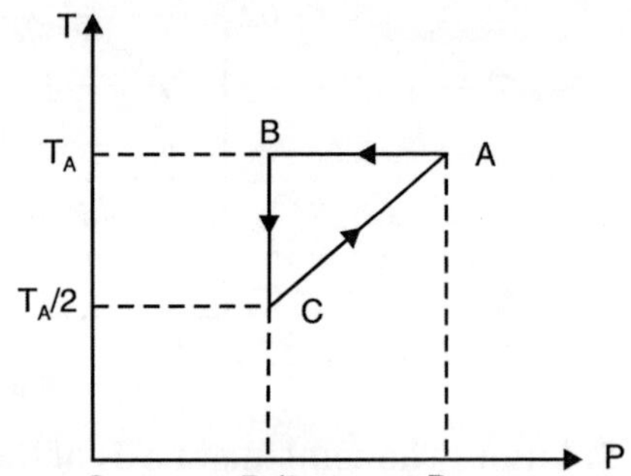

Fig. 10.24b

$$P_A \cdot V_A = P_B \cdot V_B = P_B \cdot 2V_A$$

$$\therefore P_B = \frac{P_A}{2}$$

(b) Work done in part AB

$$W_1 = 3RT\log_e \frac{V_B}{V_A}$$

$$= 3 \times 8.3 \times T_A \times 2.303 \log_{10^2}$$

$$= 3 \times 8.3 \times 2.303 \times 0.3010 \times T_A$$

$$= 17.26\,T_A$$

Work done in part *BC* is given by

$$W_2 = 3P\Delta V = 3P_B(V_C - V_B)$$

$$= \frac{3P_A}{2} \times (V_A - 2V_A)$$

$$= -\frac{3}{2}P_A \cdot V_A = -\frac{3RT_A}{2}$$

On putting the values we have

$$W_2 = -3 \times 8.3 \times \frac{T_A}{2} = -12.45\ T_A$$

Work done in part *CA* is zero since $\Delta V = 0$

W = Net work done = $17.26\ T_A - 12.45 T_A = 4.81\ T_A$

Change in the internal energy is zero.

Heat supplied $Q = W = 4.81 T_A$

10.131. How much energy is absorbed by 10 kg molecules of an ideal gas if it expands from an initial pressure of 8 atmosphere to 4 atmosphere at a constant temperature of 27°C. (Roorkee 1992)

Solution:

$$Q = \int_{V_1}^{V_2} PdV$$

$$= \int_{V_1}^{V_2} \frac{nRT}{V} \cdot dV$$

$$= nRT \cdot \log_e \frac{V_2}{V_1}$$

From Boyle's law $P_1V_1 = P_2V_2$

or $\quad \frac{V_2}{V_1} = \frac{P_1}{P_2}$ hence $Q = nRT \log_e \frac{P_1}{P_2}$

On putting the values we have

$$Q = 10 \times 1000 \times 8.3 \times 300 \times 2.303 \times 0.3010$$

$$= 1.726 \times 10^7 \text{ Joules}$$

10.132. Two moles of helium gas undergo a cyclic process as shown in the Fig. 10.25. Assuming the gas to be ideal. Calculate the following quantities in this process:

(a) The net change in the heat energy.

(b) The net work done.

(c) The net change in internal energy. (IIT 1992)

Solution:

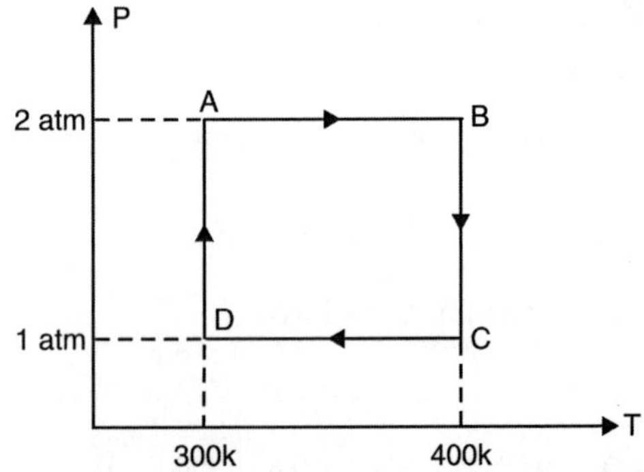

Fig. 10.25

(b) Workdone along AB is given by

$W_{AB} = 2P(V_2 - V_1) = 2R(T_2 - T_1) = 2 \times 8.3 \times (400 - 300) = 1660$ J

Work done along BC

$$W_{BC} = 2RT_C \log_e \frac{V_2}{V_1} = 2RT_C \log_e \frac{P_2}{P_1}$$

$$= 2 \times 8.3 \times 400 \times 2.303 \log_{10^2}$$

$$= 2 \times 8.3 \times 400 \times 2.303 \times 0.3010 = 4602.9 \text{ Joule}$$

Work done along path CD is given by

$$W_{CD} = 2R[T_D - T_C] = 2 \times 8.3 \times (300 - 400)$$

$$= -1660 \text{ Joule}$$

Work done along path DA is given by

$$W_{DA} = 2RT_D \log_e \frac{P_D}{P_A}$$

$$= -2 \times 8.3 \times 300 \times 2.303 \times \log_{10^2}$$

$$= -2 \times 8.3 \times 300 \times 2.303 \times 0.3010$$

$$= -3452.2 \text{ Joule}$$

Net work done $= 4602.9 - 3452.2$

$= 1150.7$ Joule

(c) Since it is a cyclic process $\Delta U = 0$

(a) $\Delta Q = \Delta U + \Delta W$

$\Delta Q = \Delta W = 1150.7$ Joule

10.133. Fig. 10.26 shows three isothermals at temperatures $T_1 = 4000$K, $T_2 = 2000$K and $T_3 = 1000$K. When one mole of an ideal monoatomic gas is taken through the paths $A \to B$, $B \to C$, $C \to D$ and $D \to A$, find the change in the internal energy ΔU, the work done by the gas W and the heat Q absorbed by the gas in each path. Also find these quantities for complete cycle $ABCDA$. Given $V_A = 1$ m³, $V_B = 2$ m³. (Roorkee 1993)

Solution:

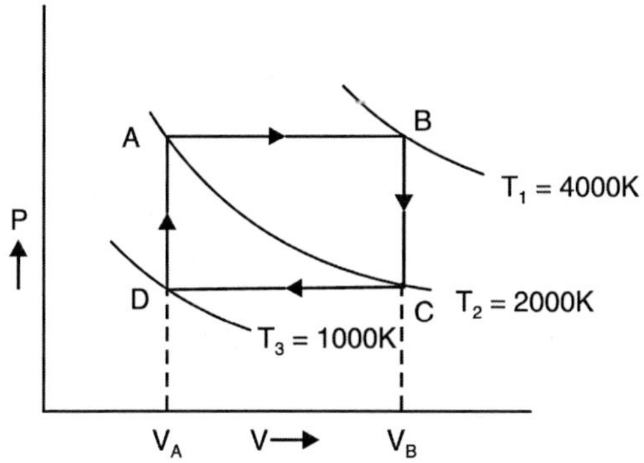

Fig. 10.26

(i) The system returns to its original state hence for this cyclic process $\Delta U = 0$

(ii) Work done along the path AB is given by

$$W_{AB} = P(V_2 - V_1) = R(T_1 - T_2) = 8.31 \times 2000$$
$$= 16.62 \times 10^3 \text{J}$$

Work done along path BC = Work done along path $DA = 0$. Since $\Delta V = 0$

Work done along path $CD = -R(T_2 - T_3) = -8.31 \times 1000$
$= -8.31 \times 10^3$J since it is compression.

Net work done = $(16.62 - 8.31) \times 10^3\text{J} = 8.31 \times 10^3\text{J}$

(iii) Heat absorbed along path $AB = C_p(T_1 - T_2)$

$$= \frac{R\gamma}{\gamma - 1} \cdot (T_1 - T_2)$$

$\gamma = \frac{5}{3}$ for monoatomic gas

$$= \frac{8.31 \times \frac{5}{3} \times 2000}{\frac{2}{3}}$$

$$= 8.31 \times \frac{5}{2} \times 2000 = 41.55 \times 10^3 \text{ J}$$

Heat released along path $BC = C_V[T_1 - T_2]$

$$= \frac{R}{\gamma - 1} \times 2000$$

$$= \frac{8.31}{\frac{5}{3}-1} \times 2000$$

$$= 24.93 \times 10^3 \text{J}$$

Heat released along path $CD = C_P (T_2 - T_3)$

$$= \frac{R\gamma}{\gamma - 1} \times 1000$$

$$= \frac{8.31 \times \frac{5}{3} \times 1000}{\frac{2}{3}}$$

$$= 20.775 \times 10^3 \text{J}$$

Heat absorbed along path $DA = C_V (T_2 - T_3)$

$$= \frac{R}{\gamma - 1} \times 1000$$

$$= 8.31 \times \frac{3}{2} \times 1000$$

$$= 12.465 \times 10^3 \text{J}$$

Net heat absorbed during the complete cycle is given by

$$= (12.465 + 41.55 - 20.775 - 24.93) \times 10^3 \text{J}$$

$$= 8.31 \times 10^3 \text{J}$$

10.134. One mole of a monoatomic ideal gas is taken through the cycle shown in Fig. 10.27. $A \to B$ adiabatic expansion. $B \to C$ cooling at constant volume. $C \to D$ adiabatic compression. $D \to A$ heating at constant volume. The pressure and temperature at A, B etc. are denoted by P_A, P_B ...., T_A, T_B .... etc. respectively. Given that

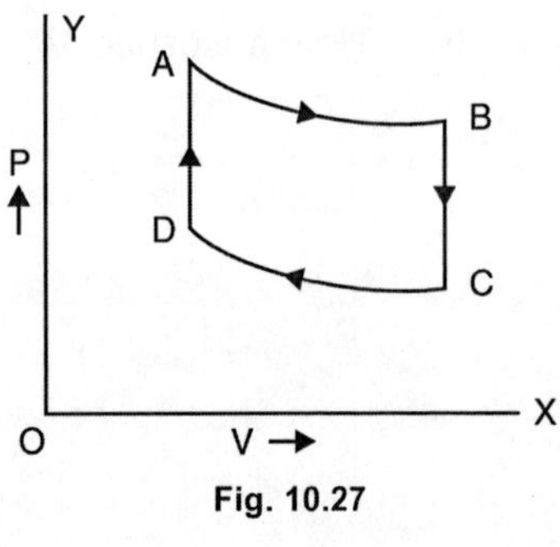

Fig. 10.27

$$T_A = 1000\text{K}, P_B = \frac{2}{3} P_A \text{ and } P_C = \frac{1}{3} P_A$$

Calculate the following quantities:

(i) The work done by the gas in process $A \to B$.

(ii) The heat lost by the gas in process $B \to C$.

(iii) The temperature T_D.

Given that $\left(\frac{2}{3}\right)^{2/5} = 0.85$ (IIT 1993)

Solution: The work done in adiabatic process $A \to B$ is given by

$$W_1 = \frac{R(T_A - T_B)}{\gamma - 1} = \frac{R(T_A - T_B)}{\left(\frac{5}{3} - 1\right)}; \quad \gamma = \frac{5}{3}$$

$$W_1 = \frac{3}{2} R(T_A - T_B)$$

For adiabatic change

$$P_A^{\gamma-1} \cdot T_A^{-\gamma} = P_B^{\gamma-1} \cdot T_B^{-\gamma}$$

$$\left(\frac{P_A}{P_B}\right)^{\frac{2}{3}} = \left(\frac{T_A}{T_B}\right)^{\frac{5}{3}}$$

$$\left(\frac{T_A}{T_B}\right) = \left(\frac{3}{2}\right)^{\frac{2}{5}}$$

$$\therefore T_B = T_A \cdot \left(\frac{2}{3}\right)^{\frac{2}{5}}$$

$$= 1000 \times 0.85 = 850\text{K}$$

$$W_1 = \frac{3}{2} \times 8.31 \times (1000 - 800) = 1869.83\text{J}$$

(ii) Heat lost by the gas in process $B \to C$ is given by

$$C_V(T_B - T_C) = \frac{R}{\gamma - 1}(T_B - T_C) = \frac{3}{2} R(T_B - T_C)$$

Now $\frac{P_D}{T_B} = \frac{P_C}{T_C}$

$$\therefore T_C = \frac{P_C}{P_B} \times T_B$$

$$= \frac{P_A/3}{2P_A/3} T_B = T_B/2 = 425\text{K}$$

$$\text{Heat lost} = \frac{3}{2} \times 8.31\,(850 - 425) = 5293.63\text{J}$$

(iii) For path $A \to B$

$$P_A^{\gamma-1} \cdot T_A^{-\gamma} = P_B^{\gamma-1} \cdot T_B^{-\gamma}$$

$$\left(\frac{P_A}{P_B}\right)^{\gamma-1} = \left(\frac{T_A}{T_B}\right)^{\gamma}$$

For path $B \to C$

$$\frac{P_B}{T_B} = \frac{P_C}{T_C} \quad \text{or} \quad \frac{P_B}{P_C} = \frac{T_B}{T_C}$$

For path $C \to D$

$$\left(\frac{P_D}{P_C}\right)^{\gamma-1} = \left(\frac{P_D}{T_C}\right)^{\gamma}$$

For path $A \to D$

$$\frac{P_A}{P_D} = \frac{T_A}{T_B}$$

$$\left(\frac{P_A}{P_B} \times \frac{P_C}{P_D}\right)^{\gamma-1} = \left(\frac{T_A \times T_C}{T_B \times T_D}\right)^{\gamma}$$

and $\frac{P_A}{P_B} \times \frac{P_C}{P_D} = \frac{T_A}{T_B} \times \frac{T_C}{T_D}$

$$\left(\frac{P_A}{P_B} \times \frac{P_C}{P_D}\right)^{\gamma-1} = \left(\frac{T_A}{T_B} \times \frac{T_C}{T_D}\right)^{\gamma-1}$$

$$\left(\frac{T_A}{T_B} \times \frac{T_C}{T_D}\right)^{\gamma} = \left(\frac{T_A}{T_B} \times \frac{T_C}{T_D}\right)^{\gamma-1}$$

$$\therefore T_A \cdot T_C = T_D \cdot T_B$$

$$1000 \times 425 = T_D \times 850$$

$$T_D = 500\text{K}$$

10.135. There are two vessels. Each of them contains one mole of a monoatomic ideal gas. Initial volume of the gas in each vessel is 8.3×10^{-3} m^3 at 27°C. Equal amount of heat is supplied to each vessel. In one of the vessels, the volume of the gas is doubled without change in its internal energy, whereas the volume of the gas is held constant in the second vessel. The vessels are now connected to allow free mixing of the gas. Find the final temperature and pressure of the combined gas system.

(Roorkee 1994)

Solution: According to first law of thermodynamics

$$\delta Q = dU + \delta W$$

For the first vessel there is no change in temperature hence $dU = 0$

$$\therefore \delta Q = \delta W = PdV$$

$$Q = \int_{V_1}^{V_2} PdV$$

$$= RT \log_e \frac{V_2}{V_1}$$

$$= RT \log_e 2 \quad \text{Q} \quad V_2 = 2V_1$$

For second vessel $\Delta W = 0$

$$Q = C_V dT = \frac{3}{2} R \cdot dT$$

$$RT \log_{e^2} = \frac{3}{2} RdT$$

$$\therefore dT = \frac{2}{3} T \times 2.303 \times \log_{10^2}$$

$$= \frac{2}{3} \times 300 \times 2.303 \times 0.3010$$

$$= 138.6 \text{ K}$$

Temperature of the gas in the second vessel

$$= T + dT = 300 + 138.6 = 438.6\text{K}$$

$$T_f = \text{final temperature} = \frac{T + (T + dT)}{2} = \frac{300 + 438.6}{2}$$

$$= 369.3\text{K}$$

$$P_f = \frac{2RT_f}{V_f} = \frac{2 \times 8.3 \times 369.3}{3 \times 8.3 \times 10^{-3}}$$

$$= 2.46 \times 10^5 \text{ N/m}^2$$

10.136. An ideal gas is taken through a cyclic thermodynamic process through four steps. The amount of heat involved in these steps are Q_1 = 5960J, Q_2 = –5585J, Q_3 = –2980J and Q_4 = 3645J respectively. The corresponding works involved are W_1 = 2200J, W_2 = – 825J, W_3 = –1100J and W_4 respectively.

(i) Find the value of W_4

(ii) What is the efficiency of the cycle? (IIT 1994)

Solution: Net heat absorbed

$$Q = Q_1 + Q_2 + Q_3 + Q_4$$

$$= 5960 - 5585 - 2980 + 3645$$

$$= 1040\text{J}$$

Net work obtained

$$W = W_1 + W_2 + W_3 + W_4$$

$$= 2200 - 825 - 1100 + W_4$$

$$= 275 + W_4$$

For the cyclic process $dU = 0$

$$\therefore 1040 = 275 + W_4$$

$$\therefore W_4 = 1040 - 275 = 765\text{J}$$

(ii) $$\eta = \frac{\text{Work obtained}}{\text{Heat absorbed}} = \frac{1040}{5960 + 3645}$$
$$= 0.1082 = 10.82\%$$

10.137. A vertical hollow cylinder contains an ideal gas. The gas is enclosed by a 5 kg movable piston with an area of cross-section 5×10^{-3} m^2. Now, the gas is heated from 300 K to 350 K and the piston rises by 0.1 m. The piston is now clamped at this position and the gas is cooled back to 300 K. Find the difference between the heat energy added during heating process and energy lost during the cooling process (1 atmospheric pressure = 10^5 N/m^2). (Roorkee 1996)

Solution:

$$\delta Q = dU + \delta W$$

$$\delta Q - dU = PdV = \left[P_0 + \frac{Mg}{A}\right] A \cdot dx$$

$$= [P_0 A + Mg]\, dx = [10^5 \times 5 \times 10^{-3} + 5 \times 10] \times 0.1 = 55\text{J}$$

10.138. A sample of 2 kg of monoatomic helium (assumed ideal) is taken through the process *ABC* and another sample of 2 kg of the same gas is taken through the process *ADC* (see Fig. 10.28). Given molecular mass of helium = 4.

(i) What is the temperature of helium in each of the states *A, B, C* & *D*?

(ii) Is there any way of telling afterwards which sample of helium went through the process *ABC* and which went through the process *ADC*? Write Yes or No.

(iii) How much is the heat involved in each of the processes *ABC* and *ADC*? (IIT 1997)

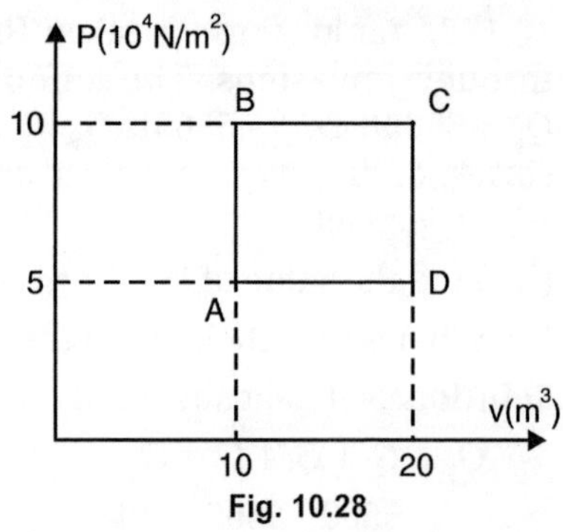

Fig. 10.28

Solution:

(i) $$n = \text{mass/mol wt}$$
$$= \frac{2 \times 10^3}{4} = 500 \text{ mol}$$

$$T_A = \frac{5 \times 10^4 \times 10}{500\, R} = \frac{1000}{R} = \frac{1000}{8.31} = 120.33 \text{ K}$$

$$T_B = \frac{10 \times 10^4 \times 10}{500 R} = \frac{2000}{R} = 240.66 \text{ K}$$

$$T_C = \frac{10 \times 10^4 \times 20}{500R} = \frac{4000}{R} = 481.32\ \text{K}$$

$$T_D = \frac{5 \times 10^4 \times 20}{500R} = \frac{2000}{R} = 240.66\ \text{K}$$

(ii) No, as change in internal energy is path independent.

(iii) For process ABC

Work done along $AB = 0$

Work done along $BC = P(V_2 - V_1) = 10^5 (20 - 10) = 10^6$ J

$\therefore$ Work done along $ABC = 10^6$ J

$$dU = nC_V dT = 500 \times \frac{3}{2} R [T_C - T_A]$$

$$= 750 R \left[\frac{4000}{R} - \frac{1000}{R}\right] = 2.25 \times 10^6\ \text{J}$$

$$\delta Q = dU + \delta W = 2.25 \times 10^6 + 10^6$$

$$= 3.25 \times 10^6\ \text{J}$$

Work done along path ADC = Work along path AD + work along path DC

$$= 5 \times 10^4 \times (20 - 10) + 0 = 0.5 \times 10^6\ \text{J}$$

$$dU = 2.25 \times 10^6\ \text{J}$$

$$\delta Q = (2.25 + 0.5) \times 10^6 = 2.75 \times 10^6\ \text{J}$$

10.139. Two moles of an ideal mono atomic gas are confined within a cylinder by a massless and frictionless spring loaded piston of cross-sectional area 4×10^{-3} m^2. The spring is, initially, in its relaxed state. Now the gas is heated by an electric heater, placed inside the cylinder, for some time. During this time the gas expands and does 50 J of work in moving the piston through a distance 0.1 m. The temperature of the gas increases by 50 K. Calculate the spring constant and the heat supplied by the heater. (Roorkee 1997)

Solution:

$$W = \text{work done} = \int P dV = \int \left(P_0 + \frac{Kx}{A}\right) A dx$$

$$W = P_0 A x + \frac{1}{2} K x^2$$

On putting the values

$$50 = 10^5 \times 4 \times 10^{-3} \times 0.1 + \frac{1}{2} Kx (0.1)^2$$

$$K = \frac{2(50 - 40)}{0.01} = 2000\ \text{N/m}$$

$$dU = 2\,C_V\,dT = 2 \times \frac{3}{2} R \times 50 = 150 \times 8.3 = 1245 \text{ J};$$

$$C_V = \frac{3}{2} R \text{ for monoatomic gas}$$

$$\delta Q = dU + \delta W = 1245 + 50 = 1295 \text{ J}$$

10.140. Two vessels *A* and *B*, thermally insulated, contain an ideal monoatomic gas. A small tube fitted with a valve connects these vessels. Initially the vessel *A* has 2 litres of gas at 300 K and 2×10^5 N/m^2 pressure while vessel *B* has 4 litres of gas at 350 K and 4×10^5 N/m^2 pressure. The valve is now opened and the system reaches equilibrium in pressure and temperature. Calculate the new pressure and temperature.

(Roorkee 1997)

Solution:

For ideal gas $U = \frac{3}{2} nRT = \frac{3}{2} PV$

$$\frac{3}{2} P_1V_1 + \frac{3}{2} P_2V_2 = \frac{3}{2} PV$$

$$\therefore P = \frac{P_1V_1 + P_2V_2}{V}$$

On putting the values we have

$$P = \frac{2 \times 2 \times 10^5 + 4 \times 4 \times 10^5}{2+4} = 3.3 \times 10^5 \text{ N/m}^2$$

$n = n_1 + n_2$ and therefore

$$\frac{P_1V_1}{RT_1} + \frac{P_2V_2}{RT_2} = \frac{PV}{RT}$$

$$\frac{2 \times 10^5 \times 2}{300} + \frac{4 \times 10^5 \times 4}{350} = \frac{10 \times 10^5 \times 6}{3T}$$

$$\frac{4}{300} + \frac{16}{350} = \frac{20}{T}$$

$$T = 338.71 \text{ K}$$

10.141. An ideal gas undergoes a cyclic operation consisting of alternate isothermals at temperatures T_1, T_2 & T_3 and adiabatics as shown in the Fig. 10.29. In each isothermal expansion the volume of the gas increases in the same proportion. Find the efficiency.

Solution: Heat absorbed at temperature T_1 is given by

$$Q_1 = RT_1 \log_e \frac{V_2}{V_1}$$

$$= RT_1 \log_e n; \qquad V_2 = nV_1$$

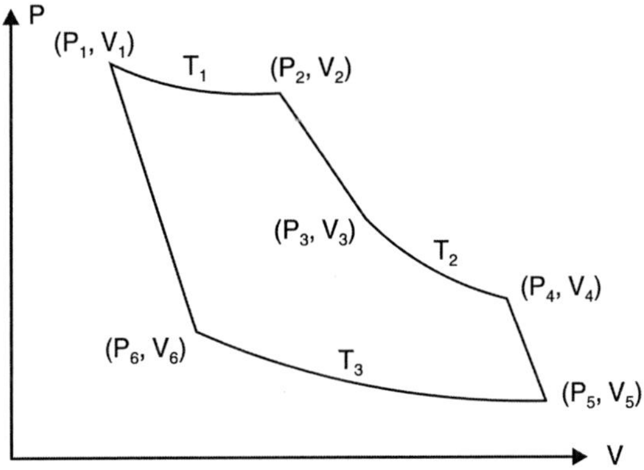

Fig. 10.29

Heat absorbed at temperature T_2 is given by

$$Q_2 = RT_2 \log_e \frac{V_4}{V_3}$$

$$= RT_2 \log_e n; \qquad V_4 = nV_3$$

Total heat absorbed $= R(T_1 + T_2) \log_e n$

Heat rejected at temperature T_3 is given by

$$Q_3 = -RT_3 \log_e \cdot \frac{V_6}{V_5}$$

Along the adiabatics we have

$$T_1 V_2^{\gamma-1} = T_2 V_3^{\gamma-1}$$

or $$V_3 = \left(\frac{T_1}{T_2}\right)^{1/\gamma-1} \cdot V_2$$

Similarly

$$T_2 V_4^{\gamma-1} = T_3 V_5^{\gamma-1}$$

$$\therefore V_5 = \left(\frac{T_2}{T_3}\right)^{1/\gamma-1} \cdot V_4$$

and $$V_6 = \left(\frac{T_1}{T_3}\right)^{1/\gamma-1} \cdot V_1$$

$$Q_3 = -RT_3 \log_e \left[\left(\frac{T_1}{T_2}\right)^{1/\gamma-1} \cdot \frac{V_1}{V_4}\right]$$

$$= -RT_3 \log_e \left[\left(\frac{T_1}{T_2}\right)^{1/\gamma-1} \cdot \frac{V_2}{n^2 V_3}\right]$$

$$= -RT_3 \log_e \left[\left(\frac{T_1}{T_2}\right)^{1/\gamma-1} \times \frac{1}{n^2}\cdot\left(\frac{T_1}{T_2}\right)^{-1/\gamma-1}\right]$$

$$= 2RT_3 \log_e n$$

$$\eta = 1 - \frac{Q_3}{Q_1+Q_2} = 1 - \frac{2T_3}{T_1+T_2}$$

10.142. A diatomic gas goes through a cycle consisting of two isochoric and two adiabatic curves. During adiabatic process the volume becomes n fold. Assuming the gas as an ideal one find the efficiency of the cycle.

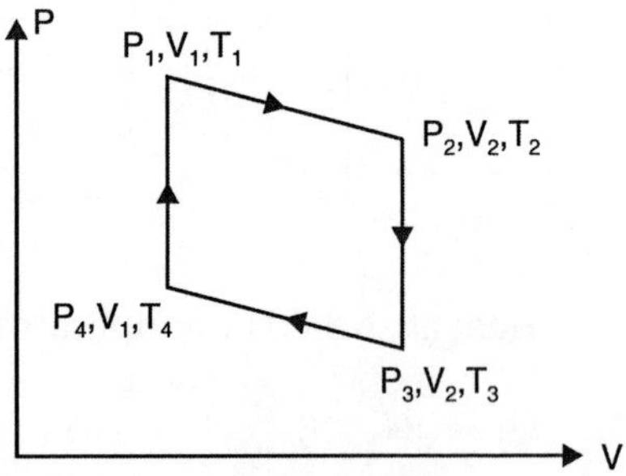

Fig. 10.30

Solution: Amount of heat absorbed during isochoric process is given by

$$Q_1 = C_V(T_1 - T_4) = \frac{C_V}{R}V_1[P_1 - P_4];$$

$P_1V_1 = RT_1$ and $P_4V_1 = RT_4$ (see Fig. 10.30)

Heat rejected at volume V_2 is given by

$$Q_2 = C_V[T_2 - T_3] = \frac{C_V}{R}V_2[P_2 - P_3]$$

$$\eta = 1 - \frac{Q_2}{Q_1} = 1 - \frac{V_2(P_2 - P_3)}{V_1(P_1 - P_4)}$$

Along the adiabatics we have

$P_1V_1^\gamma = P_2V_2^\gamma$ and $P_3V_2^\gamma = P_4V_1^\gamma$

$P_1 = P_2 n^\gamma$ and $P_4 = P_3 \cdot n^\gamma$

$$\text{Efficiency } \eta = 1 - n\frac{\left[P_1/n^\gamma - P_4/n^\gamma\right]}{(P_1 - P_4)}$$

$$= 1 - n\cdot n^{-\gamma} = 1 - n^{1-\gamma}$$ For a diatomic gas $\gamma = 7/5$

$$= 1 - n^{1-7/5} = 1 - n^{-2/5}$$

10.143. A monoatomic gas with the adiabatic exponent $\gamma = 5/3$ goes through a cycle consisting of two isobaric and two adiabatic curves as shown in the Fig. 10.31. During the adiabatic process the pressure becomes n fold. Assuming the gas to be an ideal one, find the efficiency of the cycle.

Solution: Amount of heat absorbed is given by

$$Q_1 = \frac{C_P}{R} P_1 (V_2 - V_1)$$

Amount of heat rejected is given by

$$Q_2 = \frac{C_P}{R} \cdot P_2 (V_3 - V_4)$$

Efficiency $\eta = 1 - \frac{Q_2}{Q_1} = 1 - \frac{P_2 (V_3 - V_4)}{P_1 (V_2 - V_1)}$

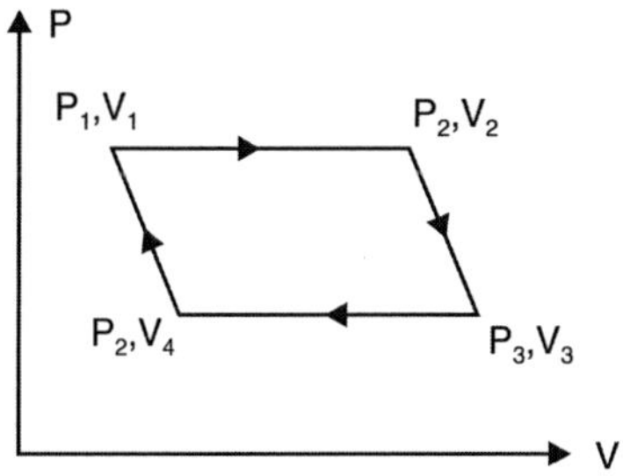

Fig. 10.31

Along the two adiabatics we have

$P_1 V_2^{\gamma} = P_2 V_3^{\gamma}$ and $P_1 V_1^{\gamma} = P_2 V_4^{\gamma}$

Further $P_1 = nP_2$

$nV_2^{\gamma} = V_3^{\gamma}$ and $nV_1^{\gamma} = V_4^{\gamma}$

$$\eta = 1 - \frac{1}{n} \frac{\left[n^{1/\gamma} \cdot V_2 - n^{1/\gamma} \cdot V_1 \right]}{V_2 - V_1}$$

$$= 1 - n^{1/\gamma - 1}$$

$$= 1 - n^{-2/5}$$

10.144. An ideal gas with the adiabatic exponent γ passes through a cycle consisting of two isobaric and two isochoric lines as shown in the Fig. 10.32. The temperature of the gas increases n times during isochoric heating and in isobaric expansion. Find the efficiency.

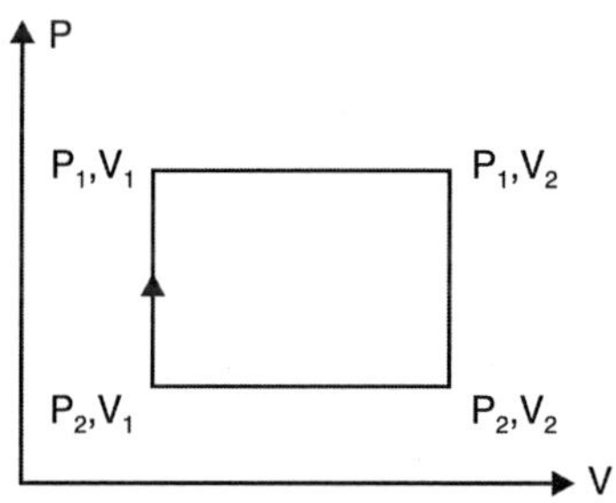

Fig. 10.32

Solution: Heat absorbed during isochoric process is given by

$$Q_1 = C_V \left[T_1 - T_1/n \right]$$

Heat absorbed during isobaric expansion is given by

$$Q_1' = C_P [nT_1 - T_1]$$

Heat rejected along the two remaining operations is given by

$$Q_2 + Q_2' = C_V T_1 (n - 1) + C_P \left(T_1 - T_1/n \right)$$

$$\text{Efficiency } \eta = 1-\frac{Q_2+Q_2'}{Q_1+Q_1'} = 1-\frac{C_V(n-1)+C_P\left(1-1/n\right)}{C_V\left(1-1/n\right)+C_P(n-1)}$$

$$= 1-\frac{(n-1)+\gamma\left(1-1/n\right)}{\left(1-1/n\right)+\gamma(n-1)}$$

$$= 1-\frac{n\left(1-1/n\right)+\gamma\left(1-1/n\right)}{\left(1-1/n\right)+n\gamma\left(1-1/n\right)}$$

$$= 1-\frac{n+\gamma}{1+n\gamma}$$

10.145. Find the amount of work done by a mole of a Vander Waal's gas upon isothermal expansion from volume V_1 to V_2. The temperature of the gas is T.

Solution: The Vander Waal's equation of state is given by

$$\left(P+\frac{a}{V^2}\right)(V-b) = RT$$

$$\text{Work done } W = \int_0^W dW = \int_{V_1}^{V_2}\left(\frac{RT}{V-b}-\frac{a}{V^2}\right)dV$$

$$= RT\log_e\frac{V_2-b}{V_1-b}+a\left(\frac{1}{V_2}-\frac{1}{V_1}\right)$$

10.146. Obtain an equation of an adiabatic in the variables V & T and P & V for a Vander Waal's gas.

Solution: In adiabatic process we have $TV^{\gamma-1}$ = Constt.

$$TV^{\gamma-1} = TV^{\frac{C_P}{C_V}-1} = TV^{\frac{R}{C_V}} = T(V-b)^{\frac{R}{C_V}} = \text{Constt.}$$

$$\text{Further } \left(P+\frac{a}{V^2}\right)(V-b)^{\gamma} = \text{Constt.}$$

$$\text{or } \quad \left(P+\frac{a}{V^2}\right)(V-b)^{\frac{C_P}{C_V}} = \text{Constt.}$$

$$\text{or } \quad \left(P+\frac{a}{V^2}\right)(V-b)^{\frac{R}{C_V}+1} = \text{Constt.}$$

10.147. One mole of a gas at temperature 273K occupies a volume = 0.55 lit. Assuming it to be an actual gas. Calculate the pressure that it exerts. $a = 0.37$ Nm^4 mol^{-2}, $b = 43 \times 10^{-6}$ m^3 mol^{-1} and $R = 8.31$ J mol^{-1} K^{-1}.

Solution:

$$P = \frac{RT}{V-b} - \frac{a}{V^2}$$

$$= \frac{8.31 \times 273}{(550-43)\times 10^{-6}} - \frac{0.37}{\left(550\times 10^{-6}\right)^2}$$

$$= 4.47 \times 10^6 - 1.22 \times 10^6$$

$$= 3.25 \times 10^6\ \text{N/m}^2$$

10.148. Two moles of hydrogen expands into an evacuated chamber as a result of which the volume of the gas increases from $V_1 = 2$ lit. to $V_2 = 10$ lit. What amount of heat Q must be imparted to the gas for its temperature to remain unchanged? $a = 0.25$ atmos · lit²/mol².

Solution: For a Vander Waal's gas the internal energy varies with volume as

$$\frac{dU}{dV} = \frac{a}{V^2}$$

$$dU = \frac{a}{V^2}\cdot dV$$

$$Q = U = \int_{V_1}^{V_2} \frac{a}{V^2}\, dV = \frac{a\left(V_2 - V_1\right)}{V_1 V_2}$$

The above equation is for one mole of a gas. For υ moles we have

$$Q = \frac{\upsilon^2 \cdot a\left(V_2 - V_1\right)}{V_1 V_2}$$

On substituting the values we have

$$Q = 2^2 \times 0.24 \times 10^5 \times 10^{-6} \times \frac{2}{5\times 10^{-3}}$$

$$= 38.4\ \text{J}$$

10.149. Two thermally insulated vessels are interconnected by a tube equipped with a valve. One vessel of volume V_1 contains υ moles of Co_2. The other vessel of volume V_2 is evacuated. The valve having been opened, the gas adiabatically expanded. The gas obeys Vander Waal's equation. Find the change in its temperature.

Solution:

$$\frac{dU}{dV} = \frac{a}{V^2}$$

$$\therefore dU = \frac{a}{V^2}\, dV$$

$$\int dU = \Delta U = \int_{V_1}^{V_1+V_2} \frac{a dV}{V^2} = \frac{aV_2}{V_1(V_1+V_2)}$$

For υ moles of the gas $\Delta U = \dfrac{\upsilon^2 aV_2}{V_1(V_1+V_2)} = -\dfrac{\upsilon R}{\gamma-1}(T_1 - T_2)$

$$\therefore T_1 - T_2 = -\frac{\upsilon a V_2(\gamma-1)}{V_1 R(V_1+V_2)}$$

10.150. One mole of nitrogen expands adiabatically into an evacuated enclosure as a result of which its volume increases from V_1 = 1 lit to V_2 = 10.0 lit. Determine the increment ΔT of the temperature of the gas. a = 1.35 atmos · lit²/mol².

Solution:

$$\frac{dU}{dV} = \frac{a}{V^2}$$

$$\int dU = \int \frac{a}{V^2} \cdot dV$$

$$\Delta U = \int_{V_1}^{V_2} \frac{a}{V^2} dV = \frac{a(V_2 - V_1)}{V_1 V_2}$$

$$\Delta U = \frac{a(V_2 - V_1)}{V_1 V_2} = -\frac{R}{\gamma-1}(T_1 - T_2)$$

$$\therefore \Delta T = T_1 - T_2 = -\frac{a(V_2 - V_1)(\gamma-1)}{RV_1 V_2}$$

$$= -\frac{1.35\times10^5\times10^{-6}}{8.31\times10\times10^{-3}}\times\frac{9}{10}\times\frac{2}{5}$$

$$= -\frac{1.35\times18\times10^2}{8.31\times50} = -5.84 \text{ K}$$

10.151. One mole of an ideal gas expands from (p_1, V_1, T) to (p_2 V_2, T) in two stages. In the first stage the opposing constant pressure is P and the volume changes from V_1 to V and in the second stage the opposing pressure is p_2 and the volume changes from V to V_2. For what value of P work is maximum? What is the maximum value of the work done?

Solution:

$W = P(V - V_1) + p_2(V_2 - V)$

Further $p_1 V_1 = p_2 V_2 = PV = RT$

$$\therefore W = P\left(\frac{RT}{P} - \frac{RT}{p_1}\right) + p_2\left(\frac{RT}{p_2} - \frac{RT}{P}\right) = RT\left[2 - \frac{P}{p_1} - \frac{p_2}{P}\right]$$

$$\frac{dW}{dP} = RT\left[-\frac{1}{p_1} + \frac{p_2}{P^2}\right] = 0$$

$$\therefore P = \sqrt{p_1 p_2}$$

$$W_{\max} = RT\left[2 - \sqrt{\frac{p_2}{p_1}} - \sqrt{\frac{p_2}{p_1}}\right]$$

$$= 2RT\left[1 - \sqrt{\frac{p_2}{p_1}}\right]$$

10.152. The critical temperature and pressure of argon is – 122°C and 48 atmosphere respectively. Calculate the radius of the argon atom.

Solution:

$$\frac{RT_C}{P_C V_C} = \frac{8}{3}$$

$$\frac{RT_C}{P_C \cdot 3b} = \frac{8}{3}$$

$$\therefore b = \frac{RT_C}{8P_C}$$

On putting the values we have

$$b = \frac{8.31 \times 151}{8 \times 48 \times 1.013 \times 10^5}$$

$$= 3.22 \times 10^{-5} \text{ m}^3/\text{mol}$$

Further $b = 4 \times \frac{4}{3}\pi r^3 \cdot N$; N = Avagadro number

$$\therefore r^3 = \frac{3b}{16\pi N} = \frac{3 \times 3.22 \times 10^{-5}}{16 \times 3.14 \times 6.02 \times 10^{23}}$$

$$= 3.2 \times 10^{-30}$$

$$\therefore r = \left(3.2 \times 10^{-30}\right)^{1/3}$$

$$= 1.475 \text{ A}$$

10.153. What pressure is required to convert Co_2 gas into liquid state at a temperature of (i) 31°C (ii) 40°C. $a = 3.62$ atmos $\cdot\, l^2/\text{mol}^2$; $b = 0.043\ l/\text{mol}$.

Solution:

(i) 31°C is the critical temperature of Co_2 hence the required pressure is the critical pressure $= \frac{a}{27b^2}$

$$p_C = \frac{3.62}{27 \times (0.043)^2} = 72.5 \text{ atmosphere}$$

(ii) At 40°C any pressure will not turn it into liquid state.

10.154. Calculate the critical temperature of hydrogen $a = 0.24$ atmos · l^2/mol^2, $b = 0.027\ l/\text{mol}$.

Solution:

$$T_C = \frac{8a}{27Rb}$$

$$= \frac{8 \times 0.24 \times 1.013 \times 10^5 \times 10^{-6}}{27 \times 8.31 \times 0.027 \times 10^{-3}} = 32.1\text{ K}$$

10.155. Find the pressure due to the forces of interaction of the molecules contained in one kilomole of a gas at N.T.P. The critical temperature and critical pressure of this gas are $T_C = 417$ K and $P_C = 76$ atmosphere respectively.

Solution:

$$P_i = \frac{a}{V^2} = \frac{aP^2}{R^2T^2}$$

$$= \frac{aP^2}{T^2} \times \frac{27^2 b^2 T_C^2}{64a^2};\ T = \frac{8a}{27Rb}$$

$$= \frac{27^2 T_C^2 P^2}{64T^2} \times \frac{b^2}{a}$$

$$= \frac{27 T_C^2 P^2}{64 p_C \cdot T^2};\ p_C = \frac{a}{27b^2}$$

On substituting the values we have

$$P_i = \frac{27 \times 417^2 \times 1.013 \times 10^5}{64 \times 76 \times 273^2} = 1.31 \times 10^3\ \text{N/m}^2$$

10.156. Find the value of $C_p - C_V$ for a Vander Waal's gas.

Solution:

$$p = \frac{RT}{V-b} - \frac{a}{V^2}$$

$$\left(\frac{\partial p}{\partial T}\right)_V = \frac{R}{V-b}$$

On differentiating the initial equation w.r.t. T keeping p constt. we have

$$0 = \frac{R}{V-b} - \frac{RT}{(V-b)^2}\left(\frac{\partial V}{\partial T}\right)_p + \frac{2a}{V^3}\left(\frac{\partial V}{\partial T}\right)_p$$

$$\left(\frac{\partial V}{\partial T}\right)_p = \frac{\dfrac{R}{V-b}}{\dfrac{RT}{(V-b)^2} - \dfrac{2a}{V^3}}$$

We know that $C_p - C_V = T\left(\frac{\partial p}{\partial T}\right)_V \cdot \left(\frac{\partial V}{\partial T}\right)_p$

On putting the values we have

$$C_P - C_V = \frac{\dfrac{T\cdot R}{V-b}\cdot\dfrac{R}{V-b}}{\dfrac{RT}{(V-b)^2} - \dfrac{2a}{V^3}} = \frac{R}{1 - \dfrac{2a(V-b)^2}{RTV^3}}$$

10.157. A process in a system is called polytropic if it does not change the systems heat capacity $C = dQ/dT$. Find the equation of such a process for a monoatomic perfect gas.

Solution:

$$dQ = C\cdot dT = \frac{C}{R}(PdV + VdP) = C_V dT + PdV$$

$$\frac{C(PdV + VdP)}{R} = C_V \cdot \frac{VdP}{R} + PdV$$

or $\quad C(PdV + VdP) - C_V VdP - RPdV = 0$

or $\quad C PdV + C VdP - C_V VdP - (C_P - C_V) PdV = 0$

or $\quad C PdV + C VdP - C_V VdP - C_P PdV = 0;\ PdV \cdot C_V = 0$

or $\quad (C_P - C) PdV + (C_V - C) VdP = 0$

or $\quad (C_P - C)\dfrac{dV}{V} + (C_V - C)\dfrac{dP}{P} = 0$

or $\quad \dfrac{dP}{P} + \left(\dfrac{C_P - C}{C_V - C}\right)\dfrac{dV}{V} = 0$

On integration we have

$$PV^n = \text{Constt. where } n = \frac{C_P - C}{C_V - C}$$

10.158. A real gas satisfies Vander Waal's equation of state. Determine the velocity of sound in such a gas.

Solution: The velocity of sound in a gas is determined by

$$v_{\text{sound}} = \sqrt{\left(\frac{dP}{d\rho}\right)_S}\ ;\ S \text{ stands for entropy}$$

We know that adiabatic elasticity = $\gamma \times$ isothermal elasticity

$$\therefore \left(\frac{dP}{dV}\right)_S = \gamma\left(\frac{dP}{dV}\right)_T$$

Further $V\rho = M$; M = mol.wt

$$\rho dV + V d\rho = 0$$

$$\therefore dV = -\frac{V d\rho}{\rho}$$

$$\rho\left(\frac{dP}{-V d\rho}\right) = \gamma\left(\frac{dP}{dV}\right)_T$$

$$\frac{dP}{d\rho} = -\frac{\gamma V}{\rho}\cdot\left(\frac{dP}{dV}\right)_T$$

$$= -\frac{\gamma V^2}{M}\cdot\left(\frac{dP}{dV}\right)_T$$

$$\therefore v_{\text{sound}} = \sqrt{\frac{dP}{d\rho}} = V\sqrt{-\frac{\gamma}{M}\left(\frac{dP}{dV}\right)_T}$$

Now $$P = \frac{RT}{V-b} - \frac{a}{V^2}$$

$$\left(\frac{dP}{dV}\right)_T = -\frac{RT}{(V-b)^2} + \frac{2a}{V^3}$$

$$\therefore v_{\text{sound}} = V\sqrt{\frac{\gamma}{M}\left[\frac{RT}{(V-b)^2} - \frac{2a}{V^3}\right]}$$

$$\cong \frac{V}{V-b}\cdot\sqrt{\frac{\gamma RT}{M}}$$

$$\cong \frac{V}{V-b}\cdot v_{\text{perfect gas}}$$

10.159. A vessel of volume V is divided into two equal parts by a partition with an opening closed by a plug. One of the halves of the vessel contains a mole of Vander Waal's gas (with a, b & C_V known) having the temperature T. The plug is now removed and the entire space is being occupied by the gas. Assuming the expansion of the gas to be an adiabatic find

(i) the increment of the internal energy

(ii) the work of the forces of inter molecular attraction

(iii) the increment of the temperature.

Solution:

(i) $\Delta U = 0$

(ii) $\dfrac{dU}{dV} = \dfrac{a}{V^2}$

$$\Delta U = \int dU = \int_{V/2}^{V} \frac{a}{V^2} dV = -\frac{a}{V} + \frac{2a}{V} = \frac{a}{V}$$

In adiabatic process $\delta Q = 0 = \Delta U + \delta W$

$$\therefore \delta W = -\Delta U = -\frac{a}{V}$$

(iii) $-\Delta U = C_V dT$

$$\therefore dT = -\frac{\Delta U}{C_V} = -\frac{a}{VC_V}$$

10.160. Find the adiabatic equation of a gas whose equation of state is $p = p_0 (1 + \alpha T - \beta V)$; C_V = Constt.

Solution: In adiabatic process $\delta Q = 0 = dU + pdV$

$$U = f(V, T)$$

$$dU = \left(\frac{\partial U}{\partial T}\right)_V dT + \left(\frac{\partial U}{\partial V}\right)_T dV$$

$$0 = \left(\frac{\partial U}{\partial T}\right)_V dT + \left(\frac{\partial U}{\partial V}\right)_T dV + pdV$$

$$= C_V dT + \left[\left(\frac{\partial U}{\partial V}\right)_T + p\right] dV$$

We know that $\left(\dfrac{\partial U}{\partial V}\right)_T = T\left(\dfrac{\partial p}{\partial T}\right)_V - p$

For the given equation

$$T\left(\frac{\partial P}{\partial T}\right)_V = T(p_0 \alpha)$$

$$\therefore \left(\frac{\partial U}{\partial V}\right)_T = Tp_0\alpha - p_0 - p_0 \alpha T + p_0 \beta V$$

$$= p_0 (\beta V - 1)$$

$$\therefore 0 = C_V dT + p_0 (\beta V - 1) dV + p_0 (1 + \alpha T - \beta V) dV$$

$$= C_V dT + p_0 \alpha T dV$$

or $\quad C_V \cdot \dfrac{dT}{T} + p_0 \alpha dV = 0$

On integration we have

$$C_V \log T + p_0 \alpha V = \text{Constt.}$$

10.161. A gas follows Berthelot's equation of state $p = \frac{RT}{V-b} - \frac{a}{TV^2}$. Find the values of the critical constants.

Solution:

$$\left(\frac{dP}{dV}\right)_T = \frac{-RT}{(V-b)^2} + \frac{2a}{TV^3} = 0$$

$$\therefore \frac{RT}{(V-b)^2} = \frac{2a}{TV^3}$$

$$\left(\frac{d^2P}{dV^2}\right)_T = \frac{2RT}{(V-b)^3} - \frac{6a}{TV^4} = 0$$

$$\therefore \frac{RT}{(V-b)^3} = \frac{3a}{TV^4}$$

On solving $V = V_C = 3b$

Now $$\frac{RT_C}{4b^2} = \frac{2a}{27b^3T_C}$$

$$\therefore T_C^2 = \frac{8a}{27Rb}$$

or $$T_C = \sqrt{\frac{8a}{27Rb}}$$

$$p_C = \frac{R}{2b}\sqrt{\frac{8a}{27Rb}} - \frac{a\sqrt{27Rb}}{\sqrt{8a}\cdot 9b^2}$$

$$= \frac{1}{12b}\sqrt{\frac{2aR}{3b}}$$

10.162. The entropy of v moles of an ideal gas increases by ΔS due to isothermal expansion. How many times should the volume of the gas be increased?

Solution:

$$\int_{S_1}^{S_2} dS = v\int\frac{\delta Q}{T} = v\int\frac{pdV}{T}$$

$$S_2 - S_1 = \Delta S = vR\int_V^{nV}\frac{dv}{v}$$

$$= vR\log n$$

$$n = e^{\Delta S/vR}$$

10.163. Find the increase in entropy of a mole of a monoatomic ideal gas when heated from 0 to 273°C if heating occurs (i) at constant volume (ii) at constant pressure.

Solution:

(i) $$\Delta S=\int_{273}^{546}\frac{3}{2}R\frac{dT}{T}=\frac{3}{2}R\log_e 2=1.5\times 8.31\times 0.693=8.6\text{ J/mol.K}$$

(ii) $$\Delta S=\int_{273}^{546}\frac{5}{2}R\frac{dT}{T}=\frac{5}{2}R\log_e 2=2.5\times 8.31\times 0.693=14.4\text{ J/mol.K}$$

10.164. The entropy of a mole of oxygen at a temperature of 25°C and a pressure of 10^5 Pa is $S_1=204.8$ J/mol.K. As a result of isothermal expansion, the volume occupied by the gas increases two-fold. Determine the entropy in the final state.

Solution:

$$S_2-S_1 = \Delta S=\int\frac{\delta Q}{T}$$

$$= \int\frac{pdV}{T}$$

$$= R\int_{v_0}^{2v_0}\frac{dv}{v}=R\log_e 2$$

$$= 8.31\times 0.693$$

$$= 5.7588$$

$$S_2 = 204.8+5.7588=210.55=210.6\text{ J/mol.K}$$

10.165. In a certain temperature region, the entropy of a thermodynamical system varies with temperature via equation $S=a+bT$ where $a=100$ J/K and $b=5$ J/K^2. What amount of heat does the system receive upon reversible heating in this region from $T_1=290$ K to $T_2=310$ K.

Solution:

$$\int_{T_1}^{T_2}dS = \int\frac{\delta Q}{T}$$

$$Q = \int_{T_1}^{T_2}TdS=b\int_{T_1}^{T_2}TdT=\frac{b}{2}\left[T_2^2-T_1^2\right]$$

On putting the values $Q=\dfrac{5}{2}\left[310^2-290^2\right]=30$ KJ

10.166. In the course of a reversible isothermal process occurring at a temperature of $T=350$ K a body does the work of 80 J while its internal energy increases by 7.5 J. Find the increase in the entropy of the body.

Solution:

$$\Delta S=\frac{\delta Q}{T}=\frac{dU+PdV}{T}=\frac{7.5+80}{350}=0.25\text{ J/mol. K}$$

10.167. One of the two thermally insulated vessels interconnected by a tube with a valve contains υ moles of an ideal gas. The other vessel is evacuated. The valve having been opened, the gas increased its volume n times. Find the increase in entropy.

Solution:

For one mole of gas $\Delta s = -\dfrac{R}{\gamma - 1}\displaystyle\int_{T_1}^{T_2} \frac{dT}{T} = \frac{R}{\gamma - 1}\log_e \frac{T_1}{T_2}$

For an adiabatic process $T_1 V^{\gamma-1} = T_2\,(nV)^{\gamma-1}$

or $\dfrac{T_1}{T_2} = n^{\gamma-1}$

$$\therefore \Delta s = \frac{R}{\gamma - 1}\log_e n^{\gamma-1} = R\log_e n$$

For υ moles the change in entropy $\Delta S = \upsilon \Delta s = \upsilon R\log_e n$

10.168. Find the entropy increment of one mole of a Vander Waal's gas due to the isothermal variation of volume from V_1 to V_2.

Solution:

$$\Delta S = \int_{V_1}^{V_2} \frac{\delta Q}{T} = \int_{V_1}^{V_2} \frac{dU + pdV}{T}$$

$$= \int_{V_1}^{V_2} \frac{\dfrac{dU}{dV}\cdot dV + pdV}{T}$$

$$= \int_{V_1}^{V_2} \frac{a}{TV^2}dV + \int_{V_1}^{V_2} \frac{RT}{T(V-b)}dV - \int_{V_1}^{V_2} \frac{a}{TV^2}dV; \quad \frac{dU}{dV} = \frac{a}{V^2}$$

$$= R\log_e\left(\frac{V_2 - b}{V_1 - b}\right)$$

10.169. One mole of a Vander Waal's gas which has initial volume V_1 and temperature T_1 was transferred to some other state where its volume is V_2 and temperature is T_2. Find the increase in entropy.

Solution:

$$\delta Q = C_V\,dT + pdV$$

$$\Delta S = \int \frac{\delta Q}{T} = \int C_V \frac{dT}{T} + \int \frac{pdV}{T}$$

$$= C_V \int_{T_1}^{T_2} \frac{dT}{T} + \int_{V_1}^{V_2} \frac{RT}{V-b}\cdot\frac{dV}{T}; \quad p = \frac{RT}{V-b} \text{ approximately}$$

$$= C_V \cdot \log_e \frac{T_2}{T_1} + R\log_e\left(\frac{V_2 - b}{V_1 - b}\right)$$

10.170. Two bodies of equal and constant thermal capacity C and at absolute temperature T_1 and T_2 ($T_1 > T_2$) respectively attains the same temperature on being placed in direct contact. Calculate the unavailable energy.

Solution: According to conservation of energy

$$C(T_1 - T_0) = C(T_0 - T_2);\ T_0 = \text{Common temperature}$$

$$T_0 = \frac{T_1 + T_2}{2}$$

$$\Delta S_1 = C\int_{T_1}^{T_0} \frac{dT}{T} = C\log_e \frac{T_0}{T_1}$$

$$\Delta S_2 = C\int_{T_2}^{T_0} \frac{dT}{T} = C\log_e \frac{T_0}{T_2}$$

$$\Delta S = \Delta S_1 + \Delta S_2 = C\log_e \frac{T_0^2}{T_1 T_2} = C\log_e \frac{(T_1 + T_2)^2}{4T_1 T_2}$$

If T_X be the lowest available temperature then the amount of unavailable energy is equal to $\Delta S \cdot T_X = T_X \cdot C\log_e \dfrac{(T_1 + T_2)^2}{4T_1 T_2}$

10.171. Two identical thermally insulated vessels interconnected by a tube with a valve contain one mole of the same ideal gas each. The gas temperature in the two vessels are T_1 and T_2 respectively. The specific heat at constant volume is C_V. The valve having been opened, the gas comes to an equilibrium state. Calculate the increase in entropy.

Solution:

$$\frac{P_1 V}{T_1} + \frac{P_2 V}{T_2} = \frac{P \cdot 2V}{T}$$

or

$$\frac{(P_1 T_2 + P_2 T_1)}{T_1 T_2} = \frac{2P}{T}$$

Initial internal energy = Final internal energy

$$C_V \cdot \frac{P_1 V}{R} + C_V \frac{P_2 V}{R} = C_V \frac{P \cdot 2V}{R}$$

$$\therefore P = \frac{P_1 + P_2}{2}$$

$$T = \frac{(P_1 + P_2) T_1 T_2}{P_1 T_2 + P_2 T_1}$$

$$= \frac{(RT_1 + RT_2) T_1 T_2}{RT_1 T_2 + RT_1 T_2} = \frac{T_1 + T_2}{2}$$

$$\Delta S = -C_V \int_T^{T_1} \frac{dT}{T} + C_V \int_{T_2}^{T} \frac{dT}{T}$$

$$= C_V \log_e \frac{T}{T_2} - C_V \log_e \frac{T_1}{T}$$

$$= C_V \log_e \frac{T^2}{T_1 T_2} = C_V \log \frac{(T_1 + T_2)^2}{4T_1 T_2}$$

10.172. Determine the entropy of a gas whose equation of state is given by $V = V_0 [1 + \alpha (T - T_0)]$; C_P = Constt.

Solution:

$$C_P = T\left(\frac{\partial S}{\partial T}\right)_P = T\left(\frac{\partial S}{\partial T}\right)_V + T\left(\frac{\partial S}{\partial V}\right)_T \cdot \left(\frac{\partial V}{\partial T}\right)_p$$

$$= T\left(\frac{\partial S}{\partial T}\right)_V + T\left(\frac{\partial p}{\partial T}\right)_V V_0 \alpha$$

$$C_P \cdot \frac{\partial T}{T} = \partial S + V_o \alpha\, \partial p$$

$$C_P \cdot \frac{dT}{T} = dS + V_o \alpha\, dp$$

On integration

$$C_P \log T = S - S_0 + V_0 \alpha p$$

or $$S = S_0\, C_P \log T - V_0 \alpha p = \text{Constt.}$$

10.173. An ideal gas, expands isothermally from state 1 to state 2. The states 1 & 2 lies on two adiabatics. The entropy corresponding to two adiabatics are S_1 & S_2 respectively. What work does the gas do in the process?

Solution: Let T be the temperature of the gas then work done by the gas

$$W = T\Delta S = T(S_2 - S_1)$$

10.174. A sufficiently long bar is heated at one end by an indirect method in order to have a unidirectional flow of heat. The bar attains the conditions of steady state. Find the temperature along the length of the bar when (i) it is exposed to atmosphere (ii) it is covered.

Solution:

(i) Consider a slice of thickness δx at a distance x from the hot end (see Fig. 10.33). Quantity of heat entering at the face P_1 per sec is given by

$$Q_1 = -KA\frac{d\theta}{dx};$$ K = thermal conductivity of the material of the bar

A = area of cross-section.

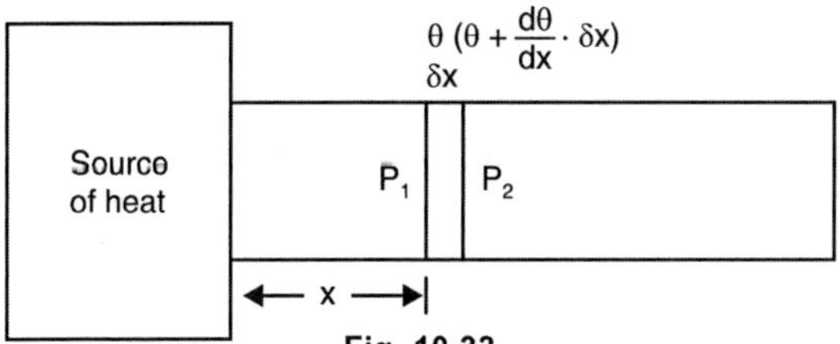

Fig. 10.33

Quantity of heat going out from the face P_2 per sec is given by

$$Q_2 = -KA\frac{d}{dx}\left[\theta + \frac{d\theta}{dx}\cdot\delta x\right];$$ θ is the excess of temperature over that of surrounding at face P_1.

Amount of heat retained by the element per sec

$$Q = Q_1 - Q_2 = -KA\cdot\frac{d\theta}{dx} - KA\frac{d\theta}{dx} + KA\frac{d^2\theta}{dx^2}\cdot\delta x$$

$$= KA\frac{d^2\theta}{dx^2}\cdot\delta x$$

Before steady state a part of it is used in heating the element and the rest is radiated to atmosphere.

Amount of heat required to heat the element

$$= A\cdot\delta x\cdot\rho\cdot S\cdot\frac{d\theta}{dt};$$ ρ = density of the material

S = sp. heat

Amount of heat radiated to the atmosphere

$$= E\cdot p\cdot\delta x\cdot\theta;$$ E = emissivity

p = perimeter

Thus $$KA\frac{d^2\theta}{dx^2}\cdot\delta x = A\cdot\delta x\cdot\rho\cdot S\cdot\frac{d\theta}{dt} + E\cdot p\cdot\delta x\cdot\theta$$

or $$\frac{d^2\theta}{dx^2} = \frac{\rho S}{K}\cdot\frac{d\theta}{dt} + \frac{Ep}{KA}\cdot\theta$$

At steady state $\frac{d\theta}{dt} = 0$

hence $$\frac{d^2\theta}{dx^2} = \frac{Ep}{KA}\cdot\theta = m^2\theta;\ m = \sqrt{\frac{Ep}{KA}}$$

This is the equation for flow of heat when the bar is at steady state and exposed to atmosphere.

Let $\theta = e^{nx}$ be the solution of the above equation

$$n^2 = m^2$$

$$n = \pm m$$

$\therefore$ general solution is given by $\theta = Be^{mx} + B'e^{-mx}$

Here B and B' are the constants whose values are to be determined with the help of boundary conditions.

At $x = 0$, $\theta = \theta_0$ and at $x = \infty$, $\theta = 0$

$$\theta_0 = B + B'$$

$$0 = Be^{+\infty} + B'e^{-\infty}$$

$$e^{-\infty} = 0 \text{ hence } B = 0 \text{ and } B' = \theta_0$$

$$\therefore \theta = \theta_0 \, e^{-mx} = \theta_0 \, e^{-\sqrt{\frac{Ep}{KA}} \cdot x}$$

Thus the distribution of temperature along the length of the bar is governed by the above stated exponential equation.

(ii) When the bar is at steady state and covered then

$$\frac{d^2\theta}{dx^2} = 0$$

or $$\frac{d\theta}{dx} = C$$

or $$d\theta = Cdx$$

further integration will yield

$$\theta = Cx + D$$

The values of constants C & D are determined with the help of two conditions.

At $x = 0$, $\theta = \theta_0$ and at $x = l$, $\theta = \theta_1$

$$\theta = Cx + D$$

$$\theta_0 = D$$

$$\theta_1 = Cl + \theta_0$$

$$\therefore C = -\frac{(\theta_0 - \theta_1)}{l}$$

$$\therefore \theta = \theta_0 - \frac{(\theta_0 - \theta_1)}{l} x$$

The distribution of temperature along the length of the bar is governed by a linear equation.

10.175. Determine the thermal conductivity of a wall which is made alternate blocks whose thermal conductivities are k_1 and k_2 as shown in the Fig. 10.34. The length and the area of cross-section of the two blocks is same.

Solution: Let S be the source of heat. Under the conditions of steady state let θ_1 and θ_2 be the temperatures at the two faces of the wall. The face of wall close to the source of heat will be at a higher temperature, i.e., $\theta_1 > \theta_2$. Let l be the length and A be the area of cross-section of each block. In the conditions of steady state, the quantity of heat passing per second through the block of conductivity k_1 is given by

$$Q_1 = \frac{k_1 A}{l}(\theta_1 - \theta_2)$$

Similarly, the quantity of heat passing through the other block of conductivity k_2 is given by

$$Q_2 = \frac{k_2 A}{l}(\theta_1 - \theta_2)$$

Total quantity of heat Q passing through the two blocks having an area of cross-section $2A$ is given by

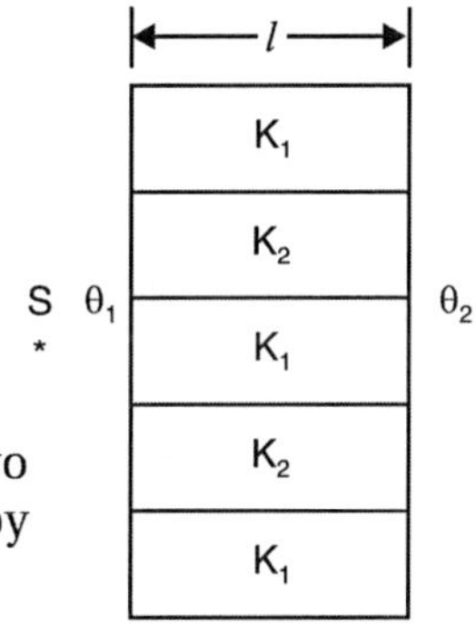

Fig. 10.34

$$Q = Q_1 + Q_2 = \frac{k_1 + k_2}{2}\frac{2A}{l}(\theta_1 - \theta_2)$$

$$= k\frac{2A}{l}(\theta_1 - \theta_2)$$

where $k = \dfrac{k_1 + k_2}{2}$ is the conductivity of the wall.

10.176. Determine the thermal conductivity of a wall which is made of two blocks of thermal conductivities k_1 and k_2 as shown in Fig. 10.35. The length and the area of cross-section of the two blocks is the same.

Solution: Let S be the source of heat. Under the conditions of steady state, let θ_1 and θ_2 be the temperatures at the two faces of the wall and θ be the temperature of the interface. The face of wall close to the source of heat will be at a higher temperature, i.e., $\theta_1 > \theta_2$. Let $2l$ be the thickness of the wall and A be its area of cross-section. The quantity of heat which will pass per second through the wall is given by

$$Q = \frac{k_1 A}{l}(\theta_1 - \theta) = \frac{k_2 A}{l}(\theta - \theta_2)$$

$$\theta = \frac{k_1\theta_1 + k_2\theta_2}{k_1 + k_2}$$

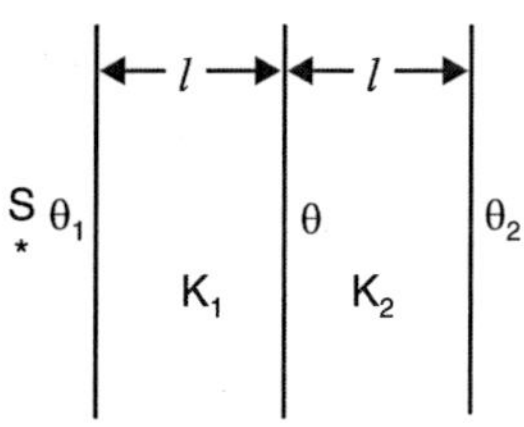

Fig. 10.35

with the above mentioned value of θ, Q takes the form.

$$Q = \frac{k_1 A}{l}\left[\theta_1 - \frac{k_1\theta_1 + k_2\theta_2}{k_1 + k_2}\right] = \frac{k_1 A}{l(k_1 + k_2)} k_2 [\theta_1 - \theta_2]$$

$$Q = \frac{2k_1 k_2}{k_1 + k_2}\frac{A}{2l}(\theta_1 - \theta_2)$$

$$= k\frac{A}{2l}(\theta_1 - \theta_2)$$

where k is the conductivity of the wall and is given by $k = \dfrac{2k_1k_2}{k_1 + k_2}$

10.177. A cubical box made of perfectly insulating material has walls of thickness 8 cm and the only way for heat to enter or leave the box is through two solid, cylindrical metallic plugs each of cross-sectional area 12 cm^2 and length 8 cm, fixed in the opposite walls of the box (see Fig. 10.36). The outer surface A is kept at 100°C while the outer surface B of the other plug is maintained at 4°C. The thermal conductivity of the material of the plug is 0.5 cal/cm/s/°C. A source of energy generating 36 cal/s is enclosed inside the box. Find the equilibrium temperature of the inner surface of the box assuming that it is the same at all points on the inner surface. (IIT 1972)

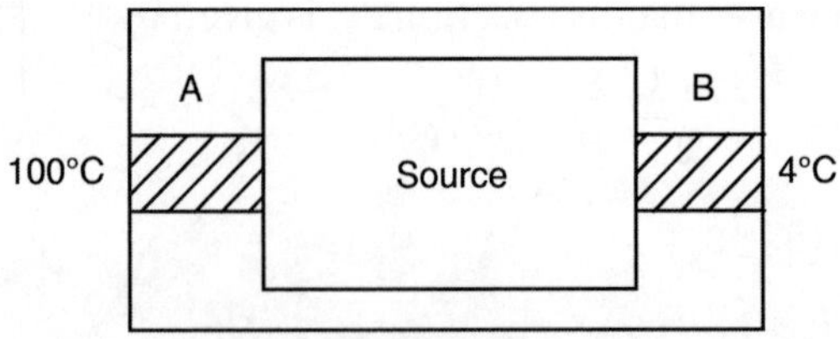

Fig. 10.36

Solution: Let the amount of heat entering per second through the plug A be Q_1 calories. The total quantity of heat inside will now be equal to $36 + Q_1$. Under the conditions of equilibrium, this quantity of heat will be conducted through the plug B.

i.e., $$36 + \frac{0.5 \times 12}{8}(100 - \theta) = \frac{0.5 \times 12}{8}(\theta - 4)$$

where θ is the temperature inside the box.

Now $$36 + 75 - \frac{3\theta}{4} = \frac{3\theta}{4} - 3$$

or $$\frac{3\theta}{2} = 114 \text{ or } \theta = 76°\text{C}$$

10.178. A metallic water tank with all-round lagging of thickness 16 mm, surface area 0.4 m^2 and thermal conductivity 0.04 w/m/K maintains water temperature at 74°C inside it with the help of an immersion heater. Heat lost per second by the lagging is 0.8 watt/degree. Determine the power of the heater. The surrounding temperature = 11°C.

Solution: Let θ be the temperature of the lagging. Heat lost per second to the surrounding = 0.8 (θ – 11)

Amount of heat conducted per second through the lagging

$$= \frac{0.04 \times 0.4}{16 \times 10^{-3}}(74 - \theta)$$

In steady state,

$$0.8(\theta - 11) = \frac{0.04 \times 0.4}{16 \times 10^{-3}}(74 - \theta)$$

or $\quad 0.8\,\theta - 8.8 = 74 - \theta$

or $\quad 1.8\,\theta = 82.8$

or $\quad \theta = \frac{82.8}{1.8} = 46°C$

Power of the heater = 0.8 × (46 – 11) = 0.8 × 35 = 28.0 watt.

10.179. Three rods of equal length and cross-section are placed end to end, one copper, one steel and one aluminium in that order. The free end of the copper rod is in contact with ice and the free end of the aluminium rod is in contact with a source of heat at 150°C. If no transfer of heat occurs at the sides, find the temperature of each junction. k_{cu} = 0.92 cgs unit, k_{steel} = 0.12 cgs unit and k_{Al} = 0.50 cgs unit.

Solution: Let l and A be the length and area of cross-section of each rod. Under the conditions of steady state, same quantity of heat will flow through each rod.

Quantity of heat flowing per second

$$= \frac{0.50 \times A}{l}(150 - \theta_1) = \frac{0.12 \times A}{l}(\theta_1 - \theta_2) = \frac{0.92 \times A}{l}(\theta_2 - 0)$$

where θ_1 is the temperature at the aluminium and steel interface and θ_2 at the steel and copper interface.

Now $\quad 25\,(150 - \theta_1) = 6\,(\theta_1 - \theta_2) = 46\,\theta_2$

$$6\,\theta_1 = 52\,\theta_2$$

Now $\quad 25 \times 150 - 25\,\theta_1 = \frac{46 \times 3}{26}\theta_1 = 53.08\,\theta_1$

$\therefore \quad \theta_1 = \frac{25 \times 150}{78.08} = 48.03°C$

and $\quad \theta_2 = \frac{3}{26} \times 48.03 = 5.54°C$

10.180. A room is maintained at 20°C by means of a heater of resistance 20Ω and connected to 200 volt mains. The temperature is uniform throughout the room and the heat from the room is going out through a glass window of area 1 m^2 and thickness 0.2 cm. Calculate the outside temperature. Coefficient of thermal conductivity of glass = 0.2 cal/m/s/°C and 1 cal = 4.2 Joule. (IIT 1978)

Solution:

Quantity of heat produced per second $= \frac{V^2}{R} = \frac{200 \times 200}{20} = 2000$ Joule

Now $\frac{2000}{4.2} = \frac{KA}{l}(\theta_1 - \theta_2)$

where l is the thickness of the glass

$$= \frac{0.2 \times 1}{0.2 \times 10^{-2}}(20 - \theta_2)$$

or $\frac{2000}{4.2} = 100(20 - \theta_2)$

$$\frac{100}{21} = 20 - \theta_2$$

or $21\theta_2 = 420 - 100$

$\therefore$ $\theta_2 = \frac{320}{21} = 15.2°C$

10.181. A lake is covered with 4cm thick ice and the temperature of air above the ice is –12°C. At what rate, expressed in cm/hr will the ice thicken? Conductivity of ice = 0.0052 c.g.s. unit, density of ice = 0.92 g/cc and latent heat of ice = 80 cal/g.

Solution: Let us assume that on the surface of the lake we already have a block of ice of thickness x cm (see Fig. 10.37) and with an area of cross-section A cm^2. The amount of heat given up when an ice layer of thickness dx is formed in time $dt = A\,dx\,\rho L$ where ρ is the density of ice.

This quantity of heat will be conducted through the ice of thickness x, i.e.,

$$A\,dx\,\rho\,L = \frac{kA\theta}{x}\,dt$$

or $\frac{dx}{dt} = \frac{k\theta}{x\rho L}$

Here $\frac{dx}{dt}$ is the rate of growth of ice

Now $\frac{dx}{dt}\frac{\text{cm}}{\text{hr}} = \frac{k\theta}{x\rho L}3600$

$$= \frac{0.0052 \times 12 \times 3600}{4 \times 0.92 \times 80}$$

$$= 0.763 \text{ cm/hr}$$

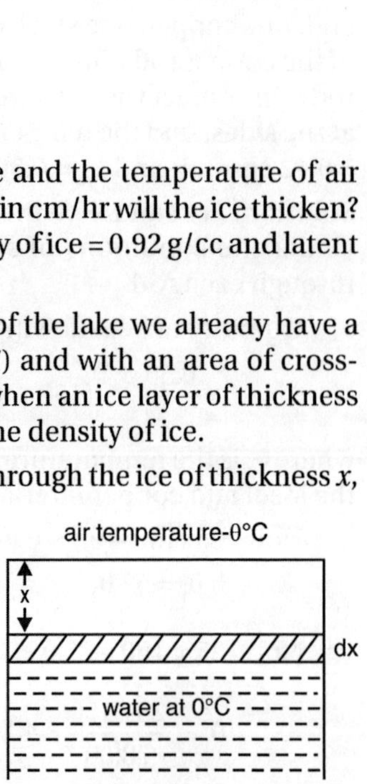

Fig. 10.37

10.182. The thickness of ice in lake is 6 cm and the temperature of air is –10°C. How long will it take for the thickness of ice to be doubled? Thermal conductivity of ice = 0.005 c.g.s. unit, density of ice = 0.92 g/cc and latent heat of ice = 80 cal/g.

Solution: Time taken for thickness x_1 to increase to x_2 is given by

$$t = \frac{\rho L}{2k\theta}\left(x_2^2 - x_1^2\right) \text{ sec}$$

where ρ is the density of ice, L = latent heat of ice, k = thermal conductivity of ice and θ is the temperature of air.

$$t = \frac{0.92 \times 80}{2 \times 0.005 \times 10}\left[12^2 - 6^2\right] \text{ sec}$$

$$= \frac{0.92 \times 80}{2 \times 0.05} \times 18 \times 6 \text{ sec}$$

$$= 22 \text{ hr } 4 \text{ min } 48 \text{ sec}$$

10.183. The ends of a rod of length l are kept at temperature T_1 & T_2 ($T_1 > T_2$). The conductivity of the material of the rod is no longer constant but varies with temperature by equation $K = \frac{\alpha}{T}$. Find (i) the temperature of the rod at a distance x from the hot end (ii) the heat flow density.

Solution: Quantity of heat flowing per sec is given by

$$Q = -KA \cdot \frac{dT}{dx}$$

$$Q = -\frac{\alpha}{T} \cdot A \cdot \frac{dT}{dx}$$

$$\int_0^x Q \cdot dx = -\alpha A \int_{T_1}^{T_x} \frac{dT}{T}$$

and $$\int_x^l Q \cdot dx = -\alpha A \int_{T_x}^{T_2} \frac{dT}{T}$$

On solving the two integrals we have

$$Q \cdot x = -\alpha A \log \frac{T_x}{T_1}$$

$$Q(l - x) = -\alpha A \log \frac{T_2}{T_x}$$

On dividing the two equation we have

$$\frac{l - x}{x} = \frac{\log T_2 - \log T_x}{\log T_x - \log T_1}$$

$$\frac{l}{x} = \frac{\log T_2 - \log T_1}{\log T_x - \log T_1}$$

$$\frac{l}{x} = \frac{\log \frac{T_2}{T_1}}{\log \frac{T_x}{T_1}}$$

$$\log\frac{T_x}{T_1} = \frac{x}{l}\cdot\log\frac{T_2}{T_1}$$

$$\therefore T_x = T_1\left(\frac{T_2}{T_1}\right)^{x/l}$$

(ii) $$Q = -KA\frac{dT}{dx}$$

$$\int_0^l Q\cdot dx = \int_{T_1}^{T_2} -\alpha A\frac{dT}{T}$$

$$Q\cdot l = \alpha A\log\frac{T_1}{T_2}$$

$$\therefore \frac{Q}{A} = \text{heat flow density} = \frac{\alpha}{l}\cdot\log\left(\frac{T_1}{T_2}\right)$$

10.184. The element of radii r_1 and r_2 and length L as shown in the Fig. 10.38 is obtained by cutting a solid cone. The two circular faces are maintained at constant temperature θ_1 and θ_2 $(\theta_2 > \theta_1)$. The conductivity of the material is K. Find the rate of flow of heat.

Solution: In view of the Fig. 10.38

$$\tan\phi = \frac{r_2 - r_1}{L}$$

Consider a slice of thickness dx at a distance x from the hot end

$$\frac{p}{x} = \tan\phi = \frac{r_2 - r_1}{L}$$

$$p = \left(\frac{r_2 - r_1}{L}\right)x$$

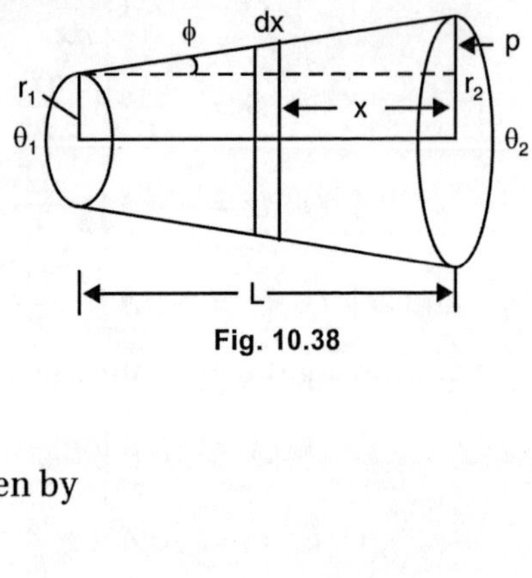

Fig. 10.38

$\therefore$ radius of the circular slice is given by

$$r_2 - \left(\frac{r_2 - r_1}{L}\right)x$$

$$Q = -K\cdot\pi\left[r_2 - \left(\frac{r_2 - r_1}{L}\right)x\right]^2\frac{d\theta}{dx}$$

$$\int_0^L \frac{-Q\,dx}{\left[r_2 - \left(\frac{r_2 - r_1}{L}\right)x\right]^2} = \int_{\theta_2}^{\theta_1} K\pi\,d\theta$$

$$\frac{LQ}{(r_2 - r_1)}\left[\frac{1}{r_1} - \frac{1}{r_2}\right] = K\pi(\theta_2 - \theta_1)$$

$$\therefore Q = \frac{K\pi r_1 r_2 (\theta_2 - \theta_1)}{L}$$

10.185. The two ends of a rod of length L is maintained at temperature T_1 and T_2 ($T_1 > T_2$). The coefficient of thermal conductivity of the material of the rod varies with temperature via equation $K = \alpha\sqrt{T}$. Find the temperature along the length of the rod.

Solution: Consider an element of thickness dx at a distance x from the hot end (see Fig. 10.39).

$$Q = -KA \cdot \frac{dT}{dx}; K = \text{Thermal conductivity of the material}$$

$$\int_0^x Q \cdot dx = \int_{T_1}^{T_x} -\alpha\sqrt{T} \cdot A \cdot dT$$

$$Q \cdot x = \frac{2}{3}\alpha \cdot A \cdot \left[T_1^{3/2} - T_X^{3/2}\right]$$

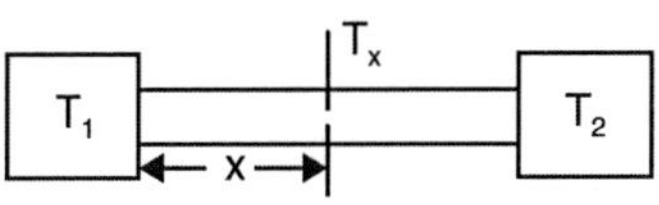

Fig. 10.39

Similarly we have

$$Q[L - x] = \frac{2}{3}\alpha A\left[T_X^{3/2} - T_2^{3/2}\right]$$

$$\frac{L - x}{x} = \frac{T_X^{3/2} - T_2^{3/2}}{T_1^{3/2} - T_X^{3/2}}$$

$$\frac{L}{x} = \frac{T_1^{3/2} \quad T_2^{3/2}}{T_1^{3/2} - T_X^{3/2}}$$

$$T_1^{3/2} - T_X^{3/2} = \left(T_1^{3/2} - T_2^{3/2}\right)\frac{x}{L}$$

$$= T_1^{3/2}\left[1 - \left(\frac{T_2}{T_1}\right)^{3/2}\right] \cdot \frac{x}{L}$$

$$T_X^{3/2} = T_1^{3/2}\left\{1 + \frac{x}{L}\left[\left(\frac{T_2}{T_1}\right)^{3/2} - 1\right]\right\}$$

$$T_x = T_1\left\{1 + \frac{x}{L}\left[\left(\frac{T_2}{T_1}\right)^{3/2} - 1\right]\right\}^{2/3}$$

10.186. A rod of length L, cross-sectional area A and thermal conductivity K is connected between two bodies of masses m_1 & m_2, and specific heats s_1 & s_2 respectivity (see Fig. 10.40). The temperature of the two bodies are T_1 & T_2 ($T_2 > T_1$) respectively. The system as a whole is thermally insulated. Calculate the temperature difference between the two bodies as a function of time.

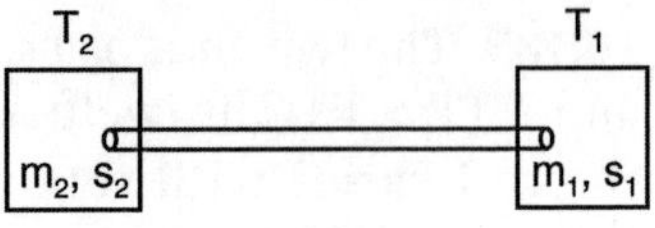

Fig. 10.40

Solution:

$$-m_2 s_2 \frac{d\theta}{dt} = \frac{KA}{l}(\theta - \phi) = m_1 s_1 \frac{d\phi}{dt}$$

or $$-m_2 s_2 \theta = m_1 s_1 \phi$$

or $$-\frac{m_2 s_2 \, d\theta}{\theta - \phi} = \frac{KA}{l} \cdot dt$$

or $$-\frac{m_2 s_2 \, d\theta}{\theta + \dfrac{m_2 s_2 \theta}{m_1 s_1}} = \frac{KA}{l} \cdot dt$$

$$\frac{m_1 s_1 m_2 s_2}{m_1 s_1 + m_2 s_2} \int_{T_2}^{\theta} \frac{d\theta}{\theta} = -\frac{KA}{l} \int_0^t dt$$

$$\log \frac{\theta}{T_2} = -\frac{KAt}{l} \cdot \frac{m_1 s_1 + m_2 s_2}{m_1 s_1 m_2 s_2}$$

$$\theta = T_2 e^{-\lambda t}; \ \lambda = \frac{KA}{l}\left(\frac{m_1 s_1 + m_2 s_2}{m_1 m_2 \, s_1 s_2}\right)$$

$$m_1 s_1 \cdot \frac{d\phi}{dt} = \frac{KA}{l}(\theta - \phi)$$

or $$\frac{m_1 s_1 \, d\phi}{\theta - \phi} = \frac{KA}{l} \cdot dt$$

or $$-\frac{m_1 s_1 d\phi}{\left(\dfrac{m_1 s_1}{m_2 s_2} + 1\right)\phi} = \frac{KA}{l} \cdot dt$$

or $$\int_{T_1}^{\phi} \frac{d\phi}{\phi} = -\left(\frac{m_1 s_1 + m_2 s_2}{m_1 m_2 \, s_1 s_2}\right) \frac{KA}{l} \int_0^t dt$$

$$\log \frac{\phi}{T_1} = -\lambda t$$

$$\phi = T_1 e^{-\lambda t}$$

$\theta - \phi$ = change of temperature in time $t = (T_2 - T_1)\, e^{-\lambda t}$

10.187. Two vessels containing liquid of mass m and specific heat s are maintained at temperature T_1 & T_2 ($T_2 > T_1$). The ends of a rod of length l, cross-sectional area A and conductivity K are inserted into the two vessels. The system is thermally insulated. Calculate the time during which the difference in temperature between the two vessels reduces to half of its original value.

Solution: In view of Q. 10.186 we have

$$\frac{T_2 - T_1}{2} = (T_2 - T_1)e^{-\lambda t}$$

$$\log_e 2 = \lambda t$$

$$\lambda = \frac{2KA}{msl}$$

$$\therefore \log_e 2 = \frac{2KA}{msl} \cdot t$$

or $$t = \frac{msl}{2KA} \cdot \log_e 2$$

10.188. The space between two coaxial cylinders of radii R_1 & R_2 is filled with a homogenous conducting material. A source of heat is placed along the axis of the cylinder. In steady state the inner and outer surfaces acquires temperatures T_1 & T_2 respectively. Find the temperature at a distance r_0 ($R_2 > r_0 > R_1$) from the axis of the cylinder.

Solution: Consider an element of thickness dr lying at a distance r from the axis of the cylinders (see Fig. 10.41). Quantity of heat flowing per sec through the element is given by

$$Q = -K \cdot 2\pi r l \cdot \frac{d\theta}{dr}$$

$$\int_{R_1}^{R_2} Q \cdot \frac{dr}{r} = -\int_{T_1}^{T_2} 2\pi k l \cdot d\theta$$

and $$\int_{R_1}^{r_0} Q \cdot \frac{dr}{r} = \int_{T_1}^{T} -2\pi k l \cdot d\theta$$

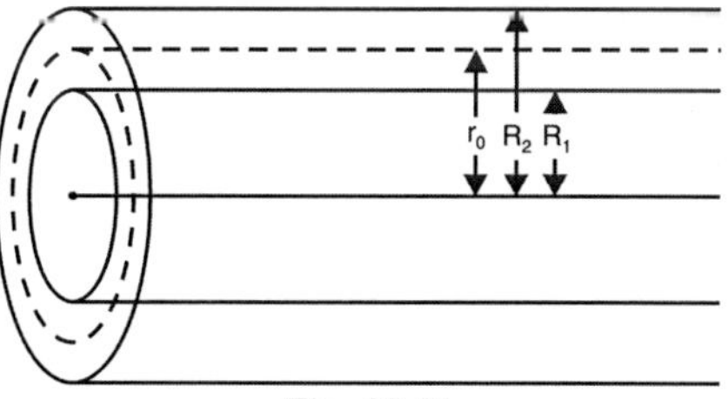

Fig. 10.41

On solving the two integrals we have

$$\frac{\log_e \frac{R_2}{R_1}}{\log_e \frac{r_0}{R_1}} = \frac{T_2 - T_1}{T - T_1}$$

$$T - T_1 = -(T_1 - T_2)\frac{\log_e \frac{r_0}{R_1}}{\log_e \frac{R_2}{R_1}}$$

or $$T = T_1 - (T_1 - T_2)\frac{\log_e \frac{r_0}{R_1}}{\log_e \frac{R_2}{R_1}}$$

10.189. The space between two concentric spheres of radii R_1 & R_2 is filled with the homogenous conducting material. A source of heat is placed at the centre of the sphere. In steady state the inner and outer surfaces acquires temperatures θ_1 and θ_2 respectively. Find the temperature at a distance r_0 from the centre of the two spheres.

Solution: Consider a spherical shell of radius r and $r + dr$ (see Fig. 10.42). The quantity of heat flowing per sec through the shell is given by

$$Q = -K \cdot 4\pi r^2 \cdot \frac{d\theta}{dt}$$

$$\int_{R_1}^{R_2} -Q\frac{dr}{r^2} = 4\pi K \cdot \int_{\theta_1}^{\theta_2} d\theta$$

$$\int_{R_1}^{r_0} -Q\frac{dr}{r^2} = 4\pi K \int_{\theta_1}^{\theta} d\theta$$

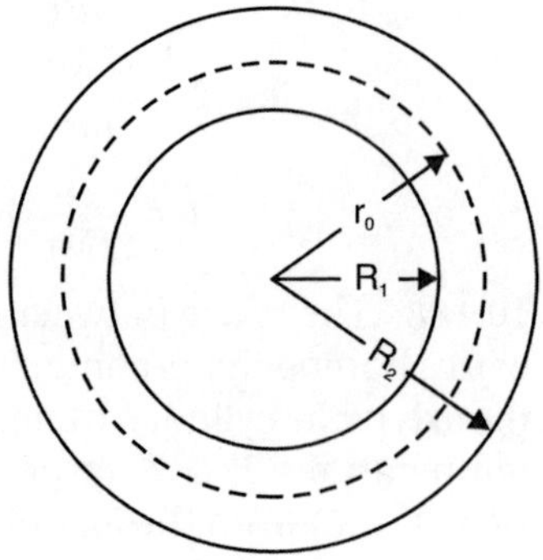

Fig. 10.42

On solving the two integrals we have

$$\frac{Q\left(\frac{1}{R_1} - \frac{1}{R_2}\right)}{Q\left(\frac{1}{r_0} - \frac{1}{R_1}\right)} = \frac{\theta_1 - \theta_2}{\theta_1 - \theta}$$

$$\theta_1 - \theta = \frac{\left(\frac{1}{r_0} - \frac{1}{R_1}\right)}{\frac{1}{R_1} - \frac{1}{R_2}}(\theta_1 - \theta_2)$$

$$\theta = \theta_1 - \frac{\left(\frac{1}{r_0} - \frac{1}{R_1}\right)}{\left(\frac{1}{R_1} - \frac{1}{R_2}\right)}(\theta_1 - \theta_2)$$

10.190. Fig. 10.43 represents a system of concentric spherical layers of materials of conductivities K_1 & K_2. The inner and the outer radii are r_1 and r_2 and the radius of the interface is r_0. A source of heat is placed at the centre of the sphere. In steady state of the inner and outer surfaces attains temperatures θ_1 and θ_2 respectively. Calculate the rate of flow of heat through the system.

Solution:

$$\frac{d\theta}{dt} = \frac{\text{Difference in temperature}}{\text{Thermal resistance}}$$

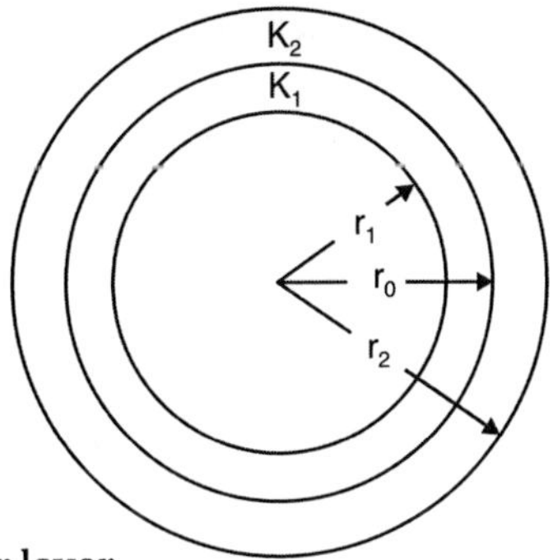

Fig. 10.43

Thermal resistance of inner layer

$$= \int_{r_1}^{r_0} \frac{dr}{K_1 \cdot 4\pi r^2}$$

$$= \frac{1}{4\pi K_1}\left[\frac{1}{r_1} - \frac{1}{r_0}\right]$$

Similarly the thermal resistance of outer layer

$$= \frac{1}{4\pi K_2}\left[\frac{1}{r_0} - \frac{1}{r_2}\right]$$

$$\text{Total thermal resistance} = \frac{1}{4\pi}\left[\frac{1}{K_1}\left(\frac{1}{r_1} - \frac{1}{r_0}\right) + \frac{1}{K_2}\left(\frac{1}{r_0} - \frac{1}{r_2}\right)\right]$$

$$\text{Rate of flow of heat} = \frac{4\pi(\theta_1 - \theta_2)}{\frac{1}{K_1}\left(\frac{1}{r_1} - \frac{1}{r_0}\right) + \frac{1}{K_2}\left(\frac{1}{r_0} - \frac{1}{r_2}\right)}$$

10.191. A uniform wire of radius R and thermal conductivity K carries a current I. A unit volume of wire produces a thermal power P. Assuming the temperature at the surface of the wire as θ_0. Find the temperature distribution across the wire.

Solution: Quantity of heat flowing per sec is given by

$$Q = -K \cdot 2\pi\, rl \cdot \frac{d\theta}{dr}$$

$$P = -\frac{K \cdot 2\pi r l}{\pi r^2 l} \cdot \frac{d\theta}{dr}$$

$$P\int_r^R r \cdot dr = -K \cdot 2\int_\phi^{\theta_0} d\theta$$

$$\frac{P}{4}\left(R^2 - r^2\right) = K\left[\phi - \theta_0\right]$$

$$\therefore \phi = \theta_0 + \frac{P}{4K}\left[R^2 - r^2\right]$$

10.192. Inside a sphere of radius R and thermal conductivity K a thermal power of density P is generated. Assuming the steady state temperature at the surface equal to θ_0. Find the temperature in the sphere.

Solution: Quantity of heat flowing per sec is given by

$$Q = -K \cdot 4\pi r^2 \frac{d\theta}{dr}$$

$$P = -\frac{K \cdot 4\pi r^2}{\frac{4}{3}\pi r^3} \cdot \frac{d\theta}{dr}$$

$$\frac{P}{3}\int_r^R r dr = -K \int_\phi^{\theta_0} d\theta$$

$$\phi = \theta_0 + \frac{P}{6K}\left(R^2 - r^2\right)$$

10.193. A cylindrical block of length 0.4 m and area of cross-section 0.04 m^2 is placed coaxially on a thin metal disc of mass 0.4 kg and of same cross-section. The upper face of the cylinder is maintained at a constant temperature of 400 K and the initial temperature of the disc is 300 K. If the thermal conductivity of the material of the cylinder is 10 watt/m-K and the specific heat of the material of the disc is 600 J/kg-K how long will it take for the temperature of the disc to increase to 350 K? Assume, for purposes of calculation, the thermal conductivity of the disc to be very high and the system to be thermally insulated except for the upper face of the cylinder. (IIT 1992)

Solution: Rate of flow of heat through the block at temperature 400 K and T is given by

$$\frac{dQ}{dt} = \frac{KA(400-T)}{L} = \frac{10 \times 0.04 \times (400-T)}{0.4}$$

$$= (400 - T)$$

Rate at which the heat is taken by the disc is given by

$$mS\frac{dT}{dt} = 0.4 \times 600 \times \frac{dT}{dt} = 240\frac{dT}{dt}$$

Now $$240\frac{dT}{dt} = 400 - T$$

$$dt = 240 \cdot \frac{dT}{400-T}$$

$$t = \int dt = 240\int_{300}^{350} \frac{dT}{400-T} = -240\left|\log_e(400-T)\right|_{300}^{350}$$

$$= 240 \log_e 2 = 240 \times 2.303 \times 0.3010$$

$$= 166.3 \text{ sec}$$

10.194. A radiation of energy 18 J falls on an element of area 1 cm^2 for 60 seconds. Determine the pressure exerted by the radiation when the surface (i) completely absorbs the radiation and (ii) completely reflects the radiation.

Solution: Pressure exerted by the radiation is given by

$$P = \frac{E}{Atc}(1+\rho)$$

Where E is the energy of the radiation, A is the area of cross-section of the element, t = time, c = velocity of light and ρ is the reflection coefficient.

(i) When the radiation is completely absorbed, then there will be no reflection i.e., $\rho = 0$

$$\therefore P = \frac{E}{Act} = \frac{18}{10^{-4} \times 3 \times 10^{8} \times 60} = 10^{-5}\,\text{Pa}$$

(ii) When the radiation is completely reflected, then $\rho = 1$

$$\therefore P = \frac{2E}{Act} = 2 \times 10^{-5}\,\text{Pa}$$

10.195. Calculate the pressure exerted by 2×10^{16} X-ray photons of frequency 7×10^{19} Hz falling every second on 1 m^2 area of a blackened element h = 6.62×10^{-34} JS.

Solution:

$$P = \frac{E}{Act} = \frac{nh\nu}{Act} = \frac{2 \times 10^{16} \times 6.62 \times 10^{-34} \times 7 \times 10^{19}}{1 \times 3 \times 10^{8} \times 1}$$

$$= 30.89 \times 10^{-7}\,\text{Pa}$$

10.196. A daylight bulb made of tungsten filament of surface area 0.5 cm^2 is heated to a temperature of 2240°K when operated at 220 volts. Calculate the wattage of the bulb. Emissivity of the filament = 0.35, $\sigma = 5.67 \times 10^{-5}$ c.g.s. units, 1 watt = 10^7 erg/s.

Solution:

$$P = eA\sigma T^4\ \text{erg/s}$$

where e is emissivity, A is surface area, σ is Stephan's constant and T is the temperature of the filament.

On substituting the values of various terms, we have

$$P = \frac{0.35 \times 0.5 \times 5.67 \times 10^{-5} \times (2240)^4}{10^7} = 24.98\ \text{watt}$$

$$= 25\ \text{watt}$$

10.197. In the above problem, if the filament temperature falls to 2000°K due to a drop of mains voltage, then what will be the wattage of the bulb.

Solution: Let P_1 and P_2 be the wattage of the bulb at temperatures T_1 and T_2 respectively. Then,

$$\frac{P_1}{P_2} = \left(\frac{T_1}{T_2}\right)^4 ; \; P_2 = \left(\frac{T_2}{T_1}\right)^4 P_1$$

$$= \left(\frac{2000}{2240}\right)^4 \times 25 = 15.88 \text{ watt}$$

10.198. A 40 watt daylight bulb has a tungsten filament of surface area 0.25 cm². What is the temperature of the filament when it reaches incandescence. Emissivity of the filament = 0.35, $\sigma = 5.67 \times 10^{-5}$ c.g.s. units; J = 4.18 J/cal.

Solution:

$$P = \frac{eA\sigma T^4}{10^7} \text{ watt}$$

or $$T = \left(\frac{40 \times 10^7}{0.35 \times 0.25 \times 5.67 \times 10^{-5}}\right)^{1/4} = 2996\text{K}$$

10.199. A blackened platinum wire of length 5 cm and perimeter = 0.02 cm is maintained at a temperature of 3000 K. At what rate is the wire losing its energy ?

$\sigma = 5.67 \times 10^{-5}$ c.g.s. units.

Solution:

$$p = \frac{eA\sigma T^4}{10^7} \text{ watt}$$

$$= \frac{1 \times 5 \times 0.02 \times 5.67 \times 10^{-5} \times 3000^4}{10^7} = 45.927 \text{ watt}$$

10.200. A blackened sphere of radius 10 cm at a temperature of 227°C is placed in a chamber with blackened wall, maintained at 27°C. Calculate its rate of loss of heat. $\sigma = 5.67 \times 10^{-5}$ c.g.s. unit; J = 4.18 J/cal.

Solution:

$$E = \frac{eA\sigma\left(T_1^4 - T_2^4\right)}{4.18 \times 10^7} \text{ cal/s.}$$

$$= \frac{1 \times 4\pi \times 10^2 \times 5.67 \times 10^{-5}}{4.18 \times 10^7} \times \left(500^4 - 300^4\right)$$

$$= 92.68 \text{ cal/s.}$$

10.201. In the above problem, if the thermal capacity of the sphere is 1000 cal/°C then what will be the rate of cooling of the sphere.

Solution:

$$-ms\frac{d\theta}{dt}=\frac{eA\sigma\left(T_1^4-T_2^4\right)}{4.18\times10^7}$$

or $$-1000\frac{d\theta}{dt}=\frac{1\times4\pi\times10^2\times5.67\times10^{-5}\times\left(500^4-300^4\right)}{4.18\times10^7}$$

or $$-1000\frac{d\theta}{dt}=92.68$$

$$\therefore\quad -\frac{d\theta}{dt}=0.09268\ \text{C/s}$$

10.202. Two copper spheres of radius r and $2r$ are placed in two chambers. For the two spheres, the excess temperature over that of surrounding is θ. Calculate (i) the ratio of rate of loss of heat and (ii) the ratio of rate of fall of temperature of the two spheres.

Solution: According to Newton's law of cooling, the rate at which a body loses heat is proportional to the excess of temperature over that of its surroundings

i.e., $$-\frac{dQ}{dt}\propto\theta$$

or $$-\frac{dQ}{dt}=KA\theta$$

where K is constant, A is the surface area

(i) For the first sphere, we have $-\dfrac{dQ_1}{dt}=K4\pi r^2\theta$

for the second sphere, we have $-\dfrac{dQ_2}{dt}=K16\pi r^2\theta$

$$\therefore\ \frac{-dQ_1}{dt}\Big/-\frac{dQ_2}{dt}=\frac{k4\pi r^2\theta}{k16\pi r^2\theta}=1:4$$

(ii) For the first sphere, we have

$$-\frac{4}{3}\pi r^3\,ds\frac{d\theta_1}{dt}=k4\pi r^2\theta;\ d=\text{density of copper}$$

s = specific heat of copper

For the second sphere, we have

$$-\frac{4}{3}\pi 8r^3ds\frac{d\theta_2}{dt}=K4\pi r^2\,4\theta$$

On dividing the two equations, we have

$$-\frac{d\theta_1}{dt}\Big/-8\frac{d\theta_2}{dt}=\frac{1}{4}$$

or $$-\frac{d\theta_1}{dt} \Big/ -\frac{d\theta_2}{dt} = \frac{2}{1}$$

10.203. A solid sphere of steel of radius $r_1 = 6$ cm and another solid sphere of copper of radius $r_2 = 12$ cm have the same emissivity and are heated to a constant temperature θ. The two spheres are allowed to cool in a chamber where the temperature is θ_0. Calculate the ratio of their rate of fall of temperature. Density of steel $d_1 = 7.8 \times 10^3$ kg/m³, density of copper $d_2 = 8.9 \times 10^3$ kg/m³, specific heat of steel $s_1 = 460$ J/kg/K and specific heat of copper $s_2 = 380$ J/kg/K.

Solution: According to Newton's law of cooling, the rate of cooling in the two cases is given by the equations

$$-\frac{4}{3}\pi r_1^3 d_1 s_1 \frac{d\theta_1}{dt} = k4\pi r_1^2 (\theta - \theta_0)$$

and $$-\frac{4}{3}\pi r_2^3 d_2 s_2 \frac{d\theta_2}{dt} = k\,4\pi\, r_2^2 (\theta - \theta_0)$$

$$\therefore \quad \frac{-d\theta_1}{dt} \Big/ -\frac{d\theta_2}{dt} = \frac{d_2\, s_2\, r_2}{d_1\, s_1 r_1}$$

$$= \frac{2d_2 s_2}{d_1 s_1}; \frac{r_2}{r_1} = \frac{12}{6} = 2$$

$$= \frac{2 \times 8.9 \times 380}{7.8 \times 460} = 1.885$$

10.204. The radiant energy received from the sun on the earth's surface is 8.2 J cm^{-2} per minute. Assuming the radiation to be black body in character, calculate the temperature of the sun. Radius of sun's sphere = 7×10^8 m, distance between the earth and sun = 1.5×10^{11} m, $\sigma = 5.67 \times 10^{-8}$ W m^{-2} k^{-4}.

Solution: $E = eA\sigma T^4$; $e = 1$ for a black body.

$$E = \text{amount of energy radiated} = \frac{8.2 \times 10^4}{60} \times 4\pi \times \left(1.5 \times 10^{11}\right)^2$$

$$A = \text{Area} = 4\pi \times (7 \times 10^8)^2$$

$$\therefore \quad T = \left(\frac{E}{A\sigma}\right)^{1/4} = \left[\frac{8.2 \times 10^4 \times 4\pi \times \left(1.5 \times 10^{11}\right)^2}{60 \times 4\pi \times \left(7 \times 10^8\right)^2 \times 5.67 \times 10^{-8}}\right]^{1/4}$$

$$= \left(\frac{8.2 \times 100 \times 2.25}{60 \times 49 \times 5.67}\right)^{1/4} \times 10^4$$

$$= 5768 \text{ K}$$

10.205. A 100 watt bulb has a tungsten filament of length 40 cm and diameter 0.05 mm. Calculate the temperature of the filament by assuming the radiation from the filament to be 85% of that of a perfectly black body maintained at that temperature $\sigma = 5.67 \times 10^{-8}\,\text{Wm}^{-2}\,\text{k}^{-4}$.

Solution:

$$E = eA\sigma T^4 = e\,2\pi rl\sigma T^4$$

$$100 = 0.85 \times 2 \times 3.14 \times 0.025 \times 10^{-3} \times 0.4 \times 5.67 \times 10^{-8} \times T^4$$

$$\therefore \quad T^4 = \frac{100 \times 10^{11}}{0.85 \times 2 \times 3.14 \times 0.025 \times 0.4 \times 5.67}$$

$$T = \left(\frac{10^{15}}{0.85 \times 2 \times 3.14 \times 5.67}\right)^{1/4} = \left(\frac{1000}{0.85 \times 2 \times 3.14 \times 5.67}\right)^{1/4} \times 10^3$$

$$= 2398\text{ K}$$

10.206. Assuming that the temperature on the surface of sun is approximately 6000°K, determine the wavelength corresponding to the maximum energy, considering that the sun is a perfectly black body. Wien's constant $b = 2.89 \times 10^{-3}$ mk.

Solution:

$$\lambda_m = \frac{b}{T} = \frac{2.89 \times 10^{-3}}{6000} = 4.82 \times 10^{-7}\,\text{m}$$

10.207. The peak emission from a black body at a certain temperature occurs at a wavelength of 9000A. On increasing the temperature, the total radiation emitted is increased 81 times. At the initial temperature when the peak radiation from the black body is incident on a metal surface it does not cause photo-emission from the surface. After the increase of temperature the peak radiation from the black body caused photo-emission. To bring these photo electrons to rest, a potential equivalent to the excitation energy between the $n = 2$ to $n = 3$ Bohr's level of hydrogen atom is required. Find the work function of the metal. (Roorkee 1993)

Solution:

$$E = \sigma T_1^4 \text{ and } 81E = \sigma T_2^4$$

$$\left(\frac{T_2}{T_1}\right)^4 = 81$$

$$T_2 = 3T_1$$

According to Wien's law we have

$$\lambda_1 T_1 = \lambda_2 T_2 = \lambda_2 \times 3T_1$$

$$9000\,T_1 = 3\lambda_2 T_1$$

$$\lambda_2 = 3000\,\text{A}°$$

$$eV_0 = \frac{hc}{\lambda_2} - W \text{ ; } V_0 = \text{cut-off potential}$$

$$eV_0 = -13.6\left(\frac{1}{9} - \frac{1}{4}\right) = \frac{17}{9}$$

$$= 1.89\, eV$$

$$\frac{hc}{\lambda_2} = \frac{6.62 \times 10^{-34} \times 3 \times 10^{8}}{3000 \times 10^{-10} \times 1.6 \times 10^{-19}} eV$$

$$= 4.14\, eV$$

$$W = 4.14 - 1.89 = 2.25\, eV$$

10.208. A wire of length 1.0 m and radius 10^{-3}m is carrying a heavy current and is assumed to radiate as a black body. At equilibrium its temperture is 900 K. While that of the surroundings is 300 K. The resistivity of the material of the wire at 300 K is $\pi^2 \times 10^{-8}\ \Omega$m and its temperature coefficient of resistance is 7.8×10^{-3}/°C. Find the current in the wire. Given Stefan's constant = $5.68 \times 10^{-8}\,\text{W} \cdot \text{m}^{-2}\,\text{k}^{-4}$. (Roorkee 1994)

Solution: According to Stefan's Law

$$E = \sigma\left(T^4 - T_0^4\right)A = 2\pi r l \sigma\left(T^4 - T_0^4\right)$$

$$= 2\pi \times 10^{-3} \times 1.0 \times 5.68 \times 10^{-8}\,[900^4 - 300^4]$$

$$= 5.68 \times 6480 \times 2\pi \times 10^{-3} \text{ watt}$$

$$\rho_{900} = \rho_{300}\,(1 + \alpha \cdot dT)$$

$$= \pi^2 \times 10^{-8}\,[1 + 7.8 \times 10^{-3} \times 600]$$

$$= \pi^2 \times 5.68 \times 10^{-8} \text{ ohm. m.}$$

$$R_{900} = \rho_{900} \times \frac{1}{\pi r^2}$$

$$= \frac{\pi^2 \times 5.68 \times 10^{-8}}{\pi \times \left(10^{-3}\right)^2}$$

$$= \pi \times 5.68 \times 10^{-2} \text{ ohm.}$$

$$P = I^2 \cdot 900 = I^2 \times \pi \times 5.68 \times 10^{-2} = 5.68 \times 6480 \times 2\pi \times 10^{-3}$$

$$I^2 = 648 \times 2 = 1296$$

$$I = 36 \text{ amp.}$$

10.209. A conductor of resistance R and heat capacity C is connected to a d.c. source of voltage V at an instant $t = 0$. Find the temperature of the conductor θ as a function of time by assuming the thermal power dissipated in the surrounding to vary as $q = k(\theta - \theta_0)$ where k is constant and θ_0 is the temperature of the surrounding.

Solution: According to conservation of energy

$$\frac{V^2}{R}\cdot t = msd\theta + E\,; \quad m = \text{mass of conductor}$$

s = sp. heat of conductor and

E = amount of energy radiated in time t.

or $$\frac{V^2}{R} = ms\frac{d\theta}{dt} + k(\theta-\theta_0);\ \frac{dE}{dt} = q = k(\theta-\theta_0)$$

$$\frac{V^2}{kR} = \frac{C}{k}\cdot\frac{d\theta}{dt} + (\theta-\theta_0)$$

$$\left[\frac{V^2}{kR} - (\theta-\theta_0)\right] = \frac{C}{k}\cdot\frac{d\theta}{dt}$$

$$\int_0^t \frac{k}{C}\cdot dt = \int_{\theta_0}^{\theta} \frac{d\theta}{\frac{V^2}{kR} - (\theta-\theta_0)}$$

$$-\frac{kt}{C} = \log\left[\frac{V^2}{kR} - (\theta-\theta_0)\right] - \log\frac{V^2}{kR}$$

$$= \log\left[1 - \frac{\theta-\theta_0}{\frac{V^2}{kR}}\right]$$

$$\frac{\theta-\theta_0}{\frac{V^2}{kR}} = 1 - e^{-kt/C}$$

$$\theta-\theta_0 = \frac{V^2}{kR}\left[1 - e^{-kt/C}\right]$$

$$\theta = \theta_0 + \frac{V^2}{kR}\left[1 - e^{-kt/C}\right]$$

10.210. A uniform wire is heated by using a d.c. source. In order to maintain its previous temperature it is required to increase the voltage by η%. Calculate the % decrease in the wire's diameter due to evaporation by assuming the amount of heat radiated to the surrounding to be proportional to the surface area of the wire.

Solution: Amount of heat generated in time t is equal to $\frac{V^2 t}{R}$

The temperature of the conductor θ is given by

$$\frac{V^2}{\rho l J}\cdot \pi r^2 t = \pi r^2 l d \cdot s(\theta - \theta_0);\; R = \frac{\rho l}{\pi r^2}$$

θ_0 = temperature of the surrounding
d = density of the material of wire
s = specific heat

or $\quad V^2 \cdot t = \rho l^2 \cdot d \cdot s\,(\theta - \theta_0)\,J$

Now the amount of heat radiated is proportional to the surface area

i.e. $$\pi r^2 l \cdot d \cdot s \cdot \frac{d\theta}{dt} = -k \cdot 2\pi r l$$

$$\therefore\; d \cdot s \cdot d\theta = -\frac{2k}{r}dt$$

or $$d \cdot s(\theta - \theta_0) = -\frac{2k}{r}\cdot t$$

or $$V^2 t = -\rho l^2 \cdot \frac{2k}{r}\cdot t J$$

or $\quad V^2 r = \text{Constt.}$

$$\log V^2 + \log r = \text{Constt.}$$

$$\frac{2dV}{V} + \frac{dr}{r} = 0$$

$$-\frac{dr}{r} = -\frac{dD}{D} = \frac{2dV}{V}$$

∴ % decrease in diameter of the wire = 2η.

10.211. A liquid at 80°C cools to 60°C in 5 minutes and to 50°C in 10 minutes. What is the temperature of the surroundings.

Solution: According to Newton's law of cooling, the rate of cooling of a body is proportional to the temperature in excess over that of the surroundings, i.e.,

$$-ms\frac{d\theta}{dt} \propto (\theta - \theta_0)$$

Where θ_0 is the temperature of the surrounding

or $$-ms\frac{d\theta}{dt} = k(\theta - \theta_0)$$

or $$\int_{\theta_2}^{\theta_1}\frac{d\theta}{\theta - \theta_0} = -\frac{k}{ms}\int_0^t dt$$

or $$\log \frac{\theta_1 - \theta_0}{\theta_2 - \theta_0} = -\frac{k}{ms} t$$

On substituting the values of θ_1, θ_2 and t we have

$$\log\left(\frac{60 - \theta_0}{80 - \theta_0}\right) = -\frac{k}{ms} \times 5$$

and $$\log\left(\frac{50 - \theta_0}{80 - \theta_0}\right) = -\frac{k}{ms} \times 10$$

or $$\log\left(\frac{50 - \theta_0}{80 - \theta_0}\right) = 2 \log\left(\frac{60 - \theta_0}{80 - \theta_0}\right)$$

or $$\log\left(\frac{50 - \theta_0}{80 - \theta_0}\right) = \log\left(\frac{60 - \theta_0}{80 - \theta_0}\right)^2$$

or $$(50 - \theta_0)(80 - \theta_0) = (60 - \theta_0)^2$$

$$4000 - 50\,\theta_0 - 80\,\theta_0 + \theta_0^2 = 3600 - 120\,\theta_0 + \theta_0^2$$

$$4000 - 3600 = 10\,\theta_0$$

$$400 = 10\,\theta_0$$

$\therefore$ $$\theta_0 = 40°C$$

10.212. A body, initially at 80°C cools to 64°C in 5 minutes and to 52°C in 10 minutes. What is the temperature of the surrounding? What will be the temperature of the body after 15 minutes?

Solution:

$$\log\left(\frac{64 - \theta_0}{80 - \theta_0}\right) = -\frac{k}{ms} \times 5$$

$$\log\left(\frac{52 - \theta_0}{80 - \theta_0}\right) = -\frac{k}{ms} \times 10$$

Here θ_0 is the temperature of the surrounding

$$\log\left(\frac{52 - \theta_0}{80 - \theta_0}\right) = \log\left(\frac{64 - \theta_0}{80 - \theta_0}\right)$$

or $$(52 - \theta_0)(80 - \theta_0) = (64 - \theta_0)^2$$

or $$52 \times 80 - 132\,\theta_0 + \theta_0^2 = 64^2 - 128\,\theta_0 + \theta_0^2$$

$$4\,\theta_0 = 4160 - 4096 = 64$$

$\therefore$ $$\theta_0 = 16°C$$

$$\log \frac{\theta - \theta_0}{80 - \theta_0} = \log\left(\frac{64 - 16}{80 - 16}\right)^2;$$ where θ is the temperature of the body after 15 minutes.

$$\frac{\theta - 16}{64} = \frac{27}{64}$$

or $\theta = 43°C$

10.213. Two identical spheres P and Q at temperatures 18 and 35°C, respectively are suspended in a room where the temperature is 25°C. After some time, the temperature of P is found to be 23°C. Calculate the temperature of the sphere Q at that instant.

Solution:

$$\log \frac{\theta_1 - \theta_0}{\theta_2 - \theta_0} = \log \frac{\theta_0 - \theta_1'}{\theta_2 - \theta_2'}$$

Let θ be the temperature of the sphere Q then, we have

$$\frac{\theta - 25}{35 - 25} = \frac{25 - 23}{25 - 18}$$

or

$$\frac{\theta - 25}{10} = \frac{2}{7}$$

$$7\theta - 175 = 20$$

$$7\theta = 195$$

$$\theta \cong 28°C$$

10.214. A calorimeter containing a hot liquid is placed inside an enclosure whose walls are at 10°C and cools from 80°C to 60°C in 10 minutes. How long will it take for the liquid to cool from 60° to 40°C if Newton's law of cooling holds?

Solution:

$$\log \frac{\theta_1 - \theta_0}{\theta_2 - \theta_0} = -\frac{k}{ms} \times t$$

Here $\theta_0 = 10°C$

$$\log \frac{60 - 10}{80 - 10} = -\frac{k}{ms} \times 10$$

$$\log \frac{40 - 10}{60 - 10} = -\frac{k}{ms} \times t$$

$\therefore$ $t \log 5/7 = 10 \log 3/5$

$t = 15.18$ minutes

11

ELECTROSTATICS

11.1. Find out the potential difference $V_A - V_B$ in the arrangement shown in Fig. 11.1. $q = 1 \mu C$, $x = 2$cm, $y = 3$ cm.

Solution: Potential at A due to the two charges is given by

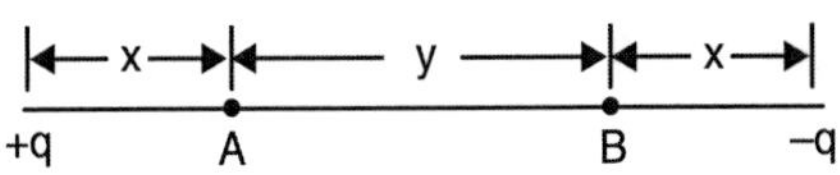

Fig. 11.1

$$V_A = \frac{q}{4\pi\varepsilon_0 x} - \frac{q}{4\pi\varepsilon_0 (x+y)}$$

Similarly potential at B due to the two charges is given by

$$V_B = -\frac{q}{4\pi\varepsilon_0 x} + \frac{q}{4\pi\varepsilon_0 (x+y)}$$

$$V_A - V_B = \frac{2q}{4\pi\varepsilon_0}\left[\frac{1}{x} - \frac{1}{(x+y)}\right] = \frac{2qy}{4\pi\varepsilon_0 x(x+y)}$$

On substituting the values of the various terms we have

$$V_A - V_B = \frac{9\times10^9 \times 2\times10^{-6} \times 3\times10^{-2}}{2\times10^{-2} \times 5\times10^{-2}} = 5.4\times10^5\ V$$

11.2. An infinite chain of charge $q = 0.5\ \mu C$ is placed at distance of 1 m, 2 m, 4 m, 8 m, etc. along the x-axis (see Fig. 11.2). Find the electric potential at the origin.

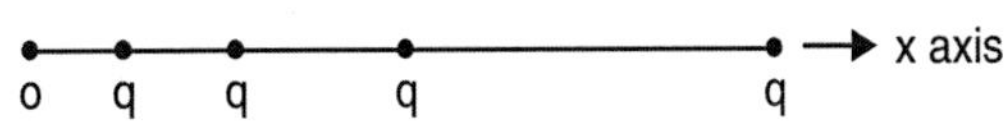

Fig. 11.2

Solution:

$$\text{Potential at } O = \frac{1}{4\pi\varepsilon_0}\left[\frac{q}{1} + \frac{q}{2} + \frac{q}{4} + \frac{q}{8} + \ldots\right]$$

$$= \frac{q}{4\pi\varepsilon_0}\left[1 + \frac{1}{2} + \frac{1}{2^2} + \frac{1}{2^3} + \ldots\right]$$

$$= \frac{q}{4\pi\varepsilon_0}\left[\frac{1}{1-1/2}\right] = \frac{2q}{4\pi\varepsilon_0}$$

On substituting the values, we have

Potential at $O = 2 \times 0.5 \times 10^{-6} \times 9 \times 10^{9} = 9{,}000$ V

11.3. There are two thin wire rings of radius $r = 4$ cm. The two rings have a common axis. The charges on the rings are $q = +10^{-8}$ and $q = -10^{-8}$ C (see Fig. 11.3). Find the potential difference between the centres of the rings separated by a distance $x = 3$ cm.

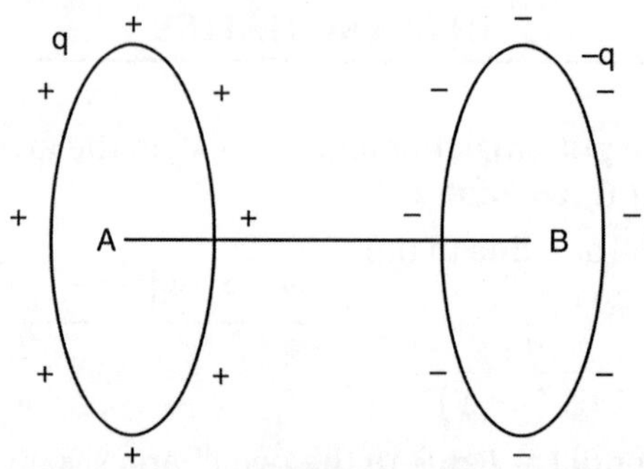

Fig. 11.3

Solution: Let A and B be the centres of the two rings. Consider the ring to be consisting of only $+q$ charge. Let us have an element on the ring carrying a charge dq on it. The potentials at A and B due to this element is given by

$$dV_A = \frac{dq}{4\pi\varepsilon_0 r}, \; dV_B = \frac{dq}{4\pi\varepsilon_0 \sqrt{x^2 + r^2}}$$

Total potential at A and B due to $+q$ charge on the ring is given by

$$V_A = \frac{1}{4\pi\varepsilon_0 r}\int dq = \frac{q}{4\pi\varepsilon_0 r}$$

$$V_B = \frac{1}{4\pi\varepsilon_0 \sqrt{x^2 + r^2}}\int dq = \frac{q}{4\pi\varepsilon_0 \sqrt{(x^2 + r^2)}}$$

Similarly total potential at A and B due to $-q$ charge on the other ring is given by

$$V_A = -\frac{q}{4\pi\varepsilon_0 \sqrt{(x^2 + r^2)}}; \; V_B = -\frac{q}{4\pi\varepsilon_0 r}$$

$$\text{Net } V_A = \left(\frac{1}{4\pi\varepsilon_0 r} - \frac{q}{4\pi\varepsilon_0 \sqrt{x^2 + r^2}}\right)$$

$$\text{Net } V_B = -\left(\frac{q}{4\pi\varepsilon_0 r} - \frac{q}{4\pi\varepsilon_0 \sqrt{x^2 + r^2}}\right)$$

$$V_A - V_B = \frac{2q}{4\pi\varepsilon_0}\left(\frac{1}{r} - \frac{1}{\sqrt{x^2 + r^2}}\right)$$

On substituting the values of the various terms, we have

$$V_A - V_B = 2 \times 10^{-8} \times 9 \times 10^9 \left(\frac{1}{4} - \frac{1}{5}\right) \times 10^{+2}$$

$$= 900 \text{ V}$$

11.4. A charge Q is distributed over two concentric hallow spheres of radius r and R ($> r$) such that the surface densities are equal. Find the potential at the common centre. (IIT 1981)

Solution: Let the charges Q_1 and Q_2 be distributed over the two spheres of radius r and R such that $Q = Q_1 + Q_2$.

$$\text{Surface charge density} = \frac{\text{charge}}{\text{area}} = \frac{Q_1}{4\pi r^2} = \frac{Q_2}{4\pi R^2}$$

$$\therefore \quad \frac{Q_1}{Q_2} = \frac{r^2}{R^2}$$

$$\text{or} \quad Q_1 + Q_2 = Q = \frac{r^2 + R^2}{R^2} \cdot Q_2$$

$$\text{or} \quad Q_2 = Q \cdot \frac{R^2}{r^2 + R^2}$$

$$\text{and} \quad Q_1 = \frac{r^2}{r^2 + R^2} \cdot Q$$

$$\text{Potential of large sphere } V_R = \frac{Q}{4\pi\varepsilon_0 R} \cdot \frac{R^2}{(r^2 + R^2)}$$

$$\text{Potential of small sphere } V_r = \frac{Q}{4\pi\varepsilon_0 r} \cdot \frac{r^2}{(r^2 + R^2)}$$

The required potential is given by

$$V = V_R + V_r = \frac{Q}{4\pi\varepsilon_0}\left(\frac{R + r}{R^2 + r^2}\right)$$

11.5. A thin rod carrying a charge q uniformly distributed over its length l is placed along x axis at a distance d from the origin O. Find the electric potential at a point P at a distance b along the y axis.

Solution: Consider an element of length dx lying at a distance x from the origin (see Fig. 11.4). The potential at P due to the element is given by

$$dV = \frac{q}{l} \cdot \frac{dx}{4\pi\varepsilon_0 \sqrt{b^2 + x^2}}$$

Net potential at P

$$= \frac{q}{4\pi\varepsilon_0 l}\int_d^{(l+d)} \frac{dx}{\sqrt{b^2+x^2}}$$

$$= \frac{q}{4\pi\varepsilon_0 l}\left|\log\left(x+\sqrt{b^2+x^2}\right)\right|_d^{l+d}$$

$$= \frac{q}{4\pi\varepsilon_0}\log\left[\frac{(l+d)+\sqrt{(l+d)^2+b^2}}{d+\sqrt{d^2+b^2}}\right]$$

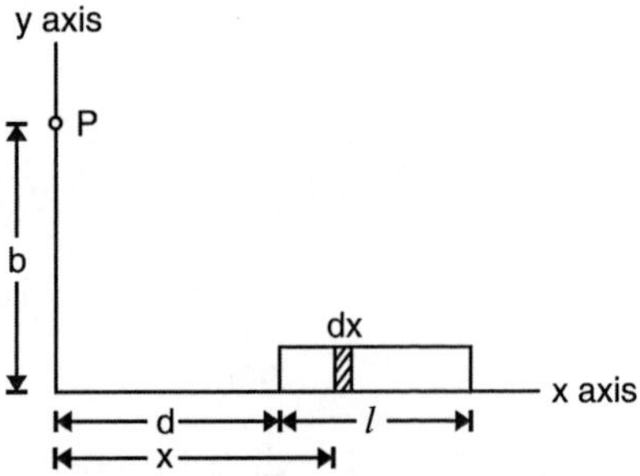

Fig. 11.4

11.6. Find out the electric potential at the rim of a charged disc of surface charge density σ.

Solution: Consider a thin wedge of length l and angular width $d\theta$ (see Fig. 11.5) charge on an element of length dr at a distance r from $P = \sigma\, r d\theta \cdot dr$

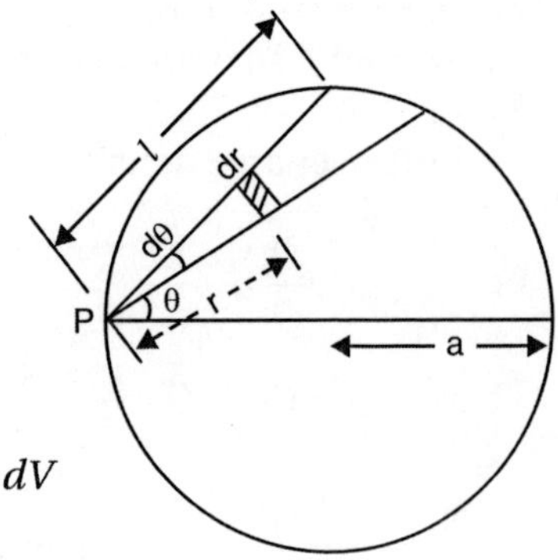

Fig. 11.5

Potential at P due to this element

$$= \frac{1}{4\pi\varepsilon_0}\cdot\frac{\sigma r d\theta\cdot dr}{r}$$

Net potential due to the entire wedge dV

$$= \frac{1}{4\pi\varepsilon_0}\cdot\sigma d\theta\int_0^l dr$$

$$dV = \frac{\sigma l d\theta}{4\pi\varepsilon_0}$$

$l = 2a\cos\theta$; a = radius of the disc

$$dV = \frac{\sigma\cdot 2a\cdot\cos\theta\cdot d\theta}{4\pi\varepsilon_0}$$

Net potential $V = \dfrac{\sigma a}{2\pi\varepsilon_0}\displaystyle\int_{-\pi/2}^{\pi/2}\cos\theta\cdot d\theta = \dfrac{\sigma a}{\pi\varepsilon_0}$

11.7. Find out the potential at the centre of a charged disc of surface charge density σ.

Solution: Consider an element of length dr at a distance r from the centre O (see Fig. 11.6). Potential due to this element at O is given by

$$\frac{1}{4\pi\varepsilon_0}\cdot\frac{\sigma r d\theta\cdot dr}{r} = \frac{\sigma\cdot d\theta\cdot dr}{4\pi\varepsilon_0}$$

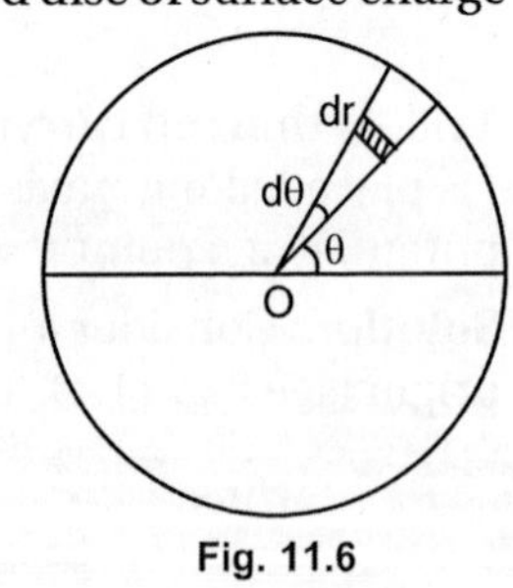

Fig. 11.6

The potential due to the entire wedge $= \dfrac{\sigma d\theta}{4\pi\varepsilon_0}\int_0^a dr$;

a = radius of the disc

$$dV = \frac{\sigma a d\theta}{4\pi\varepsilon_0}$$

Net potential V at the centre $= \dfrac{\sigma a}{4\pi\varepsilon_0}\int_0^{2\pi} d\theta = \dfrac{\sigma a}{2\varepsilon_0}$

11.8. A thin wire ring of radius a contains a charge q uniformly distributed over its length. Find the potential at a point lying on the circumference of the ring.

Solution: Consider an element on the circumference subtending an angle $d\theta$ (see Fig. 11.7).

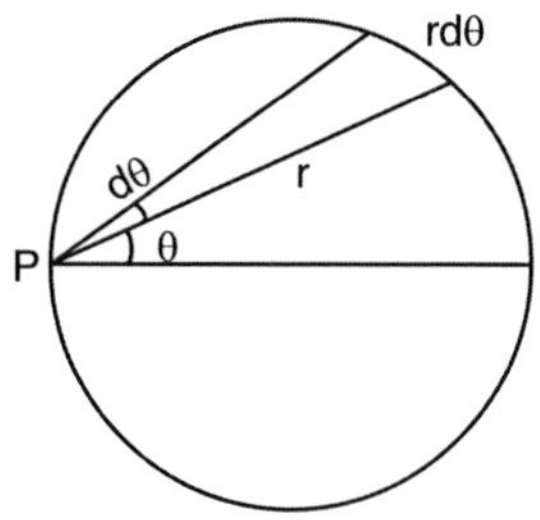

Fig. 11.7

Charge on the element $= r\,d\theta \cdot \dfrac{q}{2\pi a}$;

a = radius of the ring

Potential at P due to this element

$$dV = \frac{qr\,d\theta}{2\pi a \cdot 4\pi\varepsilon_0} \cdot \frac{1}{r} = \frac{q \cdot d\theta}{8\pi^2\varepsilon_0 a}$$

Net potential at P due to the entire ring

$$V = \int_{-\pi/2}^{\pi/2} \frac{q}{8\pi^2\varepsilon_0 a} \cdot d\theta = \frac{q}{8\pi\varepsilon_0 a}$$

11.9. A right angled sheet of base b and height h is placed in xy plane with its vertex O at the origin as shown in the Fig. 11.8. The sheet is uniformly charged and has surface charge density σ. Find the potential at the origin.

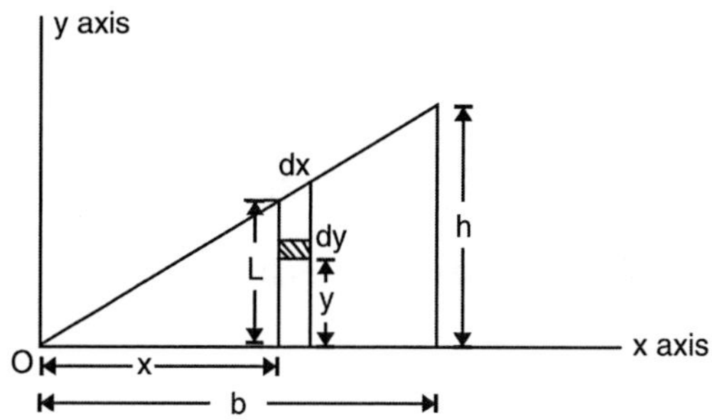

Fig. 11.8

Solution: Consider an element of area $dx \cdot dy$ at a distance x from the origin.

Charge on the element $= \sigma \cdot dx \cdot dy$

dV = Potential due to the strip of width dx

$$= \int_0^L \frac{\sigma \cdot dx \cdot dy}{4\pi\varepsilon_0 \sqrt{x^2 + y^2}}$$

$$= \frac{\sigma \cdot dx}{4\pi\varepsilon_0} \left| \log\left[y + \sqrt{x^2 + y^2} \right] \right|_0^L$$

$$= \frac{\sigma \cdot dx}{4\pi\varepsilon_0} \log\left[\frac{h + \sqrt{h^2 + b^2}}{b} \right]; \frac{L}{x} = \frac{h}{b}$$

$$\text{Net potential } V \text{ at } O = \frac{\sigma}{4\pi\varepsilon_0} \cdot \log\left[\frac{h + \sqrt{h^2 + b^2}}{b} \right] \int_0^b dx$$

$$= \frac{\sigma b}{4\pi\varepsilon_0} \log\left[\frac{h + \sqrt{h^2 + b^2}}{b} \right]$$

11.10. Calculate the amount of work done in assembling a disc of radius R having a charge density σ.

Solution: The total charge Q on thedisc $= \sigma \cdot \pi R^2$

Let us consider a disc of radius r having surface charge density σ.

$$\text{The potential due to the disc at the rim} = \frac{\sigma r}{\pi\varepsilon_0}$$

Consider a shell of radius r and $r + dr$

Area of the shell $= 2\pi\, rdr$

Charge contained in the shell $dq = \sigma \cdot 2\pi\, rdr$

$dW = V \cdot dq$

$$\text{Net work done} = W = \int dW = \int_0^R Vdq = \int_0^R \frac{\sigma r}{\pi\varepsilon_0} \cdot 2\pi r \sigma dr$$

$$= \frac{\sigma^2}{\varepsilon_0} \int_0^R r^2 dr = \frac{2\sigma^2 R^3}{3\varepsilon_0}$$

$$\text{Net work done in terms of the total charge } Q = \frac{2Q^2 R^3}{3\pi^2 R^4 \varepsilon_0}$$

$$= \frac{2Q^2}{3\pi^2 \varepsilon_0 R}$$

11.11. Calculate the potential at the centre of a spherical charge distribution given by

$\rho = \rho_0 r$ for $r \le R$

$\rho = 0$ for $r > R$

Solution: For the given distribution the charge at centre ($r = 0$) is zero. The value of the charge increases with increasing the value of r.

Consider a spherical shell of radius r and $r + dr$

Charge contained in the shell = $4\pi r^2\, dr \cdot \rho$

Potential due to the shell at the centre

$$dV = \frac{4\pi r^2 dr\rho}{4\pi\varepsilon_0 r} = \frac{\rho_0}{\varepsilon_0} r^2 \cdot dr$$

$$V = \int dV = \int_0^R \frac{\rho_0}{\varepsilon_0} \cdot r^2 dr = \frac{\rho_0 R^3}{3\varepsilon_0}$$

11.12. Calculate the potential at the centre for the charge distribution

$\rho = \rho_0$ for $r < R$

$\rho = 0$ for $r > R$

Solution: This is the case of uniform charge distribution.

Consider a shell of radius r and $r + dr$. Charge contained in this shell $= 4\,\pi\, r^2\, dr\, \rho$

Potential due to the shell at the centre

$$dV = \frac{4\pi r^2 dr\rho}{4\pi\varepsilon_0 r}$$

Net potential $V = \dfrac{\rho_0}{\varepsilon_0}\displaystyle\int_0^R r dr = \dfrac{\rho_0 R^2}{2\varepsilon_0}$

11.13. Calculate the potential at the centre for the charge distribution.

$\rho = \rho_0 \left[1 - \dfrac{r^2}{R^2}\right]$ for $r \le R$

$\rho = 0$ for $r > R$

Solution: In this case the maximum charge is at the centre ($r = 0$) and its value decreases on increasing r.

Consider a shell of radius r and $r + dr$

Charge contained in this shell = $4\pi\, r^2\, dr\, \rho$

Potential due to the shell at the centre is given by

$$dV = \frac{4\pi r^2 dr\rho}{4\pi\varepsilon_0 r}$$

$$= \frac{\rho_0}{\varepsilon_0}\left[1 - \frac{r^2}{R^2}\right] r \cdot dr$$

$$V = \int dV = \frac{\rho_0}{\varepsilon_0}\int_0^R\left[1-\frac{r^2}{R^2}\right]rdr$$

$$= \frac{\rho_0 R^2}{4\varepsilon_0}$$

11.14. A charge carrying infinitely long thread has linear charge density $\lambda = 0.3\ \mu c/m$. Calculate the potential difference between arbitrarily chosen two points 1 & 2 lying on its perpendicular bisector. Point 2 is $\eta = 3$ times farther from point 1.

Solution: Electric field at point 1 (see Fig. 11.9) due to the charge carrying thread $= \dfrac{\lambda}{2\pi\varepsilon_0 x}$

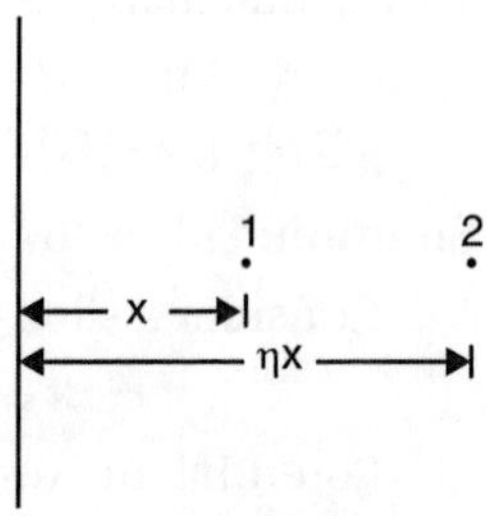

Fig. 11.9

Let the point 1 be shifted through dx

then $\dfrac{\lambda\cdot dx}{2\pi\varepsilon_0 x} = -dV$

$$\therefore V = -\int_1^2 dV = \frac{\lambda}{2\pi\varepsilon_0}\cdot\int_x^{\eta x}\frac{dx}{x}$$

$$= \frac{\lambda}{2\pi\varepsilon_0}\cdot\log_e\eta$$

$$= \frac{2.303\,\lambda}{2\pi\varepsilon_0}\log_{10}\eta$$

On substituting the values of the terms we have

$$V = 2.303\times0.3\times10^{-6}\times2\times9\times10^{9}\times\log_{10}3$$

$$= 5.93\times10^{3}\text{ volt}$$

11.15. An electric dipole of dipole moment p is placed in an external constant electric field E_0 such that vector p is parallel to the external field (see Fig. 11.10). Find the radius of a sphere which encloses the dipole and is formed by one of the equipotential surfaces.

Solution: The electric field at O due to dipole is given by

$$E = \frac{p}{4\pi\varepsilon_0 R^3}$$

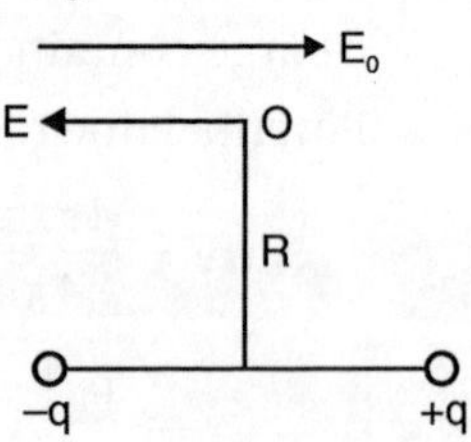

Fig. 11.10

For certain value of R the net electric field at O is equal to zero. The potential at O under this condition is constant hence

$$\frac{p}{4\pi\varepsilon_0 R^3} = E_0$$

or $$R = \left(\frac{p}{4\pi\varepsilon_0 E_0}\right)^{1/3}$$

11.16. A metallic sphere of radius r is placed concentrically inside a hollow sphere of radius R. The inner sphere is earthed by a wire through an opening of the outer sphere as shown in the Fig. 11.11. A change Q is imparted to the outer sphere. Find its potential.

Solution: The two spheres forms a capacitor and a conductor. The total capacity is given by

$$C = \frac{4\pi\varepsilon_0 Rr}{R-r} + 4\pi\varepsilon_0 R$$

$$= \frac{4\pi\varepsilon_0 R^2}{R-r}$$

$$V = \frac{Q(R-r)}{4\pi\varepsilon_0 R^2}$$

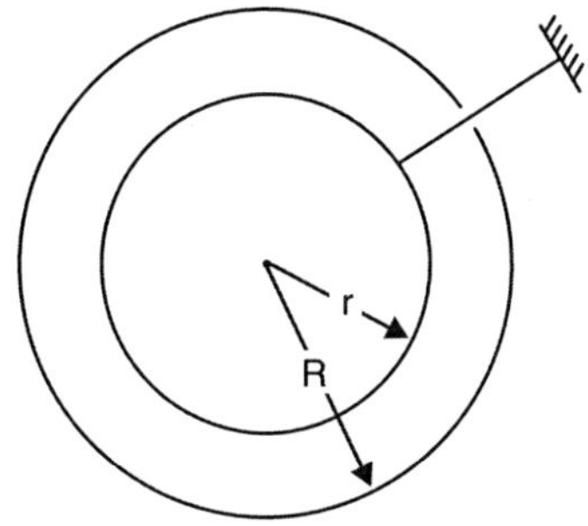

Fig. 11.11

11.17. A charge q is uniformly distributed over a length l of a conductor. Find out the potential at the point O on the perpendicular bisector at a distance d from the conductor.

Solution: Consider an element of length dx lying at a distance x from the centre of the conductor as shown in the Fig. 11.12.

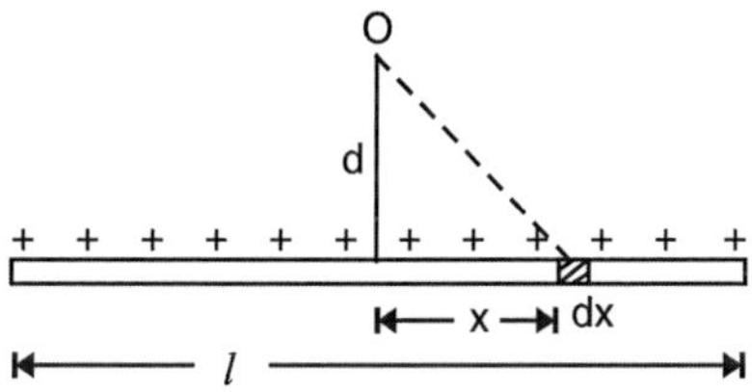

Fig. 11.12

Potential at O due to the element

$$= \frac{q}{l} \cdot \frac{dx}{4\pi\varepsilon_0 \sqrt{x^2 + d^2}}$$

Net potential at O $= \dfrac{q}{4\pi\varepsilon_0 l} \displaystyle\int_{-l/2}^{+l/2} \frac{dx}{\sqrt{x^2 + d^2}}$

$$= \frac{q}{4\pi\varepsilon_0 l}\left[\log \frac{\sqrt{4d^2 + l^2} + l}{\sqrt{4d^2 + l^2} - l}\right]$$

11.18. A rod of length L lies along the x-axis with its one end at the origin (see Fig. 11.13). The rod is charged and has linear charge density $\lambda = ax$

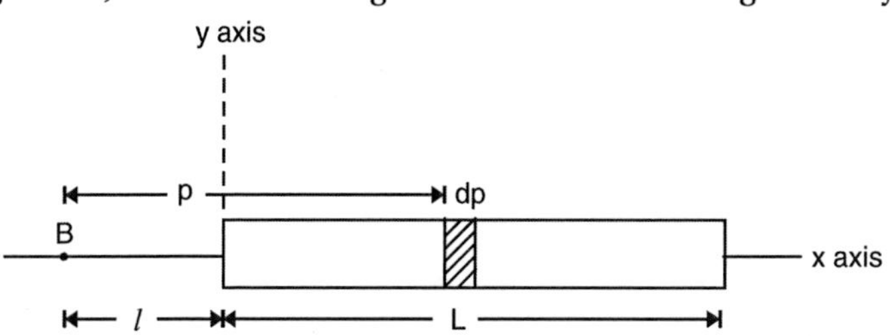

Fig. 11.13

where a is a positive constant. Find out the potential at B lying at a distance l from the origin.

Solution: Consider an element of length dp at a distance p from B

Potential at B due to this element $= \dfrac{1}{4\pi\varepsilon_0} \cdot \dfrac{a(p-l)\,dp}{p}$

Net potential at B is given by

$$V = \frac{a}{4\pi\varepsilon_0}\int_l^{(l+L)}\left[dp - \frac{l}{p}\cdot dp\right]$$

$$= \frac{a}{4\pi\varepsilon_0}\left[L - l\log\left(\frac{L+l}{l}\right)\right]$$

11.19. At what distance from the centre along the axis of a charged disc of surface charge density σ the electric potential will have the same value as at its centre. The error in this assumption is not more than 5%.

Solution: Consider a shell of radius x and $x + dx$ on the charged disc of radius R and surface charge density σ.

Potential at a point lying on the axis of the disc at a distance l from its centre is given by

$$dV = \frac{\sigma \cdot 2\pi \cdot x dx}{4\pi\varepsilon_0\sqrt{x^2 + l^2}}$$

$$\text{Net potential} = \frac{\sigma}{2\varepsilon_0}\int_0^R \frac{x dx}{\sqrt{x^2 + l^2}} = \frac{\sigma}{2\varepsilon_0}\left[\sqrt{R^2 + l^2} - l\right]$$

Potential at the centre of the disc $= \dfrac{\sigma R}{2\varepsilon_0}$

$$\text{Now} \quad \frac{\dfrac{\sigma R}{2\varepsilon_0} - \dfrac{\sigma}{2\varepsilon_0}\left[\sqrt{R^2 + l^2} - l\right]}{\dfrac{\sigma R}{2\varepsilon_0}} = \frac{1}{20}$$

$$1 - \frac{\sqrt{R^2 + l^2} - l}{R} = \frac{1}{20}$$

$$\left(\frac{19}{20}R + l\right)^2 = R^2 + l^2$$

$$\frac{19^2 R^2}{20^2} + \frac{38 Rl}{20} = R^2$$

$$\frac{38l}{20} = R\left[1 - \frac{19^2}{20^2}\right] = R \cdot \frac{39}{20^2}$$

$$\therefore\ l = \frac{39R}{20 \times 38} = 0.05R$$

11.20. Find the potential of an electrostatic field

$E = 2\,a\,x\,y\,i + a\,(x^2 - y^2)\,j$

Solution:

$$E_x = -\frac{d\phi_1}{dx};\quad E_y = -\frac{d\phi_2}{dy}$$

$$d\phi_1 = -E_x\,dx = -2ay\,xdx$$

$$\phi_1 = -ayx^2 + C_1$$

Similarly

$$d\phi_2 = -E_y\,dy = -a\,(x^2 - y^2)\,dy$$

$$\phi_2 = -a\left[x^2y - \frac{y^3}{3}\right] + C_2$$

$$\phi = -ax^2y + a\frac{y^3}{3} + \text{Constant}$$

11.21. A point charge Q is located at a distance x from the centre of an uncharged conducting sphere of radius R (see Fig. 11.14). Find the potential of the sphere.

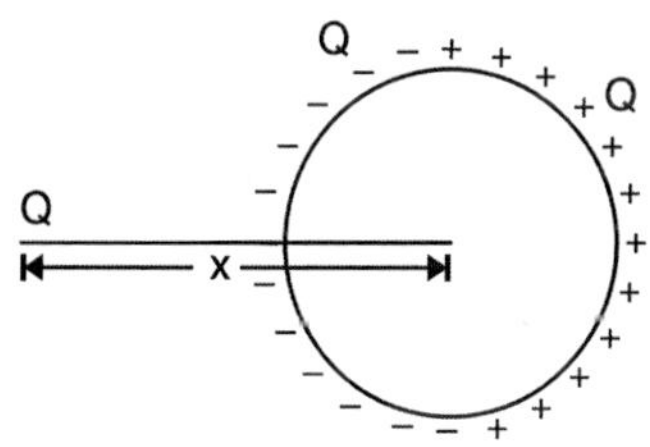

Fig. 11.14

Solution: The externally located charge will induce charges $-Q$ and $+Q$ on the sphere. The net potential of the sphere is given by

$$V = \frac{Q}{4\pi\varepsilon_0 x} + \frac{Q}{4\pi\varepsilon_0 R} - \frac{Q}{4\pi\varepsilon_0 R} = \frac{Q}{4\pi\varepsilon_0 x}$$

11.22. Find the potential at the centre of a uniformly charged hemisphere of radius r and surface charge density σ.

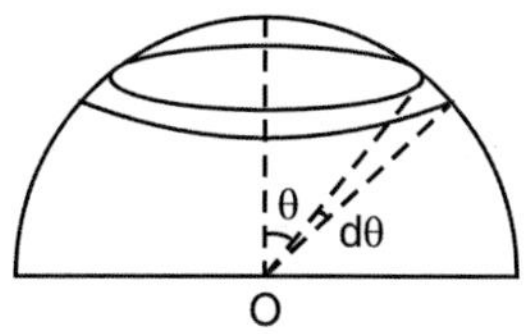

Fig. 11.15

Solution: Consider an element as shown in the Fig. 11.15. The area of the element = $2\,\pi\,r\sin\theta \cdot rd\,\theta$

Charge contained in the element

$= 2\,\pi\,r\sin\theta \cdot rd\,\theta \cdot \sigma$

$$\text{Potential due to the element at } O = \frac{2\pi r\sin\theta \cdot rd\theta \cdot \sigma}{4\pi\varepsilon_0 r}$$

$$\text{Net potential} = \frac{r\sigma}{2\varepsilon_0}\int_0^{\pi/2} \sin\theta\,d\theta = \frac{r\sigma}{2\varepsilon_0}$$

11.23. A charge Q is uniformly distributed through the volume of a solid sphere of radius R. Find the electric potential at (i) the centre of the sphere (ii) a point inside the sphere at a distance r from its centre.

Solution:

(i) Consider a spherical shell of radius x and $x + dx$ (see Fig. 11.16a)

$$\text{Volume of the shell} = \frac{4\pi}{3}\left[(x+dx)^3 - x^3\right] = \frac{4\pi}{3}\cdot 3x^2 dx$$

$$\text{Charge in the shell} = \frac{Q}{\frac{4}{3}\pi R^3}\cdot\frac{4\pi}{3}\cdot 3x^2 dx$$

$$= \frac{3Q\cdot x^2 dx}{R^3}$$

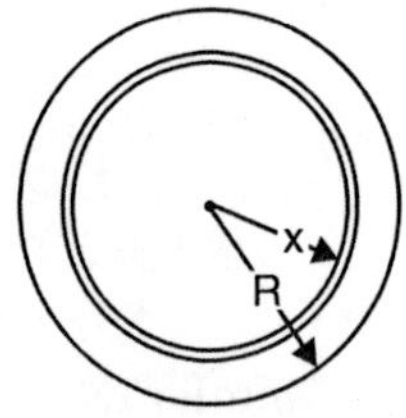

Fig. 11.16a

Potential due to the shell at the centre

$$dV = \frac{3Q\,x^2 dx}{4\pi\varepsilon_0 R^3\cdot x}$$

$$\text{Net potential at the centre} = \int_0^R \frac{3Q\,xdx}{4\pi\varepsilon_0 R^3} = \frac{3Q}{8\pi\varepsilon_0 R}$$

(ii) Potential at $C = \dfrac{Q}{4\pi\varepsilon_0 R}$ (see Fig. 11.16b)

Electric field at D inside the sphere at a distance r from the centre

$$E_D = \frac{Qr}{4\pi\varepsilon_0 R^3}$$

$$V_D - V_C = -\int_R^r E_D\cdot dr = -\frac{Q}{4\pi\varepsilon_0 R^3}\int_R^r rdr = \frac{Q}{8\pi\varepsilon_0 R^3}\left(R^2 - r^2\right)$$

$$V_D = V_C + \frac{QR^2}{8\pi\varepsilon_0 R^3} - \frac{Qr^2}{8\pi\varepsilon_0 R^3}$$

$$= \frac{Q}{4\pi\varepsilon_0 R} + \frac{Q}{8\pi\varepsilon_0 R} - \frac{Qr^2}{8\pi\varepsilon_0 R^3}$$

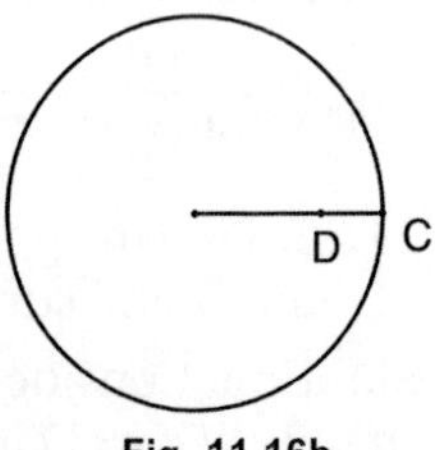

Fig. 11.16b

$$= \frac{3Q}{8\pi\varepsilon_0 R} - \frac{Qr^2}{8\pi\varepsilon_0 R^3}$$

$$= \frac{3Q}{8\pi\varepsilon_0 R}\left[1 - \frac{r^2}{3R^2}\right]$$

11.24. Calculate the amount of work required to assemble a sphere of charge Q and radius r_0.

Solution: Here we will calculate the amount of work done in assembling charges together from ∞ to form the sphere.

Consider a sphere of radius r. On this a shell of radius dr has been added. This process is to be continued till its radius becomes r_0.

Let ρ be the volume density of the charge.

Charge contained in the sphere of radius $r = \frac{4\pi}{3} r^3 \rho$

Potential of the sphere $V = \frac{\frac{4}{3}\pi r^3 \rho}{4\pi\varepsilon_0 r}$

Consider a spherical shell of radius r and $r + dr$

Charge on the shell $dq = 4\pi r^2\, dr\, \rho$

$$dW = V \cdot dq = \frac{4\pi}{3} \frac{r^3 \rho}{4\pi\varepsilon_0 r} 4\pi r^2 dr\, \rho$$

$$\text{Net work done } W = \int dW = \frac{4\pi\rho^2}{3\varepsilon_0} \int_0^{r_0} r^4 \cdot dr$$

$$= \frac{4\pi\rho^2 \cdot r_0^5}{15\varepsilon_0} = \frac{3Q^2}{20\pi\varepsilon_0 r_0}$$

11.25. N identical spherical droplets are charged to the same potential v. The charge on each droplet is q and the energy of each droplet is equal to E_0. Now the droplets are combined to form a big drop. Find the potential V and energy E of the drop formed.

Solution:

$$\text{Potential } v = \frac{q}{4\pi\varepsilon_0 r}; \; r = \text{radius of the droplet}$$

$$\text{Now} \quad N \times \frac{4}{3}\pi r^3 = \frac{4}{3}\pi R^3; \; R = \text{radius of the big drop}$$

$$R = N^{1/3} r$$

$$\text{Potential } V = \frac{Nq}{4\pi\varepsilon_0 R} = \frac{Nq}{4\pi\varepsilon_0 N^{1/3} r}$$

$$= N^{2/3} \cdot v$$

$$E_0 = \frac{1}{2} \frac{q^2}{4\pi\varepsilon_0 r}$$

$$E = \frac{1}{2} \cdot \frac{(Nq)^2}{4\pi\varepsilon_0 R}$$

$$= \frac{1}{2} \frac{q^2}{4\pi\varepsilon_0 r} \frac{N^2}{N^{1/3}}$$

$$= N^{5/3} \cdot E_0$$

11.26. A charge q is placed at the four corners of a square at a distance a from the origin O. Find out the potential at the point $P(x, y, z)$ which is close to the origin.

Solution: On accounting the distance between the charges and point P (see Fig. 11.17) the potential at P is given by

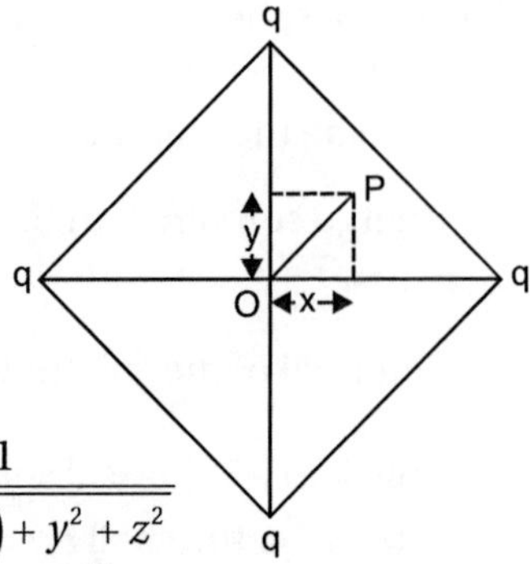

Fig. 11.17

$$V = \frac{q}{4\pi\varepsilon_0}\left[\frac{1}{\sqrt{(a-x)^2+y^2+z^2}} + \frac{1}{\sqrt{(a+x^2)+y^2+z^2}} + \frac{1}{\sqrt{x^2+(a-y)^2+z^2}} + \frac{1}{\sqrt{x^2+(a+y)^2+z^2}}\right]$$

$$= \frac{q}{4\pi\varepsilon_0 a}\left[\frac{1}{\sqrt{1-\left(\frac{2ax-x^2-y^2-z^2}{a^2}\right)}} + \frac{1}{\sqrt{1+\left(\frac{2ax+x^2+y^2+z^2}{a^2}\right)}} + \frac{1}{\sqrt{1-\left(\frac{2ay-x^2-y^2-z^2}{a^2}\right)}} + \frac{1}{\sqrt{1+\left(\frac{2ay+x^2+y^2+z^2}{a^2}\right)}}\right]$$

$$= \frac{q}{4\pi\varepsilon_0 a}\left[1+\frac{1}{2}\left(\frac{2ax-x^2-y^2-z^2}{a^2}\right)+\frac{3}{8}\times\frac{4a^2x^2}{a^4} + 1-\frac{1}{2}\left(\frac{2ax+x^2-y^2-z^2}{a^2}\right)+\frac{3}{8}\times\frac{4a^2x^2}{a^4}+1+\frac{1}{2}\left(\frac{2ay-x^2-y^2-z^2}{a^2}\right) + \frac{3}{8}\times\frac{4a^2y^2}{a^4}+1-\frac{1}{2}\left(\frac{2ay+x^2+y^2+z^2}{a^2}\right)+\frac{3}{8}\cdot\frac{4a^2y^2}{a^4}\right]$$

$$= \frac{q}{4\pi\varepsilon_0 a}\left[4-\frac{2x^2-2y^2-2z^2}{a^2}+\frac{3x^2+3y^2}{a^2}\right]$$

$$= \frac{q}{4\pi\varepsilon_0 a}\left[4+\frac{x^2+y^2-2z^2}{a^2}\right]$$

11.27. An electric quadrupole consisting of four charges is presented in Fig. 11.18. Find out the electric potential at P lying at a distance r from O. Under the assumption $r >> a$.

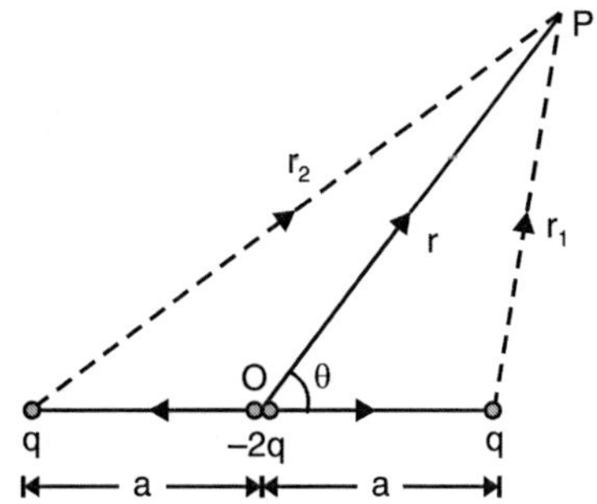

Fig. 11.18

Solution:

$$r_1 = \sqrt{r^2 + a^2 - 2ar\cos\theta} \text{ and}$$

$$r_2 = \sqrt{a^2 + r^2 + 2ar\cos\theta}$$

$$V = -\frac{2q}{4\pi\varepsilon_0 r} + \frac{q}{4\pi\varepsilon_0\sqrt{r^2 + a^2 - 2ar\cos\theta}} + \frac{q}{4\pi\varepsilon_0\sqrt{r^2 + a^2 + 2ar\cos\theta}}$$

$$= \frac{q}{4\pi\varepsilon_0 r}\left[-2 + \left[1 + \left(\frac{a^2 - 2ar\cos\theta}{r^2}\right)\right]^{-1/2} + \left[1 + \left(\frac{a^2 + 2ar\cos\theta}{r^2}\right)\right]^{-1/2}\right]$$

$$= \frac{q}{4\pi\varepsilon_0 r}\left[-2 + 1 - \left(\frac{a^2 - 2ar\cos\theta}{2r^2}\right) + \frac{3}{8}\cdot\frac{4a^2r^2\cos^2\theta}{r^4}\right.$$

$$\left. + 1 - \frac{1}{2}\left(\frac{a^2 + 2ar\cos\theta}{r^2}\right) + \frac{3}{8}\cdot\frac{4a^2r^2\cos^2\theta}{r^4} + \ldots\right]$$

$$= \frac{q}{4\pi\varepsilon_0 r}\left[-\frac{a^2}{r^2} + \frac{3a^2\cos^2\theta}{r^2}\right]$$

$$V = \frac{qa^2}{4\pi\varepsilon_0 r^3}\left[3\cos^2\theta - 1\right]$$

11.28. Three spherical shells of radii r_1, r_2 & r_3 and having surface charge densities σ, $-\sigma$ & σ are shown in Fig. 11.19. Find (i) the potential of each shell (ii) the relation between r_1, r_2 & r_3 when potential of inner and outer shell assumes the same value.

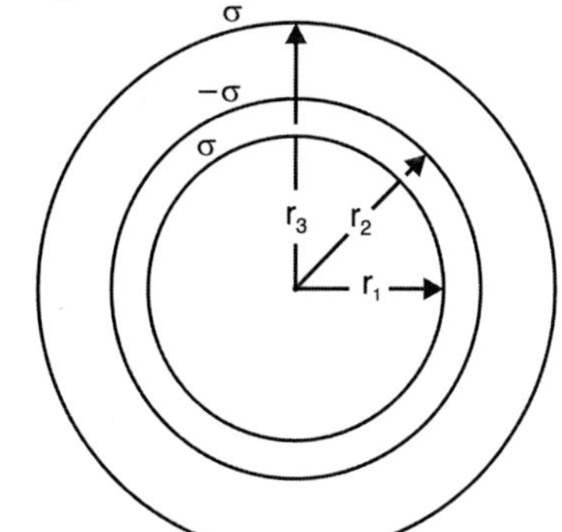

Fig. 11.19

Solution:

(i) Potential V_1 of inner shell of radius r_1 is given by

$$V_1 = \frac{4\pi r_1^2\sigma}{4\pi\varepsilon_0 r_1} - \frac{4\pi r_2^2\sigma}{4\pi\varepsilon_0 r_2} + \frac{4\pi r_3^2\sigma}{4\pi\varepsilon_0 r_3}$$

$$= \frac{\sigma}{\varepsilon_0}[r_1 - r_2 + r_3]$$

Potential V_2 of the middle shell of radius r_2 is given by

$$V_2 = \frac{4\pi r_1^2\sigma}{4\pi\varepsilon_0 r_2} - \frac{4\pi r_2^2\sigma}{4\pi\varepsilon_0 r_2} + \frac{4\pi r_3^2\sigma}{4\pi\varepsilon_0 r_3}$$

$$= \frac{\sigma}{\varepsilon_0}\left[\frac{r_1^2}{r_2} - r_2 + r_3\right]$$

Potential V_3 of the outer shell of radius r_3 is given by

$$V_3 = \frac{4\pi r_1^2\sigma}{4\pi\varepsilon_0 r_3} - \frac{4\pi r_2^2\sigma}{4\pi\varepsilon_0 r_3} + \frac{4\pi r_3^2\sigma}{4\pi\varepsilon_0 r_3}$$

$$= \frac{\sigma}{\varepsilon_0 r_3}\left[r_1^2 - r_2^2 + r_3^2\right]$$

(ii) $\frac{\sigma}{\varepsilon_0}[r_1 - r_2 + r_3] = \frac{\sigma}{\varepsilon_0 r_3}\left[r_1^2 - r_2^2 + r_3^2\right]$ given

or $r_1 r_3 - r_2 r_3 + r_3^2 = r_1^2 - r_2^2 + r_3^2$

or $r_3 (r_1 - r_2) = (r_1 - r_2)(r_1 + r_2)$

or $r_3 = r_1 + r_2$

11.29. Electric potential in the x-y plane is given by

$$V = \frac{x}{(x^2+y^2)^{3/2}} + \frac{y}{(x^2+y^2)^{5/2}}$$

Find the x and y component of electric field.

Solution:

$$E_x = -\frac{dV}{dx} = -\frac{d}{dx}\left[x(x^2+y^2)^{-3/2} + y(x^2+y^2)^{-5/2}\right]$$

$$= -\left[(x^2+y^2)^{-3/2} - x\cdot\frac{3}{2}(x^2+y^2)^{-5/2}\cdot 2x\right] + \frac{5}{2}\cdot y\cdot(x^2+y^2)^{-7/2}\cdot 2x$$

$$= -\frac{1}{(x^2+y^2)^{3/2}} + \frac{3x^2}{(x^2+y^2)^{5/2}} + \frac{5xy}{(x^2+y^2)^{7/2}}$$

$$= \frac{1}{(x^2+y^2)^{3/2}}\left[\frac{3x^2}{(x^2+y^2)} + \frac{5xy}{(x^2+y^2)^2} - 1\right]$$

$$E_y = -\frac{dV}{dy} = -\frac{d}{dy}\left[x(x^2+y^2)^{-3/2} + y(x^2+y^2)^{-5/2}\right]$$

$$= +\frac{3}{2}\cdot x(x^2+y^2)^{-5/2}\cdot 2y - (x^2+y^2)^{-5/2} + \frac{\frac{5}{2}y\cdot 2y}{(x^2+y^2)^{7/2}}$$

$$= \frac{1}{(x^2+y^2)^{5/2}}\left[3xy + \frac{5y^2}{(x^2+y^2)} - 1\right]$$

11.30. Electric charge with surface charge densities σ_1 and σ_2 respectively is uniformly spread over *xz* and *yz* planes. Calculate the potential difference between points 0, 0, 0 and 5, 7, 9.

Solution:

$$-\frac{dV}{dr} = E = \frac{\sigma}{2\varepsilon_0}$$

$$dV = -\frac{\sigma}{2\varepsilon_0} \cdot dr$$

$$V = \int -\frac{\sigma}{2\varepsilon_0} \cdot dr = -\frac{7\sigma_1}{2\varepsilon_0} - \frac{5\sigma_2}{2\varepsilon_0}$$

$$= -\frac{1}{2\varepsilon_0}[7\sigma_1 + 5\sigma_2]$$

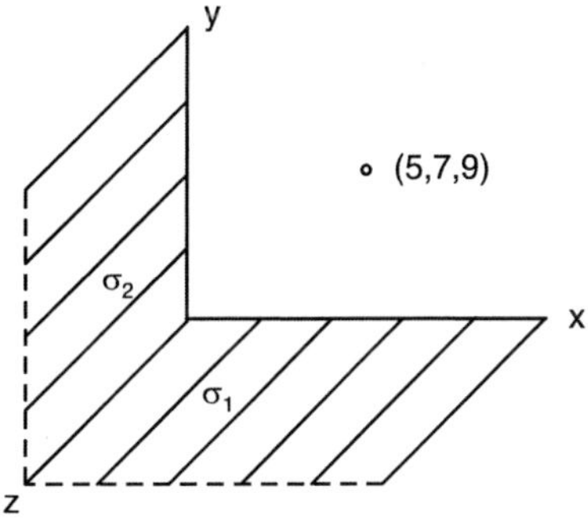

Fig. 11.20

11.31. A spherical conducting shell of radius *a*, centered at the origin has a potential field given by

$$V = V_0 \text{ when } r \le a$$

$$\text{and } V = V_0 \frac{a}{r} \text{ when } r > a$$

It has zero value at infinity. Find the energy stored.

Solution: Energy stored per unit volume = $\frac{1}{2}\varepsilon_0 E^2$ where *E* is the electric field.

$$\text{Now } E = -\frac{dV}{dr} = V_0 \frac{a}{r^2}$$

$$W = \frac{1}{2}\varepsilon_0 \int_0^{2\pi}\int_0^{\pi}\int_0^{\infty} \left(\frac{V_0 a}{r^2}\right)^2 r^2 \sin\theta \, dr \, d\theta \, d\phi = 2\pi\varepsilon_0 V_0^2 a$$

11.32. For a potential $V = 3x^2 + 4y^2$ volt. Calculate the energy stored in the volume described by $0 \le x \le 1\text{m}$, $0 \le y \le 1\text{m}$ and $0 \le z \le 1\text{m}$.

Solution:

$$E = -(6xi + 8yj)$$

$$W = \frac{1}{2}\varepsilon_0 \int_0^1\int_0^1\int_0^1 (36x^2 + 64y^2)\, dx\, dy\, dz$$

$$= \frac{1}{2}\varepsilon_0 \times \frac{100}{3} = \frac{1}{2} \times 8.85 \times 10^{-12} \times \frac{100}{3}$$

$$= 147.5 \text{ pJ}$$

11.33. An uncharged conductor is charged by repeated contact with a plate which after each touch with the conductor is filled with charge *Q* from an external source. Find the maximum charge which can be imparted to the conductor by this method if the conductor acquires a charge q at the first touch.

Solution: Let c_1 and c_2 be the capacities of the plate and conductor respectively (see Fig. 11.21). Before the first touch the plate has charge Q and the conductor is uncharged. At the first touch q charges are transferred from the plate to the conductor in order to equalise the potential, i.e.,

$$\frac{Q-q}{c_1} = \frac{q}{c_2}$$

or $$\frac{Q-q}{q} = \frac{c_1}{c_2}$$

If this process of charging is continued then the charges will add up on the conductor. After each touch the plate will have Q charges. Let Q_0 be the maximum charge which can be given to the conductor. Then at the last touch we have

$$\frac{Q}{c_1} = \frac{Q_0}{c_2}$$

or $$\frac{Q}{Q_0} = \frac{c_1}{c_2} = \frac{Q-q}{q}$$

$\therefore$ $$Q_0 = \frac{Qq}{Q-q}$$

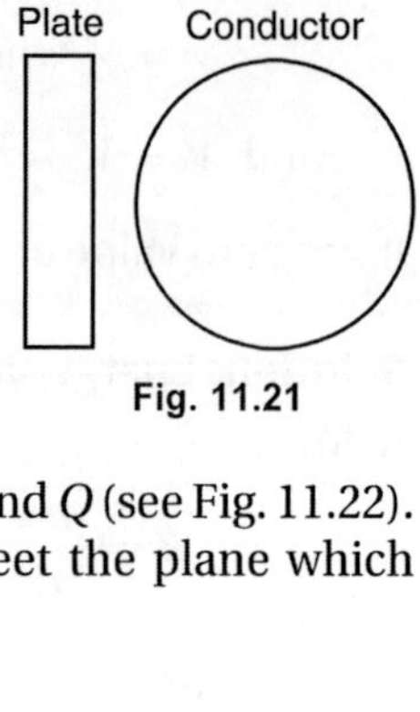

Fig. 11.21

11.34. Two charges $+q$ and $-q$ are situated at point P and Q (see Fig. 11.22). The lines of forces starting from P at an angle θ meet the plane which bisects PQ at right angles to A. Show that

$$\sin \theta/2 = \sqrt{2} \sin \frac{1}{2}(APQ)$$

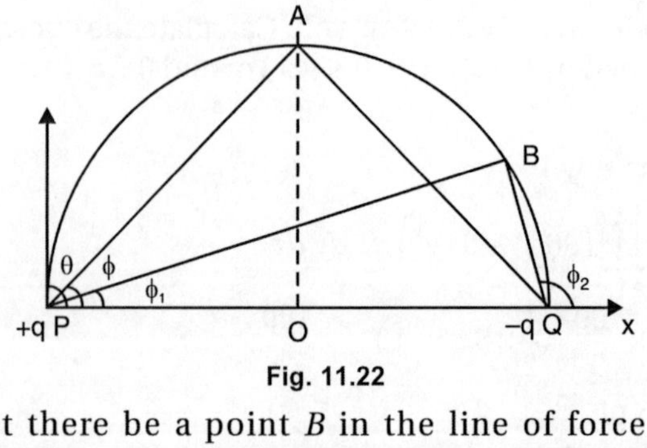

Fig. 11.22

Solution: Let there be a point B in the line of force. Then, from considerations of geometry, we have $\phi_1 = BPQ$, $\phi_2 = BQX$ and $\phi = APQ$. As B tends to P, ϕ_1 approaches θ and ϕ_2 approaches π. The equation of line of force is given by

$$q\cos\phi_1 - q\cos\phi_2 = q\cos\theta + q$$

or $\cos\phi_1 - \cos\phi_2 = \cos\theta + 1 = 2\cos^2\theta/2$

At point A, $\phi_1 = \phi = APQ$ and $\phi_2 = \pi - \phi$

$\therefore$ $\cos\phi - \cos(\pi - \phi) = 2\cos^2\theta/2$

or $\cos\phi = \cos^2\theta/2$

$$1 - 2\sin^2\phi/2 = 1 - \sin^2\theta/2$$

or $\sin\theta/2 = \sqrt{2}\sin\phi/2 = \sqrt{2}\sin\frac{1}{2}(APQ)$

11.35. Two charges $+q_1$ and $-q_2$ are placed at A & B respectively (see Fig. 11.23). A line of force starts from q_1 at an angle θ with the line AB. Find the angle ϕ at which it will terminate at $-q_2$.

Solution: Number of lines of force originating from A per unit solid angle is $\frac{q_1}{4\pi}$. Number of lines of forces through a cone of half angle θ is $\frac{q_1}{4\pi}\cdot 2\pi(1-\cos\theta)$. Number of lines of forces terminating at B at an angle ϕ is given by $\frac{q_2}{4\pi}\cdot 2\pi(1-\cos\phi)$

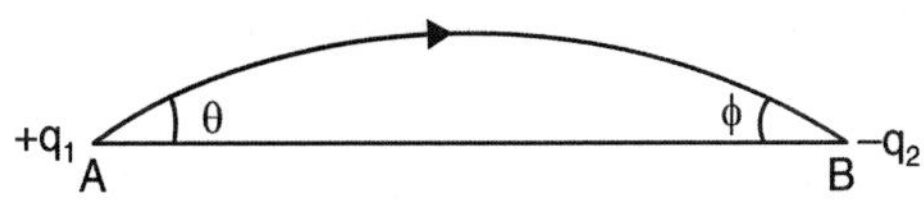

Fig. 11.23

$$\frac{q_1}{2}(1-\cos\theta) = \frac{q_2}{2}(1-\cos\phi)$$

$$q_1\sin^2\frac{\theta}{2} = q_2\sin^2\frac{\phi}{2}$$

$$\sqrt{\frac{q_1}{q_2}}\sin\frac{\theta}{2} = \sin\frac{\phi}{2}$$

$$\therefore \phi = 2\sin^{-1}\left[\sin\frac{\theta}{2}\sqrt{\frac{q_1}{q_2}}\right]$$

11.36. Find the force acting between the two dipole of dipole moments $p_1 = ql$ and $p_2 = q'l'$ respectively placed x apart along the straight line joining the dipoles.

Solution: Electric field due to the dipole of dipole moment p_1 at $-q'$ is given by

$$E = \frac{2p_1}{4\pi\varepsilon_0 x^3} \quad \text{(see Fig. 11.24)}$$

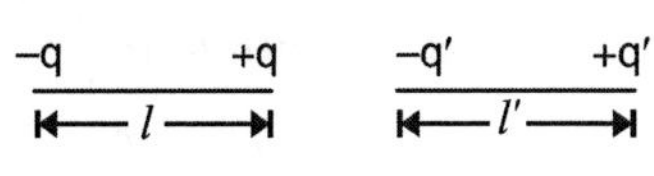

Fig. 11.24

Force F_1 at the charge $-q'$ is given by

$$F_1 = \frac{2pq'}{4\pi\varepsilon_0 x^3}$$

Rate of change of electric field along the length of dipole of dipole moment p_2 is given by

$$\frac{dE}{dx} = -\frac{2p_1}{4\pi\varepsilon_0}\times\frac{3}{x^4}$$

Change of field over a length l' is equal to

$$-\frac{2p_1 \cdot 3l'}{4\pi\varepsilon_0 x^4}$$

$$\text{Electric field at } +q' = \left(\frac{2p_1}{4\pi\varepsilon_0 x^3} - \frac{2p_1 \cdot 3l'}{4\pi\varepsilon_0 \cdot x^4}\right)$$

$$\text{Force } F_2 \text{ at } +q' = \left(\frac{2p_1}{4\pi\varepsilon_0 x^3} - \frac{2p_1 \cdot 3l'}{4\pi\varepsilon_0 \cdot x^4}\right) q'$$

Net force acting on the dipole

$$F_1 - F_2 = \frac{2p_1 \cdot 3l' \cdot q'}{4\pi\varepsilon_0 x^4} = \frac{6p_1 p_2}{4\pi\varepsilon_0 x^4}$$

11.37. Two copper spheres of radii $R_1 = 2$ cm and $R_2 = 3$ cm are connected to a source of emf $E = 2{,}000$ V. Calculate the force acting between them if their centres are $R = 1$ m apart.

Solution: Let Q_1 and Q_2 be the charges on the sphere such that $Q_1 + Q_2 = 0$. The potential difference between the spheres is E. Therefore, we have

$$\frac{Q_1}{4\pi\varepsilon_0 R_1} - \frac{Q_2}{4\pi\varepsilon_0 R_2} = E$$

On substituting $Q_1 = -Q_2$, we have

$$-\frac{Q_2}{4\pi\varepsilon_0}\left[\frac{R_1 + R_2}{R_1 R_2}\right] = E$$

$$\therefore \quad Q_1 = -Q_2 = \frac{4\pi\varepsilon_0 R_1 R_2 E}{R_1 + R_2}$$

$$\therefore \quad F = \frac{1}{4\pi\varepsilon_0}\left(\frac{4\pi\varepsilon_0 \cdot R_1 R_2 E}{R_1 + R_2}\right)^2 \times \frac{1}{R^2}$$

$$= 4\pi\varepsilon_0 \cdot \frac{R_1^2 R_2^2}{R^2} \cdot \frac{E^2}{(R_1 + R_2)^2}$$

On substituting the values of various terms we have

$$F = \frac{1}{9 \times 10^9} \times \frac{4 \times 10^{-4} \times 9 \times 10^{-4}}{1} \times \frac{4 \times 10^6}{25 \times 10^{-4}} = 6.4 \times 10^{-8}\,\text{N}$$

11.38. What would be the interaction force between two copper spheres, each of mass 2 g, separated by a distance of 1 m, if the total electronic charge in them differs from the total charge on the nuclei by 0.5%. $Z = 29$, atomic weight of copper = 64, $N = 6 \times 10^{23}$/mol.

Solution: Let q be the charge on each sphere. The force of interaction is given by

$$F = \frac{1}{4\pi\varepsilon_0} \cdot \frac{q^2}{x^2} = 9 \times 10^9 \cdot q^2; \; x = 1\,\text{m}$$

$$\text{Number of atoms in 2 g of copper} = \frac{6 \times 10^{23} \times 2}{64} = \frac{6}{32} \times 10^{23}$$

$$\text{Total positive charge (nuclear charge)} = \frac{6}{32} \times 10^{23} \times 29 \times 1.6 \times 10^{-19}$$

$$\text{Total electronic charge on each sphere} = \frac{6}{32} \times \frac{10^{23} \times 29 \times 1.6 \times 10^{-19} \times 0.5}{100}$$

$$= \frac{6 \times 29 \times 0.8 \times 100}{32} = 435\,\text{C}$$

$\therefore \quad F = 9 \times 10^9 \times (435)^2 = 1.7 \times 10^{15}$ N

11.39. It has been noticed that two similar balls containing same amount of charge suspended from the same point by two threads of same length do not deflect from the vertical in vacuum and also when immersed in kerosene of density $p_0 = 0.8 \times 10^3$ kg/m^3 and dielectric constant $K = 2$. Find the density of the material of the ball.

Solution: Let q be the charge on each ball of radius r. For no deflection from the vertical in vacuum we have

$$\frac{q^2}{4\pi\varepsilon_0 x^2} = \frac{G \cdot m^2}{x^2} = \frac{G \cdot \left(\frac{4}{3}\pi r^3 \rho\right)^2}{x^2}; \quad x = \text{distance between the centres of the ball}$$

ρ – density of the material of the ball

For no deflection when immersed in kerosene we have

$$\frac{q^2}{4\pi\varepsilon_0 K x^2} = \frac{G m'^2}{x^2} = G \cdot \left[\frac{4}{3}\pi r^3 (\rho - \rho_0)\right]^2 \Big/ x^2$$

On dividing the two we have

$$K = \frac{\rho^2}{(\rho - \rho_0)^2}$$

or $$\sqrt{K} = \frac{\rho}{\rho - \rho_0}$$

or $$\rho = \frac{\rho_0 \sqrt{K}}{\left(\sqrt{K} - 1\right)}$$

On substituting the values of the various terms we have

$$\rho = \frac{0.8\times10^3\times\sqrt{2}}{\left(\sqrt{2}-1\right)}$$

$$= \frac{0.8\times10^3\times1.414}{0.414}$$

$$= 2.74\times10^3\ \text{kg/m}^3$$

11.40. A ring of radius $r = 5$ cm, made of a thin wire, carries a charge $q = 10^{-6}$ C uniformly distributed on it. Calculate the increment of the force stretching the wire if a point charge $q_0 = 0.5\times10^{-6}$ C is placed at the centre of the ring.

Solution: Let the sought force of tension be T. Consider an element of the ring with a length $r\cdot\Delta\theta$.

$$\text{Charge on the element} = \frac{q}{2\pi r}\cdot r\Delta\theta = \frac{q\cdot\Delta\theta}{2\pi}$$

Force of repulsion

$$F = \frac{1}{4\pi\varepsilon_0}\cdot\frac{q\Delta\theta}{2\pi}\cdot\frac{q_0}{r^2}$$

Tension T will have two components. $T\cos\Delta\,\theta/2$ (see Fig. 11.25) will cancel out and the net effective force $2T\sin\Delta\theta/2$ will act downwards.

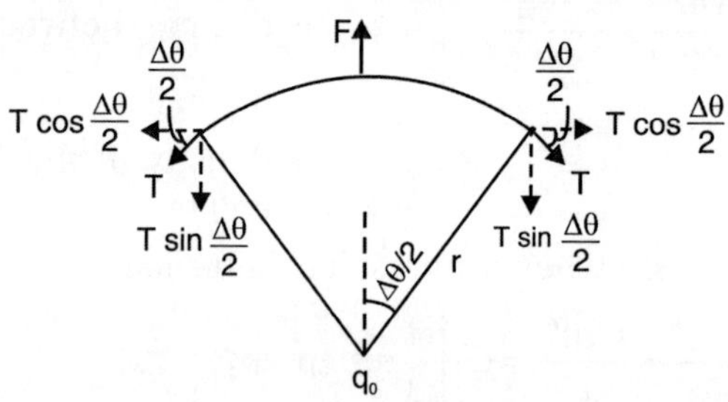

Fig. 11.25

For small angles,

$$2T\sin\Delta\theta/2 = 2T\cdot\Delta\theta/2 = T\cdot\Delta\theta$$

$$\text{Hence,}\qquad T\cdot\Delta\theta = \frac{qq_0\,\Delta\theta}{2\pi\cdot4\pi\varepsilon_0 r^2}$$

$$\therefore\qquad T = \frac{qq_0}{2\pi\cdot4\pi\varepsilon_0 r^2}$$

On substituting the values of various terms, we have

$$T = \frac{9\times10^9\times10^{-6}\times0.5\times10^{-6}}{2\times3.14\times25\times10^{-4}} = \frac{45}{50\times3.14} = 0.286\text{N}$$

11.41. A thin wire ring of mass m and radius R contains q charge uniformly distributed over its length. A charge Q ($Q >> q$) is placed at the centre of the ring. Find the total tension developed in the ring when it rotates with an angular velocity ω about an axis passing through its centre and perpendicular to its plane.

Solution: Consider an element AB on the ring as shown in Fig. 11.26. In view of the forces acting on the element we have

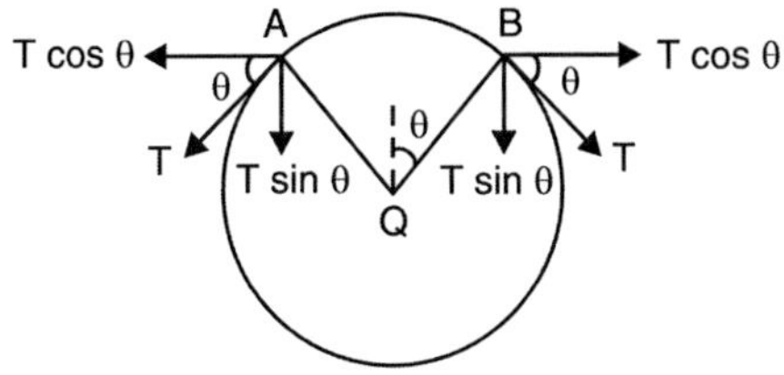

Fig. 11.26

$$\frac{Q}{4\pi\varepsilon_0 R^2}\cdot\frac{q}{2\pi R}\cdot R\cdot 2\theta+\frac{m}{2\pi R}\cdot R\cdot 2\theta\cdot R\omega^2=2T\sin\theta=T\cdot 2\theta;$$

For small values of θ, $\sin\theta = \theta$

$$\frac{Q\cdot q}{8\pi^2\varepsilon_0 R^2}+\frac{mR\omega^2}{2\pi}=T$$

11.42. A charge Q is placed at a distance $OA = d$ from the centre of a earthed conducting sphere of radius R. Find (i) the potential at a point P on the sphere, (ii) the force between the sphere and the charge.

Solution: (i) Let us chose a point B inside the sphere such that $OA \cdot OB = R^2$

Now $\quad \dfrac{OA}{R}=\dfrac{R}{OB}$ (see Fig. 11.27)

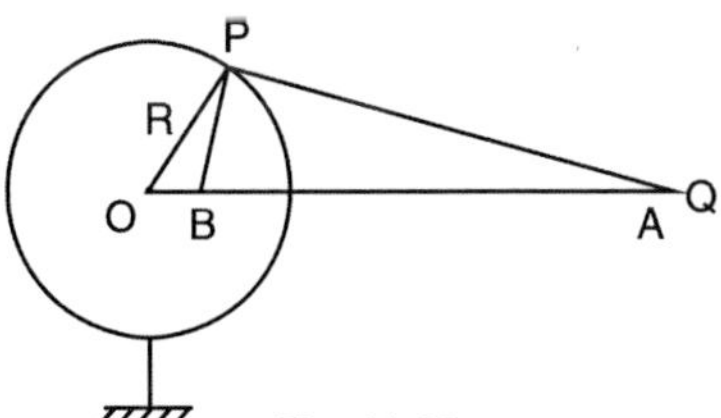

Fig. 11.27

Consider triangle OAP and OPB

$$\frac{BP}{OB}=\frac{AP}{OP}$$

or $$\frac{BP}{AP}=\frac{OB}{OP}=\frac{R^2}{OA\cdot OP}=\frac{R^2}{d\cdot R}=\frac{R}{d}$$

Now a charge $-\dfrac{R}{d}\cdot Q$ will be an electric image of the charge Q placed at A. The purpose of the sphere is to provide an electrical image of the charge Q.

$$\text{Potential at } P=\frac{1}{4\pi\varepsilon_0}\left[\frac{Q}{AP}-\frac{RQ}{d\cdot BP}\right]=\frac{1}{4\pi\varepsilon_0}\left[\frac{Q}{AP}-\frac{BP}{AP}\times\frac{Q}{BP}\right]=O$$

$$\text{(ii) Force} = \frac{1}{4\pi\varepsilon_0}\frac{Q \times -Q\,{}^R\!/_d}{AB^2} = -\frac{RQ^2}{4\pi\varepsilon_0 d\left[OA - OB\right]^2}$$

$$= -\frac{RQ^2}{4\pi\varepsilon_0 d\left[d - \dfrac{R^2}{d}\right]^2} = -\frac{RQ^2 d}{4\pi\varepsilon_0\left[d^2 - R^2\right]^2}$$

11.43. A charge Q is placed at a distance d from the centre of an uncharged insulated sphere of radius R as shown in Fig. 11.28. Find the force of the attraction between the sphere and the charge.

Solution: Let B be the electrical image of charge Q placed at A.

Charge at $B = -\dfrac{QR}{d}$

In order to have net charge equal to zero. We must have a charge $\dfrac{QR}{d}$ at the centre O of the sphere

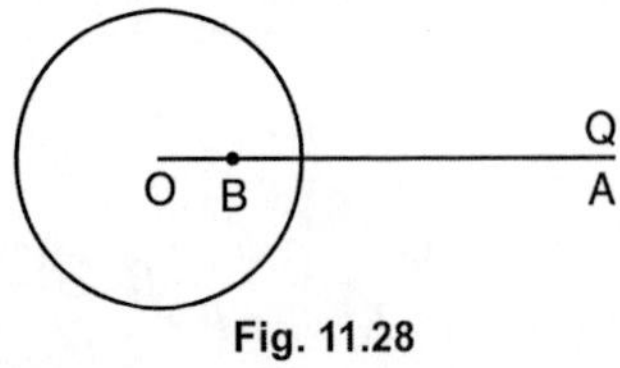

Fig. 11.28

$$F = \frac{1}{4\pi\varepsilon_0} \times -\frac{Q^2R}{d \cdot AB^2} + \frac{Q^2R}{d \cdot OA^2} \times \frac{1}{4\pi\varepsilon_0} = -\frac{Q^2R}{4\pi\varepsilon_0 d}\left[\frac{1}{\left(d - {}^{R^2}\!/_d\right)^2} - \frac{1}{d^2}\right]$$

$$= -\frac{Q^2R}{4\pi\varepsilon_0 d}\left[\frac{d^2}{\left(d^2 - R^2\right)^2} - \frac{1}{d^2}\right]$$

$$= -\frac{Q^2R^3\left[2d^2 - R^2\right]}{4\pi\varepsilon_0 d^3\left[d^2 - R^2\right]^2}$$

If $d >> R$ then

$$F = -\frac{2Q^2R^3}{4\pi\varepsilon_0 d^5}$$

11.44. An electric dipole of dipole moment $p = 2qa$ is located along y axis (see Fig 11.29). Calculate the electric field at a point P (x, y) using $r^2 = x^2 + y^2$.

Solution: Electric field E_1 due to $+q$ charge of the dipole is equal to

$$E_1 = \frac{q}{4\pi\varepsilon_0\left[x^2 + (y - a)^2\right]}$$

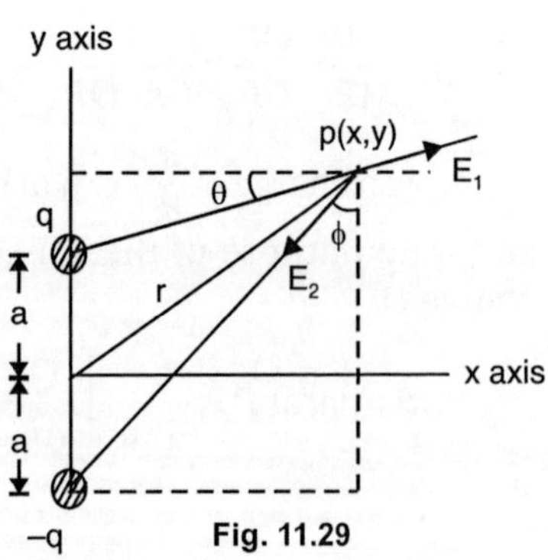

Fig. 11.29

Field E_2 due to $-q$ charge of the dipole is equal to

$$E_2 = \frac{q}{4\pi\varepsilon_0\left[x^2+(y+a)^2\right]}$$

E_x = Net electric field of the dipole along x axis and is given by $E_1\cos\theta - E_2\sin\phi$

$$E_x = \frac{qx}{4\pi\varepsilon_0\left[x^2+(y-a)^2\right]^{3/2}} - \frac{qx}{4\pi\varepsilon_0\left[x^2+(y+a)^2\right]^{3/2}}$$

$$= \frac{qx}{4\pi\varepsilon_0\left(x^2+y^2\right)^{3/2}}\left\{\left[1+\left(\frac{a^2-2ay}{x^2+y^2}\right)\right]^{-3/2} - \left[1+\frac{a^2+2ay}{x^2+y^2}\right]^{-3/2}\right\}$$

$$= \frac{qx}{4\pi\varepsilon_0\left(x^2+y^2\right)^{3/2}}\left\{1-\frac{3}{2}\left(\frac{a^2-2ay}{x^2+y^2}\right)-1+\frac{3}{2}\left(\frac{a^2+2ay}{x^2+y^2}\right)\right\}$$

$$= \frac{qx}{4\pi\varepsilon_0\left(x^2+y^2\right)^{3/2}}\times\frac{6ay}{\left(x^2+y^2\right)} = \frac{3pxy}{4\pi\varepsilon_0\left(x^2+y^2\right)^{5/2}}$$

$$E_y = E_1\sin\theta - E_2\cos\phi$$

$$= \frac{q(y-a)}{4\pi\varepsilon_0\left[x^2+(y-a)^2\right]^{3/2}} - \frac{q(y+a)}{4\pi\varepsilon_0\left[x^2+(y+a)^2\right]^{3/2}}$$

$$= \frac{q}{4\pi\varepsilon_0\left(x^2+y^2\right)^{3/2}}\left[(y-a)\left(1+\frac{a^2-2ay}{x^2+y^2}\right)^{-3/2} - (y+a)\left(1+\frac{a^2+2ay}{x^2+y^2}\right)^{-3/2}\right]$$

$$= \frac{q}{4\pi\varepsilon_0\left(x^2+y^2\right)^{3/2}}\left\{(y-a)\left[1-\frac{3}{2}\left(\frac{a^2-2ay}{x^2+y^2}\right)+\frac{15}{8}\left(\frac{4a^2y^2-4a^3y}{\left(x^2+y^2\right)^2}\right)\right] - (y+a)\left[1-\frac{3}{2}\left(\frac{a^2+2ay}{x^2+y^2}\right)+\frac{15}{8}\left(\frac{4a^2y^2+4a^3y}{\left(x^2+y^2\right)^2}\right)\right]\right\}$$

$$= \frac{q}{4\pi\varepsilon_0 \left(x^2+y^2\right)^{3/2}} \left[-2a + \frac{3}{2} \cdot \frac{a^2 \cdot 2a}{\left(x^2+y^2\right)} + \frac{3}{2} \cdot \frac{2ay \cdot 2y}{\left(x^2+y^2\right)} \right.$$

$$\left. + \frac{15}{2} \frac{a^2 y^2}{\left(x^2+y^2\right)^2} \times -2a - \frac{15}{2} \cdot \frac{a^3 y}{\left(x^2+y^2\right)^2} \times 2y \right]$$

$$= \frac{q}{4\pi\varepsilon_0 \left(x^2+y^2\right)^{3/2}} \left[-2a + \frac{3a^3}{\left(x^2+y^2\right)} + \frac{6ay^2}{x^2+y^2} \right]$$

On neglecting the second term we have

$$= \frac{2qa}{4\pi\varepsilon_0 \left(x^2+y^2\right)^{3/2}} \left[\frac{2y^2 - x^2}{x^2+y^2} \right]$$

$$= \frac{p\left(2y^2 - x^2\right)}{4\pi\varepsilon_0 \left(x^2+y^2\right)^{5/2}}$$

$$\therefore E = \frac{3pxy}{4\pi\varepsilon_0 \left(x^2+y^2\right)^{5/2}} i + \frac{p\left(2y^2-x^2\right) j}{4\pi\varepsilon_0 \left(x^2+y^2\right)^{5/2}}$$

11.45. An electric dipole of dipole moment p is placed in a non uniform electric field due to a point charge Q situated at the origin (see Fig. 11.30). Calculate the force experienced by the dipole.

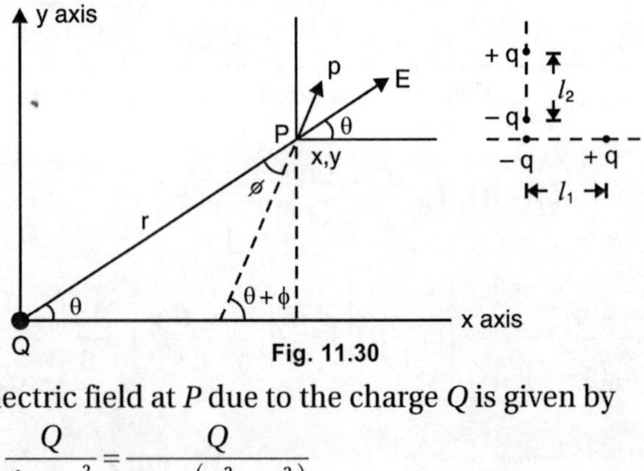

Fig. 11.30

Solution: Electric field at P due to the charge Q is given by

$$E = \frac{Q}{4\pi\varepsilon_0 r^2} = \frac{Q}{4\pi\varepsilon_0 \left(x^2+y^2\right)}$$

The x and y component of the field E is given by

$$E_x = \frac{Qx}{4\pi\varepsilon_0 \left(x^2+y^2\right)^{3/2}},\ E_y = \frac{Qy}{4\pi\varepsilon_0 \left(x^2+y^2\right)^{3/2}}$$

Dipole moment along x axis $= p\cos(\phi + \theta) = q \cdot l_1$

Dipole moment along y axis $= p\sin(\phi + \theta) = q \cdot l_2$

Force at $-q$ of dipole along x axis $= \dfrac{-Qxq}{4\pi\varepsilon_0\left(x^2+y^2\right)^{3/2}}$

Rate of change of field along x axis

$$= \frac{d}{dx}\left[\frac{Qx}{4\pi\varepsilon_0\left(x^2+y^2\right)^{3/2}}\right] = \frac{Q\left(y^2-2x^2\right)}{4\pi\varepsilon_0\left(x^2+y^2\right)^{5/2}}$$

Change of field over a length $l_1 = \dfrac{Q\left(y^2-2x^2\right)l_1}{4\pi\varepsilon_0\left(x^2+y^2\right)^{5/2}}$

Field at $+q$ charge of dipole along x axis

$$= -\frac{Qx}{4\pi\varepsilon_0\left(x^2+y^2\right)^{3/2}} - \frac{Q\left(y^2-2x^2\right)l_1}{4\pi\varepsilon_0\left(x^2+y^2\right)^{5/2}}$$

Force at $+q$ charge of the dipole along x axis

$$= -\frac{Qxq}{4\pi\varepsilon_0\left(x^2+y^2\right)^{3/2}} - \frac{Q\left(y^2-2x^2\right)l_1 q}{4\pi\varepsilon_0\left(x^2+y^2\right)^{5/2}}$$

Net force on the dipole lying along x axis due to the electric field E_x is given by

$$-\frac{Qxq}{4\pi\varepsilon_0\left(x^2+y^2\right)^{3/2}} + \frac{Qxq}{4\pi\varepsilon_0\left(x^2+y^2\right)^{3/2}} + \frac{Q\left(y^2-2x^2\right)ql_1}{4\pi\varepsilon_0\left(x^2+y^2\right)^{5/2}}$$

$$= \frac{Q\left(y^2-2x^2\right)p\cos(\theta+\phi)}{4\pi\varepsilon_0\left(x^2+y^2\right)^{5/2}}$$

Force at $-q$ of dipole along y axis $= -\dfrac{Qxq}{4\pi\varepsilon_0\left(x^2+y^2\right)^{3/2}}$

Rate of change of field E_x along y axis $= -\dfrac{3Qxy}{4\pi\varepsilon_0\left(x^2+y^2\right)^{3/2}}$

Net value of x component of the field along y axis

$$= \left(\frac{Qx}{4\pi\varepsilon_0\left(x^2+y^2\right)^{3/2}} - \frac{3Qxy \cdot l_2}{4\pi\varepsilon_0\left(x^2+y^2\right)^{5/2}}\right)$$

Force acting along x axis on the dipole placed along y axis

$$-\frac{Qxq}{4\pi\varepsilon_0\left(x^2+y^2\right)^{3/2}}+\frac{Qxq}{4\pi\varepsilon_0\left(x^2+y^2\right)^{3/2}}-\frac{3Qxy\cdot ql_2}{4\pi\varepsilon_0\left(x^2+y^2\right)^{5/2}}$$

$$=-\frac{3Qxy\,p\sin(\theta+\phi)}{4\pi\varepsilon_0\left(x^2+y^2\right)^{5/2}}$$

Net force F_x along x axis

$$=\frac{Q\left(y^2-2x^2\right)p\cos(\theta+\phi)}{4\pi\varepsilon_0\left(x^2+y^2\right)^{5/2}}-\frac{3Qxy\,p\cdot\sin(\theta+\phi)}{4\pi\varepsilon_0\left(x^2+y^2\right)^{5/2}}$$

$$F_x=\frac{Qp}{4\pi\varepsilon_0\left(x^2+y^2\right)^{5/2}}\left[\left(y^2-2x^2\right)\cos(\theta+\phi)-3xy\sin(\theta+\phi)\right]$$

On similar lines

$$F_y=\frac{Qp}{4\pi\varepsilon_0\left(x^2+y^2\right)^{5/2}}\left[-3\,xy\cos(\theta+\phi)+\left(x^2-2y^2\right)\sin(\theta+\phi)\right]$$

11.46. Two small spheres each of mass m carrying equal charge q are suspended from a point by two threads of length l. In the conditions of equilibrium the centres of the spheres are a distance x apart x = l. The entire charge leaks from each sphere. Find out the rate of leakage of the charge from the spheres if the velocity of their approach varies as $v=\alpha/\sqrt{x}$ where α is a constant.

Solution: The force of repulsion F between the two spheres is given by

$$F=\frac{1}{4\pi\varepsilon_0}\cdot\frac{q^2}{x^2}\quad\text{(Fig. 11.31)}$$

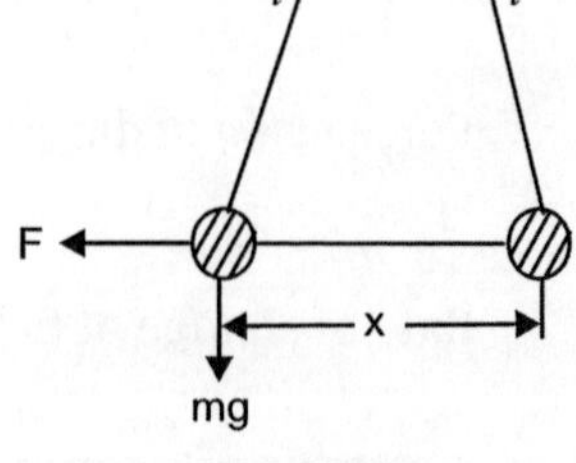

Fig. 11.31

Under the conditions of equilibrium

$$\tan\theta=\frac{F}{mg}=\frac{q^2}{4\pi\varepsilon_0 mg\,x^2}$$

If θ is small then $\sin\theta=\tan\theta=\dfrac{x}{2l}$. Hence

$$\frac{x}{2l}=\frac{q^2}{4\pi\varepsilon_0 mg\,x^2}$$

or $$x=\left(\frac{q^2 l}{2\pi\varepsilon_0 mg}\right)^{1/3}$$

As the charges leak from the spheres, in the light of the above equation we have

$$\frac{d}{dt}(x)=\frac{d}{dt}\left(\frac{q^2 l}{2\pi\varepsilon_0 mg}\right)^{1/3}$$

or $$\frac{dx}{dt}=\frac{1}{3}\left(\frac{q^2 l}{2\pi\varepsilon_0 mg}\right)^{-2/3}\times\frac{2ql}{2\pi\varepsilon_0 mg}\cdot\frac{dq}{dt}$$

or $$\frac{\alpha}{\sqrt{x}}\cdot\frac{3\pi\varepsilon_0 mg}{ql}\cdot\left(\frac{q^2 l}{2\pi\varepsilon_0 mg}\right)^{2/3}=\frac{dq}{dt}$$

or $$\frac{3\pi\varepsilon_0 mg}{ql}\cdot\alpha\cdot x^{3/2}=\frac{dq}{dt}$$

or $$\frac{3\pi\varepsilon_0 mg}{ql}\cdot\alpha\cdot q\sqrt{\frac{l}{2\pi\varepsilon_0 mg}}=\frac{dq}{dt}$$

or $$\frac{3}{2}\alpha\sqrt{\frac{2\pi\varepsilon_0 mg}{l}}=\frac{dq}{dt}$$

11.47. A thin non-conducting rod of length l = 1 m carries a charge q = 10^{-6} C spread uniformly along it. Find the electric field at the point O lying on the perpendicular by bisector at a distance y = 25 cm.

Solution: Consider an element of length dx at a distance x from the centre of the rod (see Fig. 11.32). The charge on the element $= q/l\cdot dx$. The electric field dE at the point O is given by

$$dE=\frac{q\cdot dx}{4\pi\varepsilon_0 l\left(x^2+y^2\right)}$$ Now dE has two components

$dE_x = dE\sin\theta$ and $dE_y = dE\cos\theta$

$$\text{Net } E_x=\int_{-l/2}^{l/2}\frac{1}{4\pi\varepsilon_0}\cdot\frac{q\,dx}{l\left(x^2+y^2\right)}\cdot\sin\theta \text{ and}$$

$$E_y=\int_{-l/2}^{l/2}\frac{q\,dx\cos\theta}{4\pi\varepsilon_0 l\left(x^2+y^2\right)}$$

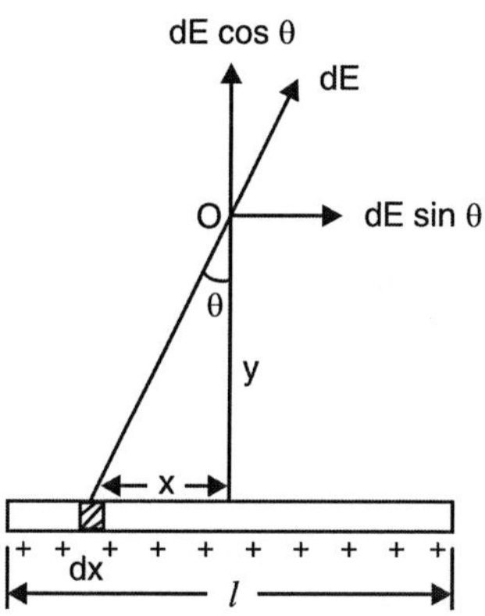

Fig. 11.32

$E_x = 0$ since every charge element on the left has a corresponding element on the right. Therefore, the total field is given by

$$E=E_y=2\int_0^{l/2}\frac{qy\cdot dx}{4\pi\varepsilon_0\, l\left(x^2+y^2\right)^{3/2}}$$

Now $y \tan\theta = x$ $\quad \therefore\ y\sec^2\theta\, d\theta = dx$

Under the above substitution, we have

$$E = 2\int \frac{qy}{4\pi\varepsilon_0 l}\cdot\frac{y\sec^2\theta\, d\theta}{y^3\sec^3\theta}$$

$$= \frac{q}{2\pi\varepsilon_0 yl}\cdot\int \cos\theta\, d\theta = \frac{q}{2\pi\varepsilon_0 yl}\sin\theta$$

$$= \frac{q}{l\cdot 2\pi\varepsilon_0 y}\left[\frac{x}{\left(x^2+y^2\right)^{1/2}}\right]_0^{l/2} = \frac{q}{2\pi\varepsilon_0 y\sqrt{l^2+4y^2}}$$

On substituting the values of various terms, we have

$$E = \frac{2\times 10^{-6}\times 9\times 10^9}{0.25\sqrt{1+4\times 0.25^2}} = \frac{18\times 10^3}{0.25\sqrt{1.25}} = 64.4\times 10^3\ \text{V/m}$$

11.48. A semi-infinite non-conducting thin rod has charge per unit length $\lambda = 10^{-6}$ C/m. Find the direction and the magnitude of electric field at the point O lying at a distance $y = 1$ m.

Solution: Consider an element of length dx situated at a distance x from one end of the rod (Fig. 11.33). Charge on this element = λdx.

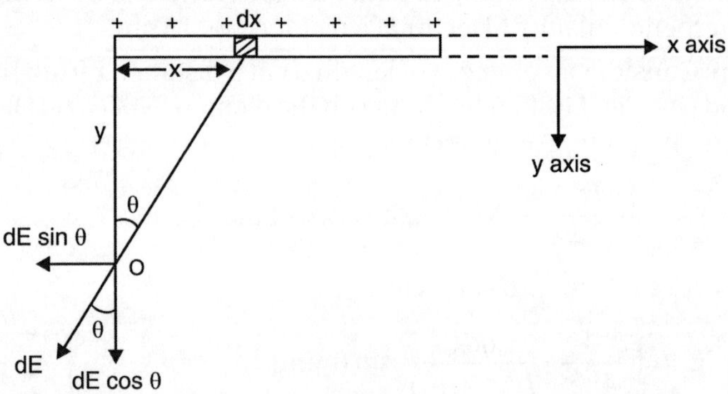

Fig. 11.33

The electric field at O due to this element is given by

$$dE = \frac{1}{4\pi\varepsilon_0}\cdot\frac{\lambda dx}{x^2+y^2}$$

x-component of this field, $dE_x = \dfrac{1}{4\pi\varepsilon_0}\cdot\dfrac{\lambda dx\cdot\sin\theta}{x^2+y^2}$

y-component of this field, $dE_y = \dfrac{1}{4\pi\varepsilon_0}\cdot\dfrac{\lambda dx\cdot\cos\theta}{x^2+y^2}$

Net field along x-axis $E_x = \int_0^\infty \frac{1}{4\pi\varepsilon_0} \cdot \frac{\lambda x dx}{\left(x^2+y^2\right)^{3/2}}$

Net field along y-axis $E_y = \int_0^\infty \frac{1}{4\pi\varepsilon_0} \cdot \frac{\lambda y dx}{\left(x^2+y^2\right)^{3/2}}$

In view of the figure, $x = y\tan\theta$ $\quad dx = y\sec^2\theta\, d\theta$

$$\therefore E_x = \frac{\lambda}{4\pi\varepsilon_0} \cdot \int_0^{\pi/2} \frac{y\sec^2\theta\, y\cdot\tan\theta d\theta}{y^3\sec^3\theta} = \frac{\lambda}{4\pi\varepsilon_0 y} \cdot \int_0^{\pi/2} \sin\theta d\theta = \frac{\lambda}{4\pi\varepsilon_0 y}$$

Similarly,

$$E_y = \frac{\lambda}{4\pi\varepsilon_0} \cdot \int_0^{\pi/2} \frac{y^2\sec^2\theta d\theta}{y^3\sec^3\theta} = \frac{\lambda}{4\pi\varepsilon_0 y} \cdot \int_0^{\pi/2} \cos\theta d\theta = \frac{\lambda}{4\pi\varepsilon_0 y}$$

$$\tan\theta = \frac{E_x}{E_y} = 1$$

$$\therefore \theta = 45°$$

i.e., the resultant yield makes an angle of 45° with the y-axis.

$$\text{Resultant } E = \sqrt{E_x^2 + E_y^2} = \frac{\lambda}{4\pi\varepsilon_0 y} \cdot \sqrt{2}$$

On substituting the values we have

$$E = \frac{10^{-6} \times 9 \times 10^9 \times 1.41}{1} = 12.69 \times 10^3 \text{ V/m}$$

11.49. Find out the intensity of the electric field at O lying on the axis of a thing uniformly charged disc of surface charge density σ.

Solution: In the figure O is a point from which the disc is seen at angle ϕ.

Let us consider an element of area dA as shown in Fig. 11.34. The charge on the element is equal to $\sigma\, dA$.

The electric field at O due to this element is given by

$$dE = \frac{1}{4\pi\varepsilon_0} \cdot \frac{\sigma dA}{r^2}$$

dE will give two components $dE\cos\theta$ along the y-axis and $dE\sin\theta$ along the x-axis. From consideration of symmetry, $dE\sin\theta$ will vanish. Thus the net field will be along y-axis. Thus

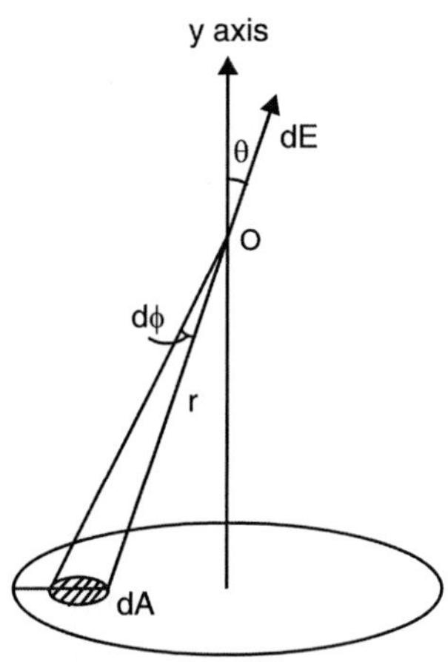

Fig. 11.34

$$dE_y = dE\cos\theta = \frac{1}{4\pi\varepsilon_0}\cdot\frac{\sigma\, dA\cos\theta}{r^2}$$

Here $\dfrac{dA\cos\theta}{r^2} = d\phi$

In the light of the above,

$$dE_y = \frac{1}{4\pi\varepsilon_0}\sigma\, d\phi$$

On integrating the above, we have the total value of the field as

$$E_y = \int dE_y = \frac{1}{4\pi\varepsilon_0}\cdot\sigma\int d\phi = \frac{\sigma\phi}{4\pi\varepsilon_0}$$

11.50. Two similar rods of length $l = 1$ m carrying equal charge $q = 10^{-8}$ C are placed as shown in Fig. 11.35. Find the electric field at the point O, taking $d = 25$ cm.

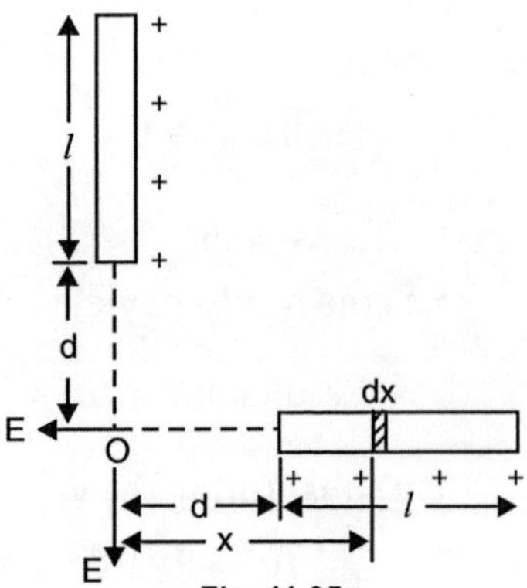

Fig. 11.35

Solution: Consider an element of length dx lying at a distance x from the point O. The charge on this element is equal to $(q/l)\,dx$. The electric field at O due to this element is given by

$$dE = \frac{1}{4\pi\varepsilon_0}\cdot\frac{q}{l}\cdot\frac{dx}{x^2}$$

Net field due to one rod

$$E = \frac{q}{4\pi\varepsilon_0 l}\int_d^{l+d}\frac{dx}{x^2} = \frac{q}{4\pi\varepsilon_0 l}\left(\frac{1}{d} - \frac{1}{l+d}\right)$$

$$= \frac{q}{4\pi\varepsilon_0\cdot d\,(l+d)}$$

The same amount of field will be obtained by the other similar rod. Hence the net field due to both the rods is given by

$$\frac{\sqrt{2}\,q}{4\pi\varepsilon_0(l+d)\,d}$$

On substituting the values of various terms, we have

$$\frac{9\times10^9\times1.41\times10^{-8}}{0.25\times1.25} = \frac{90\times1.41}{0.25\times1.25} = 406.08\ \text{V/m}$$

11.51. A ring of radius $r = 3$ cm carries a charge of $q = 10^{-10}$ C uniformly distributed on it. Find the electric field at the point O lying on the axis of the ring at a distance $x = 4$ cm from its centre.

Solution: Charge q is uniformly distributed on the circumference of the ring. Charge per unit length of the ring = $q/2\pi r$. Consider an element of length ds on the ring (Fig. 11.36). The charge on this element = $ds \cdot q / 2\pi r$.

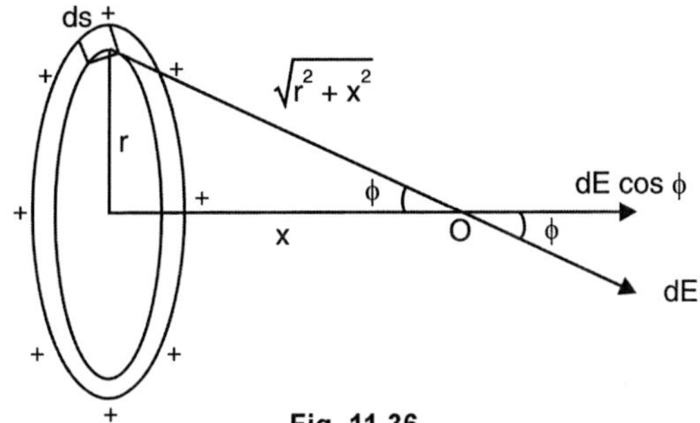

Fig. 11.36

The electric field at O due to the element is given by

$$dE = \frac{1}{4\pi\varepsilon_0} \cdot \frac{ds \cdot q}{2\pi r} \cdot \frac{1}{\left(r^2 + x^2\right)}$$

Of the two components of this field, only $dE \cos \phi$ will exist along the axis of the ring from considerations of symmetry.

$$\therefore \text{ Net field} = E = \int dE \cos \phi = \frac{q \cdot x}{4\pi\varepsilon_0 \cdot 2\pi r \cdot \left(r^2 + x^2\right)^{3/2}} \int ds$$

$$= \frac{qx}{4\pi\varepsilon_0 \cdot \left(r^2 + x^2\right)^{3/2}}$$

On substituting the values of the various terms, we have

$$E = \frac{9 \times 10^9 \times 10^{-10} \times 4 \times 10^{-2}}{125 \times 10^{-6}} = \frac{36}{125} \times 10^3 = 288 \text{ V/m}$$

11.52. A charged thin non-conducting ring of radius r = 1 m has a charge density λ which varies linearly according to equation $\lambda = 10^{-8} \cos \phi$ where ϕ is the azimuth angle. Find out the electric field at the centre of the ring.

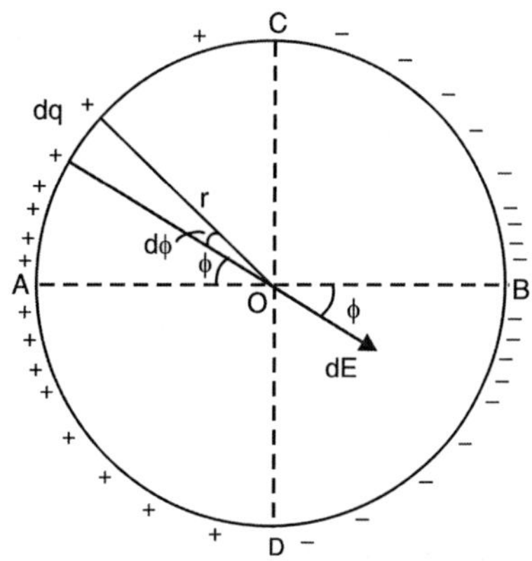

Fig. 11.37

Solution: In view of the value of $\cos \phi$ the charge distribution is shown in Fig. 11.37. At A we have a larger concentration of positive charges whereas at B we have a concentration of negative charges. As we move along

AC or *AD*, the concentration will continue to diminish. The same is true for portions *BC* and *BD* of the ring.

Consider an element of length *ds* on the circumference of the ring. The charge on the element $dq = 10^{-8}\, r\cos\phi \cdot d\phi$.

The electric field at *O* due to this element is given by

$$dE = \frac{1}{4\pi\varepsilon_0}\cdot\frac{dq}{r^2}$$

Now *dE* will have two components *dE* cos ϕ along *OB* and *dE* sin ϕ along *OD*.

From consideration of symmetry, the field along *OD* will cancel out. Hence the net field acting along *OB* is given by

$$E = \int_0^{2\pi} dE\cos\phi = \frac{10^{-8}}{4\pi\varepsilon_0 r^2}\int_0^{2\pi} r\cos\phi\cdot d\phi\cos\phi = \frac{10^{-8}}{4\pi\varepsilon_0 r}\int_0^{2\pi}\cos^2\phi\, d\phi$$

$$= \frac{10^{-8}}{4\pi\varepsilon_0 r}\cdot\pi;\quad \int_0^{2\pi}\cos^2\phi\, d\phi = \pi$$

On substituting the values of *r* and π, we have

$$E = \frac{9\times10^9\times10^{-8}\times3.14}{1} = 282.6\ \text{V/m}$$

11.53. A thin non-conducting rod carrying a charge $q = 2\times10^{-10}$ C is turned into a semi-circle of radius $r = 2$ m. The charges are uniformly distributed on its circumference. Find out the electric field at the centre of the semi-circle.

Solution: Consider an element of length *ds* (Fig. 11.38). The charge

on this element $= \dfrac{q}{\pi r}ds = \dfrac{q\,d\theta}{\pi}$

The field at *O* due to the element is given by

$$dE = \frac{1}{4\pi\varepsilon_0}\cdot\frac{q}{\pi}\cdot\frac{1}{r^2}\cdot d\theta$$

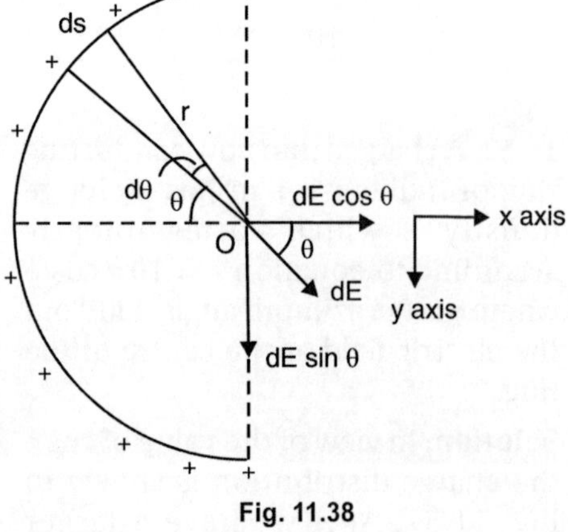

Fig. 11.38

This field has two components as *dE* cos θ and *dE* sin θ along *x* and *y* axes respectively. On symmetry considerations the effective field is 2*dE* cos θ. The net field is E_x and is equal to

$$E = E_x = 2\int_0^{\pi/2} dE\cos\theta = \frac{2q}{4\pi\varepsilon_0\cdot\pi\cdot r^2}\int_0^{\pi/2}\cos\theta\cdot d\theta$$

or $$E = \frac{2q}{4\pi\varepsilon_0\pi\cdot r^2}$$

On substituting the values, we have

$$E = \frac{2\times2\times10^{-10}\times9\times10^9}{3.14\times4} = \frac{0.9}{3.14} = 0.287\ \text{V/m}$$

11.54. A thin and uniform non-conducting wire is turned into a semi-circle of radius $r = 3$ m. On the upper and lower half of the semi-circle, charges $q = +10^{-10}$ C and $q = -10^{-10}$ C are uniformly distributed as shown in Fig. 11.39. Find the electric field at the centre of the semi-circle.

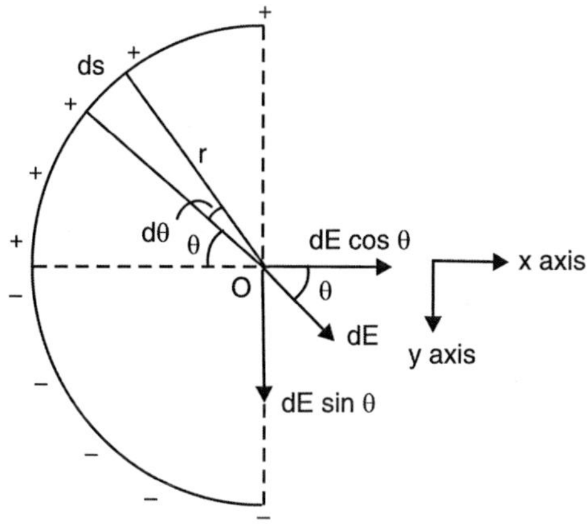

Fig. 11.39

Solution: Consider an element of length ds. Charge on this element is given by

$$\frac{2q}{\pi r}ds = \frac{2q}{\pi}\cdot d\theta;\ ds = r\cdot d\theta$$

Electric field at O due to this element is represented as

$$dE = \frac{1}{4\pi\varepsilon_0}\cdot\frac{2q\,d\theta}{\pi r^2}$$

This field will be away from the element and have two components $dE\cos\theta$ and $dE\sin\theta$ along the x-axis and the y-axis respectively. If we consider an element of length ds situated in the lower half, the magnitude of the electric field will be the same as stated above but will be directed towards the element. It will have two components. One equal the opposite to $dE\cos\theta$ and the other equal to and in the direction of $dE\sin\theta$. Therefore the net field will be

$$E = 2\int_0^{\pi/2} dE\sin\theta = \int_0^{\pi/2}\frac{4q}{4\pi^2\varepsilon_0 r^2}\sin\theta\,d\theta = \frac{4q}{4\pi^2\varepsilon_0 r^2}$$

On substituting the values of various terms, we have

$$E = \frac{4\times10^{-10}\times9\times10^9}{3.14\times9} = \frac{0.4}{3.14} = 0.127\ \text{V/m}$$

11.55. One fourth arc of a circle of radius R contains positive charge. The charge per unit length is λ. Find the electric field at the centre.

Solution: Consider an element of length $ds = R\,d\theta$ (Fig. 11.40). Charge on this element $= \lambda\, ds = \lambda R \cdot d\theta$.

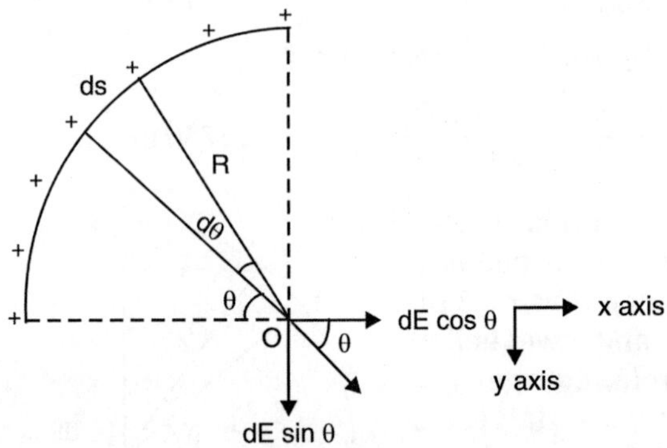

Fig. 11.40

The electric field at O due to this element is given by

$$dE = \frac{1}{4\pi\varepsilon_0} \cdot \frac{\lambda R\, d\theta}{R^2}$$

$dE \sin\theta$ and $dE \cos\theta$ are the components of this field along two mutually perpendicular axes. The net field along the x and y-axes is given by

$$E_x = \int_0^{\pi/2} dE \cos\theta = \frac{\lambda R}{4\pi\varepsilon_0 R^2} \int_0^{\pi/2} \cos\theta\, d\theta = \frac{\lambda}{4\pi\varepsilon_0 R}$$

$$\text{and } E_y = \int_0^{\pi/2} dE \sin\theta = \frac{\lambda}{4\pi\varepsilon_0 R} \int_0^{\pi/2} \sin\theta\, d\theta = \frac{\lambda}{4\pi\varepsilon_0 R}$$

$$\text{Resultant field } E = \sqrt{E_x^2 + E_y^2} = \frac{\lambda}{4\pi\varepsilon_0 R} \cdot \sqrt{2}$$

11.56. A charged thin uniform non-conducting wire *abcd* with charge per unit length λ is bent into the form shown in Fig. 11.41. Find out the intensity of the electric field at the point O assuming that the radius of the arc *bc* is less than lengths *ab* and *cd* in the wire.

Solution: The wire *abcd* is essentially made of two straight portions *ab* and *cd* and 1/4th arc of a circle *bc*. The net field at O due to *ab* and *cd* is zero (see Q. 11.48). Therefore the field at O is only due to arc *bc*. In view of Q. 11.55, the

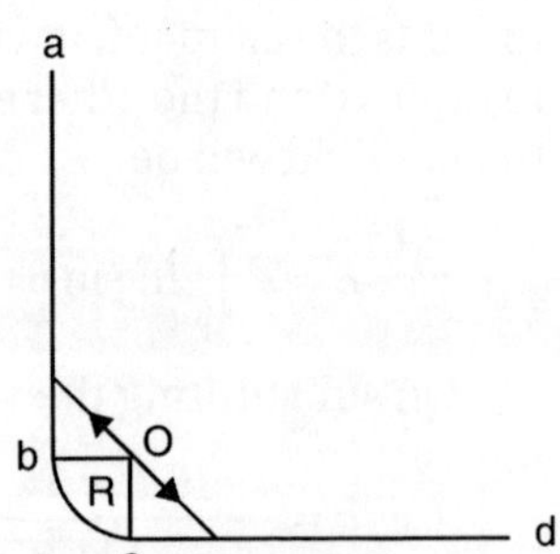

Fig. 11.41

field at O is $\frac{\lambda\sqrt{2}}{4\pi\varepsilon_0 R}$. Therefore the net field at O due to the wire *abcd* is $\frac{\lambda\sqrt{2}}{4\pi\varepsilon_0 R}$.

11.57. In Fig. 11.42, a long and uniformly charged thread with charge per unit length $\lambda = 10^{-3}$ C/m is placed on the axis of a thin ring of radius $R = 1$ m. If charges $q = 10^{-6}$ C are uniformly distributed on the circumference of the ring, find the force acting between the ring and the thread.

Solution: In Fig. 11.42 consider an element of thread of length dx located at a distance x from the centre of the ring. The electric field at P lying on the circumference of the ring due to the element is given by

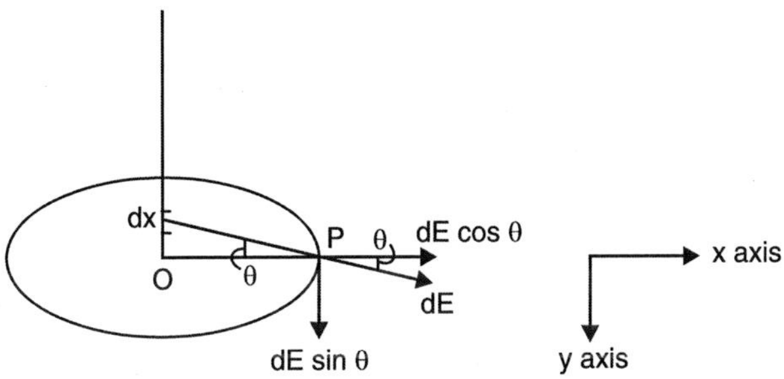

Fig. 11.42

$$dE = \frac{1}{4\pi\varepsilon_0} \cdot \frac{\lambda dx}{(x^2 + R^2)}$$

This field will have components $dE \cos\theta$ and $dE \sin\theta$ acting along the x and y-axes. From considerations of symmetry, the field along the x-axis will cancel out.

The electric field is along y-axis is given by

$$E_y = \int_0^\infty dE \sin\theta = \int_0^\infty \frac{\lambda}{4\pi\varepsilon_0} \cdot \frac{x\,dx}{(x^2 + R^2)^{3/2}}; \quad x = R\tan\theta; dx = R\sec^2\theta d\theta$$

$$= \int_0^{\pi/2} \frac{\lambda}{4\pi\varepsilon_0} \frac{R\tan\theta \cdot R\sec^2\theta d\theta}{R^3\sec^3\theta} = \frac{\lambda}{4\pi\varepsilon_0 R} \int_0^{\pi/2} \sin\theta d\theta = \frac{\lambda}{4\pi\varepsilon_0 R}$$

Now consider an element at P containing a charge dq. The force on this element is equal to

$$dF = \frac{\lambda}{4\pi\varepsilon_0 R} \cdot dq$$

$$\therefore \text{Net force } F = \frac{\lambda}{4\pi\varepsilon_0 R} \int dq = \frac{\lambda q}{4\pi\varepsilon_0 R}$$

On substituting the values, we have

$$F = 9 \times 10^9 \times 10^{-3} \times 10^{-6} = 9 \text{ N}$$

11.58. A rigid insulated wire frame in the form of a right-angled triangle *ABC*, is set in a vertical plane as shown in the Fig. 11.43a. Two beads of equal masses *m* each and carrying charges q_1 and q_2 are connected by cord of length *l* and can slide without friction on the wires. Considering the case when the beads are stationary, determine (IIT 1978)

(i) angle α,

(ii) the tension in the cord, and

(iii) the normal reaction of the beads

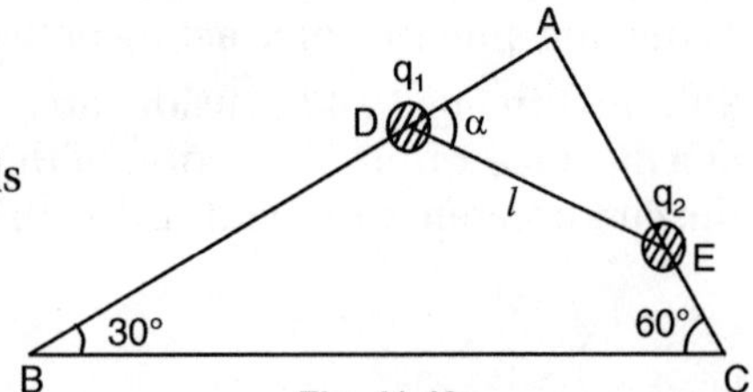

Fig. 11.43a

Solution: In this case four forces are working :

(i) *mg*, the weight of the bead, (ii) normal force *N*, (iii) Tension in the cord, *T*, and (iv) force of repulsion *F*. All the four forces are represented in Fig. 11.43b.

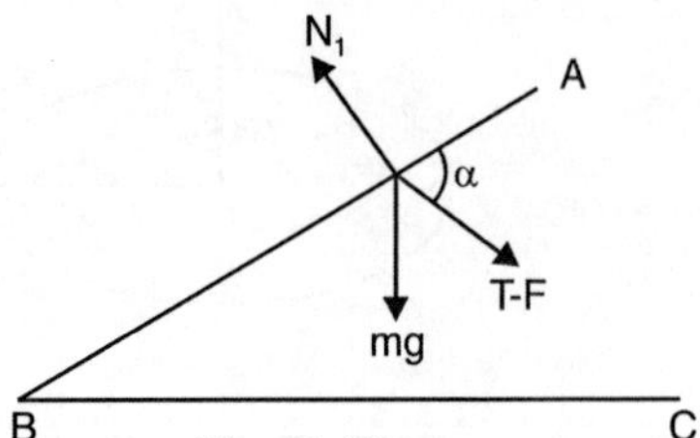

Fig. 11.43b

The net force along the plane *AB* is zero since there is no motion, i.e.,

$$mg \cos 60 - (T - F) \cos \alpha = 0$$

or $\quad mg \cos 60 = (T - F) \cos \alpha \qquad$ (i)

For the bead at *E*, we have

$$mg \sin 60 = (T - F) \sin \alpha \qquad \text{(ii)}$$

The net force perpendicular to $AB = 0$

$$N_1 = mg \cos 30 + (T - F) \sin \alpha \qquad \text{(iii)}$$

Similarly,

$$N_2 = mg \cos 60 + (T - F) \cos \alpha \qquad \text{(iv)}$$

Equation (i) and (ii) yield

$$T - F = \pm\, mg$$

or $\quad T = F + mg = \dfrac{kq_1q_2}{l^2} + mg$

Now $mg \cos 60 = mg \cos \alpha$

$\therefore \quad \alpha = 60°$

$$N_1 = mg \cos 30 + mg \sin 60$$

$$= 2\, mg \sin 60 = \sqrt{3}\, mg$$

$$N_2 = mg \cos 60 + mg \cos 60 = 2\, mg \cos 60 = mg$$

11.59. A point charge $q = 10^{-5}$ is located at a distance l = 50 cm from an infinite conducting plane. What amount of work has to be performed in order to slowly remove this charge very far from the plane?

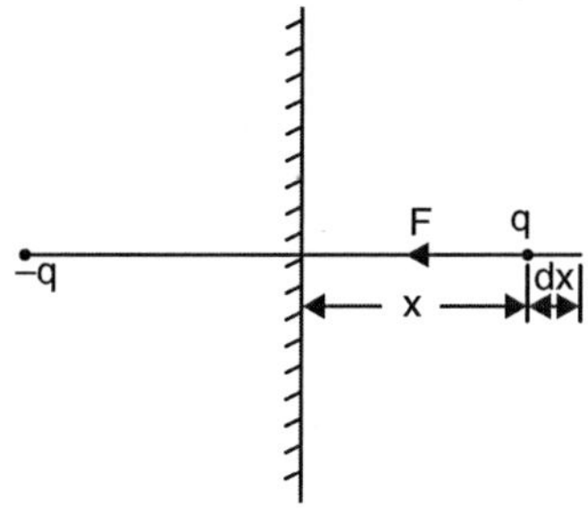

Fig. 11.44

Solution: The role of the conducting plane is to provide an electrical image of the charge q with negative sign (see Fig. 11.44). Due to this charge there will be a force of attraction. The work done is given by $dW = F \cdot dx$

$$\text{Net work done} = W = \int dW = \int F \cdot dx = \frac{q^2}{4\pi\varepsilon_0}\int_l^\infty \frac{dx}{4x^2}$$

$$= \frac{q^2}{16\pi\varepsilon_0 l}$$

On substituting the values, we have

$$W = \frac{10^{-5} \times 10^{-5} \times 9 \times 10^9}{4 \times 0.5} = 0.45 \text{ J}$$

11.60. A point charge q is located at a distance l from an infinite conducting plane (see Fig. 11.45). Find the interaction energy of that charge with chose induced on the plane.

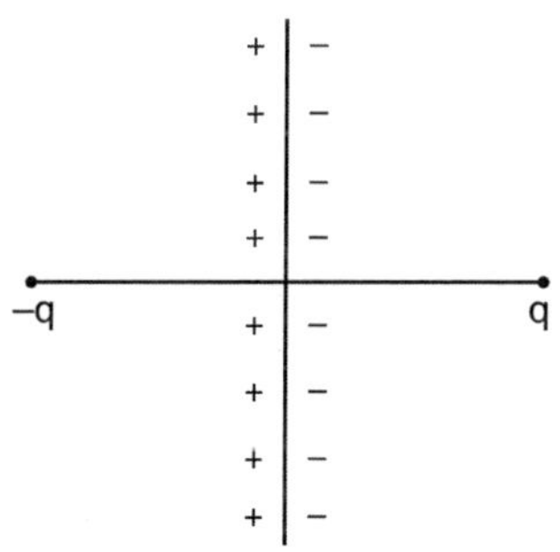

Fig. 11.45

Solution: The role of the conducting plane is to provide an electrical image of the charge with a negative sign.

$$\text{Potential energy} = -\frac{q^2}{8\pi\varepsilon_0 l}$$

11.61. There is an infinite straight chain of alternating charges q and $-q$. The distance between the neighbouring charges is equal to a. Find the interaction energy of each charge with all the others.

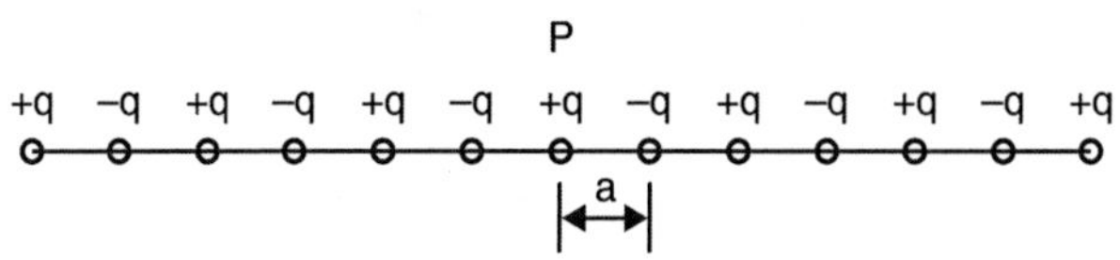

Fig. 11.46

Solution:

Potential energy of interaction

= $q \times$ potential at P due to all the charges

$$= q \times \frac{1}{4\pi\varepsilon_0}\left(\frac{-2q}{a} + \frac{2q}{2a} - \frac{2q}{3a} + \ldots\right)$$

$$= -\frac{2q^2}{4\pi\varepsilon_0 a}\left(1 - \frac{1}{2} + \frac{1}{3} - \frac{1}{4} + \ldots\right)$$

$$= -\frac{2q^2}{4\pi\varepsilon_0 a}\log_e 2$$

11.62. A charge $+q$ is freely placed at the four corners of a square of side a. Calculate the amount of charge which when placed at the centre of the square will keep the system in equilibrium.

Solution: For equilibrium the total interaction energy of the system must be equal to zero. Let the four corners be marked as 1, 2, 3 and 4 as shown in Fig. 11.47. Let a charge Q be placed at the centre of the square. The total energy is given by

$$W = 0 = W_{12} + W_{13} + W_{14} + W_{23} + W_{24} + W_{34} + W_{1Q} + W_{2Q} + W_{3Q} + W_{4Q}$$

$$= \frac{q^2}{4\pi\varepsilon_0 a} + \frac{q^2}{4\pi\varepsilon_0 a\sqrt{2}} + \frac{q^2}{4\pi\varepsilon_0 a} + \frac{q^2}{4\pi\varepsilon_0 a}$$

$$+ \frac{q^2}{4\pi\varepsilon_0 a\sqrt{2}} + \frac{q^2}{4\pi\varepsilon_0 a} + \frac{4qQ\sqrt{2}}{4\pi\varepsilon_0 a}$$

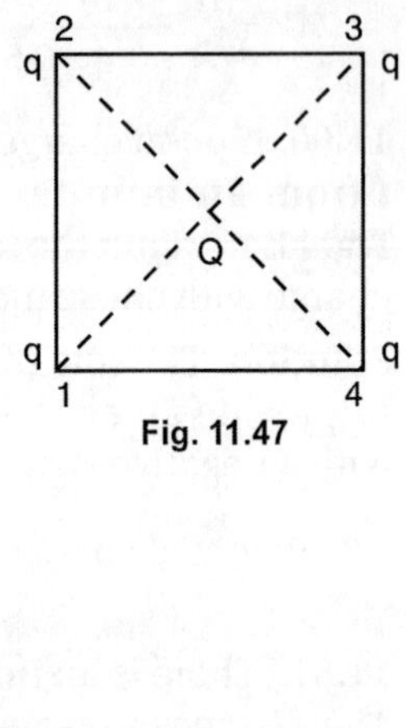

Fig. 11.47

or $$\frac{q}{4\pi\varepsilon_0 a}\left\{\left[4 + \sqrt{2}\right]q + 4Q\sqrt{2}\right\} = 0$$

$$\therefore \quad Q = -\frac{\left(4 + \sqrt{2}\right)q}{4\sqrt{2}}$$

$$= -\frac{q}{4}\left(1 + 2\sqrt{2}\right)$$

11.63. An infinitely long straight charged conductor having linear charge density λ and a point charge $-Q$ are placed as shown in the Fig. 11.48. For what value of OO' the resultant electric field at O' is zero.

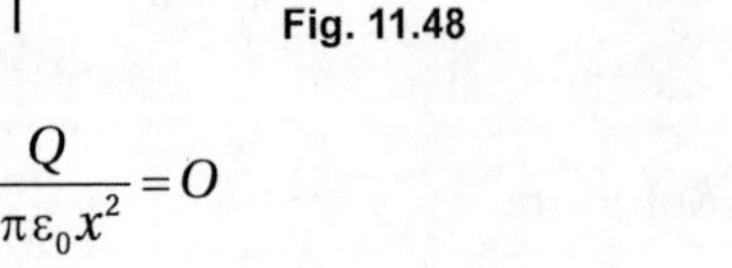

Fig. 11.48

Solution:

$$\text{Net field at } O' = \frac{\lambda}{2\pi\varepsilon_0\,(x + a)} - \frac{Q}{4\pi\varepsilon_0 x^2} = O$$

or $2\lambda x^2 - Qx - Qa = O$

$$x = \frac{Q \pm \sqrt{Q^2 + 8\lambda a Q}}{4\lambda}$$

$$= \frac{Q}{4\lambda} + \sqrt{\frac{Q^2}{16\lambda^2} + \frac{aQ}{2\lambda}}$$

11.64. A point charge q is located at the centre of a thin ring of radius R with uniformly distributed charge $-q$. Find out the magnitude of the electric field strength vector at a point lying on the axis of the ring at a distance x from its centre when $x >> R$.

Solution:

$$\text{Net electric field at } \overleftarrow{A} = \frac{q}{4\pi\varepsilon_0 x^2} - \frac{qx}{4\pi\varepsilon_0 \left(x^2 + R^2\right)^{3/2}}$$

$$= \frac{q}{4\pi\varepsilon_0 x^2} - \frac{qx}{4\pi\varepsilon_0 x^3}\left[1 - \frac{3}{2}\cdot\frac{R^2}{x^2}\right]$$

$$= \frac{3qR^2}{8\pi\varepsilon_0 x^4}$$

$$\text{Similarly net field at } \vec{B} = \frac{3qR^2}{8\pi\varepsilon_0 x^4}$$

$$\text{Vector field at } B = \frac{3qR^2}{4\pi\varepsilon_0 x^4}$$

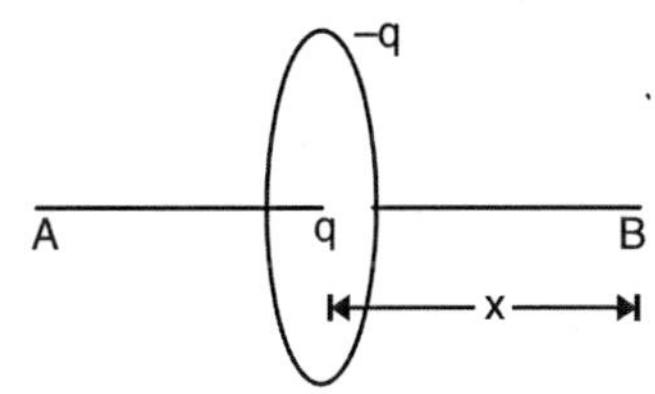

Fig. 11.49

11.65. A sphere of radius r carries a surface charge density $\sigma = \vec{ar}$ where a is a constant and r is the radius vector of a point of the sphere relative to its centre. Find the electric field strength vector at the centre of the sphere.

Solution: Consider an element at a distance x and $x + dx$

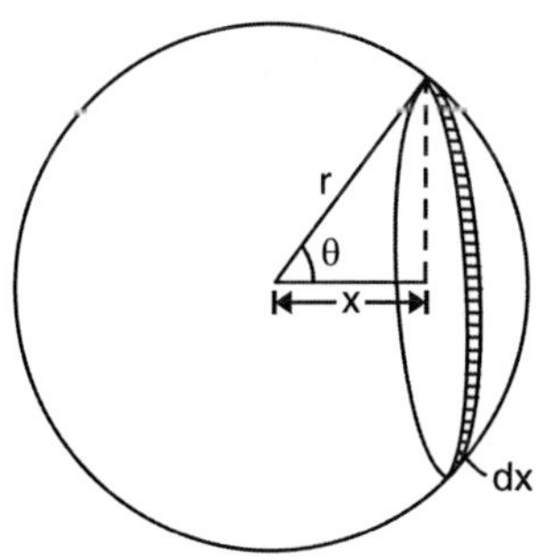

Fig. 11.50

$$\text{Area of the element} = 2\pi\sqrt{r^2 - x^2}\cdot dx$$

$$\text{Charge on the element} = 2\pi\sqrt{r^2 - x^2}\cdot dr\cdot ar$$

$$\text{Effective field at the centre} = \int_0^r \frac{2\pi\sqrt{r^2 - x^2}\, dx\cdot ar\cdot x}{4\pi\varepsilon_0 r^2\cdot r}$$

$$= \frac{ar}{6\varepsilon_0} = -\frac{a\vec{r}}{6\varepsilon_0}$$

Field due to the other half of the sphere $= -\dfrac{a\vec{r}}{6\varepsilon_0}$

Vector field $= -\dfrac{a\vec{r}}{3\varepsilon_0}$

11.66. Find the force of repulsion between the two halves of a spherical conductor charged to a potential of V volt.

Solution: Electrostatic pressure $= \dfrac{\sigma^2}{2\varepsilon_0}$

Consider an element of area ds. Force on the element $= \dfrac{\sigma^2}{2\varepsilon_0} ds$

Out of the two components of this force, $\dfrac{\sigma^2}{2\varepsilon_0} ds \sin\theta$ (see Fig. 11.51) will be parallel to the plane which is dividing the two spheres. The sum of this force will be equal to zero. The total outward force will be the force of repulsion and is equal to

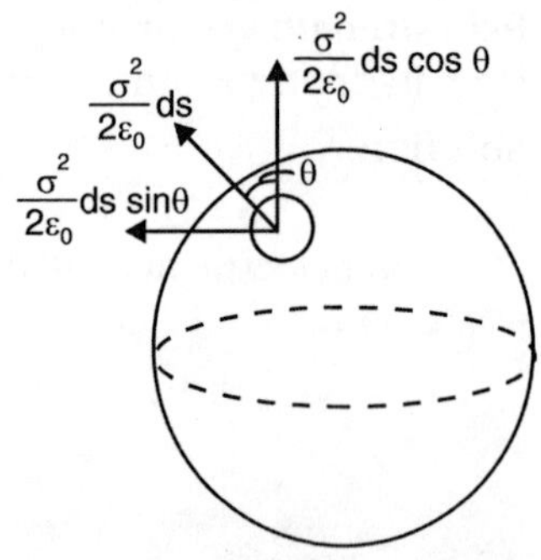

Fig. 11.51

$$F = \sum \frac{\sigma^2}{2\varepsilon_0} \cdot ds\cos\theta = \frac{\sigma^2}{2\varepsilon_0} \sum ds\cos\theta = \frac{\sigma^2}{2\varepsilon_0}\pi r^2;$$

r = radius of the sphere.

$$\text{Potential } V = \frac{Q}{4\pi\varepsilon_0 r} = \frac{\sigma \cdot 4\pi r^2}{4\pi\varepsilon_0 r}$$

$$V = \frac{r\sigma}{\varepsilon_0}$$

$$F = \left(\frac{\varepsilon_0 V}{r}\right)^2 \times \frac{1}{2\varepsilon_0} \cdot \pi r^2 = \frac{1}{2}\varepsilon_0 \pi V^2$$

11.67. Consider the classical model of an atom such that a nucleus of charge $+e$ is uniformly distributed within a sphere of radius $2A$. An electron of charge $-e$ at a radial distance $1A$ moves inside the sphere. Find`the force attracting the electron to the centre of the sphere. Calculate the frequency with which the electron would oscillate about the centre of the sphere. (Roorkee 1995)

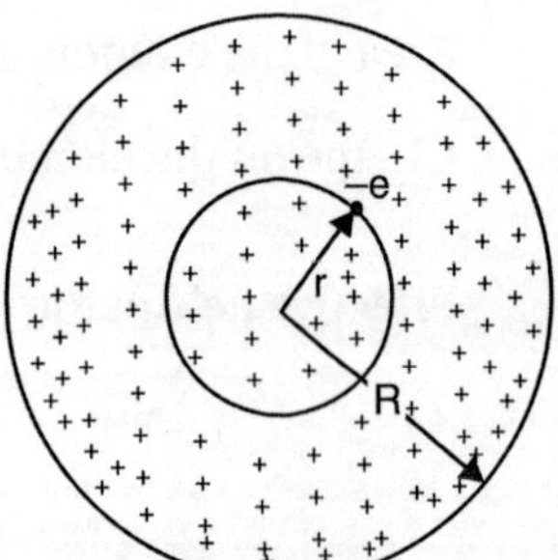

Fig. 11.52

Solution: Only those charges which lie within the sphere of radius r will exert force on the electron. Let q be the amount of charge contained in the sphere of radius r then

$$\frac{q}{\frac{4}{3}\pi r^3} = \frac{e}{\frac{4}{3}\pi R^3}$$

$$\therefore q = e \cdot \frac{r^3}{R^3}$$

$$\text{Force } F = \frac{1}{4\pi\varepsilon_0} \cdot \frac{er^3}{R^3} \times \frac{-e}{r^2} = -\frac{e^2}{4\pi\varepsilon_0 R^3} \cdot r$$

On putting the values we have

$$|F| = 9 \times 10^9 \times \frac{\left(1.6 \times 10^{-19}\right)^2 \times 10^{-10}}{8 \times 10^{-30}} = 2.88 \times 10^{-9}\,\text{N}$$

$$\text{Force} = m \times \text{acceleration} = -\frac{e^2}{4\pi\varepsilon_0 R^3} \cdot \text{displacement}$$

This indicates that the electron will execute simple harmonic motion in which

$$\omega = 2\pi n = \sqrt{\frac{e^2}{4\pi\varepsilon_0 R^3 \cdot m}}$$

or $$n = \frac{1}{2\pi} \times e \frac{1}{\sqrt{4\pi\varepsilon_0 R^3 m}}$$

$$= \frac{1.6 \times 10^{-19}}{2 \times 3.14\sqrt{\frac{1}{9 \times 10^9} \times 8 \times 10^{-30} \times 9 \times 10^{-31}}}$$

$$= 9 \times 10^{14}\,\text{Hz}$$

11.68. Two similar discs are placed parallel to each other at a separation of $x = 1$ cm. They are charged by unlike charges keeping the surface charge density same. What could the maximum value of the radii of the discs be for the field between the centres of the discs to differ from the field of the plane capacitor by not more than 5%.

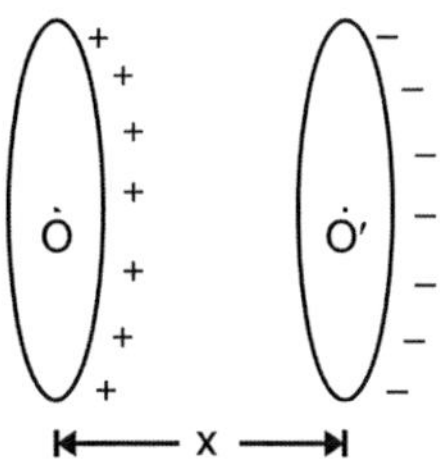

Fig. 11.53

Solution: Consider an element of radius r and $r + dr$ on the surface of the disc containing $+ve$ charges.

Electric field due to the element at $O' = \dfrac{\sigma \cdot 2\pi r dr \cdot x}{4\pi\varepsilon_0 \left(x^2 + r^2\right)^{3/2}}$;

σ = surface charge density

Net field at O' due to the disc of radius $R = \dfrac{\sigma x}{2\varepsilon_0}\displaystyle\int_0^R \frac{r dr}{\left(x^2 + r^2\right)^{3/2}}$

$$= \frac{\sigma}{2\varepsilon_0}\left[1 - \frac{x}{\sqrt{x^2 + R^2}}\right]$$

Similar calculations will give the same value of field at O due to $-ve$ charges on the other disc.

In view of the problem we have

$$\frac{\dfrac{\sigma}{2\varepsilon_0} - \dfrac{\sigma}{2\varepsilon_0} - \dfrac{\sigma x}{2\varepsilon_0\sqrt{x^2 + R^2}}}{\dfrac{\sigma}{2\varepsilon_0}} = \frac{5}{100}$$

$$20x = \sqrt{R^2 + x^2}$$

$$399x^2 = R^2$$

$$R \cong 20x = 20 \text{ cm}$$

11.69. A ball of mass 10^{-2} kg and having charges 3×10^{-6} C is tied at one end of a 1 m long thread. The other end of the thread is fixed and a charge -3×10^{-6}C is placed at this end. The ball can move in a circular orbit of radius 1 m in the vertical plane. Initially the ball is at the bottom. Find the minimum initial horizontal velocity of the ball so that it will be able to complete the full circle. $g = 10$ m/sec^2. (Roorkee 1996)

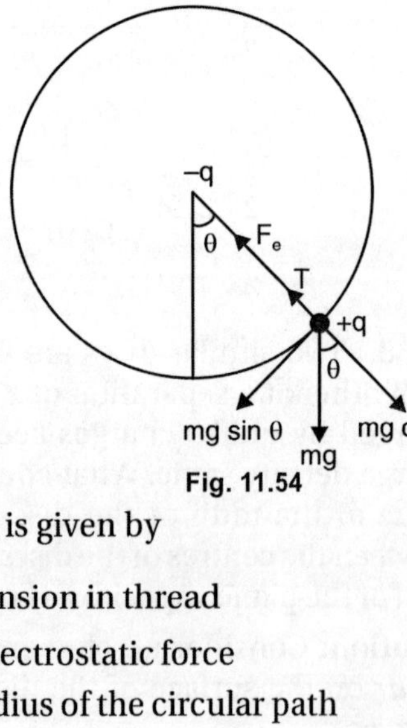

Fig. 11.54

Solution: The equation of motion of ball is given by

$$T + F_e - mg\cos\theta = \frac{mv^2}{r};$$ T = tension in thread

F_e = electrostatic force

r = radius of the circular path

$$T = \frac{mv^2}{r} - F_e + mg\cos\theta$$

T will have the minimum value as T_{min} when $\theta = 180°$

$$T_{min} = \frac{mv_H^2}{r} - F_e - mg$$

For looping the loop $T_{min} \geq 0$

$$\therefore v_H^2 \geq r\left[\frac{F_e}{m} + g\right]$$

$$v_H^2 = 1\left[\frac{9\times10^9\times\left(3\times10^{-6}\right)^2}{10^{-2}} + 10\right] = 181.1$$

In view of conservation of energy

$$\frac{1}{2}mv_0^2 = mg\cdot 2r + \frac{1}{2}mv_H^2$$

$$v_0^2 = 4gr + v_H^2$$

$$v_0 = \sqrt{18.1 + 4\times10\times1} = 7.62 \text{ m/s}$$

11.70. A charge 10^{-9} coul is located at origin in free space and another charge Q at (2, 0, 0). If the x component of electric field at (3, 1, 1) is zero calculate the value of Q. Is the y component zero at (3, 1, 1)?

(Roorkee 1995)

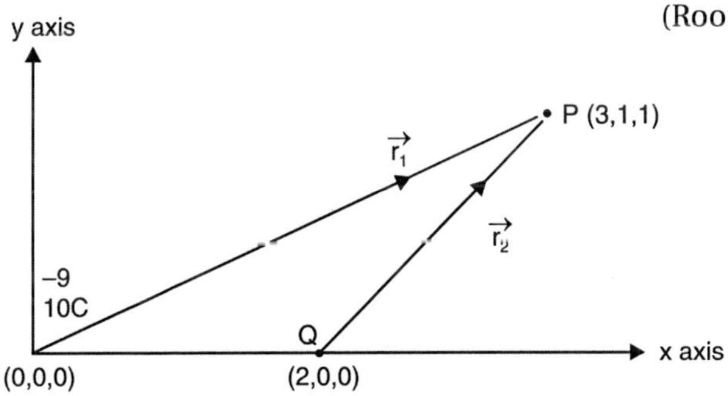

Fig. 11.55

Solution:

$$\vec{r_1} = 3i + j + k \quad \therefore r_1 = \sqrt{3^2 + 1^2 + 1^2} = \sqrt{11} \text{ m}$$

$$\vec{r_2} = (3-2)i + j + k \quad \therefore r_2 = \sqrt{1^2 + 1^2 + 1^2} = \sqrt{3} \text{ m}$$

Electric field at P due to 10^{-9} C and Q C charge is equal to

$$\frac{10^{-9}}{4\pi\varepsilon_0(11)^{3/2}}(3i + j + k) + \frac{Q}{4\pi\varepsilon_0(3)^{3/2}}(i + j + k)$$

$$x \text{ component of the field} = 0 = \left[\frac{10^9}{4\pi\varepsilon_0\,(11)^{3/2}}\times 3 + \frac{Q}{4\pi\varepsilon_0\,(3)^{3/2}}\right]$$

$$\therefore Q = -\left(\frac{3}{11}\right)^{3/2} \times 3\times 10^{-9} = -4.27\times 10^{-10}\,\text{C}$$

$$E_y = \frac{1}{4\pi\varepsilon_0}\left[\frac{10^{-9}}{(11)^{3/2}} - \frac{\left(\frac{3}{11}\right)^{3/2}\times 3\times 10^{-9}}{3^{3/2}}\right] = \frac{-2\times 10^{-9}}{(11)^{3/2}} \text{ which is not zero}$$

11.71. The field potential inside a charged ball depends only on the distance from its centre as $\phi = ar^2 + b$ where a & b are constant. Find the space charge distribution $\rho\ (r)$ inside the ball.

Solution:

Electric field $E = -\frac{d\phi}{dr} = -2ar$

Electric field at a distance r inside the ball is given by $\frac{\rho r}{3\varepsilon_0}$

On equating the two we have

$$-2ar = \frac{\rho r}{3\varepsilon_0}$$

$$\therefore \rho = -6\varepsilon_0 a$$

11.72. In the region of space between the plates of a parallel plate capacitor there is a uniformly distributed $+ve$ charge with a volume density ρ. The plates are connected electrically and their potential is set to zero. Calculate the electric field and potential between the plates.

Solution: In the space between the two plates of the parallel plate capacitor consider a thin layer of charge of thickness dx (see Fig. 11.56a). The electric field by this layer on the two sides will have the value $\rho x/2\varepsilon_0$.

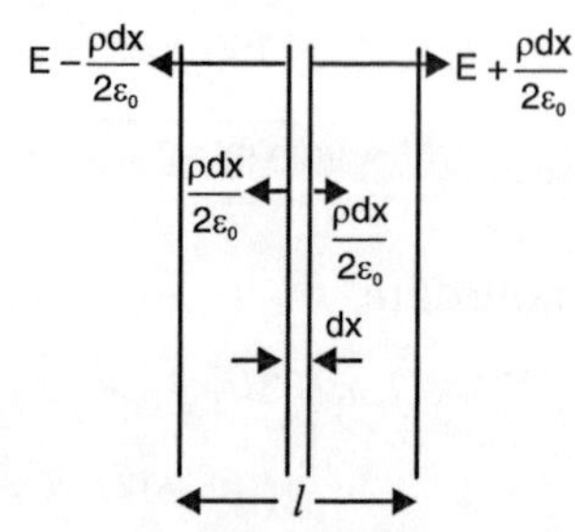

Fig. 11.56a

Let all the charges confined on the left of this layer produces an electric field E. Net electric field at the left and right boundary of the element is $E - \frac{\rho dx}{2\varepsilon_0}$ and $E + \frac{\rho dx}{2\varepsilon_0}$ respectively.

Change of field over a distance $dx = dE = \frac{\rho dx}{\varepsilon_0}$

On integration we have $E = \frac{\rho x}{\varepsilon_0} + A$. The value of A is to be determined with the help of boundary conditions. At $x = 0$, $E = E_0$ $\therefore A = E_0$.

or $$E = E_0 + \frac{\rho x}{\varepsilon_0}$$

$$-\frac{dV}{dx} = E_0 + \frac{\rho x}{\varepsilon_0}$$

On integration we have

$$V_1 - V_2 = E_0 x + \frac{\rho x^2}{2\varepsilon_0}$$

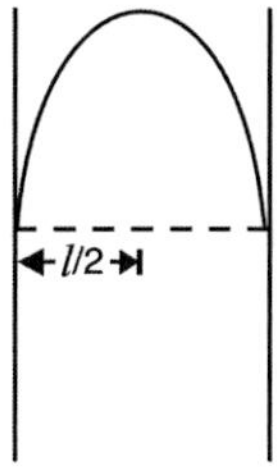

Fig. 11.56b

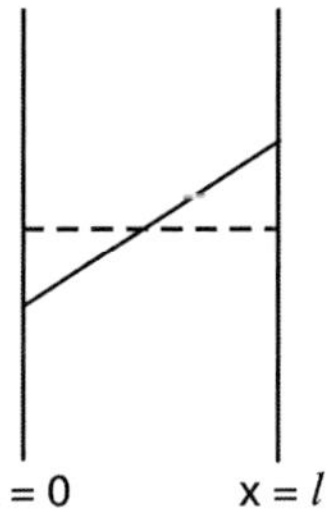

Fig. 11.56c

Now V_1 and V_2 both are zero at $x = 0$ and $x = l$

$$\therefore E_0 = -\frac{\rho l}{2\varepsilon_0}$$

$$\therefore E = -\frac{\rho l}{2\varepsilon_0} + \frac{\rho x}{\varepsilon_0}$$

E at $x = 0 = -\dfrac{\rho l}{2\varepsilon_0}$ and at $x = l$, $E = \dfrac{\rho l}{2\varepsilon_0}$ (see Fig. 11.56c)

$$E = -\frac{dV}{dx} = -\frac{\rho l}{2\varepsilon_0} + \frac{\rho x}{\varepsilon_0}$$

$$V = \frac{\rho l x}{2\varepsilon_0} - \frac{\rho x^2}{2\varepsilon_0}$$

V will be in the form of parabola with its maximum value at $x = l/2$ (see Fig. 11.56b).

11.73. Three point charges $1C$, $2C$ and $3C$ are placed at the corners of an equilateral triangle of side 1 m. Calculate the work required to move these charges to the corners of a small equilateral triangle of side 0.5 m as shown in Fig. 11.57.

(Roorkee 1997)

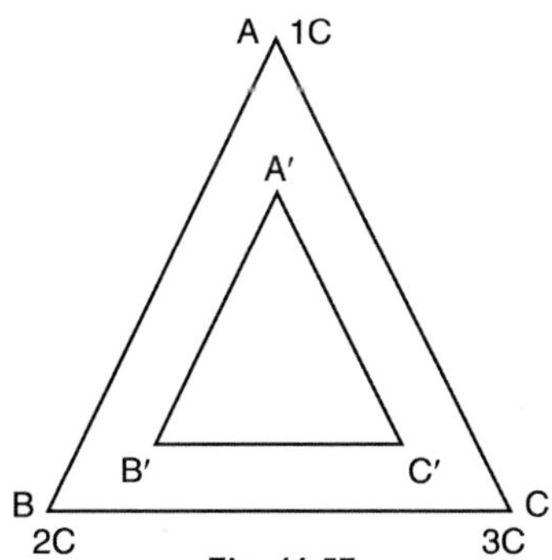

Fig. 11.57

Solution: Initial electrostatic potential energy

$$U_i = 9 \times 10^9 \left[\frac{1\times 2}{1} + \frac{2\times 3}{1} + \frac{1\times 3}{1}\right] = 99 \times 10^9 \text{ J}$$

$$U_f = 9 \times 10^9 \left[\frac{1\times 2}{0.5} + \frac{2\times 3}{0.5} + \frac{1\times 3}{0.5}\right] = 22 \times 9 \times 10^9 = 198 \times 10^9 \text{ J}$$

Amount of work done $= U_f - U_i = 99 \times 10^9$ J

11.74. Calculate the energy of the interaction between the electron cloud of a hydrogen atom and the proton (which is the nucleus of the hydrogen atom). The charge density inside the cloud is $\rho = \frac{e}{\pi a^3} e^{-2r/a}$ where a is the radius of the first Bohr orbit.

Solution: Consider a thin spherical shell of radius r and $r + dr$

Volume of the shell $= 4\pi r^2 dr$

Charge in the shell $= \rho \cdot 4\pi r^2 dr = 4\pi r^2 \cdot dr \cdot \frac{e}{\pi a^3} e^{-2r/a}$

$$\text{Interaction energy } W = \int dW = \int_0^\infty \frac{1}{4\pi\varepsilon_0} \cdot 4r^2 \cdot \frac{e}{a^3} e^{-2r/a} \cdot \frac{e}{r} dr$$

$$= \frac{e^2}{\pi\varepsilon_0 a^3} \int_0^\infty r e^{-2r/a} dr$$

$$= -\frac{e^2}{4\pi\varepsilon_0 a}$$

11.75. To what potential a soap bubble of radius r be charged in order to have pressure inside and outside same.

Solution: Total pressure inside a charged soap bubble is given by

$$P = P_0 + \frac{4T}{r} - \frac{\sigma^2}{2\varepsilon_0}$$

where P_0 = atmospheric pressure, T = surface tension of soap solution.

$\frac{\sigma^2}{2\varepsilon_0}$ = electrostatic pressure

In view of the problem

$$P_0 + \frac{4T}{r} - \frac{\sigma^2}{2\varepsilon_0} = P_0$$

$$\therefore \frac{4T}{r} = \frac{\sigma^2}{2\varepsilon_0} = \frac{Q^2}{16\pi^2 r^4} \cdot \frac{1}{2\varepsilon_0} = \frac{16\pi^2\varepsilon_0^2 r^2 V^2}{16\pi^2 r^4 \times 2\varepsilon_0} = \frac{\varepsilon_0 V^2}{2r^2}$$

$$\therefore V = \sqrt{\frac{8Tr}{\varepsilon_0}}$$

11.76. A soap bubble of radius a is formed from a soap solution of surface tension T. The bubble is charged to a potential V. As a result of repulsion its radius increases to r. Calculate the atmospheric pressure. Assume the process to be an isothermal.

Solution:

$$\left(P+\frac{4T}{a}\right)\times\frac{4}{3}\pi a^3 = \left(P+\frac{4T}{r}-\frac{\sigma^2}{2\varepsilon_0}\right)\times\frac{4}{3}\pi r^3$$

or
$$P(r^3-a^3) = \frac{\sigma^2 r^3}{2\varepsilon_0} - 4T\left(r^2-a^2\right)$$

$$= \frac{1}{2}\varepsilon_0 V^2 r - 4T\left(r^2-a^2\right)$$

$$\therefore P = \frac{1}{\left(r^3-a^3\right)}\left[\frac{1}{2}\varepsilon_0 V^2 r - 4T\left(r^2-a^2\right)\right]$$

11.77. Calculate the electric field intensity at a point inside a spherically symmetrical charge distribution of radius a. The volume density of the charge ρ (r) is given by

$$\rho(r) = \rho_0\left(1-\frac{r^3}{a^3}\right)^{1/2} \quad \text{when } r \le a$$

$$= 0 \quad \text{when } r > a$$

Solution: According to Gauss law

$$\varepsilon_0 E\cdot 4\pi r^2 = 4\pi\int_0^r \rho_0\left(1-\frac{r^3}{a^3}\right)^{1/2} r^2 dr$$

$$\varepsilon_0 E\cdot r^2 = -\frac{2\rho_0 a^3}{9}\left[1-\frac{r^3}{a^3}\right]^{3/2}$$

$$= \frac{2\rho_0 a^3}{9\varepsilon_0 r^2}\left[1-\left(1-\frac{r^3}{a^3}\right)^{3/2}\right]$$

11.78. Electrons are emitted as a result of thermal emission from a negatively charged plate into a parallel plate capacitor where field strength is $E = 10^3$ V/m. In view of electrical images calculate the distance from the emitter plate at which the velocity of the electrons will be maximum.

Solution: Force F_1 due to the electric field is given by

$$F_1 = eE$$

Force F_2 due to electrical image is given by

$$F_2 = -\frac{e^2}{4\pi\varepsilon_0\cdot(2x)^2}\ ;\ x = \text{distance}$$

Net force $F = F_1 + F_2 = 0 = eE - \dfrac{e^2}{4\pi\varepsilon_0\cdot 4x^2}$

or $x = \frac{1}{4}\sqrt{\frac{e}{\pi\varepsilon_0 E}}$

$= \frac{1}{4}\sqrt{\frac{4\times 9\times 10^{9}\times 1.6\times 10^{-19}}{10^{3}}}$

$= \frac{2\times 3\times 0.4}{4}10^{-6} = 6\times 10^{-7}\,\text{m}$

11.79. An ideal gas whose molecules have a polarizability $\alpha = 4 \times 10^{-30}$ m³ is contained in a large vessel at 300 K. A charged parallel plate capacitor with a field strength $E = 3 \times 10^{6}$ V/m is placed inside the vessel. Find the relative difference between the concentrations of the molecules inside and outside the capacitor.

Solution: In electric field the gas molecules will be polarised.

The dipole moment is given by $p = \varepsilon_0 \alpha E$.

Energy $W = -\frac{pE}{2} = -\frac{\varepsilon_0\alpha E^2}{2}$. The gas molecules will follow Maxwell Boltzman distribution. The molecular concentration is given by

$$n = n_0 e^{-W/KT} = n_0\, e^{\frac{\varepsilon_0\alpha E^2}{2KT}} = n_0\left[1 + \frac{\varepsilon_0\alpha E^2}{2KT}\right]$$

$$\frac{n-n_0}{n_0} = \frac{\varepsilon_0\alpha E^2}{2KT} = \frac{1}{4\pi\times 9\times 10^{9}}\times\frac{4\times 10^{-30}\times 9\times 10^{12}}{2\times 1.38\times 10^{-23}\times 300} = 3.85\times 10^{-8}$$

The molecular concentration and hence the pressure between the plates will be more than the remaining part.

11.80. A charge + Q is placed in front of an infinite conducting earthed plate. Find (i) electric potential and electric field at P. (ii) Surfce charge density at P. (iii) Force of attraction between the charge + Q and the plane.

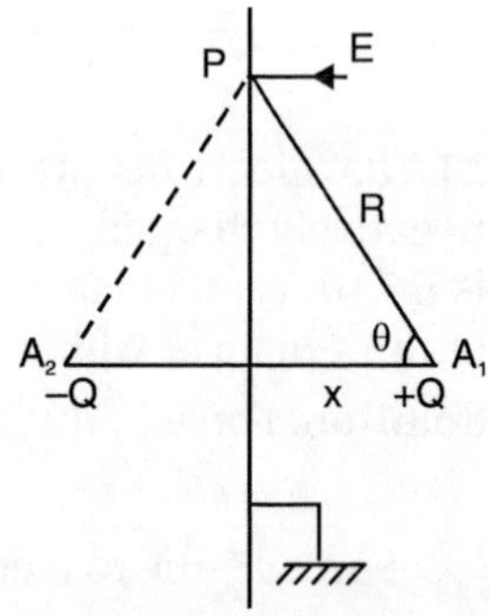

Fig. 11.58

Solution: The conducting plane will provide an electrical image of charge + Q as – Q at A_2.

(i) Electrical potential at P

$$= \frac{Q}{4\pi\varepsilon_0 A_1P} - \frac{Q}{4\pi\varepsilon_0 A_2P}$$

$= 0$ Since $A_1P = A_2P = R$

Electric field at $P = 2E\cos\theta = \frac{2Q}{4\pi\varepsilon_0 R^2}\cdot\frac{x}{R} = \frac{2Qx}{4\pi\varepsilon_0 R^3}$

(ii) Surface charge density at $P = \sigma = -\dfrac{E}{4\pi} = -\dfrac{Qx}{8\pi^2\varepsilon_0 R^3}$

(iii) Force of attraction between the plane and charge + Q

$$= -\frac{Q^2}{4\pi\varepsilon_0 \cdot 4x^2}$$

11.81. A charge q is uniformly distributed throughout the volume of a solid sphere of radius R. Find the potential difference between the centre and the surface of the sphere.

Solution: Electric field inside the sphere at a distance r from its centre is given by the relation.

$$\varepsilon_0 \cdot E \cdot 4\pi r^2 = \frac{q}{\frac{4}{3}\pi R^3} \cdot \frac{4}{3}\pi r^3$$

$$\therefore E = \frac{qr}{4\pi\varepsilon_0 R^3}$$

$$\int_{\text{Surface}}^{\text{Centre}} dV = -\int_R^0 E\,dr$$

$$V_{\text{Centre}} - V_{\text{Surface}} = \int_0^R \frac{qr\,dr}{4\pi\varepsilon_0 R^3} = \frac{q}{8\pi\varepsilon_0 R} = \frac{4\pi R^3\rho}{24\pi\varepsilon_0 R} = \frac{\rho R^2}{6\varepsilon_0};$$

$$\rho = \text{charge per unit volume} = \frac{q}{\frac{4}{3}\pi R^3}$$

11.82. A point charge Q is placed at the geometrical centre of a hollow cylinder of length l and radius r. Calculate the amount of electric flux passing through (i) the two plane surfaces (ii) the curved surface.

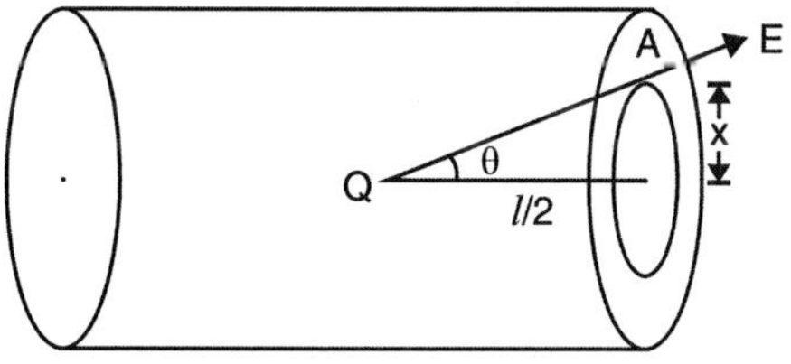

Fig. 11.59

Solution: Electric field at A due to the charge Q. (see Fig. 11.59)

$$= \frac{Q}{4\pi\varepsilon_0\left[x^2 + \frac{l^2}{4}\right]} = \frac{4Q}{4\pi\varepsilon_0\left[l^2 + 4x^2\right]}$$

$$\text{Field normal to the flat surface} = \frac{4Q}{4\pi\varepsilon_0\left[l^2 + 4x^2\right]} \cdot \cos\theta$$

$$= \frac{4Q \cdot l}{4\pi\varepsilon_0\left[l^2 + 4x^2\right]^{3/2}}$$

Consider a shell of radius x and $x + dx$

Flux through this shell $= \dfrac{4Q\cdot l\cdot 2\pi x dx}{4\pi\varepsilon_0\left[l^2+4x^2\right]^{3/2}}$

Net flux through one flat surface $= \displaystyle\int_0^R \frac{2Ql\cdot x dx}{\varepsilon_0\left(l^2+4x^2\right)^{3/2}}$

$$= \frac{Ql}{2\varepsilon_0}\left[\frac{1}{l}-\frac{1}{\sqrt{l^2+4R^2}}\right]$$

$$= \frac{Q}{2\varepsilon_0}\left[1-\frac{1}{\sqrt{1+4R^2/l^2}}\right]$$

Net flux through both the flat surfaces $= \dfrac{Q}{\varepsilon_0}\left[1-\dfrac{1}{\sqrt{1+4R^2/l^2}}\right]$

(ii) Flux through the curved surface

$$=\frac{Q}{\varepsilon_0}-\frac{Q}{\varepsilon_0}+\frac{Q}{\varepsilon_0\sqrt{1+4R^2/l^2}}=\frac{Q}{\varepsilon_0\sqrt{1+4R^2/l^2}}$$

11.83. Two point charges $+q$ and $-q$ are symmetrically placed on the axis of a circular ring of radius r at a distance l from the centre (see Fig. 11.60a). Find the electric flux passing through the ring.

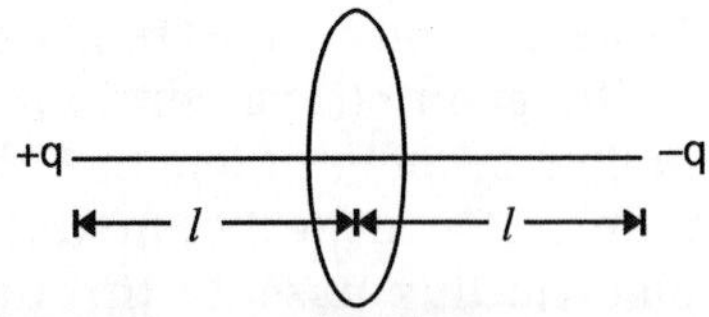

Fig. 11.60a

Solution: Consider a shell of radius x and $x + dx$.

Area of the shell $= 2\pi x dx$

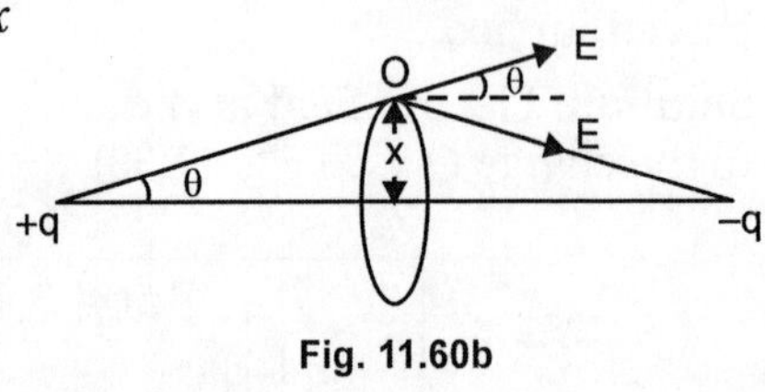

Fig. 11.60b

Flux through this shell is equal to the product of the component of electric field perpendicular to the plane of the shell and area.

$$= \frac{2q\cdot\cos\theta}{4\pi\varepsilon_0\left[l^2+x^2\right]}\cdot 2\pi x\cdot dx \text{ (see Fig. 11.60b)}$$

$$\text{Net flux through the ring} = \frac{q}{\varepsilon_0} \cdot \int_0^r \frac{l \cdot x \cdot dx}{\left(l^2 + x^2\right)^{3/2}}$$

$$= \frac{q}{\varepsilon_0}\left[1 - \frac{1}{\sqrt{1 + r^2/l^2}}\right]$$

11.84. A solid sphere of radius R is uniformly charged and has charge per unit volume σ. Find the flux passing through a plane located at a distance r_0 ($r_0 < R$) from the centre.

Solution: Electric field at a distance r inside the charged sphere is given by

$$\varepsilon_0 \cdot 4\pi r^2 \cdot E = \sigma \cdot \frac{4}{3}\pi r^3$$

$$\therefore \quad E = \frac{\sigma r}{3\varepsilon_0}$$

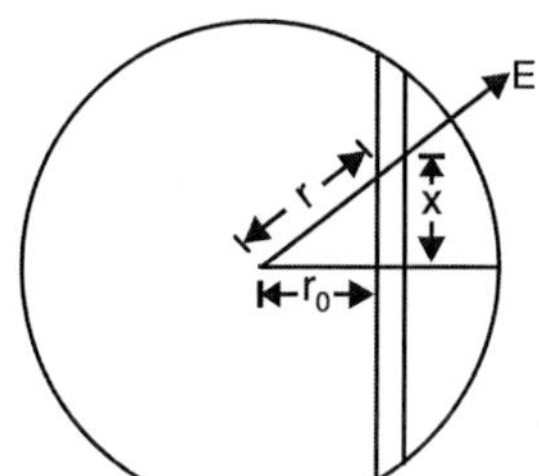

Fig. 11.61

Consider a shell of radius x and $x + dx$ located at a distance r_0 from the centre. Flux through this shell is given by

$$\frac{\sigma r}{3\varepsilon_0} \cdot \cos\theta \cdot 2\pi x dx \quad \text{(see Fig. 11.61)}$$

$$= \frac{\sigma r}{3\varepsilon_0} \cdot \frac{r_0}{r} \cdot 2\pi x dx$$

$$= \frac{\sigma r_0}{3\varepsilon_0} \cdot 2\pi x dx$$

$$\text{Net flux passing through the plane} = \int_0^{\sqrt{R^2 - r_0^2}} \frac{2\pi\sigma r_0}{3\varepsilon_0} \cdot x dx$$

$$= \frac{\pi\sigma r_0}{3\varepsilon_0}\left(R^2 - r_0^2\right)$$

11.85. A long charged thread having constant linear change density $\lambda = \frac{10^{-6}}{\pi}$ coul/m is placed along the axis of a circular loop of radius $R = 5$ cm with one of the ends at the centre (see Fig. 11.62). Calculate the total electric flux passing through the loop.

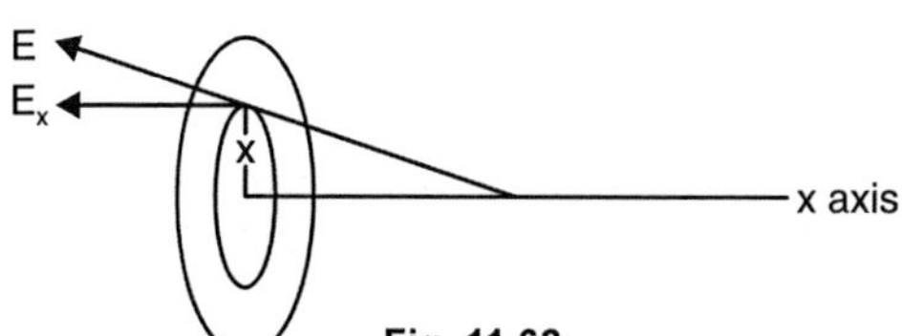

Fig. 11.62

Solution: The electric field E at one end of an infinitely long conductor at a distance x is given by

$$E = \frac{\lambda\sqrt{2}}{4\pi\varepsilon_0 x}$$

Consider a shell of radius x and $x + dx$

Area of the shell = $2\pi\, xdx$

Flux passing through the shell = $2\pi\, xdx \cdot E_x$

where E_x is the x component of the electric field E

$$E_x = \frac{\lambda}{4\pi\varepsilon_0 x}$$

$$d\phi = \frac{\lambda}{4\pi\varepsilon_0 x} \cdot 2\pi x dx$$

$$\text{Total flux } \phi = \frac{\lambda}{2\varepsilon_0}\int_0^R dx = \frac{\lambda R}{2\varepsilon_0}$$

On substituting the values of the various terms we have

$$\phi = \frac{10^{-6}}{\pi} \times \frac{5\times10^{-2}\times4\pi\times9\times10^{9}}{2}$$

$$= 900$$

11.86. A spherical shell of radius r carries a charge q uniformly distributed over its surface. Calculate the mechanical stress on the shell.

Solution: Let r' be the new radii as a result of the expansion of the shell. According to conservation of energy

$$\frac{q^2}{8\pi\varepsilon_0 r} - \frac{q^2}{8\pi\varepsilon_0 r'} = p\cdot\frac{4\pi}{3}\left(r'^3 - r^3\right)$$

$$\frac{3q^2\,(r'-r)}{32\pi^2\varepsilon_0 rr'} = p\left(r'^3 - r^3\right)$$

$$\frac{3q^2}{32\pi^2\varepsilon_0} = p\cdot rr'\left(r'^2 + rr' + r^2\right)$$

$$= 3pr^4$$

$$\therefore p = \frac{q^2}{32\pi^2\varepsilon_0 r^4}$$

11.87. The radius of a soap bubble is 10 mm. What charge should be imparted to it to make it begin to swell. Surface tension of soap solution $T = 3\times10^{-2}$ N/m.

Solution: Excess of pressure inside a soap bubble is equal to $\frac{4T}{R}$. When the electrostatic pressure exceeds the pressure due to surface tension the bubble begins to swell.

i.e. $$\frac{q^2}{32\pi^2\varepsilon_0 R^4} > \frac{4T}{R}$$

or $$q^2 > 32\pi^2\varepsilon_0 \cdot 4T \cdot R^3$$

$$q > \sqrt{32\pi \cdot 4\pi\varepsilon_0 \cdot T \cdot R^3}$$

On substituting the values of various terms we have

$$q > \sqrt{\frac{32 \times 3.14 \times 3 \times 10^{-2} \times 10^{-6}}{9 \times 10^9}}$$

$$q > 1.83 \times 10^{-8}\text{C}$$

11.88. A charge of 5×10^{-8} C is imparted to a soap bubble of radius R_0 = 5 cm. Find the increase in the radius due to the mutual repulsion of the surface charges. Neglect the effect of surface tension. Atmospheric pressure $p = 10^5$ N/m^2.

Solution:

$$p \cdot \frac{4}{3}\pi R_0^3 = \left(p - \frac{\sigma^2}{2\varepsilon_0}\right)\frac{4}{3}\pi R^3; \frac{\sigma^2}{2\varepsilon_0} = \text{electrostatic pressure}$$

$$\left(\frac{p}{p - \frac{\sigma^2}{2\varepsilon_0}}\right)^{1/3} = \frac{R}{R_0}$$

$$1 + \frac{\sigma^2}{3 \times 2\varepsilon_0 p} = \frac{R}{R_0}$$

$$\Delta R = R - R_0 = \frac{\sigma^2 R_0}{6\varepsilon_0 \cdot p} = \frac{q^2}{4\pi\varepsilon_0 p \times 24\pi R_0^3}$$

On substituting the values of various terms we have

$$\Delta R = \frac{25 \times 10^{-16} \times 9 \times 10^9}{10^5 \times 24 \times 3.14 \times 125 \times 10^{-6}}$$

$= 2.39 \times 10^{-8}$m

11.89. A charge Q given to a soap bubble of radius R_0 is uniformly distributed over its surface. Due to the mutual repulsion of the surface charges the radius of the bubble increases and as a result of this the air pressure inside the bubble drops. Neglecting the effect of surface tension, calculate the optimum value of the charge Q at which the pressure inside the bubble becomes $p\left(\frac{R_0}{R}\right)^3$ where p is the atmospheric pressure and R is the final radius acquired by the bubble.

Solution: Let us consider a charged spherical surface of density σ. The electric field near the surface $= \frac{\sigma}{2\varepsilon_0}$

Consider an element of area ds.

$$\text{Force acting on the element} = \sigma \cdot ds \cdot \frac{\sigma}{2\varepsilon_0}$$

$$\therefore \text{ Pressure} = \frac{\sigma^2}{2\varepsilon_0}$$

This is known as the electrostatic pressure and it always acts outwards.

In the light of this electrostatic pressure the net pressure inside the charged soap bubble is given by

$$p + \frac{4T}{R} - \frac{\sigma^2}{2\varepsilon_0}; \quad p = \text{atmospheric pressure}$$

$$T = \text{Surface tension}$$

On neglecting the effect of surface tension we have

$$p - \left(\frac{Q}{4\pi R^2}\right)^2 \cdot \frac{1}{2\varepsilon_0} = p \cdot \left(\frac{R_0}{R}\right)^3$$

or

$$p\left[1 - \frac{R_0^3}{R^3}\right] = \frac{Q^2}{32\pi^2\varepsilon_0 R^4}$$

$$\therefore Q = \sqrt{32\pi^2 \varepsilon_0 R p\left(R^3 - R_0^3\right)}$$

11.90. Two infinitely long parallel threads carrying charge with linear charge densities $+\lambda$ and $-\lambda$ are placed l apart at points A & B perpendicular to the plane of the paper (see Fig. 11.63). Find the electric potential and electric field at the point P lying at a distance $r >> l$ at an angle θ to the vector l.

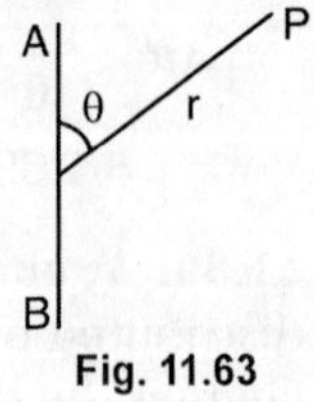

Fig. 11.63

Solution: Let $AP = r_1$ and $BP = r_2$

Electric field due to an infinitely long charged conductor at a distance r is given by

$$E = \frac{\lambda}{2\pi\varepsilon_0 r}$$

$$dV = -E \cdot dr = -\frac{\lambda dr}{2\pi\varepsilon_0 r}$$

Potential at P due to conductor at $A = V_1 = -\frac{\lambda}{2\pi\varepsilon_0} \cdot \int_0^{r_1} \frac{dr}{r} = \frac{\lambda}{2\pi\varepsilon_0} \cdot \log_e r_1$

Potential at P due to conductor at $B = V_2 = +\frac{\lambda}{2\pi\varepsilon_0} \cdot \int_0^{r_2} \frac{dr}{r} = \frac{\lambda}{2\pi\varepsilon_0} \cdot \log_e r_2$

Net potential $V = V_1 + V_2$

$$= \frac{\lambda}{2\pi\varepsilon_0} \cdot \log_e \frac{r_2}{r_1}$$

$$= \frac{\lambda}{2\pi\varepsilon_0}\left[\log\left(r + \frac{l}{2}\cos\theta\right) - \log\left(r - \frac{l}{2}\cos\theta\right)\right]$$

$$= \frac{\lambda l \cdot \cos\theta}{2\pi\varepsilon_0 r}$$

Let us represent this potential in Cartesian coordinate system

$$V = \frac{\lambda \cdot ly}{2\pi\varepsilon_0 \left(x^2 + y^2\right)}$$

$$E_x = -\frac{dV}{dx} = \frac{\lambda l}{2\pi\varepsilon_0} \cdot \frac{2xy}{\left(x^2 + y^2\right)^2}$$

$$E_y = -\frac{dV}{dy} = \frac{\lambda l}{2\pi\varepsilon_0} \frac{\left(x^2 - y^2\right)}{\left(x^2 + y^2\right)^2}$$

$$E = \sqrt{E_x^2 + E_y^2} = \frac{\lambda l}{2\pi\varepsilon_0}\left[\frac{4x^2y^2 + x^4 + y^4 - 2x^2y^2}{\left(x^2 + y^2\right)^4}\right]^{1/2}$$

$$= \frac{\lambda l}{2\pi\varepsilon_0 \left(x^2 + y^2\right)}$$

$$= \frac{\lambda l}{2\pi\varepsilon_0 r^2}$$

11.91. Protons accelerated through a potential difference V travels in the form of a beam of radius r and constitutes a current I. Calculate the electric field at the surface of the beam.

Solution:

$I = ne\pi r^2 v$; v = velocity of proton

n = number of protons per unit volume

$$\varepsilon_0 \cdot 2\pi r v \cdot E = I$$

$$\therefore\ E = \frac{I}{2\pi r \varepsilon_0 v}$$

$$\frac{1}{2} mv^2 = eV$$

$$\therefore\ v = \sqrt{\frac{2eV}{m}}$$

or
$$E = \frac{I}{2\pi r \varepsilon_0}\sqrt{\frac{m}{2eV}}$$

11.92. There is an infinite long straight uniformly charged wire having charge per unit length λ. Find out the electric field E at a distance r from the wire using Gauss's law.

Solution: According to Gauss's law, we have

$$\varepsilon_0 \oint E \cdot ds = q$$

In order to apply Gauss's law we have to construct a hypothetical surface called Gaussian surface in the form of a circular cylinder of radius r and length l taking the wire at its axis (see Fig. 11.64). E is constant over the cylindrical surface. Both E and ds at any point on the Guassian surface are directed radially outward. The angle between them is zero and therefore $E \cdot ds = E\, ds$. Thus,

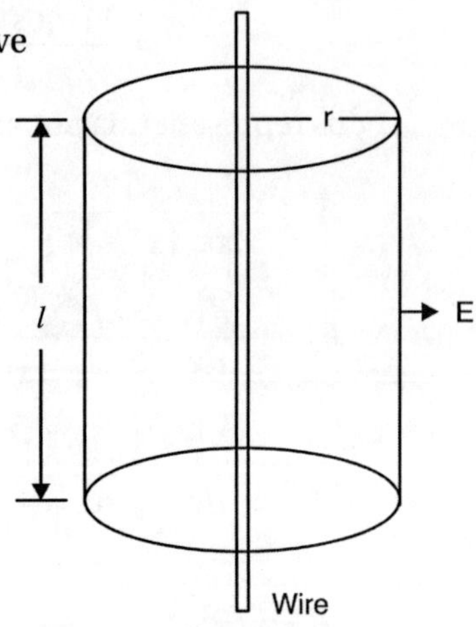

Fig. 11.64

$$\oint E \cdot ds = \oint E\, ds = E \oint ds = E \cdot 2\pi r l$$

There is no flux through the circular caps of the cylinder, i.e., $E \cdot ds = 0$.

Total flux is therefore equal to $E \cdot 2\,\pi\, r\, l$.

Charge enclosed by the Gaussian surface = λl

$$\therefore q = \lambda l$$

Therefore, $$E = \frac{\lambda l}{\varepsilon_0 2\pi r l} = \frac{\lambda}{2\pi \varepsilon_0 r}$$

11.93. There is an infinite long uniformly charged cylinder of radius R, having charge per unit volume ρ. Using Gauss's theorem, find the electric field E at a distance r ($r < R$) from the axis of the cylinder.

Solution: Consider a Gaussian surface in the form of cylinder of radius r and length l (see Fig. 11.65).

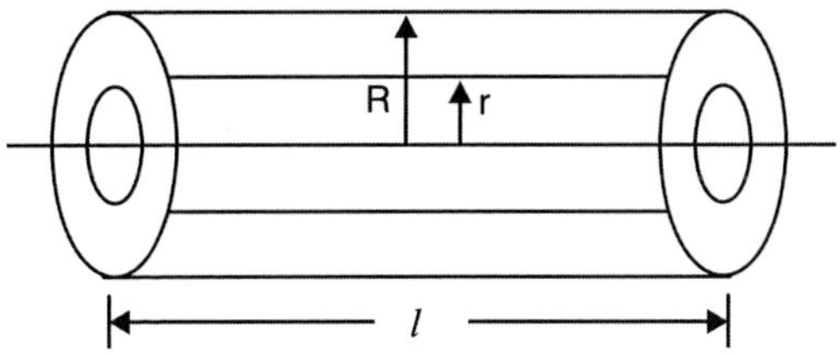

Fig. 11.65

$$\oint E \cdot ds = \oint E\, ds = E \oint ds = E \cdot 2\pi r l$$

There is no flux through the circular caps of the cylinder.

Total flux $= E \cdot 2\pi r l$

Volume of the Gaussian surface $= \pi r^2 l$

Charge enclosed by the Gaussian surface $= q = \rho \cdot \pi r^2 l$

$$\therefore \quad \varepsilon_0 \cdot 2\pi r l \cdot E = \rho \cdot \pi r^2 l$$

$$\therefore \quad E = \frac{\rho r}{2\varepsilon_0}$$

11.94. A section of two long concentric uniformly charged cylinders of radius r_1 and r_2 carrying equal and opposite charges is shown in Fig. 11.66. The charge per unit length in each case is λ. Using Gauss's theorem, find the electric field at a distance r from the axis of the cylinder when (i) $r > r_2$, (ii) $r < r_1$, and (iii) $r_1 < r < r_2$.

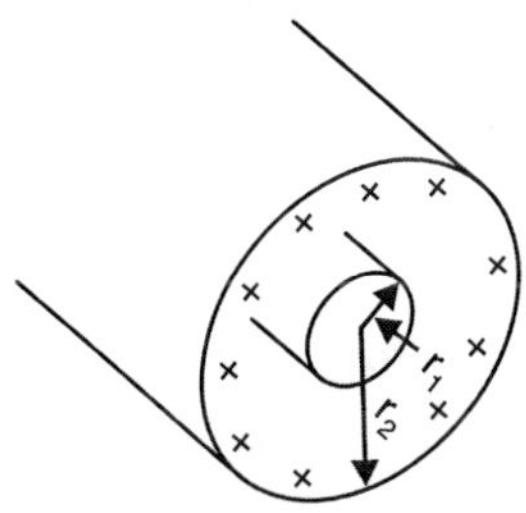

Fig. 11.66

Solution:

(i) when $r > r_2$: Consider a Gaussian surface in the form of a cylinder of radius r. This cylinder encloses both the cylinders. The net charge enclosed is equal to zero. The electric field will therefore be equal to zero.

(ii) when $r < r_1$: Consider a Gaussian surface in the form of cylinder of radius r. The cylinder will not enclose any charge. Hence the electric field will be equal to zero.

(iii) when $r_1 < r < r_2$: Consider a Gaussian surface in the form of a cylinder of radius r and length l. The cylinder will enclose charge $q = \lambda l$.

Now $\oint E \cdot ds = \oint E\, ds = E \oint ds = E \cdot 2\pi r l$

There is no flux through the circular caps (i.e., the ends of the cylinder).

Therefore total flux $= E \cdot 2\pi r l$.

$$\therefore E = \frac{\lambda l}{2\pi r l \cdot \varepsilon_0} = \frac{\lambda}{2\pi \varepsilon_0 r}$$

11.95. Two concentric uniformly charged spherical shells of radii r_1 and r_2 carry charges Q_1 and Q_2 respectively. Using Gauss's theorem, find the electric field at a distance r from the centre of the spherical shell when (i) $r < r_1$, (ii) $r > r_2$, and (iii) $r_1 < r < r_2$.

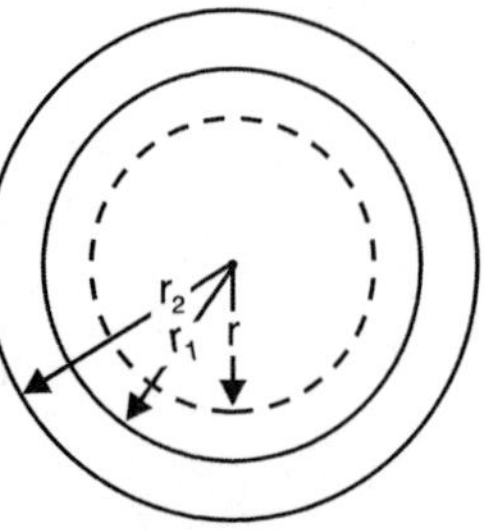

Fig. 11.67a

Solution:

(i) When $r < r_1$: Consider a Gaussian surface in the form of sphere of radius r as shown in Fig. 11.67a. This sphere will not enclose any charge. Hence the electric field will be equal to zero.

(ii) When $r > r_2$: Consider a Gaussian surface of radius r as shown in Fig. 11.67b. The sphere will enclose total charge = $Q_1 + Q_2$.

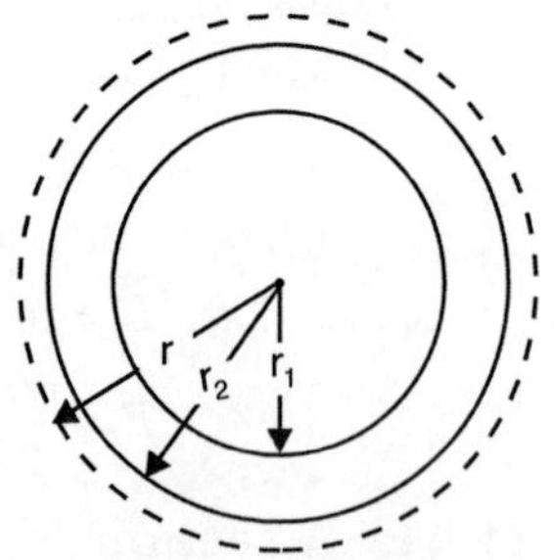

Fig. 11.67b

Now

$$\oint E \cdot ds = \oint E\, ds = E \oint ds = E \cdot 4\pi r^2$$

$$\therefore E = \frac{Q_1 + Q_2}{4\pi\varepsilon_0 r^2}$$

(iii) When $r_1 < r < r_2$: Consider a Gaussian surface of radius r as shown in Fig. 11.67c. The sphere will enclose a charge Q_1. Now

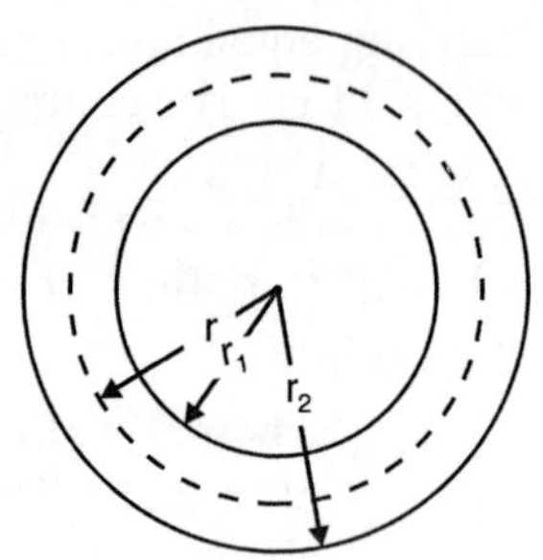

Fig. 11.67c

$$\oint E \cdot ds = \oint E\, ds = E \oint ds = E \cdot 4\pi r^2$$

$$\therefore E = \frac{Q_1}{4\pi\varepsilon_0 r^2}$$

11.96. Using Gauss's theorem, find the intensity of the electric field inside a uniformly charged sphere of volume density ρ.

Solution: Construct a Gaussian surface in the form of a sphere of radius r concentric to the given sphere. The charge enclosed by the sphere = $(4/3)\,\pi\, r^3\, \rho$.

Now $\oint E \cdot ds = E \oint ds = E \cdot 4\pi r^2$

Therefore $$E = \frac{\frac{4}{3}\pi r^3 \rho}{4\pi\varepsilon_0 r^2} = \frac{r\rho}{3\varepsilon_0}$$

11.97. Find the electric field intensity E in the region of intersection of two spheres uniformly charged by unlike charges with volume densities ρ and $-\rho$, if the distance between the centres of the spheres is determined by vector l (Fig. 11.68).

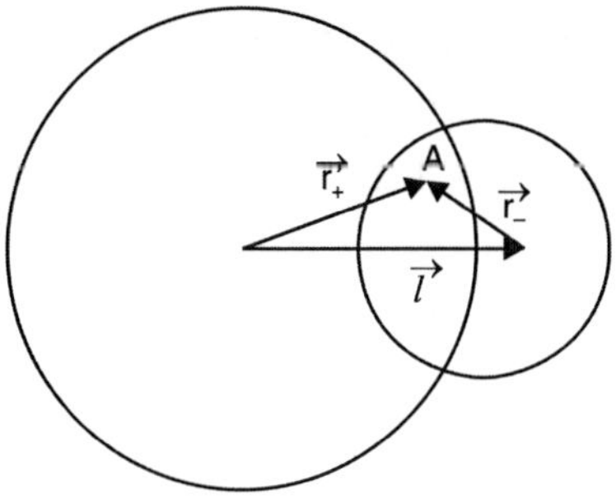

Fig. 11.68

Solution: The electric field within a uniformly charged sphere is given by

$$E = \frac{\rho r}{3\varepsilon_0} \qquad \text{(see Q. 11.96)}$$

The electric field in the region of intersection of the two spheres will be due to superposition of the fields due to two uniformly charged spheres. The field at a point A is given by

$$E = E_+ + E_- = \rho\,(r_+ - r_-)/3\varepsilon_0 = \rho l/3\varepsilon_0$$

11.98. In a charged sphere of radius R the volume density of charge at a distance r from the centre is defined by $\rho = \rho_0\,(1 - r/R)$ where ρ_0 is constant and represents the charge density at the centre of the sphere. Find the electric field strength both inside and outside the sphere.

Solution: For points outside the sphere, let us first find out the total charge.

$$Q = \int_0^R \frac{4\pi}{3} \cdot \rho \cdot 3r^2 dr$$

$$= 4\pi\rho_0 \int_0^R (1 - r/R)\, r^2 dr$$

$$= 4\pi\rho_0 \left[R^3/3 - R^3/4 \right]$$

$$= \frac{\pi\rho_0 R^3}{3}$$

According to Gauss theorem

$$\varepsilon_0 \cdot E \cdot 4\pi r^2 = \frac{\pi\rho_0 R^3}{3}$$

$$\therefore E = \frac{\rho_0 R^3}{12\varepsilon_0 r^2}$$

For points inside the sphere

$$4\pi r^2 \cdot E = \frac{1}{\varepsilon_0} \int \rho\, dV$$

or $$4\pi r^2 E = \frac{4\pi}{3\varepsilon_0} \int \rho_0 \left(1 - r/R\right) 3r^2 dr$$

$$E = \frac{\rho_0 r}{3\varepsilon_0}\left(1 - \frac{3r}{4R}\right)$$

11.99. A ball of radius R having spherical symmetric charge distribution is located in a medium where the volume density of charge varies inversely with the distance r measured from the centre of the ball via relation $\rho = \frac{a}{r}$. At what value of the charge on the ball the electric field strength outside the ball is independent of the distance r?

Solution: Let the charge on the sphere be q. Construct a spherical Gaussian surface of radius r outside the sphere then according to Gauss law

$$E \cdot 4\pi r^2 = \frac{q}{\varepsilon_0} + \frac{1}{\varepsilon_0}\int_R^r \frac{a}{r} \cdot 4\pi r^2 dr$$

$$E \cdot 4\pi r^2 = \frac{q - 2\pi a R^2}{\varepsilon_0} + \frac{4\pi a r^2}{2\varepsilon_0}$$

$$E = \frac{q - 2\pi a R^2}{4\pi r^2 \varepsilon_0} + \frac{a}{2\varepsilon_0}$$

Now E is independent of r only when the first term is zero. Hence

$$q = 2\pi a R^2$$

11.100. The volume density of charge in a medium is given by $\sigma = \sigma_0 e^{-ar^3}$ where σ_0 and a assumes positive constant values. Find the electric field strength as a function of r. Examine the result under the two conditions $ar^3 << 1$ and $ar^3 >> 1$.

Solution: Using Gauss theorem

$$\varepsilon_0 \cdot E \cdot 4\pi r^2 = \frac{4\pi}{3}\int_0^r \sigma_0 e^{-ar^3} \cdot 3r^2 dr$$

$$\varepsilon_0 \cdot E \cdot 4\pi r^2 = \frac{4\pi}{3}\frac{\sigma_0}{a}\int e^{-x} dx \,;\ ar^3 = x$$

$$= -\frac{4\pi\sigma_0}{3a}\left| e^{-ar^3} \right|_0^r$$

$$E = \frac{4\pi\sigma_0}{4\pi\varepsilon_0 r^3 \cdot 3a}\left[1 - e^{-ar^3}\right]$$

When $ar^3 << 1$

$$e^{-ar^3} = 1 - ar^3$$

and the charge distribution takes the form $\sigma = \sigma_0$

$$\varepsilon_0 \cdot 4\pi r^2 \cdot E = \frac{4\pi}{3}\sigma_0 \int 3r^2 dr$$

$$E = \frac{\sigma_0 r}{3\varepsilon_0}$$

When $ar^3 >> 1$ then the equation obtained for electric field reduces to

$$E = \frac{\sigma_0}{3\varepsilon_0 a r^2}$$

11.101. For a given charge distribution

$$\rho = \rho_0 \left(1 - r^2\!\!\middle/\!a^2\right) \text{ for } r \le a$$

$$\rho = 0 \text{ for } r > a$$

Find (i) the electric field inside the distribution

(ii) the distance from the centre at which the electric field has maximum value

(iii) the electric field outside the distribution.

Solution:

(i) According to Gauss Law

$$\varepsilon_0 \cdot E \cdot 4\pi r^2 = \frac{4\pi}{3}\int \rho_0 \left(1 - r^2\!\!\middle/\!a^2\right) 3r^2 dr$$

$$E = \frac{\rho_0}{\varepsilon_0 r^2}\left[\frac{r^3}{3} - \frac{r^5}{5a^2}\right]$$

$$= \frac{\rho_0}{\varepsilon_0}\left[\frac{r}{3} - \frac{r^3}{5a^2}\right]$$

Now $$\frac{dE}{dr} = \frac{\rho_0}{\varepsilon_0}\left[\frac{1}{3} - \frac{3r^2}{5a^2}\right] = 0$$

$$\therefore r = \frac{a}{3}\sqrt{5}$$

(ii) Amount of charge contained is given by

$$q = \frac{4\pi}{3}\int_0^a \rho_0 \left(1 - r^2\!\!\middle/\!a^2\right) 3r^2 dr$$

$$= 4\pi\rho_0 \left[a^3\!/3 - a^3\!/5\right] = \frac{8\pi\rho_0 a^3}{15}$$

For points outside the distribution the electric field is given by

$$\varepsilon_0 \cdot E \cdot 4\pi r^2 = \frac{8\pi}{15}\rho_0 a^3$$

$$\therefore E = \frac{2a^3\rho_0}{15\varepsilon_0 r^2}$$

11.102. A long cylinder of radius $r = 1.5$ cm has some charge uniformly distributed over its surface. These charges produces an electric field $E = 81$ KV/m at the surface of the cylinder. Calculate the amount of current produced when the cylinder starts moving with a constant uniform velocity of 8 m/sec.

Solution: According to Gauss theorem electric field at the surface of the charged cylinder of length l and radius r is given by

$$\varepsilon_0 \cdot E \cdot 2\pi r l = Q$$

$$\therefore E = \frac{Q}{2\pi\varepsilon_0 rl}$$

Consider a plane perpendicular to the axis of the cylinder. Let the cylinder moves through a distance *dx* along its axis in time *dt*.

$$I = \frac{dq}{dt} = \frac{Q}{2\pi\varepsilon_0 rl} \cdot 2\pi r \cdot \frac{dx}{dt} = \frac{Q}{l} \cdot v$$

On substituting the values of Q/l we have

$$I = 2\pi r \varepsilon_0 E \cdot v$$

On putting the values of various terms we have

$$I = \frac{1}{2 \times 9 \times 10^9} \times 1.5 \times 10^{-2} \times 81 \times 10^{+3} \times 8 = 0.54 \mu a$$

11.103. Calculate the amount of charge which, when placed on the surface of an uncharged soap bubble of radius R_0 = 1 cm, increases its radius by 0.05 cm through mutual repulsion of the charged surfaces. Atmospheric pressure = 10^5 N/m^2.

Solution: Let R_0 and R be the radii of the soap bubble before and after imparting the charge and p be the atmospheric pressure.

From the principle of conservation of energy, we have

$$p\,dv = \frac{q^2}{2c_1} - \frac{q^2}{2c_2}$$

where dv is the change in the volume of the soap bubble = $\frac{4}{3}\pi\left(R^3 - R_0^3\right)$

$c_1 = 4\pi\varepsilon_0 R_0$, $c_2 = 4\pi\varepsilon_0 R$ and q is the amount of charge imparted to the bubble. On substituting these values, we have

$$p \cdot \frac{4}{3}\pi\left(R^3 - R_0^3\right) = \frac{q^2}{2 \cdot 4\pi\varepsilon_0}\left(\frac{1}{R_0} - \frac{1}{R}\right)$$

or $$p \cdot \frac{4\pi}{3}(R - R_0)\left(R^2 + RR_0 + R_0^2\right) = \frac{q^2}{8\pi\varepsilon_0} \cdot \frac{R - R_0}{RR_0}$$

$$\therefore \quad q = \left[p \cdot \frac{32}{3}\pi^2 R_0 R \varepsilon_0 \left(R^2 + RR_0 + R_0^2\right)\right]^{1/2}$$

On substituting the values of the various terms we have

$$q = \left[10^5 \cdot \frac{32}{3} \cdot \pi^2 \times 10^{-2} \times 1.05 \times 10^{-2} \times 8.85 \times \right.$$

$$\left. 10^{-12}\left(10^{-4} + 1.05 \times 10^{-4} + 1.1025 \times 10^{-4}\right)\right]^{1/2}$$

$$= 10^{-7}\left(\frac{3.2}{3}\cdot\pi^2 \times 1.05 \times 8.85 \times 3.1525\right)^{1/2}$$

$$q = 1.755 \times 10^{-6}\,\text{C}$$

11.104. Calculate the force with which the plates of a charged parallel-plate capacitor attract each other.

Solution: Consider a parallel-plate capacitor in which the plates carry a charge q and are separated by a distance x (see Fig. 11.69). The energy of the capacitor is given by $\frac{q^2 \cdot x}{2\varepsilon_0 A}$, where A is the surface area of the plates. On displacing one of the plates through the distance dx, the work done is equal to $F \cdot dx$ where F is the force of attraction between the plates. The final energy of the capacitor is $\frac{q^2}{2\varepsilon_0 A}(x+dx)$. From the law of conservation of energy, we have

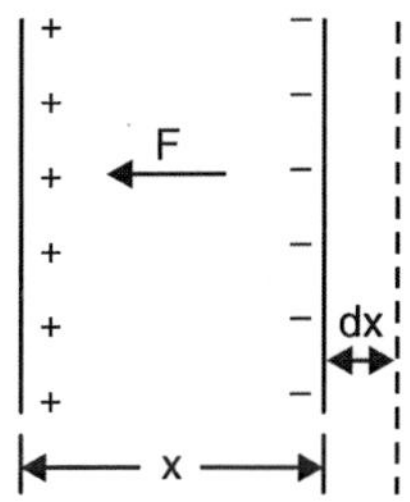

Fig. 11.69

$$F \cdot dx = \frac{q^2}{2\varepsilon_0 A}\left[(x+dx)-x\right]$$

or $$Fdx = \frac{q^2}{2\varepsilon_0 A}dx \text{ or } F = \frac{q^2}{2\varepsilon_0 A}$$

11.105. A parallel-plate capacitor is made of square plates each of length l separated by distance t. Calculate its capacity when plates are making a small angle ϕ with each other as shown in Fig. 11.70.

Solution: Consider an element of thickness dx lying at a distance x from one end of the condenser. The area of the elementary condenser thus formed $= l \cdot dx$ and the separation between the plates $= t + x\phi$. The capacity of the elementary capacitor is given by

$$dc = \frac{\varepsilon_0 l\, dx}{t + x\phi}$$

The total capacity $C = \int dc = \int_0^l \frac{\varepsilon_0 l\, dx}{t + x\phi}$

or $$C = \frac{\varepsilon_0 l}{t}\int_0^l \left(1 + \frac{x\phi}{t}\right)^{-1} dx$$

$$\cong \frac{\varepsilon_0 l}{t}\int_0^l \left(1 - \frac{x\phi}{t}\right) dx$$

$$\cong \frac{\varepsilon_0 l^2}{t}\left(1 - \frac{l\phi}{2t}\right)$$

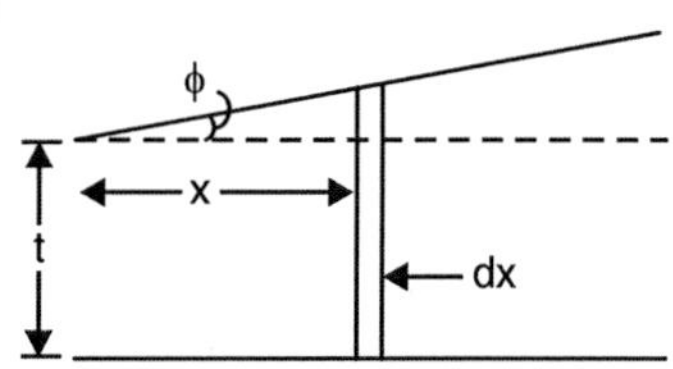

Fig. 11.70

11.106. Four metallic square plates *a, b, c* and *d*, each of length l and separated by distance t, are located in air. Find the capacity between the points P and Q when the plates are interconnected in the two different arrangements shown in Fig. 11.71.

Solution: In Fig. 11.71a plates *a, b* and *c* will form two capacitors which are in parallel and will be in series with a third capacitor formed by the plates *c* and *d*.

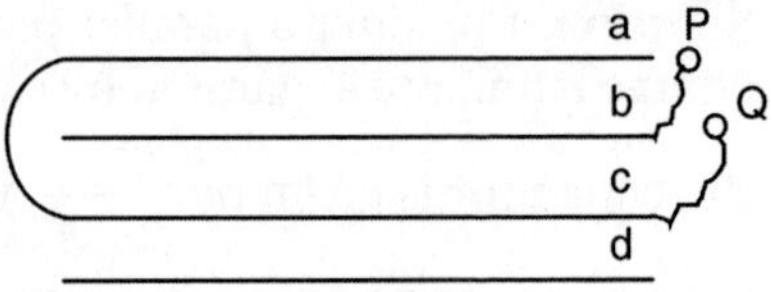

Fig. 11.71a

Now

$$C = \frac{\varepsilon_0 l^2}{t} + \frac{\varepsilon_0 l^2}{t} = \frac{2\varepsilon_0 l^2}{t}$$

$$\frac{1}{C'} = \frac{1}{C} + \frac{t}{\varepsilon_0 l^2} = \frac{t}{2\varepsilon_0 l^2} + \frac{t}{\varepsilon_0 l^2} = \frac{3t}{2\varepsilon_0 l^2}$$

or $$C' = \frac{2\varepsilon_0 l^2}{3t}$$

In Fig. 11.71b, plates *a* and *b*, and *c* and *d*, will form two capacitors in series and these will be in parallel with a third capacitor formed by the plates *b* and *c*.

Fig. 11.71b

Now $$\frac{1}{C} = \frac{t}{\varepsilon_0 l^2} + \frac{t}{\varepsilon_0 l^2} = \frac{2t}{\varepsilon_0 l^2}$$

$\therefore$ $$C = \frac{\varepsilon_0 l^2}{2t}$$

$$C' = C + \frac{\varepsilon_0 l^2}{t} = \frac{\varepsilon_0 l^2}{2t} + \frac{\varepsilon_0 l^2}{t} = \frac{3\varepsilon_0 l^2}{2t}$$

11.107. An infinite circuit (Fig. 11.72a) is formed by a capacitor of capacity $C = 2\mu F$. Find the capacitance between the points A and B.

Solution: Let the total capacitance between the two points A and B be C_x. The infinite network can be replaced by a simple network as shown in Fig. 11.72b. Total capacity between A and B can be represented as

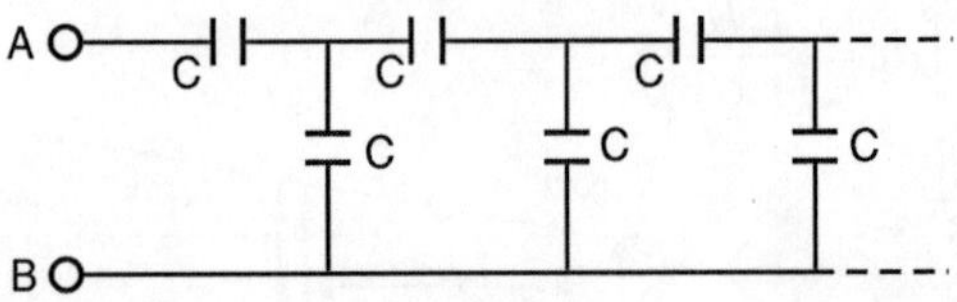

Fig. 11.72a

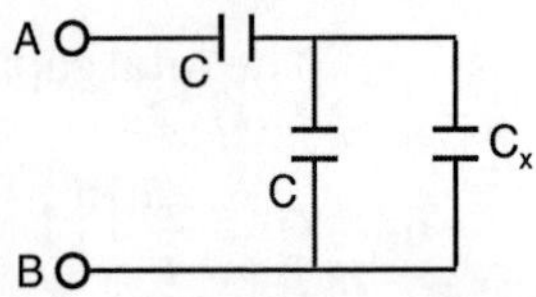

Fig. 11.72b

$$\frac{1}{C_x} = \frac{1}{C} + \frac{1}{C + C_x} = \frac{2C + C_x}{C(C + C_x)} \quad \text{or} \quad C^2 + CC_x = 2CC_x + C_x^2$$

or $$C_x^2 + CC_x - C^2 = 0$$

or $$C_x = -C \pm \sqrt{\frac{C^2 + 4C^2}{2}} = \frac{C}{2}\left[\sqrt{5} - 1\right]$$

On substituting $C = 2\mu F$,

$$C_x = \sqrt{5} - 1 = 2.236 - 1 = 1.236\ \mu F$$

11.108. A parallel-plate capacitor of area A is filled with two materials of having dielectric constant k_1 and k_2 as shown in Figs. 11.73a and 11.73b. Find its equivalent capacity.

Solution: In Fig. 11.73a each part represents a capacitor whose area is $A/2$ and thickness x. The two capacitors are in parallel. The total capacity is given by

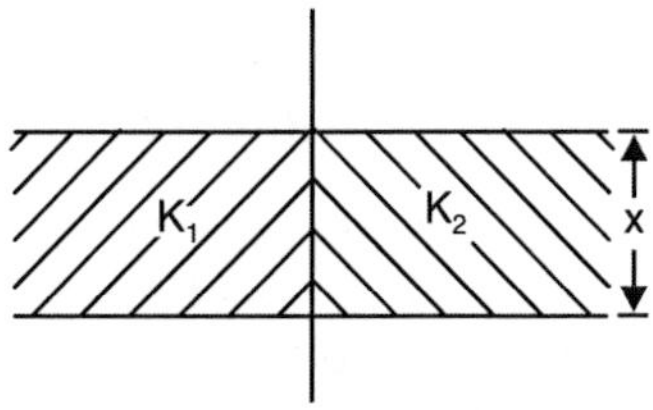

Fig. 11.73a

$$C = \frac{k_1 \varepsilon_0 A}{2x} + \frac{k_2 \varepsilon_0 A}{2x} = \frac{\varepsilon_0 A}{x}\left(\frac{k_1 + k_2}{2}\right)$$

In Fig. 11.73b, each part represents a capacitor whose thickness is $x/2$. The two capacitors are in series. The total capacity is represented by

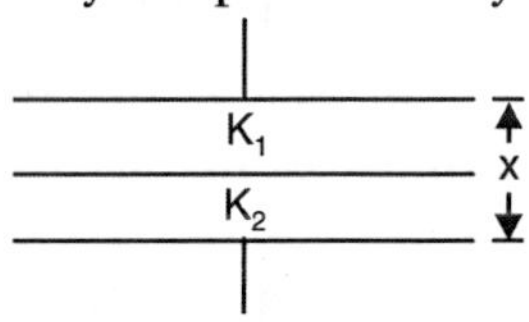

Fig. 11.73b

$$\frac{1}{C} = \frac{x}{2k_1 \varepsilon_0 A} + \frac{x}{2k_2 \varepsilon_0 A} = \frac{x}{2\varepsilon_0 A}\left(\frac{k_1 + k_2}{k_1 k_2}\right)$$

$$\therefore \quad C = \frac{2\varepsilon_0 A}{x}\left(\frac{k_1 k_2}{k_1 + k_2}\right)$$

11.109. A charge of $Q = 20 \times 10^{-3}$ C has been imparted to two similar parallel plate capacitors connected in parallel. In both the capacitors one plate moves with a velocity $v = 0.05$ mm/s such that the distances between the plates of the first capacitor increases uniformly according to equation $d_1 = (2 + vt)$ mm and the distance between the plates of the second decreases according to the equation $d_2 = (2 - vt)$ mm. Calculate the current in the circuit.

Solution: Let at an instant t the two capacitors contains q_1 and q_2 charges respectively such that $Q = q_1 + q_2$.

Now $$\frac{q_1}{C_1} = \frac{q_2}{C_2}$$

where $C_1 = \dfrac{\varepsilon_0 A}{d_1 \times 10^{-3}}$ and $C_2 = \dfrac{\varepsilon_0 A}{d_2 \times 10^{-3}}$ where A stands for the area of each capacitor, represented in square metres. In view of the identity stated above, $\dfrac{q_1}{q_2} = \dfrac{d_2}{d_1}$

or $$q_1 = q_2 \frac{(2 - vt)}{2 + vt}$$

On solving the equation $q_1 = Q\cdot\frac{(2-vt)}{4}$ and $q_2 = Q\cdot\frac{(2+vt)}{4}$

Current $I = -\frac{dq_1}{dt} = \frac{dq_2}{dt} = \frac{Qv}{4}$

On substituting the values of Q and v, we have

$$I = \frac{20\times10^{-3}\times0.05}{4} = 250\mu A$$

The current $I = 250\ \mu A$ will flow from the positively charged plate of the first capacitor to the positively charged plate of the second capacitor.

11.110. A condenser of capacity $C_1 = 1\ \mu F$ can be charged to a maximum voltage of $V_1 = 4.5$ KV; while another capacitor of capacity $C_2 = 1.5\ \mu F$ to maximum voltage of $V_2 = 4$ KV. Calculate the voltage which these two capacitors withstand when they are connected in series.

Solution: The charges on the two capacitors are

$$Q_1 = C_1V_1 = 10^{-6}\times4.5\times10^3 = 4.5\times10^{-3}\text{ Coul}$$
$$Q_2 = C_2V_2 = 1.5\times10^{-6}\times4\times10^3 = 6\times10^{-3}\text{ Coul}$$

In view of the above a charge $Q_1 = 4.5\times10^{-3}$ Coul can safely pass through both the capacitors.

$\therefore$ Max voltage across the combination $= \frac{Q_1}{C_1}+\frac{Q_1}{C_2}$

$$= \frac{4.5\times10^{-3}}{10^{-6}}+\frac{4.5\times10^{-3}}{1.5\times10^{-6}}$$
$$= 4.5\times10^3+3\times10^3\text{ Volt}$$
$$= 7.5\text{ KV}$$

11.111. The two plates of parallel plate capacitor separated by distance d has length l, breadth b. Find the capacitance when the upper plate is tilted through an angle $\theta = \frac{2\Delta d}{l}$.

Solution: Consider an element of length dx located at a distance x from one end (see Fig. 11.74). The capacity of the elementary capacitor is given by

$$dc = \frac{\varepsilon_0 b\cdot dx}{d-\Delta d+x\cdot\frac{2\Delta d}{l}}$$

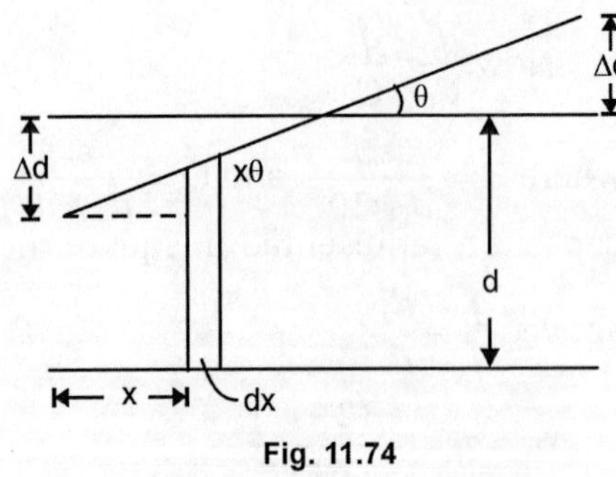

Fig. 11.74

$$\therefore C = \int dc = \int_0^l \frac{\varepsilon_0 b\,dx}{d - \Delta d + x \cdot \frac{2\Delta d}{l}}$$

$$= \frac{\varepsilon_0 bl}{2\Delta d}\left|\log\left(d - \Delta d + x\cdot\frac{2\Delta d}{l}\right)\right|_0^l$$

$$= \frac{\varepsilon_0 A}{2\Delta d}\log\left[\frac{\left(1+\Delta d/d\right)}{\left(1-\Delta d/d\right)}\right]$$

$$= \frac{\varepsilon_0 A}{2\Delta d}\left[\frac{\Delta d}{d} - \frac{(\Delta d)^2}{2d^2} + \frac{(\Delta d)^3}{3d^3} + \frac{\Delta d}{d} + \frac{(\Delta d)^2}{2d^2} + \frac{(\Delta d)^3}{3d^3}\right]$$

$$= \frac{\varepsilon_0 A}{d}\left[1 + \frac{1}{3}\left(\frac{\Delta d}{d}\right)^2\right]$$

11.112. The two square plates of length a and separated by distance d forms a parallel plate capacitor. Two materials of respective dielectric constants K_1 & K_2 are filled in the space between the plates as shown in the figure. Find the capacity of such capacitor.

Solution: Consider an element of thickness dx. (See Fig. 11.75)

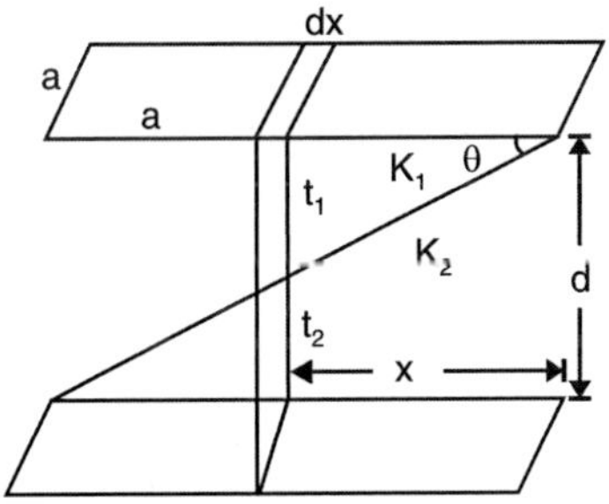

Fig. 11.75

$$\tan\theta = \frac{t_1}{x} = \frac{d}{a}$$

$$\therefore \quad t_1 = \frac{d\cdot x}{a}$$

$$t_2 = d - \frac{d\cdot x}{a} = \frac{d}{a}(a-x)$$

$$\frac{1}{dc} = \frac{d\cdot x}{\varepsilon_0 a^2 K_1 dx} + \frac{d(a-x)}{\varepsilon_0 a^2 K_2 dx}$$

$$= \frac{d}{\varepsilon_0 a^2 dx}\left[\frac{x}{K_1} - \frac{x}{K_2} + \frac{a}{K_2}\right]$$

$$= \frac{d}{\varepsilon_0 a^2 dx}\left[\frac{aK_1 + x(K_2 - K_1)}{K_1 K_2}\right]$$

$$\therefore\ dc = \frac{\varepsilon_0 a^2\cdot dx\cdot K_1 K_2}{d\left[aK_1 + x(K_2 - K_1)\right]}$$

$$C = \int dc = \frac{\varepsilon_0 a^2 K_1 K_2}{d}\int_0^a \frac{dx}{aK_1 + x(K_2 - K_1)} = \frac{\varepsilon_0 a^2 K_1 K_2}{d(K_2 - K_1)}\log_e\frac{K_2}{K_1}$$

11.113. A parallel plate capacitor of plate length l, breadth b and plate separation d is constructed by using three materials of dielectric constant K_1, K_2 & K_3 as shown in the Fig. 11.76. Find the capacitance of such a capacitor in terms of known parameters.

Solution: The two dielectrics K_2 and K_3 covers half of the area hence their equivalent capacity is

$$\frac{1}{C_1} = \frac{\frac{d}{2}}{K_2\,\varepsilon_0 \frac{l}{2}\cdot b} + \frac{\frac{d}{2}}{K_3\,\varepsilon_0 \cdot \frac{l}{2}\cdot b}$$

$$\frac{1}{C_1} = \frac{d}{\varepsilon_0 lb}\left[\frac{1}{K_2} + \frac{1}{K_3}\right]$$

$$C_1 = \frac{K_2 K_3\,\varepsilon_0 lb}{d\,(K_2 + K_3)}$$

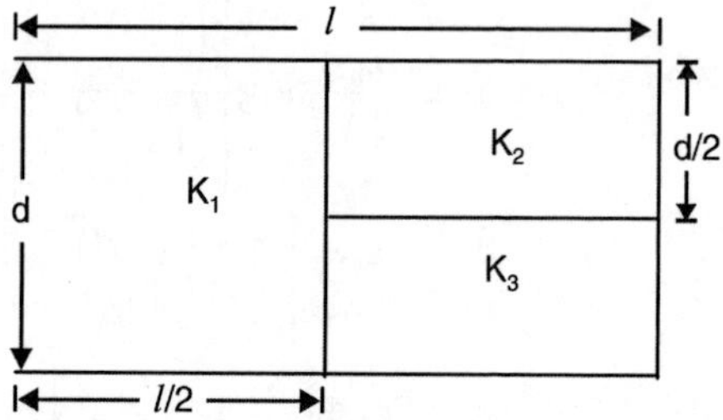

Fig. 11.76

Capacity of the capacitor formed by dielectric K_1

$$C_2 = \frac{K_1\,\varepsilon_0 \frac{l}{2}\cdot b}{d}$$

$\therefore$ Total capacity $C = C_1 + C_2 = \dfrac{\varepsilon_0 lb}{d}\left[\dfrac{K_1}{2} + \dfrac{K_2 K_3}{K_2 + K_3}\right]$

11.114. Find the equivalent capacitance between x & y for the given arrangement of the capacitors.

Solution: In view of the symmetry the charge distribution in the capacitors is shown in the Fig. 11.77.

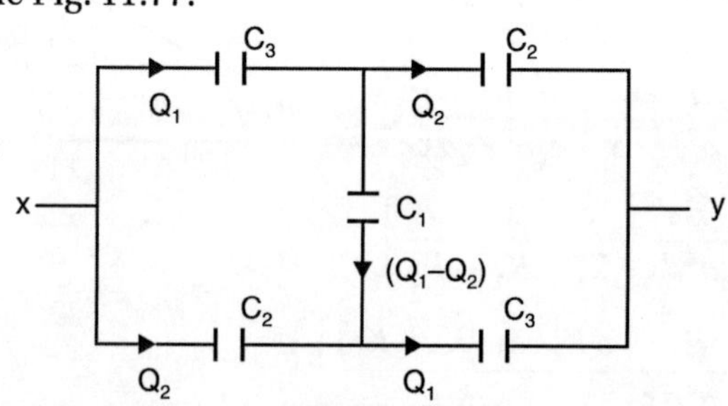

Fig. 11.77

In any mesh we have

$$\frac{Q_2}{C_2} = \frac{Q_1}{C_3} + \frac{Q_1 - Q_2}{C_1}$$

$$Q_2\left[\frac{1}{C_1} + \frac{1}{C_2}\right] = Q_1\left[\frac{1}{C_1} + \frac{1}{C_3}\right]$$

or $$\frac{Q_2}{C_2}[C_1+C_2] = \frac{Q_1}{C_3}[C_1+C_3]$$

$$\text{Total charge } Q = Q_1+Q_2 = Q_1\left[1+\frac{C_2(C_1+C_3)}{C_3(C_1+C_2)}\right]$$

or $$Q_1 = \frac{Q\cdot C_3(C_1+C_2)}{C_1C_2+C_1C_3+2C_2C_3}$$

$$V = \frac{Q_1}{C_3}+\frac{Q_2}{C_2}$$

$$= \frac{Q_1}{C_3}+\frac{Q_2(C_1+C_3)}{C_3(C_1+C_2)}$$

$$= Q_1\left[\frac{2C_1+C_2+C_3}{C_3(C_1+C_2)}\right]$$

or $$Q_1 = \frac{QC_3(C_1+C_2)}{C_1C_2+C_1C_3+2C_2C_3} = \frac{C_3(C_1+C_2)V}{2C_1+C_2+C_3}$$

$\therefore$ $$Q = \left(\frac{C_1C_2+C_1C_3+2C_2C_3}{2C_1+C_2+C_3}\right)V$$

$$C_{eq} = \frac{C_1C_2+C_1C_3+2C_2C_3}{2C_1+C_2+C_3}$$

11.115. (a) The plates of a parallel plate capacitor are connected to a cell. How the various parameters will change when a material of dielectric constant K is introduced.

Solution: Energy supplied by the battery is QV. Energy stored in the capacitor is $1/2\ QV$. After introducing the dielectric energy supplied by the battery is KQV. Energy stored in the capacitor = $1/2\ KQV$.

Additional energy supplied by the battery = $QV(K-1)$.

(b) After charging a parallel plate capacitor it is disconnected from the battery and the material of dielectric constant K is introduced. Discuss how the various parameters will change.

Solution: The amount of charge Q will have the same value as it was earlier. The capacity will increase K times. The potential across the plates will now $\frac{V}{K}$. Energy retained in the capacitor = $\frac{1}{2}\frac{QV}{K}$

$$\text{Loss in energy} = \frac{1}{2}QV - \frac{1}{2}\frac{QV}{K} = \frac{1}{2}QV\left[1-\frac{1}{K}\right]$$

Due to this loss in energy the dielectric will be accelerated between the plates of the capacitor.

11.116. Find the mutual capacitance per unit length of pair of parallel wires located in air. The cross-sectional radii of each wire is r and the axis of the two wires is separated by S.

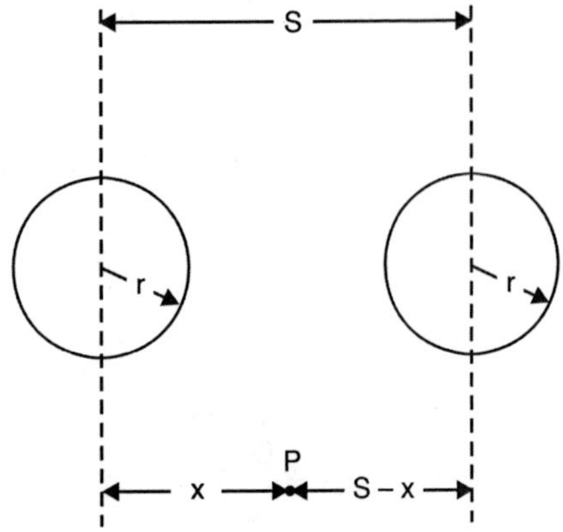

Fig. 11.78

Solution: Let q be the charge per unit length on each wire.

Net electric field at P is given by (see Fig. 11.78)

$$E = \frac{q}{2\pi\varepsilon_0 x} + \frac{q}{2\pi\varepsilon_0 (S-x)}$$

$$V = \text{pot. diff.} = -\int E \cdot dx$$

$$= -\frac{q}{2\pi\varepsilon_0}\left[\int_{S-r}^{r}\left(\frac{1}{x}+\frac{1}{S-x}\right)\right]dx$$

$$= -\frac{q}{2\pi\varepsilon_0}\left[\log_e \frac{r}{S-r} - \log_e \frac{S-r}{r}\right]$$

$$= \frac{q}{\pi\varepsilon_0}\log_e \frac{S-r}{r}$$

$$\cong \frac{q}{\pi\varepsilon_0}\log_e \frac{S}{r}$$

$$C = \frac{ql}{V};\ l = \text{length of wire}$$

$$= \frac{ql\pi\varepsilon_0}{q\log_e \frac{S}{r}} = \frac{\pi\varepsilon_0 l}{\log_e \frac{S}{r}}$$

$$\therefore \text{ Mutual capacitance per unit length} = \frac{\pi\varepsilon_0}{\log_e \frac{S}{r}}$$

11.117. In the arrangement shown in the Fig. 11.79 the switch S_1 is closed and condenser of capacity C_1 is charged to a potential V_0. Now the switch S_1 is opened and switch S_2 is closed. The charge accumulated in capacitor C_1 passes through the two capacitors C_2 & C_3. Calculate the amount of charge that will flow through the connecting wires.

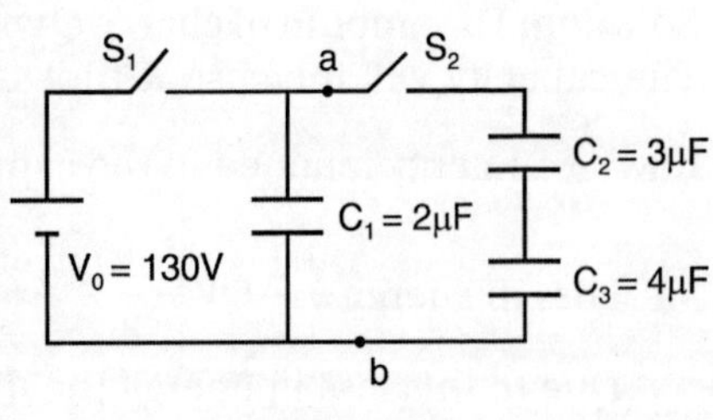

Fig. 11.79

Solution: Total charge accumulated in condenser of capacity $C_1 = CV_0$. After closing the switch S_2 the common potential across a & b is given by

$$V = \frac{C_1 V_0}{C_1 + \dfrac{C_2 C_3}{C_2 + C_3}}$$

Charge retained by condenser of capacity C_1 is equal to

$$C_1 V = \frac{C_1^2 V_0}{C_1 + \dfrac{C_2 C_3}{C_2 + C_3}}$$

Charge flown through the wires

$$Q = C_1 V_0 - \frac{C_1^2 V_0}{C_1 + \dfrac{C_2 C_3}{C_2 + C_3}}$$

$$= \frac{V_0}{\dfrac{1}{C_1} + \dfrac{1}{C_2} + \dfrac{1}{C_3}} = \frac{C_1 C_2 C_3 V_0}{C_1 C_2 + C_2 C_3 + C_1 C_3}$$

On substituting the values of various terms we have

$$Q = \frac{2 \times 3 \times 4 \times 130 \times 10^{-6}}{6 + 12 + 8} = 120 \mu C$$

11.118. Two identical parallel plate capacitors A & B of capacity C are connected as shown in the Fig. 11.80. The capacitors are fully charged by closing the switch. The switch is now open and the space between the plates is filled with a dielectric material. If the ratio of total initial energy stored and total final energy stored in the two capacitors is 2/5, find the value of the dielectric constant of the material filled between the plates of the two capacitors.

Solution: Total initial energy $= \frac{1}{2} CV^2 + \frac{1}{2} CV^2 = CV^2$; V = charging potential

Final energy of $A = \frac{1}{2} KCV^2$

Final energy of $B = \frac{1}{2} CV^2/K$

Now

$$\frac{CV^2}{CV^2 \left[\dfrac{K}{2} + \dfrac{1}{2K} \right]} = \frac{2}{5}$$

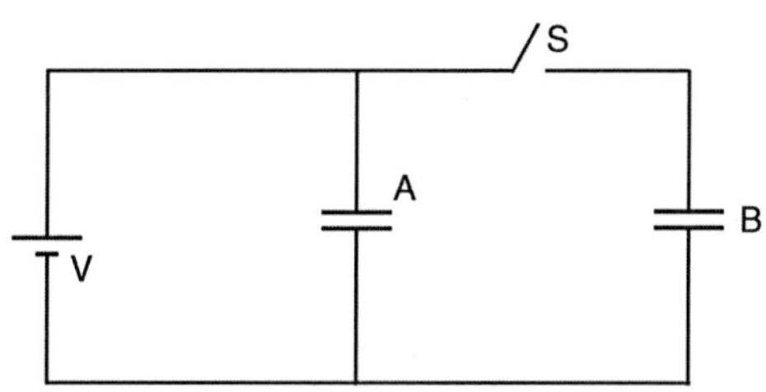

Fig. 11.80

$$\frac{2K}{K^2+1}=\frac{2}{5}$$

$$K = 4.79$$

11.119. Two parallel plate capacitors of capacities C_1 and C_2 are charged with a charging potential V_0 but with opposite polarity. The two capacitors are removed from the charging set up and are connected as shown in the Fig. 11.81. Find (i) the potential difference between points a & b after the switch S is closed, (ii) the total energy of the capacitors before and after closing the switch.

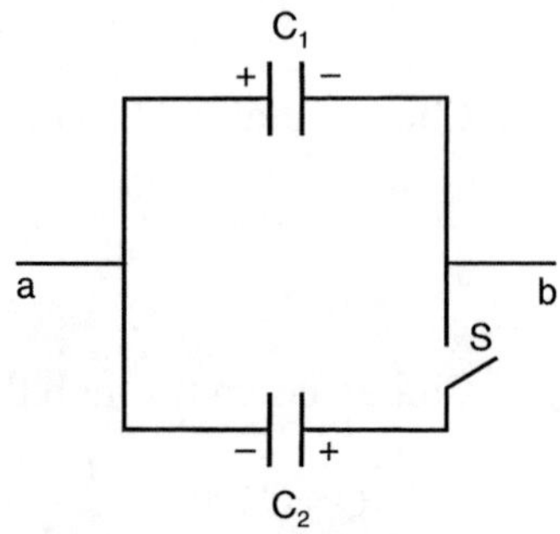

Fig. 11.81

Solution:

(i) Charge on the two capacitors are

$Q_1 = C_1V_0$ and $Q_2 = -C_2V_0$

On closing the switch S total charge $Q = (C_1 - C_2)V_0$

This charge has to be distributed between the two capacitors in order to obtain a common potential V given by

$$V=\frac{Q}{C_1+C_2}=\frac{(C_1-C_2)V_0}{C_1+C_2}$$

(ii) Total energy before closing the switch $=\frac{1}{2}(C_1+C_2)V_0^2$

Total energy after closing the switch $=\frac{1}{2}(C_1+C_2)\left[\frac{(C_1-C_2)V_0}{(C_1+C_2)}\right]^2$

11.120. **(a)** Find the capacity of a spherical capacitor of inner radii R_1 and outer radii R_2. When the outer sphere is charged and the inner sphere is earthed.

Solution: The inner sphere is at zero potential and the outer is having a potential V. This arrangment is equivalent to a spherical capacitor of radii R_1 and R_2 + spherical conductor of radii R_2

$$\text{Total capacity } C=\frac{4\pi\varepsilon_0 R_1R_2}{R_2-R_1}+4\pi\varepsilon_0 R_2=\frac{4\pi\varepsilon_0 R_2^2}{R_2-R_1}$$

(b) In the above problem what will be the capacity if the outer sphere is a thick one and having inner and outer radii as R_2 & R_3 respectively.

Solution: In this case the arrangement is equivalent to a spherical capacitor of radii R_1 & R_2 + spherical conductor of radii R_3.

$$\text{Total capacity } C = \frac{4\pi\varepsilon_0 R_1 R_2}{R_2 - R_1} + 4\pi\varepsilon_0 R_3$$

(c) Find the capacity if the outer sphere of problem 11.120(b) is surrounded by another sphere of radii R_4 and is earthed.

Solution: This arrangement is equivalent to the spherical capacitors of radii R_1 & R_2 and R_3 & R_4 respectively.

$$\text{Total capacity } C = \frac{4\pi\varepsilon_0 R_1 R_2}{R_2 - R_1} + \frac{4\pi\varepsilon_0 R_3 R_4}{R_4 - R_3}$$

11.121. The two plates of a parallel plate capacitor having area $A = 100\ \text{cm}^2$ are $t = 3$ mm apart. On introducing a $d = 2$mm thick dielectric it has been found that the capacity increases by an appreciable amount. In order to restore the same value of capacity the distance between the plates has been increased by h = 1.6 mm. Find the dielectric constant of the material.

Solution: The capacity of the parallel plate capacitor without dielectric is given by

$$C = \frac{\varepsilon_0 A}{t}$$

Capacity C' with dielectric inside the plates is given by

$$C' = \frac{\varepsilon_0 A}{t - d\left(1 - \frac{1}{K}\right)}$$

On increasing the plate separation by h we have

$$C = \frac{\varepsilon_0 A}{t + h - d\left(1 - \frac{1}{K}\right)} = \frac{\varepsilon_0 A}{t}$$

In the light of above equation we have

$$h = d\left(1 - \frac{1}{K}\right)$$

On putting the values we have

$$1.6 = 2\left(1 - \frac{1}{K}\right)$$

$$\frac{2}{K} = 2 - 1.6 = 0.4$$

$$\therefore K = \frac{2}{0.4} = 5$$

11.122. Two parallel plate capacitors each of capacity C are connected in series to a cell of emf E. Now one of the capacitor is filled with

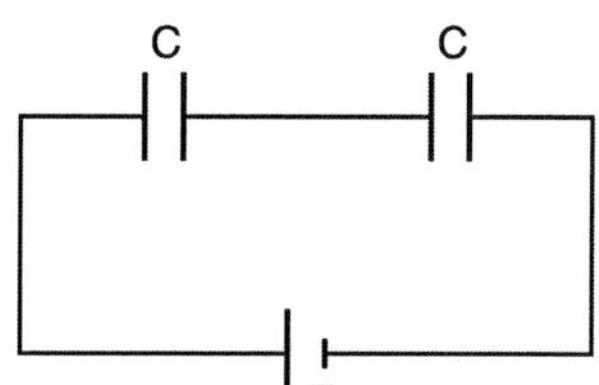

Fig. 11.82

a material of dielectric constant K (see Fig. 11.82). Find (i) the amount of charge flows through the cell (ii) how many times the electric field in that capacitor will decrease.

Solution:

(i) Q = charge through the capacitor $= \dfrac{EC}{2}$

Potential difference across each capacitor $= \dfrac{E}{2}$

E_1 = electric field across each capacitor $= \dfrac{E}{2x}$;

x = separation between the plates

$$\frac{1}{C_{eq}} = \frac{1}{KC} + \frac{1}{C}$$

$$C_{eq} = \frac{KC}{1+K}$$

Charge flowing through the capacitor $Q' = C_{eq} \cdot E = \dfrac{KCE}{1+K}$

Amount of charge flows through the cell $= Q' - Q$

$$= \frac{KCE}{1+K} - \frac{EC}{2} = \frac{EC(K+1)}{2(1+K)}$$

(ii) Potential difference V' across the first capacitor

$$= \frac{Q'}{2C} = \frac{E}{1+K}$$

Electric field $= E_2 = \dfrac{V'}{x} = \dfrac{E}{(1+K)x}$

$$E_2 = \frac{2E_1}{1+K}$$

11.123. A potential difference of V = 100 volt is applied across a spherical capacitor of inner and outer radii as R_1 = 3 cm and R_2 = 3.5 cm respectively. What velocity will be required by an electron while moving towards the centre of the spheres into the space between the two spheres from r_2 = 3.4 cm to r_1 = 3.1 cm.

Solution: Electric field at a distance r is given by

$$E = \frac{Q}{4\pi\varepsilon_0 r^2}$$

$$= \frac{CV}{4\pi\varepsilon_0 r^2}$$

$$= \frac{R_1 R_2 V}{(R_2 - R_1) r^2}$$

$$\int_{V_2}^{V_1} dV = -\int_{r_1}^{r_2} E \cdot dr$$

$$V_1 - V_2 = \frac{R_1 R_2 V}{R_2 - R_1}\left[\frac{1}{r_1} - \frac{1}{r_2}\right]$$

$$\text{Pot diff} = \frac{R_1 R_2 V\,(r_2 - r_1)}{(R_2 - R_1)\, r_1 r_2}$$

$$\frac{1}{2}mv^2 = \frac{eV \cdot R_1 R_2\,(r_2 - r_1)}{(R_2 - R_1)\, r_1 r_2}$$

$$v = \sqrt{\frac{2eV\, R_1\, R_2\,(r_2 - r_1)}{m\,(R_2 - R_1)\, r_1 r_2}}$$

On substituting the values of the various terms we have

$$v = \sqrt{\frac{2 \times 1.6 \times 10^{-19} \times 100 \times 3 \times 10^{-2} \times 3.5 \times 10^{-2} \times 0.3 \times 10^{-2}}{9 \times 10^{-31} \times 0.5 \times 10^{-2} \times 3.1 \times 10^{-2} \times 3.4 \times 10^{-2}}}$$

$$= 4.61 \times 10^{6}\ \text{m/sec.}$$

11.124. In a cylinderical capacitor the radii of the inner and outer cylinders are $r = 3$ cm and $R = 3.6$ cm respectively. The space between the two cylinders is filled with two layers of insulating material. The layer adjoining the inner cylinder is $d = 2$ mm thick. Calculate the ratio of the voltage drop in the two layers.

Solution: Electric field inside a cylinderical capacitor of length l at a distance x is given by

$$E = \frac{Q}{2\pi\varepsilon_0 x l}$$

$$V_0 = \frac{Q}{2\pi\varepsilon_0 l} \cdot \log_e \frac{R}{r}$$

$$\therefore\ E = \frac{V_0}{x \log_e \frac{R}{r}}$$

$$dV = -E \cdot dx$$

$$V_1 = -\int_{r+d}^{d} \frac{V_0}{\log_e \frac{R}{r}} \cdot \frac{dx}{x} = \frac{V_0}{\log_e \frac{R}{r}} \cdot \log_e\left(\frac{r+d}{d}\right)$$

$$V_2 = -\int_{R}^{r+d} \frac{V_0}{\log_e \frac{R}{r}} \cdot \frac{dx}{x} = \frac{V_0}{\log_e \frac{R}{r}} \cdot \log_e\left(\frac{R}{r+d}\right)$$

$$\therefore \frac{V_1}{V_2} = \frac{\log_e\left(\frac{r+d}{r}\right)}{\log_e\left(\frac{R}{r+d}\right)}$$

On substituting the values we have

$$\frac{V_1}{V_2} = \frac{\log_{10}\frac{3.2}{3}}{\log_{10}\frac{3.6}{3.2}} = 0.547$$

11.125. A potential difference of V' volt is maintained across the two electrodes of a cylindrical capacitor having inner and outer radii as r & R respectively. Calculate the velocity acquired by an electron while moving from point A to point B lying at a distance of l_1 and l_2 respectively from the axis of the capacitor.

Solution: Electric field inside a cylindrical capacitor at a distance x from the axis is given by

$$E = \frac{V'}{x\log_e\frac{R}{r}}$$

$$dV = -E\,dx$$

or $$V = -\int E\,dx = -\int_{l_1}^{l_2}\frac{V'}{\log_e\frac{R}{r}}\cdot\frac{dx}{x}$$

$$= \frac{V'\log_e\frac{l_1}{l_2}}{\log_e\frac{R}{r}}$$

$$\frac{1}{2}mv^2 = eV = \frac{eV'\log_e\frac{l_1}{l_2}}{\log_e\frac{R}{r}}$$

$$\therefore v = \sqrt{\frac{2eV'\log_e\frac{l_1}{l_2}}{m\cdot\log_e\frac{R}{r}}}$$

11.126. A mica sheet of dielectric constant $K = 7.5$ is fixed between the square plates of a parallel plate capacitor. The two plates each of area 81 cm^2 attracts each other by a force of 4.5×10^{-3} N. Calculate the surface density of the bound charges on the sheet.

Solution: Force between the plates is given by

$$F = \frac{q^2}{2\varepsilon_0 AK}$$

$$F = \left(\frac{K\varepsilon_0 A}{d}\right)^2 \cdot \frac{V^2}{2\varepsilon_0 AK}$$

$$= \frac{\varepsilon_0 AKV^2}{2d^2}$$

$$\frac{\varepsilon_0 V}{d} = \sqrt{\frac{2F\varepsilon_0}{KA}}$$

σ = Surface charge density of bound charges

$$= \frac{\varepsilon_0 V(K-1)}{d} = \sqrt{\frac{2F\varepsilon_0}{KA}} \cdot (K-1)$$

On substituting the values we have

$$\sigma = \sqrt{\frac{2\times 4.5\times 10^{-3}\times 8.85\times 10^{-12}}{7.5\times 81\times 10^{-4}}}\times 6.5$$

$$= 7.44\times 10^{-6}\,\text{Coul/m}^2$$

11.127. The space between the two cylinders of a concentric cylindrical capacitor is filled with two cylindrical layers of dielectric constant K_1 and K_2 respectively. The inner radii of the two layers are R_1 & R_2 respectively (see Fig. 11.83). The breakdown value of the electric field for the two dielectrics are E_1 and E_2 respectively. Calculate the breakdown voltage for the capacitor under the condition $K_1R_1E_1 < K_2R_2E_2$.

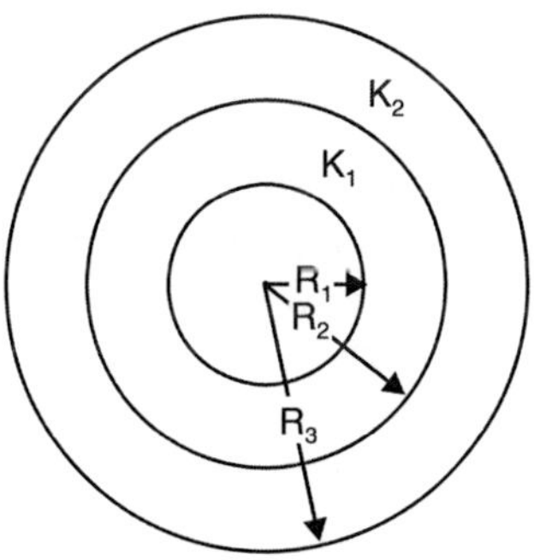

Fig. 11.83

Solution: Let l be the length of the capacitor. The voltage applied across the two electrodes is given by

$$V_0 = \frac{Q}{2\pi\varepsilon_0 l}\log_e \frac{R_3}{R_1}$$

Electric field inside the first dielectric at a distance x from the axis is given by

$$\varepsilon_0 K_1 \cdot 2\pi xl \cdot E = Q$$

or $$E = \frac{Q}{K_1 \cdot 2\pi\varepsilon_0 xl}$$

or $\dfrac{V_0}{E} = K_1\, x \log_e \dfrac{R_3}{R_1}$

Now $\dfrac{V_0}{E_1} = K_1\, R_1 \log_e \dfrac{R_3}{R_1}$

Similarly $\dfrac{V_0}{E_2} = K_2\, R_2 \log_e \dfrac{R_3}{R_1}$

Using these relations we have

$$\frac{K_1}{K_2} = \frac{E_2 R_2}{E_1 R_1}$$

$$V_1 = -\int E\,dx = \int_{R_2}^{R_1} -\frac{V_0}{K_1 \log_e \dfrac{R_3}{R_1}} \cdot \frac{dx}{x} = \frac{V_0}{K_1 \log_e \dfrac{R_3}{R_1}} \cdot \log_e \frac{R_2}{R_1}$$

Similarly $V_2 = \dfrac{V_0}{K_2 \log_e \dfrac{R_3}{R_1}} \cdot \log_e \dfrac{R_3}{R_2}$

$$\therefore \quad V = V_1 + V_2 = \left[E_1 R_1 \log_e \frac{R_2}{R_1} + E_2 R_2 \log_e \frac{R_3}{R_2} \right]$$

$$= E_1 R_2 \left[\log_e \frac{R_2}{R_1} + \frac{K_1}{K_2} \cdot \log_e \frac{R_3}{R_2} \right]$$

11.128. Find the capacity of a spherical capacitor in which the inner and outer spheres has radii $r_1 = 8.00$ cm and $r_2 = 10.00$ cm respectively and the space between the two spheres is filled with a material whose dielectric constant varies inversely with the distance r measured from the centre via relation $K = 2.00\, r_1 / r$.

Solution: Electric field inside the dielectric at a distance r is given by

$$E = \frac{Q}{4\pi\varepsilon_0 K r^2}$$

$$E = -\frac{dV}{dr}$$

$$\therefore \quad dV = -E \cdot dr$$

$$\int_V^0 dV = \int_{r_1}^{r_2} -\frac{Q}{4\pi\varepsilon_0 \cdot 2\dfrac{r_1}{r_2} \cdot r^2}\, dr$$

$$V = \frac{Q}{8\pi\varepsilon_0 r_1} \log_e \frac{r_2}{r_1}$$

$$\therefore C = \frac{8\pi\varepsilon_0 r_1}{2.303 \log_{10} \frac{r_2}{r_1}}$$

On substituting the values we have

$$C = \frac{2}{9\times 10^9} \times \frac{8\times 10^{-2}}{2.303 \log_{10} \frac{10}{8}}$$

$$= \frac{2\times 8\times 10^{-11}}{9\times 2.303\times 0.0969}$$

$$= 79.6 \text{ pf}$$

11.129. A capacitor of capacity C_1 is fully charged by using a cell of emf E. At an instant $t = 0$ it is connected to an uncharged capacitor of capacity C_2 through a resistance R. In what way the charge in the capacitor of capacity C_2 grows with time.

Solution: Let the total charge on capacitor C_1 be $Q = EC_1$

Let q be the charge on the second capacitor then we have

$$\frac{q}{C_2} + Ri = \frac{Q-q}{C_1}$$

$$R\frac{dq}{dt} = \frac{Q}{C_1} - q\left[\frac{1}{C_1} + \frac{1}{C_2}\right]$$

$$\int \frac{C_1 C_2 dq}{QC_2 - q(C_1 + C_2)} = \int \frac{dt}{R}$$

$$\frac{C_1 C_2}{C_1 + C_2} \log\left[QC_2 - q(C_1 + C_2)\right] = -\frac{t}{R} + A$$

At $t = 0$, $q = 0$ $\therefore A = \frac{C_1 C_2}{C_1 + C_2} \log QC_2$

$$\frac{C_1 C_2}{C_1 + C_2} \log\left[1 - \frac{q(C_1 + C_2)}{QC_2}\right] = -\frac{t}{R}$$

$$1 - \frac{q(C_1 + C_2)}{QC_2} = e^{-\left(\frac{C_1 + C_2}{C_1 C_2}\right)\frac{t}{R}}$$

$$q = \left(\frac{QC_2}{C_1 + C_2}\right)\left[1 - e^{-\left(\frac{C_1 + C_2}{C_1 C_2}\right)\frac{t}{R}}\right]$$

$$= \left(\frac{EC_1 C_2}{C_1 + C_2}\right)\left[1 - e^{-\left(\frac{C_1 + C_2}{C_1 C_2}\right)\frac{t}{R}}\right]$$

11.130. In the given circuit (see Fig. 11.84) find the voltage across the capacitor as a function of time after shorting the switch S at an instant $t=0$.

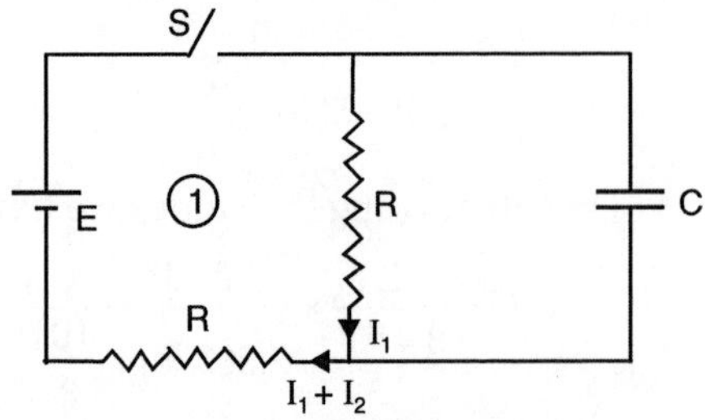

Fig. 11.84

Solution:

$$E-\frac{q}{C}-R(I_1+I_2)=0$$

$$E=\frac{q}{C}+R(I_1+I_2) \quad \text{... (i)}$$

In the first mesh

$$E-2RI_1-RI_2=0$$

$$\therefore I=\frac{E-RI_2}{2R} \quad \text{... (ii)}$$

On substituting the values of I_1 in equation 1 we have

$$E=\frac{q}{C}+\frac{E}{2}-\frac{RI_2}{2}+RI_2$$

$$\frac{E}{2}-\frac{q}{C}-\frac{RI_2}{2}=0$$

or $$EC-2q=CR\frac{dq}{dt}$$

$$\int-\frac{dq}{EC-2q}=\int-\frac{dt}{RC}$$

$$\log(EC-2q)=-\frac{2t}{RC}+A$$

At $t=0$, $q=0$ $\therefore$ $A=\log EC$

or $$\log\left(1-\frac{2q}{EC}\right)=-\frac{2t}{RC}$$

$$q=\frac{EC}{2}\left[1-e^{-\frac{2t}{RC}}\right]$$

or $$V=\frac{E}{2}\left[1-e^{-\frac{2t}{RC}}\right]$$

11.131. Two dielectric materials of thickness d_1 and d_2 and dielectric constants K_1 and K_2 respectively are filled in the space between the two plates of parallel plate capacitor without creating any air gap (see Fig. 11.85). A *dc* voltage V is applied across the two plates provides an electric field in the direction shown in the Fig. 11.85. Find the charge density of the bound charges on the boundary of the two materials.

Solution: Let A be the area of the plate then

$$\frac{1}{C}=\frac{d_1}{K_1\varepsilon_0 A}+\frac{d_2}{K_2\varepsilon_0 A}$$

$$\therefore C=\frac{K_1K_2\varepsilon_0 A}{d_1K_2+K_1d_2}$$

$$\frac{q}{A}=\frac{CV}{A}=\frac{K_1K_2\varepsilon_0 V}{d_1K_2+K_1d_2}$$

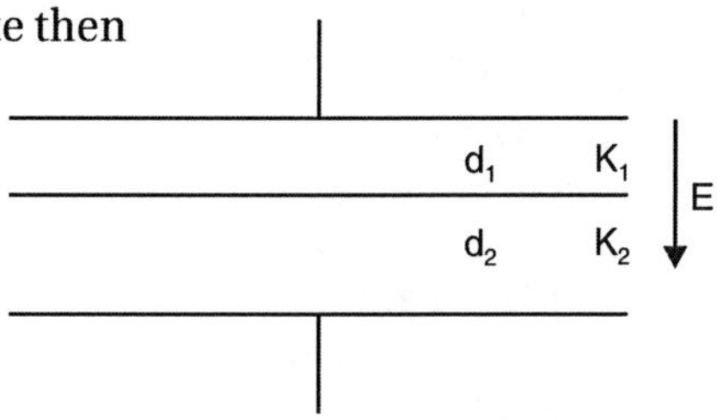

Fig. 11.85

Now $q\left(1-\frac{1}{K_1}\right)=q_1$

and $q\left(1-\frac{1}{K_2}\right)=q_2$

$$\therefore \frac{q_1-q_2}{A}=\frac{q}{A}\left[\frac{K_1-K_2}{K_1K_2}\right]=\frac{K_1K_2\varepsilon_0 V(K_1-K_2)}{(d_1K_2+K_2d_1)K_1K_2}$$

$$=\frac{\varepsilon_0 V(K_1-K_2)}{d_1K_2+K_1d_2}$$

11.132. Two materials of thickness d_1 & d_2, dielectric constant K_1 and K_2 and resistivities ρ_1 and ρ_2 are filled by putting one above the other between the two plates of parallel plate capacitor of area A. An electric field is established across the plates in a direction as shown in the Fig. 11.86 by applying a d.c. voltage V. Under what condition the surface charge density of extraneous charges at the boundary of the two materials is zero?

Solution: The resistances of the two materials are

$$R_1=\frac{\rho_1 d_1}{A} \text{ and } R_2=\frac{\rho_2 d_2}{A}$$

Total resistance $R=\frac{\rho_1 d_1+\rho_2 d_2}{A}$

Current $I=\frac{VA}{\rho_1 d_1+\rho_2 d_2}$

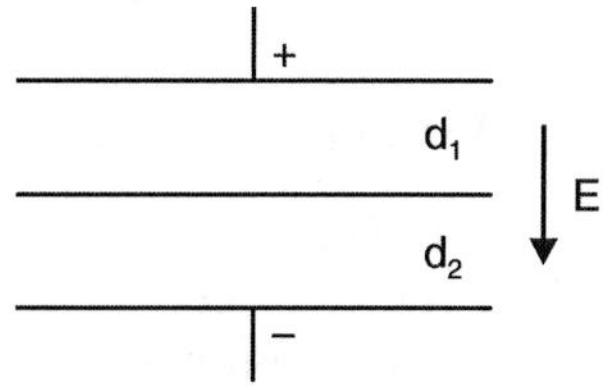

Fig. 11.86

Electric field in one material $=\frac{V'}{d_1}=\frac{IR_1}{d_1}=\frac{I\rho_1}{A}=\frac{\sigma_1}{K_1}$

Similarly electric field in other material $=\frac{V''}{d_2}=\frac{IR_2}{d_2}=\frac{I\rho_2}{A}=\frac{\sigma_2}{K_2}$

$$\sigma_1 - \sigma_2 = \frac{I}{A}[\rho_1 K_1 - \rho_2 K_2] = 0$$

or $\rho_1 K_1 = \rho_2 K_2$

11.133. The space between the plates of a parallel plate capacitor is filled with an in homogenous material whose dielectric constant and resistivity varies uniformly from one plate to the other with their values as K_1 and ρ_1 at one plate *a* and K_2 & ρ_2 at the other plate *b* respectively. On connecting the two electrodes of the capacitor to a dc source a current *I* flows through it in the direction from *a* to *b* (see Fig. 11.87). Find the total extraneous charge in the material.

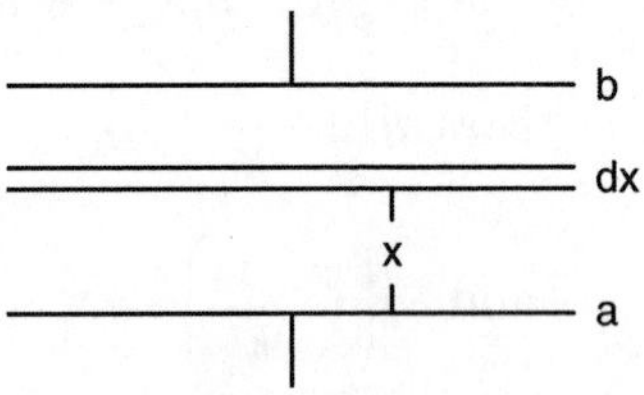

Fig. 11.87

Solution: Let *A* be area of the plate and *d* be the separation between the plates. Let

$$\alpha = \frac{K_2 - K_1}{d} \text{ and } \alpha' = \frac{\rho_2 - \rho_1}{d}.$$

Consider an element of thickness dx at a distance x from the lower end. The

$$\text{capacity dc} = \varepsilon_0 A \frac{[K_1 + \alpha x]}{dx}$$

$$\text{Resistance of the element} = (\rho_1 + \alpha' x)\frac{dx}{A}$$

$$\text{Voltage across the element} = I(\rho_1 + \alpha' x)\frac{dx}{A}$$

$$q = \frac{\varepsilon_0 A[K_1 + \alpha x]}{dx} \times \frac{I(\rho_1 + \alpha' x)\,dx}{A}$$

$$= \varepsilon_0 I(K_1 + \alpha x)[\rho_1 + \alpha' x]$$

At $x = 0$ let $q = Q_1$ and at $x = d$ $q = Q_2$

$$Q_1 = I\varepsilon_0 K_1 \rho_1,\ Q_2 = I\varepsilon_0 K_2 \rho_2$$

Extraneous charge $Q = Q_2 - Q_1 = \varepsilon_0 I\,[K_2\rho_2 - K_1\rho_1]$

11.134. The diameter of the outer cylinder of a cylindrical capacitor is D_2. Find the diameter D_1 of the core of this capacitor so that for a given potential difference across the capacitor, the electric field strength at the core is minimum.

Solution: The electric field at a distance *r* inside a cylindrical capacitor is given by

$$E = \frac{Q}{2\pi\varepsilon_0 r l} = -\frac{dV}{dr}$$

$$V = \frac{Q}{2\pi\varepsilon_0 l}\log_e \frac{r_2}{r_1}$$

$$\therefore E = \frac{2\pi\varepsilon_0 l \cdot V}{\log_e \frac{r_2}{r_1}} \cdot \frac{1}{2\pi\varepsilon_0 r l}$$

$$= \frac{V}{r \log_e \frac{r_2}{r_1}}$$

Electric field at the core

$$E = \frac{V}{\frac{D_1}{2} \log_e \frac{D_2}{D_1}} = \frac{2V}{D_1 \log_e \frac{D_2}{D_1}}$$

Now $$\frac{dE}{dD_1} = \left[-\frac{2V}{D_1^2 \log_e \frac{D_2}{D_1}} + \frac{2V}{D_1^2 (\log_e D_2 - \log_e D_1)^2} \right]$$

$$0 = \frac{2V\left[-(\log_e D_2 - \log_e D_1) + 1\right]}{D_1^2 (\log_e D_2 - \log_e D_1)^2}$$

or $$\log_e \frac{D_2}{D_1} = 1$$

or $$D_1 = \frac{D_2}{e}$$

11.135. Find the capacitance of a capacitor formed by two identical spheres of radius r placed in a medium of dielectric constant K. The distance between the centres of the sphere is L ($L >> r$).

Solution: Electric field at O due to two charged spheres is given by

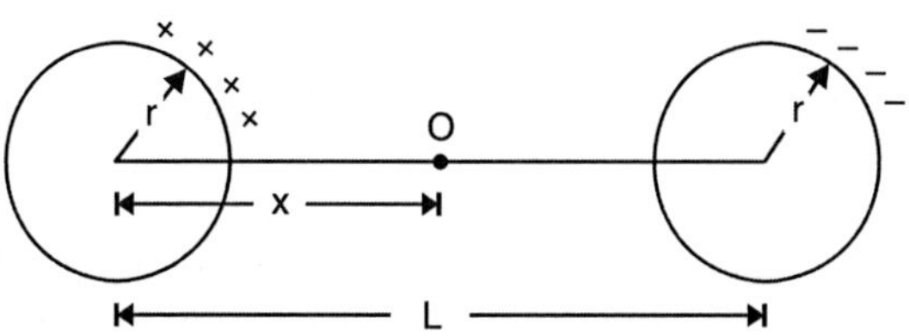

Fig. 11.88

$$E = -\frac{dV}{dx} = \frac{Q}{4\pi\varepsilon_0 x^2 K} + \frac{Q}{4\pi\varepsilon_0 (L-x)^2 K}$$

$$-\int_V^0 dV = \frac{Q}{4\pi\varepsilon_0 K} \int_r^{L-r} \left[\frac{dx}{x^2} + \frac{dx}{(L-x)^2} \right]$$

$$V = \frac{Q}{4\pi\varepsilon_0 K}\left[-\frac{1}{x}+\frac{1}{(L-x)}\right]_r^{L-r}$$

$$= \frac{Q}{4\pi\varepsilon_0 K}\left[\frac{2}{r}-\frac{2}{L-r}\right]$$

$$V = \frac{Q}{2\pi\varepsilon_0 r K}$$

$$\therefore C = 2\pi\varepsilon_0 r K$$

11.136. A parallel plate capacitor of capacity C is connected to the battery of emf E in parallel with a resistor. The capacitor plates are rapidly brought together so that the separation distance is halved. It is assumed that during the motion of the plates the charge on the capacitor remains same. Find the joule heat dissipated by the resistor.

Solution:

$Q = W_1 - W_2$

where Q = heat librated

W_1 = work done by the extraneous forces in the battery

W_2 = change in the electrostatic energy of the battery

Let dq be the change in the charge on the capacitor, then

$$dq = (2C - C)\,E = CE$$

$$W_1 = E \cdot dq = CE^2$$

When the separation between the plates is reduced to half of its value

Capacity = $2C$ and $E = \frac{E}{2}$ (since the charges remains the same)

$$W_2 = \frac{1}{2}\cdot 2C \cdot \left(\frac{E}{2}\right)^2$$

$$\therefore Q = W_1 - W_2 = CE - \frac{CE^2}{4} = \frac{3CE^2}{4}$$

11.137. A metallic plate is placed between the plates of a parallel plate capacitor parallel to them. The thickness of the plate is 1/3 of the separation between the plates. The capacitor is charged to a potential of $V = 100$ volt. Taking the capacity of the capacitor as 0.025 μF, find out the amount of work done in extracting the plate between the plates of the capacitor.

Solution: Amount of work done W = change in energy = $E_1 - E_2$

where E_2 is the final energy and E_1 is the initial energy

$$W = \frac{1}{2}\cdot\frac{3\varepsilon_0 A}{2t}\cdot V^2 - \frac{1}{2}\cdot\frac{\varepsilon_0 A}{t}\cdot V^2$$

$$= \frac{CV^2}{4};\; C = \frac{\varepsilon_0 A}{t}$$

On putting the values we have

$$W = \frac{0.025 \times 10^{-6} \times 10^{4}}{4}$$
$$= 62.5\ \mu J$$

11.138. A parallel plate capacitor is charged to a potential of 120 volt and then connected to a resistance. After a lapse of one second the potential difference across the capacitor is 90 volt. Find the potential difference across the plates after a lapse of 2 sec.

Solution: In the process of discharge of a capacitor through a resistance

$$Q = Q_0 e^{-t/RC}$$

or $$CV = CV_0\, e^{-t/RC}$$

or $$V = V_0\, e^{-t/RC}$$

$$90 = 120\, e^{-1/RC}$$

or $$e^{-1/RC} = \frac{3}{4}$$

After a lapse of 2 sec

$$V = 120\, e^{-2/RC}$$
$$= 120 \times \left(\frac{3}{4}\right)^2$$
$$= 67.5 \text{ volt}$$

11.139. The two plates of a parallel plate capacitor are $t = 3$ mm apart. A $d - 2$ mm thick material of dielectric constant $K = 5$ is introduced between the plates by keeping it parallel to the plates. The capacity will now be greater than the capacity of air capacitor in which the plates are 3 mm a part. In order to have the same value of capacity the distance between the plates is to be changed. Calculate the new separation between the plates.

Solution:

$$\frac{\varepsilon_0 A}{t - d\left(1 - \frac{1}{K}\right) + t'} = \frac{\varepsilon_0 A}{t}$$ here t' = increase in the distance

or $$t' = d\left(1 - \frac{1}{K}\right)$$
$$= 2\left(1 - \frac{1}{5}\right) = 1.6 \text{ mm}$$

The new separation between the plates = 3 + 1.6 = 4.6 mm

11.140. A potential difference of 100 volt is applied across the plates of a parallel plate capacitor. The two plates are 5 mm apart. A plane parallel plate of porcelain of dielectric constant 6 and thickness 4 mm is placed in contact with one of the plates. After charging the capacitor the supply is removed and the dielectric is also removed. Find the potential difference across the plates.

Solution:

$$Q = \frac{\varepsilon_0 A \times 100}{5\times10^{-3} - 4\times10^{-3}\left(1-\frac{1}{6}\right)} = \frac{\varepsilon_0 AV}{5\times10^{-3}};$$

V = required potential difference

A = area of cross-section

$$\frac{100}{5-4\left(1-\frac{1}{6}\right)} = \frac{V}{5}$$

$$V = \frac{5\times100\times6}{10} = 300 \text{ volt}$$

11.141. A capacitor of capacity 10μF is periodically charged through a potential difference of 120 volt. The capacitor is discharged through a 10 cm long solenoid with 200 number of turns. The magnetic induction at the centre of the solenoid is 3.02 × 10^4 w/m^2. How many times is the capacitor switched over a second?

Solution:

$$\mu_0 Ni = 3.02\times10^{-4}$$

$$4\pi\times10^{-7}\times\frac{200}{0.1}\times10\times10^{-6}\times n\times120 = 3.02\times10^{-4}$$

$$\therefore n = \frac{3.02\times10^4}{96\,\pi} = 100$$

11.142. A potential of 100 volt is applied to the plates of a parallel plate capacitor spaced 2 mm apart. The space between the plates is filled with a dielectric of dielectric constant $K = 6$. Find (a) the surface charge density of the polarization charges and (b) the change in the surface charge density on the plates when the capacitor is filled with dielectric when (1) the dielectric is introduced when the source is connected (2) the dielectric is introduced when the source is disconnected from the capacitor.

Solution: Let σ_0 and σ_d be the surface charge density of the capacitor without and with dielectric and σ_b the surface charge density of bound charges.

The effective charge density $\sigma' = \sigma_d - \sigma_b$

$$E_0 = \frac{V_1}{d} = \frac{\sigma_0}{\varepsilon_0}$$

$$E = \frac{\sigma_d}{K\varepsilon_0} = \frac{\sigma'}{\varepsilon_0} = \frac{V_2}{d}$$

$$\sigma_b = \sigma_d - \sigma' = (K-1)\,\varepsilon_0 E = (K-1)\,\varepsilon_0 \frac{V_2}{d}$$

When $V_1 = V_2 = V$

$$(1)\ (a)\ \ \sigma_b = (K-1)\,\varepsilon_0 \frac{V}{d} = \frac{(6-1)\times 8.85\times 10^{-12}\times 100}{2\times 10^{-3}} = 22.1\times 10^{-7}\,\text{C/m}^2$$

$$(b)\ \ \sigma_d - \sigma_0 = -\frac{V\varepsilon_0}{d} + KE\varepsilon_0 = (K-1)\frac{\varepsilon_0 V}{d} = 22.1\times 10^{-7}\,\text{C/m}^2$$

$$(2)\ \ V_2 = \frac{V_1}{K}$$

$$(a)\ \ \sigma_b = \varepsilon_0\,(K-1)\frac{V_2}{d} = \frac{\varepsilon_0\,(K-1)V_1}{Kd} = \frac{8.85\times 10^{-12}\times 5\times 100}{6\times 2\times 10^{-3}}$$

$$= \frac{8.85\times 5}{12}\times 10^{-7} = 3.68\times 10^{-7}\,\text{C/m}^2$$

$$(b)\ \ \sigma_b = \sigma_0$$

11.143. A material of dielectric constant K is completely filled between the two metallic plates of radii r_1 and r_2 respectively. The height of the plates is equal to h. Find the capacitance.

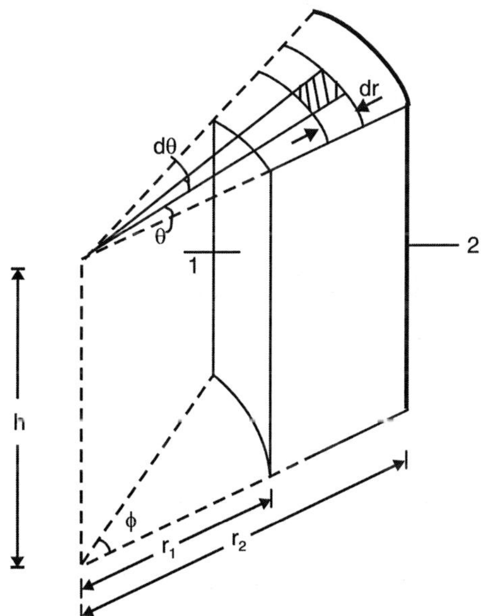

Fig. 11.89

Solution: Consider an element of height dh lying between the radii r and $r + dr$ and between angle θ and $\theta + d\theta$

The area of elementary capacity $= rd\,\theta \cdot dh$

Capacity of the element

$$= \frac{K\varepsilon_0\, rd\theta \cdot dh}{dr}$$

Capacity of the dielectric layer of height dh over an angle ϕ is given by

$$dc = K\varepsilon_0 \frac{r}{dr}\, dh \int_0^{\phi} d\theta = K\varepsilon_0\,\phi \frac{r}{dr}\cdot dh$$

$$\frac{1}{dc} = \frac{dr}{rK\varepsilon_0\,\phi\, dh}$$

Capacity of the complete dielectric layer of height dh lying between the two radii r_1 & r_2 is given by

$$\frac{1}{dc'} = \sum \frac{1}{dc} = \frac{1}{K\varepsilon_0\phi dh}\int_{r_1}^{r_2}\frac{dr}{r} = \frac{2.303\log_{10}\dfrac{r_2}{r_1}}{K\varepsilon_0\phi dh}$$

$$dc' = \frac{K\varepsilon_0\phi\, dh}{2.303 \log_{10}\frac{r_2}{r_1}}$$

$$\text{Total capacity } C = \int dc' = \frac{K\varepsilon_0\phi}{2.303 \log_{10}\frac{r_2}{r_1}} \cdot \int_0^h dh$$

$$= \frac{K\varepsilon_0\,\phi\cdot h}{2.303 \log_{10}\frac{r_2}{r_1}}$$

11.144. Two materials of dielectric constant, K_1 and K_2 are completely filled in the space between the two metallic plates of height h and radii r_1 and r_3 respectively (see Fig. 11.90). The two plates marked as 1 & 2 are connected across a d.c source of emf E. Calculate the voltage drop across each dielectric.

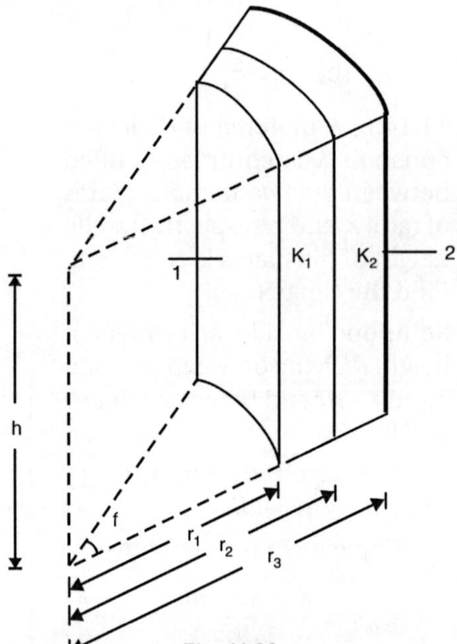

Fig. 11.90

Solution: It is a system of two capacitors of capacities C_1 & C_2 connected in series.

$$C_1 = \frac{K_1\varepsilon_0\phi h}{2.303 \log_{10}\frac{r_2}{r_1}},$$

$$C_2 = \frac{K_2\varepsilon_0\phi h}{2.303 \log_{10}\frac{r_3}{r_2}}$$

Let V_1 and V_2 be the voltage drop across the two dielectrics then

$$Q = C_1V_1 = C_2V_2;\ V_1 + V_2 = E$$

$$V_1 = \frac{C_2}{C_1 + C_2}\cdot E \text{ and } V_2 = \frac{C_1E}{C_1 + C_2}$$

$$V_1 = \frac{\dfrac{E\cdot K_2\varepsilon_0\phi h}{2.303\log_{10}\frac{r_3}{r_2}}}{\dfrac{K_1\varepsilon_0\phi h}{2.303\log_{10}\frac{r_2}{r_1}} + \dfrac{K_2\varepsilon_0\phi h}{2.303 \log_{10}\frac{r_3}{r_2}}} = \frac{E\cdot K_2\cdot\log_{10}\frac{r_2}{r_1}}{K_1 \log_{10}\frac{r_3}{r_2} + K_2 \log_{10}\frac{r_2}{r_1}}$$

Similarly

$$V_2 = \frac{E \cdot K_1 \cdot \log_{10} \frac{r_3}{r_2}}{K_1 \log_{10} \frac{r_3}{r_2} + K_2 \log_{10} \frac{r_2}{r_1}}$$

11.145. In the space between two metallic plates *ABCD* and *EFGH* a homogenous material of dielectric constant *K* is completely filled in a way in which the two surfaces *HADG* and *EBCF* attains a curvature of radii r_1 & r_2 respectively. Taking the height of the capacitor formed as *h* find its capacitance.

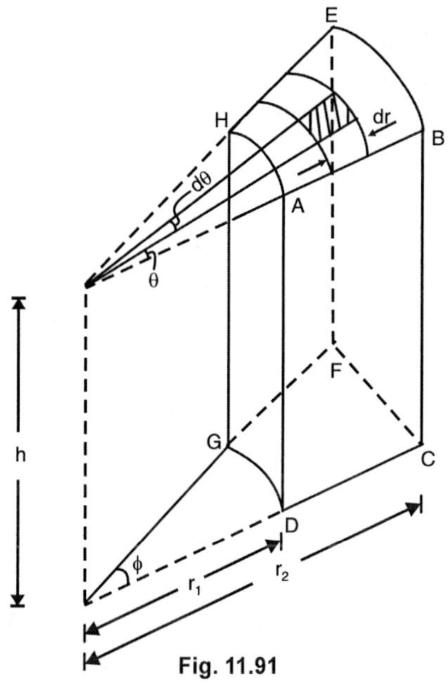

Fig. 11.91

Solution: Consider an element of height *dh* lying between radii *r* and $r + dr$ between the angles θ and $\theta + d\theta$ (see Fig. 11.91).

Area of the elementary capacitor $= dr \cdot dh$

Capacity of the elementary capacitor $= \dfrac{K\varepsilon_0 dr\, dh}{r d\theta}$

Capacity of the dielectric layer upto height *h* and also between the two radii

$$dc = \frac{K\varepsilon_0}{d\theta} \cdot \int_{r_1}^{r_2} \frac{dr}{r} \cdot \int_0^h dh = 2.303 \frac{K\varepsilon_0 h}{d\theta} \cdot \log_{10} \frac{r_2}{r_1}$$

$$\frac{1}{C} = \sum \frac{1}{dc} = \frac{1}{2.303\, K\varepsilon_0 h \log_{10} \frac{r_2}{r_1}} \cdot \int_0^{\phi} d\theta$$

$$= \frac{\phi}{2.303\, K\varepsilon_0 h \cdot \log_{10} \frac{r_2}{r_1}}$$

or $$C = \frac{2.303\, K\varepsilon_0 h \cdot \log_{10} \frac{r_2}{r_1}}{\phi}$$

11.146. A ring capacitor is formed by filling the space between the two rings of radii r_1 and r_2 and height h with a material of dielectric constant K. Find the capacitance.

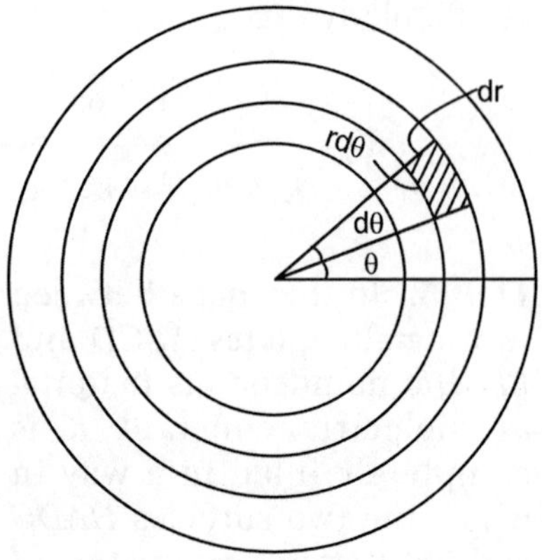

Fig. 11.92

Solution: Consider an element of the dielectric of radius r and $r + dr$ at an angle θ and $\theta + d\theta$. The capacity of the element is given by

$$dc = \frac{K\varepsilon_0 hr\, d\theta}{dr}$$

or $$\frac{1}{dc} = \frac{dr}{K\varepsilon_0 hd\theta \cdot r}$$

Capacity of all such capacitors lying between θ and $\theta + d\theta$

$$\sum \frac{1}{dc} = \frac{1}{dc'} = \frac{1}{K\varepsilon_0 hd\theta}\int_{r_1}^{r_2}\frac{dr}{r} = \frac{\log_e \frac{r_2}{r_1}}{K\varepsilon_0 h\, d\theta}$$

or $$dc' = \frac{K\varepsilon_0 hd\theta}{\log_e \frac{r_2}{r_1}}$$

$$\text{Net capacity } C = \int dc' = \frac{K\varepsilon_0 h}{\log_e \frac{r_2}{r_1}} \cdot \int_0^{2\pi} d\theta$$

$$= \frac{2\pi K\varepsilon_0 h}{\log_e \frac{r_2}{r_1}}$$

11.147. Two flat rings of radii r_1 and r_2 were prepared by a wire of diameter d. The rings were placed concentrically and the space between them is filled with a material of dielectric constant K. Find the capacitance of the capacitor formed.

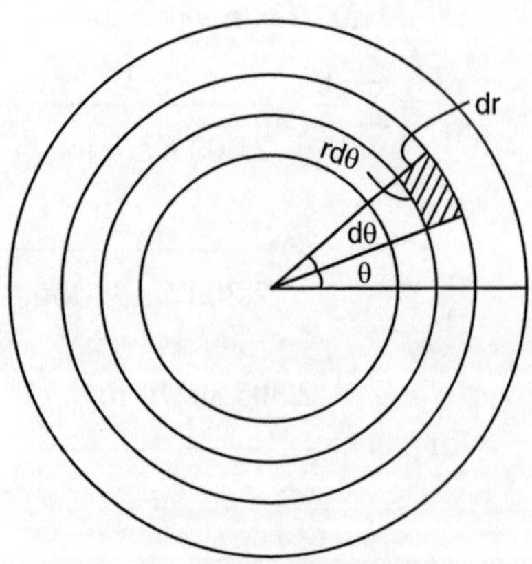

Fig. 11.93

Solution: Consider an element of the dielectric lying between radii r and $r + dr$ at an angle θ and $\theta + d\theta$. The capacity of the element is given by

$$dc = \frac{K\varepsilon_0 d \cdot r d\theta}{dr}$$

or $$\frac{1}{dc} = \frac{dr}{K\varepsilon_0 d \cdot d\theta \cdot r}$$

The capacity of all capacitors lying between angle θ and $\theta + d\theta$

$$\sum \frac{1}{dc} = \frac{1}{dc'} = \frac{1}{K\varepsilon_0 d \cdot d\theta} \cdot \int_{r_1}^{r_2} \frac{dr}{r} = \frac{\log_e \frac{r_2}{r_1}}{K\varepsilon_0 d \cdot d\theta}$$

$$\therefore dc' = \frac{K\varepsilon_0 d \cdot d\theta}{\log_e \frac{r_2}{r_1}}$$

Net capacity $C = \int dc' = \frac{K\varepsilon_0 d}{\log_e \frac{r_2}{r_1}} \cdot \int_0^{2\pi} d\theta$

$$C = \frac{2\pi K\varepsilon_0 \cdot d}{\log_e \frac{r_2}{r_1}}$$

11.148. In the circuit shown in Fig. 11.94, find the potential difference between points x and y if $E = 190$ V and $C_2/C_1 = 3$.

Solution: In the second mesh, we have

$$\frac{Q_2}{C_1} + \frac{Q_2}{C_2} = \frac{Q_1}{C_2}$$

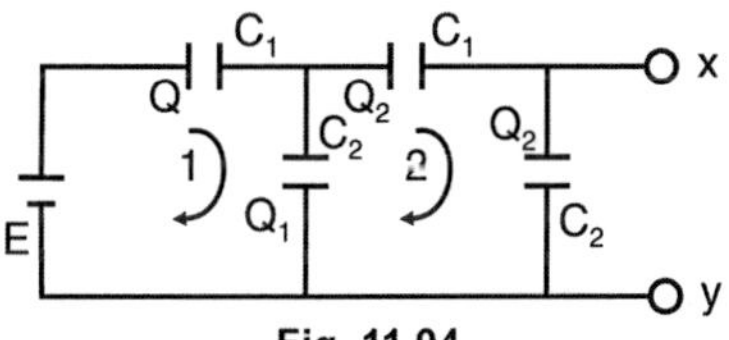

Fig. 11.94

or $$\frac{Q_2}{C_1} + \frac{Q_2}{3C_1} = \frac{Q_1}{3C_1}; \; C_2 = 3C_1$$

or $4Q_2 = Q_1$

In the first mesh, we have

$$\frac{Q_1 + Q_2}{C_1} + \frac{Q_1}{C_2} = E = 190$$

$$\frac{4Q_2}{C_1} + \frac{Q_2}{C_1} + \frac{4Q_2}{3C_1} = E = 190$$

or $$\frac{19Q_2}{3C_1} = 190$$

or $$\frac{Q_2}{3C_1} = V_{xy} = \frac{190}{19} = 10 \text{ V}$$

11.149. A section of a circuit is shown in Fig. 11.95. The potential difference between x and y is equal to 4 V. Find the voltage across each capacitor. $C_1 = 2\ \mu F$, $C_2 = 1.5\ \mu F$ and $E = 10$ V.

Solution:

$$V_x - V_y = 4 = \frac{Q}{C_1} - 10 + \frac{Q}{C_2}$$

(the same charge of $Q\ \mu C$ will flow through each capacitor).

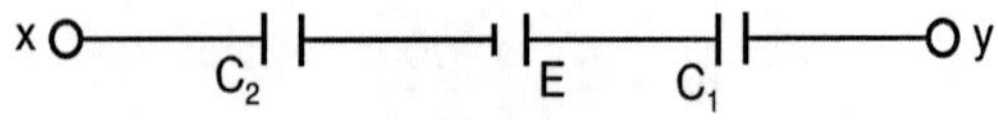

Fig. 11.95

or $$14 = \frac{Q}{2} + \frac{2Q}{3}$$

$$14 = \frac{3Q + 4Q}{6}$$

$$Q = 12\ \mu C$$

Voltage across 1.5 μF capacitor = 8 V

Voltage across 2 μF capacitor = 6 V

11.150. In the circuit shown in Fig. 11.96, find the potential difference across the two capacitors if $E_1 = 4$ V, $E_2 = 1$ V and $C_1/C_2 = 2$.

Solution:

$$E_1 - E_2 = \frac{Q}{C_1} + \frac{Q}{C_2} = \frac{Q(C_1 + C_2)}{C_1 C_2}; Q \text{ is the charge in the circuit.}$$

$$\text{Potential difference across } C_1 = \frac{Q}{C_1} = \frac{(E_1 - E_2)C_2}{C_1 + C_2} = \frac{E_1 - E_2}{1 + C_1/C_2}$$

$$\text{Potential difference across } C_2 = \frac{Q}{C_2} = \frac{(E_1 - E_2)C_1}{(C_1 + C_2)} = \frac{E_1 - E_2}{1 + (C_2/C_1)}$$

On substituting the values, we have

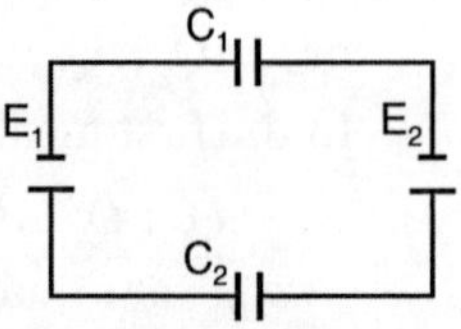

Fig. 11.96

$$\text{Potential difference across } C_1 = \frac{3}{1+2} = 1\text{V}$$

$$\text{Potential difference across } C_2 = \frac{3}{1+\frac{1}{2}} = 2\text{V}$$

11.151. Determine the potential difference between the points A and B of the circuit shown in Fig. 11.97, if $E = 21$ V, $C_1 = 1\ \mu F$, $C_2 = 2\ \mu F$, $C_3 = 3\ \mu F$ and $C_4 = 4\ \mu F$.

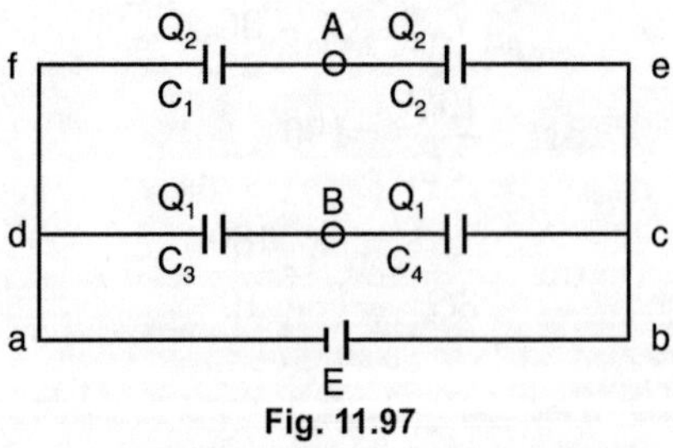

Fig. 11.97

Solution: In the mesh *abcd*, we have

$$Q_1\left(\frac{1}{C_3} + \frac{1}{C_4}\right) = E$$

or $\quad Q_1 = \left(\dfrac{E \cdot C_3 C_4}{C_3 + C_4}\right)$

Similarly in the mesh *abef* we have

$$Q_2\left(\frac{1}{C_1}+\frac{1}{C_2}\right)=E$$

or $\quad Q_2 = \dfrac{E \cdot C_1 C_2}{C_1 + C_2}$

Now $V_A - V_B = \dfrac{Q_2}{C_1} - \dfrac{Q_1}{C_3} = \dfrac{E \cdot C_2}{(C_1+C_2)} - \dfrac{E \cdot C_4}{(C_3+C_4)}$

$$= \frac{E[C_2C_3 + C_2C_4 - C_1C_4 - C_2C_4]}{(C_1+C_2)(C_3+C_4)}$$

$$= \frac{E[C_2C_3 - C_1C_4]}{(C_1+C_2)(C_3+C_4)}$$

On substituting the values of various terms, we have

$$V_{AB} = V_A - V_B = \frac{21(2\times3-1\times4)}{(1+2)(3+4)} = \frac{21\times2}{3\times7} = 2\text{ V}$$

11.152. Determine the potential of points 1, 2 and 3 of the circuit shown in Fig. 11.98 assuming C_1 = 1 μF, C_2 = 2 μF, C_3 = 3 μF and E_1 = 1V, E_2 = 2V, E_3 = 3V and the potential at point *O* to be zero.

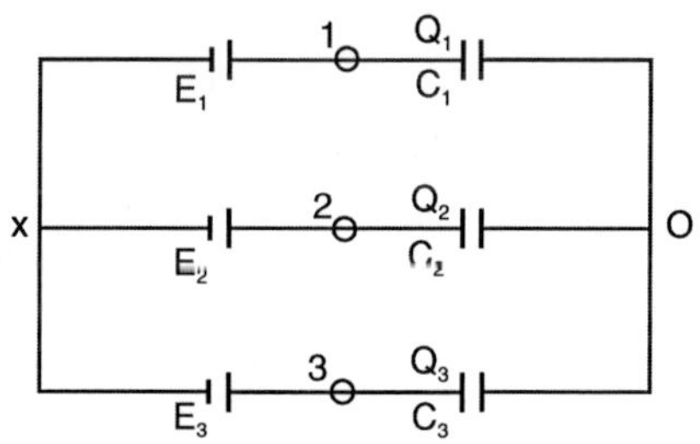

Fig. 11.98

Solution: Let us choose a reference point *x*. In view of the circuit, we have

$$V_x - O = E_1 - \frac{Q_1}{C_1} = E_2 - \frac{Q_2}{C_2} = E_3 - \frac{Q_3}{C_3}$$

or $\quad V_xC_1 = E_1C_1 - Q_1, V_xC_2 = E_2C_2 - Q_2, V_xC_3 = E_3C_3 - Q_3$

$\therefore \quad V_x = \dfrac{E_1C_1 + E_2C_2 + E_3C_3 - (Q_1+Q_2+Q_3)}{C_1+C_2+C_3}$

or $\quad V_x = \dfrac{E_1C_1 + E_2C_2 + E_3C_3}{C_1+C_2+C_3}$

since at the point *O*, $Q_1 + Q_2 + Q_3 = 0$. Let ϕ_1, ϕ_2 and ϕ_3 be the potentials at points 1, 2 and 3. Then we have

$$V_x - \phi_1 = E_1 \text{ or } \phi_1 = V_x - E_1 = \frac{E_1C_1 + E_2C_2 + E_3C_3}{C_1 + C_2 + C_3} - E_1$$

$$= \frac{E_2C_2 + E_3C_3 - E_1(C_2 + C_3)}{C_1 + C_2 + C_3}$$

Similarly

$$\phi_2 = V_x - E_2 = \frac{E_1C_1 + E_3C_3 - E_2(C_1 + C_3)}{C_1 + C_2 + C_3}$$

and $$\phi_3 = V_x - E_3 = \frac{C_1E_1 + C_2E_2 - E_3(C_1 + C_2)}{C_1 + C_2 + C_3}$$

On substituting the values, we have

$$\phi_1 = \frac{4 + 9 - 5}{6} = \frac{4}{3}\text{V}$$

$$\phi_2 = \frac{1 + 9 - 2 \times 4}{6} = \frac{1}{3}\text{V}$$

$$\phi_3 = \frac{1 + 4 - 3 \times 3}{6} = -\frac{2}{3}\text{V}$$

11.153. A condenser of capacity, $C = 2\ \mu\text{F}$ is connected to two cells of emf $E_1 = 3\text{V}$ and $E_2 = 2\text{V}$ through a switch S as shown in Fig. 11.99. Calculate the amount of heat generated in the circuit when the switch S is shifted from position a to position b.

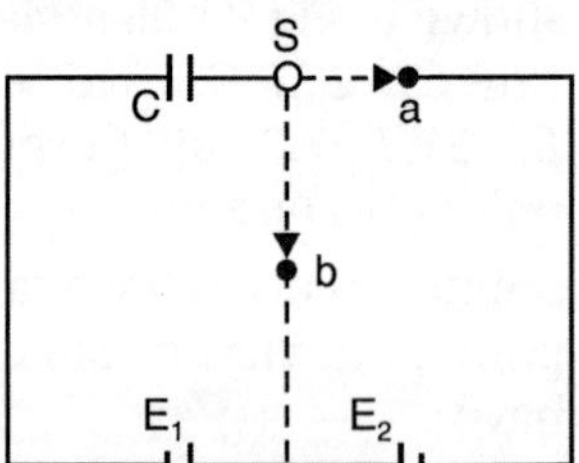

Fig. 11.99

Solution: When the switch S is in position a, the condenser will be in the charging state and the total charge on it = $C(E_1 - E_2)$. On turning the switch S to position b the charges left on capacitor equal to CE_1. The amount of the charge displaced = $CE_1 - C(E_1 - E_2) = CE_2$.

$$\text{Amount of heat generated} = \frac{1}{2}\frac{C^2E_2^2}{C} = \frac{1}{2}CE_2^2$$

On substituting the values, we have

$$\text{Heat generated} = \frac{1}{2} \times 2 \times 10^{-6} \times 4 = 4 \times 10^{-6}\text{J}$$

11.154. Two condensers of capacity $C = 2\mu\text{F}$ and $C_0 = 1\mu\text{F}$ are connected to a cell of emf $E = 10$ volt through a switch S as shown in Fig. 11.100a. Calculate the amount of heat generated in the circuit when the switch S is shifted from position a to position b.

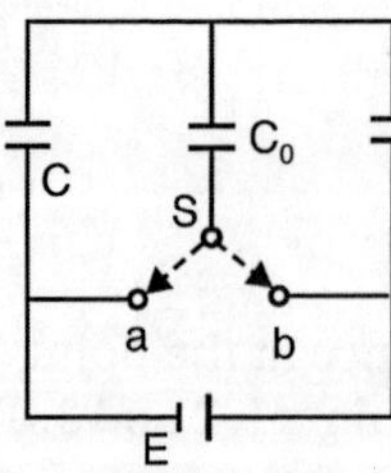

Fig. 11.100a

Solution: When the switch S is in positions a and b respectively we come across to circuits as shown in Figs. 11.100b and c.

In the first case we have

$$\frac{Q_1}{C} = \frac{Q_2}{C_0} \text{ or } Q_1 = \frac{C}{C_0} Q_2$$

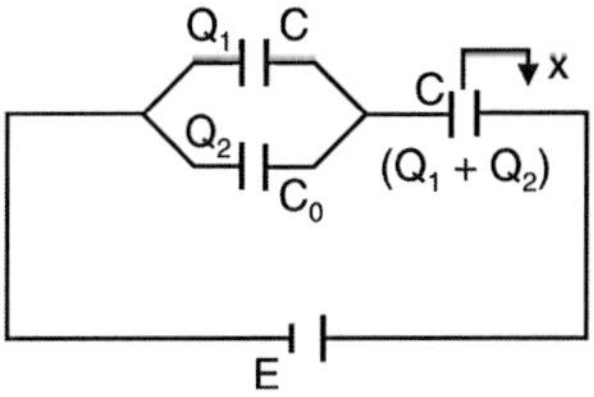

Fig. 11.100b

$$\text{Total capacity} = \frac{C(C+C_0)}{2C+C_0} \text{ Total}$$

$$\text{Charge } Q_1 + Q_2 = \frac{C(C+C_0)}{2C+C_0} E$$

and $Q_1 = C^2 E / 2C + C_0$

Initial energy of the capacitor marked x is given by

$$\frac{1}{2}\frac{(Q_1+Q_2)^2}{C} = \frac{1}{2}\frac{C^2(C+C_0)^2 E^2}{C(2C+C_0)^2}$$

$$\text{Final energy of the capacitor marked } x = \frac{1}{2} Q_1^2 / C = \frac{1}{2}\frac{C^4 E^2}{C(2C+C_0)^2}$$

(see Fig. 11.100c).

$$\text{Change in the energy} = \frac{1}{2}\frac{CE^2}{(2C+C_0)^2}\left[(C+C_0)^2 - C^2\right]$$

$$= \frac{1}{2}\frac{CE^2C_0(2C+C_0)}{(2C+C_0)^2} = \frac{1}{2}\frac{CC_0E^2}{2C+C_0}$$

Similarly the other capacitor of capacity C will also involve a change of energy equal to $\dfrac{CC_0E^2}{2(2C+C_0)}$

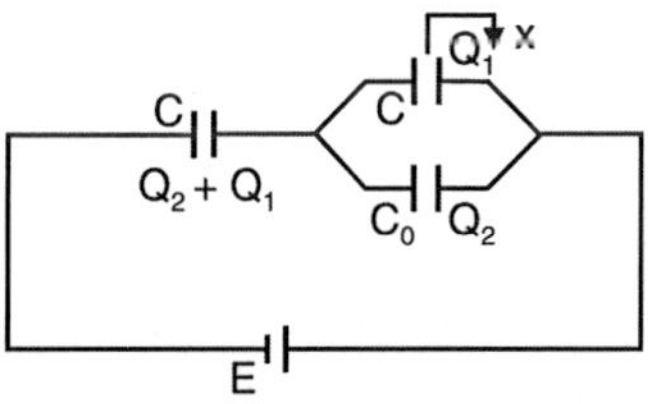

Fig. 11.100c

Hence the net change in the energy H which will be in the form of heat is given by $H = \dfrac{CC_0E^2}{(2C+C_0)}$

On substituting the values of the various terms, we have

$$H = \frac{2\times10^{-6}\times10^{-6}\times10^2}{5\times10^{-6}} = 40\times10^{-6}\text{ J}$$

11.155. At what value of the capacitance C_x in the circuit shown in Fig. 11.101a will the total capacitance between points P and Q be independent of the number of cells. Assume $C = 5$ μF.

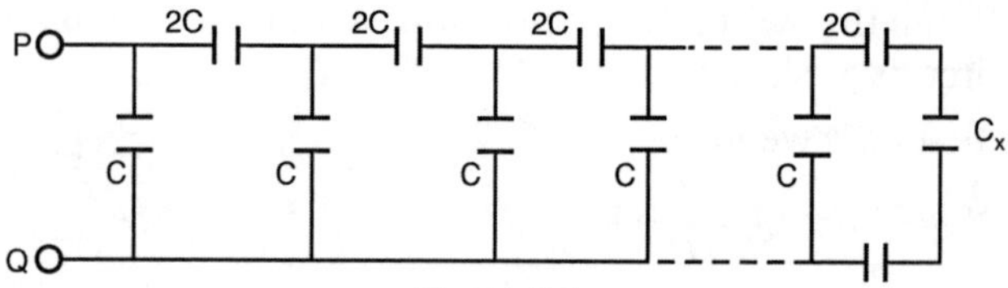

Fig. 11.101a

Solution: Form two circuits consisting of one cell and two cells as shown in Figs. 11.101b and c. Equivalent capacitance between P and Q of

Fig. 11.101b $= \dfrac{3CC_x + 2C^2}{2C + C_x}$

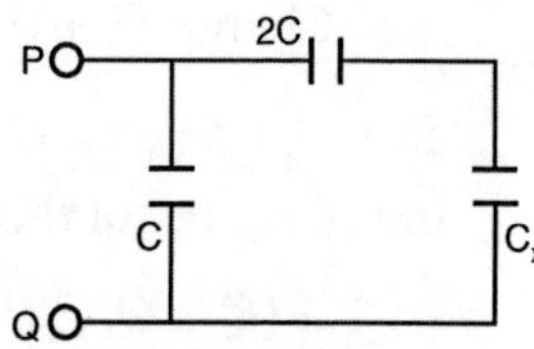

Fig. 11.101b

Equivalent capacitance between P and Q of Fig. 11.101c

$$= \frac{10C^2 + 11CC_x}{6C + C_x}$$

These two capacitance must be equal, i.e.,

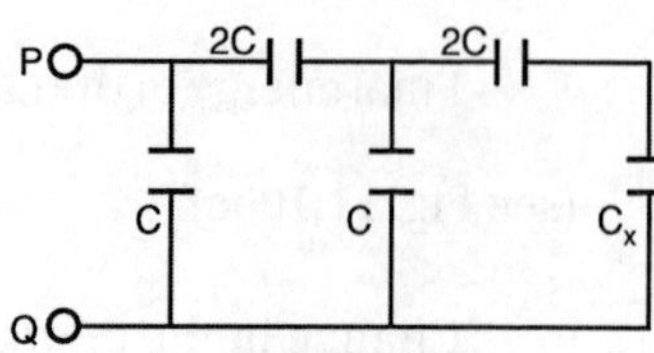

Fig. 11.101c

$$\frac{3CC_x + 2C^2}{2C + C_x} = \frac{10C^2 + 11CC_x}{6C + 5C_x}$$

or $$\frac{3C_x + 2C}{2C + C_x} = \frac{10C + 11C_x}{6C + 5C_x}$$

or $$18CC_x + 15C_x^2 + 12C^2 + 10CC_x = 20C^2 + 22CC_x + 10CC_x + 11C_x^2$$

or $$C_x^2 - CC_x - 2C^2 = 0$$

The only permissible value of $C_x = 2C$.

On substituting $C = 5$ μF, we have

$C_x = 2 \times 5 \times 10^{-6} = 10 \times 10^{-6}$ F.

11.156. A parallel-plate capacitor has square plates of length $a = 2$ cm. The separation $t = 2$ mm between the plates is filled with a material whose dielectric constant varies uniformly from one plate to the other. Find the capacity of the parallel-plate capacitor if $K_1 = 2$ and $K_2 = 2.24$ are the values of dielectric constant at the two plates.

Solution: The electric field inside a parallel-plate capacitor filled with dielectric is given by

$$E = \frac{\sigma}{\varepsilon_0 K}$$

where σ = charge per unit area and K is dielectric constant.

Change of the dielectric constant per unit length $= \dfrac{K_2 - K_1}{t} = \alpha$

The dielectric constant K at a distance x from the plate where the dielectric constant is K_1 is given by $K = K_1 + \alpha x$.

$$\therefore E = \frac{\sigma}{\varepsilon_0 (K_1 + \alpha x)}$$

Potential difference between the plates $= \int_0^t E \cdot dx$

$$\therefore V = \frac{\sigma}{\varepsilon_0} \int_0^t \frac{dx}{(K_1 + \alpha x)}$$

$$= \frac{\sigma}{\alpha \varepsilon_0} \log_e \frac{K_1 + \alpha t}{K_1}$$

or $$V = \frac{\sigma}{\alpha \varepsilon_0} \cdot \log_e \frac{K_2}{K_1}$$

$$= \frac{\sigma t}{(K_2 - K_1)\varepsilon_0} \cdot \log_e \frac{K_2}{K_1}$$

$$V = \frac{Qt}{(K_2 - K_1)\varepsilon_0 A} \cdot \log_e \frac{K_2}{K_1};$$ where Q is the charge and A is the area of the plate.

$$\therefore C = \frac{(K_2 - K_1)\varepsilon_0 A}{t \cdot \log_e \frac{K_2}{K_1}} = \frac{(K_2 - K_1)\varepsilon_0 A}{2.303 \cdot t \cdot \log_{10} \frac{K_2}{K_1}}$$

On substituting the values of the various terms, we have

$$C = \frac{(2.24 - 2) \times 8.85 \times 10^{-12} \times 4 \times 10^{-4}}{2.303 \times 2 \times 10^{-3} \times \log_{10} 1.12}$$

$$= \frac{0.24 \times 8.85 \times 2 \times 10^{-13}}{2.303 \times 0.0492} = 3.75 \times 10^{-12}\,\text{F}$$

11.157. In a spherical capacitor the radii a of the inner and b of the outer sphere are 1 and 1.2 cm respectively. Find its capacity when the first half of the radial separation between them is filled with material whose dielectric constant is $k_1 = 2$ and the second half with another material of dielectric constant $k_2 = 3$ as shown in Fig. 11.102.

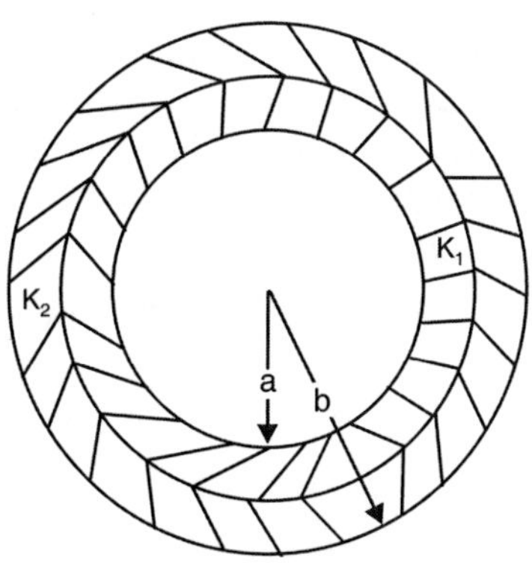

Fig. 11.102

Solution: The potential difference between the inner and the outer sphere of the capacitor is given by

$$V = \int_a^b \frac{Q}{4\pi\varepsilon_0 kr^2} dr$$

$$= \frac{Q}{4\pi\varepsilon_0}\left[\int_a^r \frac{dr}{k_1 r^2} + \int_r^b \frac{dr}{k_2 r^2}\right]$$

$$= \frac{Q}{4\pi\varepsilon_0}\left[-\frac{1}{k_1 r} + \frac{1}{k_1 a} - \frac{1}{k_2 b} + \frac{1}{k_2 r}\right]$$

or $$Q = \frac{4\pi\varepsilon_0 V}{\left[\frac{1}{r}\left(\frac{1}{k_2} - \frac{1}{k_1}\right) + \left(\frac{1}{k_1 a} - \frac{1}{k_2 b}\right)\right]}$$

or $$C = \frac{4\pi\varepsilon_0}{\left[\frac{1}{r}\left(\frac{1}{k_2} - \frac{1}{k_1}\right) + \left(\frac{1}{k_1 a} - \frac{1}{k_2 b}\right)\right]}$$

On substituting the values, we have

$$C = \frac{1}{9\times 10^9\left[\frac{1}{1.1}\times 10^{+2}\left(\frac{1}{3} - \frac{1}{2}\right) + \left(\frac{1}{2\times 10^{-2}} - \frac{1}{3\times 1.2\times 10^{-2}}\right)\right]}$$

$$= \frac{1}{9\times 10^9\left(\frac{100\times -1}{6.6} + 50 - \frac{100}{3.6}\right)}$$

$$= \frac{1}{9\times 10^9\times 7.0708}$$

$$= 15.7\times 10^{-12}\,\text{F}$$

11.158. A parallel-plate condenser consists of two plates each with area A separated by distance l_1. Calculate the amount of work done against electrostatic force in increasing the separation between the plates from l_1 to l_2 when: (i) the charge Q on the capacitor remains constant, and (ii) the voltage V cross the capacitor remains constant.

Solution:

(i) The required work is given by

$$W = \int dw = \int_{l_1}^{l_2} F\,dx = \frac{1}{2}\frac{Q^2}{C_2} - \frac{1}{2}\frac{Q^2}{C_1}$$

$$= \frac{1}{2}\left(\frac{Q^2 l_2}{\varepsilon_0 A} - \frac{Q^2}{\varepsilon_0 A} l_1\right)$$

$$W = \frac{Q^2}{2\varepsilon_0 A}(l_2 - l_1)$$

(ii) $$dw = F \cdot dx = Q \cdot E \cdot dx = CV \cdot \frac{V}{2x} \cdot dx$$

$$= \frac{\varepsilon_0 A}{x} \cdot \frac{V^2}{2x} \cdot dx; \; C = \frac{\varepsilon_0 A}{x} \text{ and } E = \frac{V}{2x}$$

$$= \frac{\varepsilon_0 A V^2}{2} \frac{dx}{x^2}$$

$$\therefore W = \int dw = \int_{l_1}^{l_2} F \cdot dx = \frac{\varepsilon_0 A V^2}{2} \int_{l_1}^{l_2} \frac{dx}{x^2} = \frac{\varepsilon_0 A V^2}{2}\left(\frac{1}{l_1} - \frac{1}{l_2}\right)$$

$$= \frac{\varepsilon_0 A V^2}{2 l_1 l_2}(l_2 - l_1)$$

11.159. Two condensers of capacity $C_1 = 2\mu F$ and $C_2 = 3\mu F$ are connected to two identical cells of emf $E = 5V$ as shown in Fig. 11.103. What charges q_1 and q_2 will flow after the shorting of the switch S in the circuit.

Solution: When the switch S is open, then

$$q_2 = \frac{EC_1C_2}{C_1 + C_2}$$

Now when the switch S is closed, the charge q_2 will flow in the opposite direction and the charge $q_1 = EC_2$ will flow in the circuit in the direction shown in the figure.

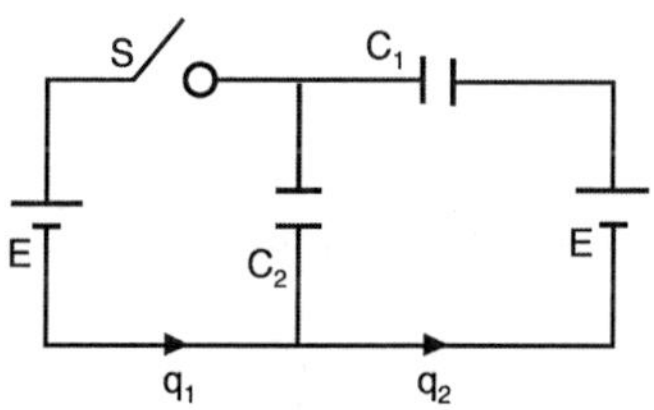

Fig. 11.103

$$q_2 = \frac{5 \times 2 \times 10^{-6} \times 3 \times 10^{-6}}{5 \times 10^{-6}} = 6 \times 10^{-6} = 6\mu C$$

and $q_1 = EC_2 = 5 \times 3 \times 10^{-6} = 15\mu C$

11.160. Two capacitors of capacities $C_1 = 2\mu F$ and $C_2 = 4\mu F$ are connected to two identical cells of emf $E = 45V$ as shown in Fig. 11.104. What charges will flow after the shorting of the switch S through sections 1, 2 and 3 in the direction shown by arrows?

Solution: Before the switch is closed, the total charge flowing in the circuit is determined by the equation.

$$2E = \frac{Q}{C_1} + \frac{Q}{C_2}$$

$$\therefore \quad Q = \frac{2EC_1C_2}{C_1 + C_2}$$

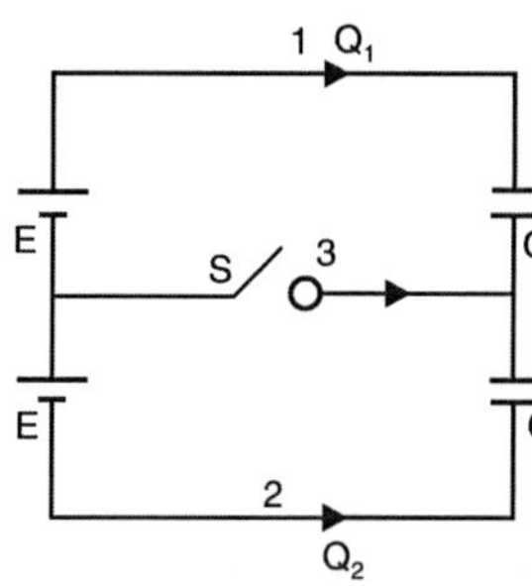

Fig. 11.104

On closing S, the potential difference across $C_1 = E$ and the charge retained by it $= EC_1$.

$\therefore$ Charge Q_1 flowing through section 1 = $EC_1 - \dfrac{2EC_1C_2}{C_1+C_2}$

$$= EC_1\left(\frac{C_1+C_2-2C_2}{C_1+C_2}\right) = -\frac{EC_1(C_2-C_1)}{C_1+C_2}$$

On substituting the values of C_1, C_2 and E, we have

$$Q_1 = \frac{-45\times2\times10^{-6}\times2\times10^{-6}}{6\times10^{-6}} = -30\mu C$$

Similarly charge Q_2 flowing through section 2 = $-\left(EC_2 - \dfrac{2EC_1C_2}{C_1+C_2}\right)$

$$= \frac{-EC_2(C_2-C_1)}{C_1+C_2}$$

On substituting the values of the various terms we have

$$Q_2 = -\frac{45\times4\times10^{-6}\times2\times10^{-6}}{6\times10^{-6}} = -60\ \mu C$$

Charge flowing through section 3 = $E(C_2 - C_1) = 45\times2\times10^{-6}$

$= 90\ \mu C$

11.161. A parallel-plate capacitor of area $A = 200\text{ cm}^2$ is immersed in water of dielectric constant $k = 81$ by keeping it in a horizontal position such that the gap $t = 2$ mm between the plates is filled by it. Now the capacitor is connected to a source of constant voltage $V = 1000$ V. Calculate the force acting on the plate from the dielectric.

Solution: Required force $F' = F_0 - F$

where F is the resultant force = F_0/K and F_0 is the force acting on a plate when there is no dielectric.

$$\therefore \quad F' = F_0\left(1-\frac{1}{k}\right) = F_0\frac{(k-1)}{k}$$

$$= \frac{Q^2}{2\varepsilon_0 A}\left(\frac{k-1}{k}\right) = \frac{C^2V^2(k-1)}{2\varepsilon_0 A\cdot k} = \frac{\varepsilon_0^2A^2k^2V^2(k-1)}{t^2\cdot 2\varepsilon_0 A\cdot k}$$

$$= \frac{\varepsilon_0 Ak(k-1)\cdot V^2}{2t^2}$$

On substituting the values, we have

$$F' = \frac{8.85\times10^{-12}\times200\times10^{-4}\times81\times80\times10^{6}}{2\times4\times10^{-6}} = 143.37\text{ N}$$

11.162. A parallel-plate capacitor with square plates of length $a = 10$ cm separated by distance $d = 2$ mm touches the surface of water contained in a vessel (Fig. 11.105). The capacitor is fully charged by connecting its plates to a $v = 1{,}000$ V source and then disconnected from it. To what height will the liquid rise? Dielectric constant of water $k = 81$.

Solution: The liquid close to the edges of the two plates is polarised. As a consequence of this it will be pulled into the space between the two plates. The capacity of the capacitor thus formed will increase which in turn will be accompanied by decrease in energy. According to conservation of energy the loss in energy is balanced by the potential energy of the liquid column.

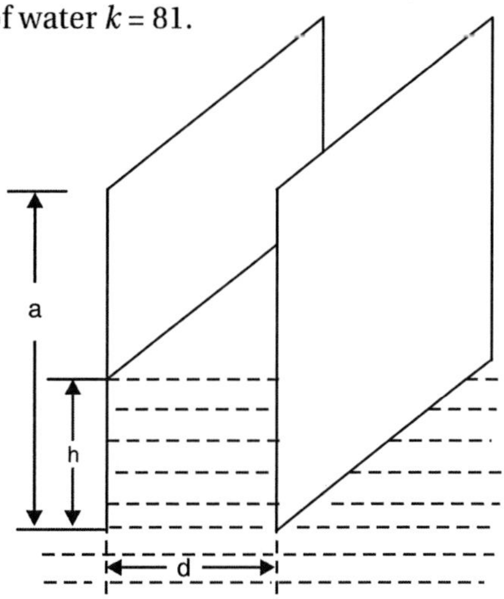

Fig. 11.105

Now $\dfrac{q^2}{2C_1} = \dfrac{q^2}{2C_2} + \dfrac{mgh}{2}$

where $C_1 = \dfrac{\varepsilon_0 a^2}{d}, C_2 = \dfrac{\varepsilon_0 a}{d}\left[a + (k-1)h\right]$

C_2 is the parallel combination of two capacitors

$$C_2 = \frac{\varepsilon_0 a(a-h)}{d} + \frac{k\varepsilon_0 ah}{d}$$

$m = \rho\, ahd$, where ρ = density of water

Now $\dfrac{q^2 d}{2\varepsilon_0 a^2} = \dfrac{q^2 d}{2\varepsilon_0 a\left[a + (k-1)h\right]} + \dfrac{\rho ahd \cdot gh}{2}$

or $\dfrac{q^2}{\varepsilon_0 a}\left[\dfrac{1}{a} - \dfrac{1}{\left[a + (k-1)h\right]}\right] = \rho ah^2 g$ or $\dfrac{q^2 (k-1)h}{\varepsilon_0 a^2\left[a + (k-1)h\right]} = \rho ah^2 g$

Now, substituting $q = \dfrac{\varepsilon_0 a^2 v}{d}$ in the above equation we have,

$$\frac{\varepsilon_0^2 a^4 v^2 \cdot (k-1)h}{d^2 \cdot \varepsilon_0 a^2\left[a + (k-1)h\right]} = \rho ah^2 g$$

or $$\frac{\varepsilon_0 av^2 (k-1)}{d^2} = \rho hg\left[a + (k-1)h\right]$$

or $$\frac{\varepsilon_0 av^2 (k-1)}{d^2} = \rho hga + \rho gh^2 (k-1)$$

or $$h^2+\frac{ah}{k-1}-\frac{\varepsilon_0 a v^2}{\rho g d^2}=0$$

$$h=-\frac{a}{2(k-1)}\pm\sqrt{\frac{a^2}{4(k-1)^2}+\frac{\varepsilon_0 a v^2}{\rho g d^2}}$$

$$=-\frac{a}{2(k-1)}+\sqrt{\frac{a^2}{4(k-1)^2}+\frac{\varepsilon_0 a v^2}{\rho g d^2}}$$

On substituting the values of various terms, we have

$$h=-\frac{0.1}{2\times 80}+\sqrt{\frac{10^{-2}}{4\times 80\times 80}+\frac{8.85\times 10^{-12}\times 0.1\times 10^{6}}{10^{3}\times 9.8\times 4\times 10^{-6}}}$$

$$=-\frac{0.1}{2\times 80}+\sqrt{\frac{1}{4\times 64}+\frac{8.85}{9.8\times 4}}\times 10^{-2}$$

$$=-\frac{0.1}{2\times 80}+0.4791\times 10^{-2}$$

$= 0.4166 \times 10^{-2}$ m

$= 0.42$ cm

11.163. In a cylindrical condenser of mean radius $R = 3$ cm the gap of width $t = 3$ mm between the two coaxial cylinders is filled with glass of dielectric constants $k = 6$. The capacitor is connected to a 1000 V supply. Calculate the force pulling the dielectric inside the capacitor.

Solution:

$$\text{Force} = -\frac{d}{dx}\,(w) \text{ keeping } q \text{ constant.}$$

$$=-\frac{d}{dx}\left(\frac{q^2}{2C}\right)$$

$$=\frac{q^2}{2C^2}\frac{dC}{dx}=\frac{v^2}{2}\cdot\frac{dC}{dx}$$

If the depth of the dielectric be x and the length of the capacitor be l, then

$$C=\frac{k\cdot\varepsilon_0\, x\cdot 2\pi R}{t}+\frac{\varepsilon_0 (l-x)\cdot 2\pi R}{t}=\frac{\varepsilon_0 2\pi R}{t}\left[l+(k-1)x\right]$$

$$\therefore \quad \frac{dC}{dx}=\frac{\varepsilon_0\cdot 2\pi R}{t}(k-1)$$

$$\therefore \quad F=\frac{\varepsilon_0\cdot 2\pi R v^2\,(k-1)}{2t}=\frac{\varepsilon_0\,(k-1)\pi R v^2}{t}$$

On substituting the values of the various terms, we have

$$F = \frac{8.85\times 10^{-12}\times 5\times 3.14\times 3\times 10^{-2}\times 10^{6}}{3\times 10^{-3}}$$

$$= 8.85\times 5\times 3.14\times 10^{-5}\text{ N}$$

$$= 1.389\times 10^{-3}\text{ N}$$

11.164. A capacitor is formed by using two fixed semicircular plates of radius $R = 5$ cm and a movable semicircular sheet of glass of thickness $t = 2$ mm and dielectric constant $k = 6$. The capacitor is connected to a 1000 V supply. Calculate the moment of the force acting on the glass plate about an axis passing through the centre of the semicircular plate O, when it makes an angle θ as shown in Fig. 11.106.

Solution: On rotating the plate through an angle $d\theta$ the energy of the system will decrease. When the charge q is constant, then we have

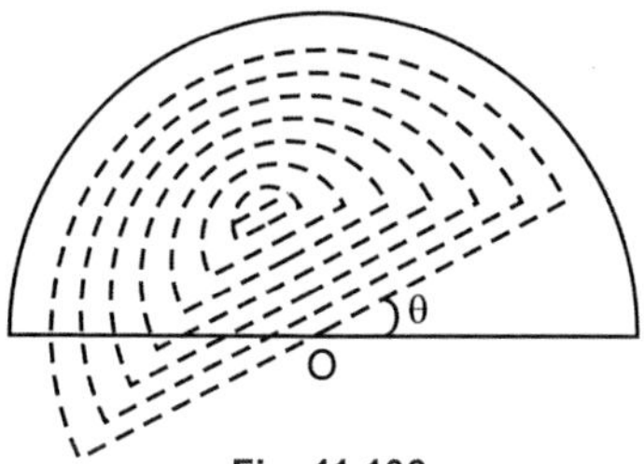

Fig. 11.106

$$\tau \cdot d\theta = -w$$

or $$\tau = -\frac{dw}{d\theta} = -\frac{d}{d\theta}\left(\frac{q^2}{2c}\right) = \frac{q^2}{2c^2}\cdot\frac{dc}{d\theta}$$

where c is the total capacity equal to $c_1 + c_2$, c_1 is the capacity with dielectric and c_2 = capacity of the capacitor without dielectric. Area described with angle $\theta = \theta R^2/2$.

$$\therefore \quad c = \frac{k\varepsilon_0(\pi-\theta)R^2}{2t} + \frac{\varepsilon_0\theta R^2}{2t};\; c_1 = \frac{k\varepsilon_0(\pi-\theta)R^2}{2t} \text{ and } c_2 = \frac{\varepsilon_0\theta R^2}{2t}$$

$$\therefore \quad \frac{dc}{d\theta} = \frac{\varepsilon_0 R^2}{2t}(1-k)$$

$$\therefore \quad \tau = -\frac{c^2V^2}{2c^2}\cdot\frac{\varepsilon_0 R^2}{2t}(k-1)$$

$$= -\frac{(k-1)V^2\cdot\varepsilon_0 R^2}{4t}$$

On substituting the values of the various terms, we have

$$\tau = -\frac{5\times 10^{6}\times 8.85\times 10^{-12}\times 25\times 10^{-4}}{8\times 10^{-3}} = -138.28\times 10^{-7}\text{ N·m}$$

11.165. An isotropic dielectric is filled in the gap between the two plates a and b of a parallel plate capacitor of area A (see Fig. 11.107). The dielectric constant at the two plates a and b is K_1 and K_2 respectively and in the inbetween space it increases uniformly in the direction a to b. Find the space density of the bound charges as a function of dielectric constant when an electric field E is established across the plates in the direction as shown in the Fig. 11.107.

Solution: The dielectric constant of the material at a distance x from the lower plate is given by

$$K = K_1 + \frac{K_2 - K_1}{d}x;\ d = \text{separation between the plates.}$$

$$\frac{dK}{dx} = \frac{K_2 - K_1}{d}$$

We know that $q\left(1 - \frac{1}{K}\right) = q'$; q being the charge on the capacitor

$$+\frac{q}{K^2}\cdot\frac{dK}{dx} = \frac{dq'}{dx}$$

$$\frac{q}{K^2}\frac{(K_2 - K_1)}{d} = \frac{dq'}{dx}$$

$$\sigma' = \frac{dq'}{dx\cdot A} = \frac{q}{K^2}\frac{(K_2 - K_1)}{d\cdot A}$$

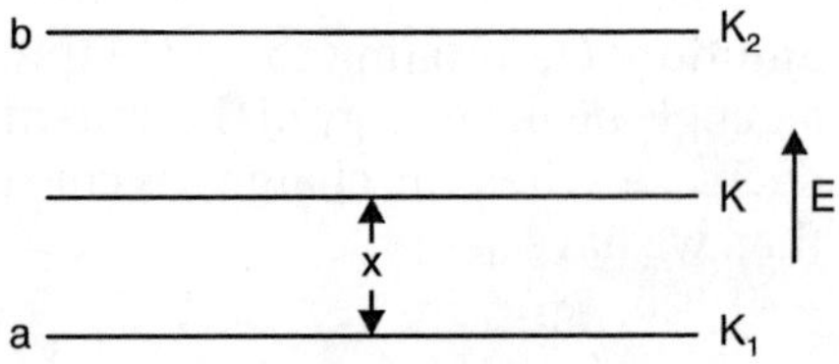

Fig. 11.107

11.166. A capacitor of capacity C has plates separated by a distance. The shape of the plates is arbitrary. A poorly conducting medium of resistivity ρ and dielectric constant K is filled up in between the plates. Calculate the value of the leakage current when a voltage V is applied across it.

Solution:

$$Q = CVe^{-t/RC}$$

$$I = \frac{CV}{RC} = \frac{V}{R};\quad R = \frac{\rho d}{A}\quad A = \text{area of the plate}$$

d = separation between the plates

$$= \frac{V\cdot A}{\rho\cdot d} = \frac{CV}{K\rho\varepsilon_0};\ C = \frac{K\varepsilon_0 A}{d}$$

11.167. The space between the two plates of a parallel plate capacitor is filled with glass of dielectric constant $K = 6$ and resistivity $\rho = 100$ G. Ω. m. The capacity of the capacitor with dielectric is $C = 5$ nF. Calculate the leakage current of the capacitor when a dc voltage of 3×10^3 volt is applied across it.

Solution: The system presents a RC circuit in which the leakage current is observed during discharge

$$Q = CVe^{-t/RC}$$

$$I = \frac{dQ}{dt} = \frac{V}{R} e^{-t/RC}$$

$$\text{at } t = 0, I = \frac{V}{R} = \frac{V}{\rho \frac{d}{A}};$$ d = separation between the plates
A = area of the plates

$$= \frac{VA}{\rho d}$$

$$I = \frac{CV}{K\varepsilon_0 \rho}; \quad C = \frac{K\varepsilon_0 A}{d}$$

On substituting the values we have

$$I = \frac{3\times 10^3 \times 5\times 10^{-9} \times 9\times 10^9 \times 4\pi}{6\times 100\times 10^9}$$

$$= 2.82\ \mu\text{A}$$

11.168. A cylindrical capacitor with mean curvature radius of electrodes equal to $r = 60$ mm and an air gap of $d = 2$ mm between the electrodes is submerged vertically in a container filled with water with a velocity of $v = 9$ m/sec. Calculate the amount of current along the wires when a dc voltage of $V = 250$ volt is maintained across the two electrodes of the capacitor. Dielectric constant of water 81.

Solution: Length covered by the liquid inside the capacitor in time

$$dt = v \cdot dt$$

$$\text{Total capacity} = \frac{K\varepsilon_0 \cdot 2\pi r v dt}{d} + \frac{\varepsilon_0 (l - v dt)\cdot 2\pi r}{d};$$ l = length of the capacitor

$$C = \frac{2\pi\varepsilon_0 r}{d}\left[(K-1) v dt + l\right]$$

$$Q = CV = \frac{2\pi\varepsilon_0 r}{d}\left[(K-1) v dt + l\right] V$$

$$I = \frac{dQ}{dt} = \frac{2\pi\varepsilon_0 r}{d}(K-1) v V$$

$$= \frac{1}{2\times 9\times 10^9} \times 60\times 10^{-3} \times 80\times 9\times 250$$

$$= 0.6\ \mu\text{a}$$

11.169. A cylindrical capacitor having mean curvature radius of the electrodes $R = 5$mm and an air gap of $d = 2$ mm touches the surface of water of dielectric constant $K = 81$ with its end. Calculate the height upto which water will rise into the gap between the two cylinders when a dc voltage of $V = 210$ volt is maintained across the electrodes of the capacitor.

Solution:

$$F = -\left(\frac{dW}{dx}\right)_q = -\frac{d}{dh}\left(\frac{q^2}{2C}\right) = \frac{q^2}{2C^2}\cdot\frac{dC}{dh} = \frac{V^2}{2}\frac{dC}{dh}$$

Let the length of water column be h and the length of capacitor be l then

$$C = \frac{K\varepsilon_0\cdot 2\pi Rh}{d} + \frac{\varepsilon_0(l-h)2\pi R}{d} = \frac{\varepsilon_0 2\pi R}{d}[Kh+l-h]$$

$$\frac{dC}{dh} = \frac{2\pi R\varepsilon_0(K-1)}{d}$$

$$\text{Force } F = \frac{\varepsilon_0(K-1)\pi RV^2}{d}$$

$$\text{Pressure} = \frac{F}{2\pi Rd} = \frac{\varepsilon_0(K-1)\pi RV^2}{2\pi Rd^2} = h\rho g$$

$$\therefore h = \frac{\varepsilon_0(K-1)V^2}{2\rho g\cdot d^2}$$

On substituting the values of various terms occurring in the above equation

$$h = \frac{80\times 210^2}{4\pi\times 9\times 10^9\times 2\times 10^3\times 9.8\times 4\times 10^{-6}}$$

$$= 0.4\text{ mm}$$

11.170. A poorly conducting medium is filled in the space between the two spheres of a spherical condenser of capacity C with inner and outer radii of the spheres as r_1 and r_2 respectively. The capacitor is fully charged through an external dc source. After the charging is complete it is disconnected from the source. For what value of the resistivity of the material the potential difference across the capacitor reduces to half of its value in an interval of time Δt?

Solution:

$$Q = Q_0 e^{-t/RC}$$

$$CV = Q_0 e^{-t/RC}$$

$$CdV = Q_0 e^{-t/RC}\times -\frac{dt}{RC} = -\frac{CVdt}{RC}$$

$$\int_V^{V/2}\frac{dV}{V} = \int_0^{\Delta t} -\frac{dt}{RC}$$

$$\log_e 2 = \frac{\Delta t}{RC}$$

$$\therefore R = \frac{\Delta t}{C \log_e 2}$$

Consider a spherical shell of radius x and $x + dx$.

$$\text{Resistance of the shell } dR = \frac{\rho dx}{4\pi x^2}$$

$$R = \int dR = \rho \int_{r_1}^{r_2} \frac{dx}{4\pi x^2}$$

$$R = \frac{\rho}{4\pi}\left[\frac{r_2 - r_1}{r_1 r_2}\right]$$

$$\frac{\rho}{4\pi}\left[\frac{r_2 - r_1}{r_1 r_2}\right] = \frac{\Delta t}{C \log_e 2}$$

$$\therefore \rho = \frac{4\pi r_1 r_2 \Delta t}{(r_2 - r_1) C \log_e 2}$$

11.171. The square plates of a parallel plate capacitor of length l are clamped by keeping the plates d apart. A potential difference of V volt is applied across the plates with the help of a battery and a dielectric of thickness d and dielectric constant K is inserted into the capacitor (see Fig. 11.108). Calculate the force with which the dielectric will be attracted into the capacitor.

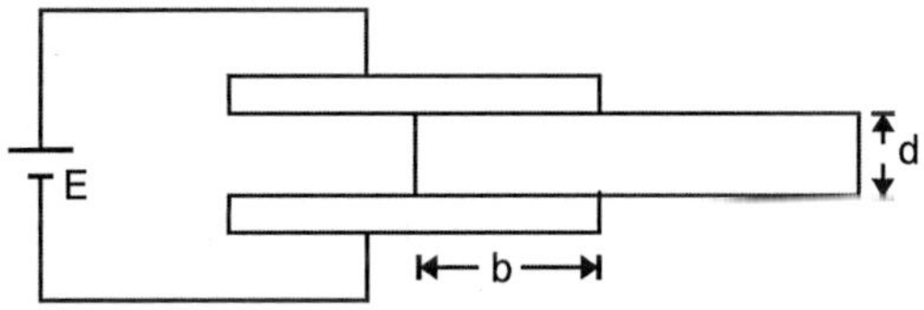

Fig. 11.108

Solution: Let the slab occupies a length x inside the capacitor. Let us displace the slab by dx. The capacity will change and due to this there will be a change in the charge of the capacitor by an amount dq.

Work done by the battery $= Vdq$

Total change in stored energy $= dU = F \cdot dx + Vdq$

$$U = \text{Total energy} = \frac{1}{2}\varepsilon_0 E^2 \cdot l \cdot dx + \frac{1}{2}K\varepsilon_0 E^2 l \cdot d(l - x);$$

$$E = \text{electric field} = \frac{V}{d}$$

$$U = \frac{1}{2}\varepsilon_0 E^2 \left[Kl - (K - l)x\right] l \cdot d$$

$$dU = -\frac{1}{2}\varepsilon_0 E^2 (K - l) l \cdot d \cdot dx$$

$$q = \sigma_1 A_1 + \sigma_2 A_2 = \varepsilon_0 E \cdot x \cdot l + K\varepsilon_0 El(l-x)$$
$$dq = \varepsilon_0 El\,dx - K\varepsilon_0 El \cdot dx$$
$$= -(K-1)\,\varepsilon_0 E \cdot l \cdot dx$$
$$F \cdot dx = -\frac{1}{2}\varepsilon_0 E^2 (K-1)\, l \cdot d \cdot dx + V(K-1)\,\varepsilon_0 E \cdot l \cdot dx$$
$$F = V(K-1)\varepsilon_0 \frac{V}{d} \cdot l - \frac{1}{2}\varepsilon_0 \frac{V^2}{d}(K-1)l$$
$$= \frac{1}{2}\frac{V^2 (K-1)\varepsilon_0 l}{d}$$

11.172. The square plates of a parallel plate capacitor of length l are properly clamped and are connected to a cell of emf E. A slab of mass m and thickness d equal to plate separation and dielectric constant K is introduced between the plates such that its length b is inside the capacitor. The slab is released from rest. Show that its motion is not simple harmonic. Calculate the time period of its motion.

Solution: The force with which the dielectric will now move inside the capacitor is

$$F = \frac{1}{2}\frac{E^2\varepsilon_0 l(K-1)}{d}$$
$$\frac{md^2x}{dt^2} = \frac{1}{2}\frac{E^2\varepsilon_0 l(K-1)}{d}$$
$$\frac{d^2x}{dt^2} = \frac{1}{2}\frac{E^2\varepsilon_0 l^2(K-1)}{ml \cdot d}$$

Here the acceleration is not proportional to displacement hence it will not execute simple harmonic motion.

On integrating the above equation we have

$$\frac{dx}{dt} = \frac{1}{2}\frac{E^2\varepsilon_0 l(K-1)t}{md} + C$$

at $t = 0$, $\frac{dx}{dt} = 0 \therefore C = 0$

$$\therefore \frac{dx}{dt} = \frac{1}{2}\frac{E^2\varepsilon_0 l(K-1)t}{m \cdot d}$$

A further integration shows

$$x = \frac{1}{4}\frac{E^2\varepsilon_0 l(K-1)t^2}{m \cdot d} = (l-b)$$
$$\therefore t = \sqrt{\frac{4 \cdot m \cdot d(l-b)}{E^2\,\varepsilon_0 l(K-1)}}$$

11.173. A leaky parallel plate capacitor is filled completely with a material having dielectric constant $K = 5$ and electrical conductivity $\sigma = 7.4 \times 10^{-12}\ \Omega^{-1}\ m^{-1}$. If the charge on the plate at the instant $t = 0$ is $q = 8.85\ \mu C$, then calculate the leakage current at the instant $t = 12$ sec.

Solution: In case of discharge of a capacitor through a resistance R.

$$Q = Q_0 e^{-t/RC}$$

$$i = -\frac{dQ}{dt} = \frac{Q_0}{RC} e^{-t/RC}$$

$$RC = \frac{\varepsilon_0 KA}{d} \times \frac{1}{\sigma} \cdot \frac{d}{A} = \frac{\varepsilon_0 K}{\sigma}$$

On putting the values

$$RC = \frac{5 \times 8.85 \times 10^{-12}}{7.4 \times 10^{-12}} = 6$$

$$i = \frac{8.85 \times 10^{-6}}{6} \times e^{-12/6} = \frac{8.85 \times 10^{-6}}{6 \times 7.39} = 0.2\,\mu A$$

12

Simple Circuit and Electrolysis

12.1. A truncated right circular conical conductor as shown in the Fig. 12.1 has length l, and resistivity ρ. The end radii of the circular sections are r_1 and r_2 respectively. Find the resistance of the conductor across the two ends.

Solution: Consider an elementary disc of thickness dx at a distance x from one end

$$\tan\theta = \frac{r_2 - r_1}{l}$$

$$\text{Radius of the disc} = r_1 + x\tan\theta$$

$$= \left[r_1 + \frac{x(r_2 - r_1)}{l}\right]$$

Resistance of the element is given by

$$dR = \frac{\rho\, dx}{\pi\left[r_1 + \frac{x(r_2 - r_1)}{l}\right]^2}$$

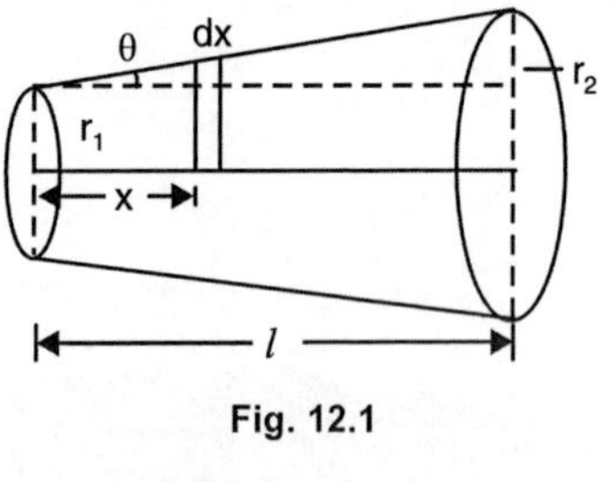

Fig. 12.1

Total resistance R is given by

$$R = \int dR + \frac{\rho}{\pi}\int_0^l \frac{dx}{\left[r_1 + \frac{x(r_2 - r_1)}{l}\right]^2}$$

$$= \frac{\rho l}{\pi r_1 r_2}$$

12.2. Calculate the insulation resistance of a cable of specific resistance ρ, length l having inner and outer radii as r_1 and r_2 respectively.

Solution: Let us consider a thin concentric layer of radius r and $r + dr$.

Surface area = $2\pi rl$

Resistance of the element $dR = \dfrac{\rho\, dr}{2\pi rl}$

Total resistance R is given by

$$R = \int dR = \frac{\rho}{2\pi l}\int_{r_1}^{r_2}\frac{dr}{r} = \frac{\rho}{2\pi l}\cdot\log_e\frac{r_2}{r_1}$$

12.3. A conductor of resistivity ρ has been shaped as a hollow cylinder of length L and inner and outer radii as r_1 and r_2. Find the current flowing through the conductor in a direction parallel to its length when a potential difference V is maintained across the two ends.

Solution: Consider a shell of radius r and $r + dr$

Resistance of the element is given by

$$dR = \frac{\rho L}{2\pi r\,dr}$$

$$\sum\frac{1}{dR} = \frac{2\pi}{\rho L}\int_{r_1}^{r_2} r\,dr$$

$$R = \frac{\rho L}{\pi\left(r_2^2 - r_1^2\right)}$$

$$I = \frac{V\pi\left(r_2^2 - r_1^2\right)}{\rho L}$$

12.4. The opposite sides of a conducting sphere of resistivity ρ and radius r has been grounded in order to obtain plane parallel surfaces separated by a distance L (see Fig. 12.2). Find the resistance of the sphere between the two faces. What will be the resistance of the sphere when L approach $2r$?

Solution: Consider a strip of thickness dx located at a distance x from the centre. The resistance of the strip is given by

$$dR' = \frac{\rho\,dx}{\pi\left(r^2 - x^2\right)}$$

Resistance of half of the sphere

$$R' = \int dR' = \int_0^{L/2}\frac{\rho\,dx}{\pi\left(r^2 - x^2\right)}$$

Let us substitute $x = r\tan\theta$

$$R' = \frac{\rho}{\pi r}\int \sec\theta = \frac{\rho}{\pi r}\log\left(\sec\theta + \tan\theta\right)$$

$$= \frac{\rho}{\pi r}\log_e\left[\frac{r + x}{\sqrt{r^2 - x^2}}\right]_0^{L/2}$$

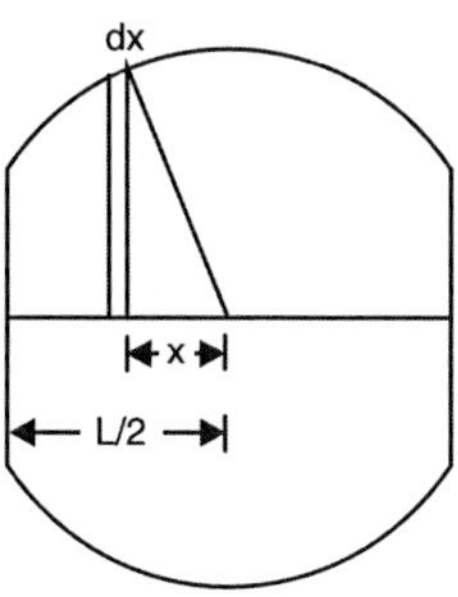

Fig. 12.2

$$= \frac{\rho}{2\pi r}\left|\log_e\left(\frac{r+x}{r-x}\right)\right|_0^{L/2}$$

$$= \frac{\rho}{2\pi r}\log_e\left(\frac{2r+L}{2r-L}\right)$$

The resisance of the full sphere $R = 2R' = \frac{\rho}{\pi r}\log_e\left(\frac{2r+L}{2r-L}\right)$

When $L \to 2r$ then resistance becomes ∞

12.5. A metallic quarter ring has rectangular cross-sectional area as shown in the Fig. 12.3. The inner and outer radii of the ring is r_1 and r_2 respectively. The thickness of the ring is p and the resistivity of the material is ρ. Find the resistance between the face A & B.

Solution: Consider an element of radius r and $r + dr$. The resistance of the element is given by

$$dR = \rho \cdot \frac{2\pi r}{4 \cdot p \cdot dr}$$

$$\sum \frac{1}{dR} = \frac{2p}{\pi\rho}\int_{r_1}^{r_2}\frac{dr}{r}$$

$$\frac{1}{R} = \frac{2p}{\pi\rho}\cdot\log_e\frac{r_2}{r_1}$$

$$\therefore R = \frac{\pi\rho}{2p\log_e\frac{r_2}{r_1}}$$

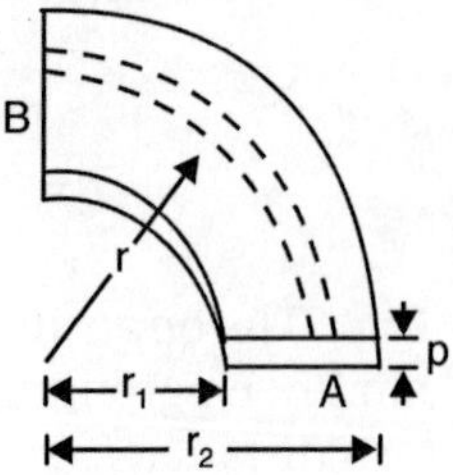

Fig. 12.3

12.6. A wedge shaped material of resistivity ρ is shown in the Fig. 12.4a. Find its resistance between the two faces A & B.

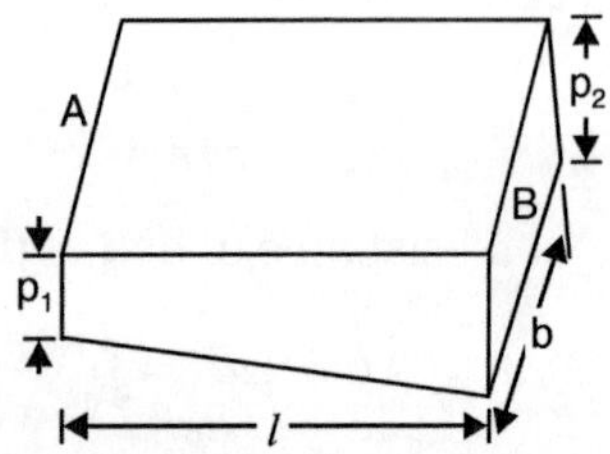

Fig. 12.4a

Solution: Consider an element of thickness dx at a distance x as shown in Fig. 12.4b.

$$\tan\theta = \left(\frac{p_2 - p_1}{l}\right)$$

Resistance of the element is given by

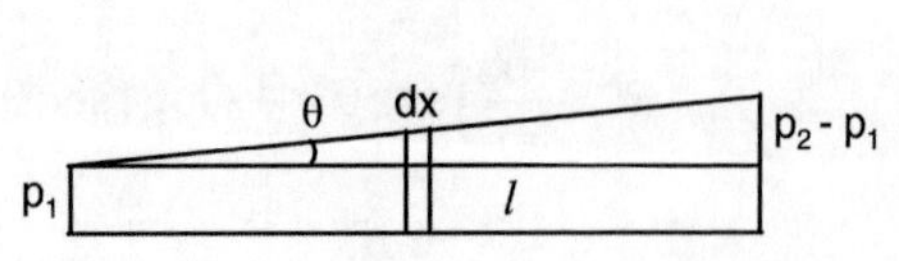

Fig. 12.4b

$$dR = \frac{\rho \cdot dx}{\left[p_1 + \frac{x(p_2 - p_1)}{l}\right]b}$$

Total resistance R is given by

$$R = \int dR = \int_0^l \frac{\rho\, dx}{\left[p_1 + \frac{x(p_2 - p_1)}{l}\right] b}$$

$$= \frac{\rho l}{b(p_2 - p_1)} \log_e \frac{p_2}{p_1}$$

12.7. Two electrodes in the form of metal spheres of radius a are placed in a medium with a resistivity ρ (see Fig. 12.5a). The distance between the sphere centres is b ($b >> a$). Find the resistance between the electrodes.

Solution: The electric field at O due to the two balls is

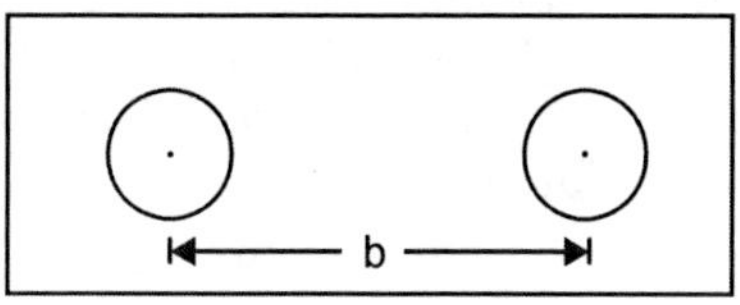

Fig. 12.5a

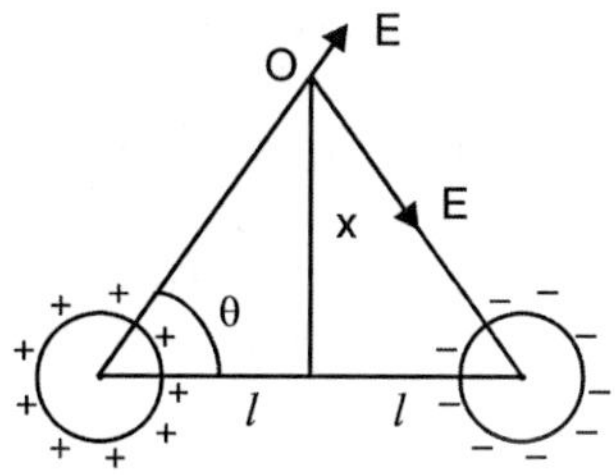

Fig. 12.5b

$$E = \frac{q}{4\pi\varepsilon_0 (x^2 + l^2) K}; \quad q = \text{charge on the sphere}$$

Net field at $O = E_0 = 2\, E \cos\theta = \dfrac{q}{2\pi\varepsilon_0 K} \cdot \dfrac{l}{(x^2 + l^2)^{3/2}}$

According to ohm's law

$$\frac{E_0}{\rho} = j$$

$$j = \frac{ql}{2\pi\varepsilon_0 K\rho (x^2 + l^2)^{3/2}}$$

Consider a ring of radius x and $(x + dx)$, area of the ring $= 2\pi x dx$

$$\text{Current } i = \frac{ql}{2\pi\varepsilon_0 K\rho (x^2 + l^2)^{3/2}} \cdot 2\pi x dx = \frac{ql}{\varepsilon_0 K\rho}\left[-\frac{1}{\sqrt{l^2 + x^2}}\right]_0^\infty = \frac{q}{\varepsilon_0 K\rho}$$

V = Potential difference between the two balls $= \dfrac{q}{4\pi\varepsilon_0 a K} + \dfrac{q}{4\pi\varepsilon_0 a K}$

$$= \frac{q}{2\pi\varepsilon_0 a K}$$

$$R = \frac{V}{i} = \frac{q}{2\pi\varepsilon_0 K a} \times \frac{\varepsilon_0 K\rho}{q} = \frac{\rho}{2\pi a}$$

12.8. A flat ring of thickness d inner radii r_1 and outer radii r_2 is prepared with a material of resistivity ρ. A potential difference V is maintained between the outer and inner surfaces of the ring. Find the resistance of the ring.

Solution: Consider a ring of radius r and width dr. The resistance of the element is given by

$$dR = \frac{\rho \cdot dr}{2\pi r d}$$

$$\text{Net resistance } R = \int_{r_1}^{r_2} \frac{\rho dr}{2\pi d \cdot r} = \frac{\rho}{2\pi d} \log_e \frac{r_2}{r_1}$$

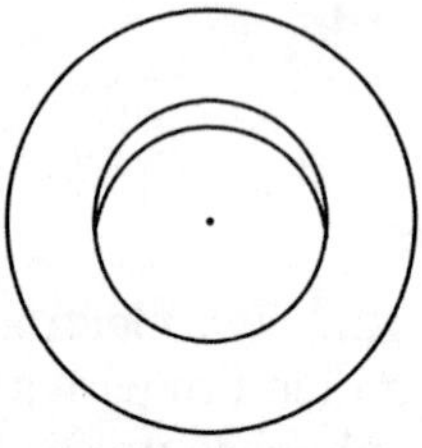

Fig. 12.6

12.9. In a spherical capacitor the outer sphere has potentialities to contract by keeping its spherical shape and also remaining concentric to the inner sphere. A charge q is imparted to such a capacitor. The outer sphere is compressed under the action of the electrical force and as a consequence of that the outer radius a diminishes and attains a value b $(b < a)$. Calculate the amount of work done by the electrical forces.

Solution: Amount of work done by the electrical forces is given by

$$W = \frac{1}{2}\frac{q^2}{C_1} - \frac{1}{2}\frac{q^2}{C_2}$$

where $$C_1 = \frac{4\pi\varepsilon_0 r_1 a}{a - r_1}$$

and $$C_2 = \frac{4\pi\varepsilon_0 r_1 b}{b - r_1} \quad \text{(see Fig. 12.7)}$$

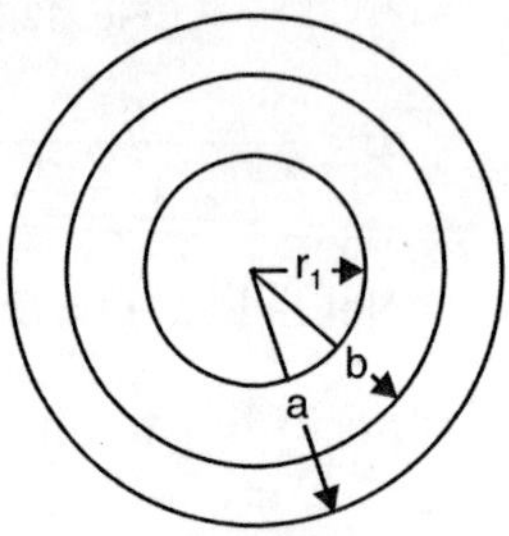

Fig. 12.7

$$W = \frac{q^2}{8\pi\varepsilon_0 r_1 a}(a - r_1) - \frac{q^2 (b - r_1)}{8\pi\varepsilon_0 r_1 b}$$

$$= \frac{q^2 [ab - br_1 - ab + ar_1]}{8\pi\varepsilon_0 r_1 ab}$$

$$= \frac{q^2 (a - b)}{8\pi\varepsilon_0 ab}$$

12.10. A regular star with 5 verticles is formed with the help of uniform resistor wire. The resistance of the part $AC = r\,\Omega$ and part $BD = \frac{2r}{3}\,\Omega$. Find the equivalent resistance between A & B.

Solution: The current distribution is shown in the Fig. 12.8. In the first mesh we have

$$rI_1 + \frac{2r}{3} I_4 = r I_2$$

or $\quad 3I_1 + 2I_4 = 3I_2 \qquad \text{... (1)}$

In the second mesh we have

$$2rI_6 = \frac{2r}{3}[I_2 + I_4 - I_6]$$

or $\quad 4I_6 = I_2 + I_4 \qquad \text{... (2)}$

In the third mesh we have

$$2rI_7 = \frac{2r}{3}[I_2 + I_4 - I_7]$$

or $\quad 4I_7 = I_2 + I_4 \qquad \text{... (3)}$

or $\quad I_6 = I_7 \qquad \text{... (4)}$

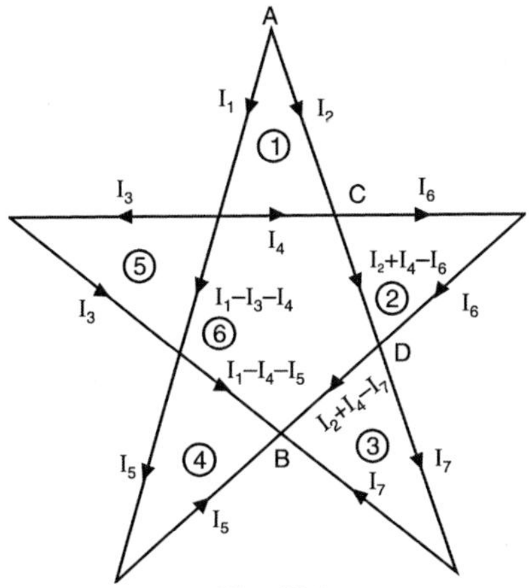

Fig. 12.8

From the 4th & 5th mesh we have

$$4I_5 = I_1 - I_4$$

and $\quad 4I_3 = I_1 - I_4$

or $\quad I_3 = I_5$

In the 6th mesh we have

$$I_4 + I_2 + I_4 - I_6 + I_2 + I_4 - I_7 = I_1 - I_3 - I_4 + I_1 - I_4 - I_5$$

or $\quad 3I_4 + 2I_2 - 2I_6 = 2I_1 - 2I_3 - 2I_4$

or $\quad 2I_1 - 2I_2 - 2I_3 - 5I_4 + 2I_6 = 0$

or $\quad 2I_1 - 2I_2 - 2I_3 - 5I_4 + I_2/2 + I_4/2 = 0$

$$4I_1 - 3I_2 - 4I_3 - 9I_4 = 0$$

$$4I_1 - 3I_2 - 9I_4 = I_1 - I_4$$

$$3I_1 - 3I_2 = 8I_4 - 12I_2 - 12I_1$$

$$15I_1 = 15I_2$$

$\therefore \quad I_1 = I_2, I_4 = 0, I_3 = I_1/4 = I_5, I_6 = I_7 = I_2/4$

For equivalent resistance between A and B we have

$$V = rI_2 + \frac{2r}{3}(I_2 + I_4 - I_6) + \frac{2r}{3}(I_2 + I_4 - I_7)$$

$$= r\left[I_1 + \frac{4}{3}\left(I_1 - \frac{I_1}{4}\right)\right]$$

$$= r[I_1 + I_1] = r \cdot 2I_1$$

$$= rI$$

$\therefore \quad R_{eq} = r\,\Omega$

12.11. Twelve resistor each of resistance r are connected so that each is along one edge of a cube as shown in the Fig. Find the equivalent resistance across a and d.

Solution: The current distribution in the resistances is shown in the Fig. 12.9. Now following the path *abcd* we have

$$V = \frac{I}{3}r + \frac{I}{6}r + \frac{I}{3}r$$

$$= \frac{5}{6}rI$$

$\therefore$ Requirement resistance between a & $d = \frac{5r}{6}\Omega$

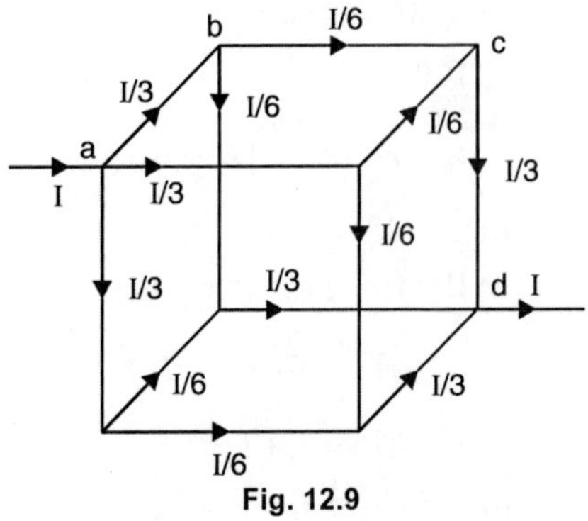

Fig. 12.9

12.12. Twelve resistor each of resistance r are connected so that each is along one edge of a cube as shown in the Fig. 12.10. Find the equivalent resistane across a and b.

Solution: The current distribution is shown in Fig. 12.10. In the first mesh we have

$$rI_2 + rI_3 + rI_2 = rI_1$$

or $\quad 2I_2 + I_3 = I_1 \quad$ ... (1)

In the second mesh we have

$$4r(I_2 - I_3) = rI_3$$

or $\quad 4I_2 = 5I_3 \quad$ ... (2)

or $\quad I_3 = \frac{4}{5}I_2$

Now $I_1 = 2I_2 + \frac{4}{5}I_2 = \frac{14I_2}{5}$

Main current $I = I_1 + 2I_2 = I_1 + \frac{5I_1}{7} = \frac{12I_1}{7}$

Now $V = I_1 r$

$$= \frac{7r}{12}I$$

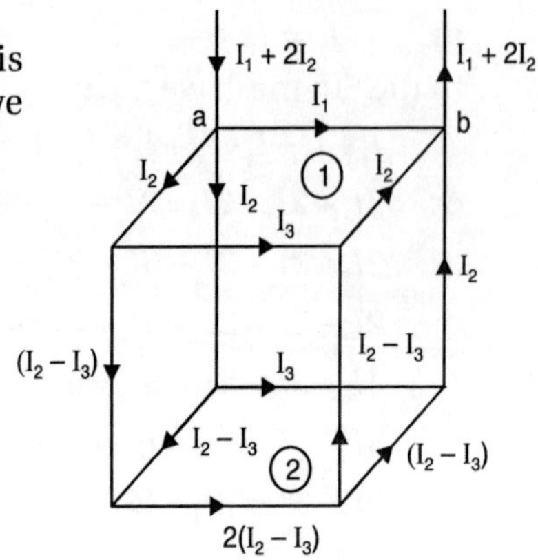

Fig. 12.10

$\therefore$ Equivalent resistance between a & $b = \frac{7r}{12}\Omega$

12.13. Eleven wires each of resistance 5Ω are connected as shown in the Fig. 12.11. Find the equivalent resistance between a and b.

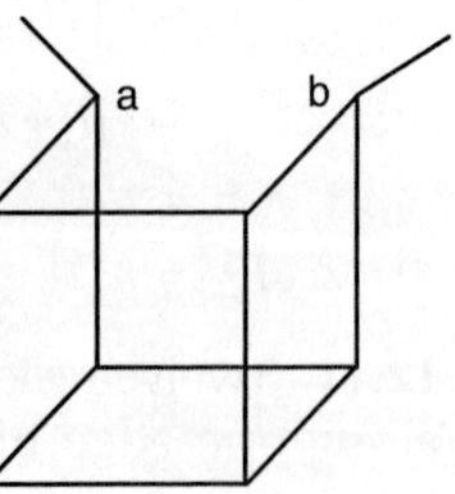

Fig. 12.11

Solution: Let the equivalent resistance be x then in view of Q. No. 12.12 we have

$$\frac{1}{x}+\frac{1}{r}=\frac{12}{7r}$$

$$x=\frac{7r}{5}$$

On putting $r = 5\Omega$

$$x = 7\Omega$$

12.14. Four wires each of resistance $r = 3\Omega$ are connected as shown in the Fig. 12.12. Find the equivalent resistance between points 1 and 2.

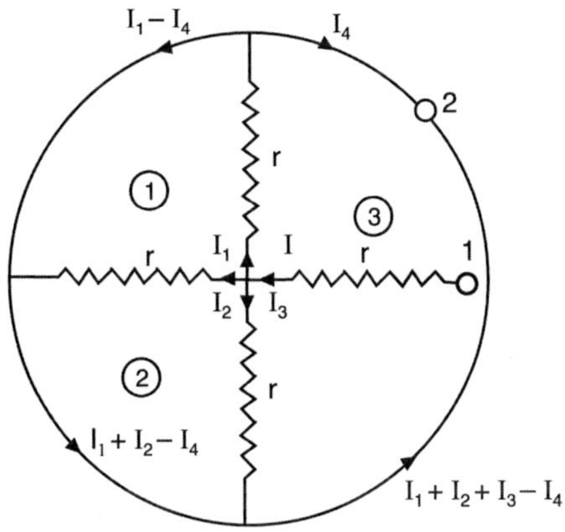

Fig. 12.12

Solution: The current distribution is shown in the Fig. 12.12. In the first mesh we have

$$I_1 r - I_2 r = 0$$

$$\therefore \quad I_1 = I_2$$

From the second and third mesh we have

$$I_2 = I_3 \text{ and } I_1 = I_3$$

Main current $I = I_1 + I_2 + I_3 = 3I_1$

$$\text{Now } V = Ir + I_1 r$$

$$= 4I_1 r$$

$$= \frac{4r}{3}I$$

$\therefore$ Equivalent resistance $= \dfrac{4r}{3}$

On putting $r = 3\Omega$

Equivalent resistance $= \dfrac{4}{3} \times 3 = 4\Omega$

12.15. Twelve wires each of resistance $r = 8\Omega$ are connected as shown in the Fig. 12.13. Find the equivalent resistance between points 1 and 2.

Solution: The current distribution in the various resistances are shown in the Fig. 12.13.

In the first mesh we have

$$rI_2 + rI_5 + rI_5 = rI_1$$

or

$$I_1 = I_2 + 2I_5 \qquad \text{...(1)}$$

In the second mesh we have

$$rI_1 + 2r(I_1 + I_5) = rI_4$$

or $3I_1 + 2I_5 = I_4$ ... (2)

In the third mesh we have

$$rI_2 + 2r(I_2 - I_5) = rI_3$$

or $3I_2 - 2I_5 = I_3$ ... (3)

In the fourth mesh we have

$$rI_3 + r(I_2 + I_3 - I_5) = rI_4 + r(I_1 + I_4 + I_5)$$

or $I_2 + 2I_3 = I_1 + 2I_4 + 2I_5$ ... (4)

or $I_2 + 2I_3 = I_2 + 2I_4 + 4I_5$

or $2I_3 = 6I_1 + 8I_5$

or $2I_3 = 6I_2 + 20I_5$

or $2I_3 = 6I_2 + 30I_2 - 10I_3$

$I_3 = 3I_2$ ... (5)

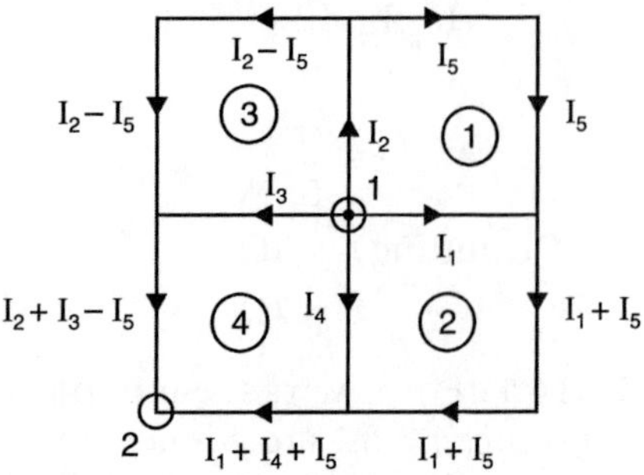

Fig. 12.13

In view of this $I_5 = 0$ and $I_1 = I_2$, $3I_2 = I_3$ and $3I_1 = I_4$

Main current $I = I_1 + I_1 + 3I_1 + 3I_1 = 8I_1$

Now $V = rI_4 + r(I_1 + I_4 + I_5)$

$= r[I_1 + 2I_4]$

$= r[I_1 + 6I_1]$

$= 7r \cdot I_1$

$= \dfrac{7r}{8} I$

or Equivalent resistance $= \dfrac{7r}{8}\,\Omega$

On putting $r = 8\Omega$ we have

Equivalent resistance $= 7\Omega$

12.16. The resistivity of a cylindrical conductor of cross-sectional area A is no longer constant but varies inversely as square of distance r from the axis of conductor as $\rho = \dfrac{a}{r^2}$, where a is a positive constant. Calculate (i) the resistance per unit length of the conductor (ii) the electric field strength in the conductor due to which a current I flows through it.

Solution:

(i) Consider a cylindrical shell of radius r and $r + dr$ and of unit length.

$$\text{Its resistance } dx = \frac{\rho}{2\pi r\,dr} = \frac{a}{2\pi r^3 dr}$$

$$\sum \frac{1}{dx} = \frac{1}{x} = \int_0^R \frac{2\pi r^3 dr}{a};\ R = \text{radius of the conductor}$$

$$\frac{1}{x} = \frac{\pi R^4}{2a} = \frac{A^2}{2\pi a}$$

$$\text{or} \quad x = \frac{2\pi a}{A^2}$$

(ii) $Ix = V(\text{potential per unit length}) = E = \dfrac{2\pi a I}{A^2}$

12.17. A metallic sphere of radius r surrounded by a thin metallic shell of radius R forms the two electrodes. The space between them is filled with a homogenous material of resistivity ρ. Find the resistance of the gap between the two electrodes.

Solution: Consider a shell of radius x and $x + dx$

$$\text{Resistance of the element} = \frac{\rho\,dx}{4\pi x^2}$$

$$\text{Net resistance } R' = \frac{\rho}{4\pi}\int_r^R \frac{dx}{x^2}$$

$$= \frac{\rho}{4\pi}\left[\frac{R-r}{R\cdot r}\right]$$

12.18. A inhomogenous poorly conducting material is filled in the space between the square plates of a parallel plate capacitor. The conductivity of the material varies linearly in a direction perpendicular to its length from σ_1 to σ_2 such that $\sigma_2 > \sigma_1$. Assuming the length of the plate as l and plate separation as d. Find (i) the resistance of the material filled between the two plates (ii) the current through the capacitor due to a potential difference V applied across the plates.

Solution:

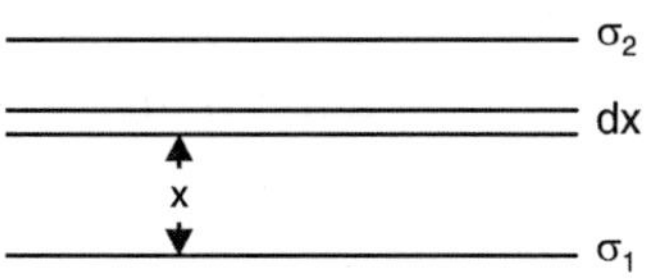

Fig. 12.14

$$\text{Conductivity per unit length} = \frac{\sigma_2 - \sigma_1}{d}$$

Consider a layer of thickness dx at a distance x from the lower surface where the conductivity is σ_1 (See Fig. 12.14).

$$\text{Conductivity at a distance } x = \sigma_1 + \frac{\sigma_2 - \sigma_1}{d}\cdot x$$

$$\text{Resistivity} = \frac{1}{\sigma_1 + \frac{\sigma_2 - \sigma_1}{d} \cdot x}$$

$$\text{Resistance of the element } dR = \frac{dx}{\sigma_1 + \left(\frac{\sigma_2 - \sigma_1}{d}\right) x} \cdot \frac{1}{l^2}$$

Net resistance of the material between the plates

$$R = \int dR = \int_0^d \frac{dx}{l^2 \left[\sigma_1 + \left(\frac{\sigma_2 - \sigma_1}{d}\right) x\right]}$$

$$= \frac{d}{l^2 (\sigma_2 - \sigma_1)} \log_e \frac{\sigma_2}{\sigma_1}$$

(ii) $$I = \frac{V}{R} = \frac{V \cdot l^2 (\sigma_2 - \sigma_1)}{d \cdot \log_e \frac{\sigma_2}{\sigma_1}}$$

12.19. Find the equivalent resistance between the points A & B of the circuit shown in Fig. 12.15.

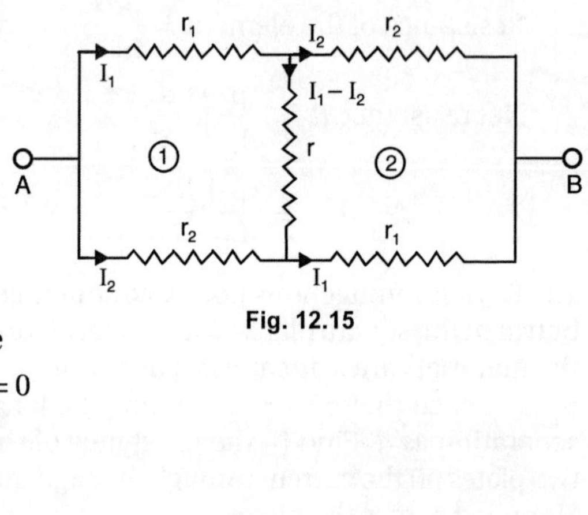

Fig. 12.15

Solution: On the basis of the symmetry considerations the current distribution is shown in the Fig. 12.15.

In the first mesh we have

$$-I_1 r_1 - (I_1 - I_2) r + I_2 r_2 = 0$$

or $$I_2 (r + r_2) = I_1 (r + r_1)$$

or $$I_2 = \frac{I_1 (r + r_1)}{(r + r_2)}$$

$$\text{Main current } I = I_1 + I_2$$

$$= I_1 + \frac{I_1 (r + r_1)}{r + r_2}$$

$$= \frac{I_1 [2r + r_1 + r_2]}{r + r_2}$$

Now $V = I_2 r_2 + I_1 r_1$

$$= \frac{I_1 (r + r_1) r_2}{r + r_2} + I_1 r_1$$

$$= \frac{I_1 [r r_2 + r_1 r_2 + r r_1 + r_1 r_2]}{r + r_2}$$

$$= \frac{I_1 [r(r_1 + r_2) + 2r_1 r_2]}{(r + r_2)}$$

$$= \frac{[r(r_1 + r_2) + 2r_1 r_2]}{(r + r_2)} \frac{(r + r_2) I}{(2r + r_1 + r_2)}$$

$$\therefore \quad \text{Equivalent resistance } R = \frac{[r(r_1 + r_2) + 2r_1 r_2]}{(2r + r_1 + r_2)}$$

12.20. The current I flowing through and the potential difference V across a non ohmic resistor is given by the relation $I = 0.2\ V^2$. This resistor is connected in series to an ohmic resistor and a 6 volt d.c source. For what value of the non ohmic resistor the current in the circuit is 0.4 amp?

Solution: Let R and R' be the values of the resistance of non ohmic and ohmic resistors respectively.

$$R + R' = \frac{6}{0.4} = 15\Omega$$

$$0.4 = 0.2\ V^2$$

$$\therefore V = \sqrt{\frac{0.4}{0.2}} = \sqrt{2} = 0.4 R'$$

$$\therefore R' = \frac{1.414}{0.4} = 3.535\ \Omega$$

$$\therefore R = 15 - 3.535 = 11.465\ \Omega$$

12.21. A material of thickness t and resistivity ρ is enclosed between the two plates AB and CD of a parallel plate capacitor as shown in the Fig. 12.16. The two plates are at an angle ϕ. The curved surfaces AC and BD are having radii r_1 and r_2 respectively. Calculate the current flowing through the material when a constant voltage V is applied across the two plates AB and CD of the capacitor.

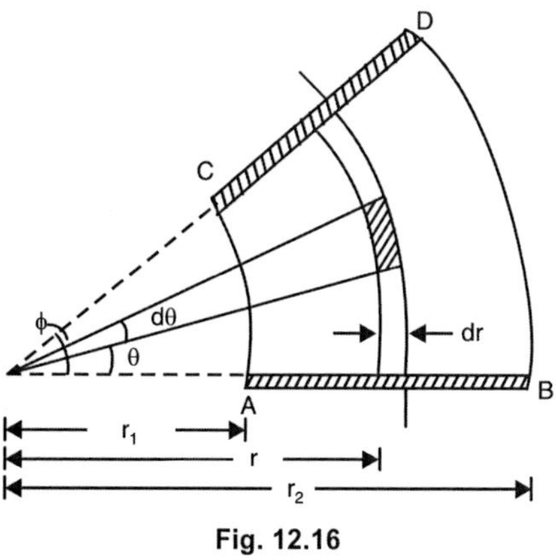

Fig. 12.16

Solution: Consider an element represented by the shaded portion in the Fig 12.16. The resistance dR' of the element is given by

$$dR' = \frac{\rho r d\theta}{t \cdot dr}$$

or $$\frac{1}{dR'} = \frac{t}{\rho d\theta} \cdot \frac{dr}{r}$$

or $$\frac{1}{dR} = \sum \frac{1}{dR'} = \frac{t}{\rho d\theta} \int_{r_1}^{r_2} \frac{dr}{r} = \frac{t}{\rho d\theta} \cdot \log_e \frac{r_2}{r_1}$$

$$\therefore dR = \frac{\rho \cdot d\theta}{t \cdot \log_e \frac{r_2}{r_1}}$$

$$R = \int dR = \int_0^{\phi} \frac{\rho d\theta}{t \cdot \log_e \frac{r_2}{r_1}} = \frac{\rho \phi}{t \cdot \log_e \cdot \frac{r_2}{r_1}}$$

$$I = \frac{V}{R} = \frac{V \cdot t}{\rho \phi} \log_e \frac{r_2}{r_1}$$

12.22. In the above problem if *AC* and *BD* are taken as the plates instead of *AB* and *CD* then what will be the resistance of the material enclosed between the plates.

Solution: In view of the Fig. 12.16 the resistance of the element is given by

$$dR' = \frac{\rho dr}{r d\theta \cdot t}$$

$$dR = \int dR' = \frac{\rho}{d\theta \cdot t} \int_{r_1}^{r_2} \frac{dr}{r}$$

$$dR = \frac{\rho}{d\theta \cdot t} \cdot \log_e \frac{r_2}{r_1}$$

$$\frac{1}{R} = \sum \frac{1}{dR} = \int_0^{\phi} \frac{t \cdot d\theta}{\rho \log_e \frac{r_2}{r_1}}$$

$$\frac{1}{R} = \frac{t \cdot \phi}{\rho \log_e \frac{r_2}{r_1}}$$

or $$R = \frac{\rho \log_e \frac{r_2}{r_1}}{t \cdot \phi}$$

12.23. A homogenous poorly conductive medium of resistivity ρ fills up the space between two thin coaxial ideally conducting cylinders. The radii of cylinders are equal to r_1 and r_2 with $r_1 < r_2$. The length of each cylinder is l. Find the resistance of the medium between the two cylinders.

Solution: Consider a thin cylindrical shell of radius r and $r + dr$. The resistance of the shell is given by

$$dR = \frac{\rho \cdot dr}{2\pi r l}$$

$$\text{Net resistance } R = \frac{\rho}{2\pi l}\int_{r_1}^{r_2}\frac{dr}{r} = \frac{\rho}{2\pi l}\cdot \log_e \frac{r_2}{r_1}$$

12.24. Determine the equivalent resistance between points A and B as shown in Fig. 12.17.

Solution: Let us assume that points A and B are connected to a hypothetical cell. The current distribution in the various resistance is shown in Fig. 12.17.

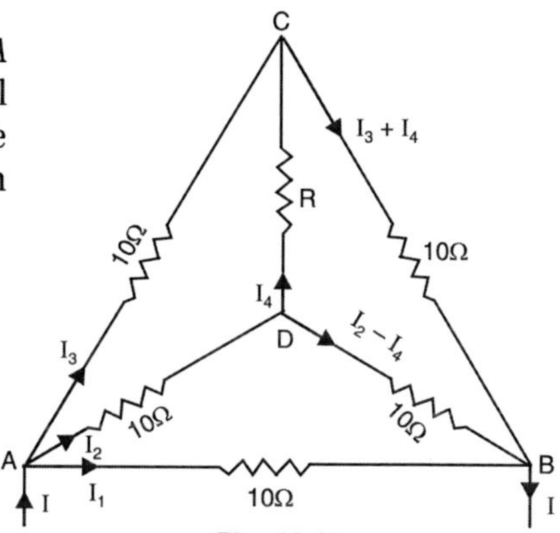

Fig. 12.17

In the mesh ABD we have

$$-10\,I_1 + 10\,(I_2 - I_4) + 10\,I_2 = 0$$

or $\quad 2I_2 - I_4 = I_1 \qquad \ldots \text{(i)}$

In the mesh ADC, we have

$$-10\,I_2 - I_4 R + 10\,I_3 = 0$$

or $\quad 10\,I_3 = 10\,I_2 + I_4 R \qquad \ldots \text{(ii)}$

In the mesh BCD, we have

$$-10\,(I_2 - I_4) + 10\,(I_3 + I_4) + R\,I_4 = 0$$

or $\quad -10\,I_2 + 10\,I_4 + 10\,I_3 + 10\,I_4 + R\,I_4 = 0$

or $\quad (20 + 2R)\,I_4 = 0 \qquad \ldots$ (iii), in view of Eq. (ii)

since $20 + 2R \neq 0$ hence $I_4 = 0$

Now, from Eq. (i) we have $2\,I_2 = I_1$ and from Eq. (ii) $I_2 = I_3$

Along the path AB, we have

$10\,I_1 = V$; V is the potential difference across the cell

Main current $I = I_1 + I_2 + I_3 = I_1 + 2\,I_2 = 2\,I_1$

$\therefore \quad 10\,I_1 = 5 \times 2\,I_1 = 5\,I = V$

$\therefore \quad$ Equivalent resistance $= 5\ \Omega$

12.25. A part of circuit is shown in Fig. 12.18. Determine the equivalent resistance between points A and B.

Solution: Let us assume that the points A and B are connected to a hypothetical cell. The current distribution in various resistances is shown in Fig. 12.18.

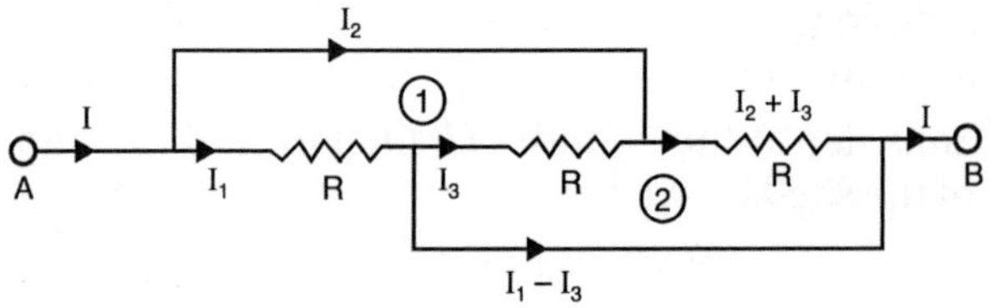

Fig. 12.18

In the first mesh, we have

$$-I_1R - I_3R = 0$$

or $\quad I_1 = -I_3 \quad$ ... (i)

In the second mesh, we have

$$-I_3R - I_2R - I_3R = 0$$

or $\quad I_2 = -2\,I_3 \quad$ ... (ii)

$\therefore \quad I_2 = 2\,I_1$

Let us follow a straight path between A and B.

$-I_1R - I_3R - I_2R - I_3R + V = 0$; V = potential difference across the cell

or $\quad V = R\,[I_1 + I_2 + 2\,I_3] = R\,I_1$ in view of Eq. (ii)

Main current $I = I_1 + I_2 = I_1 + 2\,I_1 = 3\,I_1$

Now $V = R\,I_1 = \dfrac{R}{3}\,3\,I_1 = \dfrac{R}{3}\,I$

$\therefore \quad$ equivalent resistance between A and $B = \dfrac{R}{3}\,\Omega$

12.26. Two conductors with temperature coefficient of resistance α_1 and α_2 have resistances R_{01} and R_{02} at 0°C. Find the temperature coefficient of a circuit consisting of these conductors if they are connected in (i) series and (ii) parallel.

Solution: In series combination the total resistance of the circuit at a temperature t is given by

$$R = R_{01} + R_{02} + \alpha_1\,R_{01}\,t + \alpha_2\,R_{02}\,t$$

On comparing it with equation, $R = R_0\,(1 + \alpha' t)$ we have

$$R_0 = R_{01} + R_{02} \text{ and } \alpha' = \frac{R_{01}\cdot\alpha_1 + R_{02}\cdot\alpha_2}{R_{01} + R_{02}}$$

In parallel combination, $R = \dfrac{R_{01}\cdot R_{02}\cdot(1+\alpha_1 t)(1+\alpha_2 t)}{R_{01}\,(1+\alpha_1 t) + R_{02}\,(1+\alpha_2 t)} = R_0\,(1+\alpha'' t)$

Where $R_0 = \dfrac{R_{01}\cdot R_{02}}{R_{01} + R_{02}}$

and $\alpha'' \cong \dfrac{R_{02}\alpha_1 + R_{01}\alpha_2}{R_{01} + R_{02}}$

12.27. There is an infinite chain of two resistances $R_1 = 2\,\Omega$ and $R_2 = 3\,\Omega$, as shown in Fig. 12.19a. Determine the equivalent resistance between A and B.

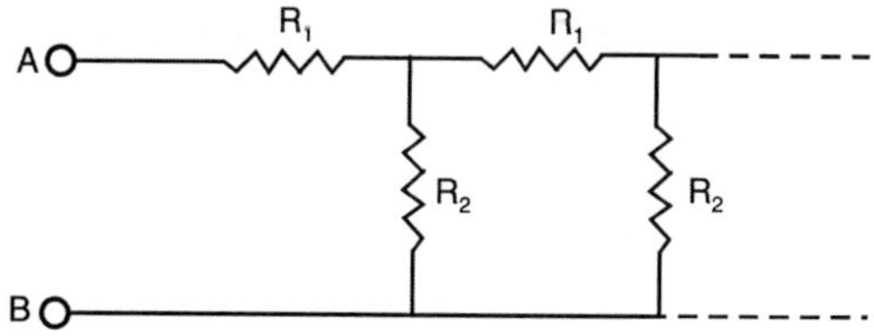

Fig. 12.19a

Solution: Let the equivalent resistance between A and B be R. The infinite mesh can be reduced to a simple mesh, as shown in Fig. 12.19b.

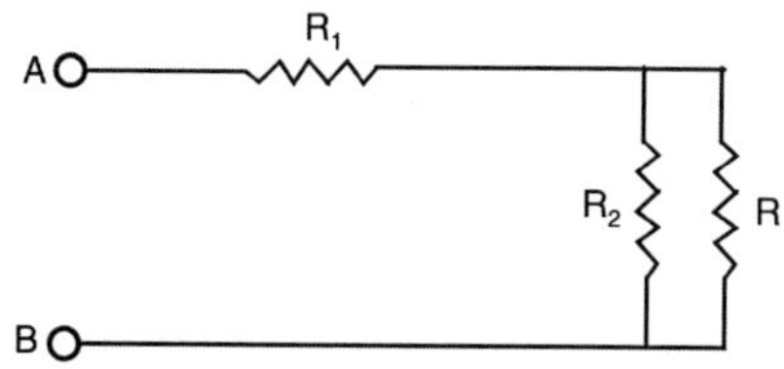

Fig. 12.19b

The total resistance between A and B is given by

$$R = R_1 + \frac{R_2 R}{R + R_2}$$

or $$R^2 + RR_2 = R_1 R + R_1 R_2 + R_2 R$$

or $$R^2 - R_1 R - R_1 R_2 = 0$$

or $$R = \frac{R_1 \pm \sqrt{R_1^2 + 4R_1 R_2}}{2}$$

On substituting the values of R_1 and R_2, we have

$$R = \frac{2 \pm \sqrt{4 + 24}}{2} = \sqrt{7} + 1 = 3.645\,\Omega$$

12.28. At what value of the resistance x in the circuit shown in Fig. 12.20 will the total resistance between points A and B be independent of the number of cells? Assume $R = 2\,\Omega$.

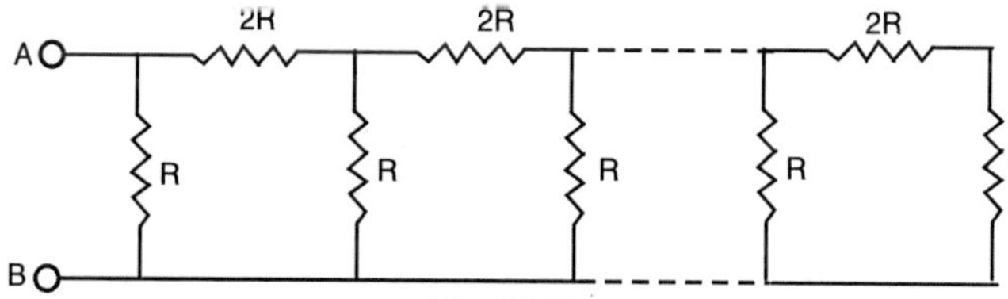

Fig. 12.20a

Solution: Consider two circuits consisting of one cell and two cells, as shown in Figs. 12.20b and c. Equivalent resistance between AB of Fig. 12.20b is given by

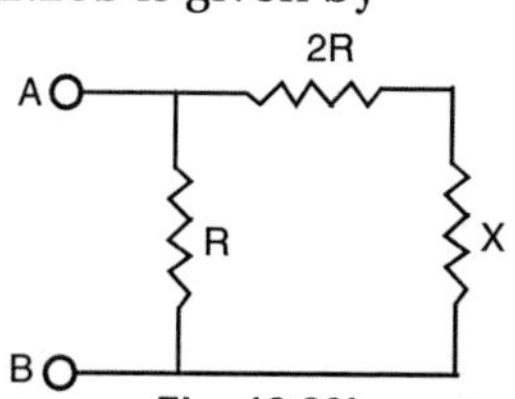

Fig. 12.20b

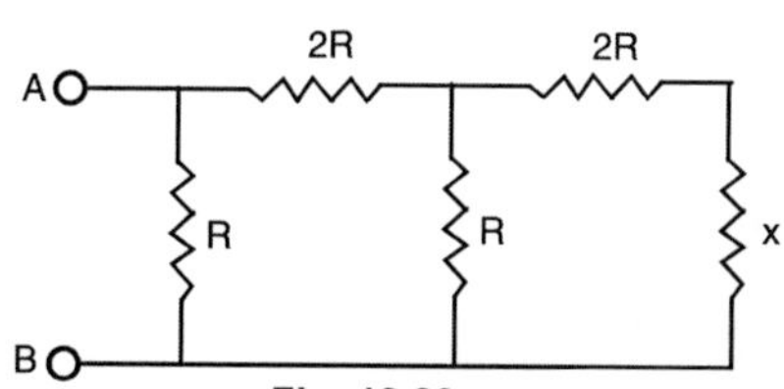

Fig. 12.20c

$$R' = \frac{R(2R+x)}{3R+x} \quad \text{... (i)}$$

Equivalent resistance between AB of Fig. 12.20c is given by

$$R'' = \frac{R(8R+3x)}{11R+4x} \quad \text{... (ii)}$$

These two resistances must be equal, i.e.,

$$\frac{2R+x}{3R+x} = \frac{8R+3x}{11R+4x} \quad \text{... (iii)}$$

or $\quad 22R^2 + 8Rx + 11Rx + 4x^2 = 24R^2 + 9Rx + 8Rx + 3x^2$

or $\quad x^2 + 2Rx - 2R^2 = 0$

or $\quad x = -\dfrac{2R \pm \sqrt{4R^2 + 8R^2}}{2} = \left(\sqrt{3} - 1\right)R$

On putting $R = 2\,\Omega$, we have

$$x = 2\left(\sqrt{3} - 1\right) = 1.464\,\Omega$$

12.29. In the circuit shown in Fig. 12.21, under what condition will the maximum thermal power be generated?

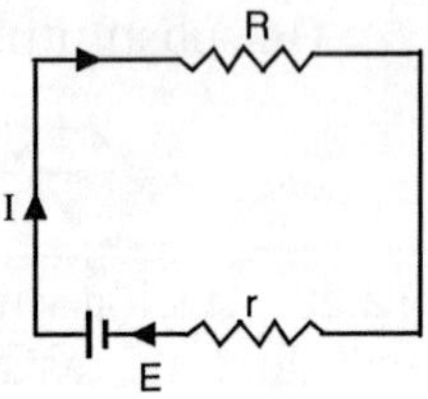

Fig. 12.21

Solution: According to Kirchhoff's law, we have

$$E - Ir - IR = 0 \quad \text{... (i)}$$

where I is the current in the circuit.

Now $I = \dfrac{E}{R+r}$ $\quad$ ... (ii)

Thermal power generated in resistance R is given by

$$P = I^2 R = \frac{E^2}{(R+r)^2} R = E^2 R (R+r)^{-2} \quad \text{... (iii)}$$

In the above equation P is a function of R. Now find the value of $\dfrac{dP}{dR}$ and equate it to zero.

$$\text{Now } \frac{dP}{dR} = 0 = E^2 \left[\frac{1}{(R+r)^2} - \frac{2R}{(R+r)^3}\right]$$

$$= \frac{E^2}{(R+r)^3}[r - R]$$

$\therefore \quad R = r$

On differentiating $\dfrac{dP}{dR}$ once again with respect to R, we have

$$\frac{d^2P}{dR^2} = E^2 \left[\frac{-2}{(R+r)^3} - \frac{2}{(R+r)^3} + \frac{6R}{(R+r)^4}\right]$$

On substituting $R = r$, we have

$$\frac{d^2P}{dR^2} = E^2\left[-\frac{-2}{8R^3} - \frac{2}{8R^3} + \frac{6}{16R^3}\right] = -\frac{E^2}{8R^3} \text{ a negative quantity}$$

Hence the maximum power will be generated under the condition, when $R = r$.

12.30. Two identical cells of emf E and internal resistance r are connected in parallel across a resistor R, as shown in Fig. 12.22. For what value of R is the power delivered to the resistor maximum?

Solution: On applying Kirchhoff's loop theorem to the mesh *abcd*, we have

$$-I_1r + E - I_1R - I_2R = 0$$

$$\therefore \quad I_1(R+r) + I_2R = E \qquad \text{... (i)}$$

Similarly, for the mesh *efcd*, we have

$$-I_2r + E - I_1R - I_2R = 0$$

$$\therefore \quad I_1R + I_2(R+r) = E \qquad \text{... (ii)}$$

Multiplying Eq. (i) by R and Eq. (ii) by $(R + r)$, we have

$$I_1R(R+r) + I_2R^2 = ER \qquad \text{... (iii)}$$

$$I_1R(R+r) + I_2(R+r)^2 = E(R+r) \qquad \text{... (iv)}$$

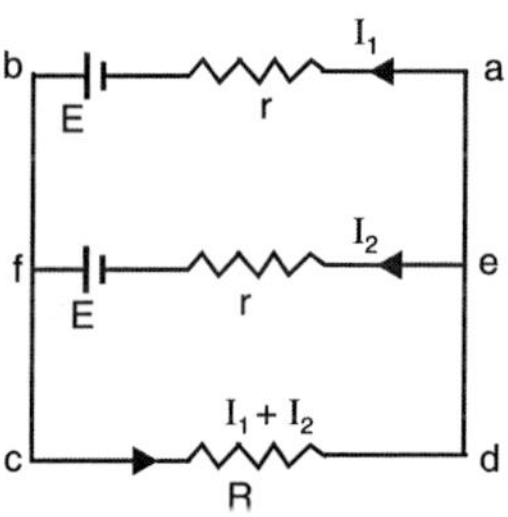

Fig. 12.22

On substracting the two, we have

$$I_2r(r+2R) = Er$$

or $\quad I_2 = E/(2R+r)$

and $\quad I_1 = \dfrac{E}{r+2R}$

$$\text{Total current in the resistor} = \frac{2E}{r+2R}$$

$$\text{Power } P = \frac{4E^2}{(r+2R)^2}R$$

On differentiating P with respect to R, we have

$$\frac{dP}{dR} = 4E^2\left[\frac{1}{(r+2R)^2} - \frac{4R}{(r+2R)^3}\right]$$

$$= \frac{4E^2}{(r+2R)^2}[r+2R-4R]$$

On equating it to zero, we have

$$r = 2R$$

or $\quad R = r/2$

Thus maximum power will be obtained when $R = r/2$.

This can be verified by substituting $R = r/2$ in the expressions for $\frac{d^2P}{dR^2}$ where the entire quantity comes out to be negative.

12.31. A cell of emf E and internal resistance r is connected across the point AB containing three identical resistance inter-connected as shown in Fig. 12.23. At what value of R will the thermal power generated in the circuit be maximum.

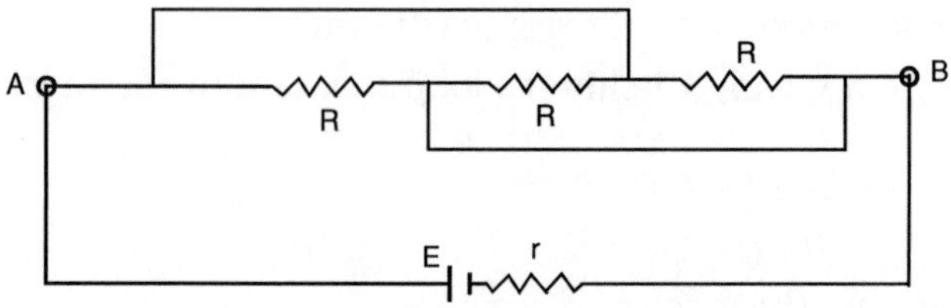

Fig. 12.23

Solution: In the light of Q. 12.25, the equivalent resistance between A and B is $R/3$. The above circuit now reduces to a circuit in which a cell of emf E and internal resistance r is connected to a resistance $R/3$.

According to Q. 12.29, the maximum power will be generated under the condition when $R/3 = r$ or $R = 3r$.

12.32. A dc source of emf E with negligible internal resistance is connected to three resistances R_1, R_2 and R_x as shown in Fig. 12.24. At what value of the resistance R_x will the thermal power generated be independent of small variations of that resistance. When (i) $R_1 = R_2 = 10\ \Omega$, (ii) $R_1 = 10\ \Omega$ and $R_2 = 15\ \Omega$.

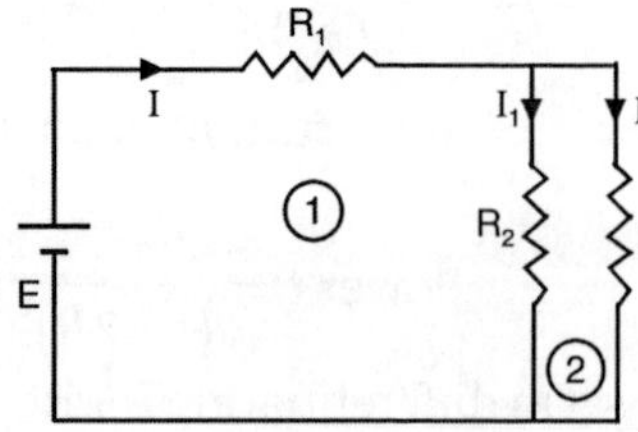

Fig. 12.24

Solution: According to Kirchhoff's loop theorem, in the first mesh we have

$$E - R_1 I - I_1 R_2 = 0$$

or $\quad E = (R_1 + R_2)\, I_1 + R_1 I_2 \quad$ ... (i);

$$I = I_1 + I_2$$

According to Kirchhoff's loop theorem, in the second mesh we have

$$-I_1 R_2 + I_2 R_x = 0$$

$\therefore \quad I_1 = \dfrac{I_2 R_x}{R_2} \quad$ ... (ii)

On substituting the values of I_1 in Eq. (i), we have

$$E = \left[\frac{(R_1 + R_2)\, R_x}{R_2} + R_1\right] I_2$$

or $$\frac{ER_2}{R_1R_2+(R_1+R_2)R_x}=I_2 \qquad \text{... (iii)}$$

Thermal power in R_x is given by

$$P=\frac{E^2R_2^2R_x}{\left[(R_1+R_2)R_x+R_1R_2\right]^2} \qquad \text{... (iv)}$$

Differentiating P with respect to R_x and equate it to zero

$$\frac{dP}{dR_x}=E^2R_2^2\left[\frac{1}{\left[(R_1+R_2)R_x+R_1R_2\right]^2}-\frac{2R_x(R_1+R_2)}{\left[(R_1+R_2)R_x+R_1R_2\right]^3}\right]$$

$$=\frac{E^2R_2^2}{\left[(R_1+R_2)R_x+R_1R_2\right]^3}\left[(R_1+R_2)R_x+R_1R_2-2R_x(R_1+R_2)\right]$$

On equating it to zero, we have

$$R_x=\frac{R_1R_2}{R_1+R_2}$$

On differentiating $\frac{dP}{dR_x}$ once again with respect to R_x and then putting

$R_x=\frac{R_1R_2}{R_1+R_2}$, negative quantity will be obtained.

(i) On putting $R_1=R_2=10\,\Omega$

$$R_x=\frac{10\times 10}{20}=5\,\Omega$$

(ii) On putting $R_1=10\Omega, R_2=15\,\Omega$

$$R_x=\frac{10\times 15}{25}=6\,\Omega$$

12.33. In the circuit shown in Fig. 12.25, determine the potentials of the terminals of the battery. The internal resistance is zero.

Solution: According to Kirchhoff's loop theorem, we have

$$50-400\,I-600\,I=0$$

or $I=1/20$ amp

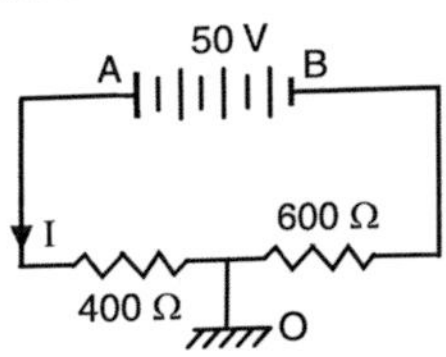

Fig. 12.25

Now $V_{A0}=V_A-V_0=400\times\frac{1}{20}=20\text{V}$

Since $V_0=0, V_A=+20\text{V}$

Now $V_{B0}=V_B-V_0=-600\times\frac{1}{20}=-30\text{V}$

In view of the two equations, we have

$$2I_1 + 3I_2 = 4I_1 + 2I_2$$

or $I_2 = 2I_1$

On substituting this identity in Eq. (i), we have

$$2I_1 + 6I_1 = 1$$

or $8I_1 = 1$

$$I_1 = \frac{1}{8} A = 0.125 A$$

$$I_2 = \frac{1}{4} A = 0.25 A$$

$$I_1 + I_2 = \frac{3}{8} A = 0.375 A$$

12.37. A part of a circuit is shown in Fig. 12.29, determine the current in the 10 Ω resistance. Potential at $A = 20$V, potential at B = potential at $C = 9$V.

Solution:

$$V_{AB} = V_A - V_B = 11 = 20I_1 + 10I_1 + 10I_2$$

or $30I_1 + 10I_2 = 11$ ... (i)

$$V_{AC} = V_A - V_C = 11 = 30I_2 + 10I_1 + 10I_2$$

or $10I_1 + 40I_2 = 11$ ... (ii)

In view of the two equations, we have

$$30I_1 + 10I_2 = 10I_1 + 40I_2$$

or $20I_1 = 30I_2$

or $I_1 = \frac{3}{2} I_2$

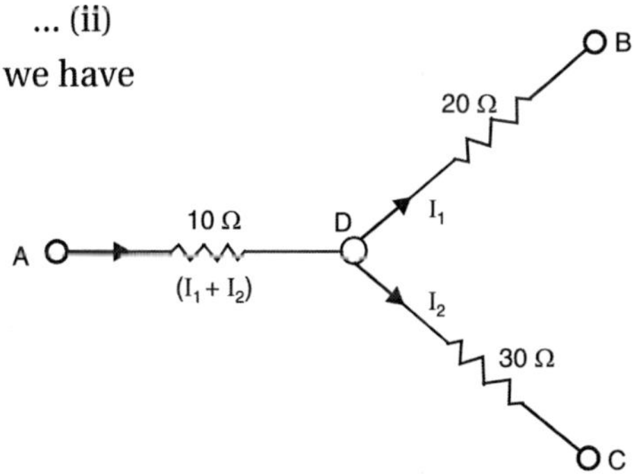

Fig. 12.29

On substituting this identity in Eq. (i), we have

$$10I_2 + 45I_2 = 11$$

$$55I_2 = 11$$

or $I_2 = 11/55 = 0.2$ amp

$$I_1 = \frac{3}{2} \times 0.2 = 0.3 \text{ amp}$$

Total current through the 10 Ω resistance $= I_1 + I_2 = 0.2 + 0.3 = 0.5$ A

12.38. A straight wire of length l = 500 m carries a current I = 80 amp. Calculate the total momentum of the electrons in the wire.

Solution:

$I = neA\,v_d$; n = number of electrons per unit volume

A = area of cross-section of the wire

v_d = drift velocity

In view of the two equations, we have

$$2I_1 + 3I_2 = 4I_1 + 2I_2$$

or $\quad I_2 = 2I_1$

On substituting this identity in Eq. (i), we have

$$2I_1 + 6I_1 = 1$$

or $\quad 8I_1 = 1$

$$I_1 = \frac{1}{8} A = 0.125 A$$

$$I_2 = \frac{1}{4} A = 0.25 A$$

$$I_1 + I_2 = \frac{3}{8} A = 0.375 A$$

12.37. A part of a circuit is shown in Fig. 12.29, determine the current in the 10 Ω resistance. Potential at $A = 20$ V, potential at B = potential at $C = 9$ V.

Solution:

$$V_{AB} = V_A - V_B = 11 = 20I_1 + 10I_1 + 10I_2$$

or $\quad 30I_1 + 10I_2 = 11 \qquad \text{... (i)}$

$$V_{AC} = V_A - V_C = 11 = 30I_2 + 10I_1 + 10I_2$$

or $\quad 10I_1 + 40I_2 = 11 \qquad \text{... (ii)}$

In view of the two equations, we have

$$30I_1 + 10I_2 = 10I_1 + 40I_2$$

or $\quad 20I_1 = 30I_2$

or $\quad I_1 = \frac{3}{2} I_2$

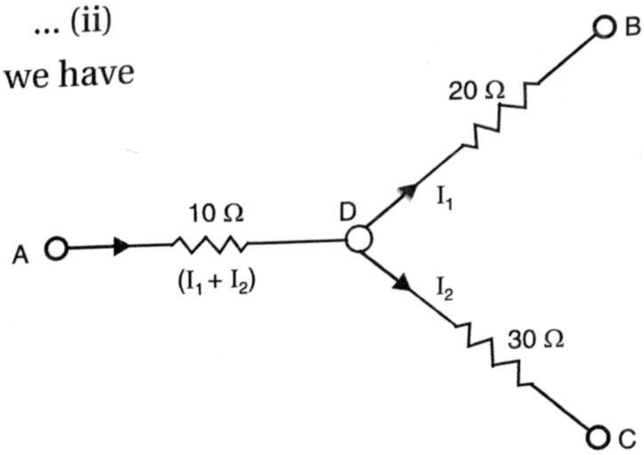

Fig. 12.29

On substituting this identity in Eq. (i), we have

$$10I_2 + 45I_2 = 11$$

$$55I_2 = 11$$

or $\quad I_2 = 11/55 = 0.2$ amp

$$I_1 = \frac{3}{2} \times 0.2 = 0.3 \text{ amp}$$

Total current through the 10 Ω resistance $= I_1 + I_2 = 0.2 + 0.3 = 0.5$ A

12.38. A straight wire of length l = 500 m carries a current I = 80 amp. Calculate the total momentum of the electrons in the wire.

Solution:

$I = neA\,v_d$; $\quad n$ = number of electrons per unit volume

A = area of cross-section of the wire

v_d = drift velocity

$$\frac{R_1R_2 + R_1R_3 + R_2R_3}{R_1 + R_2}, \frac{R_2R_3 + R_1R_2 + R_1R_3}{R_2 + R_3}, \frac{R_1R_3 + R_1R_2 + R_2R_3}{R_1 + R_3}$$

Let V be the supply voltage then

$$\frac{20V^2(R_1 + R_2)}{R_1R_2 + R_1R_3 + R_2R_3} = \frac{40V^2(R_2 + R_3)}{R_1R_2 + R_1R_3 + R_2R_3} = \frac{16V^2(R_1 + R_3)}{R_1R_2 + R_1R_3 + R_2R_3}$$

$$= Q = V^2t\left(\frac{R_1R_2 + R_2R_3 + R_1R_3}{R_1R_2R_3}\right)$$

Now $\quad 20\,(R_1 + R_2) = 40\,(R_2 + R_3)$

or $\quad R_1 = R_2 + 2R_3$

Further $\quad 5\,(R_2 + R_3) = 2R_1 + 2R_3$

$5\,(R_2 + R_3) = 2R_2 + 4R_3 + 2R_3$

$3R_2 = R_3$

$R_1 = R_2 + 6R_2 = 7R_2$

Now $$\frac{20V^2 \times 8R_2}{7R_2^2 + 21R_2^2 + 3R_2^2} = \frac{V^2 \cdot t \times 31R_2^2}{21R_2^3}$$

$$\therefore \quad t = \frac{160 \times 21}{31^2} = 3'\,29.8 \text{ sec.}$$

12.41. An electric discharge in air occurs at an electric field of 3×10^6 V/m. Assuming the mean free path λ of an electron equal to 6×10^{-6}m. Find out the velocity of an electron which can ionise air molecules.

Solution:

Ionisation energy $= e\,E \cdot \lambda$

$= 1.6 \times 10^{-19} \times 3 \times 10^6 \times 6 \times 10^{-6}$

$= 1.6 \times 18 \times 10^{-19}$

$= 28.8 \times 10^{-19}$ J

Now $$\frac{1}{2}mv^2 = 28.8 \times 10^{-19}$$

$$\therefore \quad v = \sqrt{\frac{2 \times 28.8 \times 10^{-19}}{9 \times 10^{-31}}}$$

$$= \sqrt{\frac{2 \times 28.8}{9}} \times 10^6$$

$$= 2.53 \times 10^6 \text{ m/sec.}$$

$$\frac{R_1R_2 + R_1R_3 + R_2R_3}{R_1 + R_2}, \frac{R_2R_3 + R_1R_2 + R_1R_3}{R_2 + R_3}, \frac{R_1R_3 \ldots R_2 +}{R_1 \ldots}$$

Let V be the supply voltage then

$$\frac{20V^2(R_1 + R_2)}{R_1R_2 + R_1R_3 + R_2R_3} = \frac{40V^2(R_2 + R_3)}{R_1R_2 + R_1R_3 + R_2R_3} = \frac{16V^2(R_1 \; R_3)}{R_1R_2 + R_1R_3 + \; R_3}$$

$$= Q = V^2t\left(\frac{R_1R_2 + R_2R_3 + R_1R_3}{R_1R_2R_3}\right)$$

Now $\quad 20\,(R_1 + R_2) = 40\,(R_2 + R_3)$

or $\quad R_1 = R_2 + 2R_3$

Further $\quad 5\,(R_2 + R_3) = 2R_1 + 2R_3$

$$5\,(R_2 + R_3) = 2R_2 + 4R_3 + 2R_3$$

$$3R_2 = R_3$$

$$R_1 = R_2 + 6R_2 = 7R_2$$

Now $\quad \dfrac{20V^2 \times 8R_2}{7R_2^2 + 21R_2^2 + 3R_2^2} = \dfrac{V^2 \cdot t \times 31R_2^2}{21R_2^3}$

$\therefore \quad t = \dfrac{160 \times 21}{31^2} = 3'\,29.8$ sec.

12.41. An electric discharge in air occurs at an electric field of 3×10^6 V/m. Assuming the mean free path λ of an electron equal to 6×10^{-6}m. Find out the velocity of an electron which can ionise air molecules.

Solution:

Ionisation energy $= e\,E \cdot \lambda$

$$= 1.6 \times 10^{-19} \times 3 \times 10^6 \times 6 \times 10^{-6}$$

$$= 1.6 \times 18 \times 10^{-19}$$

$$= 28.8 \times 10^{-19}\,\text{J}$$

Now $\quad \dfrac{1}{2}mv^2 = 28.8 \times 10^{-19}$

$\therefore \quad v = \sqrt{\dfrac{2 \times 28.8 \times 10^{-19}}{9 \times 10^{-31}}}$

$$= \sqrt{\frac{2 \times 28.8}{9}} \times 10^6$$

$$= 2.53 \times 10^6 \text{ m/sec.}$$

$$\frac{d_1 l_1 A_1 s_1 \Delta t}{d_2 l_2 A_2 [s_2 \Delta t_1 + L]} = \frac{\rho_1 l_1 A_2}{A_1 \rho_2 l_2}$$

$$\Delta t = \frac{\rho_1 d_2 A_2^2 [s_2 \cdot \Delta t_1 + L]}{d_1 \rho_2 A_1^2 s_1}$$

$$\frac{1.7 \times 10^{-8} \times 11300 \times 1 \left[126 \times (327 - 27) + 2.26 \times 10^4\right]}{8600 \times 2.2 \times 10^{-7} \times 4 \times 395}$$

$$\frac{1.7 \times 11.3 [6.04] \times 10^4}{86 \times 2.2 \times 4 \times 395} = 3.9°\text{C}$$

12.45. An electric kettle with 0.5 lit of water at 25°C is connected to a source of emf of 150 V in series with a resistance of 25 Ω. A current of 2 amp flows in the circuit. Taking the efficiency of the kettle to be 80%. Calculate the time in which the water will begin to boil.

Solution: Potential drop across the 25 Ω resistance = 50 volt. Potential difference across the kettle = 100 volt. Resistance R of the kettle = 50 Ω.

$$\text{Now } \frac{V^2}{R} t \times \frac{0.8}{4.2} = 0.5 \times 1000 \times 1 \times (100 - 25)$$

$$\therefore \quad t = \frac{50 \times 500 \times 75 \times 4.2}{0.8 \times 100^2} \text{ sec}$$

$$= 16.4 \text{ min}$$

12.46. In an ionisation chamber the two electrodes each of area 100 cm^2 are placed 6 cm apart. An electric field produces 10^9 pair of monovalent ions/sec/c.c. The mobility of each ion is 3.5×10^{-4}m^2/v.s. Calculate the value of the saturation current and the applied electric field.

Solution:

Saturation current $I = neAv$

$$n = 2 \times \frac{10^9}{10^{-6}} = 2 \times 10^{15} /\text{m}^3; \; v = \frac{6 \times 10^{-2}}{1} = 6 \times 10^{-2}$$

$$\therefore \quad I = 2 \times 10^{15} \times 1.6 \times 10^{-19} \times 100 \times 10^{-4} \times 6 \times 10^{-2}$$

$$= 12 \times 1.6 \times 10^{-8}$$

$$= 0.192 \ \mu\text{A}$$

$$\text{Electric field } E = \frac{v}{\text{mobility}} = \frac{l}{\text{mobility}}$$

$$= \frac{6 \times 10^{-2}}{3.5 \times 10^{-4}} = \frac{6 \times 100}{3.5} = 171.4 \text{ V/m}$$

$$\frac{d_1 l_1 A_1 s_1 \Delta t}{d_2 l_2 A_2 \left[s_2 \Delta t_1 + L\right]} = \frac{\rho_1 l_1 A_2}{A_1 \rho_2 l_2}$$

$$\Delta t = \frac{\rho_1 d_2 A_2^2 \left[s_2 \cdot \Delta t_1 + L\right]}{d_1 \rho_2 A_1^2 s_1}$$

$$\frac{1.7 \times 10^{-8} \times 11300 \times 1 \left[126 \times (327 - 27) + 2.26 \times 10^4\right]}{8600 \times 2.2 \times 10^{-7} \times 4 \times 395}$$

$$\frac{1.7 \times 11.3 \left[6.04\right] \times 10^4}{86 \times 2.2 \times 4 \times 395} = 3.9°\text{C}$$

12.45. An electric kettle with 0.5 lit of water at 25°C is connected to a source of emf of 150 V in series with a resistance of 25 Ω. A current of 2 amp flows in the circuit. Taking the efficiency of the kettle to be 80%. Calculate the time in which the water will begin to boil.

Solution: Potential drop across the 25 Ω resistance = 50 volt. Potential difference across the kettle = 100 volt. Resistance R of the kettle = 50 Ω.

$$\text{Now } \frac{V^2}{R} t \times \frac{0.8}{4.2} = 0.5 \times 1000 \times 1 \times (100 - 25)$$

$$\therefore \quad t = \frac{50 \times 500 \times 75 \times 4.2}{0.8 \times 100^2} \text{ sec}$$

$$= 16.4 \text{ min}$$

12.46. In an ionisation chamber the two electrodes each of area 100 cm² are placed 6 cm apart. An electric field produces 10^9 pair of monovalent ions/sec/c.c. The mobility of each ion is 3.5×10^{-4} m²/v.s. Calculate the value of the saturation current and the applied electric field.

Solution:

Saturation current $I = neAv$

$$n = 2 \times \frac{10^9}{10^{-6}} = 2 \times 10^{15}\,/\text{m}^3; \; v = \frac{6 \times 10^{-2}}{1} = 6 \times 10^{-2}$$

$$\therefore \quad I = 2 \times 10^{15} \times 1.6 \times 10^{-19} \times 100 \times 10^{-4} \times 6 \times 10^{-2}$$

$$= 12 \times 1.6 \times 10^{-8}$$

$$= 0.192\ \mu\text{A}$$

$$\text{Electric field } E = \frac{v}{\text{mobility}} = \frac{l}{\text{mobility}}$$

$$= \frac{6 \times 10^{-2}}{3.5 \times 10^{-4}} = \frac{6 \times 100}{3.5} = 171.4\,\text{V/m}$$

12.49. Three cells of emf E_1, E_2 & E_3 and internal resistance r_1, r_2 & r_3 respectively are connected to an external resistance R as shown in Fig. 12.31. Under what condition maximum thermal power will be generated in the external resistance R?

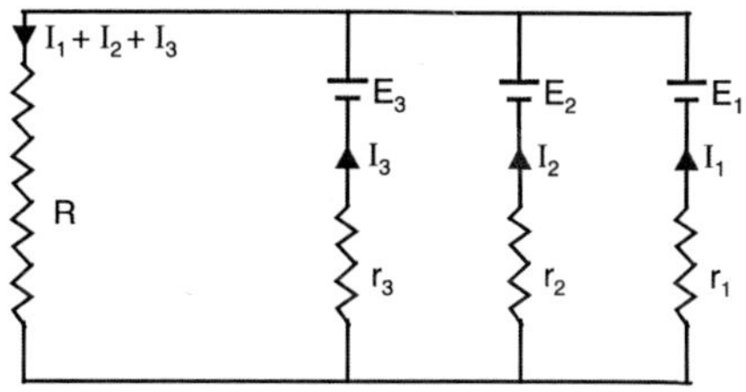

Fig. 12.31

Solution: We can have three mesh equation with the three cells and external resistance R as

$$E_1 - I_1 r_1 - R\,(I_1 + I_2 + I_3) = 0$$
$$E_2 - I_2 r_2 - R\,(I_1 + I_2 + I_3) = 0$$
$$E_3 - I_3 r_3 - R\,(I_1 + I_2 + I_3) = 0$$

From these equations we have

$$r_2 r_3 E_1 = I_1 r_1 r_2 r_3 + R\, r_2 r_3\,(I_1 + I_2 + I_3)$$
$$r_1 r_3 E_2 = I_2 r_1 r_2 r_3 + R\, r_1 r_3\,(I_1 + I_2 + I_3)$$
$$r_1 r_2 E_3 = I_3 r_1 r_2 r_3 + R\, r_1 r_2\,(I_1 + I_2 + I_3)$$

On adding these three equations we have

$$E_1 r_2 r_3 + E_2 r_1 r_3 + E_3 r_1 r_2 = r_1 r_2 r_3 I + R\,(r_1 r_2 + r_1 r_3 + r_2 r_3)\, I;\ I = I_1 + I_2 + I_3$$

or
$$\frac{E_1 r_2 r_3 + E_2 r_1 r_3 + E_3 r_1 r_2}{r_1 r_2 + r_1 r_3 + r_2 r_3} - \frac{r_1 r_2 r_3 I}{r_1 r_2 + r_1 r_3 + r_2 r_3} - IR$$

The above equation represents that the equivalent emf of the three cells is $\dfrac{E_1 r_2 r_3 + E_2 r_1 r_3 + E_3 r_1 r_2}{r_1 r_2 + r_1 r_3 + r_2 r_3}$ and equivalent resistance is $\dfrac{r_1 r_2 r_3}{r_1 r_2 + r_1 r_3 + r_2 r_3}$. This equivalent cell is connected to an external resistance R. For maximum thermal power generated in the external resistance R we must have

$$R = \frac{r_1 r_2 r_3}{r_1 r_2 + r_1 r_3 + r_2 r_3}$$

12.50. Two similar bulbs of rated voltage V and rated power P are connected in series across a cell of emf E and internal resistance r. Determine the internal resistance of the cell if the voltage drop across the internal resistance is 2% of the emf of the cell.

Solution:

Current through each bulb $i = \dfrac{P}{V}$

$$\text{Now } \frac{P}{V}\cdot\frac{r}{E} = \frac{2}{100} = \frac{1}{50} = \frac{ir}{E}$$

$$i = \frac{E}{2\dfrac{V^2}{P} + r} = \frac{E}{50r}$$

$$r = \frac{2V^2}{49P}$$

12.51. A uniform copper wire of mass 2.23×10^{-3} kg carries a current of 1 A when 1.7 volt is applied across it. Calculate the length and area of cross-section. If the wire is uniformly stretched to double its length; calculate the new resistance. Density of copper = 8.92×10^3 kg/m³ and resistivity is 1.7×10^{-8} Ω.m. (Roorkee 1995)

Solution: Let L be the length and S be the area of cross-section then

$$L \times S = \frac{m}{d} = \frac{2.23 \times 10^{-3}}{8.92 \times 10^{3}} = \frac{1}{4} \times 10^{-6}$$

$$R = \frac{V}{I} = 1.7\Omega = \rho \frac{L}{S}$$

$$\therefore \quad \frac{L}{S} = \frac{1.7}{\rho} = 10^8$$

$$\frac{L \times L}{10^8} = \frac{1}{4} \times 10^{-6}$$

or $\quad L = 5$ m and $S = 5 \times 10^{-8}$ m²

$$R' = \frac{\rho L}{S} = \frac{\rho L^2 \cdot d}{LSd} = \frac{\rho L^2 d}{m}$$

$$= \frac{1.7 \times 10^{-8} \times 100 \times 8.92 \times 10^{3}}{2.23 \times 10^{-3}}$$

$$= 1.7 \times 4 = 6.8\ \Omega$$

12.52. One m long metallic wire is broken into two unequal parts P and Q. P part of the wire is uniformly extended into another wire R. Length of R is twice the length of P and the resistance of R is equal to that of Q. Find the ratio of the length of the resistance of P and R and also the ratio of lengths of P and Q. (Roorkee 1996)

Solution: Let l be the length of P and $(1 - l)$ be the length of Q.

Length of $R = 2l$ area of $R = \frac{S}{2}$ where S = area of P & Q both.

$$R_R = R_Q \quad \therefore \frac{\rho \cdot 2l}{\frac{S}{2}} = \frac{\rho(1-l)}{S}$$

$$4l = 1 - l$$

$$l = \frac{1}{5} = 0.2 \text{ m}$$

$$\frac{R_P}{R_R} = \frac{\rho \frac{l}{S}}{4\rho \frac{l}{S}} = \frac{1}{4}, \quad \frac{L_P}{L_Q} = \frac{0.2}{0.8} = \frac{1}{4}$$

12.53. A capacitor filled with dielectric of dielectric constant $K = 2.1$ loses half of the charges imparted to it in time $t = 5$ min. Assuming the leakage of the charge through dielectric, calculate the resistivity of the dielectric.

Solution:

$$Q = Q_0 e^{-t/RC}$$

or $$\frac{1}{2} Q_0 = Q_0 e^{-t/RC}$$

or $$\frac{t}{RC} = \log_e 2$$

or $$\frac{t}{\rho \frac{d}{A} \cdot \frac{K\varepsilon_0 A}{d}} = \log_e 2$$

$\therefore$ $$\rho = \frac{t}{K\varepsilon_0 \log_e 2}$$

On putting the values we have

$$\rho = \frac{5 \times 60 \times 4\pi \times 9 \times 10^9}{2.1 \times 0.693} = 2.33 \times 10^{13}\, \Omega.\text{m}$$

12.54. In the circuit shown in Fig. 12.32 find the values of the currents I_1, I_2, I_3 & I_4.

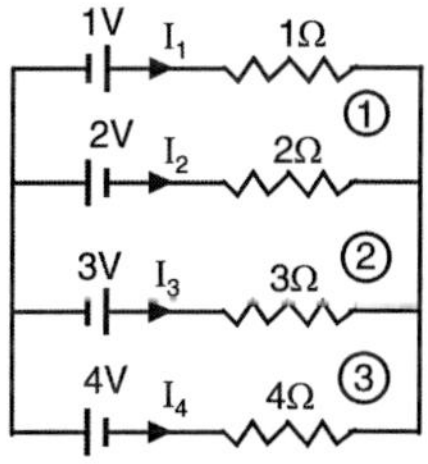

Fig. 12.32

Solution: In mesh 1 we have

$$1 - I_1 + 2I_2 + 2 = 0$$

or $$I_1 - 2I_2 = 3 \qquad \text{... (1)}$$

In mesh 2 we have

$$-2 - 2I_2 + 3I_3 - 3 = 0$$

or $$3I_3 - 2I_2 = 5 \qquad \text{... (2)}$$

In mesh 3 we have

$$3 - 3I_3 + 4I_4 + 4 = 0$$

or $$7 = 3I_3 - 4I_4 \qquad \text{... (3)}$$

On moving along the periphery we have

$$5 = I_1 - 4I_4 \qquad \text{... (4)}$$

In view of the circuit diagram $I_1 + I_2 + I_3 = -I_4$. On putting this in equation 4 we have

$$5I_1 + 4I_2 + 4I_3 = 5$$

or $$5I_1 + 4I_2 + \frac{4}{3}(2 + I_1) = 5$$

$$15I_1 + 12I_2 + 8 + 4I_1 = 15$$

or $19I_1 + 12I_2 = 7$

From equation 1 we have $I_1 - 2I_2 = 3$

On solving the two equations we have $I_1 = 1$ amp

From equation 1, $I_2 = -1$ amp

From equation 2, $I_3 = 1$ amp

From equation 3, $I_4 = -1$ amp

$\therefore \quad I_1 = I_3 = 1$ amp and $I_2 = I_4 = -1$ amp

12.55. A group of N cells, whose emf varies directly with their internal resistance via equation $E_N = 1.5\ r_N$ are connected as shown in Fig. 12.33. Find (i) the current in the circuit, (ii) the potential difference across the points P and Q.

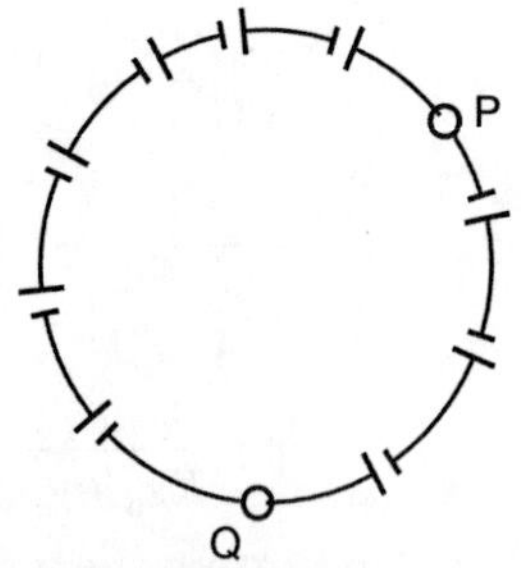

Fig. 12.33

Solution:

(i) According to Kirchhoff's loop theorem, we have

$$E_1 - Ir_1 + E_2 - Ir_2 + \ldots N \text{ terms} = 0$$

$$\therefore \quad I = \frac{E_1 + E_2 + \ldots N \text{ terms}}{r_1 + r_2 + \ldots N \text{ terms}}$$

On substituting the values of E_1, E_2, etc., we have

$$I = \frac{1.5(r_1 + r_2 + \ldots N \text{ terms})}{r_1 + r_2 + \ldots N \text{ terms}}$$

$$\therefore \quad I = 1.5 \text{ amp}$$

(ii) Let us divide the circuit into say 3 and $N - 3$ cells, i.e., there are 3 cells in between P and Q through one path and $N - 3$ cells through another path.

$$V_{PQ} = V_P - V_Q = -E_3 + Ir_3 - E_2 + Ir_2 - E_1 + Ir_1$$
$$= -1.5\,r_3 + 1.5\,r_3 - 1.5\,r_2 + 1.5\,r_2 - 1.5\,r_1 + 1.5\,r_1$$
$$= 0$$

12.56. In the circuit shown in Fig. 12.34, determine the potential difference $\phi_A - \phi_B$ between the plates A and B of the capacitor C.

Solution: On applying Kirchhoff's loop theorem to the mesh *abcd*, we have

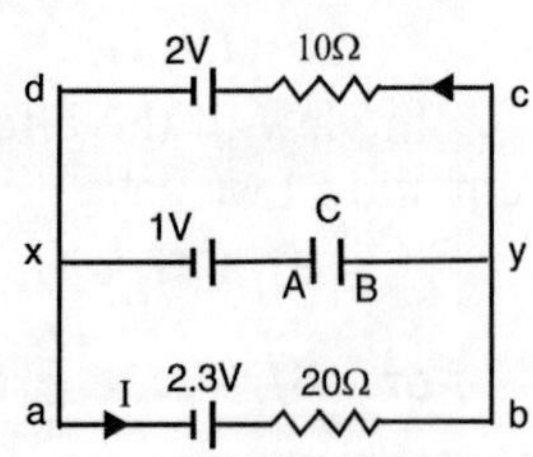

Fig. 12.34

$$2.3 - 20I - 10I - 2 = 0$$

or $0.3 = 30I$

or $I = \dfrac{0.3}{30} = 0.01$ A

Potential difference between *ab*, *xy* and *dc* is the same.

or $19I_1$ ... we have $I_1 - 2I_2 = 3$

From eq... two equations we have $I_1 = 1$ amp

On sol...ion 1, $I_2 = -1$ amp

From ...ation 2, $I_3 = 1$ amp

Fro... quation 3, $I_4 = -1$ amp

Fr...

∴ $I_1 = I_3 = 1$ amp and $I_2 = I_4 = -1$ amp

12... A group of N cells, whose emf varies directly ...th their internal resistance via equation $E_N = 1.5\, r_N$ are connected as shown in Fig. 12.33. Find (i) the current in the circuit, (ii) the potential difference across the points P and Q.

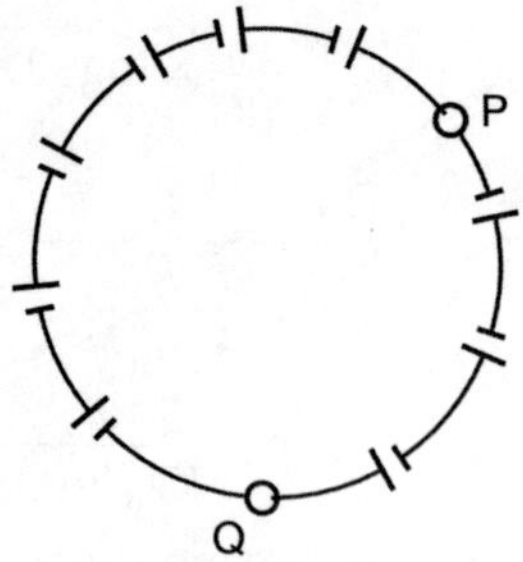

Fig. 12.33

Solution:

(i) According to Kirchhoff's loop theorem, we have

$$E_1 - Ir_1 + E_2 - Ir_2 + \ldots N \text{ terms} = 0$$

$$\therefore \quad I = \frac{E_1 + E_2 + \ldots N \text{ terms}}{r_1 + r_2 + \ldots N \text{ terms}}$$

On substituting the values of E_1, E_2, etc., we have

$$I = \frac{1.5\,(r_1 + r_2 + \ldots N \text{ terms})}{r_1 + r_2 + \ldots N \text{ terms}}$$

$$\therefore \quad I = 1.5 \text{ amp}$$

(ii) Let us divide the circuit into say 3 and $N - 3$ cells, i.e., there are 3 cells in between P and Q through one path and $N - 3$ cells through another path.

$$V_{PQ} = V_P - V_Q = -E_3 + Ir_3 - E_2 + Ir_2 - E_1 + Ir_1$$

$$= -1.5\,r_3 + 1.5\,r_3 - 1.5\,r_2 + 1.5\,r_2 - 1.5\,r_1 + 1.5\,r_1$$

$$= 0$$

12.56. In the circuit shown in Fig. 12.34, determine the potential difference $\phi_A - \phi_B$ between the plates A and B of the capacitor C.

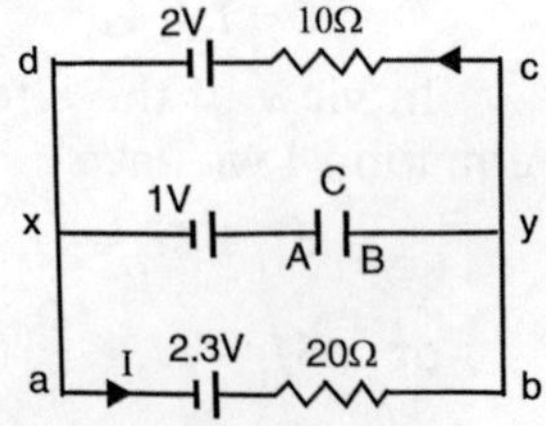

Fig. 12.34

Solution: On applying Kirchhoff's loop theorem to the mesh *abcd*, we have

$$2.3 - 20I - 10I - 2 = 0$$

or $\quad 0.3 = 30I$

or $\quad I = \dfrac{0.3}{30} = 0.01$ A

Potential difference between *ab*, *xy* and *dc* is the same.

Now $V_{ab} = V_a - V_b = 20 \times 0.01 - 2.3 = -2.1\text{V}$

$\therefore \quad V_{xy} = V_x - V_y = -2.1 = -1 + (\phi_A - \phi_B)$

or $\quad \phi_A - \phi_B = -2.1 + 1 = -1.1\text{ V}$

12.57. In the given circuit (see Fig. 12.35), determine the current in the two resistors and the potential difference between P and Q. The internal resistance of the cells is negligible.

Solution: On applying Kirchhoff's loop theorem in the first mesh, we have

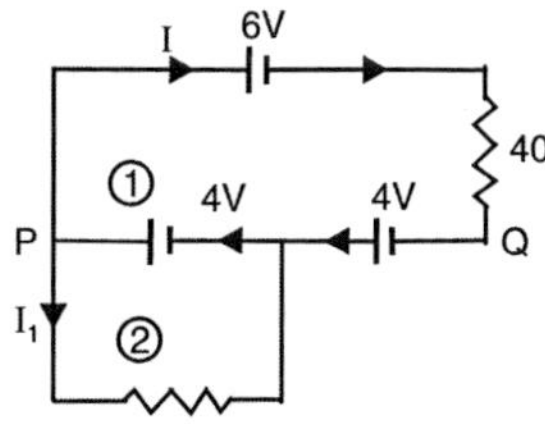

Fig. 12.35

$$4 + 4 - 6 - 40\,I = 0$$

or $\quad 40\,I = 2$

$$I = \frac{1}{20}\text{ A} = 0.05\text{ A}$$

In the second mesh, we have

$$4 - 80\,I_1 = 0$$

or $\quad I_1 = \dfrac{4}{80} = \dfrac{1}{20} = 0.05\text{ A}$

$$V_{PQ} = V_P - V_Q = 4 + 4 = 8\text{ V}$$

The same value of V_{PQ} can also be verified by the two other available paths as

$$4 + 80 \times 0.05 = 8\text{ volt}$$

$$40 \times 0.05 + 6 = 8\text{ volt}$$

12.58. In the given circuit (Fig. 12.36), calculate (a) the heat produced per second in each resistor and (b) the power supplied by the cell of emf 4 V.

Solution:

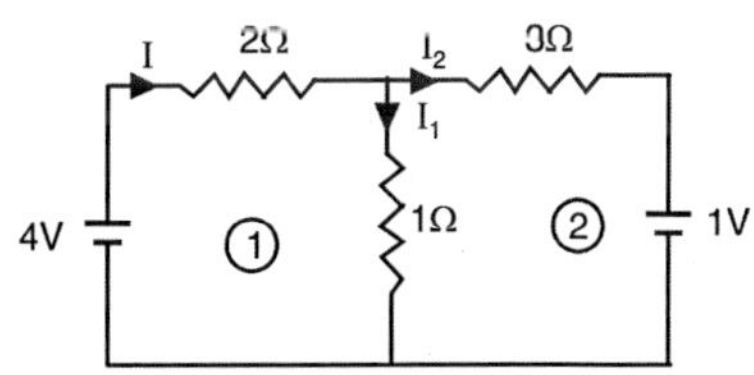

Fig. 12.36

(a) On applying Kirchhoff's loop theorem in the first mesh, we have

$$4 - 2\,(I_1 + I_2) - I_1 = 0;\ I = I_1 + I_2$$

or $\quad 3I_1 + 2I_2 = 4 \qquad \text{... (i)}$

Similarly, in the second mesh, we have

$$-3I_2 - 1 + I_1 = 0$$

or $\quad I_1 - 3I_2 = 1 \qquad \text{... (ii)}$

On solving the two equations

$$I_1 = \frac{14}{11}\text{ amp, } I_2 = \frac{1}{11}\text{ A and } I = \frac{15}{11}\text{ A}$$

Heat produced per second in the 1 Ω resistance $= \left(\dfrac{14}{11}\right)^2 \times 1$

$= 1.62\text{ W}$

$$\text{Heat produced per second in the } 2\,\Omega \text{ resistance} = \left(\frac{15}{11}\right)^2 \times 2$$
$$= 3.72 \text{ W}$$

$$\text{Heat produced per second in the } 3\,\Omega \text{ resistance} = \left(\frac{1}{11}\right)^2 \times 3$$
$$= 0.0248 \text{ W}$$

(b) Power supplied by cell of emf 4 V $= \dfrac{4 \times 15}{11} = 5.45$ W

12.59. Calculate the three current I_1, I_2 and $I_1 + I_2$ and also the potential difference between A and B in the circuit shown in the Fig. 12.37.

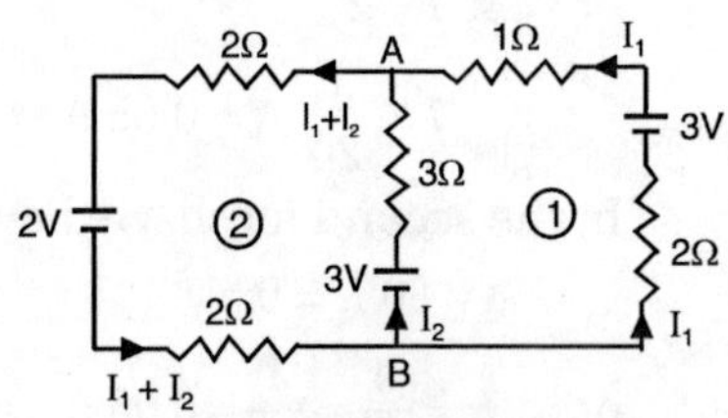

Fig. 12.37

Solution: On applying Kirchhoff's loop theorem in the first mesh, we have

$$-2I_1 + 3 - I_1 + 3I_2 - 3 = 0$$

or $\quad I_1 = I_2 \qquad \text{... (i)}$

Similarly, in the second mesh, we have

$$3 - 3I_2 - 2(I_1 + I_2) - 2 - 2(I_1 + I_2) = 0$$

or $\quad 1 = 4I_1 + 7I_2 \qquad \text{... (ii)}$

On solving the two equations, and $I_1 = I_2 = \dfrac{1}{11}$ A and $I_1 + I_2 = \dfrac{2}{11}$ A

$$V_{AB} = V_A - V_B = 3 - \frac{3}{11} = \frac{30}{11} = 2.73 \text{ V}$$

12.60. Find the emf and the internal resistance of a source which is equivalent to two cells connected in parallel (Fig. 12.38) whose emf's are equal to E_1 and E_2 and internal resistance are r_1 and r_2.

Solution: On applying Kirchhoff's loop theorem in the mesh *abcd*, we have

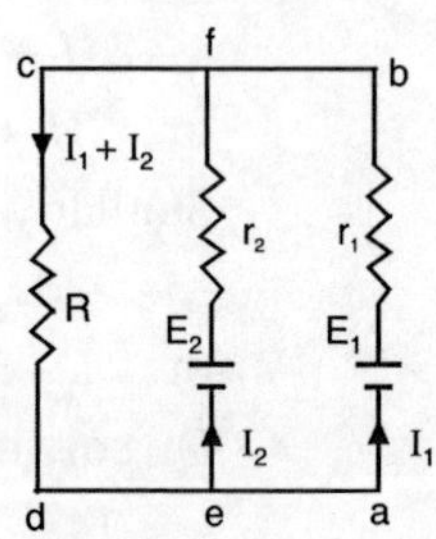

Fig. 12.38

$$E_1 - I_1 r_1 - (I_1 + I_2) R = 0$$

or $\quad E_1 = I_1 (R + r_1) + I_2 R \qquad \text{... (i)}$

Similarly in the mesh *efcd*, we have

$$E_2 - I_2 r_2 - (I_1 + I_2) R = 0$$

or $\quad E_2 = I_1 R + I_2 (r_2 + R) \qquad \text{... (ii)}$

On multiplying Eq. (i) by r_2 and Eq. (ii) by r_1 and then on adding them, we have

$$E_1 r_2 + E_2 r_1 = I_1 [r_1 r_2 + R(r_1 + r_2)] + I_2 [r_1 r_2 + R(r_1 + r_2)] \qquad \text{... (iii)}$$

On dividing the above equation by $r_1 + r_2$, we have

$$\frac{E_1 r_2 + E_2 r_1}{r_1 + r_2} = (I_1 + I_2)\frac{r_1 r_2}{r_1 + r_2} + R(I_1 + I_2)$$

On comparing it with the equation

$$E = (I_1 + I_2)\, r_0 + (I_1 + I_2)\, R$$

We have the equivalent emf $E = \dfrac{E_1 r_2 + E_2 r_1}{r_1 + r_2}$

and internal resistance $r_0 = \dfrac{r_1 r_2}{r_1 + r_2}$

12.61. Find the potential difference $\phi_A - \phi_B$ between the plates of a capacitor C in the circuit diagram shown in Fig. 12.39. The internal resistance of the cells are negligible.

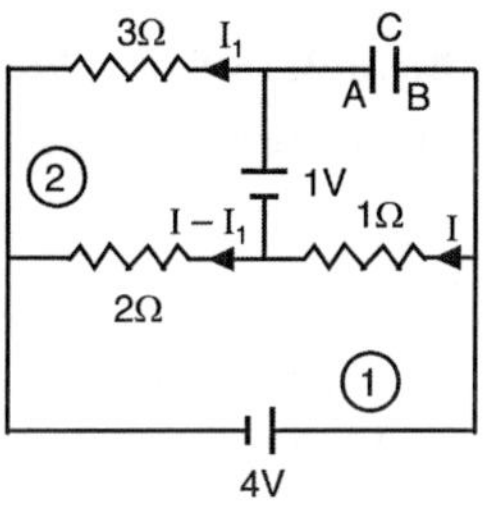

Fig. 12.39

Solution: On applying Kirchhoff's loop theorem in the first mesh, we have

$$4 - I - 2(I - I_1) = 0$$

or $\quad 3I - 2I_1 = 4 \qquad \text{... (i)}$

Similarly in the second mesh, we have

$$1 - 3I_1 + 2(I - I_1) = 0$$

or $\quad 5I_1 - 2I = 1 \qquad \text{... (ii)}$

On multiplying the Eq. (i) by 5 and Eq. (ii) by 2, we have

$$20 = 15I - 10I_1 \qquad \text{... (iii)}$$

$$2 = -4I + 10I_1 \qquad \text{... (iv)}$$

On adding the two, we have

$$11I = 22 \text{ or } I = 2\text{A}$$

$$5I_1 = 1 + 2I = 5 \quad \therefore I_1 = 1\text{ A}$$

$$\phi_{AB} = \phi_A - \phi_B = -1 \times 2 + 1 = -1\text{V}$$

12.62. In the given circuit, three cells with negligible internal resistance are connected as shown in the Fig. 12.40. Find (i) the current through the 1 Ω resistance and the potential difference between the points A and B.

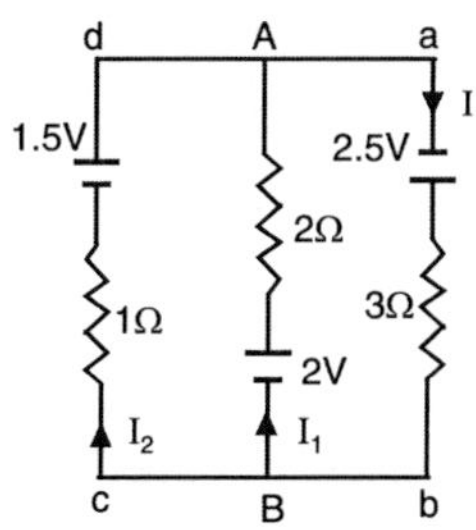

Fig. 12.40

Solution: On applying Kirchhoff's loop theorem in the mesh *abBA*, we have

$$2.5 - 3I + 2 - 2I_1 = 0$$

or $\quad 4.5 - 3I_1 - 3I_2 - 2I_1 = 0; I = I_1 + I_2$

or $\quad 5I_1 + 3I_2 = 4.5 \qquad \text{... (i)}$

Similarly in the mesh *abcd*, we have

$$2.5 - 3I - I_2 + 1.5 = 0$$

or $4 = 3I_1 + 4I_2$ ... (ii)

On solving the two equations

$$I_1 = \frac{6}{11} = 0.545 \text{ A}$$

$$I_2 = \frac{6.5}{11} = 0.59 \text{ A}$$

$$V_{AB} = V_A - V_B = 2 - 2\times\frac{6}{11} = \frac{10}{11} = 0.909\ V$$

12.63. A constant voltage of $V = 50$ V from a dc source is applied across the point *C* and *D* of the given circuit (Fig. 12.41). Find the current flowing through the arm *AB*.

Solution: On applying Kirchhoff's loop theorem in mesh 2, we have

$$-I_1 + 3I_2 = 0$$

or $I_1 = 3I_2$ ... (i)

Potential difference between points *C* and *D* through the two paths is the same.

V_{CD} through path $CAD = 2\,(I_1 - I_3) + I_1 = 4\,(I_2 + I_3) + 3I_2$

(V_{CD} through path *CBD*)

or $3I_1 - 2I_3 = 7I_2 + 4I_3$

or $9I_2 - 7I_2 = 6I_3$

or $I_2 = 3I_3$ ... (ii)

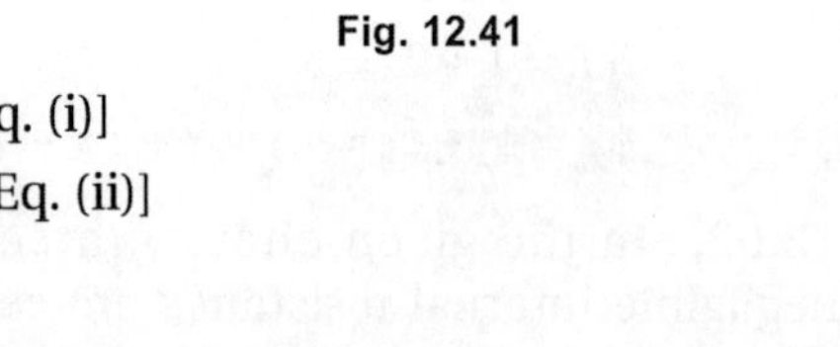

Fig. 12.41

Now $V_{CD} = 50 = 2\,(I_1 - I_3) + I_1$

$= 3I_1 - 2I_3$

$= 9I_2 - 2I_3$ [in view of Eq. (i)]

$= 27I_3 - 2I_3$ [in view of Eq. (ii)]

or $50 = 25I_3$

$\therefore\ I_3 = 2$A

Thus, a current of 2 amp will flow from *A* to *B*.

12.64. In the series circuit shown in Fig. 12.42a *E, F, G* and *H* are cells of emf 2, 1, 3 and 1V, respectively and their internal resistances are 2, 1, 3 and 1 Ω respectively. Calculate (i) the potential difference between *B* and *D* and (ii) the potential difference across the terminals of each of the cells *G* and *H*. (IIT 1984)

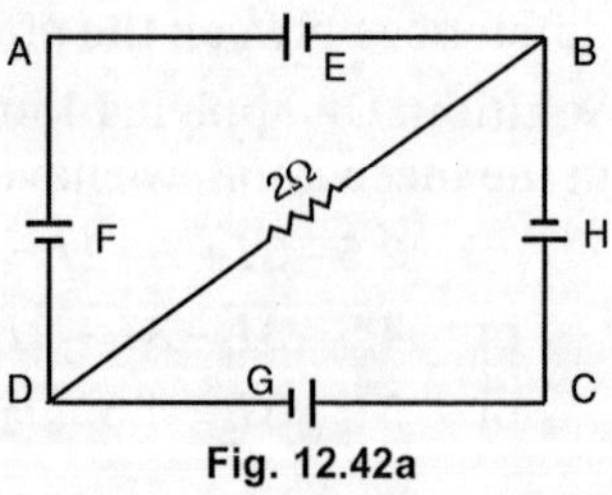

Fig. 12.42a

Solution: The current flowing in the circuit is shown in Fig. 12.42b. On applying Kirchhoff's loop theorem in the mesh *DCB*, we have

$$-3I_1 + 3 - 1 - I_1 - 2(I_1 - I_2) = 0$$

or $\quad 2 = 6I_1 - 2I_2$

or $\quad 1 = 3I_1 - I_2 \qquad \text{... (i)}$

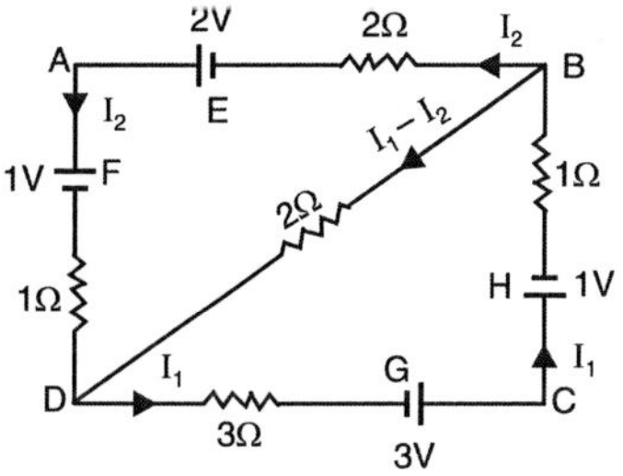

Fig. 12.42b

Similarly in the loop *BAD*, we have

$$-2I_2 + 2 - 1 - I_2 + 2(I_1 - I_2) = 0$$

or $\quad 1 = 5I_2 - 2I_1 \qquad \text{... (ii)}$

In view of Eqs. (i) and (ii), we have

$$3I_1 - I_2 = 5I_2 - 2I_1$$

or $\quad 5I_1 = 6I_2$

$\therefore \quad I_1 = \frac{6}{5}I_2$

With this identity, we have

$$1 = 3I_1 - \frac{5}{6}I_1$$

$$6 = 13I_1$$

or $\quad I_1 = \frac{6}{13}\text{A}$

$$I_2 = \frac{5}{6}I_1 = \frac{5}{6} \times \frac{6}{13}\text{A} = \frac{5}{13}\text{A}$$

$$V_{BD} = V_B - V_D = (I_1 - I_2) \times 2 = 2 \times \frac{1}{13} = \frac{2}{13} = 0.154\text{V}$$

Potential difference across the terminals of $G = 3 - \frac{6}{13} \times 3$

$$= \frac{21}{13} = 1.615 \text{ V}$$

Potential difference across the terminals of $H = 1 + \frac{6}{13}$

$$= \frac{19}{13} = 1.46 \text{ V}$$

12.65. In the given circuit (Fig. 12.43), determine (a) the charge on each capacitor and (b) the potential difference between *a* and *b*.

Solution: On applying Kirchhoff's law to the resistive circuit, we have

$$12 - I - 6I - 3I = 0$$

or $\quad 10I = 12$

or $\quad I = 1.2 \text{ A}$

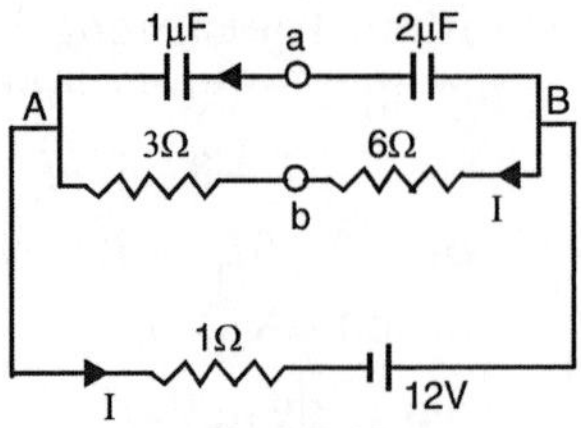

Fig. 12.43

$$V_{BA} = V_B - V_A = 9 \times 1.2 = 10.8\,\text{V}$$

$$\text{Total capacity} = \frac{2}{3}\,\mu\text{F}$$

$$\therefore \quad Q = \frac{2}{3} \times 10.8 = 2 \times 3.6 = 7.2\,\mu\text{C}$$

$$V_{Aa} = V_A - V_a = -7.2\text{V}$$

$$V_{Ab} = V_A - V_b = -3 \times 1.2 = -3.6\,\text{V}$$

$$V_{ab} = V_a - V_b = V_A - V_b - (V_A - V_a) = -3.6 + 7.2 = 3.6\,\text{V}$$

12.66. Calculate the steady current in 2 Ω resistor shown in Fig. 12.44. The internal resistance of the battery is negligible and the capacitance of the condenser is 0.2 μF. (IIT 1982)

Solution: On applying Kirchhoff's law to the mesh *abcd*, we have

$$6 - 3I_2 - 2.8I = 0$$

$$\text{or} \quad 6 = 2.8I_1 + 5.8I_2; \quad I = I_1 + I_2$$

In the mesh formed by 2 Ω and 3 Ω resistance, we have $2I_1 = 3I_2$

$$\text{or} \quad I_2 = \frac{2}{3} I_1$$

With this value of I_2, we have

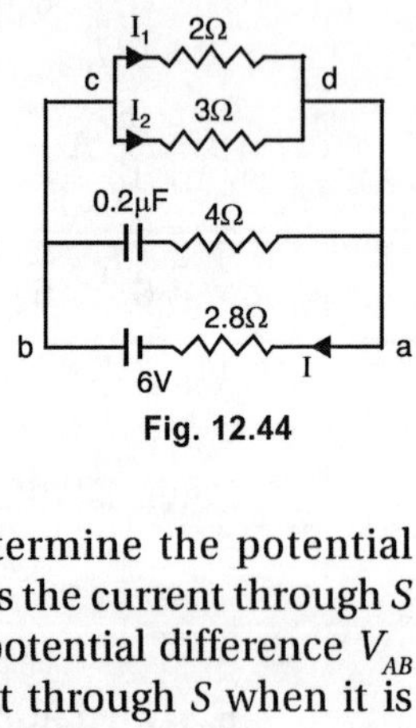

Fig. 12.44

$$6 = 2.8\,I_1 + 5.8 \times \frac{2}{3} I_1$$

$$\text{or} \quad 18 = 8.4\,I_1 + 11.6I_1$$

$$\text{or} \quad 20I_1 = 18$$

$$\text{or} \quad I_1 = \frac{18}{20} = 0.9\text{ A}$$

12.67. In the circuit shown in Fig. 12.45a, determine the potential difference V_{AB} when (a) switch *S* is open, (b) what is the current through *S* when it is closed? (c) In Fig. 12.45b what is the potential difference V_{AB} when switch is open, and (d) what is the current through *S* when it is closed?

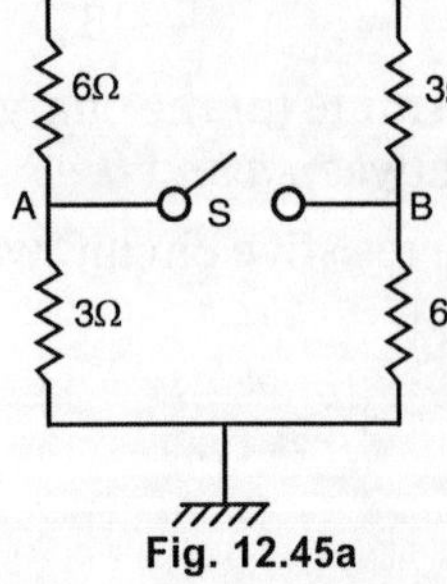

Fig. 12.45a

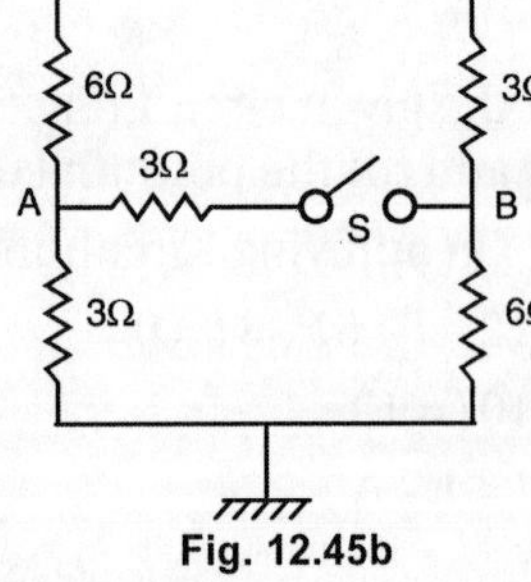

Fig. 12.45b

Solution:

(a) When S is open, the current distribution is shown in Fig. 12.45c. Potential difference across the terminals of the battery through the two paths is given by

$$9I_1 = 9I_2 = 36$$

or $I_1 = I_2 = 4$ amp ... (i)

$V_{BO} = V_B - V_O = V_B$ since V_O is the potential of the point which is earthed and is equal to zero

$\therefore$ $V_B = 6 \times 4 = 24$ V ... (ii)

$V_{AO} = V_A - V_O = 3 \times 4 = 12$ V ... (iii)

$\therefore$ $V_A = 12$ V

$\therefore$ $V_{AB} = V_A - V_B = 12 - 24 = -12$ V ... (iv)

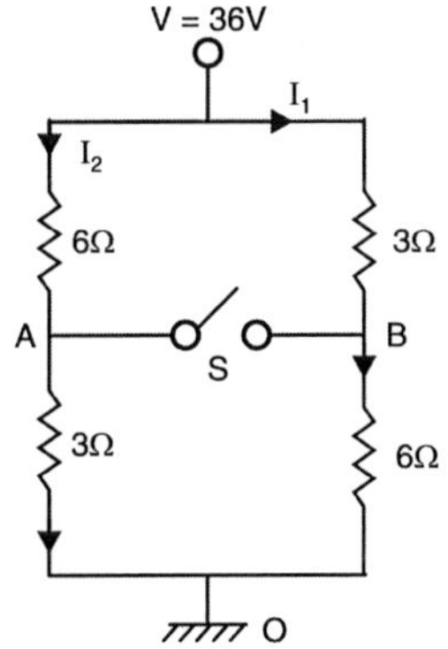

Fig. 12.45c

(b) When the switch S is closed, the current distribution is shown in Fig. 12.45d. The potential difference across the terminals of the battery through the two paths is given by

$36 = 3I_1 + 6I_1 - 6I_3$ ... (v)

$36 = 6I_2 + 3I_2 + 3I_3$ ... (vi)

or $9I_1 - 6I_3 = 36 = 9I_2 + 3I_3$ ... (vii)

In the first mesh, we have

$$3I_1 = 6I_2 \text{ or } I_1 = 2I_2$$

On substituting the values of I_1 in Eq. (vii), we have

$$18I_2 - 6I_3 = 36 = 9I_2 + 3I_3$$

On solving it $I_3 = 3$ A

Exactly in the similar way, proceed for the second circuit

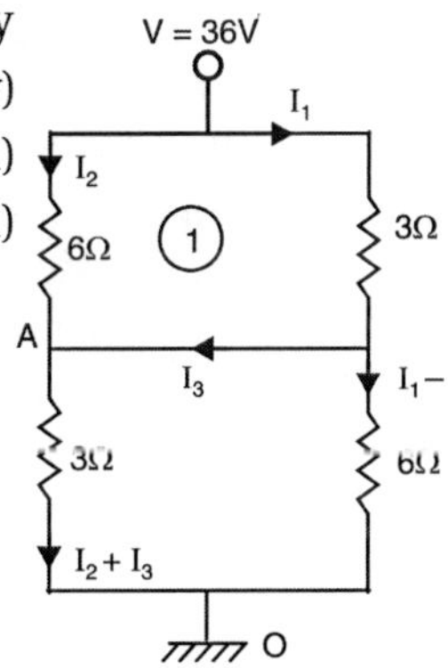

Fig. 12.45d

(c) $V_{AB} = -12$ V

(d) $I_3 = \dfrac{12}{7}$ A

12.68. In the circuit shown in Fig. 12.46, find the potential difference between points A and B when the (a) switch S is open, (b) what is the final potential of point B when the switch S is closed and (c) how much does the charge on each capacitor change when S is closed?

Solution:

(a) When the switch S is open, the point B is at zero potential whereas A is at 18 V. Hence the potential difference between A and B = 18 V.

(b) When the switch is closed, current passes only through the resistances. Let this current be I. According to Kirchhoff's law, we have

$18 = 3I + 6I$ or $I = 2$ A

Now $V_{BO} = V_B - V_O = 3 \times 2 = 6$ V

Since $V_O = 0$

$\therefore \quad V_B = 6$ V

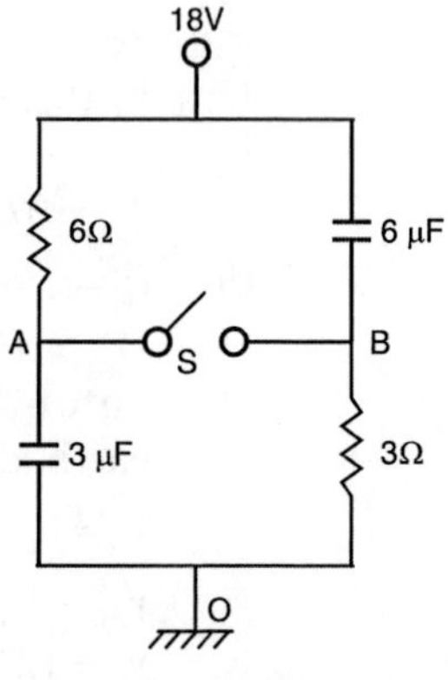

Fig. 12.46

(c) When the switch S is open, the 6 μF capacitor will contain 108 μC charge and the 3 μF capacitor will have a charge of 54 μF. When the switch is closed, the potential difference across the 6 μF and the 3 μF capacitors is 12 volt and 6 volt, respectively. The two capacitors will now have 72 μF and 18 μC of charge respectively. The change in charge in each case = 36 μC.

12.69. Determine the potential difference between points A and B in the circuit shown in the Fig. 12.47a. If A and B are connected, calculate the current in the 12 volt cell.

Solution: Since AB is open, no current will flow along *ef*. Consider a mesh *abcd*. Let us assume that in this mesh, I current is flowing. Now according to Kirchhoff's loop theorem, we have

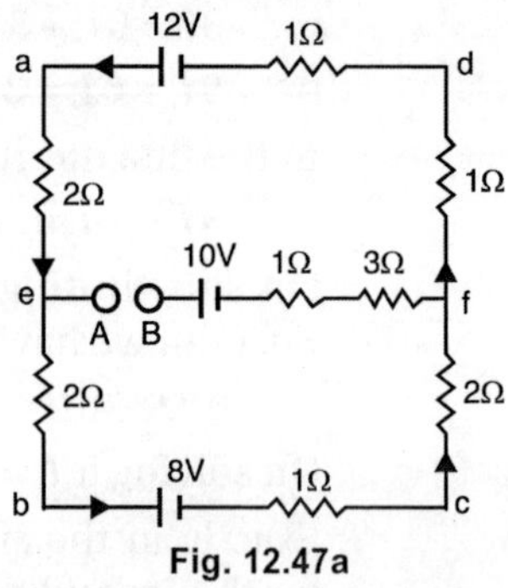

Fig. 12.47a

$$-I + 12 - 2I - 2I - 8 - I - 2I - I = 0$$

$$4 = 9I$$

or $\quad I = \dfrac{4}{9}$ A

The potential difference across *ef*

$$= 12 - 4 \times \frac{4}{9} = 10.22 \text{ V}$$

Now $V_{AB} + 10 = 10.22$

$\therefore \quad V_{AB} = 0.22$ V

On applying Kirchhoff's law in mesh *abcd* of Fig. 12.47b, we have

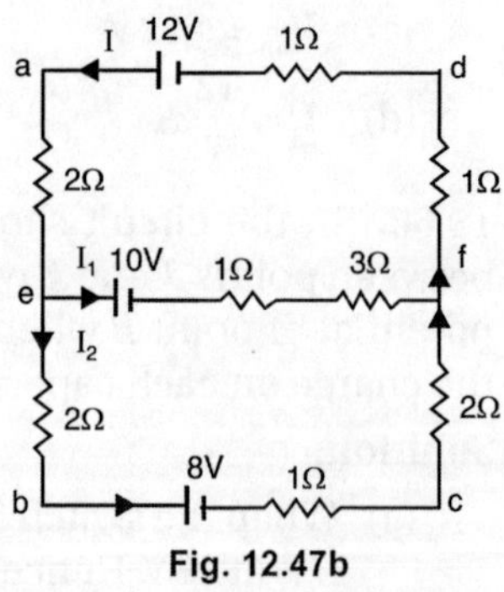

Fig. 12.47b

$$12 - 2I - 2I_2 - 8 - I_2 - 2I_2 - I - I = 0$$

$$4 = 4I + 5I_2$$

or $\quad 4 = 4I_1 + 9I_2 \quad \text{... (i)}$

Similarly in the mesh *aefd*, we have

$$12 - 2I - 10 - I_1 - 3I_1 - I - I = 0$$

$$2 - 4I + 4I_1$$

or $$2 = 8I_1 + 4I_2 \quad \text{... (ii)}$$

On solving the two equations, we have

$$I_1 = \frac{1}{28}\text{ A and } I_2 = \frac{12}{28}\text{ A}$$

$$\therefore \quad I = \frac{1}{28} + \frac{12}{28} = \frac{13}{28}\text{ A}$$

12.70. In the circuit shown in Fig. 12.48a, $E_1 = 3$ V, $E_2 = 2$ V and $E_3 = 1$ V, $R = r_1 = r_2 = r_3 = 1\ \Omega$. (i) Find the potential difference between the points A and B and current in each arm. (ii) If r_2 is short circuited and the point A is connected to point B, then find the currents through E_1, E_2 and E_3 and resistor R. (IIT 1981)

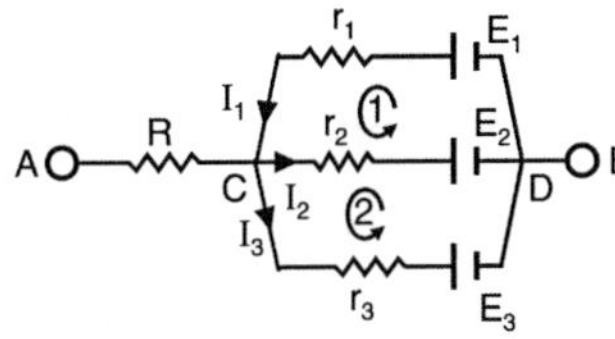

Fig. 12.48a

Solution:

(i) On applying Kirchhoff's loop theorem in the first mesh, we have

$3 - I_1 - I_2 - 2 = 0$ or $I_1 + I_2 = 1$ ... (i)

Similarly in the second mesh, we have

$2 + I_2 - I_3 - 1 = 0$ or $I_3 - I_2 = 1$ ... (ii)

Further in view of the current distribution, we have $I_1 = I_2 + I_3$

On substituting the value of I_1 in Eq. (i), we have

$2I_2 + I_3 = 1 = I_3 - I_2$; $I_2 = 0$ and $I_1 = I_3 = 1$ A

Potential difference across AB = Potential difference across CD, since no current is flowing through the resistance R.

Therefore

$$V_{AB} = V_{CD} = V_C - V_D = 2\text{ V}$$

(ii) In the first mesh of Fig. 12.48b, we have

$3 - I_1 - 2 = 0$ or $I_1 = 1$ A

In the second mesh, we have

$-I_3 - 1 + 2 = 0$ or $I_3 = 1$ A

In the third mesh

$-I_4 + 1 + I_3 = 0 \ \therefore I_4 = 2$ A

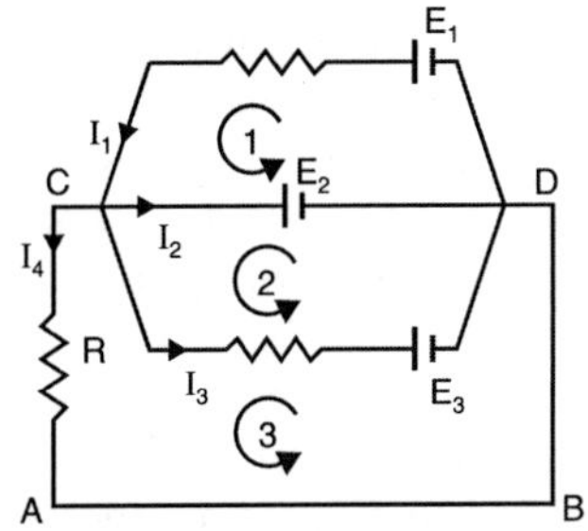

Fig. 12.48b

According to the current distribution in the circuit, we have

$$I_1 = I_2 + I_3 + I_4$$

$$\therefore \quad I_2 = I_1 - I_3 - I_4 = 1 - 1 - 2 = -2 \text{ A}$$

12.71. A condenser of capacity $C = 4.00$ μF is connected to a d.c. source of emf $E = 160$ V as shown in Fig. 12.49. After charging the condenser, switch S was thrown over from contact 1 to contact 2. Determine the amount of heat generated in the resistance $R_1 = 500\ \Omega$ if $R_2 = 300\ \Omega$.

Solution: When the switch S is in position 1, the condenser will be in the charging condition. The energy stored in the capacitor is equal to $\frac{1}{2}CE^2$. On transferring the switch S to position 2, the energy thus stored will pass through the two resistors R_1 and R_2, i.e.,

$$\frac{1}{2}CE^2 = I^2(R_1 + R_2)t$$

$$\therefore \ I^2 t = \frac{1}{2}\frac{CE^2}{R_1 + R_2}$$

Heat generated in the resistor

$$R_1 = I^2 R_1 t = \frac{1}{2}\frac{CE^2 R_1}{R_1 + R_2}$$

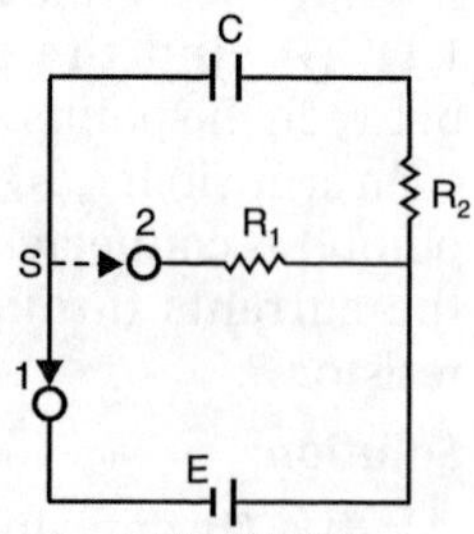

Fig. 12.49

On substituting the values of various terms, the heat generated is given by

$$\frac{\frac{1}{2} \times 4 \times 10^{-6} \times 160^2 \times 500}{800} = 32.0 \text{ mJ}$$

12.72. A part of a circuit in a steady state along with the currents flowing in its branches and the resistances is shown in the Fig. 12.50. Calculate the energy stored in the capacitor C (4 μF). (IIT 1986)

Solution: Total current at A = 3 amp. This current will flow in the part AD and DE of the circuit. At E, a current of 2 amp will pass along 4 Ω resistance and 1 amp along EB. At B, the total current is 2 amp and this will pass through the 1 Ω resistance and the cell of emf 3 V. According to Kirchhoff's loop theorem, the potential difference V_{AB} across the capacitor C is given by

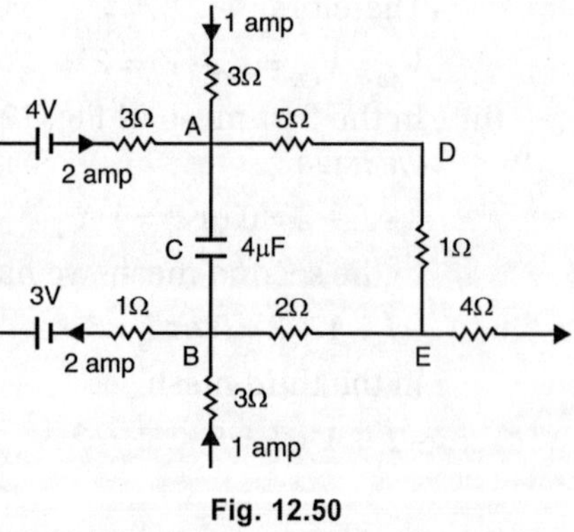

Fig. 12.50

$V_{AB} = V_A - V_B = 2 \times 1 + 1 \times 3 + 5 \times 3 = 20$ volt

Energy stored in the capacitor $= \frac{1}{2}CV^2 = \frac{1}{2} \times 4 \times 10^{-6} \times 400$

$= 8 \times 10^{-4}$ Joule

12.73. The potentiometric arrangement shown in the Fig. 12.51 can be used to apply potential V across a device having resistance R. The length of the wire of the potentiometer is l and its resistance is R_0. A voltage V_0 is applied at its terminals. Find the voltage across the device as a function of distance x. Investigate the case when $R >> R_0$.

Solution: When $R >> R_0$ the voltage across is given by

$$V_R = \frac{V_0}{R_0} \cdot \frac{xR_0}{l} = \frac{V_0 \cdot x}{l}$$

When R is not very large then

$$I_1 \cdot \frac{x}{l} \cdot R_0 = I_2 R$$

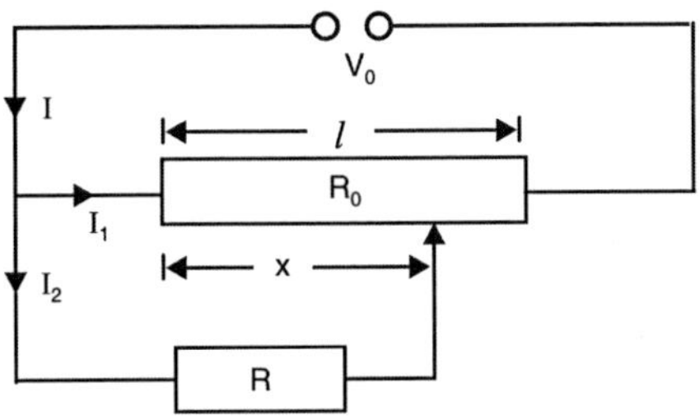

Fig. 12.51

$$V_0 = I_2 R + \frac{(l-x)}{l} \cdot R_0 (I_1 + I_2)$$

$$V_0 = V_R + \frac{l-x}{l} \cdot R_0 (I_1 + I_2)$$

$$= V_R + \left(\frac{l-x}{l}\right) R_0 \left(\frac{V_R \cdot l}{xR_0} + \frac{V_R}{R}\right)$$

$$V_R = \frac{V_0 Rxl}{Rxl + (l-x)lR + (R-x)R_0 x}$$

$$= \frac{V_0 Rx}{Rl + R_0 x\left(1 - \frac{x}{l}\right)}$$

12.74. In the circuit shown in Fig. 12.52, each battery is of 5 V and has an internal resistance of 0.2 ohm. What is the reading of an ideal voltmeter.

Solution: Let I be the current in the circuit then according to Kirchhoff's law

$$8 \times 5 - 8 \times 0.2\, I = 0$$

or $I = 25$ A

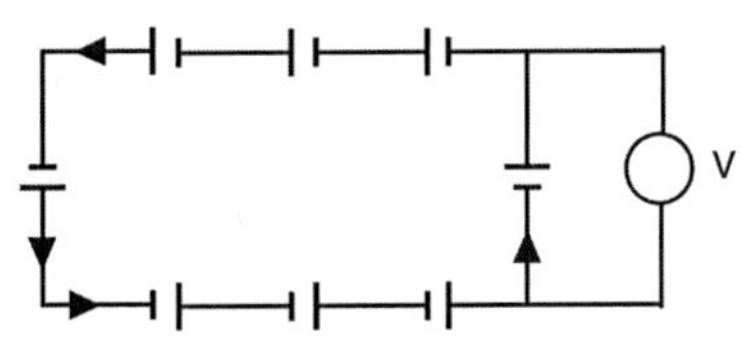

Fig. 12.52

Potential difference across the cell across which the voltmeter is connected is given by

$$V = 5 - 25 \times 0.2 = 0 \text{ volt}$$

12.75. 24 identical current sources have an emf of $E = 1.00$ V and an internal resistance of $r = 0.2\ \Omega$. The sources are connected so as to form a battery of n series connected sections each of which consists of m sources connected in parallel. An element of resistance $R = 0.3\ \Omega$ is connected across the battery. For what value of n will the power in the element will be maximum? What is the maximum value of power?

Solution: Current for a mixed grouping is given by

$$I = \frac{mnE}{mR + nr}$$

The current assumes a maximum value when $mR = nr$

or $$\frac{m}{n} = \frac{0.2}{0.3} = \frac{2}{3}$$

$$mn = 24$$

$$\frac{2}{3}n^2 = 24$$

$$n = 6$$

$$I_{max} = \frac{mnE}{2\sqrt{mnRr}} = \sqrt{\frac{mn}{Rr}} \cdot \frac{E}{2}$$

Maximum power $P = I_{max}^2 R = \dfrac{mnE^2}{4r} = \dfrac{24 \times 1}{4 \times 0.2} = 30$ watt

12.76. A battery E_2 of emf 6 volt and internal resistance 0.4 Ω has been charged by a battery of emf $E_1 = 27$ volt and internal resistance 0.6 Ω. For this a resistance of 6 Ω is connected to the battery as shown in the Fig. 12.53. Total time of charging is 3.5 min. Find (i) total energy given to the circuit (ii) energy given to the outer circuit (iii) total energy given to the 6 volt battery (iv) changed chemical energy by this battery (v) total dropped energy in the circuit in the form of heat.

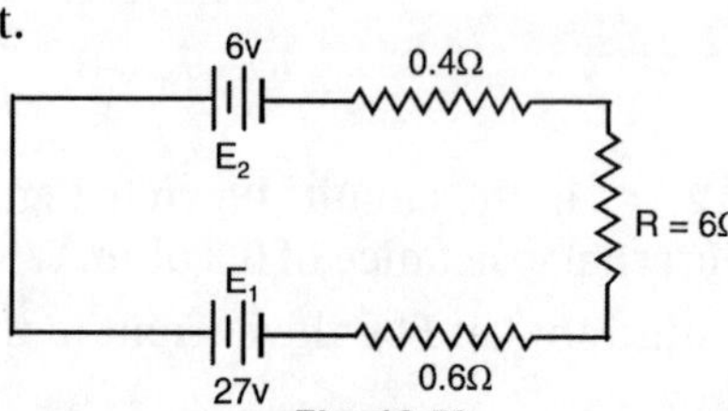

Fig. 12.53

Solution: According to Kirchhoff's law, we have

$$27 - 6 - 0.4\,I - 6\,I - 0.6\,I = 0$$

$$I = 3.0 \text{ amp}$$

$$Q = I \cdot t = 3 \times 3.5 \times 60 = 630 \text{ coul}$$

(i) $W_1 = E_1 Q = 27 \times 630 = 17010$ J

(ii) V_1 = voltage across the cell $= 6 + 0.4 \times 3 = 7.2$ volt

V_2 = voltage across the resistance $R = 6 \times 3 = 18$ volt

$V = V_1 + V_2 = 25.2$ volt

$W_2 = VQ = 25.2 \times 630 = 15876$ Joule

(iii) $W_3 = V_1 Q = 7.2 \times 630 = 4536$ J

(iv) $W_4 = E_2Q = 6 \times 630 = 3780$ J

(v) $W_5 = W_1 - W_4 = 13230$ J

12.77. A battery of internal resistance 4 Ω is connected to the network of resistances as shown in the Fig. 12.54. What must be the value of R so that maximum power is delivered to the network? What is this maximum power? (IIT 1995[s])

Solution: A study of section *ACDH* shows that it is a balanced wheatstone bridge and therefore no current will pass through 6 R resistance connected in *BF* arm.

Equivalent resistance between *AD* is

$$\frac{1}{R'} = \frac{1}{3R} + \frac{1}{6R} = \frac{1}{2R}$$

$$R' = 2R$$

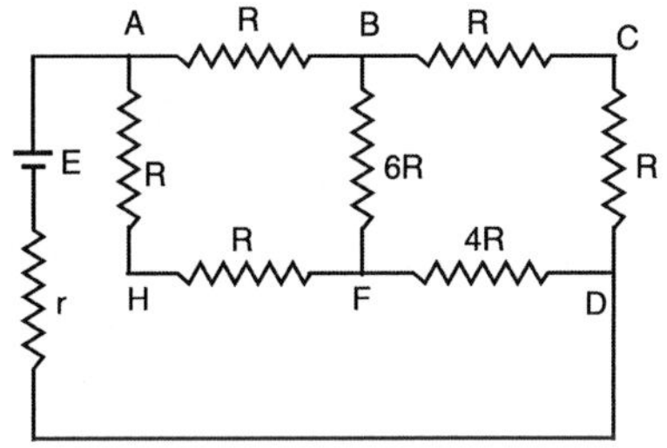

Fig. 12.54

For maximum power

$$2R = r = 4\ \Omega$$

$$\therefore \quad R = 2\ \Omega$$

$$I = \frac{E}{r + 2R} = \frac{E}{8}$$

$$\text{Maximum power} = I^2 \cdot 2R = \frac{E^2}{64} \times 4 = \frac{E^2}{16}$$

12.78. An infinite ladder network of resistances is connected with 1 Ω and 2 Ω resistances as shown in Fig. 12.55a. The 6 V battery between A and B has negligible resistance (a) Show that the effective resistance between A and B is 2 Ω (b) What is the current that passes through 2 Ω resistance nearest to the battery? (IIT 1987; Roorkee 1989)

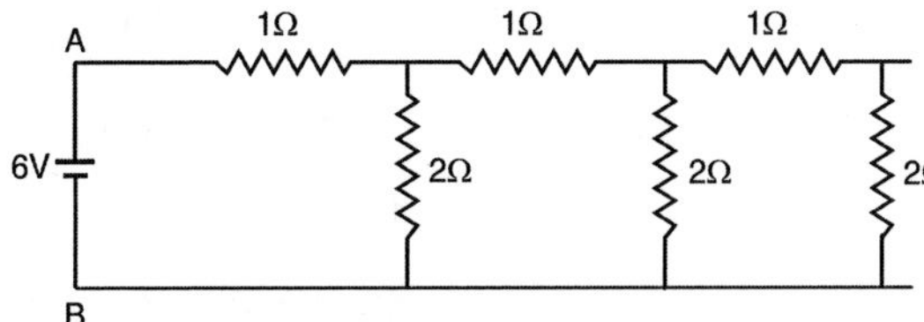

Fig. 12.55a

Solution: In view of Fig. 12.55b R_{eq} is given by

$$R_{eq} = 1 + \frac{2R_{eq}}{2 + R_{eq}}$$

or $$R_{eq}^2 - R_{eq} - 2 = 0$$

or $$R_{eq} = \frac{1}{2}\left[1 \pm \sqrt{1+8}\right] = 2\ \Omega$$

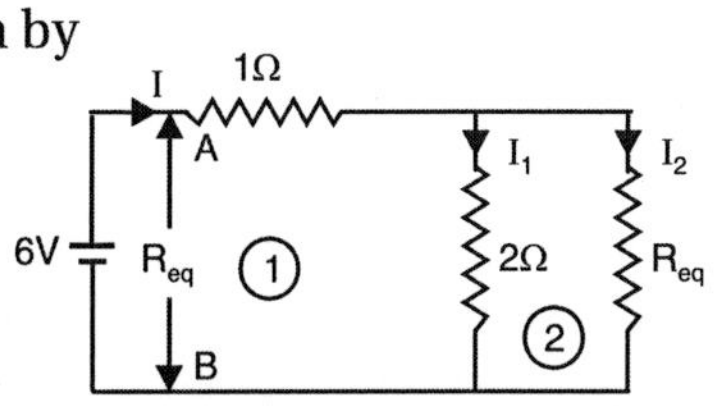

Fig. 12.55b

In the second mesh

$$2I_1 = R_{eq} \cdot I_2 = 2I_2$$

$\therefore \quad I_1 = I_2$

In the first mesh we have

$$6 - 1 \times 2I_1 - 2I_1 = 0$$

$$4I_1 = 6$$

$$I_1 = 1.5 \text{ amp}$$

12.79. In the given circuit

$$E_1 = 3\text{V}, E_2 = 2\text{V}, E_3 = 6\text{V}$$

$$R_1 = 2R_4 = 6\,\Omega$$

$$R_3 = 2R_2 = 4\,\Omega$$

and $C = 5\,\mu\text{F}$

Find the current in R_3 and energy stored in the capacitor. (IIT 1988)

Solution: In mesh 1 we have

$$6 - 4I_1 = 0$$

$\therefore \quad I_1 = 1.5$ amp

In mesh 2 we have

$$-2 - 2I_2 + 4I_1 - 3I_2 - 3 = 0$$

$$-5 - 5I_2 + 6 = 0$$

$$5I_2 = 1$$

$\therefore \quad I_1 = \dfrac{1}{5} = 0.2$ amp

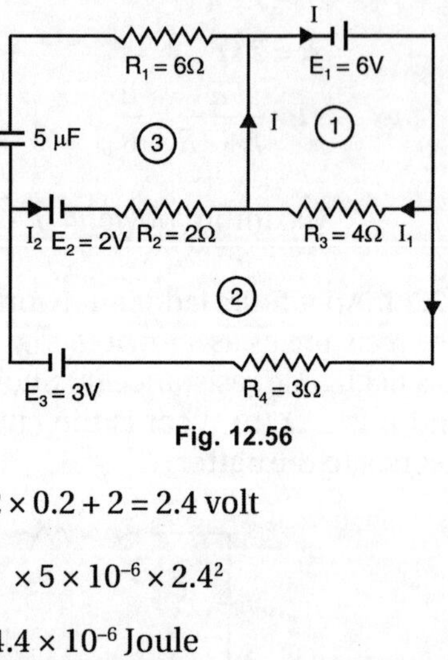

Fig. 12.56

$\therefore$ current in $R_3 = 1.5$ amp

Voltage across the capacitor $= 2 \times 0.2 + 2 = 2.4$ volt

$$\text{Energy stored in capacitor} = \frac{1}{2} \times 5 \times 10^{-6} \times 2.4^2$$

$$= 14.4 \times 10^{-6} \text{ Joule}$$

12.80. Two cells of emf E and internal resistances r_1 & r_2 are connected as shown in the Fig. 12.57. For what value of R the voltmeter reading is zero?

Solution: Current in the circuit is given by

$$I = \frac{2E}{R + r_1 + r_2}$$

Now $V = E - Ir_1 = 0$

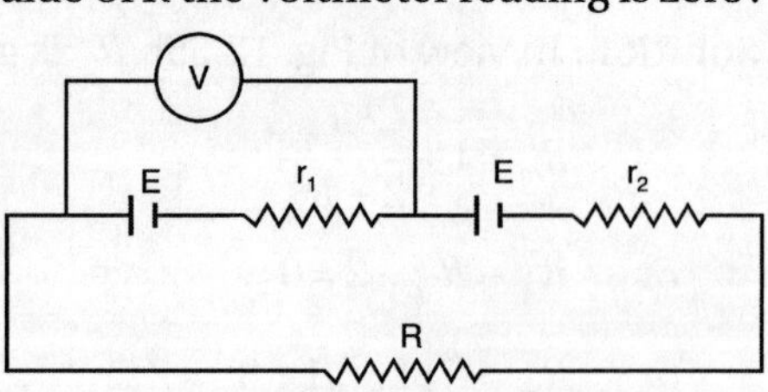

Fig. 12.57

$$\therefore \qquad E = Ir_1 = \frac{2Er_1}{R + r_1 + r_2}$$

$$R + r_1 + r_2 = 2r_1$$

$$\therefore \qquad R = r_1 - r_2$$

12.81. An ammeter and a voltmeter are connected in a series to a battery of emf E and having negligible internal resistance. On putting a wire of resistance R across the voltmeter it has been found that its value decreases η times whereas the ammeter reading increases η times. Find the reading of the voltmeter after the introduction of the resistance.

Solution: Let R_A and R_V be the resistances of the ammeter and voltmeter respectively.

$$\text{Current in the circuit } I = \frac{E}{R_A + R_V}$$

$$\text{Voltmeter reading} = \frac{ER_V}{R_A + R_V}$$

After introducing a resistance across the voltmeter we have current I' as

$$I' = \frac{E}{R_A + \dfrac{R \cdot R_V}{R + R_V}} = \frac{(R + R_V)E}{R_A(R + R_V) + RR_V}$$

Further $I_1R_V = I_2R$; I_1 and I_2 are the currents through voltmeter and resistance

$$\therefore \quad I_2 = I_1 \frac{R_V}{R}$$

$$I' = I_1 + I_1 \frac{R_V}{R} = I_1 \frac{(R + R_V)}{R}$$

$$I_1 = \frac{ER}{R_A(R + R_V) + RR_V}$$

$$\text{Voltmeter reading} = \frac{ERR_V}{R_A(R + R_V) + RR_V}$$

$$\frac{ERR_V}{R_A(R + R_V) + RR_V} = \frac{ER_V}{\eta(R_A + R_V)} \quad \text{(given)}$$

$$R\eta(R_A + R_V) = R_AR + R_AR_V + RR_V$$

$$RR_V(\eta - 1) = R_A[R + R_V - \eta R]$$

$$\therefore \quad R_A = \frac{RR_V(\eta - 1)}{R_V - (\eta - 1)R}$$

$$\eta \cdot \frac{E}{R_A + R_V} = \frac{(R + R_V)E}{R_A(R + R_V) + RR_V}$$

$$R_A\,\eta\,(R+R_V)+\eta\,R\,R_V=R_A\,(R+R_V)+R\,R_V+R_V^{\,2}$$

$$R_A\,(R+R_V)\,(\eta-1)+(\eta-1)\,R\,R_V=R_V^{\,2}$$

$$\frac{R(\eta-1)}{R_V-(\eta-1)R}(R+R_V)(\eta-1)+(\eta-1)R=R_V$$

$$R(\eta-1)^2(R+R_V)+(\eta-1)RR_V-(\eta-1)^2R^2=R_V^2-(\eta-1)RR_V$$

$$+RR_V(\eta-1)^2+2(\eta-1)RR_V=R_V^2$$

$$R\left[\eta^2-2\eta+1+2\eta-2\right]=R_V$$

$$R\left[\eta^2-1\right]=R_V$$

$$R_A=\frac{R\cdot R\left(\eta^2-1\right)(\eta-1)}{R\left(\eta^2-1\right)-(\eta-1)R}$$

$$=\frac{R\left(\eta^2-1\right)}{\eta}$$

$$\text{Voltmeter reading}=\frac{E\cdot R\cdot R\left(\eta^2-1\right)}{\dfrac{R\left(\eta^2-1\right)}{\eta}\left[R+R\eta^2-R\right]+R\left(\eta^2-1\right)R}$$

$$=\frac{ER^2\left(\eta^2-1\right)}{R^2\left[\eta\left(\eta^2-1\right)+\left(\eta^2-1\right)\right]}$$

$$=\frac{E}{\eta+1}$$

12.82. In the circuit shown in the Fig. 12.58 find the potential difference between point *a* & *b*.

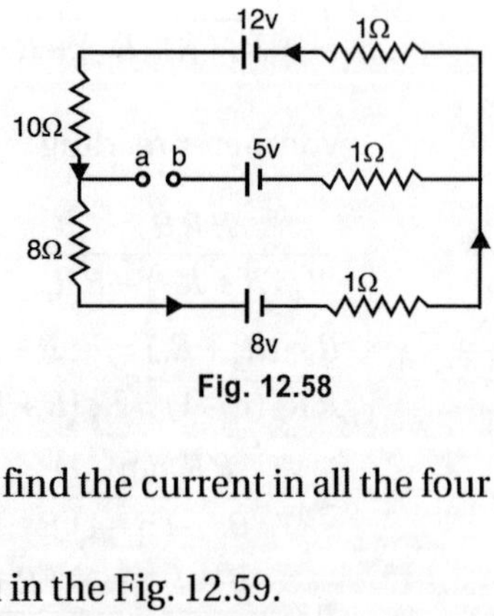

Fig. 12.58

Solution:

$$12-10\,I-8\,I-8-I-I=0$$

$$4=20\,I$$

$$I=\frac{4}{20}=0.2\text{ amp}$$

$$V_{ab}=V_a-V_b\;=-5-0.2+12-10\times0.2$$

$$=4.8\text{ volt}$$

12.83. In the circuit shown in the Fig. 12.59 find the current in all the four resistors.

Solution: The current distribution is shown in the Fig. 12.59.

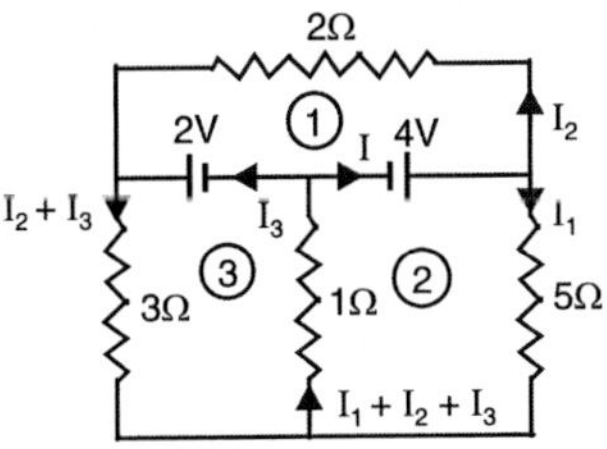

Fig. 12.59

In the first mesh we have

$$4 - 2I_2 - 2 = 0$$

$$\therefore \quad I_2 = 1 \text{ amp} \qquad \text{... (1)}$$

In the second mesh we have

$$4 - 5I_1 - I_1 - I_2 - I_3 = 0$$

$$4 = 6I_1 + I_2 + I_3$$

$$\text{or} \quad 6I_1 + I_3 = 3 \qquad \text{... (2)}$$

In the third mesh we have

$$2 - 3I_2 - 3I_3 - I_1 - I_2 - I_3 = 0$$

$$2 = I_1 + 4I_2 + 4I_3$$

$$\text{or} \quad I_1 + 4I_3 = -2 \qquad \text{... (3)}$$

$$6I_1 + I_3 = 2$$

$$24I_1 + 4I_3 = 8$$

$$I_1 + 4I_3 = -2$$

$$23I_1 = 10$$

$$\therefore \quad I_1 = \frac{10}{23} \text{ amp}$$

$$4I_3 = -2 - \frac{10}{23}$$

$$= -\frac{56}{23}$$

$$\therefore \quad I_3 = -\frac{14}{23}$$

Current in 5 Ω resistor = $\frac{10}{23}$ amp

Current in 2 Ω resistor = 1 amp

Current in 3 Ω resistor = $\frac{9}{23}$ amp

Current in 1 Ω resistor = $\frac{19}{23}$ amp

12.84. Find the emf E of the cell shown in the Fig. 12.60.

Solution: In the first mesh we have

$$E - (I_1 + I_2) - 3(I_1 + I_2) - 3I_1 + 4 = 0$$

$$\text{or} \quad E + 4 = 7I_1 + 4I_2$$

In the second mesh we have

$$-3I_1 + 4 + 5I_2 = 0$$

$$-3(2-I_2)+4+5I_2=0$$
$$-6+3I_2+4+5I_2=0$$
$$8I_2=2$$
$$I_2=\frac{2}{8}=0.25\text{ amp}$$
$$I_1=2-0.25=1.75\text{ amp}$$
$$E=7\times1.75+4\times0.25-4$$
$$=12.25+1-4$$
$$=9.25\text{ volt}$$

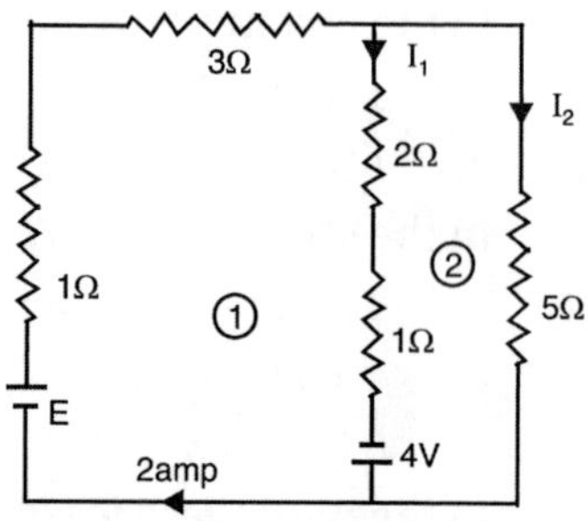

Fig. 12.60

12.85. A cell of emf 3.4 volt and internal resistance 3 Ω is connected to an ammeter having resistance 2 Ω and to an external resistance of 100 Ω. When a voltmeter is connected across the 100 Ω resistance the ammeter reading is 0.04 amp. Find the voltage read by the voltmeter and its resistance. Had the voltmeter been an ideal one what would have been its reading? (Roorkee 1990)

Solution: Let R be resistance of the voltmeter. Then

$$I=0.04=\frac{3.4}{3+2+\dfrac{100R}{100+R}}\quad\text{(see Fig. 12.61)}$$

On solving, we get

$$R=400\ \Omega$$

Now $100\times I_1=400\,(I-I_1)$

$$500\,I_1=400\,I$$

$$\therefore\quad I_1=\frac{4}{5}\times0.4=0.032\text{ amp}$$

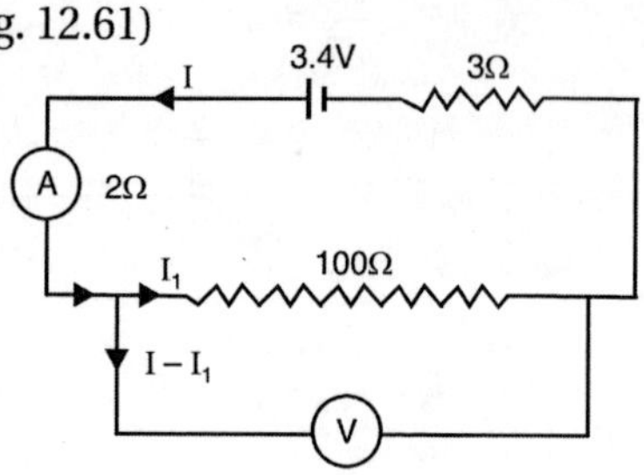

Fig. 12.61

Reading of the voltmeter = 100 × 0.032 = 3.2 volt

When the voltmeter is an ideal one then

$$I=\frac{3.4}{3+2+100}=0.03238\text{ amp}$$

Reading of the voltmeter = 3.238 volt

12.86. A circuit containing a two position switch S as shown in the Fig. 12.62a.

(a) The switch S is in position 1. Find the potential difference V_A-V_B and the rate of production of Joule heat in R_1.

(b) If now the switch S is put in position 2 at $t=0$ find

(i) steady current in R_4 and

(ii) the time when the current in R_4 is half of the steady value. Also calculate the energy stored in the inductor L at that time.

Solution: Applying Kirchhoff's law in 1st mesh of Fig. 12.62b

$$-2I_1 + 12 - 2(I_1 + I_2) = 0$$

$$12 = 4I_1 + 2I_2$$

or $\quad 2I_1 + I_2 = 6 \qquad \text{... (i)}$

In the second mesh we have

$$-2I_1 + 12 + 2I_2 - 3 = 0$$

$$9 = 2I_1 - 2I_2 \qquad \text{... (2)}$$

$$6 = 2I_1 + I_2$$

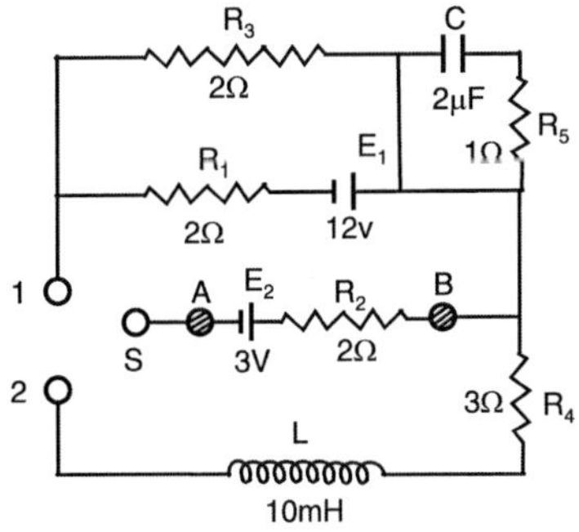

Fig. 12.62a

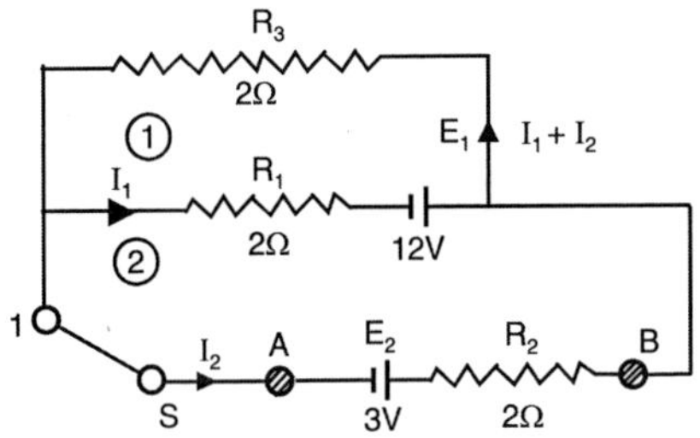

Fig. 12.62b

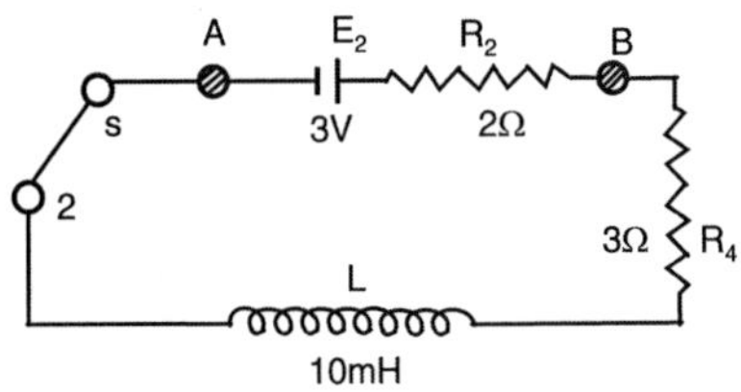

Fig. 12.62c

On solving the two equations we have

$$I_1 = 3.5 \text{ amp}, I_2 = -1 \text{ amp}$$

$$V_{AB} = V_A - V_B = 2 \times -1 - 3 = -5 \text{ volt}$$

$$\text{Joule heat in } R_1 = I_1^2 R_1 = 3.5^2 \times 2 = 24.5 \text{ J/sec}$$

(b) When the switch is in position 2 we have (see Fig. 12.62c)

Total resistance = 5 Ω

$$\text{Steady current } I_0 = \frac{E}{R} = \frac{3}{5} = 0.6 \text{ amp}$$

Now $\quad I = I_0\left[1 - e^{-\frac{Rt}{L}}\right]$

or $\quad \frac{1}{2} = \left[1 - e^{-\frac{Rt}{L}}\right]$

or $\quad e^{-\frac{Rt}{L}} = \frac{1}{2}$

or $\quad t = \frac{L}{R}\log_e 2$

On putting the values we have

$$t = \frac{10 \times 10^{-3}}{5} \times 2.303 \log_{10} 2$$

$$= 1.386 \times 10^{-3} \text{ sec}$$

Energy stored in the inductor

$$E=\frac{1}{2}LI^2 \text{ here } I=\frac{I_0}{2}=0.3\text{ amp}$$

$$=\frac{1}{2}\times 10\times 10^{-3}\times 0.09$$

$$=4.5\times 10^{-4}\text{ J}$$

12.87. In Fig. 12.63, find the value of I_1 and I_2 (a) immediately after S is closed; (b) a long time later, (c) immediately after S is opened again and (d) a long time later.

Solution:

(a) Immediately after S is closed, no current will pass through the inductance. Therefore,

$$I_1 = I_2 = \frac{9}{3} = 3\text{ A}$$

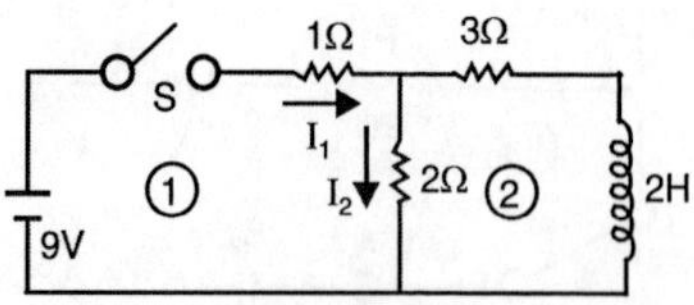

Fig. 12.63

(b) In the first mesh, we have

$9 - I_1 - 2I_2 = 0$

In the second mesh, we have

$3\,(I_1 - I_2) = 2I_2$ (resistance of the inductance is equal to zero)

or $\quad 3I_1 = 5I_2$

$$9 = I_1 + 2\times\frac{3}{5}I_1$$

$$45 = (5+6)\,I_1$$

$$\therefore \quad I_1 = \frac{45}{11} = 4.09\text{ A}$$

$$I_1 = \frac{3}{5}I_1 = \frac{3}{5}\times\frac{45}{11} = \frac{27}{11} = 2.45\text{ A}$$

(c) Immediately after S is opened again, there will be a current equal to $I_1 - I_2$ in the second mesh. This is due to the presence of inductance.

$$\therefore \quad I_2' = (I_1 - I_2) = -(4.09 - 2.45) = -1.64\text{ A}$$

(d) $I_2 = I_2' = 0$

12.88. A potentiometer wire of length 100 cm has a resistance of 10 Ω. It is connected in series with a resistance and an accumulator of emf 2 V and of negligible internal resistance. A source of emf 10 mV is balanced against a length of 40 cm of the potentiometer wire. What is the value of the external resistance? (IIT 1976)

Solution: Resistance per unit length of the wire $AB = \dfrac{10}{100} = 0.1\ \Omega$

Resistance of 40 cm length of the wire = $40 \times 0.1 = 4\ \Omega$

Current I in wire = $\dfrac{10 \times 10^{-3}}{4} = 2.5 \times 10^{-3}$A

Now $I = 2.5 \times 10^{-3} = \dfrac{2}{R+10}$

or $2.5\,R + 25 = 2000$

or $2.5\,R = 1975$

$$R = \frac{19750}{25} = 790\ \Omega$$

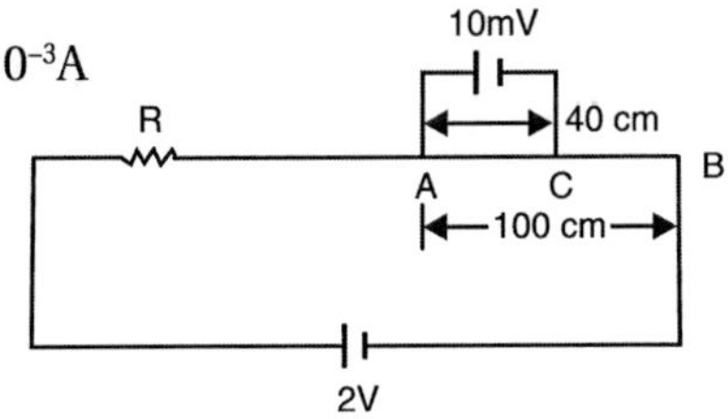

Fig. 12.64

12.89. An electrical circuit is shown in Fig. 12.65. Calculate the potential difference across the resistor of 400 Ω, as will be measured by the voltmeter V of resistance 400 Ω either by applying Kirchhoff's rule or otherwise.

(IIT 1996)

Solution: In the first mesh we have

$$-400 \times \frac{I_3}{2} + 200\,(I_1 + I_2 - I_3) + 100\,(I_2 - I_3) = 0$$

$$-2I_3 + 2I_1 + 2I_2 - 2I_3 + I_2 - I_3 = 0$$

$$2I_1 + 3I_2 = 5I_3 \qquad \dots (1)$$

In the second mesh we have

$$-100\,I_2 - 100\,(I_2 - I_3) + 100\,I_1 = 0$$

or $I_1 = I_2 + I_2 - I_3$

$= 2I_2 - I_3 \qquad \dots (2)$

or $4I_2 - 2I_3 + 3I_2 = 5I_3$

$7I_2 = 7I_3$

$\therefore \quad I_2 = I_3$

$I_1 = I_3$

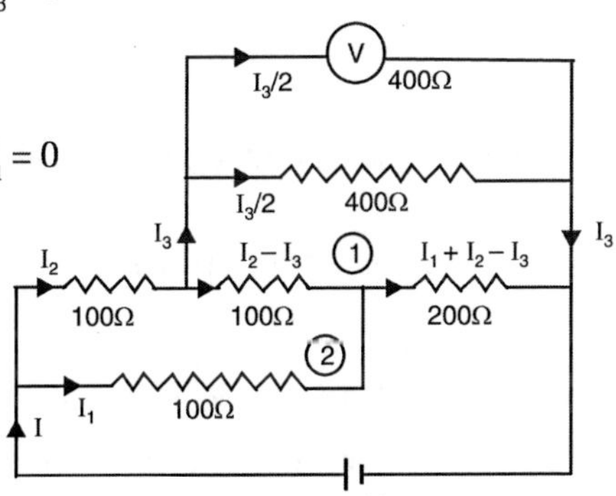

Fig. 12.65

In the third mesh we have

$$10 - 100\,I_1 + 200\,(I_1 + I_2 - I_3) = 0$$

or $10 = 100\,I_3 + 200\,I_3$

$$I_3 = \frac{10}{300} = \frac{1}{30}$$

$\therefore$ Reading of voltmeter $= \dfrac{I_3}{2} \times 400$

$$= \frac{1}{2} \times \frac{1}{30} \times 400 = \frac{20}{3} \text{ volt}$$

12.90. Find the emf (V) and internal resistance r of a single battery which is equivalent to a parallel combination of two batteries of emfs V_1 and V_2

and internal resistances r_1 and r_2 respectively with polarities as shown in Fig. 12.66a. (IIT 1997)

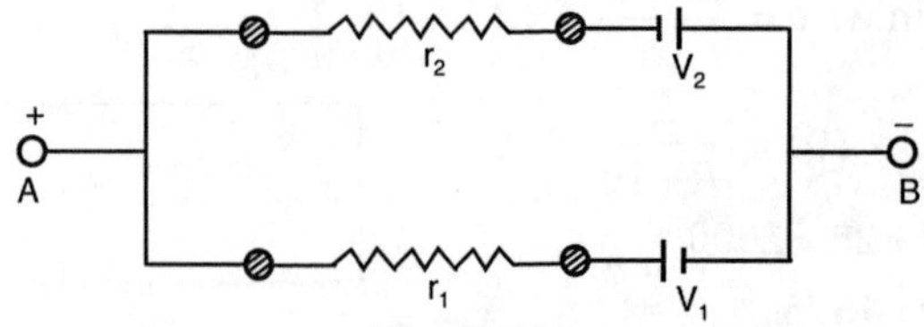

Fig. 12.66a

Solution: Applying Kirchhoff's law, we have (see Fig. 12.66b)

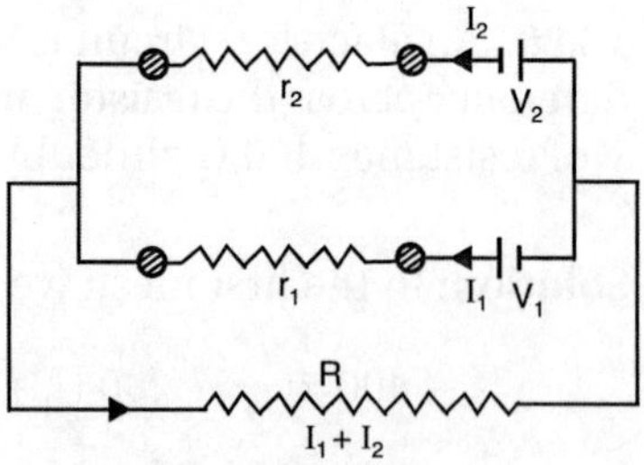

Fig. 12.66b

$$V_1 - I_1 r_1 - (I_1 + I_2) R = 0$$

$$-V_2 - I_2 r_2 - (I_1 + I_2) R = 0$$

or $$r_2 V_1 = I_1 r_1 r_2 + (I_1 + I_2) R r_2$$

$$-r_1 V_2 = I_2 r_2 r_1 + (I_1 + I_2) R r_1$$

On adding

$$V_1 r_2 - V_2 r_1 = (I_1 + I_2)\, r_1 r_2 + (I_1 + I_2)\, R\,(r_1 + r_2)$$

or $$\frac{V_1 r_2 - V_2 r_1}{r_1 + r_2} = (I_1 + I_2)\frac{r_1 r_2}{r_1 + r_2} + (I_1 + I_2) R$$

In view of the above the equivalent emf $= \dfrac{V_1 r_2 - V_2 r_1}{r_1 + r_2}$

and equivalent internal resistance $= \dfrac{r_1 r_2}{r_1 + r_2}$

12.91. Determine the mass of a silver ion when the charge on it is 1.6×10^{-19} C. Given the electrochemical equivalent of silver $= 1.118 \times 10^{-6}$ kg/C.

Solution: Electrochemical equivalent is defined as

Z = mass of one ion/charge of one ion.

$\therefore$ Mass of one ion $= Z \times$ charge on ion

$$= 1.118 \times 10^{-6} \times 1.6 \times 10^{-19}$$

$$= 1.788 \times 10^{-25} \text{ kg}$$

12.92. Using the data given, calculate the electronic charge e. Given Faraday number $= 9.65 \times 10^7$ C. Avogadro's number $= 6.02 \times 10^{26}$ kg/mol.

Solution: Faraday number f = Avogadro's number $\times\, e$

$$\therefore\ e = \frac{\text{Faraday number}}{\text{Avogadro number}} = \frac{f}{N} = \frac{9.65 \times 10^7}{6.02 \times 10^{26}} = 1.60 \times 10^{-19}\,\text{C}$$

12.93. How much current should be passed through a copper voltameter for a period of 30 minutes in order to deposit 1 g of copper? Atomic weight of silver = 107.9 g, Atomic weight of copper = 63.4 g, Electrochemical equivalent of silver = 1.118×10^{-6} kg/C.

Solution:

$$\text{Equivalent weight } W = \frac{\text{Atomic weight } A}{\text{Valency } V}$$

$$\frac{\text{Equivalent weight of silver } W_{Ag}}{\text{Equivalent weight of copper } W_{Cu}} = \frac{\text{Electrochemical equivalent of silver } Z_{Ag}}{\text{Electrochemical equivalent of copper } Z_{Cu}}$$

$$\therefore \quad Z_{Cu} = Z_{Ag} \times \frac{W_{Cu}}{W_{Ag}}$$

$$= Z_{Ag} \times \frac{A_{Cu}}{2 \times A_{Ag}} = \frac{1.118 \times 10^{-6}}{2} \times \frac{63.4}{107.9}$$

We know that $m = Zit$

$$\therefore \quad i = \frac{m}{Zt} = \frac{10^{-3} \times 2 \times 107.9}{1.118 \times 10^{-6} \times 63.4 \times 30 \times 60} = 1.691 \text{ A}$$

12.94. One gm of hydrogen, on burning produces 34,000 calories of heat. Find the minimum potential difference required for the electrolysis of water. Given the electrochemical equivalent of hydrogen = 1.044×10^{-8} kg/C, $J = 4.18$ J/cal.

Solution: 1.044×10^{-8} kg of hydrogen will be liberated by 1 C of charge

$\therefore$ Charge required for the liberation of 10^{-3} kg of hydrogen

$$= \frac{10^{-3}}{1.044 \times 10^{-8}} \text{ C}$$

Now $W = VQ$; V is the required potential

$$\therefore \quad V = \frac{W}{Q} = \frac{34000 \times 4.18 \times 1.044 \times 10^{-8}}{10^{-3}}$$

$$= 0.34 \times 4.18 \times 1.044 = 1.484 \text{ volt}$$

12.95. One ampere current flows for one minute through a silver voltameter. It deposits m = 0.067 g of silver on the cathode. Calculate, (i) the electrochemical equivalent of silver and (ii) how much charge should flow to deposit $M = 108$ g of silver?

Solution:

(i) Electrochemical equivalent of silver

$$= Z_{Ag} = \frac{m}{I \cdot t} = \frac{0.067 \times 10^{-3}}{1 \times 60} = 1.117 \times 10^{-6} \text{ kg/C}$$

(ii) $q = \frac{M}{Z_{Ag}} = \frac{108 \times 10^{-3}}{1.117 \times 10^{-6}} = 9.67 \times 10^4 \text{C}$

12.96. A piece of metal weighs of 200 g. It has to be gold plated through 5% of its weight. How much time is required for gold plating with a current of 2 amp? Electrochemical equivalent of hydrogen = 0.1044×10^{-4} g/C, atomic weight of gold = 197.1 g, atomic weight of hydrogen = 1.008 g.

Solution:

$$\text{Chemical equivalent of gold} = \frac{197.1}{3} = 65.7$$

$$\text{Chemical equivalent of hydrogen} = 1.008$$

$$\text{Electrochemical equivalent of gold} = \frac{0.1044 \times 10^{-4} \times 65.7}{1.008}$$

$$= 6.805 \times 10^{-4} \text{ g/C}$$

$$\text{Mass to be deposited} = \frac{200 \times 5}{100} = 10\text{g}$$

$$t = \frac{m}{ZI} = \frac{10}{6.805 \times 10^{-4} \times 2} = \frac{5 \times 10^4}{6.805} = 2.04 \text{ hours}$$

12.97. In the electrolysis of sulphuric acid solution, 100 mg of hydrogen is liberated in a period of 20 minutes. The resistance of the electrolyte is 0.5 Ω. Calculate the power consumed. Electrochemical equivalent of hydrogen is 1.044×10^{-8} kg/C.

Solution: Current through the electrolyte is given by

$$I = \frac{m}{Zt}$$

$$\text{Power consumed} = I^2R = \frac{m^2}{Z^2t^2} \times R W$$

$$= \frac{(100 \times 10^{-6})^2 \times 0.5}{(1.044 \times 10^{-8})^2 \times (20 \times 60)^2}$$

$$= \frac{10^{-8} \times 0.5 \times 10^{12}}{1.044^2 \times 12 \times 12} = \frac{5000}{144 \times 1.044^2}$$

$$= 31.86 \text{ W}$$

12.98. In the electrolysis of a zinc sulphate solution, the current in the bath increases linearly given by the relation $I = (2 + 0.04\ t)$. Calculate the amount of zinc deposited at the cathode 5 minutes after the current started to vary. Electrochemical equivalent of zinc = 3.4×10^{-7} kg/C.

Solution: Initially at $t = 0$, current is 2 amp. After a time of 300 s, the current attains a constant value of 14 amp. The total charge passing through the solution can be obtained by drawing a graph between I and t, keeping t along the x-axis.

$$\text{Total charge } Q = \frac{12 \times 300}{2} + 2 \times 300 = 2{,}400 \text{ C}$$

$$\text{Mass of zinc liberated} = 3.4 \times 10^{-7} \times 2400 = 81.6 \times 10^{-5} \text{ kg}$$

12.99. Two electrolytic baths, containing solutions of copper sulphate and auric chloride are connected in series and to a current source. In the copper sulphate bath, 1.5 g of copper was deposited at the cathode. Calculate, (i) the mass of trivalent gold deposited in the other bath and (ii) the number of copper and gold atoms deposited on the cathodes. Electrochemical equivalent of copper = 3.29×10^{-7} kg/C, Electrochemical equivalent of gold = 6.81×10^{-7} kg/C.

Solution:

(i) The two electrolytic baths are connected in series, therefore same amount of charge will flow through both of them.

The amount of charge flowing through the copper voltameter

$$= \frac{1.5 \times 10^{-3}}{3.29 \times 10^{-7}} \text{C}$$

Mass of gold deposited = charge × Z_{Au}

$$= \frac{1.5 \times 10^{-3}}{3.29 \times 10^{-7}} \times 6.81 \times 10^{-7}$$

$$= \frac{1.5 \times 6.81}{3.29} \times 10^{-3} \text{ kg} = 3.1 \times 10^{-3} \text{kg}$$

(ii) Number of copper atoms deposited

$$= \frac{1.5 \times 10^{-3}}{3.29 \times 10^{-7} \times 2 \times 1.6 \times 10^{-19}} = 1.42 \times 10^{22}$$

Number of gold atoms deposited

$$= \frac{1.5 \times 10^{-3}}{3.29 \times 10^{-7} \times 3 \times 1.6 \times 10^{-19}} = 9.5 \times 10^{21}$$

12.100. On applying a voltage of 5 volts across the terminals of a nickel sulphate voltameter, 3.25 g of nickel is deposited on the cathode in a period of 60 minutes. Assuming the resistance of the solution to be equal to 1.5 Ω, find the polarization emf. Electrochemical equivalent of nickel = 3.04×10^{-7} kg/C.

Solution: The passage of current through an electrolyte may cause the electrodes to polarize, which induces a counter emf so as to decrease the current in the circuit. The same is given by the relation.

$$I = \frac{V - E_{pol}}{R}$$

$$m = ZIt$$

$$m = \frac{Z(V - E_{pol})}{R} t$$

or $$V - E_{pol} = \frac{m \cdot R}{Z \times t} = \frac{3.25 \times 10^{-3} \times 1.5}{3.04 \times 10^{-7} \times 60 \times 60} = \frac{3.25 \times 1.5}{3.04 \times 0.36} = 4.454$$

$\therefore$ $$5 - 4.454 = E_{pol}$$

or $$E_{pol} = 0.546\,V$$

12.101. A water voltameter is in series with a copper sulphate voltameter and in 45 minutes a current produces a volume of $2 \times 10^{-4} m^3$ of oxygen at pressure of 700 mm of Hg and a temperature of 27°C. Calculate the current flowing in the circuit. Mass per coulomb of hydrogen liberated = 1.05×10^{-8} kg/C, the density of oxygen at s.t.p. is 1.4 kg/m^{-3} and the relative atomic masses of oxygen and copper are 16.0 and 63.6, respectively.

Solution:

Mass per Coulomb of oxygen = $1.05 \times 10^{-8} \times 8 = 8.4 \times 10^{-8}$ kg/C

Mass of oxygen at s.t.p. = $2 \times 10^{-4} \times \frac{70}{76} \times \frac{273}{300} \times 1.4$ kg = $It \times 8.4 \times 10^{-8}$

$\therefore$ $$I = \frac{2 \times 10^{-4} \times 70 \times 273 \times 1.4}{76 \times 300 \times 45 \times 60 \times 8.4 \times 10^{-8}} = 1.03\text{ A}$$

12.102. During the electrolysis of weakly acidified water, 0.5 litre of the hydrogen is obtained at a pressure of 0.13 MPa over a period of 50 minutes. Determine the temperature of hydrogen if the current in the circuit is 1.6 A. Electrochemical equivalent of hydrogen = 1.045×10^{-8} kg/C.

Solution:

Mass of H_2 = $1.045 \times 10^{-8} \times 50 \times 60 \times 1.6$ kg

= $48 \times 1.045 \times 10^{-3}$ g

The number of molecules in the above mass

= $24 \times 1.045 \times 10^{-3} \times 6.02 \times 10^{23}$

We know that $PV = nKT$

$\therefore$ $$T = \frac{PV}{nK} = \frac{0.13 \times 10^{6} \times 0.5 \times 1000 \times 10^{-6}}{24 \times 1.045 \times 6.02 \times 10^{20} \times 1.38 \times 10^{-23}}$$

$$= \frac{65 \times 10^{3}}{24 \times 1.045 \times 6.02 \times 1.38}$$

$$= 312\text{ K}$$

12.103. How long does it take for a copper anode 50 × 10 × 1 mm^3 in size to completely dissolve if a copper sulphate solution is electrolysed at a current of 0.3 A? Density of copper = 8.9 × 10^{+3} kg/m^3, electrochemical equivalent of copper = 3.29 × 10^{-7} kg/C.

Solution:

Mass of anode = 500 × (10^{-3})^3 × 8.9 × 10^3 kg

$$t = \frac{500 \times 10^{-9} \times 8.9 \times 10^3}{3.29 \times 10^{-7} \times 0.3 \times 3600}$$

$$= 12.52 \text{ hours}$$

12.104. The nickel plating of an article is carried with the current density of 0.8 A/dm^2. Assuming nickel to be divalent. Determine the rate of growth of the nickel layer. Electrochemical equivalent of nickel = 3.04 × 10^{-7} kg/C. Density of nickel = 8.9 × 10^3 kg/m^3.

Solution:

$$m = ZIt$$

$A \cdot x \cdot d = ZIt$; A = area; d = density & x = thickness

$$\frac{dx}{dt} = \frac{ZI}{dA}$$

On putting the values

$$\frac{dx}{dt} = \frac{3.04 \times 10^{-7}}{8.9 \times 10^3} \times \frac{0.8}{10^{-2}} = 2.73 \times 10^{-9} \text{ m/sec}$$

12.105. At what current density does the thickness of the silver layer in the electrolysis of silver nitrate solution grows at a rate of 4 × 10^{-3} μm/sec. Electrochemical equivalent of silver = 1.118 × 10^{-6} kg/C. Density of silver = 10.5 × 10^3 kg/m^3.

Solution:

$$m = ZIt$$

$$A \cdot x \cdot d = ZIt$$

$$\frac{dx}{dt} = \frac{Z}{d}\frac{I}{A} = \frac{Z}{d} j$$

$$\therefore \quad j = \frac{d}{Z} \cdot \frac{dx}{dt}$$

$$= \frac{10.5 \times 10^3 \times 4 \times 10^{-3} \times 10^{-6}}{1.118 \times 10^{-6}}$$

$$= \frac{10.5 \times 4}{1.118} = 37.567 \text{ A/m}^2$$

12.106. An electrolysis is carried for a period of 10 min at a current density of 5 A/m^2. Calculate the number of monovalent atoms deposited per square meter of cathode.

Solution:

$$M = ZIt$$

or $\quad m \cdot n = \frac{m}{e} \cdot I \cdot t;$ $\qquad Z = \frac{\text{mass of one ion}}{\text{charge of one ion}}$

$$\frac{n}{A} = \frac{I}{A} \cdot \frac{t}{e} \qquad m = \text{mass of one ion}$$

$$= j \cdot \frac{t}{e} \qquad n = \text{number of ions}$$

On putting the values of various terms we have

$$= \frac{5 \times 10 \times 60}{1.6 \times 10^{-19}}$$

$$= \frac{30}{16} \times 10^{22}$$

$$= 1.875 \times 10^{22}$$

12.107. A plate of area 10 cm^2 is to be electroplated with copper (density 9 gm/cm^3) to a thickness of 0.001 cm on both sides using a battery of 12 volts. Calculate the energy spent by the cell in the process of deposition. In this energy is used to heat 100 gm of water, calculate the rise in temperature of water. Electrochemical equivalent of copper = 0.0003 gm/C. (IIT 1971)

Solution: Mass of copper deposited

$$m = \rho \cdot A \cdot d = 9 \times 20 \times 0.001 = 0.18 \text{ gm}$$

$$m = ZIt$$

$$\therefore \quad It = \frac{m}{Z} = \frac{0.18}{0.0003} = 600 \text{ C}$$

Energy given by the cell = $VIt = 12 \times 600 = 7200$ J

Now $ms\,\Delta\theta = VIt$

$$\therefore \quad \Delta\theta = \frac{VIt}{m \times s} = \frac{7200}{4.2 \times 100} = 17.14°\text{C}$$

12.108. A silver and copper voltameter are connected in parallel to a 12 volt battery of negligible internal resistance. If in 30 min 1.0 gm of silver and 1.8 gm of copper are deposited. At what rate energy is being delivered by the battery. Electrochemical equivalent of silver and copper are respectively 11.2×10^{-4} gm/C and 6.6×10^{-4} gm/C. (IIT 1975)

Solution: The current supplied by the battery will be distributed in the two voltameter. Using the relation $m = ZIt$ we have

$$I_1 = \text{current through silver voltameter} = \frac{1}{11.2\times10^{-4}\times30\times60}$$
$$= 0.496 \text{ amp}$$

$$I_2 = \text{current through copper voltameter} = \frac{1.8}{6.6\times10^{-4}\times30\times60}$$
$$= 1.515 \text{ amp}$$

$$I = I_1 + I_2 = 2.011 \text{ amp}$$
$$W = V\times I = 12\times2.011 = 24.132 \text{ watt}$$

12.109. One ampere current flows for one minute through a silver voltameter. It deposits 0.067 gm of silver on the cathode. Calculate (a) the electrochemical equivalent of silver (b) how much charge will flow to deposit 108 gm of silver? (c) how much charge is carried by each silver ion? Avogadro's no = 6.024×10^{23}. (Roorkee 1982)

Solution:

(a) $m = ZIt$

$$\therefore \quad Z = \frac{m}{It} = \frac{0.067}{1\times60} = 1.117\times10^{-3}\text{ gm/C}$$

(b) $m = ZQ$

$$\therefore \quad Q = \frac{m}{Z} = \frac{108}{1.117\times10^{-3}} = 9.67\times10^{4}\text{ C}$$

(c) $Z = \dfrac{\text{mass of one ion}}{\text{charge of one ion}}$

$$\therefore \quad \text{Charge of one ion} = \frac{\text{mass of one ion}}{Z}$$
$$= \frac{108}{6.024\times10^{23}\times1.117\times10^{-3}}$$
$$= \frac{10.8\times10^{-19}}{6.024\times1.117} = 1.60\times10^{-19}\text{ C}$$

13

ELECTROMAGNETISM

13.1. In a hydrogen atom the electron moves in the first orbit around the nucleus under the action of electrostatic forces at a frequency of 6 × 10^{15} rev/s. Calculate the magnetic field *B* at the centre of the orbit. Radius of the first orbit = 0.5 A.

Solution: The motion of the electron in the circular orbit will give rise to a current i = charge × frequency of rev/s = $1.6 \times 10^{-19} \times 6 \times 10^{15}$

$$= 9.6 \times 10^{-4} \text{ A}$$

Magnetic field at the centre of the orbit

$$= \frac{\mu_0}{4\pi}\frac{2\pi i}{r} = \frac{10^{-7} \times 2 \times 3.14 \times 9.6 \times 10^{-4}}{5 \times 10^{-11}}$$

$$= 0.4 \times 3.14 \times 9.6$$

$$= 12.06 \text{ W/m}^2$$

13.2. A current of magnitude I = 5 amp flows through a segment of wire of length l = 3 cm (see Fig. 13.1). Calculate the magnetic field at the point *P* at a distance R = 1 cm from the wire along its perpendicular bisector.

Solution: Consider an element of length dx lying at a distance x from the middle point of the wire. According to Biot-Savart's law, dB at the point *P* is given by

$$dB = \frac{\mu_0}{4\pi}\frac{I\,dx\sin\theta}{\left(x^2 + R^2\right)} \quad \text{... (i)}$$

We know that $\sin\theta = \sin(\pi - \theta)$

From the figure

$$\sin(\pi - \theta) = \frac{R}{\left(x^2 + R^2\right)^{1/2}}$$

$$\therefore \quad dB = \frac{\mu_0}{4\pi}\frac{IR\,dx}{\left(x^2 + R^2\right)^{3/2}}$$

or $$B = \frac{\mu_0}{4\pi}2IR\int_0^{l/2}\frac{dx}{\left(x^2 + R^2\right)^{3/2}}$$

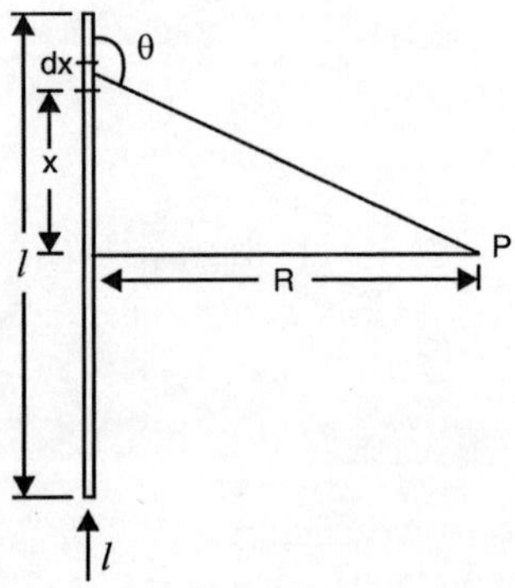

Fig. 13.1

For solving this integral, let us assume that $x = R\tan\theta$; $dx = R\sec^2\theta\, d\theta$

$$\therefore \quad B = \frac{\mu_0}{4\pi} 2IR \int \frac{R\sec^2\theta\, d\theta}{R^3\sec^3\theta} = \frac{\mu_0}{4\pi}\frac{2I}{R}\int \cos\theta\, d\theta = \frac{\mu_0}{4\pi}\frac{2I}{R}\sin\theta$$

$$= \frac{\mu_0}{4\pi}\frac{2I}{R}\left[\frac{x}{\left[x^2+R^2\right]^{1/2}}\right]_0^{l/2} = \frac{\mu_0}{4\pi}\frac{2Il}{R\left(l^2+4R^2\right)^{1/2}}$$

On substituting the values of various terms, we have

$$B = 10^{-7} \times \frac{2\times 5}{10^{-2}} \times \frac{3\times 10^{-2}}{\sqrt{13}\times 10^{-2}}$$

$$= \frac{3}{\sqrt{13}} \times 10^{-4} = 8.32\times 10^{-5}\ \text{W/m}^2$$

13.3. A rectangle of length $l = 4$ cm and width $d = 3$ cm carries a current $I = 12$ amp (Fig. 13.2). Calculate the magnetic field at its centre.

Solution: The net field at the centre of the rectangle $= B_1 + B_2 + B_3 + B_4$, where $B_1 = B_3$ are the fields obtained at a distance $d/2$ and the current is flowing along the length of the rectangle. Similarly, $B_2 = B_4$ are the fields obtained at a distance $l/2$ and the current is flowing along the width of the rectangle. Now in view of the above problem,

$$B_1 = B_3 = \frac{\mu_0 I}{2\pi d/2}\frac{l}{\sqrt{l^2+d^2}}$$

$$\text{and } B_2 = B_4 = \frac{\mu_0 I}{2\pi l/2}\frac{d}{\sqrt{d^2+l^2}}$$

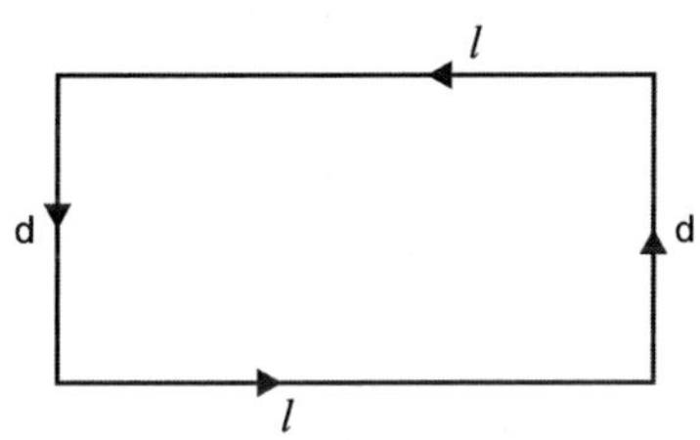

Fig. 13.2

$$\text{Net field } B = \frac{2\mu_0 I}{\pi\sqrt{d^2+l^2}}\left[\frac{l}{d}+\frac{d}{l}\right]$$

$$= \frac{\mu_0 2}{\pi}\frac{I\left(d^2+l^2\right)^{1/2}}{dl}$$

$$= \frac{\mu_0}{4\pi} 8I \frac{\left(d^2+l^2\right)^{1/2}}{dl}$$

On substituting the values of various terms, we have

$$B = 10^{-7} \times 8 \times 12 \times \frac{5\times 10^{-2}}{12\times 10^{-4}}$$

$$= 4 \times 10^{-4}\ \text{W/m}^2$$

13.4. Calculate the magnetic field at the centre of a square loop of wire of length $l = 5$ cm carrying a current $I = 5$ amp (Fig. 13.3).

Solution:

Net field at the centre is given by $B = B_1 + B_2 + B_3 + B_4$

where $B_1 = B_2 = B_3 = B_4 = \dfrac{\mu_0 I}{2\pi l/2} \dfrac{l}{\sqrt{l^2 + l^2}}$

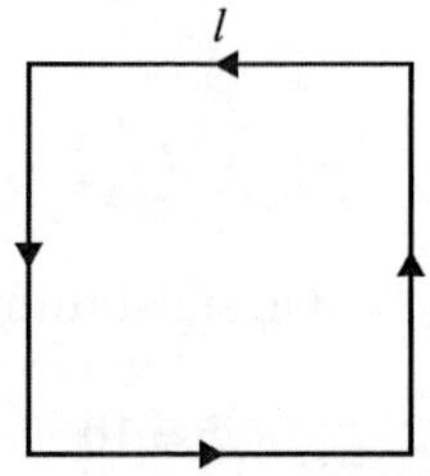

Fig. 13.3

$$\therefore \quad B = 2\sqrt{2}\,\frac{\mu_0 I}{\pi l}$$

$$= \frac{\mu_0}{4\pi}\,\frac{8\sqrt{2}\,I}{l}$$

On substituting the values of I and l, we have

$$B = 10^{-7} \times \frac{8 \times 1.41 \times 5}{5 \times 10^{-2}} = 11.28 \times 10^{-5}\ \text{W/m}^2$$

13.5. A square loop of wire of side a carries a current i as shown in Fig. 13.4. Find the magnetic field at a distance x from the centre of the loop.

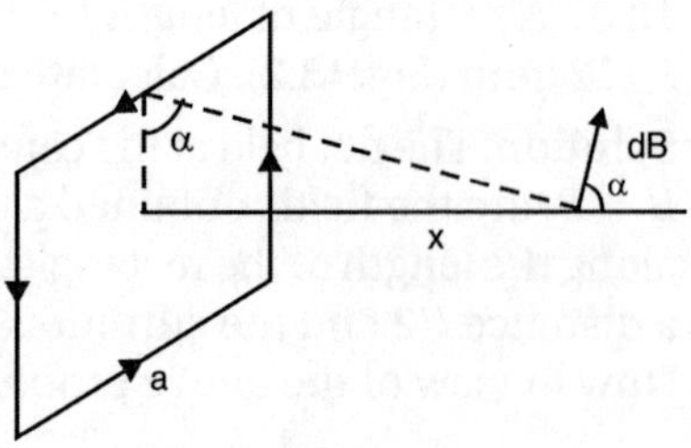

Fig. 13.4

Solution: Magnetic field due to a current i in a conductor of length l at a distance R from its perpendicular bisector is given by

$$B = \frac{\mu_0 i l}{2\pi R\left(l^2 + 4R^2\right)^{1/2}} \qquad \text{... (i)}$$

In the light of this equation the magnetic field B_1 due to one side of the loop is given by

$$B_1 = \frac{\mu_0 i a}{2\pi\left(x^2 + a^2/4\right)^{1/2}} \cdot \frac{1}{\left[a^2 + 4\left(x^2 + a^2/4\right)\right]^{1/2}} \cdot \cos\alpha$$

$$= \frac{\mu_0 i a}{\pi\left(a^2 + 4x^2\right)^{1/2}} \cdot \frac{1}{\left(2a^2 + 4x^2\right)^{1/2}} \cdot \frac{a/2}{\left(x^2 + a^2/4\right)^{1/2}}$$

$$= \frac{\mu_0 i a^2}{\pi\left(a^2 + 4x^2\right)\left(2a^2 + 4x^2\right)^{1/2}}$$

Net magnetic field due to all the four sides is equal to B.

$$B = 4B_1 = \frac{4\mu_0 i a^2}{\pi\left(a^2 + 4x^2\right)\left(2a^2 + 4x^2\right)^{1/2}}$$

For points $x >> a$ it will act like a magnetic dipole

$$B = \frac{4\mu_0 i a^2}{\pi \cdot 4x^2 \cdot 2x} = \frac{\mu_0}{4\pi} \cdot \frac{2ia^2}{x^3}$$

13.6. An infinitely long straight conductor carries a current as shown in the Fig. 13.5. Discuss the field around the conductor.

Solution: According to the amperes law the magnetic field at a distance r inside a current carrying wire of radius R is given by

$$B = \frac{\mu_0 I r}{2\pi R^2}$$; r is measured from the centre of the conductor.

The field at the centre is zero and is maximum at the surface of the conductor. The variation of the field with distance outside the conductor is represented by rectangular hyperbola. The variation of the magnetic field around the conductor in all the four quadrants is shown in the Fig. 13.5.

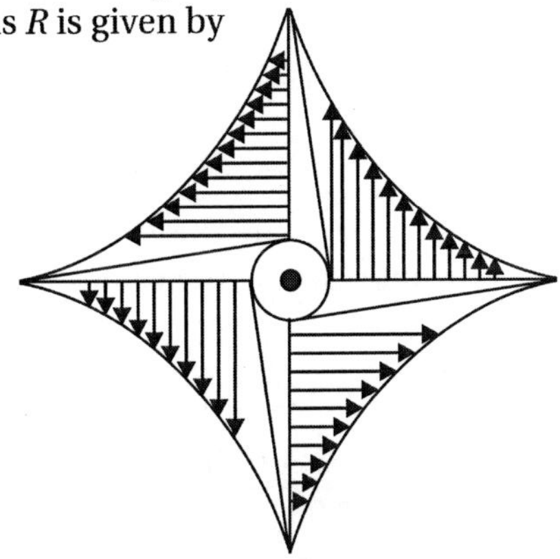

Fig. 13.5

13.7. Two infinitely long straight conductors carrying different current are shown in the Fig. 13.6a & b. Using lines of forces show that the net field in the region between the two conductors is either as difference or sum of the two fields.

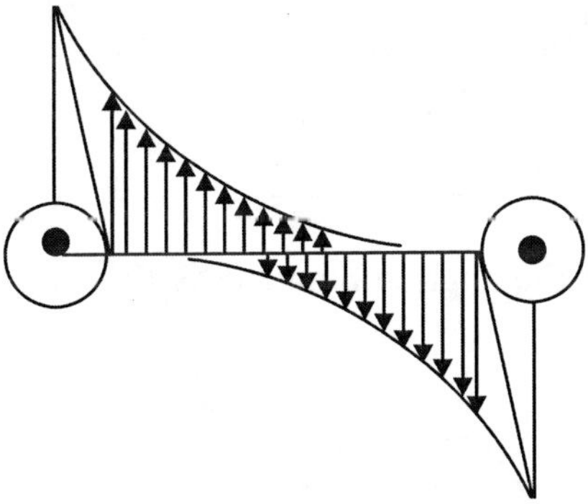

Fig. 13.6a

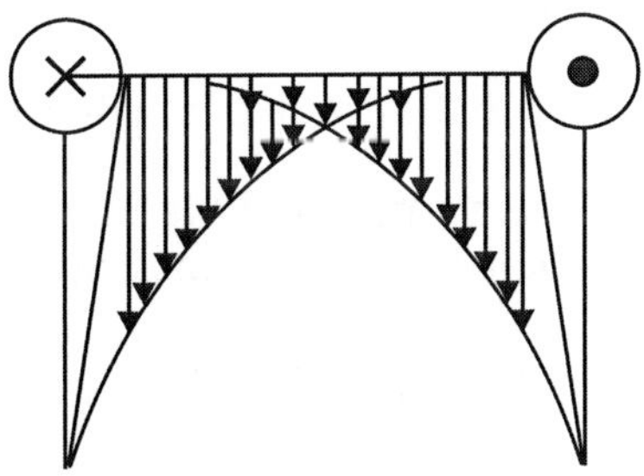

Fig. 13.6b

Solution: In Fig. 13.6a the current in the two conductors are in the same sense. The lines of forces for the two field are shown. It is evident from the figure that in the space between the two conductors the net field will be in the form of difference of the two fields. In Fig. 13.6b the sense of current in the two conductors is opposite. In the region of space between the two conductors the field will additive.

13.8. A current of $I = 6.28$ A flows through an isoceles tapezium whose dimensions are shown in the Fig. 13.7. Taking the value of $b = 5$ cm and $l = 10$ cm calculate the magnetic induction at the point O.

Solution: The magnetic field at O will be only due to conductor 2 and conductor 4.

$$B \text{ at } O = \frac{\mu_0}{4\pi}\left[\frac{2Il}{b\sqrt{(l^2+4b^2)}} - \frac{2I\cdot 2l}{2b\sqrt{4l^2+16b^2}}\right]$$

$$= \frac{\mu_0 Il}{4\pi b\sqrt{l^2+4b^2}}$$

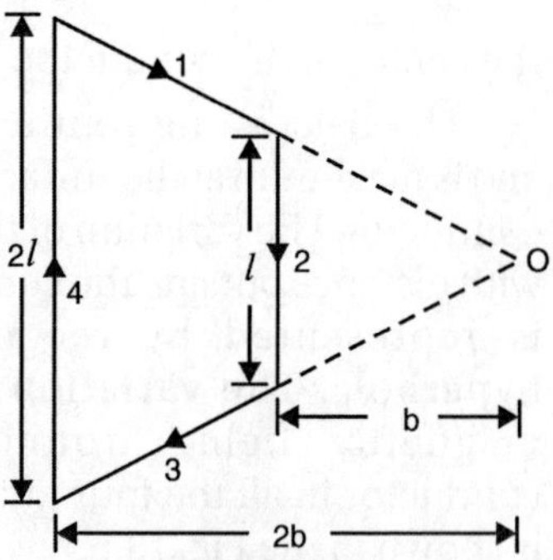

Fig. 13.7

On substituting the values we have

$$B = \frac{10^{-7}\times 6.28\times 0.1}{0.05\sqrt{0.01+4\times 0.0025}}$$

$$= \frac{6.28\times 10^{-7}}{0.05\times 0.1414}$$

$$= 8.88\times 10^{-5}\,\text{W/m}^2$$

13.9. A circuit with a steady current includes a homogenous ring and two very long radial conductors connected to it. Find the magnetic induction at the centre of the ring.

Solution: Divide the current I into I_1 & I_2 as shown in Fig. 13.8. Let the resistance of the two portions of the ring be R_1 and R_2 respectively.

Now $V_{AB} = I_1R_1 = I_2R_2$

$$B \text{ at the centre of the ring} = \frac{\mu_0}{4\pi}\left[\frac{2I_1 r\theta}{r^2} - \frac{2I_2 r(2\pi-\theta)}{r^2}\right]$$

where r = radius of the ring and θ is the angle subtended by the small section of the wire at the centre.

$$B = \frac{\mu_0}{4\pi}\left[\frac{2I_1\theta}{r} - \frac{2I_2(2\pi-\theta)}{r}\right]$$

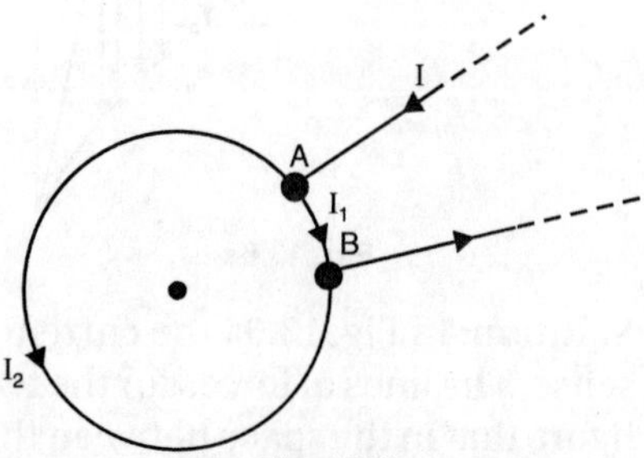

Fig. 13.8

Now $R_1 = \dfrac{\rho\cdot r\theta}{A}$, $R_2 = \dfrac{\rho(2\pi-\theta)r}{A}$

In view of $I_1R_1 = I_2R_2$ we have

$$\frac{I_1\rho r\theta}{A} = \frac{I_2\rho(2\pi-\theta)r}{A}$$

or $\quad I_1\theta = I_2(2\pi-\theta)$

In the light of this the magnetic induction B at the centre of the ring is equal to zero.

13.10. Two long straight wires, separated by a distance $l = 5$ cm carry equal anti parallel currents $I = 6.1$ amp (Fig. 13.9). Calculate the magnetic field at a point P at a distance $R = 3$ cm from the mid-point of the distance between the two wires.

Solution: The magnetic field B at the point P due to the current I in conductor 1 is given by

$$B_1 = \frac{\mu_0}{4\pi}\frac{4I}{\sqrt{l^2 + 4R^2}}$$

Similarly, the magnetic field B_2 at the point P due to the current I in the conductor 2 is given by

$$B_2 = \frac{\mu_0}{4\pi}\frac{4I}{\sqrt{l^2 + 4R^2}}$$

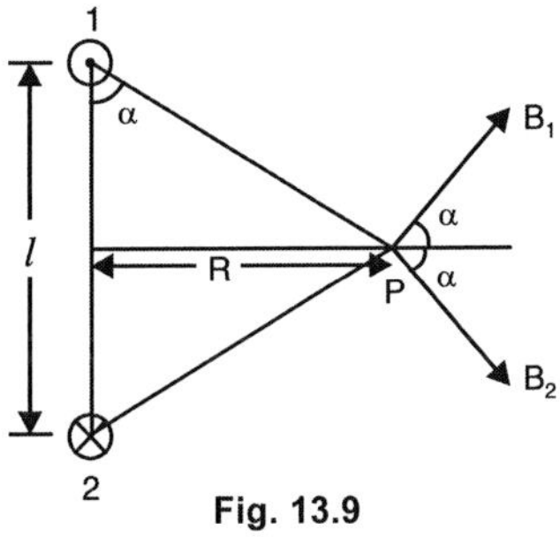

Fig. 13.9

The directions of the two fields are shown in the Fig. 13.9. Amongst the two components of B_1 and B_2 each, only the $\cos\alpha$ component will contribute to the net field. Hence

$$B = (B_1 + B_2)\cos\alpha = \frac{2\mu_0 I}{\pi\sqrt{l^2 + 4R^2}} \times \frac{l}{\sqrt{l^2 + 4R^2}}$$

or $$B = \frac{2\mu_0 I l}{\pi(l^2 + 4R^2)}$$

$$= \frac{\mu_0}{4\pi}\frac{8Il}{(l^2 + 4R^2)}$$

On substituting the values of I, l and R, we have

$$B = \frac{10^{-7} \times 8 \times 6.1 \times 5 \times 10^{-2}}{(25 + 36) \times 10^{-4}}$$

$$= \frac{4 \times 61}{61} \times 10^{-5} = 4 \times 10^{-5}\ \text{W/m}^2$$

13.11. A long wire *abcd* is bent in the form of a hair pin as shown in Fig. 13.10. The radius of the semi-circular part of the wire is 25 cm. Calculate the magnetic field at point A when a current $I = 5$ amp passes through it.

Solution: Net field at the point A = $B_1 + B_2 + B_3$, where B_1 and B_3 are the field due to the straight part *ab* and *cd* of the wire and B_2 is the field due to the semi-circular part *bc*.

For evaluating B_3, consider an element of length dx placed at a distance x from c as shown in Fig. 13.10. The magnetic field at A due to this element is given by

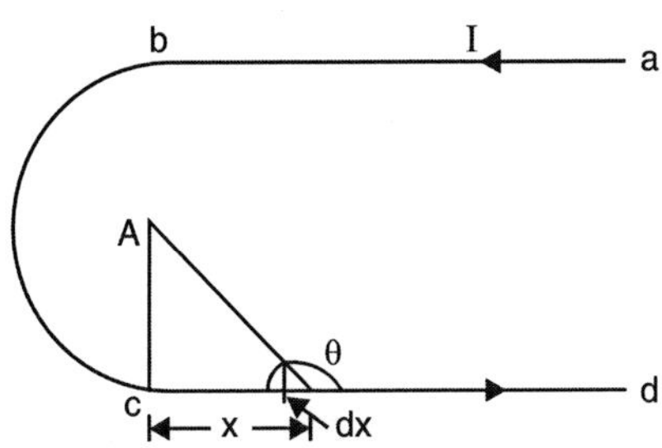

Fig. 13.10

$$dB_3 = \frac{\mu_0}{4\pi}\frac{I\,dx\sin\theta}{x^2+R^2}$$ where R is the radius of the semi-circular part of the wire

$$\sin\theta = \frac{R}{\left(x^2+R^2\right)^{1/2}}; \quad \sin\theta = \sin(\pi-\theta)$$

$$\therefore \quad dB_3 = \frac{\mu_0}{4\pi} IR \frac{dx}{\left(x^2+R^2\right)^{3/2}}$$

$$\therefore \quad B_3 = \frac{\mu_0 IR}{4\pi}\int_0^{\infty}\frac{dx}{\left(x^2+R^2\right)^{3/2}}$$

$$= \frac{\mu_0 I}{4\pi R}$$

If we proceed in a similar way, B_1 will have the same value as B_3. For evaluating B_2 consider an element of length dx situated on the circumference of the semicircular wire. The magnetic field dB_2 at the point A due to this element is given by

$$dB_2 = \frac{\mu_0}{4\pi}\frac{I\,dx}{R^2}\sin 90$$

$$\therefore \quad B_2 = \frac{\mu_0}{4\pi}\frac{I}{R^2}\int dx = \frac{\mu_0 I\,\pi R}{4\pi R^2} = \frac{\pi\mu_0 I}{4\pi R}$$

$$\text{Net field at } A = \frac{\mu_0 I}{4\pi R}(\pi+2)$$

On substituting the values of I and R, we have

$$\text{Net field} = \frac{10^{-7}\times 5}{25\times 10^{-2}}[5.14] \cong 1.03\times 10^{-5}\,\text{W/m}^2$$

13.12. A steel sphere of radius r is uniformly magnetised with magnetisation M parallel to x axis as shown in Fig. 13.11. Find the magnetic induction at the centre of the sphere.

Solution: Consider a ring of width dx and radius $y = r\sin\theta$ situated at a distance $x = r\cos\theta$ from the centre O. We know that a magnetised body of magnetisation M is equivalent to a surface current having linear dimensions of M.

dB due to the element is given by

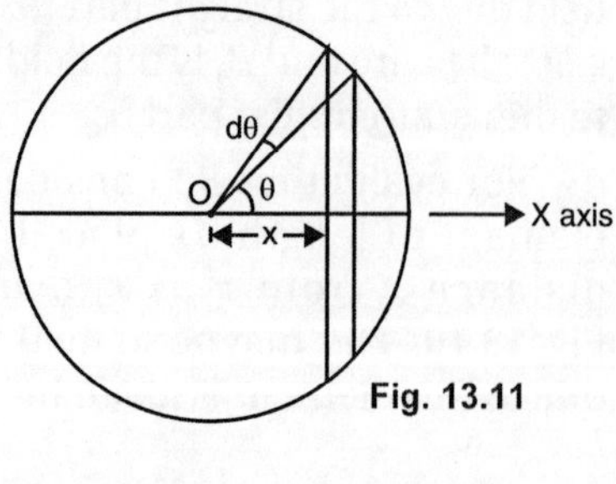

Fig. 13.11

$$dB = \frac{\mu_0}{2} j \frac{dx\cdot y^2}{\left(x^2+y^2\right)^{3/2}}$$

$$= -\frac{\mu_0}{2} j \frac{r\sin\theta\cdot d\theta\cdot r^2\sin^2\theta}{r^3}$$

$$B = -\frac{\mu_0 j}{2}\int_0^{\pi} \sin^3\theta \cdot d\theta$$

$$= -\frac{\mu_0 j}{2}\int_0^{\pi}\left[\frac{3}{4}\sin\theta - \frac{1}{4}\sin 3\theta\right]d\theta$$

$$= -\frac{\mu_0 j}{2}\left[-\frac{3}{4}\cos\theta + \frac{1}{12}\cos 3\theta\right]_0^{\pi}$$

$$= \frac{2}{3}\mu_0 j$$

$$= \frac{2}{3}\mu_0 M$$

13.13. An elliptical wire with semi major axis a and minor axis b contains a current i. Find the magnetic field at its focus.

Solution: For point P, $\theta = 180°$ and $SP = r = a(1-\varepsilon)$; a = semi major axis and ε = eccentricity

For point A in Fig. 13.12, $\theta = 360°$ and $SA = r = a(1+\varepsilon)$

Most general equation of ellipse

$$\frac{1}{r} - C_1 + C_2 \cos\theta$$

here C_1 and C_2 are the constants whose values are to be determined in the light of the above two conditions.

$$\frac{1}{r} = \frac{1-\varepsilon\cos\theta}{a(1-\varepsilon^2)}$$

$$\tan\beta = \frac{r}{dr/d\theta}$$

$$-\frac{1}{r^2}\cdot\frac{dr}{d\theta} = \frac{\varepsilon\sin\theta}{a(1-\varepsilon^2)}$$

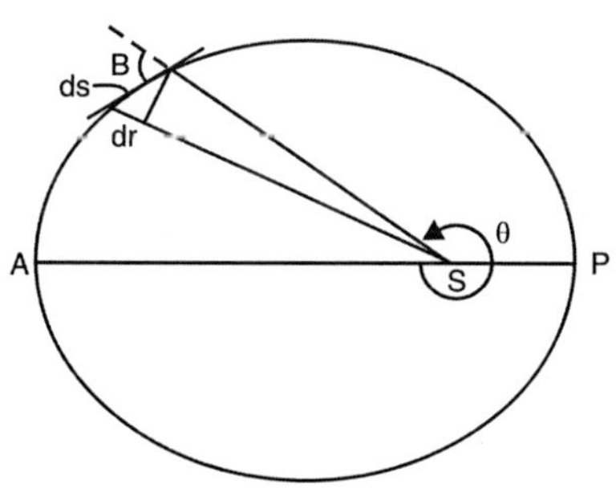

Fig. 13.12

$$\therefore \quad \frac{dr}{d\theta} = \frac{-\varepsilon\sin\theta\cdot r^2}{a(1-\varepsilon^2)}$$

$$\tan\beta = \frac{r}{dr/d\theta} = \frac{-a(1-\varepsilon^2)}{\varepsilon\sin\theta}\times\frac{1-\varepsilon\cos\theta}{a(1-\varepsilon^2)} = \frac{\varepsilon\cos\theta - 1}{\varepsilon\sin\theta}$$

$$\cos\beta = \frac{dr}{ds} \;\therefore ds = \frac{dr}{\cos\beta}$$

Magnetic field due to element of length ds is given by

$$dB = \frac{\mu_0}{4\pi} \cdot i\,ds \cdot \frac{\sin(\pi - \beta)}{r^2}$$

$$= \frac{\mu_0}{4\pi} \cdot i \cdot \frac{dr}{r^2} \cdot \tan\beta = \mu_0 i \frac{dr}{r^2} \cdot \frac{\varepsilon\cos\theta - 1}{\varepsilon\sin\theta}$$

$$B = \frac{\mu_0 i}{4\pi(1-\varepsilon^2)a} \int_0^{2\pi} (1 - \varepsilon\cos\theta)\, d\theta$$

$$= \frac{\mu_0 i\, 2\pi}{4\pi a(1-\varepsilon^2)} = \frac{\mu_0 i}{2a(1-\varepsilon^2)}; \; b = a\sqrt{1-\varepsilon^2}$$

where b = semi minor axis

$$= \frac{\mu_0 i}{2a \cdot b^2} \cdot a^2 = \frac{\mu_0 i a}{2b^2}$$

13.14. Three infinitely long straight conductors A, B & C carrying the same current I in a direction as shown in Fig. 13.13a. Find the resultant magnetic field at O when $AB = BC = BO = l$.

Solution: The magnetic field at O due to current in conductor A is given by

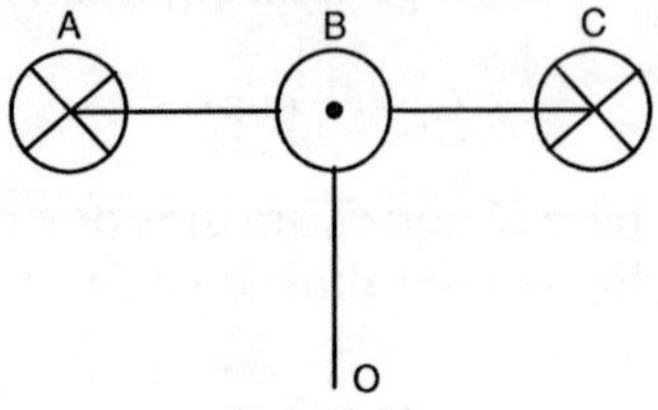

Fig. 13.13a

$$B_1 = \frac{\mu_0}{4\pi} \cdot \frac{2I}{l\sqrt{2}} = \frac{\mu_0}{4\pi} \cdot \sqrt{2}\,\frac{I}{l}$$

The magnetic field at O due to current in conductor C is given by

$$B_3 = \frac{\mu_0}{4\pi} \cdot \frac{2I}{l\sqrt{2}} = \frac{\mu_0}{4\pi} \cdot \sqrt{2}\,\frac{I}{l}$$

Both the field are perpendicular to each other (see Fig. 13.13b).

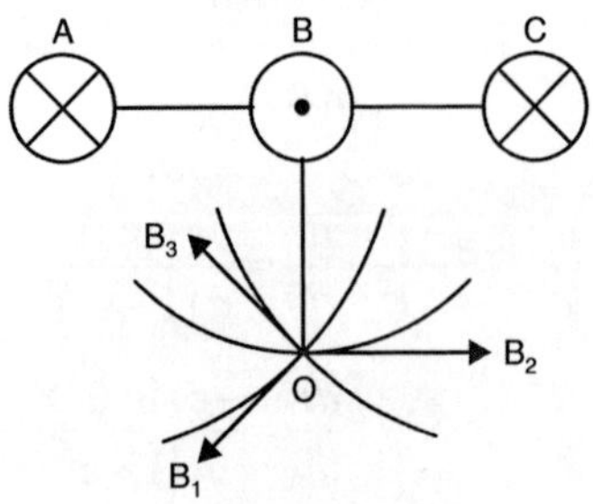

Fig. 13.13b

$$\text{Resultant of } B_1 \,\&\, B_3 = B = \frac{\mu_0}{4\pi} \cdot \frac{2I}{l}$$

The magnetic field at O due to current in conductor B is given by

$$B_2 = \frac{\mu_0}{4\pi} \cdot \frac{2I}{l} = B$$

B and B_2 are equal and opposite to each other hence the resultant field at O due to current in all the three conductors is equal to zero.

13.15. Four infinitely long straight conductors A, B, C & D carrying same current I in the direction as shown in Fig. 13.14a are placed at the corners of a square loop of side l. Find the resultant magnetic field at the centre of the loop.

Solution: The magnetic field due to all the four conductors are B_1, B_2, B_3 & B_4 respectively and

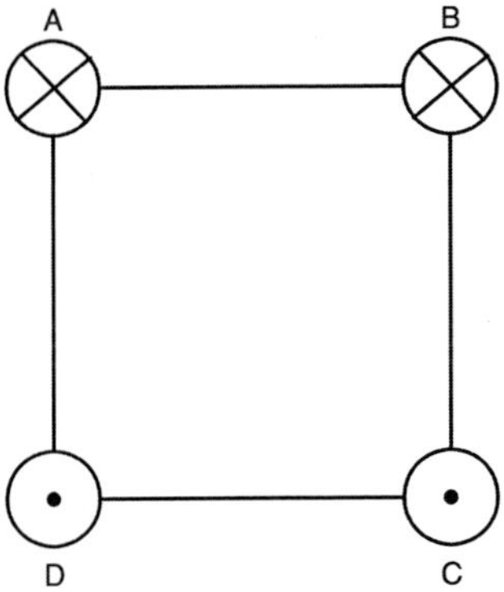

Fig. 13.14a

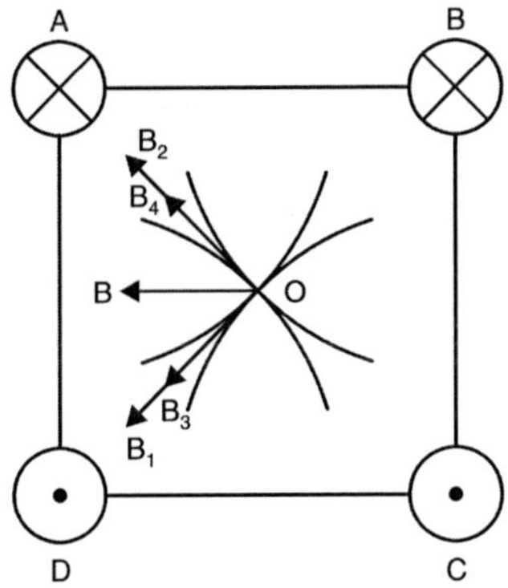

Fig. 13.14b

$$B_1 = B_2 = B_3 = B_4 = \frac{\mu_0}{4\pi} \cdot \frac{2\sqrt{2}\,I}{l}$$

In view of the directions of the current in the conductors net field due to conductors B & D will be along

$$OA = B_2 + B_4 = \frac{\mu_0}{4\pi} \cdot \frac{4\sqrt{2}\,I}{l}$$

Similarly net field due to conductors A & C will be along

$$OD = B_1 + B_3 = \frac{\mu_0}{4\pi} \cdot \frac{4\sqrt{2}\,I}{l}$$

Resultant of all the four field will be equal to $B = \frac{\mu_0}{4\pi} \cdot \frac{8I}{l}$ and will be parallel to CD and towards D.

13.16. Two infinitely long rectilinear conductors A and B, placed 8 cm apart carry currents $I_1 = 10$ amp and $I_2 = 20$ amp, respectively as shown in Fig. 13.15. Find the magnetic field at points P_1, P_2 and P_3.

$AP_1 = AP_2 = BP_3 = 2$ cm

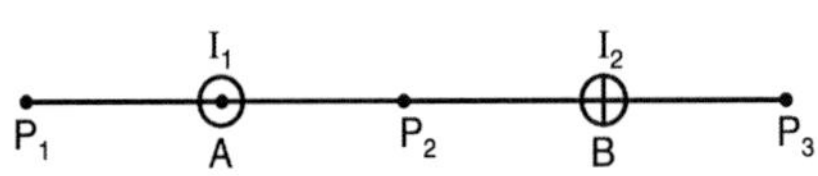

Fig. 13.15

Solution: Net field at

$$P_1 = \frac{\mu_0}{4\pi}\frac{2I_1}{AP_1} - \frac{\mu_0}{4\pi}\frac{2I_2}{BP_1}$$

$$= 10^{-7}\left[\frac{2\times 10}{2\times 10^{-2}} - \frac{2\times 20}{10\times 10^{-2}}\right]$$

$$= 10^{-5}\,[10 - 4] = 6\times 10^{-5}\ \text{W/m}^2$$

The same result can be expressed in amperes per metre by dividing with $4\pi \times 10^{-7}$.

Thus magnetic field at

$$P_1 \text{ in } A/m = \frac{6 \times 10^{-5}}{4 \times 3.14 \times 10^{7}} = \frac{600}{4 \times 3.14} = 47.78 \text{ A/m}$$

$$\text{Net field at } P_2 = \frac{\mu_0}{4\pi}\frac{2I_1}{AP_2} + \frac{\mu_0}{4\pi}\frac{2I_2}{BP_2}$$

$$= 10^{-7}\left[\frac{20}{2 \times 10^{-2}} + \frac{40}{6 \times 10^{-2}}\right] = 16.67 \times 10^{-5} \text{ W/m}^2$$

$$= 132.7 \text{ A/m}$$

$$\text{Net field at } P_3 = \frac{\mu_0}{4\pi}\left[\frac{2I_2}{BP_3} - \frac{2I_1}{AP_3}\right]$$

$$= 10^{-7}\left[\frac{40}{2 \times 10^{-2}} - \frac{2 \times 10}{10 \times 10^{-2}}\right] = 18 \times 10^{-5} \text{ W/m}^2$$

$$= 143.3 \text{ A/m}$$

13.17. Fig. 13.16 represents a cross-section of three infinitely long rectilinear conductors A, B and C carrying currents I_1, I_2 and I_3, respectively. Locate a point on the line AC at which the net field is zero. $I_1 = I_2 = 5$ amp and $I_3 = 10$ amp, $AB = BC = 10$ cm.

Solution: Let P be the point at a distance x from A at which the magnetic intensity is zero.

$$\text{Net field at } P = \frac{\mu_0}{4\pi}\left[\frac{2I_1}{x} - \frac{2I_2}{PB} + \frac{2I_3}{PC}\right] = 0$$

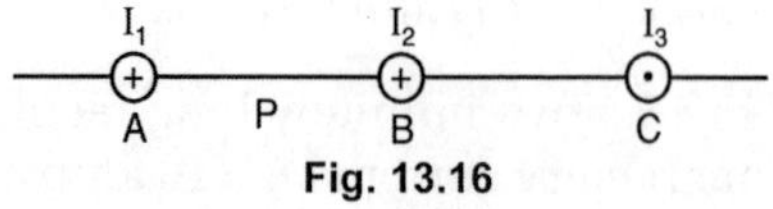

Fig. 13.16

or $$\frac{2 \times 5}{x} - \frac{2 \times 5}{10 - x} + \frac{2 \times 10}{(20 - x)} = 0$$

or $$\frac{1}{x} + \frac{2}{20 - x} = \frac{1}{10 - x}$$

or $$\frac{20 + x}{x(20 - x)} = \frac{1}{10 - x}$$

$$200 + 10x - 20x - x^2 = 20x - x^2$$

$$30\,x = 200$$

$$x = \frac{20}{3} = 6.6 \text{ cm}$$

13.18. Two infinitely long rectilinear conductors are arranged perpendicular to each other in one plane as shown in Fig. 13.17. Calculate the magnetic field at points P_1 and P_2 if $I_1 = 3$ amp $I_2 = 4$ amp. The distances $AP_1 = AP_2 = 2$ cm and $BP_1 = CP_2 = 4$ cm.

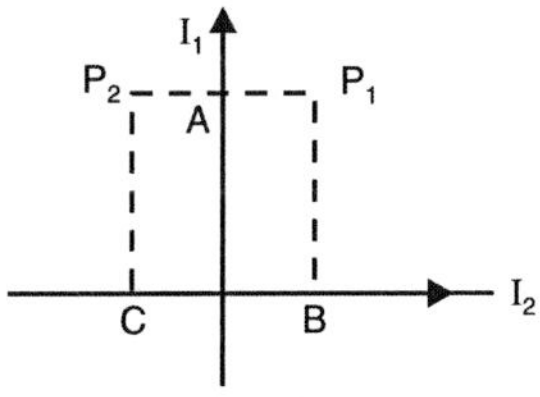

Fig. 13.17

Solution: Net field at P_1 is given by

$$\frac{\mu_0}{4\pi}\left[\frac{2I_1}{AP_1}-\frac{2I_2}{BP_1}\right]=10^{-7}\left[\frac{2\times 3}{2\times 10^{-2}}-\frac{2\times 4}{4\times 10^{-2}}\right]$$

$$=10^{-5}\,\text{W/m}^2$$

$$\cong 8\ \text{A/m}$$

Net field at P_2 is given by

$$\frac{\mu_0}{4\pi}\left[\frac{2I_1}{AP_2}+\frac{2I_2}{CP_2}\right]=10^{-7}\left[\frac{2\times 3}{2\times 10^{-2}}+\frac{2\times 4}{4\times 10^{-2}}\right]=5\times 10^{-5}\ \text{W/m}^2$$

$$=39.8\ \text{A/m}$$

13.19. The mutually perpendicular conductors carrying currents I_1 and I_2 as shown in Fig. 13.18. Assuming the conductors to be infinitely long, find the locus of points at which the magnetic field is zero.

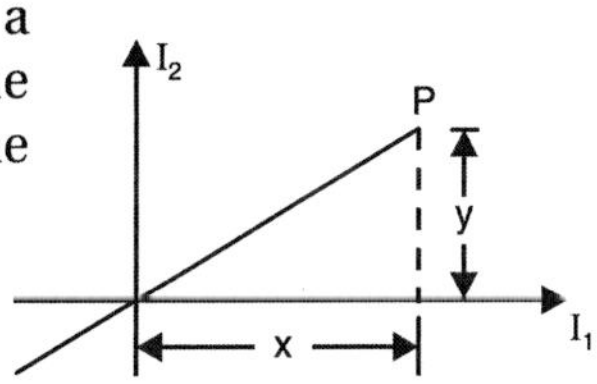

Fig. 13.18

Solution: Let y be the distance from a certain point on the conductor carrying current I_1 to a point P where the net field is zero and x is the distance from point P to a certain point on the conductor carrying current I_2. Hence

$$\frac{\mu_0}{4\pi}\cdot\frac{2I_2}{x}=\frac{\mu_0}{4\pi}\cdot\frac{2I_1}{y}$$

or $$y=\frac{I_1}{I_2}\cdot x$$

This is the equation of a straight line passing through the origin. Therefore, the locus of points where the net field is zero is a straight line passing through the origin, with a gradient equal to $\frac{I_1}{I_2}$.

13.20. A 20 cm long straight conductor carries a current. At what maximum distance from it along its perpendicular bisector can the magnetic field be considered as due to an infinitely long straight conductor carrying current? The maximum permissible error in this assumption is 5%.

Solution: The magnetic field at a distance x along the perpendicular bisector of a segment of length l and carrying a current I is given by

$$B_1=\frac{\mu_0}{2\pi x}\frac{Il}{\sqrt{l^2+4x^2}}$$ (see Q. No. 13.2)

The magnetic field at a distance x due to current in an infinitely long conductor is given by $B_2=\frac{\mu_0}{4\pi}\frac{2I}{x}$

The percentage error $= \dfrac{B_2 - B_1}{B_2} \times 100$

According to the problem, this should be up to a maximum of 5%.

Hence $\dfrac{B_2 - B_1}{B_2} \times 100 = 5$ or $\dfrac{B_2 - B_1}{B_2} = \dfrac{1}{20}$

On substituting the values of B_1 and B_2, we have

$$\frac{\dfrac{\mu_0}{4\pi} \cdot \dfrac{2I}{x} - \dfrac{\mu_0}{2\pi x} \dfrac{Il}{\sqrt{l^2 + 4x^2}}}{\dfrac{\mu_0}{4\pi} \dfrac{2I}{x}} = \frac{1}{20} \text{ or } 1 - \frac{l}{\sqrt{\left(l^2 + 4x^2\right)^{1/2}}} = \frac{1}{20}$$

or $\dfrac{19}{20} = \dfrac{l}{\left(l^2 + 4x^2\right)^{1/2}}$

On substituting $l = 20$ cm, we have

$$\left(400 + 4x^2\right) = \left(\frac{400}{19}\right)^2; \; x = \frac{10}{19}\sqrt{39} = 3.29 \text{ cm}$$

13.21. Two long parallel rectilinear conductors are 10 cm apart. Current of $I_1 = 20$A and $I_2 = 40$ flows through the conductors in one direction. What work is required (per unit length of the conductors) to move them apart to a distance of 30 cm.

Solution: Let the two conductors be x apart. Force per unit length of the conductor is given by

$$F = \frac{\mu_0}{4\pi} \cdot \frac{2I_1 I_2}{x}$$

Work done $W = \displaystyle\int_{x_1}^{x_2} F \cdot dx = \frac{\mu_0}{4\pi} \cdot 2I_1 I_2 \times 2.303 \log_{10} \frac{x_2}{x_1}$

On putting the values we have

$$W = 10^{-7} \times 2 \times 20 \times 40 \times 2.303 \log_{10} 3$$
$$= 16 \times 10^{-5} \times 2.303 \times 0.4771$$
$$= 17.58 \times 10^{-5} \text{ J/m}$$

13.22. A current $I = 2.0$ amp flows in the circuit shown in Fig. 13.19. Find out the magnetic induction B at point O assume $r_1 = 5.0$ cm and $r_2 = 20.0$ cm.

Solution: Magnetic induction at O is given by

$$B = \frac{3}{4} \cdot \frac{\mu_0}{4\pi} \cdot \frac{2\pi I}{r_1} + \frac{1}{4} \cdot \frac{\mu_0}{4\pi} \cdot \frac{2\pi I}{r_2}$$

$$= \frac{\mu_0}{4\pi} \cdot \frac{2\pi I}{4} \left[\frac{3r_2 + r_1}{r_1 r_2} \right]$$

$$= 10^{-7} \times 3.14 \frac{[0.6 + 0.05]}{0.05 \times 0.2}$$

$$= \frac{3.14 \times 10^{-7} \times 0.65}{10^{-2}}$$

$$= 2.04 \times 10^{-5} \text{W/m}^2$$

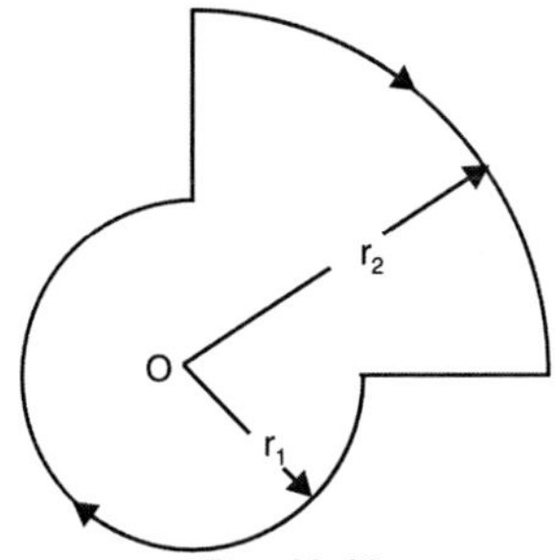

Fig. 13.19

13.23. A current I flows through a wire bent in the form of a regular polygon with n sides inscribed in a circle of radius r. Find out the magnetic induction B at the centre of the polygon.

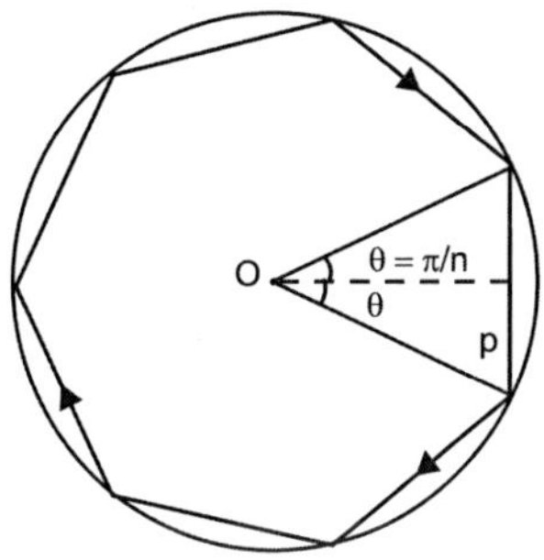

Fig. 13.20

Solution: From Fig. 13.20,

$$\frac{p}{r} = \tan \frac{\pi}{n}$$

$$\therefore \quad p = r \tan \frac{\pi}{n}$$

Length of each straight portion is given by $2p = 2\,r \tan \dfrac{\pi}{n}$

dB at O due to one straight portion is equal to

$$dB = \frac{\mu_0}{4\pi} I \cdot \frac{2r}{r^2} \tan\left(\frac{\pi}{n}\right)$$

$$= \frac{\mu_0}{4\pi} \cdot I \cdot \frac{2}{r} \cdot \tan\left(\frac{\pi}{n}\right)$$

$$\text{Net } B \text{ at the centre } O = \int dB = \frac{\mu_0}{4\pi} \cdot 2I \cdot \frac{n}{r} \cdot \tan\left(\frac{\pi}{n}\right)$$

$$= \frac{\mu_0 n I}{2\pi r} \tan\left(\frac{\pi}{n}\right)$$

13.24. A circular conductor of radius a = 5 cm. carries a current I = 2.0 amp. Find the magnetic induction at the point P lying at a distance of b = 5 mm from the centre.

Solution: Let the point P be chosen as origin coordinate of point A is (r, θ).

$$a^2 = r^2 + b^2 - 2br \cos \theta$$

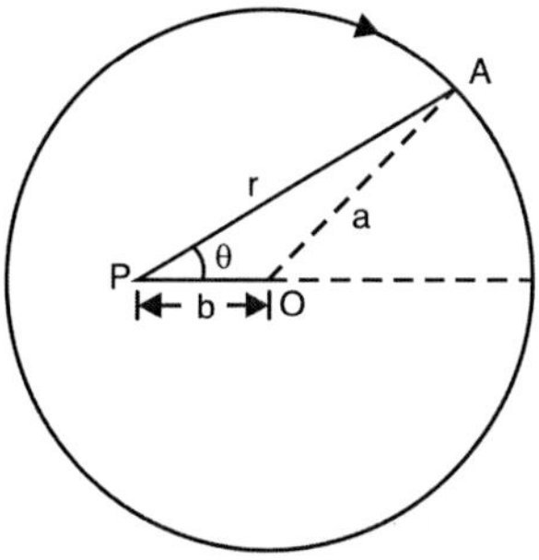

Fig. 13.21

or $\quad \dfrac{1}{r} = \dfrac{1}{b\cos\theta + \sqrt{a^2 - b^2\sin^2\theta}}$

Consider an element at A subtending angle $d\theta$ at P.

Magnetic induction dB due to the element at P is given by

$$dB = \frac{\mu_0}{4\pi}\cdot I\cdot\frac{d\theta}{r}$$

$$\text{Net induction at } P = B = \int dB = \frac{\mu_0}{4\pi}\cdot I\int_0^{2\pi}\frac{d\theta}{b\cos\theta + \sqrt{a^2 - b^2\sin^2\theta}}$$

$$= \frac{\mu_0}{4\pi}\cdot\frac{I}{a^2 - b^2}\int_0^{2\pi}\left[\sqrt{a^2 - b^2\sin^2\theta} - b\cos\theta\right]d\theta$$

$$= \frac{\mu_0}{4\pi}\cdot\frac{Ia}{(a^2 - b^2)}\int_0^{2\pi}\left[1 - \frac{b^2}{2a^2}\sin^2\theta\right]d\theta$$

Here $\dfrac{b}{a} << 1$ see the values of a & b.

$$B = \frac{\mu_0 Ia}{4\pi(a^2 - b^2)}\int_0^{2\pi}\left[1 - \frac{b^2}{4a^2} + \frac{b^2}{4a^2}\cos 2\theta\right]d\theta$$

$$= \frac{\mu_0 Ia}{4\pi(a^2 - b^2)}\cdot\frac{2\pi\cdot(4a^2 - b^2)}{4a^2}$$

$$= \frac{\mu_0}{4\pi}\cdot\frac{I\pi}{2a}\frac{(4a^2 - b^2)}{a^2 - b^2}$$

On substituting the values we have

$$B = 10^{-7}\times\frac{2\times 3.14}{2\times 5\times 10^{-2}}\left[\frac{4\times 25\times 10^{-4} - 25\times 10^{-6}}{25\times 10^{-4} - 25\times 10^{-6}}\right]$$

$$= 6.28\times 10^{-6}\,\frac{[4 - 0.01]}{1 - 0.01}$$

$$= \frac{6.28\times 3.99\times 10^{-6}}{0.99}$$

$$= 25.3\times 10^{-6}\,\text{W/m}^2$$

13.25. A long conductor bent in the form of letter L carries a current $I = 10$ amp. Find the magnetic field at a point O lying on the bisector of the angle PQR at a distance of $l = 10$ cm from the vertex Q.

Solution: Consider an element of length dy, at a distance y from the point A (Fig. 13.22). The magnetic field due to the element at O is given by

$$dB_1 = \frac{\mu_0}{4\pi} \frac{I\,dy \cdot \sin\theta}{\left(x^2 + y^2\right)}$$

Net field due to the part PQ of the conductor is equal to

$$B_1 = \int_{-x}^{\infty} \frac{\mu_0}{4\pi} \cdot \frac{I \cdot dy \cdot x}{\left(x^2 + y^2\right)^{3/2}}$$

Let $y = x \tan\phi$

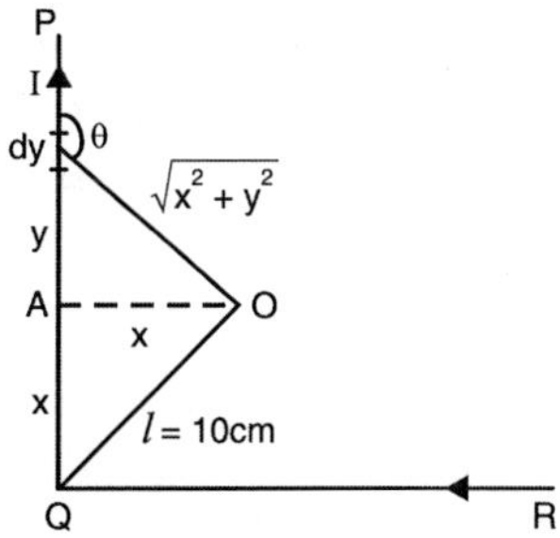

Fig. 13.22

$$B_1 = \int_{-\pi/4}^{\pi/2} \frac{\mu_0}{4\pi} \cdot \frac{I \cdot x \cdot x \cdot \sec^2\phi \cdot d\phi}{x^3 \sec^3\phi} = \frac{\mu_0}{4\pi} \cdot \frac{I}{x} \int_{-\pi/4}^{\pi/2} \cos\phi\, d\phi$$

$$= \frac{\mu_0 I}{4\pi x}\left[\frac{\sqrt{2}+1}{\sqrt{2}}\right] = \frac{\mu_0 I \cdot \left(\sqrt{2}+1\right)}{4\pi l}$$

Net magnetic field at O due to both the sections of the conductor is equal to

$$B = 2B_1 = \frac{2\mu_0 I}{4\pi l}\left(\sqrt{2}+1\right)$$

On substituting the values of various terms we have

$$B = \frac{2\times 10^{-7} \times 10 \times 2.414}{10 \times 10^{-2}} = 4.828 \times 10^{-5}\ \text{W/m}^2$$

13.26. What should be the ratio of the length and diameter of a short solenoid which will provide a magnetic field at its centre equal to that of an infinitely long solenoid with a maximum error 2%?

Solution: Magnetic field B_2 due to an infinitely long solenoid having n turns per unit length and carrying current $I = \mu_0 nI$.

Magnetic field B_1 due to short solenoid having the same number of turns and carrying same current $I = \frac{\mu_0 nI}{2}\left[\cos\theta_1 - \cos\theta_2\right]$

In view of Fig. 13.23

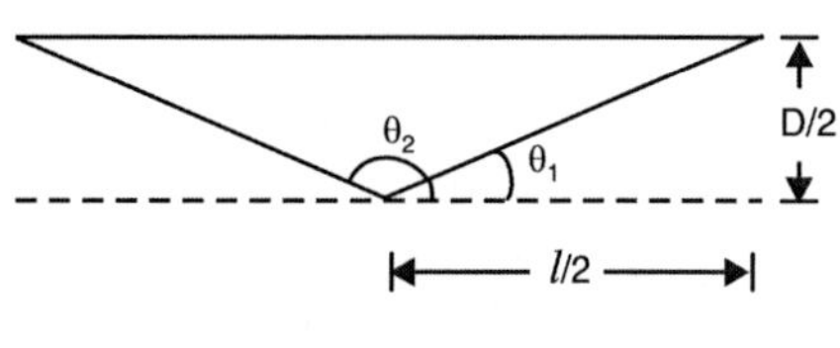

Fig. 13.23

$$B_1 = \frac{\mu_0 nIl}{\sqrt{l^2 + D^2}}$$

$$\frac{B_2 - B_1}{B_2} = \frac{1}{50}$$

$$1 - \frac{1}{50} = \frac{B_1}{B_2} = \frac{l}{\sqrt{l^2 + D^2}}$$

$$\frac{49^2}{50^2} = \frac{l^2}{\left(l^2 + D^2\right)}$$

$$49^2 \cdot l^2 + 49^2 D^2 = 50^2 l^2$$

$$49^2 D^2 = 99 l^2$$

$$\therefore \quad \frac{l}{D} = \frac{49}{\sqrt{99}} = 4.92$$

13.27. In each of the three given figures a single long straight wire has been turned into an arc of radius $R = 50$ mm and two long straight portions. Find out the magnetic induction at O when a current $I = 5$ amp flows through the wire.

Solution: In Fig. 13.24a

$$\text{Net field} = -\frac{\mu_0}{4\pi}\cdot\frac{Ik}{R} - \frac{\mu_0}{4\pi}\cdot\frac{\pi I i}{R} - \frac{\mu_0}{4\pi}\cdot\frac{Ik}{R}$$

$$= -\frac{\mu_0}{4\pi}\cdot\frac{I}{R}[\pi i + 2k]$$

$$B = \frac{\mu_0}{4\pi}\cdot\frac{I}{R}\sqrt{\pi^2 + 4}$$

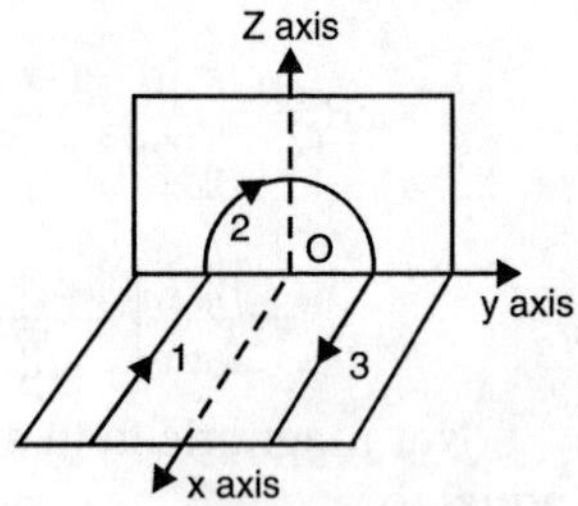

Fig. 13.24a

On putting the values

$$B = \frac{10^{-7} \times 5}{50 \times 10^{-3}}\sqrt{3.14^2 + 4}$$

$$= 37.2\ \mu\text{T}$$

In Fig. 13.24b

$$\text{Net field } B = \frac{\mu_0}{4\pi}\frac{I}{R}[-k - \pi i - i]$$

$$= -\frac{\mu_0}{4\pi}\frac{I}{R}\left[(\pi + 1)i + k\right]$$

$$B = \frac{\mu_0 I}{4\pi R}\left[(\pi + 1)^2 + 1\right]^{1/2}$$

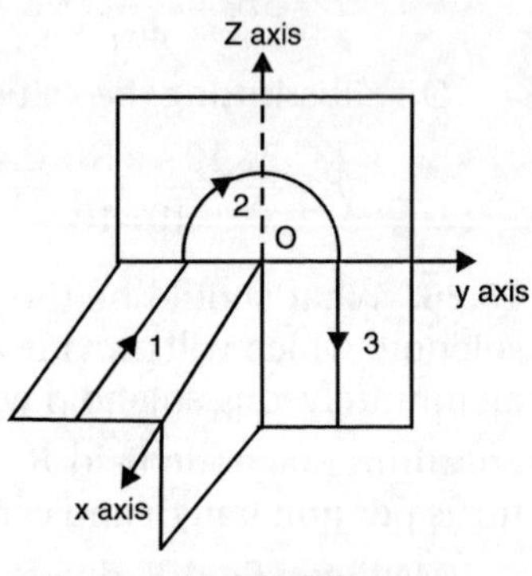

Fig. 13.24b

On putting the values

$$B = \frac{10^{-7} \times 5}{50 \times 10^{-3}}\left[4.14^2 + 1\right]^{1/2}$$

$$= 42.6\mu\text{T}$$

In Fig. 13.24c

$$\text{Net field } B = \frac{\mu_0}{4\pi}\cdot\frac{I}{R}\left[-k - \frac{3\pi}{2}i - j\right]$$

$$= -\frac{\mu_0 I}{4\pi R}\left[\frac{3\pi}{2}i + j + k\right]$$

$$B = \frac{\mu_0}{4\pi}\cdot\frac{I}{R}\left[\frac{9\pi^2}{4} + 1 + 1\right]^{1/2}$$

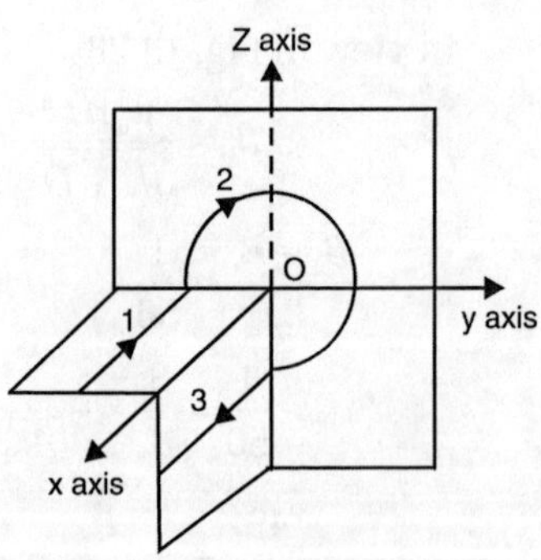

Fig. 13.24c

On putting the values

$$B = \frac{10^{-7} \times 5}{50 \times 10^{-3}} \left[\frac{9 \times 3.14^2}{4} + 1 + 1 \right]^{1/2} = 49.2 \,\mu\text{T}$$

13.28. A non conducting disc of radius r carries Q charge distributed uniformly over its surface. The disc rotates with an angular velocity ω about an axis passing through its centre and perpendicular to its plane. Find (a) the magnetic field at the centre of the disc (b) the magnetic dipole moment of the disc.

Solution: (a) Consider a shell of radius x and $x + dx$.

The area of the shell $= 2\pi\, xdx$

$$\text{Charge contained in this shell} = \frac{Q}{\pi r^2} \cdot 2\pi xdx$$

$$\text{Current } i \text{ due to this charge} = \frac{\omega}{2\pi} \cdot \frac{Q \cdot 2\pi xdx}{\pi r^2}$$

Magnetic field at the centre of the disc due to this current

$$= \frac{\mu_0}{4\pi} \cdot \frac{2\pi i}{x} = \frac{\mu_0}{4\pi} \cdot \frac{2\pi}{x} \cdot \frac{\omega Q xdx}{\pi r^2}$$

$$= \frac{\mu_0 \omega Q dx}{2\pi r^2}$$

$$\text{Net magnetic field} = \frac{\mu_0 \omega Q}{2\pi r^2} \int_0^r dx$$

$$= \frac{\mu_0 \omega Q}{2\pi r}$$

(b) Magnetic dipole moment of the shell $dM = \dfrac{\omega}{2\pi} \cdot \dfrac{Q \cdot 2\pi\, xdx \cdot \pi x^2}{\pi r^2}$

$$\text{Net magnetic dipole moment M} = \int dM = \int_0^r \frac{\omega}{2\pi} \cdot \frac{Q \cdot 2\pi xdx \cdot \pi x^2}{\pi r^2}$$

$$= \frac{\omega Q r^2}{4}$$

13.29. A sphere of radius $R = 40$ mm carries a charge $Q = 3\ \mu$C uniformly distributed over its surface. Find the magnetic induction at the centre of the sphere if it rotates with an angular velocity $\omega = 80$ rad/sec. about an axis passing through its centre.

Solution: Let us consider a disc of thickness $Rd\theta$ (see Fig. 13.25)

Area of the disc $= 2\pi R \sin\theta \cdot Rd\theta$

$$\text{Charge on the disc} = 2\pi R^2 \sin\theta \cdot d\theta \cdot \frac{Q}{4\pi R^2}$$

$$= \frac{Q}{2} \sin\theta \cdot d\theta$$

Current due to the rotation of this charge

$$I = \frac{Q}{2}\sin\theta \cdot d\theta \cdot \frac{\omega}{2\pi}$$

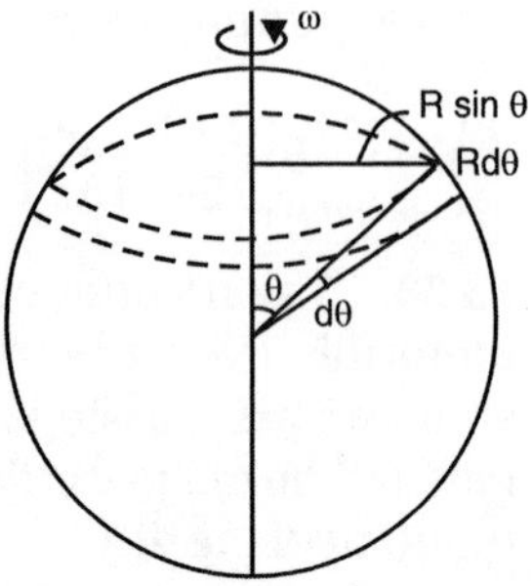

Fig. 13.25

Magnetic field due to this current

$$dB = \frac{\mu_0 I R^2 \sin^2\theta}{2\left[R^2\sin^2\theta + R^2\cos^2\theta\right]^{3/2}}$$

$$= \frac{\mu_0}{2R}\cdot\frac{Q}{2}\sin^3\theta \cdot d\theta \cdot \frac{\omega}{2\pi}$$

$$= \frac{\mu_0\,\omega Q\sin^3\theta\, d\theta}{8\pi R}$$

Net magnetic field at the centre of the sphere

$$B = \frac{\mu_0 Q\omega}{8\pi R}\int_0^{\pi}\sin^3\theta\, d\theta$$

$$= \frac{\mu_0\,\omega Q}{6\pi R}$$

On substituting the values we have

$$B = \frac{4\pi\times 10^{-7}\times 80\times 3\times 10^{-6}}{6\pi\times 40\times 10^{-3}}$$

$$= 4\times 10^{-10}\ \text{W/m}^2$$

13.30. A solid sphere of radius R and mass m contains a charge Q which is uniformly distributed over its volume. Find the magnetic moment of the sphere if it rotates about a diameter with an angular velocity ω.

Solution: Charge per unit volume of the sphere

$$\rho = \frac{Q}{\frac{4}{3}\pi R^3}$$

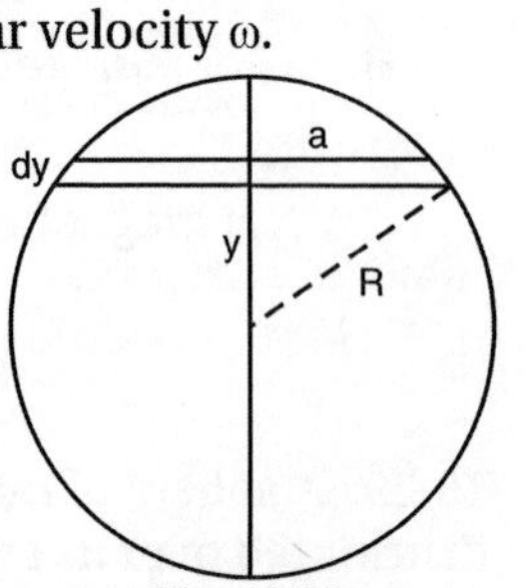

Fig. 13.26a

Consider a disc of radius a and thickness dy at a distance y from the centre of the sphere (see Fig. 13.26a). This disc is full of charge. On this disc consider a shell of radius r and $r + dr$ (see Fig. 13.26b).

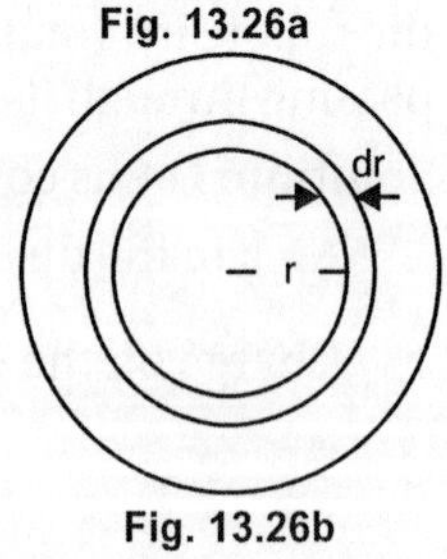

Fig. 13.26b

Area of the shell $= 2\pi\, r dr$

Volume of the shell $= 2\pi\, r dr \cdot dy$

Charge on the shell $= 2\pi\, r dr \cdot dy \cdot \rho$

Current produced due to the rotation of this charge with an angular velocity ω

$$= 2\pi\, rdr \cdot dy \cdot \rho \cdot \frac{\omega}{2\pi}$$

$$= \rho\, \omega\, rdr \cdot dy$$

$$\text{Magnetic moment of the disc} = \int_0^a \rho\omega dy \cdot rdr \cdot \pi r^2$$

$$= \frac{\rho\omega \cdot dy \pi a^4}{4}$$

$$\text{Net magnetic moment of the sphere} = \frac{\rho\omega\pi}{4}\int_{-R}^{R} a^4 \cdot dy$$

$$= \frac{\rho\omega\pi}{4}\left[\int_{-R}^{R}\left(R^2 - y^2\right)^2 \cdot dy\right]$$

$$= \frac{4}{15}\rho\omega\pi R^5$$

$$= \frac{4}{15} \cdot \frac{Q}{4/3\, \pi R^3} \cdot \omega\pi R^5$$

$$= \frac{\omega Q R^2}{5}$$

$$= \frac{Q\omega}{2m} \cdot \frac{2}{5} \cdot mR^2$$

$$= I\omega \cdot \frac{Q}{2m}$$

$$= \frac{Q}{2m} \cdot L;$$

where L = angular momentum of the sphere.

13.31. A long strip of width a carries a current I. Find out the magnetic field at point O lying in the plane of the strip at a distance b from its edge.

Solution: Consider a strip of thickness dx as shown in Fig. 13.27. Current in the ship is equal to $\frac{I}{a} \cdot dx$.

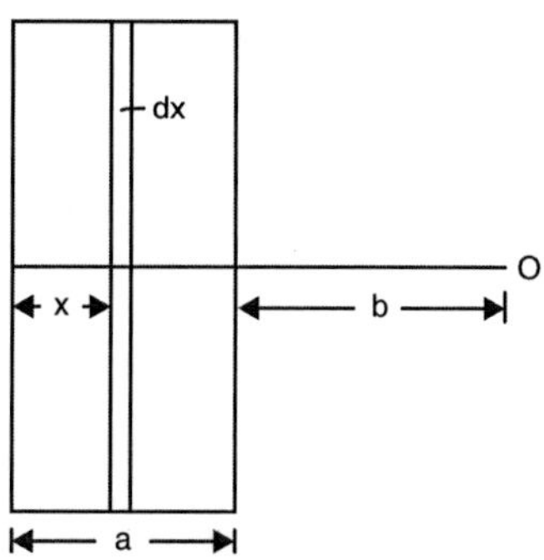

Fig. 13.27

Magnetic field due to infinitely long strip of thickness dx at a point O is given by

$$dB = \frac{\mu_0}{4\pi} \cdot \frac{2I\,dx}{a[a+b-x]}$$

Net magnetic field at O is equal to

$$B = \frac{\mu_0 I}{2\pi a} \cdot \int_0^a \frac{dx}{(a+b-x)}$$

$$= \frac{\mu_0 I}{2\pi a} \cdot \left|\log_e (a+b-x)\right|_a^0$$

$$= \frac{\mu_0 I}{2\pi a} \log_e \left(\frac{a+b}{b}\right)$$

13.32. A long straight wire of radius R carries a current I. Find out the magnetic field inside the conductor at a distance r from its centre.

Solution: Current I is uniformly distributed through the cross-section of the wire. Current I' enclosed by the radius r is given by

$$\frac{I'}{I} = \frac{\pi r^2}{\pi R^2} \text{ (see Fig. 13.28)}$$

$$\therefore \quad I' = \frac{Ir^2}{R^2}$$

According to Amperes law

$$\oint B \cdot dl = B \cdot 2\pi r = \mu_0 I'$$

$$\therefore \quad B = \frac{\mu_0}{2\pi r} \cdot \frac{Ir^2}{R^2} = \frac{\mu_0 Ir}{2\pi R^2}$$

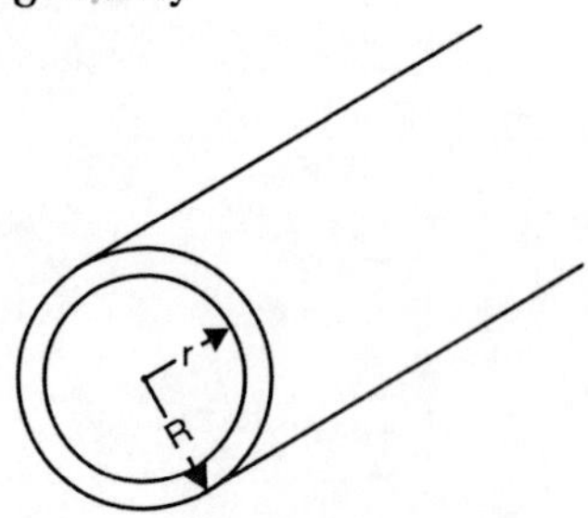

Fig. 13.28

13.33. A hollow cylindrical tube of a metal with inner and outer radii as r_1 and r_2 respectively carries a current I. Find the magnetic field inside the metal at a distance r from the centre.

Solution: In this case the current I is uniformly distributed over the surface of the metal.

$$\text{Current per unit area} = \frac{I}{\pi\left(r_2^2 - r_1^2\right)}$$

According to Ampere's Law

$$B \cdot 2\pi r = \mu_0 I'$$

I' is the current enclosed by the dotted path (see Fig. 13.29). I' is less than I and is given by

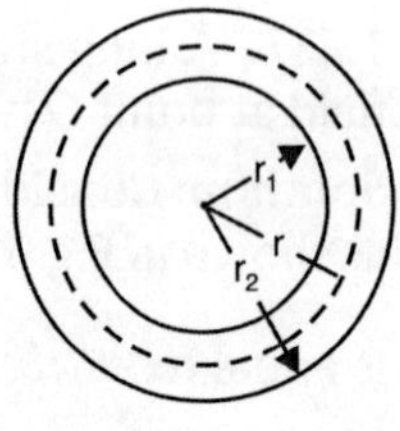

Fig. 13.29

$$I' = \frac{I\pi\left(r^2 - r_1^2\right)}{\pi\left(r_2^2 - r_1^2\right)}$$

$$\therefore \quad B = \frac{\mu_0 I\left(r^2 - r_1^2\right)}{2\pi r\left(r_2^2 - r_1^2\right)}$$

13.34. A circular coil of radius $R = 10$ cm and having number of turns $n = 203$, carrying a current $I_1 = 0.5$ amp and an infinitely long straight wire, carrying a current $I_2 = 30$ amp are arranged as shown in Fig. 13.30. Determine the net magnetic field at the point $P \cdot OP = 1$ cm and $AP = 2$ cm.

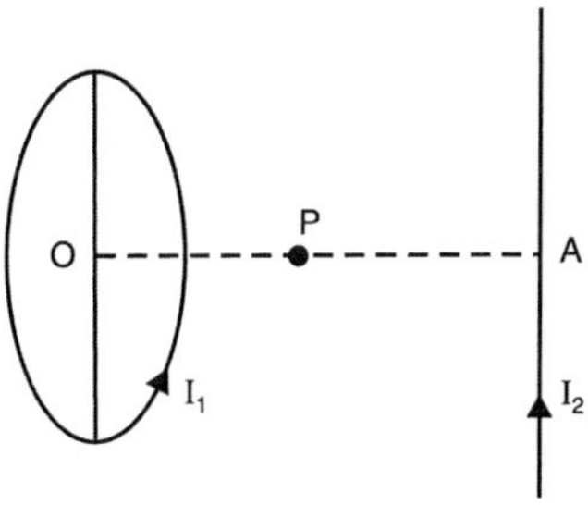

Fig. 13.30

Solution: The magnetic field at a distance x_1 from the centre of the coil in a direction perpendicular to its plane is given by

$$B_1 = \frac{\mu_0}{4\pi} \times \frac{2\pi I_1 R^2 n}{\left(x_1^2 + R^2\right)^{3/2}}$$

On substituting the values of the various terms, we have

$$B_1 = \frac{10^{-7} \times 2 \times 3.14 \times 0.5 \times 10^{-2} \times 203}{1015 \times 10^{-6}}; \; R = 10 \text{ cm}, x_1 = OP = 1 \text{ cm}, I_1 = 0.5\,A$$

$$= 6.28 \times 10^{-4} \text{W/m}^2$$

The magnetic field at a distance x_2 due to I_2 in a long straight conductor is given by

$$B_2 = \frac{\mu_0}{4\pi} \times \frac{2I_2}{x_2}$$

$$B_2 = \frac{10^{-7} \times 2 \times 30}{2 \times 10^{-2}}; \; x_2 = AP = 2 \text{ cm}, I_2 = 30 \text{ A}$$

$$= 3 \times 10^{-4} \text{ W/m}^2$$

Resultant field $B = \sqrt{B_1^2 + B_2^2} = 6.96 \times 10^{-4} \text{ W/m}^2$

13.35. Two identical coaxial coil of radius r and having the same number of turns n are placed with their centres x apart. A current I is allowed to flow through the two coils in the same direction. Determine (i) the magnetic field at the mid-point of the distance between the centres of the two coils and (ii) the condition in which the rate of change of the field along the axis becomes constant.

Solution:

(i) The magnetic field due to current in a coil at a distance l from the centre of the coil $= \dfrac{\mu_0}{4\pi} \dfrac{2\pi n I r^2}{\left(l^2 + r^2\right)^{3/2}}$

We will have the same amount of field due to the current in the other coil. The two fields will be in the same direction. Net field B_1 is given by

$$B_1 = \frac{\mu_0}{4\pi} \frac{4\pi n I r^2}{\left(l^2 + r^2\right)^{3/2}}$$

On substituting $l = x/2$ we have

$$B_1 = \frac{\mu_0}{4\pi} \frac{32\pi n I r^2}{\left(x^2 + 4r^2\right)^{3/2}}$$

(ii) The magnetic field due to current in a coil at a distance x from the centre of the coil is given by

$$B = \frac{\mu_0}{4\pi} \frac{2\pi n I r^2}{\left(x^2 + r^2\right)^{3/2}}$$

This can be represented as

$$B = A\,(x^2 + r^2)^{-3/2}$$

Differentiating the above equation with respect to x

$$\frac{dB}{dx} = \frac{-3}{2} A\left(x^2 + r^2\right)^{-5/2} \cdot 2x = -3Ax\left(x^2 + r^2\right)^{-5/2}$$

On differentiating the above equation once again,

$$\frac{d^2B}{dx^2} = -\,3A\left[\left(x^2 + r^2\right)^{-5/2} - 5x^2\left(x^2 + r^2\right)^{-7/2}\right]$$

$$= -3A\left(x^2 + r^2\right)^{-7/2}\left(r^2 - 4x^2\right)$$

On equation $\dfrac{d^2B}{dx^2}$ to zero, we find

$$x = \pm\frac{r}{2}$$

Thus, under this condition, the rate of change of magnetic field becomes constant.

13.36. Two identical coaxial coils of radius $r = 10$ cm and having the same number of turns, $n = 50$ are placed with their centres 10 cm apart. A current $I_1 = 2\,A$ flows through the two coils in the same direction. A small piece of wire of length 1 cm, carrying a current $I_2 = 3$ amp is placed perpendicular to the axis of the two coils at the mid-point of the distance between them. Find the magnetic force experienced by the wire.

Solution: Net magnetic field at mid-point of the distance between the two coils

$$B = 2\frac{\mu_0}{4\pi} \frac{2\pi n I_1 r^2}{\left(x^2 + r^2\right)^{3/2}}$$

On substituting the values of the various terms, we have

$$B = \frac{2\times10^{-7}\times2\times3.14\times50\times2\times10^{-2}}{1397.5\times10^{-6}}$$

$$= 8.987\times10^{-4}\,\text{W/m}^2$$

Force experienced by the wire $= BI_2l$

$$= 8.987\times10^{-4}\times3\times10^{-2}$$

$$= 2.696\times10^{-5}\,\text{N}$$

13.37. Calculate the amount of force acting per unit length of the wire *ab* at the point *A* of the Figs. 13.31a and b when a current I = 10 amp flows through it. R = 20 cm, l = 10 cm.

Solution: In Fig. 13.31a, there is a semi-circular wire carrying a current I. The magnetic field at the centre is given by $\frac{\mu_0 I}{4R}$. Force on the length *ab* $\frac{\mu_0 I}{4R}\cdot I\cdot 2R$.

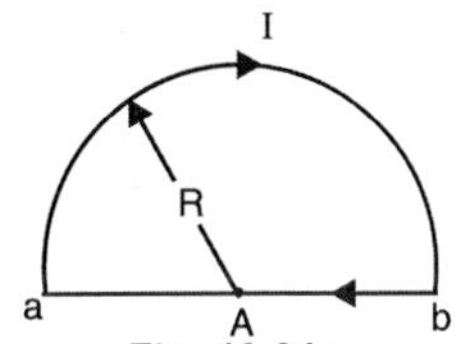

Fig. 13.31a

Hence, force per unit length

$$\frac{\mu_0 I^2}{4R} = \frac{4\times3.14\times10^{-7}\times10^2}{4\times0.2} = 0.157\,\text{mN/m}$$

In Fig. 13.31b, the magnetic field at A is the net field due to the current in part *ad* and *bc* of the conductor. Hence the net field at A is given by

$$\frac{\mu_0 I}{2\pi l}+\frac{\mu_0 I}{2\pi l}=\frac{\mu_0 I}{\pi l}$$

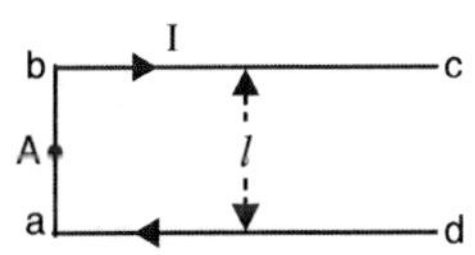

Fig. 13.31b

Force on the length $ab = \frac{\mu_0 I}{\pi l}\cdot I\cdot l$

Hence force per unit length $= \frac{\mu_0 I^2}{\pi l}$

$$= \frac{4\times3.14\times10^{-7}\times10^2}{3.14\times10\times10^{-2}}$$

$$= 0.4\,\text{mN/m}$$

13.38. Two long thin parallel conductors in the form of a straight wire and a strip of width d_2, separated by a distance d_1 as shown in Fig. 13.32 carry currents I_1 and I_2 respectively. Calculate the magnetic force of interaction acting unit length between them.

Solution: On the right hand, there is a conductor of width d_2 carrying a current I_2. Current per unit width $= \frac{I_2}{d_2}$. Consider a small strip of thickness dx lying at a distance x from the first conductor.

The current in this strip = $\dfrac{I_2 dx}{d_2}$

The magnetic force per unit length = $\dfrac{\mu_0}{4\pi}\dfrac{2I_1 I_2}{d_2}\dfrac{dx}{x}$

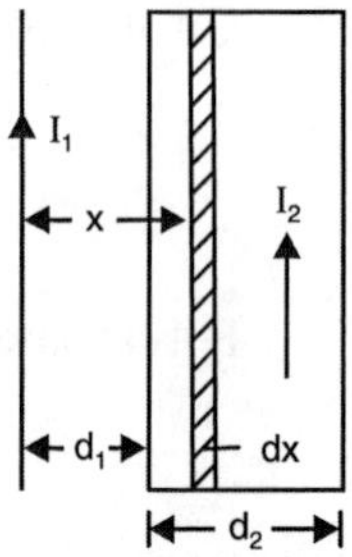

Fig. 13.32

Net magnetic foce per unit length

$$= \frac{\mu_0}{4\pi}\frac{2I_1 I_2}{d_2}\int_{d_1}^{d_1+d_2}\frac{dx}{x}$$

$$= \frac{\mu_0}{4\pi}\cdot\frac{2I_1 I_2}{d_2}\log_e\left(\frac{d_1+d_2}{d_1}\right)$$

13.39. A thin wire of length l carrying a current I has been turned into a circular coil. If the coil is placed in a magnetic field of induction B, then under what condition will the coil experience the maximum torque? Calculate also the value of the maximum torque.

Solution: Consider a coil of N turns, with area A, placed in a uniform magnetic field of induction B. On establishing a current I in the coil, it experiences a torque T equal to $BINA$.

For a circular coil $A = \pi r^2$

$\therefore \quad T = BIN\pi r^2 \quad$ ... (i)

Now $l = 2\pi rN \quad$ ... (ii)

On putting the value of r from Eq. (ii), we have

$$T = BIN\pi\frac{l^2}{4\pi^2 N^2}$$

$$\therefore \quad T = \frac{BIl^2}{4\pi N}$$

Now T will be maximum when $N = 1$

The maximum value of the torque $T = \dfrac{BIl^2}{4\pi}$

13.40. A plane spiral of n turns, having the radii of internal and external loops as r_1 and r_2 carries a current I. Calculate the magnetic induction at the centre of the spiral.

Solution: Let us consider a turn of the spiral of radius r. The magnetic induction B_1 is given by

$$B_1 = \frac{\mu_0}{4\pi}\frac{2\pi I}{r} = \frac{\mu_0 I}{2r}$$

The total magnetic induction $B = \int B_1 dN$

where dN is the number of turns lying between the radius r and $r + dr$

Now $\dfrac{dN}{dr} = \dfrac{n}{r_2 - r_1}$

$$\therefore \quad B = \int_{r_1}^{r_2} \frac{\mu_0 I}{2r} \frac{n}{r_2 - r_1} dr = \frac{\mu_0 nI}{2(r_2 - r_1)} \log \frac{r_2}{r_1}$$

13.41. A current I flows in a long straight wire with cross-section in the form of a thin half-ring of radius R as shown in Fig. 13.33a. Find the induction of the magnetic field at the point O.

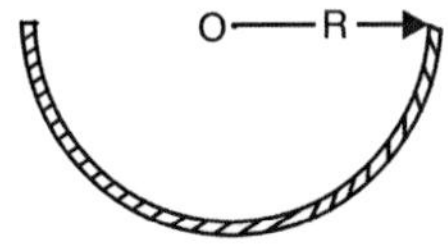

Fig. 13.33a

Solution: Divide the entire wire into elementary filaments carrying a current dI. The field due to an elementary filament is given by

$$dB = \frac{\mu_0}{4\pi} \frac{2dI}{R} = \frac{\mu_0}{2\pi R} dI$$

Now $dI = \dfrac{I}{\pi} d\theta$ (see Fig. 13.33b)

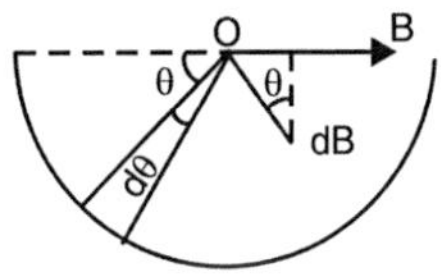

Fig. 13.33b

$$\text{Total field} = B = \int_0^{\pi} dB \sin\theta = \frac{\mu_0}{2\pi R} \frac{I}{\pi}$$

$$\int_0^{\pi} \sin\theta \, d\theta = \frac{\mu_0 I}{\pi^2 R}$$

13.42. A current $I = 10.0$ amp flows along a thin wire bent as shown in Fig. 13.34a. The radius of the curved path of the wire is equal to R = 100 mm and the angle $2\phi = 60°$. Find the magnetic induction at the point O.

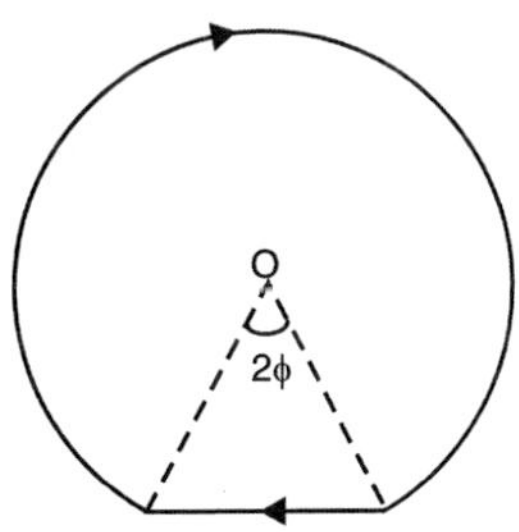

Fig. 13.34a

Solution: The total magnetic field at O is made of two parts (i) field B_1 due to the straight part of the conductor and (ii) field B_2 due to the curved part of the conductor.

Consider Fig. 13.34b we have

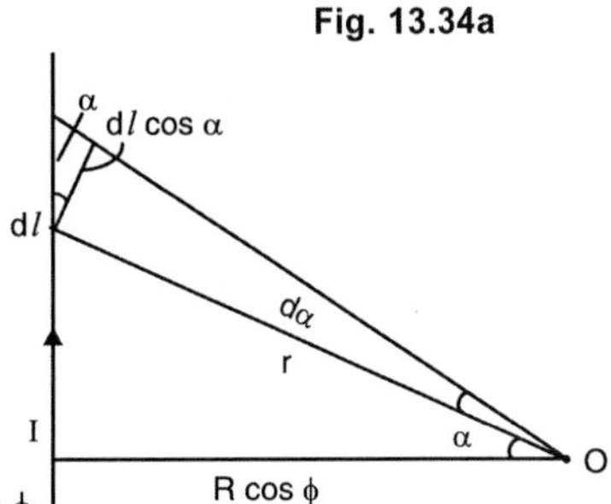

Fig. 13.34b

$$dB_1 = \frac{\mu_0}{4\pi} \frac{I\,dl \cos\alpha}{r^2}$$

$$dl \cos\alpha = r d\alpha$$

$$\therefore \quad dB_1 = \frac{\mu_0}{4\pi} \frac{I r d\alpha}{r^2} = \frac{\mu_0 I}{4\pi} \frac{d\alpha}{r}$$

$$= \frac{\mu_0 I}{4\pi} \frac{d\alpha \cos\alpha}{R\cos\phi}; \cos\alpha = \frac{R\cos\phi}{r}$$

$$= \frac{\mu_0 I}{4\pi R\cos\phi} \cos\alpha \, d\alpha$$

$$B_1 = 2\int dB_1 = \frac{2\mu_0 I}{4\pi R\cos\phi}\int_0^{\phi}\cos\alpha\, d\alpha = \frac{\mu_0 I}{2\pi R}\tan\phi$$

Second part: The field due to the curved path of the conductor is given by

$$B_2 = \frac{\mu_0 I}{4\pi}\frac{(2\pi - 2\phi)R}{R^2}$$

$$= \frac{\mu_0 I}{2\pi R}(\pi - \phi)$$

Net field $B = \dfrac{\mu_0 I}{2\pi R}[\pi - \phi + \tan\phi]$

On substituting the values we have

$$B = \frac{4\pi\times 10^{-7}\times 10}{2\pi\times 100\times 10^{-3}}\left[\pi - \frac{\pi}{6} + 0.57731\right]$$

$$= 2\times 10^{-5}\times 3.194$$

$$= 63.9\ \mu\text{T}$$

13.43. The two ends of a U-shaped wire of mass m = 10 g and length l = 20 cm are immersed in mercury, as shown in Fig. 13.35. The wire is placed in a uniform magnetic field of induction $B = 5\times 10^{-3}$ W/m². Calculate the amount of charge which upon flowing through the wire will cause it to jump up to the height of h = 10 cm.

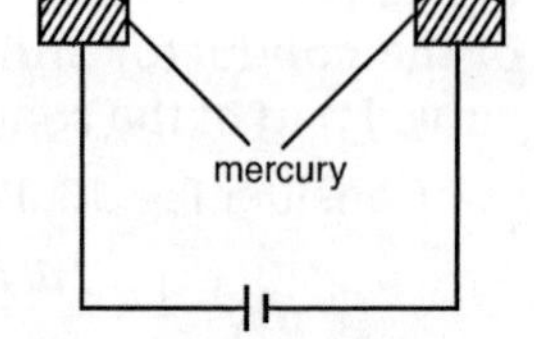

Fig. 13.35

Solution: Magnetic force on the wire = Bil

Further, we know that $F = m\dfrac{dv}{dt}$

or $\quad m dv = F dt$ or $mv = \int F dt = \int Bil\, dt$

$$= Bl\int dq = Blq$$

or $\quad v = Blq/m$

We also know that $v^2 = 2gh$ or $v = \sqrt{2gh}$

$\therefore \quad Blq = m\sqrt{2gh}$

or $\quad q = \dfrac{m}{lB}\sqrt{2gh}$

On substituting the values of the various terms, we have

$$q = \frac{10\times 10^{-3}\times\sqrt{2\times 9.8\times 0.1}}{0.2\times 5\times 10^{-3}}$$

$$= \frac{10\times 10^{-3}\times 2\times 0.7}{0.2\times 5\times 10^{-3}}$$

$$= 14\ \text{C}$$

13.44. In a radially symmetrical parallel beam of electrons the magnetic induction within the beam at a distance r from the axis is given by $B = br^{\alpha}$ where b and α are positive constant. Find the current density as a function of distance r.

Solution: Consider a shell of radius r and $r + dr$.

Area of the shell = $2\pi\, rdr$

According to Amperes Law

$$B \cdot 2\pi r = \mu_0 I = \mu_0 \int jds = \mu_0 \int j 2\pi rdr$$

$$br^{\alpha} \cdot r = \mu_0 \int jrdr$$

$$b(1+\alpha) r^{\alpha} = \mu_0 jr$$

$$\therefore \quad j = \frac{b(1+\alpha) r^{\alpha - 1}}{\mu_0}$$

13.45. A current $I = 5$ amp flows through a straight wire of radius R. Find out the magnetic flux per unit length through half the cross-section of the wire.

Solution: According to Ampere's law the magnetic field at a distance r from the axis of the wire $= \frac{\mu_0}{2\pi} \cdot \frac{Ir}{R^2}$

Magnetic flux through half the cross-section $= \phi = \int_0^R B \cdot dr$

$$= \frac{\mu_0 I}{2\pi R^2} \int_0^R rdr$$

$$= \frac{\mu_0 \cdot I}{4\pi}$$

On substituting the values we have

$$\phi = 10^{-7} \times 5$$

$$= 0.5\ \mu\text{W/m}$$

13.46. From a long thin walled cylindrical conducting tube of radius R, a longitudinal slit of width a has been taken out such that $a << R$. Find the magnetic induction inside the cylinder carrying a current I at the distance r from the place at which the slit has been taken.

Solution: Current in the portion of the cylinder left $= \frac{I}{2\pi R - a} \cong \frac{I}{2\pi R}$

Current in the portion taken out $= \frac{Ia}{2\pi R}$

The magnetic field B_1 due to the strip + magnetic field B_2 due to the remaining part of the cylinder at any point inside the cylinder is equal to zero.

$$\therefore \quad B_1 - B_2 = 0$$

or $$B_1 = B_2$$

$$B_1 = B_2 = \frac{\mu_0}{2\pi r} \cdot \frac{Ia}{2\pi R} = \frac{\mu_0 Ia}{4\pi^2 Rr}$$

13.47. Find the pressure exerted on the walls of a long thin walled cylinder when a current I flows through it.

Solution: Consider a current element. Magnetic field of induction B_1 created by the element at a point close to the surface is given by

$$B_1 = \frac{\mu_0 I}{2}$$

The magnetic induction outside the cylinder and close to its surface is equal to $B = \dfrac{\mu_0 I}{2\pi R}$

Let B_2 be the magnetic induction due to the remaining current elements excluding the element under consideration.

Inside the cylinder the magnetic induction is zero.

Due to this

$$B_1 = B_2$$

Magnetic induction outside the cylinder and close to its surface

$$B = \frac{\mu_0 I}{2\pi R} = B_1 + B_2 = 2B_2$$

or $$B_2 = \frac{B}{2}$$

$$\text{Pressure} = p = \frac{\text{force}}{\text{area}} = IB_2 = \frac{2B_2}{\mu_0} \cdot B_2 = \frac{2B_2^2}{\mu_0} = \frac{B^2}{2\mu_0}$$

$$= \frac{\mu_0^2 I^2}{4\pi^2 R^2} \times \frac{1}{2\mu_0} = \frac{\mu_0 I^2}{8\pi^2 R^2}$$

13.48. A piece of copper wire has been shaped into the form of letter M consisting of four segments. The length of each segment is represented in the Fig. 13.36. The wire is placed in the xy plane where a magnetic field of induction $B = 2\,(i + j)$ W/m² exists. Find the magnitude and the direction of the force acting on the wire when a current $I = 10$ amp flows through it.

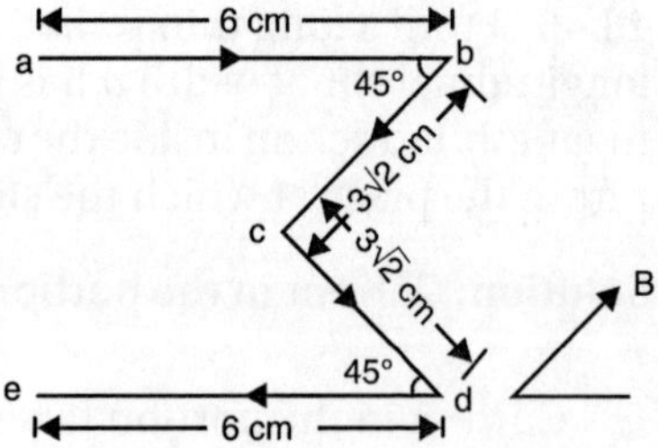

Fig. 13.36

Solution: The force acting on segment ab will be equal and opposite the force acting on segment *de*. No force will act on segment *bc*. The net force on the wire is due to the force on segment *cd* and is given by

$$2\sqrt{2}\times 10\times 3\sqrt{2}\times 10^{-2}\text{k} = 1.2\text{ N along z axis}$$

13.49. A thin copper wire has been bent in the form of a ring of radius r = 5 cm. The ring supported by two flexible conductors is placed in a uniform magnetic field of induction $B = 1\text{ W/m}^2$ in such a way so that the normal to the plane of the ring makes an angle of 90° with the lines of forces. Calculate the tension in the ring when a current I = 2 amp flows through it.

Solution: Under the action of the magnetic field the ring will turn from its initial position to final position as shown in the Figs. 13.37a & 13.37b.

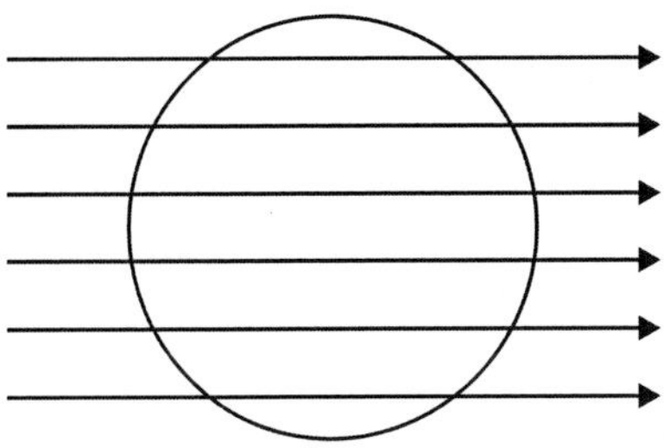

Initial Position

Fig. 13.37a

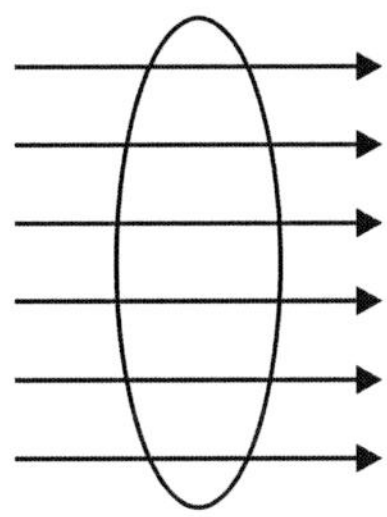

Final Position

Fig. 13.37b

Let T be the tension in the ring then (see Fig. 13.37c)

$$2T\sin\theta = 2T\theta = BI\cdot r\cdot 2\theta$$

$$\therefore \quad T = BIr$$

On substituting the values of various terms we have

$$T = 1\times 2\times 5\times 10^{-2} = 0.1\text{ N}$$

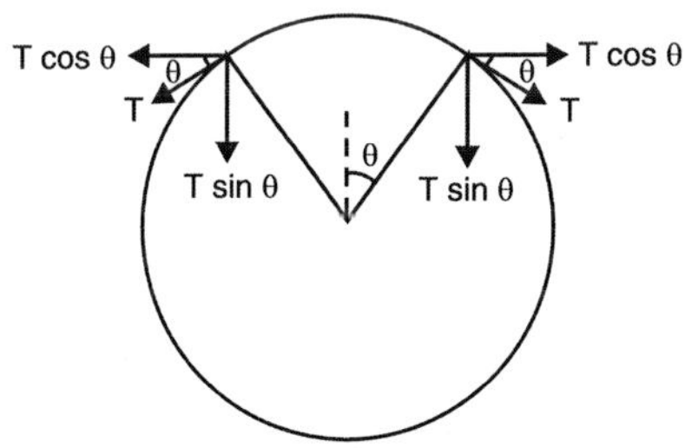

Ring with tension

Fig. 13.37c

13.50. A thin copper wire of diameter d = 1.00 mm is bent to form a ring of radius r = 5 cm. The ring carrying a current of I = 1 amp is placed in a uniform magnetic field of induction $B_1 = 1000\text{ w/m}^2$. Keeping its axis parallel to the lines of force. By what amount the magnetic field be increased in order to rupture the ring? The ultimate tensile strength of copper is $\sigma = 3000\text{ kgf/cm}^2$.

Solution: Let the required magnetic field be B_2.

The magnetic force = $B_2 Ir$; r = radius of the ring

$$\sigma = \frac{4B_2 Ir}{\pi d^2}$$

$$\therefore \quad B_2 = \frac{\sigma \pi d^2}{4Ir}$$

On substituting the values of various quantities we have

$$B_2 = \frac{3000 \times 9.8 \times 3.14 \times 10^{-6}}{10^{-4} \times 4 \times 1 \times 5 \times 10^{-2}}$$

$$= 4615.8 \text{ w/m}^2$$

$$B_2 - B_1 = 3615.8 \text{ w/m}^2$$

13.51. Two straight wires of length l & L carrying currents I_1 & I_2 respectively are placed symmetrically parallel to each other at a distance x as shown in Fig. 13.38. Calculate the magnetic force on wire due to the other.

Solution: Consider elements of length dy and dy_1 on the two wires respectively. Force on the element dy_1 due to the field by the element dy is given by

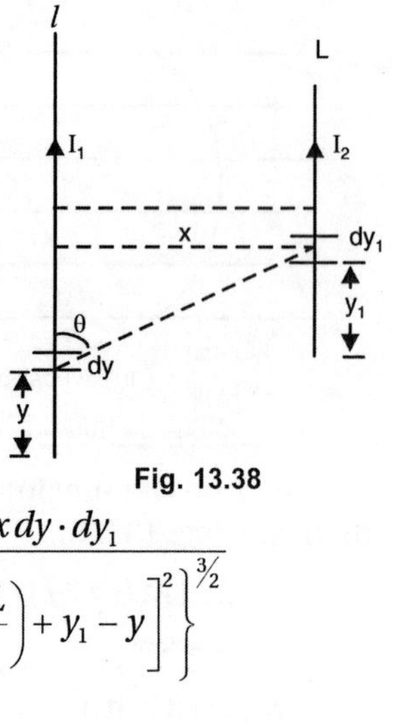

Fig. 13.38

$$dF = \frac{\mu_0}{4\pi} \cdot \frac{I_1\, dy \cdot \sin\theta \cdot I_2 dy_1}{x^2 + \left[\left(\frac{l-L}{2}\right) + y_1 - y\right]^2};$$

$$\sin\theta = \frac{x}{\sqrt{x^2 + \left[\left(\frac{l-L}{2}\right) + y_1 - y\right]^2}}$$

$$\text{Net magnetic force } F = \int_0^l \int_0^L \frac{\mu_0}{4\pi} \cdot \frac{I_1 I_2 \cdot x\, dy \cdot dy_1}{\left\{x^2 + \left[\left(\frac{l-L}{2}\right) + y_1 - y\right]^2\right\}^{3/2}}$$

$$\text{Let us substitute } \frac{l-L}{2} + y_1 - y = x \tan\phi$$

$$F = \iint \frac{\mu_0}{4\pi} \cdot \frac{I_1 I_2\, x \cdot dy \cdot x \sec^2\phi \cdot d\phi}{x^3 \sec^3\phi}$$

$$= \left| \int_0^l \frac{\mu_0}{4\pi} \cdot \frac{I_1 I_2\, dy \cdot \left[\left(\frac{l-L}{2}\right) + y_1 - y\right]}{\sqrt{x^2 + \left[\left(\frac{l-L}{2}\right) + y_1 - y\right]^2}} \right|_0^L$$

$$= \frac{\mu_0 I_1 I_2}{4\pi x} \left[\int_0^l \frac{\left[\left(\frac{l+L}{2}\right) - y\right] dy}{\sqrt{x^2 + \left(\frac{l+L}{2} - y\right)^2}} - \int_0^l \frac{\left[\left(\frac{l-L}{2}\right) - y\right] dy}{\sqrt{x^2 + \left(\frac{l-L}{2} - y\right)^2}} \right]$$

$$= \frac{\mu_0 I_1 I_2}{4\pi x}\left[-\sqrt{x^2+\left[\left(\frac{l+L}{2}\right)-y\right]^2}+\sqrt{x^2+\left[\left(\frac{l-L}{2}\right)-y\right]^2}\right]_0^l$$

$$= \frac{\mu_0}{4\pi}\cdot\frac{2I_1I_2}{x}\left[\sqrt{\left(\frac{l+L}{2}\right)^2+x^2}-\sqrt{\left(\frac{l-L}{2}\right)^2+x^2}\right]$$

13.52. How many ampreturns are required to produce a magnetic field of induction $B = 9.42 \times 10^{-4}$ w/m² in a solenoid 21 cm long and 3 cm in diameter? What potential difference has to be applied across the ends of such solenoid if the wire wound over is of 1 mm diameter and have resistivity 1.7×10^{-8} Ωm?

Solution: Magnetic field due to a solenoid of length l and diameter D is given by

$$B = \frac{\mu_0 Ni}{l} \times \frac{l}{\sqrt{l^2+D^2}}$$

$$\therefore \quad Ni = \frac{B\cdot\sqrt{l^2+D^2}}{\mu_0}$$

On substituting the values of various terms we have

$$Ni = \frac{9.42\times10^{-4}\times\sqrt{21^2+3^2}\times10^{-2}}{4\pi\times10^{-7}}$$

$$= 157.5$$

$$V = iR = \frac{4i\rho N\pi D}{\pi d^2}; \; d = \text{diameter of the wire}$$

On putting the values we have

$$V = \frac{4\times157.5\times1.7\times10^{-8}\times3\times10^{-2}}{10^{-6}}$$

$$= 0.32 \text{ volt}$$

13.53. A metallic sheet of thickness b and width l carrying a current I is placed in a magnetic field of induction B as shown in the Fig. 13.39. What will be the potential difference across the two points P & Q of the sheet.

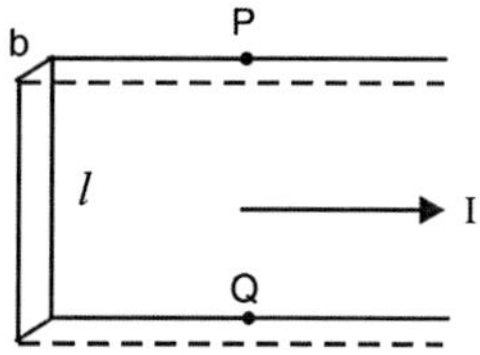

Fig. 13.39

Solution:

$I = neblv_d$; n = number of electrons per unit volume

v_d = drift velocity

Under the action of magnetic field the electron will experience a lorentz force equal to $ev_d B$. Due to this force one edge of the plate will have –ve charges and the other will have +ve charges. In this way there will be an electric field. The electron will continue to move till the electric and magnetic forces are equal.

$$\frac{eV}{l} = Bev_d$$

$$\therefore \quad V = Blv_d = \frac{Bl \cdot I}{nebl} = \frac{BI}{neb}$$

13.54. A copper wire of cross-sectional area $A = 0.5$ mm² has been turned to form a single turn ring. A current of 25 amp induces a magnetic field of induction $B = 2.24 \times 10^{-4}$ w/m² at the centre of the ring. Calculate the potential difference applied across the ends of the wire forming the ring. Resistivity of copper $\rho = 1.7 \times 10^{-8}\ \Omega$m.

Solution:

$$V = IR = I\rho \cdot \frac{2\pi r}{A}; r = \text{radius of the ring}$$

$$B = \frac{\mu_0}{2}\frac{I}{r}$$

$$\therefore \quad V = I\rho \cdot \frac{2\pi}{A} \times \frac{\mu_0 I}{2B}$$

$$= \frac{\mu_0 \rho I^2 \pi}{BA}$$

On substituting the values of the various terms we have

$$V = \frac{4\pi \times 10^{-7} \times 1.7 \times 10^{-8} \times 25^2 \times 3.14}{2.24 \times 10^{-4} \times 0.5 \times 10^{-6}}$$

$$= 0.374 \text{ volt}$$

13.55. A cylindrical conductor of cross sectional radius R carries a current I. Find the potential difference between the axis of the conductor and its surface. The concentration of the electron is n.

Solution:

$I = neAv_d$; v_d = drift velocity

A = area of cross-section = πR^2

$$v_d = \frac{I}{\pi R^2 ne}$$

Magnetic field inside the conductor at a distance r from its centre is given by

$$B = \frac{\mu_0}{2\pi}\frac{Ir}{R^2}$$

Work done in displacing a unit +ve charge through dr is given by

$$dV = v_d \cdot B \cdot dr$$

$$= \frac{I}{\pi R^2 ne} \cdot \frac{\mu_0 Ir}{2\pi R^2} \cdot dr$$

$$V = \frac{\mu_0 I^2}{2\pi^2 R^4 ne} \int_0^R rdr$$

$$= \frac{\mu_0 I^2}{2\pi^2 R^2 ne}$$

13.56. Two circular coils having current in the same direction constitutes magnetic dipoles of dipole moments $M_1 = 2$ mA·m² and $M_2 = 4$ mA·m². They are placed in such a way so that they have a collinear axis. Find the force of interaction between them if the distance between their centres $l = 20$ cm is much greater than their dimensions.

Solution: Let r be the radius of the first coil, Magnetic field due to this coil

$$B = \frac{\mu_0}{4\pi} \cdot \frac{2\pi i r^2}{\left(r^2 + l^2\right)^{3/2}}$$

when $l >> r$ then

$$B = \frac{\mu_0}{4\pi} \cdot \frac{2M_1}{l^3}$$

$$\text{Force of interaction} = \frac{dB}{dl} \cdot M_2$$

$$= -\frac{\mu_0}{4\pi} \cdot \frac{6M_1 M_2}{l^4}$$

On substituting the values we have

$$= \frac{10^{-7} \times 6 \times 2 \times 10^{-3} \times 4 \times 10^{-3}}{(0.2)^4}$$

$$= 3 \text{ nN}$$

13.57. A loop consisting of two arc of radii r_1 & r_2 ($r_2 > r_1$) respectively and two straight segments carries a current I. It subtends an angle 2θ at their common centre as shown in Fig. 13.40. Find the moment of the force acting on the loop under the action of magnetic field due to long straight wire carrying current I_0 and placed at the common centre of the two arc perpendicular to the plane of the loop.

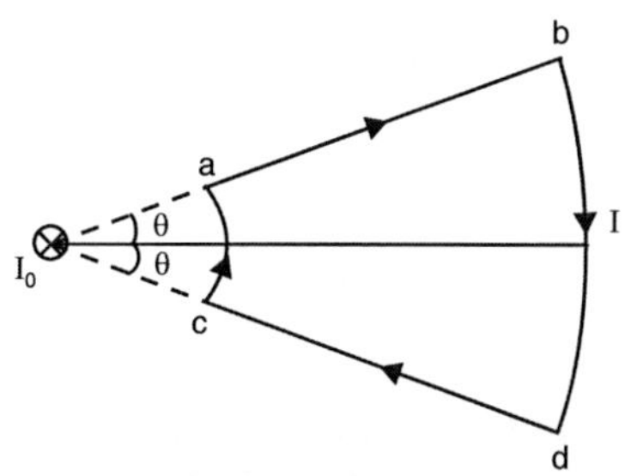

Fig. 13.40

Solution: No force will act on the curved part of the loop. We have magnetic force only on the straight segments which will constitute a torque.

Consider an element of length dr in the segment ab located at a distance r from the centre.

The magnetic force on the element is given by

$$dF = BI \cdot dr$$

$$\text{Torque on the element } = dF \cdot r \sin\theta$$

$$= \frac{\mu_0}{2\pi} \cdot \frac{I_0}{r} \cdot I \cdot r\, dr \sin\theta$$

$$\text{Total torque } = \frac{\mu_0 I I_0}{2\pi} \cdot \sin\theta \int_{r_1}^{r_2} dr$$

$$= \frac{\mu_0 I I_0}{2\pi} (r_2 - r_1) \sin\theta$$

We have the same value of the torque on other straight segment.

$$\therefore \quad \text{Net torque} = \frac{\mu_0 I I_0}{\pi} \cdot (r_2 - r_1) \sin\theta$$

13.58. A long straight wire carrying current I_0 and a square frame with sides b carrying current I are located in the same plane. The axis of the coil is parallel to the long straight wire and the two are $\eta\, b$ distance apart. Find the net magnetic force acting on the frame.

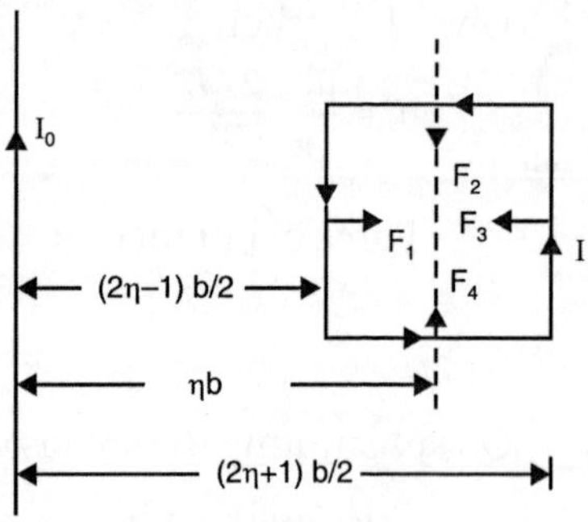

Fig. 13.41

Solution: All the four forces acting on the sides of the frame are shown in the Fig. 13.41. Net force acting on the frame is $\vec{F_1} + \vec{F_3}$

$$\vec{F_1} = \frac{\mu_0}{2\pi} \cdot \frac{I I_0 b}{(2\eta - 1)^{b}\!/_{2}}; \quad \vec{F_3} = \frac{\mu_0}{2\pi} \cdot \frac{I I_0 b}{(2\eta + 1)^{b}\!/_{2}}$$

$$\text{Net magnetic force on the frame } = \Delta\vec{F} = \vec{F_1} - \vec{F_3}$$

$$= \frac{\mu_0 I I_0}{\pi} \left[\frac{1}{2\eta - 1} - \frac{1}{2\eta + 1} \right]$$

$$= \frac{2\mu_0 I I_0}{\pi(4\eta^2 - 1)}$$

13.59. An insulated wire is wound so that it forms a flat coil with N turns. The radius of the inner and outer most turns are r_1 and r_2 respectively. Find the magnetic moment of the coil when a current I flows through it.

Solution: Consider a turn of the flat coil of radius r. The magnetic moment dM is given by

$$dM = I\pi r^2$$

Net magnetic moment M is given by

$$M = \int I\pi r^2 dN$$

where dN = number of turns lying between the radius r and $r + dr$

Now $\frac{dN}{dr} = \frac{N}{r_2 - r_1}$

$$\therefore \quad M = \int_{r_1}^{r_2} \frac{\pi I \cdot N}{r_2 - r_1} r^2 dr$$

$$= \frac{\pi N I}{3(r_2 - r_1)}\left(r_2^3 - r_1^3\right)$$

$$= \frac{1}{3}\pi NI\left(r_1^2 + r_1 r_2 + r_2^2\right)$$

13.60. In hydrogen atom an electron moves in the first orbit. Due to the motion of the electron the induction of the magnetic field generated at the centre of the orbit is B. Find the induction of the magnetic field when the electron moves in the second orbit.

Solution: The motion of the electron in the first orbit is given by equation

$$\frac{e^2}{4\pi\varepsilon_0 r^2} = mr\omega^2$$

$$\therefore \quad \omega = \sqrt{\frac{e^2}{4\pi\varepsilon_0 mr^3}} = 2\pi\upsilon \text{; where } \upsilon = \text{frequency of revolution}$$

$$= \text{current } i = e\upsilon = \frac{e^2}{2\pi\sqrt{4\pi\varepsilon_0 mr^3}}$$

$$B = \frac{\mu_0}{4\pi}\cdot\frac{2\pi i}{r} = \frac{\mu_0}{4\pi}\cdot\frac{e^2}{r\sqrt{4\pi\varepsilon_0 mr^3}} = \frac{\mu_0}{4\pi}\cdot\frac{e^2}{\sqrt{4\pi\varepsilon_0 m}}\cdot r^{5/2}$$

The radius of the second orbit $= 4r$

$$\therefore \quad B' = \frac{\mu_0}{4\pi}\cdot\frac{e^2}{\sqrt{4\pi\varepsilon_0 m}}\cdot\frac{1}{r^{5/2}}\cdot\frac{1}{4^{5/2}} = \frac{B}{32}$$

13.61. A current of 1 amp is flowing in the sides of an equilateral triangle of side 4.5×10^{-2}m. Find the magnetic field at the centroid of the triangle. (Roorkee 1991)

Solution: In Fig. 13.42

$$OP = OQ = OR = \frac{1}{3}AR = \frac{1}{3} \times 4.5 \times 10^{-2} \cdot \sin 60$$

$$= \frac{4.5 \times 10^{-2}}{2\sqrt{3}} \text{ m}$$

Net field at O = 3 times field due to one side

$$= 3 \times \frac{\mu_0}{4\pi} \times \frac{2Il}{OP\sqrt{l^2 + 4OP^2}}; \text{here } l = 4.5 \times 10^{-2} \text{ m}$$

$$= \frac{3 \times 10^{-7} \times 2 \times 1 \times 4.5 \times 10^{-2} \times 2\sqrt{3}}{4.5 \times 10^{-2}\sqrt{\left(4.5 \times 10^{-2}\right)^2 + 4 \times \frac{\left(4.5 \times 10^{-2}\right)^2}{4 \times 3}}}$$

$$= \frac{12\sqrt{3} \times 10^{-7}}{4.5 \times 10^{-2}\sqrt{1 + \frac{1}{3}}}$$

$$= \frac{12\sqrt{3} \times 10^{-7} \times \sqrt{3}}{4.5 \times 10^{-2} \times 2}$$

$$= 4 \times 10^{-5} \text{ w/m}^2$$

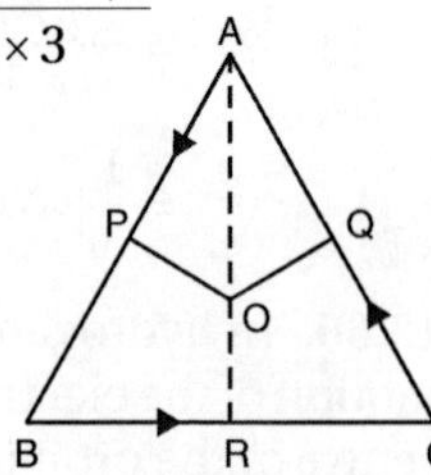

Fig. 13.42

13.62. A long horizontal wire AB which is free to move in a vertical plane and carries a steady current of 20 A, is in equilibrium at a height of 0.01 m over another parallel long wire CD, which is fixed in a horizontal plane and carries a steady current of 30 A as shown in Fig. 13.43. Show that when AB is slightly depressed and released it executes simple harmonic motion. Find the period of oscillations. (IIT 1994)

Solution: The current in the two conductors are in opposite direction. Under the conditions of equilibrium

$$\frac{mg}{L} = \frac{\mu_0}{4\pi} \cdot \frac{2I_1 I_2}{h}$$

Now if the upper wire is depressed through y downwards then the magnetic force will increase. The restoring force is given by

$$F = \frac{\mu_0}{4\pi} \cdot 2I_1 I_2 \left[\frac{1}{h-y} - \frac{1}{h}\right] L$$

$$= -\frac{\mu_0}{4\pi} \cdot \frac{2I_1 I_2 Ly}{h(h-y)}$$

$$= -\frac{\mu_0}{4\pi} \cdot \frac{2I_1 I_2 Ly}{h^2}$$

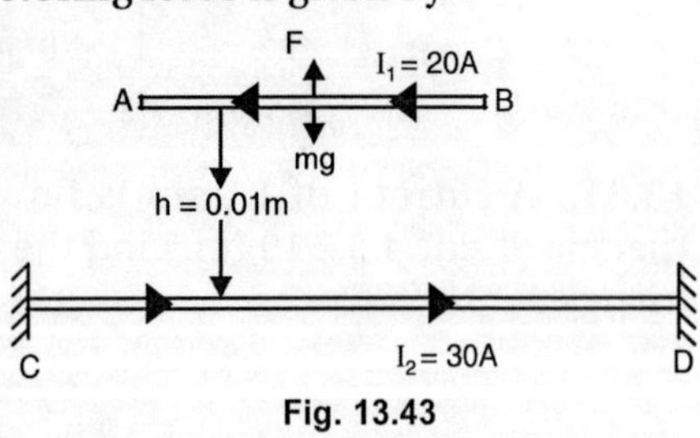

Fig. 13.43

$$m \times \text{acceleration} = -\frac{\mu_0}{4\pi} \cdot \frac{2I_1I_2L}{h^2} \cdot \text{displacement}$$

$$\text{acceleration} = -\frac{mgh}{mh^2} \text{displacement}$$

$$\omega = \frac{2\pi}{T} = \sqrt{\frac{g}{h}}$$

or $$T = 2\pi\sqrt{\frac{h}{g}}$$

13.63. An electron in the ground state of hydrogen atom is revolving in anticlock-wise direction in a circular orbit of radius r. (a) Obtain an expression for the orbital magnetic moment of the electron. (b) The atom is placed in a uniform magnetic induction $\vec{B}$ such that the plane normal of the electron orbit makes an angle of 30° with the magnetic induction as shown in Fig. 13.44. Find the torque experienced by the orbiting electron. (IIT 1996)

Solution:

(a) Current constituted as a result of the motion of electron is equal to

$$I = -e\frac{v}{2\pi r}$$

Magnetic moment $\vec{M} = I\vec{A} = -\frac{ev}{2\pi r} \cdot \pi r^2 \vec{n} = -\frac{evr}{2}\vec{n}$

Angular momentum $\vec{L} = mvr\vec{n}$

$$\therefore \quad \vec{M} = -\frac{e}{2m}\vec{L}$$

According to the second postulate of Bohr's theory we have

$$L = mvr = \frac{\eta h}{2\pi}$$

$$\vec{M} = \frac{eh}{4\pi m}(-\vec{n}); \eta = 1$$

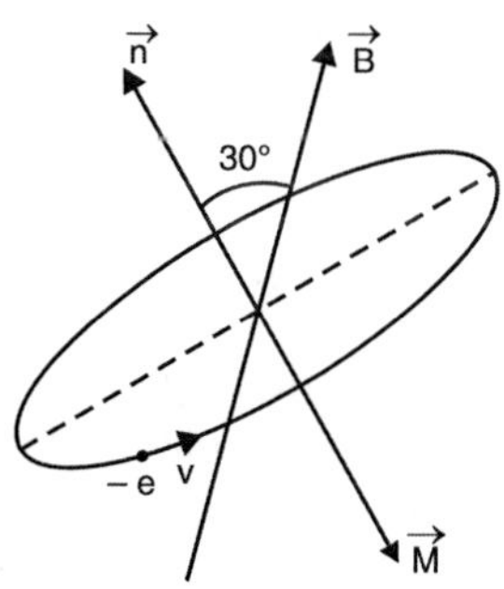

Fig. 13.44

(b) $$\vec{\tau} = \vec{M} \times \vec{B} = \frac{eh}{4\pi m}(-\vec{n}) \times \vec{B} = -\frac{ehB\sin 30}{4\pi m}k$$

$$= \frac{ehB}{8\pi m}k$$

k is unit vector ⊥ to the plane of the paper

13.64. A charge is imparted to a hollow conical surface whose base radius is 'a' and height is h. The charge is uniformly distributed over its surface and acquires a surface charge density σ. Find the magnetic moment when the cone rotates with an angular velocity ω about its own axis.

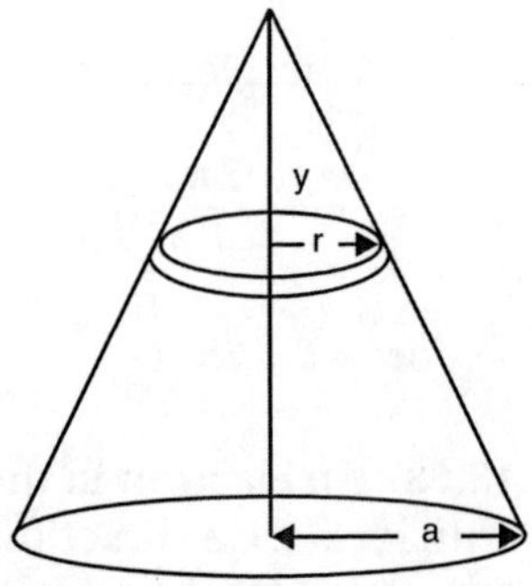

Fig. 13.45a

Solution: In this case there will be magnetic moments due to the rotation of the charges of the curved surface and base respectively. Consider an element of radius r and width dy.

The area of the element $= 2\pi r \cdot dy$

$$= \frac{2\pi a y\, dy}{h};$$

$$\frac{y}{r} = \frac{h}{a}$$

Charge on the element $= \sigma \cdot \dfrac{2\pi a y\, dy}{h}$

Current due to rotation of the charge $i = \sigma \cdot \dfrac{2\pi a y\, dy}{h} \cdot \dfrac{\omega}{2\pi} = \dfrac{\omega\sigma\, a y\, dy}{h}$

Magnetic moment dM due to the element $= \dfrac{\omega\sigma\, a y\, dy}{h} \cdot \pi r^2$

$$dM = \frac{\pi\omega\sigma a y\, dy}{h} \cdot \frac{a^2 y^2}{h^2}$$

$$= \frac{\pi\omega a^3\sigma}{h^3} \cdot y^3 dy$$

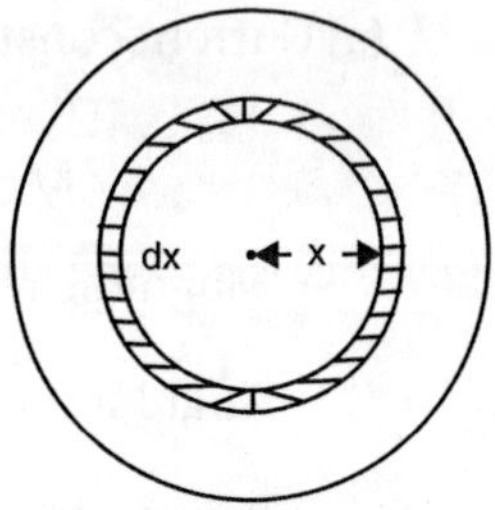

Fig. 13.45b

Net magnetic moment due to the charges lying on the curved surface

$$= m_1 = \int dM = \frac{\pi\omega a^3\sigma}{h^3} \int_0^h y^3 dy$$

$$m_1 = \frac{\pi\omega a^3\sigma h}{4}$$

For finding the magnetic moment due to the charges lying on the base consider an element of thickness dx at a distance x from the centre of the base.

Area of the element $= 2\pi\, x dx$

Charge on the element $= \sigma \cdot 2\pi\, x dx$

Current due to rotation $i_0 = \sigma \times 2\pi\, x dx \cdot \dfrac{\omega}{2\pi}$

$$= \sigma\,\omega \cdot x dx$$

Magnetic moment due to the element $= \sigma\,\omega\, x dx \cdot \pi x^2$

Net magnetic moment m_2 due to the charges lying on the base

$$= \pi\,\sigma\,\omega \int_0^a x^3 dx$$

$$m_2 = \frac{\pi\sigma\omega a^4}{4}$$

$$\text{Total magnetic moment} = m_1 + m_2 = \frac{\pi\omega a^3\sigma h}{4} + \frac{\pi\omega\sigma a^4}{4}$$

$$= \frac{\pi\omega\sigma a^3 (a+h)}{4}$$

13.65. A charge Q is uniformly distributed within the volume of a solid cone of base radius a and height h. If the cone rotates with an angular velocity ω about its own axis find the magnetic moment.

Solution: Consider a shell of radius R at a depth y below the apex (see Fig. 13.46a). The shell will be in the form of disc. On this disc consider an element of radius x and $x + dx$ (see Fig. 13.46b).

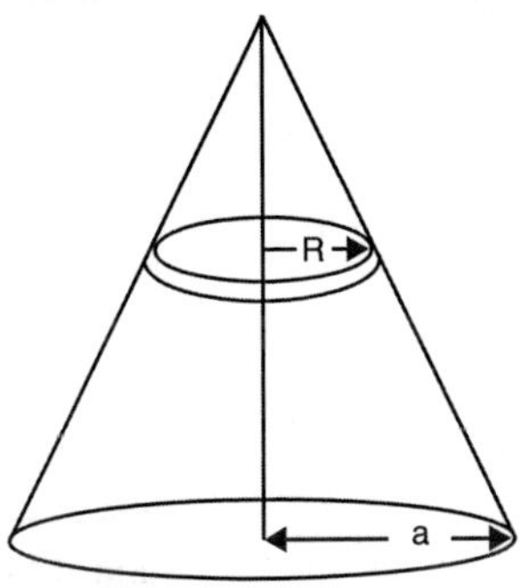

Fig. 13.46a

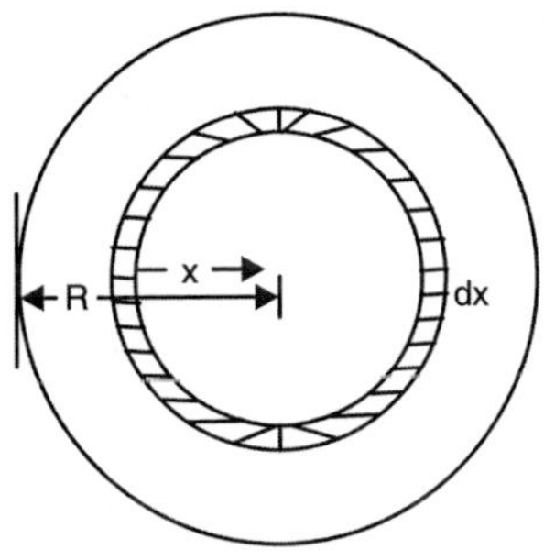

Fig. 13.46b

Volume of the element $= 2\pi x dx \cdot dy$; dy = width of the shell.

Charge on the element $= 2\pi x dx \cdot dy\,\rho$; $\rho = \dfrac{3Q}{\pi a^2 h}$

$$\text{Magnetic moment of the element} = \rho \cdot \cancel{2}\pi dy\, x \cdot dx \cdot \pi x^2 \cdot \frac{\omega}{\cancel{2}\pi}$$

$$= \pi\rho\omega dy \cdot x^3 dx$$

$$\text{Magnetic moment } dM \text{ of the disc of radius } R = \pi\rho\omega dy \cdot \int_0^R x^3 dx$$

$$= \frac{\pi\rho\omega R^4}{4} \cdot dy$$

$$= \frac{\pi\rho\omega a^4 y^4}{4h^4} \cdot dy; \frac{y}{R} = \frac{h}{a}$$

$$\text{Net magnetic moment } M = \int dM = \frac{\pi\rho\omega a^4}{4h^4} \int_0^h y^4 dy$$

$$= \frac{\pi\rho\omega a^4 \cdot h}{20}$$

$$= \frac{\pi\omega h \cdot a^4}{20} \times \frac{3Q}{\pi a^2 h}$$

$$= \frac{3Q\omega a^2}{20}$$

13.66. Inside a straight wire with a round cross-section there is a round cylindrical space whose axis is parallel to that of the wire. The displacement of the space's axis relative to that of the wire is determined by the vector *a*. A current with the identical density *j* over the entire cross-section flows through the wire. Find the magnetic induction inside the space.

Solution: Magnetic induction at a distance $(a + x)$ from axis 1 is given by

$$B_1 \cdot 2\pi(a+x) = \mu_0 \pi (a+x)^2 j$$

$$\therefore \quad B_1 = \frac{\mu_0}{2}(a+x)j$$

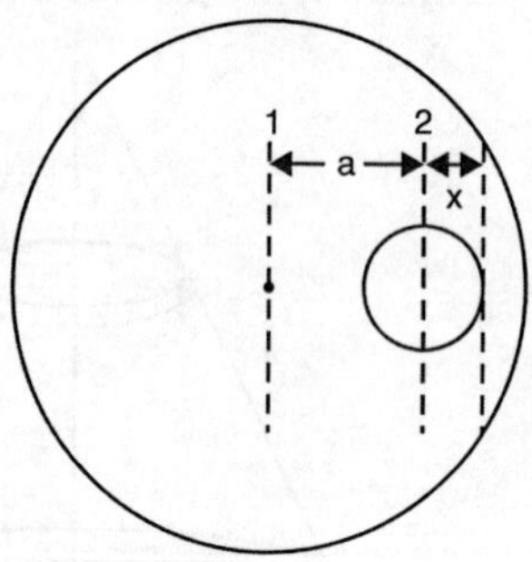

Fig. 13.47

Consider a current in the negative direction for creating a cylindrical cavity. For induction at a distance *x* from axis 2 we have

$$B_2 \cdot 2\pi x = \mu_0 j \pi x^2$$

$$\therefore \quad B_2 = \frac{\mu_0}{2} jx$$

Net magnetic induction in the cylindrical space $= \frac{\mu_0}{2}[a + x - x]j$

$$= \frac{\mu_0 a j}{2}$$

13.67. Two long parallel wires carry currents of equal magnitude but in opposite directions. These wires are suspended from rod *PQ* by four cords of same length *L* as shown in Fig. 13.48a. Determine the value of θ assuming it to be small.

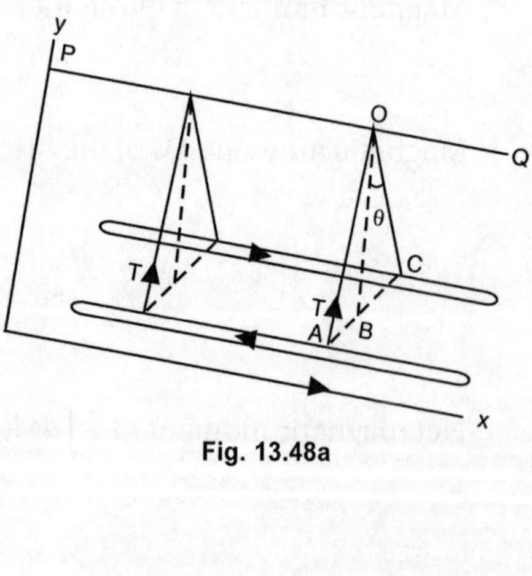

Fig. 13.48a

Solution: Repulsive force per unit length acting between the two wires which are *d* apart is given by

$$\frac{dF}{dl} = \frac{\mu_0}{4\pi}\frac{2I^2}{d}; \qquad I = \text{current in the two wires}$$

d = separation between the two wires

In view of Fig. 13.48b

$$2T\cos\theta = l\lambda g \text{ and } 2T\sin\theta = F$$

$$\therefore \quad \tan\theta = \frac{F}{l\lambda g}; \; l = \text{length of the wire}$$

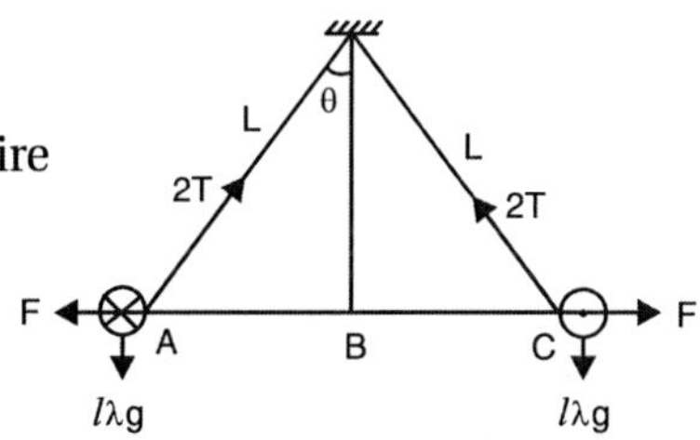

Fig. 13.48b

$$F = \frac{dF}{dl}\cdot l = \frac{\mu_0}{4\pi}\left(\frac{2I^2}{2L\sin\theta}\right)l$$

$$= \frac{\mu_0 I^2 l}{4\pi L\sin\theta}$$

$$\tan\theta = \frac{F}{l\lambda g} = \frac{\mu_0 I^2}{4\pi L\lambda g\sin\theta}$$

For small angles $\sin\theta = \theta = \tan\theta$

$$\therefore \quad \theta^2 = \frac{\mu_0 I^2}{4\pi L\lambda g}$$

$$\theta = I\sqrt{\frac{\mu_0}{4\pi\lambda L g}}$$

13.68. A straight segment OC (of length L) of a circuit carrying current I amp is placed along the x axis. Two infinitely long straight wires A and B each extending from $Z = -\infty$ to $Z = +\infty$ are fixed at $y = -a$ and $+a$ respectively as shown in Fig. 13.49. If the wires A and B each carry a current I amp into the plane of the paper, obtain the expressions for the force acting on the segment OC. What will be the force on OC if the current in the wire B is reversed? (IIT 1992)

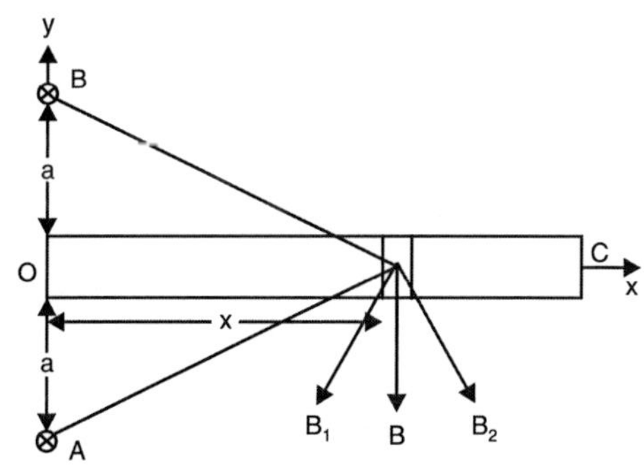

Fig. 13.49

Solution: Consider an element of length dx at a distance x from O. The effective magnetic field on the element is

$$B = \frac{2\mu_0}{4\pi}\cdot\frac{2I}{\sqrt{(a^2+x^2)}}\times\frac{x}{\sqrt{a^2+x^2}}(-j) = \frac{4\mu_0}{4\pi}\cdot\frac{Ix}{(a^2+x^2)}(-j)$$

Magnetic force on the element

$$dF = I\,dx \cdot \frac{\mu_0}{4\pi} \cdot \frac{4Ix}{\left(a^2 + x^2\right)}(i \times -j)$$

$$\text{Net force } F = \frac{\mu_0}{4\pi} \cdot 4I^2 \int_0^L \frac{x\,dx}{\left(a^2 + x^2\right)}(-k)$$

$$= \frac{\mu_0}{4\pi} \cdot 2I^2 \log_e\left(1 + \frac{L^2}{a^2}\right)(-k)$$

On reversing the current in B the net field will be along the length of the conductor OC. The magnetic force under this condition will be zero.

13.69. The gap between two parallel round plates is filled with a homogenous weakly conducting substance having a conductivity σ and dielectric constant K. The magnitude of the gap d is much smaller than the radius R of the plates. A voltage is applied across the plates that varies according to the law $U = U_m \cos \omega t$ (ω is sufficiently small for the conditions of a quasi static nature to be observed). Find an expression for the magnetic induction in the gap at a distance r from the axis of the plates that is much smaller than R.

Solution: The resistance of the material

$$x = \frac{1}{\sigma} \cdot \frac{d}{\pi R^2}$$

Current due to this resistance

$$I_1 = \frac{U_m \sigma \pi R^2}{d} \cos \omega t$$

$$\text{Charge in the capacitor } q = \frac{K \cdot \varepsilon_0 \pi R^2}{d} \cdot U_m \cos \omega t$$

$$\text{Current } I_2 \text{ due to this charge} = \frac{-K \varepsilon_0 \pi R^2 \omega}{d} \cdot U_m \sin \omega t$$

$$\text{Net current } I = \frac{U_m \pi R^2}{d}\left[\sigma \cos \omega t - K \varepsilon_0 \omega \sin \omega t\right]$$

$$= \frac{U_m \pi R^2}{d}\sqrt{\sigma^2 + K^2 \varepsilon_0^2 \omega^2}$$

$$\left[\cos \omega t \cdot \frac{\sigma}{\left(\sigma^2 + K^2 \varepsilon_0^2 \omega^2\right)^{1/2}} - \sin \omega t \cdot \frac{K \varepsilon_0 \omega}{\left(\sigma^2 + K^2 \varepsilon_0^2 \omega^2\right)^{1/2}}\right]$$

$$= \frac{U_m \pi R^2}{d}\sqrt{\sigma^2 + K^2 \varepsilon_0^2 \omega^2} \cdot \cos\left[\omega t + \tan^{-1} \frac{K \varepsilon_0 \omega}{\sigma}\right]$$

According to Amperes Law

$$B \cdot 2\pi r = \frac{\mu_0 I \pi r^2}{\pi R^2} = \frac{\mu_0 I r}{2\pi R^2}$$

$$B = \mu_0 U_m \sqrt{\sigma^2 + K^2\varepsilon_0^2\omega^2} \cdot \frac{r}{2d} \cdot \cos\left[\omega t + \tan^{-1}\frac{K\varepsilon_0\omega}{\sigma}\right]$$

13.70. Magnetic mirror is the term for the region of the magnetic field in which there is an intense concentration of lines of induction. Suppose a charged particle approaches a magnetic mirror as shown in Fig. 13.50a. What will happen to it.

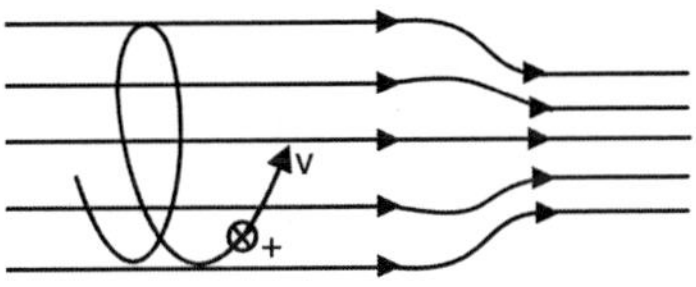

Fig. 13.50a

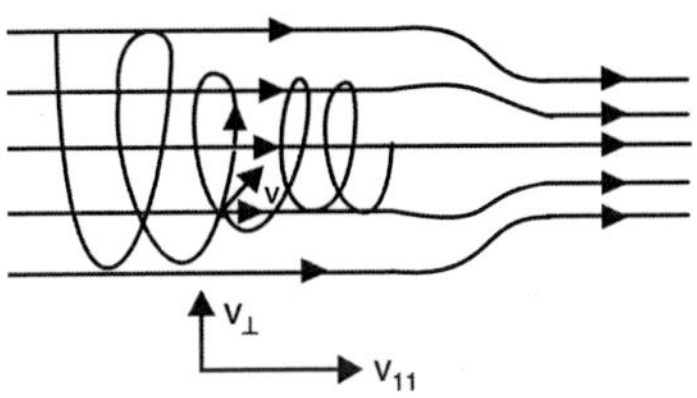

Fig. 13.50b

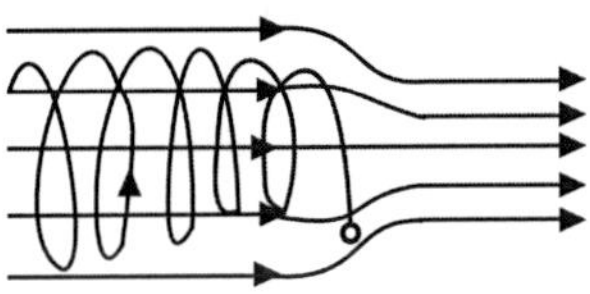
Fig. 13.50c

Solution: The particle will move along a helical winding around the lines of induction. Its velocity has two components: the lateral $v_{\perp}$ (the orbital velocity) and the longitudinal v_{11} (the drift velocity) (see Fig. 13.50b). The dipole moment of the circulating charge is directed against the field. The magnetic force tend to push the charge out into the region of weak magnetic field. As the charges approaches the magnetic mirror, its drift velocity decreases, becoming zero in the field of high gradient and from this it begins to drift in the opposite direction towards the weak field (see Fig. 13.50c).

13.71. A charged particle entering a strong uniform magnetic field experiences specular reflection, if its speed is below some limiting value. Find the kinetic energy of the electrons which experiences specular reflection if the electron beam is perpendicular to magnetic mirror.

Solution: Equation of motion of particle is given by

$$Bev = \frac{mv^2}{R}$$

or $$Be = \frac{p}{R};\ p = \text{momentum of the particle}$$

or $$R = \frac{p}{Be}$$

Electron has been chosen as the charged particle in the present case. The electron moving perpendicular to the magnetic mirror will be reflected back words if

$$\frac{p}{Be} < d$$

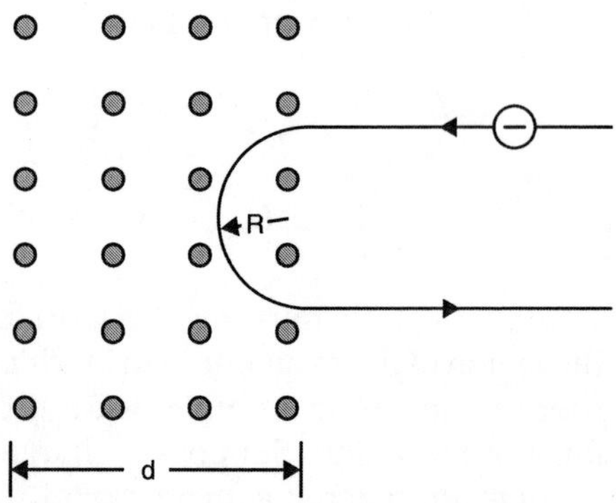

Fig. 13.51

The mass of the electron in view of its very high velocity is given by

$$m = \frac{mo}{\sqrt{1 - v^2/c^2}}$$

$$m^2c^2 - m^2v^2 = m_0^2c^2$$

$$\frac{E^2}{c^2} = \frac{E_0^2}{c^2} + p^2; \qquad E = mc^2$$

$$E_0 = m_0c^2$$

$$E = \sqrt{E_0^2 + p^2c^2} \qquad p = mv$$

Total kinetic energy $E = \sqrt{E_0^2 + p^2c^2} < \sqrt{E_0^2 + B^2e^2d^2c^2}$

Kinetic energy $K = E - E_0$

$$\text{or} \quad K < E_0\left[\sqrt{1 + \left(\frac{Bed}{m_0c}\right)^2} - 1\right]$$

13.72. A wire loop carrying a current I is placed in the $x - y$ plane as shown in Fig. 13.52. If a particle with charge $+Q$ and mass m is placed at the centre P and given a velocity v along NP. Find (a) its instantaneous acceleration (b) if an external uniform magnetic induction $B = \vec{B}i$ is applied, find the force and torque acting on the loop.

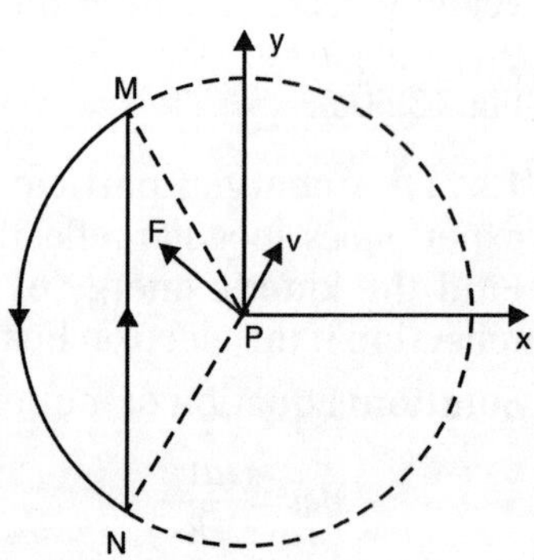

Fig. 13.52

Solution: (a) Magnetic induction due to straight portion of the wire

$$B_1 = \frac{\mu_0}{4\pi}\cdot\frac{2I}{a}\sqrt{3}\cdot(-k)$$

Magnetic induction due to curved portion of the wire

$$B_2 = \frac{\mu_0}{4\pi} \cdot \frac{2I\pi}{3a}(k)$$

Net magnetic induction $B = \frac{\mu_0}{4\pi} \cdot \frac{2I}{a}\left[\sqrt{3} - \frac{\pi}{3}\right](-k)$

Force acting on the charged particle

$$F = q\,(v \times B)$$

$$= q\,vB \text{ along } PF$$

$$F = \frac{\mu_0}{4\pi} \cdot \frac{2qvI}{a}\left[\sqrt{3} - \frac{\pi}{3}\right] \text{along } PF$$

Acceleration $A = \frac{F}{m} = 10^{-7}\,\frac{2qvI}{ma}\left[\sqrt{3} - \frac{\pi}{3}\right]$ along PF

(b) Force $F = \int I\,\vec{dL} \times \vec{B}$

$$= I \oint \vec{dL} \times \vec{B} = 0 \text{ Since } \oint \vec{dL} = 0$$

Area of the coil

$$S = \left[\frac{1}{3}\pi a^2 - \frac{1}{2} \cdot 2a \sin 60 \times a \cos 60\right] k = a^2 \left[\frac{\pi}{3} - \frac{\sqrt{3}}{4}\right] k$$

$$\vec{M} = I\vec{S} = I a^2 \left[\frac{\pi}{3} - \frac{\sqrt{3}}{4}\right] k$$

$$\vec{\tau} = \vec{M} \times \vec{B} = Ia^2 B \left[\frac{\pi}{3} - \frac{\sqrt{3}}{4}\right](k \times i)$$

$$= Ia^2 B \left[\frac{\pi}{3} - \frac{\sqrt{3}}{4}\right] j\ N.m$$

13.73. A vibration magnetometer consists of two identical bar magnets placed one over the other such that they are mutually perpendicular and bisect each other. The time period of oscillation in a horizontal magnetic field is 4 sec. If one of the magnets is taken away, find the period of oscillation of the other in the same field.

Solution: Period of oscillation is given by

$$T = 2\pi\sqrt{\frac{I}{M_0 B}}$$

Here there are two magnetic moments $\perp$ to each other hence

$M_0 = \sqrt{M^2 + M^2} = M\sqrt{2}$; M = magnetic moment of a magnet

$$4 = 2\pi\sqrt{\frac{2I}{\sqrt{2}\,MB}}$$

$$T = 2\pi\sqrt{\frac{I}{MB}}$$

On dividing $\frac{4}{T} = [2]^{1/4}$

or $T = \frac{4}{(2)^{1/4}} = 3.36$ sec

13.74. A bar magnet suspended by a horse hair lies in the magnetic meridian when there is no twist in the hair. On turning the upper end of the hair through 150° from the meridian the magnet is deflected through 30° from the meridian. Through what angle the upper end of the hair to be twisted in order to deflect the magnet through 90° from the meridian?

Solution: Let α be the angle of rotation of the hair and β be the angle of rotation of the magnet then

$C(\alpha - \beta) = MB \sin \beta$; C = twisting couple per unit angle of twist

On substituting the values of α & β we have

$$C(150 - 30) = MB \sin 30$$

$$C(\alpha - 90) = MB \sin 90$$

On dividing the two we have

$$240 = \alpha - 90$$

or $\alpha = 330°$

13.75. Find the force acting between the two magnets in the two arrangements shown in Fig. 13.53.

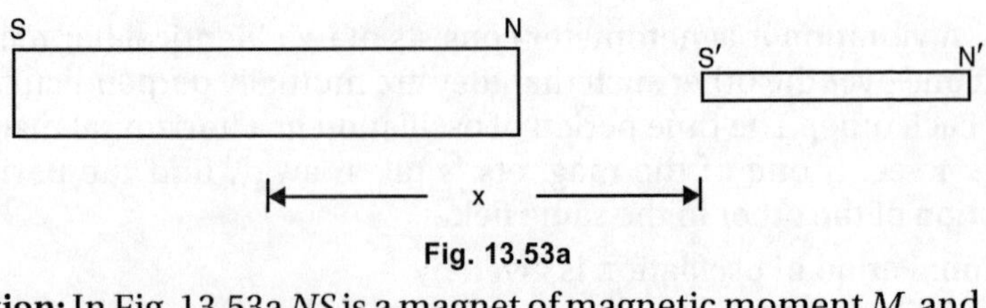

Fig. 13.53a

Solution: In Fig. 13.53a NS is a magnet of magnetic moment M_1 and $N'S'$ is a magnet of length $2\,l'$ and magnetic moment M_2.

Field at S' due to magnet $NS = \frac{\mu_0}{4\pi} \cdot \frac{2M_1}{x^3}$

Force acting at S' due to magnet NS

$= \frac{\mu_0}{4\pi} \cdot \frac{2M_1 \cdot m'}{x^3}$; m' is the pole strength of magnet $N'S'$

Rate of change of magnetic field $= \frac{\mu_0}{4\pi} \cdot \frac{d}{dx}\left(\frac{2M_1}{x^3}\right) = -\frac{\mu_0}{4\pi} \cdot \frac{6M_1}{x^4}$

Change of field over length $2l' = -\frac{\mu_0}{4\pi} \cdot \frac{6M_1}{x^4} \cdot 2l'$

Field at $N' = \left(\frac{\mu_0}{4\pi} \cdot \frac{2M_1}{x^3} - \frac{\mu_0}{4\pi} \cdot \frac{6M_1}{x^4} \cdot 2l'\right)$

Force at $N' = \left(\frac{\mu_0}{4\pi} \cdot \frac{2M_1}{x^3} - \frac{\mu_0}{4\pi} \cdot \frac{6M_1}{x^4} \cdot 2l'\right) m'$

Net force on $N'S' = \frac{\mu_0}{4\pi} \cdot \frac{2M_1}{x^3} \cdot m' - \frac{\mu_0}{4\pi} \cdot \frac{2M_1 m'}{x^3} + \frac{\mu_0}{4\pi} \cdot \frac{6M_1}{x^4} m' \cdot 2l'$

$$= \frac{\mu_0}{4\pi} \cdot \frac{6M_1M_2}{x^4}; M_2 = m' \cdot 2l$$

In Fig. 13.53b field at $N' = \frac{\mu_0}{4\pi} \cdot \frac{M_1}{y^3}$

Force acting at N' due to the magnet $= \frac{\mu_0}{4\pi} \cdot \frac{M_1 m'}{y^3}$;

m' is the pole strength of magnet N′S′.

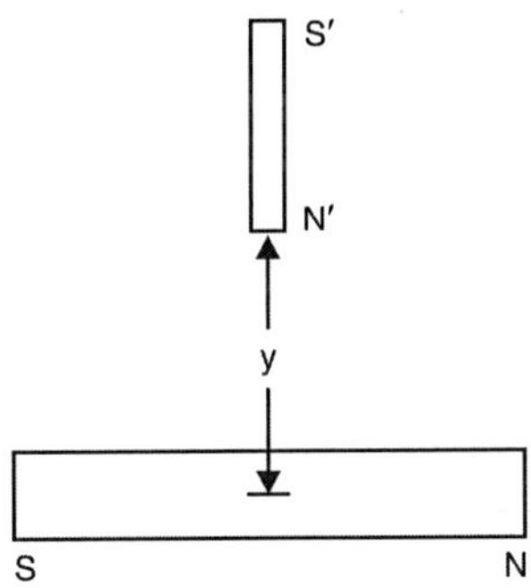

Fig. 13.53b

Rate of change of magnetic field

$$= \frac{\mu_0}{4\pi} \cdot \frac{d}{dy}\left(\frac{M_1}{y^3}\right) = -\frac{\mu_0}{4\pi} \cdot \frac{3M_1}{y^4}$$

Change of field over a length $2l'$

$$= -\frac{\mu_0}{4\pi} \cdot \frac{3M_1}{y^4} \cdot 2l'$$

Field at $S' = \left(\frac{\mu_0}{4\pi} \cdot \frac{M_1}{y^3} - \frac{\mu_0}{4\pi} \cdot \frac{3M_1}{y^4} \cdot 2l'\right)$

Force at $S' = \left(\frac{\mu_0}{4\pi} \cdot \frac{M_1}{y^3} - \frac{\mu_0}{4\pi} \cdot \frac{3M_1}{y^4} \cdot 2l'\right) m'$

Net force at $N'S' = \frac{\mu_0}{4\pi} \cdot \frac{M_1 m'}{y^3} - \frac{\mu_0}{4\pi} \cdot \frac{M_1 m'}{y^3} + \frac{\mu_0}{4\pi} \cdot \frac{3M_1}{y^4} m' \cdot 2l'$

$$= \frac{\mu_0}{4\pi} \cdot \frac{3M_1M_2}{y^4}$$

13.76. The rectangular coil of a moving coil galvanometer of area 1.5 cm^2 has 100 turns of wire. The coil is situated in a radial field of 0.2 T. The constant of the spring is 10^{-4} N/degree. Calculate the deflection of coil when a current of 2 A flows through it.

Solution: $BINA = C\theta$

$$\therefore \quad \theta = \frac{BINA}{C}$$

$$= \frac{0.2 \times 2 \times 100 \times 1.5 \times 10^{-4}}{10^{-4}} = 60°$$

13.77. Two coils, each of 100 turns are held such that one lies in the vertical plane and the other in the horizontal plane with their centres coinciding. The radii of the vertical and the horizontal coils are respectively 20 cm and 30 cm. If the direction of the currents in them are such that the earth's magnetic field at the centre of the coil is exactly neutralised, then calculate the current in each coil. (Horizontal component of the earth's field = 27.8 Am^{-1}, angle of dip = 30°.) (Roorkee 1988)

Solution:

$$\text{Angle of dip } \theta = \tan^{-1}\frac{B_V}{B_H}$$

$$\therefore \quad B_V = B_H \tan\theta$$

$$= 0.35 \times 10^{-4} \tan 30; \; 27.8\,\text{A/m} = 0.35 \times 10^{-4}\, T$$

$$= 0.35 \times 10^{-4} \times 0.577$$

$$= 0.202 \times 10^{-4}\, T$$

$$B_H = \frac{\mu_0}{4\pi}\frac{2\pi n i_1}{r} = 0.35 \times 10^{-4}$$

$$\therefore \quad i_1 = \frac{0.35 \times 10^{-4} \times r}{2\pi n} \times \frac{4\pi}{\mu_0}$$

$$= \frac{0.35 \times 10^{-4} \times 0.20 \times 10^{7}}{2 \times 3.14 \times 100} = 0.11\text{ A}$$

$$B_V = \frac{\mu_0}{4\pi}\frac{2\pi n i_2}{r} = 0.202 \times 10^{-4}$$

$$i_2 = \frac{0.202 \times 10^{-4} \times r}{2\pi n} \times \frac{4\pi}{\mu_0}$$

$$\therefore \quad i_2 = \frac{0.202 \times 10^{-4} \times 0.30 \times 10^{7}}{2 \times 3.14 \times 100}$$

$$= 0.096\text{ A}$$

13.78. If ϕ_1 and ϕ_2 be the angles of dip observed in two vertical planes at right angle to each other and ϕ be the true angle of dip, show that

$$\cot^2\phi = \cot^2\phi_1 + \cot^2\phi_2$$

Solution: Let one plane be at an angle α from the meridian

$$\tan\phi_1 = \frac{V}{H_1} = \frac{V}{H\cos\alpha}$$

$$\cos\alpha = \frac{V}{H}\cdot\cot\phi_1$$

Similarly $\tan\phi_2 = \dfrac{V}{H_2} = \dfrac{V}{H\sin\alpha}$

$\therefore \quad \sin\alpha = \dfrac{V}{H}\cdot\cot\phi_2$

$$\sin^2\alpha + \cos^2\alpha = \tan^2\phi\,(\cot^2\phi_1 + \cot^2\phi_2)$$

or $\quad \cot^2\phi = \cot^2\phi_1 + \cot^2\phi_2$

13.79. At a place the angle of dip is θ_1 and the dip needle lies in the magnetic meridian. When the dip circle is rotated through an angle ϕ in the horizontal plane then the angle of dip is θ_2. Calculate the angle ϕ.

Solution:

$$H_1 = H\cos\phi$$

$$\tan\theta_2 = \frac{V}{H_1} = \frac{V}{H\cos\phi} = \frac{\tan\theta_1}{\cos\phi}$$

$$\cos\phi = \frac{\tan\theta_1}{\tan 0_2}$$

or $\quad \phi = \cos^{-1}\left[\dfrac{\tan\theta_1}{\tan\theta_2}\right]$

13.80. If a dip needle is suspended at an angle 30° to the magnetic meridian it makes an angle of 45° with the horizontal. Calculate the angle of dip.

Solution:

$$\tan 45 = \frac{V}{H\cos 30}$$

$$\tan\theta = \tan 45\cdot\cos 30 = \frac{\sqrt{3}}{2}$$

$$\therefore\ \theta = \text{angle of dip} = \tan^{-1}\left(\frac{\sqrt{3}}{2}\right)$$

13.81. A dip circle is so set that it moves freely in the magnetic meridian. In this position the angle of dip is 40°. The dip circle is now rotated so that the plane in which the needle moves makes an angle of 30° with the magnetic meridian. Find the angle of dip in this position.

Solution: Let the new angle be θ'

$$\tan\theta' = \frac{V}{H'} = \frac{V}{H\cos 30} = \tan 40\times\frac{2}{\sqrt{3}} = 1.031$$

$$\theta' = 46°$$

13.82. What is a magnetic map? How you will define (i) isogonic lines (ii) isoclinic lines (iii) isodynamic lines?

Solution: A magnetic map shows the values of magnetic element throughout the world. In this map lines can be drawn in order to join those places which are having the same value of a magnetic element.

(i) *isogonic lines*: A line joining the places of equal declination is known as isogonic line. A line joining the places of zero declination is known as agonic line.

(ii) *isoclinic lines*: A line joining the places of equal dip is known as isoclinic lines. A line joining the places of zero dip is known as aclinic line.

(ii) *isodynamic lines*: A line joining the places of equal horizontal components is known as isodynamic line.

13.83. Two voltmeters A & B of resistances 20,000 Ω and 130,000 Ω respectively are connected across a 120 volt d.c source. Find the readings of each voltmeter.

Solution:

$$\text{Current through each voltmeter } I = \frac{120}{130{,}000 + 20{,}000}$$

$$\text{Reading of voltmeter } A = V_A = \frac{20{,}000 \times 120}{150{,}000} = 16.0 \text{ volt}$$

$$\text{Reading of voltmeter } B = V_B = \frac{130{,}000 \times 120}{150{,}000} = 104 \text{ volt}$$

13.84. A 150 V voltmeter has a resistance of 20,000 Ω. When it is connected in series with a resistance R, across a 110 volt d.c source it reads 5 V. Calculate the value of resistance R.

Solution:

$$\text{Current } I = \frac{110}{R + 20{,}000}$$

$$\text{Now} \qquad 5 = \frac{20{,}000 \times 110}{R + 20{,}000}$$

$$5R + 100{,}000 = 2{,}200{,}000$$

$$5R = 2{,}200{,}000 - 100{,}000$$

$$R = \frac{2{,}100{,}000}{5}$$

$$= 420 \text{ K}\Omega$$

13.85. A 100 volt battery has an internal resistance of 5 Ω (a) what is the reading of a voltmeter having resistance of 500 Ω when placed across the battery? (b) what maximum value may the ratio of $\frac{r}{R_V}$ have if the error in the reading of the emf of the battery is not to exceed 5%?

Solution:

(a) $V = E - Ir = 100 - \frac{5 \times 100}{500 + 5} = 99$ volt

(b) $\frac{E - V}{E} \times 100 = 5$

$20E - 20V = E$

$19E = 20V = 20\,(E - Ir) = 20\,E - \frac{20E \cdot r}{R_V + r}$

$19 = 20 - \frac{20r}{R_V + r}$

$\frac{r}{R_V} = \frac{1}{19} = 0.0526$

13.86. The resistance of a galvanometer coil is 50 Ω and the current required for full scale deflection is 500 μa (a) Find the value of the shunt resistance required to convert the galvanometer into an ammeter having a full scale reading of 5 A (b) Find the value of the shunt required to convert it into voltmeter reading 150 V.

Solution:

(a) Let S be the shunt resistance, G be the resistance of the galvanometer, I_g and I_s be the current through galvanometer and shunt respectively.

Now $G \cdot I_g = S \cdot I_s = S\,(I - I_g)$

$(S + G)\,I_g = S\,I$

$(50 + S) \times 5 \times 10^{-4} = 5\,S$

$50 + S = 10{,}000\,S$

$S = \frac{50}{(10{,}000 - 1)} \cong 5\ \text{m}\Omega$

(b) $I = 5 \times 10^{-4} = \frac{150}{50 + R}$

$R = 299{,}950\ \Omega$

13.87. The resistance of a galvanometer is 10 Ω and a current of 0.02 A causes it to deflect full scale. It is required to convert it into an ammeter reading 10 A. The only shunt available is 0.03 Ω. Find the value of R that must be connected in series.

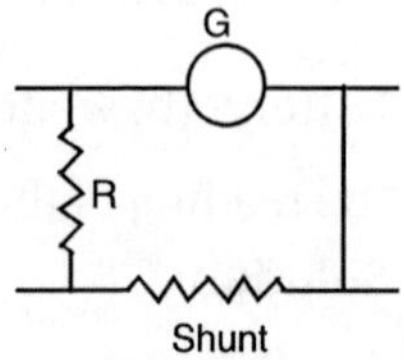

Fig. 13.54

Solution: Let G be the resistance of the galvanometer and S be the shunt resistance

$$(R+G+S)\,I_g = S\,I$$

$$(R+10+0.03)\times 2\times 10^{-2} = 0.03\times 10$$

$$R+10.03 = 15$$

$$R = 4.97\ \Omega$$

13.88. The resistance of the galvanometer is G = 50 Ω. The meter shows a full scale deflection with a current of 1 mA. Calculate the three resistances R_1, R_2 & R_3 required to convert it into a multirange ammeter deflecting full scale with currents 10 A, 1 A and 0.1 A.

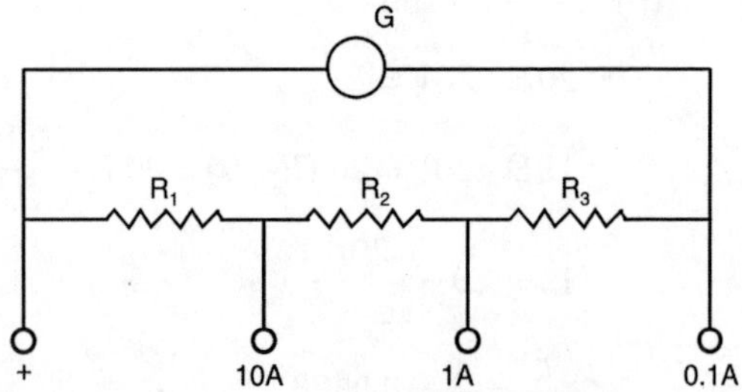

Fig. 13.55

Solution: For 10 amp current R_2 & R_3 will be in series with G, then we have

$$(G+R_2+R_3)I_g = R_1 I_S = R_1\,(I-I_g)$$

or $$(G+R_1+R_2+R_3)\,I_g = R_1 I$$

or $$(G+R_1+R_2+R_3) = R_1\,\frac{10}{10^{-3}} = 10{,}000\,R_1$$

For 1 amp current R_3 will be in series with G, then we have

$$(G+R_3)I_g = (R_1+R_2)I_S = (R_1+R_2)\,(I-I_g)$$

or $$(G+R_1+R_2+R_3)I_g = (R_1+R_2)\,I$$

or $$(G+R_1+R_2+R_3) = \frac{I}{I_g}(R_1+R_2) = 10^3\,(R_1+R_2)$$

For 0.1 amp current

$$G\cdot I_g = (R_1+R_2+R_3)I_s = (R_1+R_2+R_3)\,(I-I_g)$$

$$(G+R_1+R_2+R_3) = (R_1+R_2+R_3)\,\frac{I}{I_g} = 100\,(R_1+R_2+R_3)$$

$$10^4\,R_1 = 10^3\,(R_1+R_2)$$

$$9R_1 = R_2$$

$$10^3\,(R_1+R_2) = 100\,(R_1+R_2+R_3)$$

$$9R_1+9R_2 = R_3$$

$$90R_1 = R_3$$

$$(50 + R_1 + 9R_1 + 90R_1) = 10^4 R_1$$
$$50 = (10{,}000 - 100)\, R_1$$
$$R_1 = \frac{50}{9900} = 5 \times 10^{-3}\, \Omega$$
$$R_2 = 9R_1 = 45 \times 10^{-3}\, \Omega$$
$$R_3 = 90R_1 = 45 \times 10^{-2}\, \Omega$$

13.89. The Fig. 13.56 shows a multi range voltmeter. The resistance of the galvanometer is 20 Ω. A current of 1 mA produces a full scale deflection. Calculate the values of the resistances R_1, R_2 & R_3.

Solution:

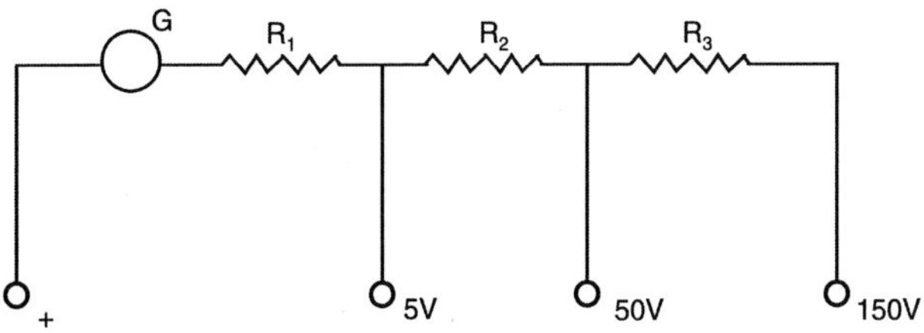

Fig. 13.56

$$10^{-3} = \frac{5}{G + R_1} = \frac{5}{20 + R_1}$$
$$20 + R_1 = 5000$$
$$\therefore \quad R_1 = 4980\, \Omega$$
$$10^{-3} = \frac{50}{20 + R_1 + R_2}$$
$$20 + R_1 + R_2 = 50{,}000$$
$$R_2 = 50{,}000 - 5{,}000$$
$$= 45{,}000\, \Omega$$
$$10^{-3} = \frac{150}{20 + R_1 + R_2 + R_3}$$
$$20 + R_1 + R_2 + R_3 = 150{,}000$$
$$R_3 = 150{,}000 - 50{,}000\, \Omega$$
$$= 100{,}000\, \Omega$$

13.90. In the two arrangements the resistance of voltmeter and ammeter is R_V and R_A respectively and their readings are V and I respectively. Calculate the value of R.

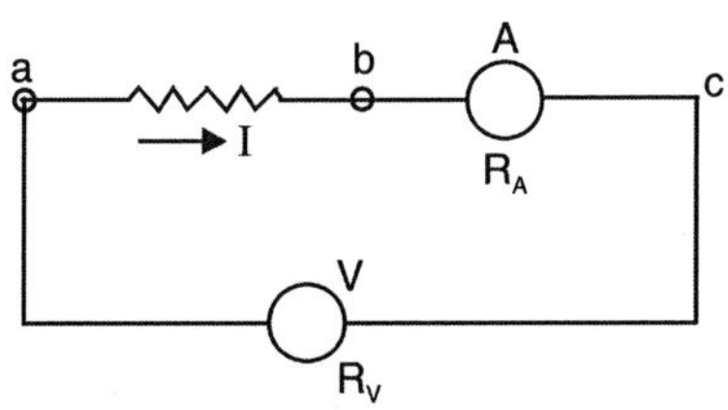

Fig. 13.57a

Solution: In Fig. 13.57a

$$(R_A + R)I = V$$

$\therefore \quad R = \dfrac{V}{I} - R_A$

In Fig. 13.57b

$$I_R \cdot R = I_V \cdot R_V$$

$$I_R = I - I_V$$

$$R \cdot I_R = RI - R \cdot I_V = V$$

$$\therefore \quad R = \frac{V}{I - I_V} = \frac{V}{I - V/R_V}$$

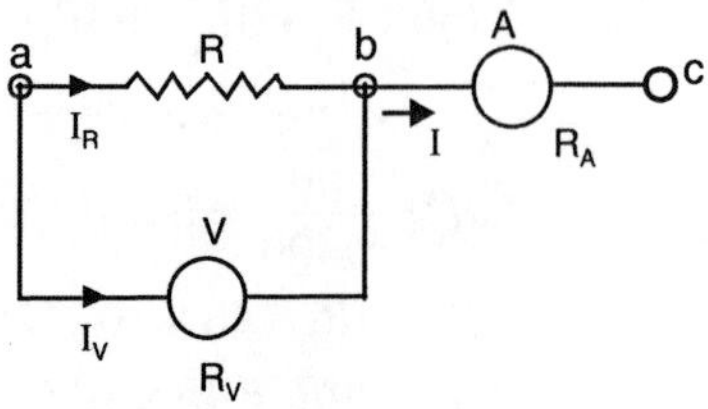

Fig. 13.57b

13.91. In the given Fig. 13.58 the resistance of the galvanometer is 100 Ω. The full scale deflection is obtained with a current of 1 mA in the coil. The emf of the cell = 2 volt. Calculate (i) the value of the series resistance R_0 (ii) the value of R corresponding to the galvanometer deflection of $\frac{1}{4}, \frac{1}{2}$ & $\frac{3}{4}$ of the full scale.

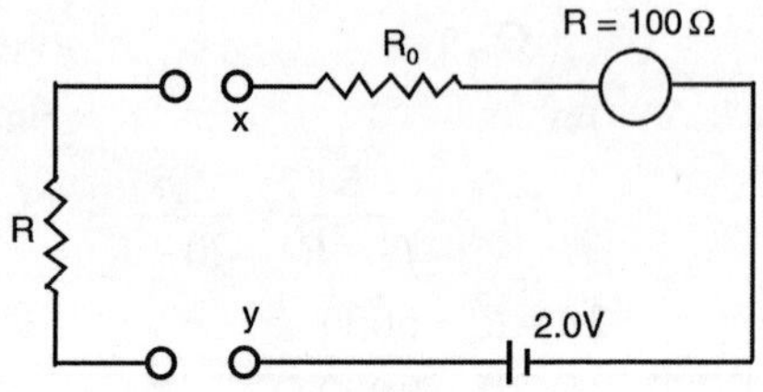

Fig. 13.58

Solution:

(i) For full scale deflection x & y are to be short circuited.

$$10^{-3} = \frac{2}{R_0 + 100}$$

$$R_0 + 100 = 2000$$

$$\therefore \quad R_0 = 1900\ \Omega$$

(ii) $0.25 \times 10^{-3} = \dfrac{2}{R + 1900 + 100} = \dfrac{2}{R + 2000}$

$$25R + 25 \times 2000 = 2 \times 10^5$$

$$25R + 0.5 \times 10^5 = 2 \times 10^5$$

$$R = \frac{1.5 \times 10^5}{25} = 6000\ \Omega$$

$$0.5 \times 10^{-3} = \frac{2}{R + 2000}$$

$$5R + 10{,}000 = 20{,}000$$

$$5R = 10{,}000$$

$$R = 2000\ \Omega$$

$$0.75 \times 10^{-3} = \frac{2}{R + 2000}$$

$$75R + 150{,}000 = 200{,}000$$

$$75R = 50{,}000$$

$$R = 666.67\ \Omega$$

13.92. A battery of emf E is connected in series with three resistances R, $2R$ & $3R$. The voltage across $2R$ is measured with a voltmeter whose resistance is $10R$. Calculate the %error.

Solution: The voltage across $2R$ without voltmeter $= \frac{E}{6R} \times 2R = \frac{E}{3}$

On introducing the voltmeter of resistance $10R$.

The total resistance $= \frac{17R}{3}$

Current in the circuit $= \frac{3E}{17R}$

Let I_1 be the current through voltmeter and I_2 be the current through $2R\Omega$.

$$I_1 \times 10R = I_2 \times 2R$$

$$5I_1 = I_2$$

$$I_1 + I_2 = \frac{3E}{17R}$$

$$6I_1 = \frac{3E}{17R}$$

$$\therefore \quad R_1 = \frac{E}{34R}$$

Reading of the voltmeter $= \frac{10ER}{34R} = \frac{5E}{17}$

$$\% \text{ error} = \frac{\frac{E}{3} - \frac{5E}{17}}{\frac{E}{3}} = 11.76\%$$

13.93. A galvanometer having 30 div has a current sensitivity of 20 μA/div. It has a resistance of 25 Ω. How will you convert it into an ammeter measuring upto 1 amp? How will you convert this ammeter into a voltmeter reading up to 1V?

Solution:

$I_g = 30 \times 20\ \mu A = 600\ \mu A$

$I_g = \frac{SI}{S+G}$; S = shunt resistance and G = galvanometer resistance

$6 \times 10^{-4} (S + 25) = S \times 1$

$6S + 150 = 10{,}000\ S$

$150 = 9994\ S$

$\therefore \quad S = \dfrac{150}{9994} \cong 0.015\,\Omega$ in parallel

$$\text{Resistance of ammeter} = \frac{0.015 \times 25}{25.015} = 0.015\,\Omega$$

Now $S + 0.015 = 1$; here full scale current = 1 amp

$\therefore \quad S = 0.985\,\Omega$ in series

13.94. In the given circuit a voltmeter reads 30 volt when it is connected across 400 Ω resistance. Calculate its reading when it is connected across 300 Ω.

Solution: Potential difference across 300 Ω = 30 volt

$\therefore \quad$ Main current $I = \dfrac{30}{300} = \dfrac{1}{10}$ amp

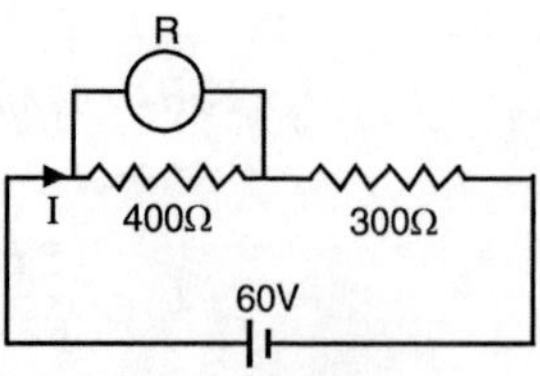

Fig. 13.59

Let I_1 be the current through 400 Ω

$\therefore \quad 400\, I_1 = 30$

$\therefore \quad I_1 = \dfrac{3}{40}$ amp

$$\text{Current through voltmeter} = \frac{1}{10} - \frac{3}{40} = \frac{1}{40} \text{ amp}$$

$$\text{Resistance of voltmeter} = \frac{30}{1/40} = 1200\,\Omega$$

When voltmeter is connected across 30 Ω the equivalent resistance

$$R_{eq} = \frac{1200 \times 300}{1500} = 240\,\Omega$$

$$I = \frac{60}{400 + 240} = \frac{60}{640} = \frac{3}{32} \text{ amp}$$

Now $I_1' \times 1200 = \left(\dfrac{3}{32} - I_1'\right) \times 300$; I_1' = current through voltmeter

$$1500\, I_1' = \frac{3 \times 300}{32}$$

$$\therefore \quad I_1' = \frac{3 \times 300}{1500 \times 32} = \frac{3}{160} \text{ amp}$$

$$\text{Reading of voltmeter} = \frac{3}{160} \times 1200 = 22.5 \text{ volt}$$

14

ELECTROMAGNETIC INDUCTION AND ALTERNATING CURRENT

14.1. A rod of length $L = 30$ cm rotates at an angular velocity $\omega = 60$ rad/s in a uniform magnetic field of induction $B = 2 \times 10^{-3}$ w/m^2 (see Fig. 14.1). Calculate the emf E developed across the ends of the rod.

Solution: Consider an element of length dl situated a distance l from one end of the rod, as shown in the Fig. 14.1. Let v be the velocity at right angles to the field B by which the element is moving. Due to this motion, the emf developed across the element is given by

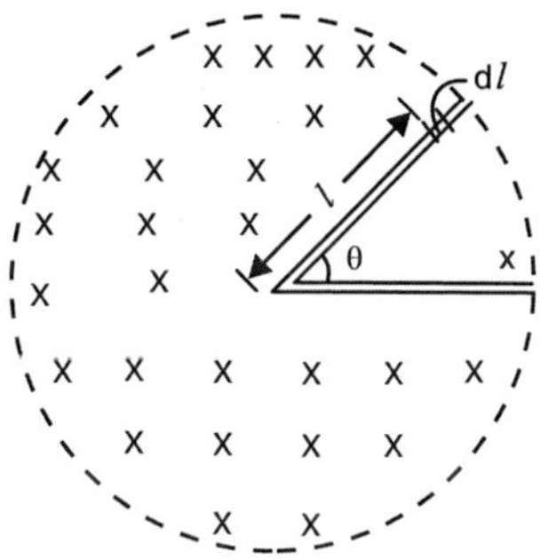

Fig. 14.1

$$dE = Bvdl;\ v = l\omega$$

$$\therefore \quad dE = B\omega l dl$$

$$\therefore \quad E = \int dE = B\omega \int_0^L l dl = \frac{1}{2} B\omega L^2$$

On substituting the values of B, ω and L, we have

$$E = \frac{1}{2} \times 2 \times 10^{-3} \times 60 \times 9 \times 10^{-2}$$

$$= 5.4 \times 10^{-3}\text{V}$$

14.2. A circular disc of radius $R = 15$ cm, located in a uniform magnetic field of induction $B_1 = 4 \times 10^{-3}$ w/m^2, rotates at an angular velocity $\omega = 50$ rad/s about an axis passing through its centre O and perpendicular to its plane (Fig. 14.2). Calculate the potential difference developed between the axis of the disc and its rim.

Solution: Let us choose a point A on the rim of the disc which is rotating about an axis passing through O. In time t the point A shifts to B and makes an angle θ with the reference. The area of the triangle

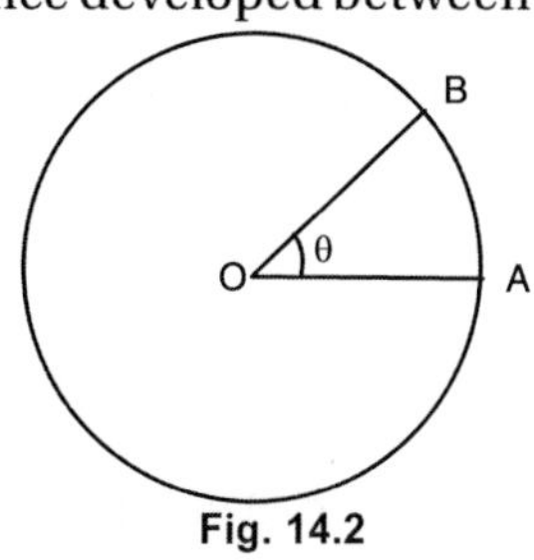

Fig. 14.2

$$AOB = \frac{R \times R\theta}{2}$$

$$\text{Flux } \phi = \frac{B_1 R^2 \theta}{2}$$

$$\therefore\ E = \frac{d\phi}{dt} = \frac{1}{2} B_1 R^2 \frac{d\theta}{dt} = \frac{1}{2} B_1 R^2 \omega$$

On substituting the values of B_1, R and ω, we have

$$E = \frac{1}{2} \times 4 \times 10^{-3} \times 0.15^2 \times 50$$

$$= 2.25 \text{ mV}$$

14.3. A superconducting ring of radius r = 10 cm and inductance L = 0.03 H is located in a magnetic field which grows from zero to $B = 2.1 \times 10^{-3}$ wb/m². The plane of the ring is perpendicular to the magnetic lines of force. Calculate the current in the ring.

Solution: As the ring is made of super-conductor its resistance is zero. Its total emf is also zero. The change in the external magnetic flux ϕ is equal to the change of magnetic flux due to induced current.

Therefore, $d\phi = L\, di$

$d\phi = \pi r^2 B$ since flux grows from zero to $\pi r^2 B$

Induced current changes from 0 to I therefore $di = I$

$$\therefore\ \pi r^2 B = LI$$

$$\text{or} \qquad I = \pi r^2 \frac{B}{L}$$

On substituting the values of r, B and L, we have

$$I = \frac{3.14 \times 10^{-2} \times 2.1 \times 10^{-3}}{0.03}$$

$$= 21.98 \times 10^{-4}$$

$$= 2.198 \times 10^{-3}$$

$$= 2.2 \text{ mA}$$

14.4. A circular loop made of a thin copper wire of mass m = 4 g is located in a uniform magnetic field keeping the plane of the loop perpendicular to the field. Calculate the induced current in the loop if the magnetic field varies at a constant rate, $\frac{dB}{dt} = 10^{-5}$ T/s. Density of copper $d = 8.9 \times 10^3$ kg/m³, resistivity of copper $\rho = 1.68 \times 10^{-8}\ \Omega \cdot$ m.

Solution: Let r_1 be the radius of the wire and r_2 be the radius of the loop then

$$m = \pi r_1^2 \times 2\pi r_2 d$$

$$\text{Induced emf in the loop } E = \pi r_2^2 \frac{dB}{dt}$$

Induced current in the loop $I = \frac{\pi r_2^2}{R}\frac{dB}{dt}$ where R is the resistance of the loop.

Further $R = \rho\frac{2\pi r_2}{\pi r_1^2} = \frac{2\rho r_2}{r_1^2}$

On substituting the values of R in the above equation, we have

$$I = \frac{r_1^2\,\pi r_2^2}{2\rho r_2}\frac{dB}{dt}$$

$$= \frac{\pi r_1^2 \cdot 2\pi r_2 d}{4\pi\rho d}\frac{dB}{dt} = \frac{m}{4\pi\rho d}\frac{dB}{dt}$$

On substituting the values of the various terms, we have

$$I = \frac{4\times 10^{-3}\times 10^{-5}}{4\times 3.14\times 1.68\times 10^{-8}\times 8.9\times 10^{3}}$$

$$= \frac{1}{3.14\times 1.68\times 8.9}\times 10^{-3} = 21.3\ \mu\text{A}$$

14.5. When magnetic lines of forces are allowed to pass through a thin wire ring of resistance r in a direction perpendicular to the plane of ring, it has been found that the flux varies as $\phi = bt\,(\tau - t)$ during the interval τ. Neglecting the inductance of the ring calculate the amount of heat generated in the loop during that time.

Solution:

$$e = -\frac{d\phi}{dt} = -\,b\tau + 2bt$$

$$\text{Amount of heat generated} = \int_0^{\tau}\frac{e^2}{r}dt$$

$$= \frac{1}{r}\left[\int_0^{\tau} b^2\tau^2\cdot dt - 4b^2\tau\int_0^{\tau} t\,dt + 4b^2\int_0^{\tau} t^2 dt\right]$$

$$= \frac{1}{3}\frac{b^2\tau^3}{r}$$

14.6. A thin wire ring of radius r is located in a time varying magnetic field such that the field is perpendicular to the plane of the ring and varies as $B = at$. Find the intensity of the induced electric field in the turn.

Solution:

Flux $\phi = at\cdot\pi\,r^2$

Induced emf $e = \frac{d\phi}{dt} = \pi\cdot a\cdot r^2$

The induced emf is numerically equal to the work performed by the electric field when a +ve charge moves in the complete one turn of the ring

$$e = a\pi r^2 = E \cdot 2\pi r;\ E = \text{induced electric field}$$

$$\therefore \quad E = \frac{ar}{2}$$

14.7. A solenoid of radius a having N turns of wire wound over a length l carries a current $I = I_0 \sin 2\pi\, nt$. Find the induced electric field at a point at a distance r when (i) $r > a$ (ii) $r < a$.

Solution:

(i) When $r > a$

$$\text{Magnetic field inside the solenoid} = \frac{\mu_0 NI}{l}$$

$$\text{Magnetic flux } \phi = \frac{\mu_0 NI}{l} \cdot N\pi a^2 = \frac{\mu_0 N^2 \pi a^2}{l} I_0 \sin 2\pi nt$$

$$\text{Induced emf} = \frac{\mu_0 N^2}{l} \pi a^2 I_0 \cdot 2\pi n \cos 2\pi nt$$

$$\text{Induced electric field } E = \frac{\mu_0 N^2\, \pi a^2 I_0 \cdot 2\pi n \cos 2\pi nt}{2\pi rl}$$

$$= \frac{\mu_0 N^2 \pi a^2 I_0 n \cos 2\pi nt}{rl}$$

(ii) when $r < a$

$$\text{Magnetic flux } \phi = \frac{\mu_0 N}{l} \cdot N\pi r^2 \cdot I_0 \sin 2\pi nt$$

$$\text{Induced emf} = \frac{\mu_0 N^2 \pi r^2\, I_0 \cdot 2\pi n \cos 2\pi nt}{l}$$

$$\text{Induced electric field } E' = \frac{\mu_0 N^2 \pi r^2 I_0 \cdot 2\pi n \cdot \cos 2\pi nt}{2\pi r l}$$

$$= \frac{\mu_0 N^2 \pi I_0\, nr \cos 2\pi nt}{l}$$

14.8. A thin wire ring of radius r_1 and resistance R_0 is placed coaxially inside a solenoid of length l radius r_2 and resistance R. The solenoid is connected to a constant voltage source. Neglecting the inductance of the ring and the resistance of the connecting wire find the maximum radial force acting per unit length of the ring.

Solution: The solenoid has finite inductance L and resistance R. On connecting it to a source of constant voltage V it will be basically a LR circuit in which current is given by

$$I = \frac{V}{R}\left[1 - e^{-Rt/L}\right]$$

Magnetic field inside the solenoid

$$B = \frac{\mu_0 NV}{lR}\left[1 - e^{-Rt/L}\right];\ N = \text{total number of turns of the solenoid.}$$

Magnetic flux through the ring

$$\phi = \frac{\mu_0 NV}{lR}\left[1 - e^{-Rt/L}\right]\pi r_1^2$$

$$\text{Induced emf } e = \frac{\mu_0 NV}{lL}\pi r_1^2 e^{-Rt/L}$$

$$\text{Current in the ring } i = \frac{e}{R_0} = \frac{\mu_0 NV \pi r_1^2 e^{-Rt/L}}{lLR_0}$$

Radial force per unit length of the ring

$$F = B \times i = \frac{\mu_0 NV}{lR} \cdot \frac{\mu_0 NV \pi r_1^2}{lLR_0}\left[1 - e^{-Rt/L}\right]e^{-Rt/L}$$

$$\text{On putting } L = \frac{\mu_0 N^2 \pi r_2^2}{l} \text{ we have}$$

$$F = \frac{\mu_0 V^2 \pi r_1^2}{lR_0 R\pi r_2^2} \cdot e^{-Rt/L}\left(1 - e^{-Rt/L}\right)$$

$$\text{On putting } \frac{dF}{dt} = 0 = \frac{\mu_0 V^2 r_1^2}{lR_0 Rr_1^2}\left[\left(1 - e^{-Rt/L}\right)e^{-Rt/L} \times \frac{-R}{L} + e^{-2Rt/L} \times \frac{R}{L}\right]$$

$$\text{We get } e^{-Rt/L} = \frac{1}{2}$$

$$F_{max} = \frac{\mu_0 V^2 r_1^2}{4RR_0 lr_2^2}$$

14.9. A horizontally placed flat circular coil of radius R and having n_1 number of turns carries a current I. Another coaxial circular coil of radius r and of n_2 number of turns is located above it at a height h as shown in the Fig. 14.3. The upper coil is allowed to fall freely with a velocity v. Assuming the velocity to be uniform throughout the course of motion. Determine the emf developed in it as a function of time.

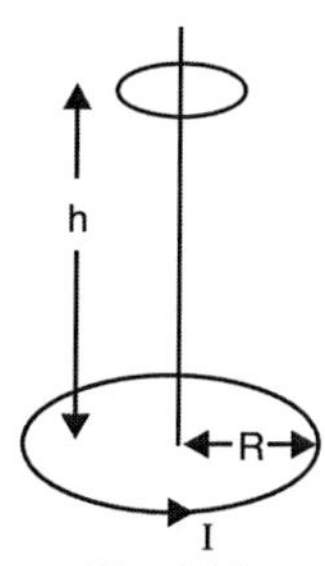

Fig. 14.3

Solution: Let the upper coil falls through a time t such that its position from the lower coil carrying I is $h - vt$

Magnetic field along the axis of the lower coil at a distance $h - vt$ is equal to

$$B = \frac{\mu_0}{4\pi} \cdot \frac{2\pi n_1 I R^2}{\left[R^2 + (h - vt)^2\right]^{3/2}}$$

Flux through the upper coil $\phi = \frac{\mu_0}{4\pi} \cdot \frac{2\pi n_1 I R^2 \cdot n_2 \pi r^2}{\left[R^2 + (h - vt)^2\right]^{3/2}}$

Induced emf $e = -\frac{d\phi}{dt} = \frac{\mu_0}{2} \cdot \frac{n_1 I R^2 \cdot n_2 \pi r^2 \cdot \frac{3}{2} \cdot 2(h - vt)v}{\left[R^2 + (h - vt)^2\right]^{5/2}}$

$$= \frac{3}{2} \cdot \frac{\mu_0 \pi I n_1 n_2 r^2 R^2 (h - vt) v}{\left[R^2 + (h - vt)^2\right]^{5/2}}$$

14.10. A straight wire of cross-sectional area A and resistivity ρ is bent to form an angle 2α. Another rod of resistance per unit length R_1 placed perpendicular to the bisector of the angle 2α forms a triangular loop *abc*. A magnetic field of induction B is applied perpendicular to the plane of the loop. Find the induced current in the loop when the rod moves with a constant speed v.

Solution: Let the rod initially be in the position shown in the Fig. 14.4. The length of the portion *ac* and *bc* each is equal to l. The length of the rod = $2\,l \sin\alpha$.

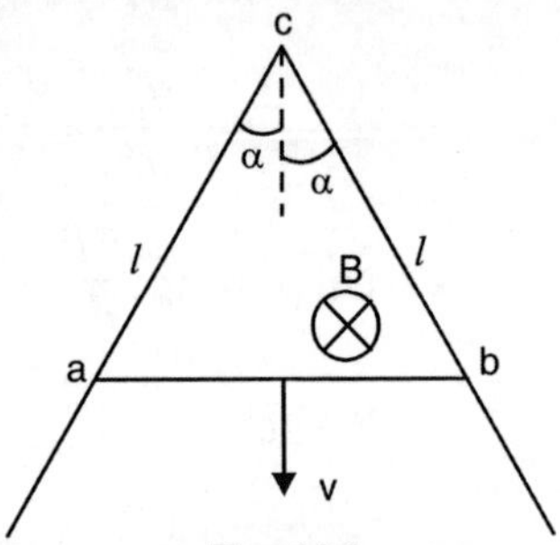

Fig. 14.4

Let the rod moves through dx in time dt

Magnetic flux $d\phi = B \cdot 2l \sin\alpha \cdot dx$

Induced emf $= B \cdot 2l \sin\alpha \cdot \frac{dx}{dt}$

$= B \cdot 2l \sin\alpha \cdot v$

Induced current in the loop $= \frac{B \cdot 2l \sin\alpha \cdot v}{\rho \cdot \frac{2l}{A} + 2l \sin\alpha \cdot R_1}$

$= \frac{B \cdot A \cdot \sin\alpha \cdot v}{\rho + AR_1 \sin\alpha}$

14.11. A frame of wire is placed in a uniform magnetic field by keeping its plane perpendicular to the field (Fig. 14.5). Determine the current in the conductors if the magnetic field varies with time according to the equation, $B = 2.2 \times 10^{-3}\, t$, $l = 22$ cm, resistance per cm length of the wire $r = 0.5\ \Omega$.

Solution: We know that induced emf $= \frac{d}{dt}(BA) = \frac{AdB}{dt}$ where A is the area.

In view of this, induced emf in the frame $AEFD = E_1 = 2.2 \times 10^{-3} \times l^2$ and in the frame $EBCF = E_2 = \frac{1}{2}\, 2.2 \times 10^{-3} \times l^2$.

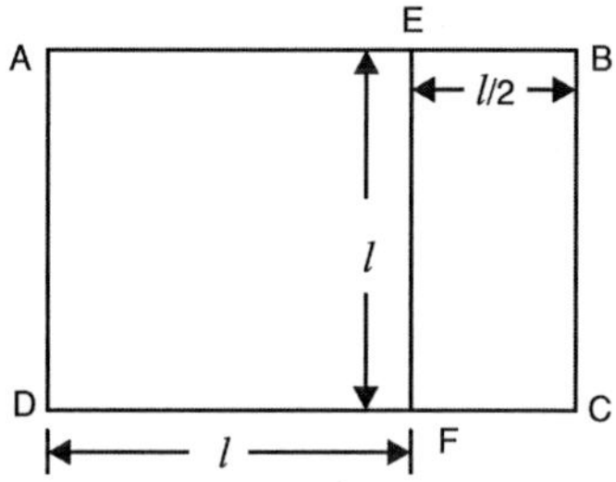

Fig. 14.5a

On applying Kirchhoff's loop theorem in the first mesh, we have

$$E_1 - 3lr\,I_1 - lr\,I_3 = 0$$

In the second mesh, we have

$$-2lr\,I_2 + E_2 + lr\,I_3 = 0$$

and $I_1 = I_2 + I_3$

On solving these equations, we get

$$I_1 = \frac{7}{22} \times 2.2 \times 10^{-3} \times \frac{l}{r}$$

$$= \frac{7}{22} \times \frac{2.2 \times 10^{-3} \times 22 \times 10^{-2} \times 10^{-2}}{0.5}$$

$$= 3.08\ \mu a$$

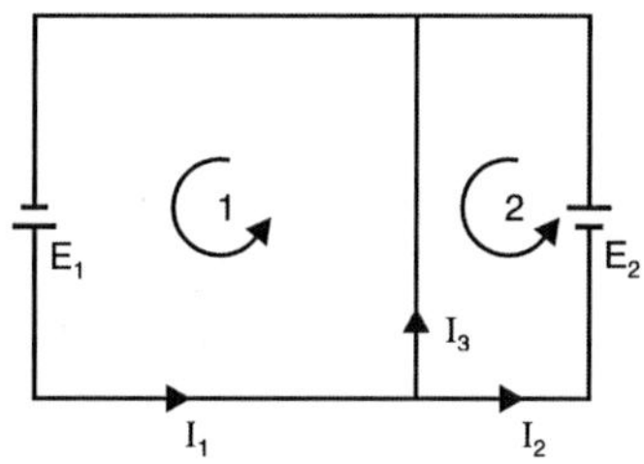

Fig. 14.5b

$$I_2 = \frac{3}{11} \times 2.2 \times 10^{-3} \times \frac{l}{r}$$

$$= \frac{3}{11} \times \frac{2.2 \times 10^{-3} \times 22 \times 10^{-2} \times 10^{-2}}{0.5}$$

$$= 2.64\ \mu a$$

and $I_3 = \frac{1}{22} \times 2.2 \times 10^{-3} \times \frac{l}{r}$

$$= \frac{1}{22} \times \frac{2.2 \times 10^{-3} \times 22 \times 10^{-2} \times 10^{-2}}{0.5}$$

$$= 0.44\ \mu a$$

14.12. Two vertical rods constituting a frame are connected through a resistance $R = 50\ \Omega$. A conductor of length $l = 10$ cm and mass $m = 10$ g can slide over the frame without friction. The system as a whole is located in a magnetic field whose intensity $B = 100\ \text{W/m}^2$ is perpendicular to the plane of the Fig. 14.6. Neglecting the resistance of the conductor and the frame, calculate the velocity of the conductor.

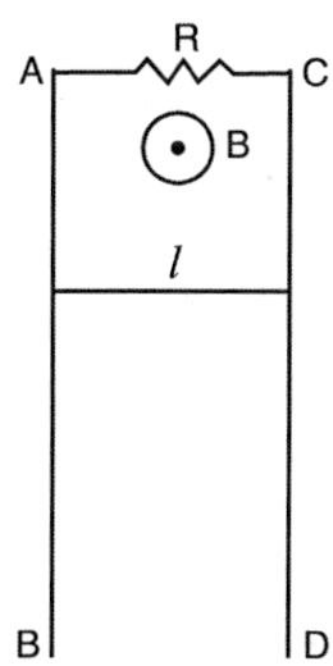

Fig. 14.6

Solution: Let the conductor acquire a velocity v in time dt. Then the area described by the conductor = $vl\,dt$ and the change of flux = $d\phi = Bvl\,dt$. The induced emf = Bvl and the induced current = $i = \frac{Bvl}{R}$.

This induced current will set up a force which will prevent the conductor from free-fall.

$$F = B\frac{Bvl}{R}l = \frac{B^2l^2v}{R}$$

The equation of motion of the rod $= mg - \dfrac{B^2l^2v}{R} = ma$

As v increases, a reduces and when $F = mg$, $a = 0$. From this condition, the conductor will move with a constant velocity.

Now $mg = \dfrac{B^2l^2v}{R}$

$\therefore \quad v = \dfrac{mgR}{B^2l^2}$

On substituting the values of various terms, we have

$$v = \frac{10\times10^{-3}\times9.8\times50}{100^2\times10^{-2}} = 49.0\times10^{-3}\text{ m/s}$$

$$= 4.9\text{ cm/s}$$

14.13. Two horizontally placed rods of negligible resistance constitute a frame on which a conductor of length l = 20 cm and mass m = 10 g can slide without friction. One end of each of the two rods is connected to a capacitor of capacity C = 5 μF. The entire system is placed in a magnetic field of induction B = 100 w/m² directed upwards (see Fig. 14.7). A force F = 6×10^{-3} N is applied at the middle of the conductor in a direction parallel to the frame. Calculate the acceleration of conductor after neglecting its resistance.

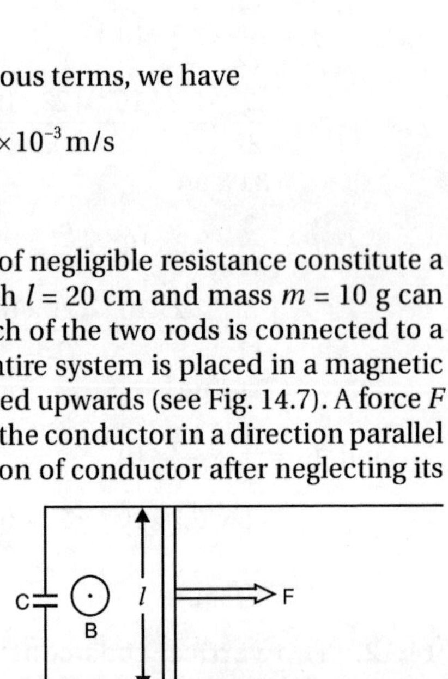

Fig. 14.7

Solution: Let the conductor acquires a velocity v in time dt. Change of flux in time dt = $d\phi$ = $B\,vldt$. The induced emf = Bvl. The charge on the plates of the capacitor = $EC = BvlC$.

The current flowing in the circuit $= I = \dfrac{dQ}{dt} = BlC\dfrac{dv}{dt} = BlCa$

where a is accleration. Due to this current, the conductor will experience a force F_1 given by $F_1 = BIl = B \cdot BlCa \cdot l = B^2l^2Ca$

This force is opposite to the force F which has been applied. Hence the equation of motion is $F - F_1 = ma$

or $\quad F = (B^2l^2C + m)\,a$

or $\quad a = \dfrac{F}{m + B^2l^2C}$

On substituting the values of the various terms, we have

$$a = \frac{6\times10^{-3}}{10\times10^{-3} + 10^{4}\times4\times10^{-2}\times5\times10^{-6}}$$

$$= \frac{6\times10^{-3}}{12\times10^{-3}} = 0.5\ \text{m/s}^2$$

14.14. A capacitor of capacity C is connected to one end of two vertical and parallel bars ab and cd of length H. A connector of length l and mass m slides over it without friction (see Fig. 14.8). A uniform magnetic field of induction B is applied perpendicular to the plane of the loop formed. Find the charge in the capacitor as a function of h.

Solution: Let the connector falls through dx in time dt.

Magnetic flux $d\phi = B\cdot l\cdot dx$

Induced emf $= Bl\cdot\dfrac{dx}{dt}$

$= BVl$

Now $\dfrac{dv}{dt} = g$

$\therefore \quad \dfrac{dv}{dx}\cdot\dfrac{dx}{dt} - g$

$\int_0^V v\,dv = g\int_0^{H-h} dx$

$V = \sqrt{2g(H-h)}$

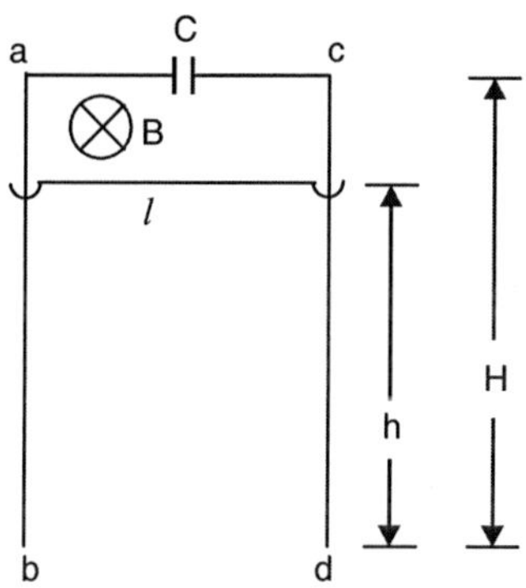

Fig. 14.8

Induced emf $= Bl\sqrt{2g(H-h)}$

Charge on the capacitor $= CBl\sqrt{2g(H-h)}$

14.15. A connector of length $l = 20$ cm and resistance $R = 38\ \Omega$ slides freely on the two horizontaly placed rods ad and bc (see Fig. 14.9). The ends of the two rods are connected to two resistance $R_1 = 30\ \Omega$ and $R_2 = 20\ \Omega$, respectively. The entire system is placed in a magnetic field of induction $B = 5\ \text{w/m}^2$ directed upwards and perpendicular to the plane of the loop. Find the current flowing in the connector during its motion with a constant velocity $v = 30$ cm/s.

Solution: Let the sliding connector move with a velocity v for a small interval of time dt. The change of flux in the same time $= Bvldt$.

The induced emf $E = Bvl$.

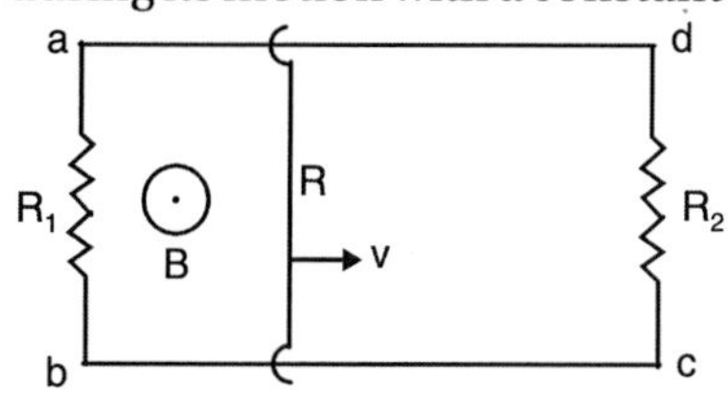

Fig. 14.9

The total resistance = $R + R_\mu$ where

$$\frac{1}{R_\mu} = \frac{1}{R_1} + \frac{1}{R_2}$$

$$\therefore \; I = \frac{Bvl}{R + R_\mu}$$

On substituting the value of various terms, we have

$$I = \frac{5 \times 0.3 \times 0.2}{38 + 12} = 6 \times 10^{-3}\,\text{A}$$

14.16. A long straight wire carrying a current I and π shaped frame with a sliding rod of length L and resistance R are placed as shown in Fig. 14.10. Neglecting the resistance of the frame, find the current induced in the sliding rod as a function of r, the separation between the wire and the rod if it moves with a constant velocity v towards the right.

Solution: Let us assume that the sliding rod is initially at OO'. Let the rod be at a distance dx from OO' in time dt and let it be at a distance r from the wire carrying current I.

The area described by the rod = $L\,dx$

Change of flux $d\phi = \dfrac{\mu_0}{2\pi}\dfrac{IL dx}{r}$

Induced emf

$$E = \frac{\mu_0}{2\pi}\frac{IL}{r}\frac{dx}{dt} = \frac{\mu_0}{2\pi}\frac{IL}{r}v$$

Induced current $I = \dfrac{E}{R} = \dfrac{\mu_0 ILv}{2\pi Rr}$

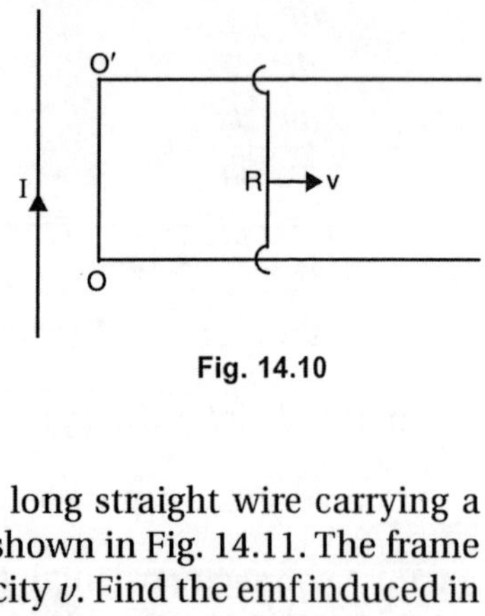

Fig. 14.10

14.17. A square frame with side a and a long straight wire carrying a current I are located in the same plane as shown in Fig. 14.11. The frame translates to the right with a constant velocity v. Find the emf induced in the frame as a function of distance r.

Solution: The magnetic field acting on the frame is given by

$$dB = \frac{\mu_0}{4\pi}2I\left[\frac{1}{r} - \frac{1}{r+a}\right] = \frac{\mu_0}{4\pi}\frac{2Ia}{r(r+a)}$$

Let the frame move through a distance dx in time dt.

The area described = adx

Change of flux $d\phi = \dfrac{\mu_0}{4\pi}\dfrac{2Ia^2 dx}{r(r+a)}$

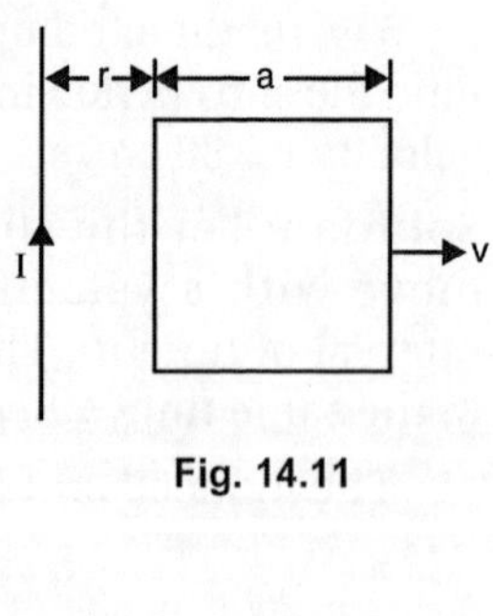

Fig. 14.11

$$E = \frac{d\phi}{dt} = \frac{\mu_0}{4\pi}\frac{2Ia^2}{r(r+a)}\frac{dx}{dt} = \frac{\mu_0}{4\pi}\frac{2Ia^2 v}{r(r+a)}$$

14.18. A square frame of side x_1 having inductance L and resistance R and a long straight wire carrying current I are placed in the same plane (see Fig. 14.12). Calculate the amount of charge that will flow through the frame if it is rotated through 180° about and axis yy' situated at a distance x_2 from the wire.

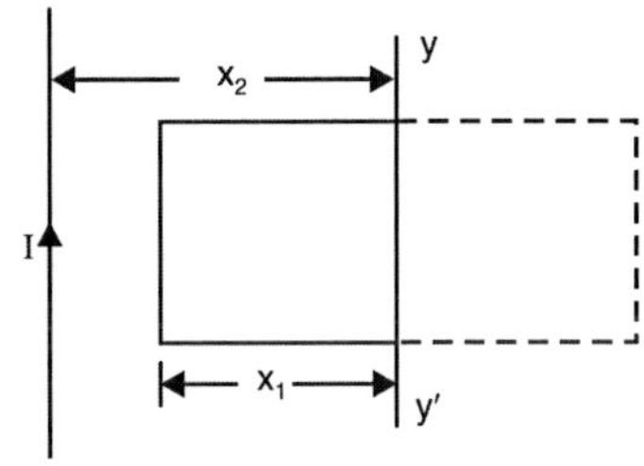

Fig. 14.12

Solution: According to Ohm's law, current in the frame during rotation is given by

$$RI = \frac{-d\phi}{dt} - \frac{LdI}{dt}$$

Amount of charge $q = \int Idt = -\frac{1}{R}\int(d\phi + LdI) = -\frac{1}{R}[\Delta\phi + L\Delta I]$

The frame has been stopped after rotation hence $\Delta I = 0$

$\therefore \quad q = \frac{\Delta\phi}{R}$ (on leaving –v sign)

Let the frame be at a distance r from the wire. Let dr be the displacement during rotation. Then $d\phi = Bx_1 dr$ and

$$\Delta\phi = \int_{x_2 - x_1}^{x_1 + x_2} Bx_2 dr = \frac{\mu_0}{4\pi}\cdot 2Ix_1 \int_{x_2 - x_1}^{x_1 + x_2} \frac{dr}{r} = \frac{\mu_0}{2\pi} Ix_1 \log_e \frac{x_2 + x_1}{x_2 - x_1}$$

$$\therefore \quad q = \frac{\Delta\phi}{R} = \frac{\mu_0}{2\pi R} Ix_1 \log_e \frac{x_2 + x_1}{x_2 - x_1}$$

14.19. Fig. 14.13 shows a current loop and a movable jumper. The inductance L of the loop is a function of x. Assuming the current I to be constant, calculate the force acting on the jumper.

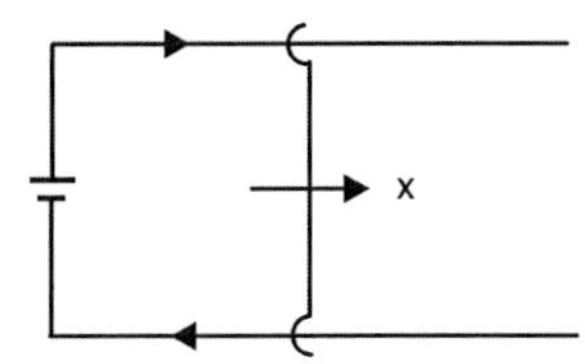

Fig. 14.13

Solution: Energy in the inductor = $\frac{1}{2}LI^2$

Let us displace the jumper to the right by dx, then

$$dw = F\,dx$$

or $$F = \frac{dw}{dx} = \frac{d}{dx}\left(\frac{1}{2}LI^2\right) = \frac{1}{2}I^2\frac{dL}{dx}$$

14.20. A rod ab of mass $m = 10$ g and length $l = 20$ cm can slide freely on two frictionless bars placed horizontally. One end of each of the two bars is connected to a resistance $R = 50\,\Omega$ as shown in Fig. 14.14. The system as a whole is placed in a magnetic field of induction $B = 10$ w/m², directed downwards. The rod starts moving towards right with an initial velocity $v_0 = 20$ cm/s. Neglecting the resistance of the horizontal bars and the rod *ab*, calculate (i) the distance travelled by the rod *ab* until it comes to a standstill and (ii) the heat developed in the resistor *R*.

Solution: (i) Let the rod move through a time *dt*.

Then its displacement $= v_0 dt$.

Change of flux $d\phi$ in time $dt = Blv_0 dt$

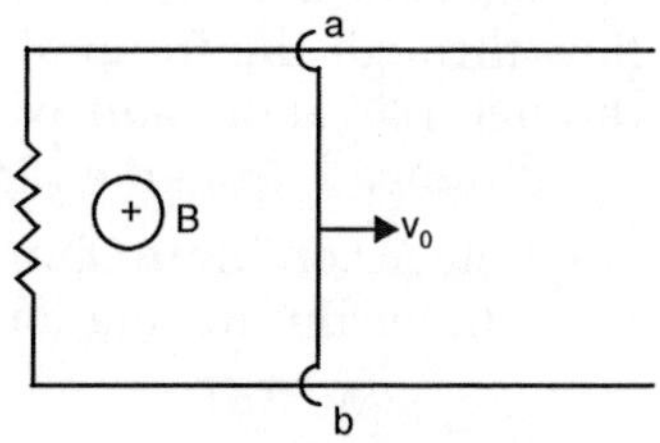

Fig. 14.14

$$\text{Induced emf} = E = \frac{d\phi}{dt} = Blv_0$$

$$\text{Induced current} = I = \frac{E}{R} = \frac{Blv_0}{R}$$

$$\text{Force acting on the rod} = BIl = \frac{B^2l^2v_0}{R}$$

According to Newton's law $\dfrac{B^2l^2v_0}{R} = ma$ where *a* is acceleration

$$\therefore \quad a = \frac{B^2l^2v_0}{mR}$$

$v_0^2 = 2as = aS$, where 2 *s* is the distance covered before the rod comes to a standstill.

$$\therefore \quad S = \frac{v_0^2}{B^2l^2v_0} mR = \frac{mv_0R}{B^2l^2}$$

On substituting the values of the various terms, we have

$$S = \frac{10^{-2} \times 0.2 \times 50}{10^2 \times 4 \times 10^{-2}}$$

$$= 2.5 \text{ cm}$$

(ii) Amount of heat developed $= \dfrac{1}{2} mv_0^2 = \dfrac{1}{2} \times 10^{-2} \times 4 \times 10^{-2} = 0.2$ mJ

14.21. A long thin wire, bent in the form of parabola $y = \alpha x^2$ has been placed in a homogenous magnetic field of induction *B*. The field is perpendicular to the plane of the parabola and directed downwards as shown in Fig. 14.15. A connector which can slide freely on the frame starts moving from the apex of the parabola along the *y*-axis with a constant acceleration *a*. Calculate the emf developed as a function of *y*.

Solution: Let the connector move through a distance *dy* along the *y*-axis in time *dt*. The area described by it is given by $dA = 2x\,dy$

Change of flux $= d\phi = B\,2x\,dy$

$$E = \frac{d\phi}{dt} = 2\,Bx\frac{dy}{dt}$$

From the equation of parabola $x = \sqrt{\frac{y}{\alpha}}$

$$\therefore \quad E = 2B\sqrt{\frac{y}{\alpha}}\frac{dy}{dt}$$

Now $v^2 = 0 + 2ay$ or $v = \frac{dy}{dt} = \sqrt{2ay}$

$$\therefore \quad E = 2B\sqrt{\frac{y}{\alpha}}\sqrt{2ay} = By\sqrt{\frac{8a}{\alpha}}$$

The emf acts in the clockwise direction.

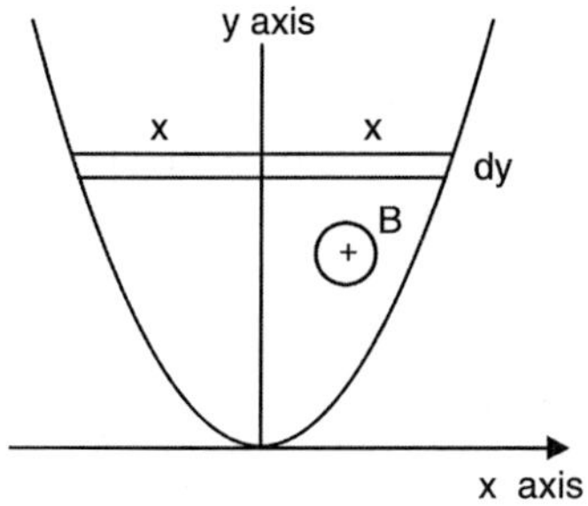

Fig. 14.15

14.22. A conducting rod AB of mass m and length l slides over two rails which are connected through a resistance R at the left end as shown in Fig. 14.16. The system is located in a homogenous magnetic field of induction B directing downwards and perpendicular to the plane of the loop. At $t = 0$, the rod starts moving towards right with an initial velocity v_0. Neglecting the resistance of the rails, find the velocity of the rod as a function of time.

Solution: Let the rod AB move through a time dt with a velocity v. Change of flux = $Blvdt$

Induced emf = $-Blv$ and is equal to RI

$$\therefore \quad I = -\frac{Blv}{R}$$

$$\text{Magnetic force} = BIl = -\frac{B^2l^2v}{R}$$

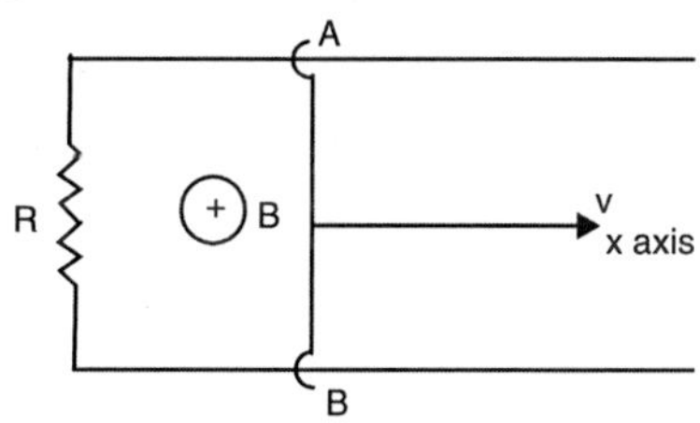

Fig. 14.16

According to Newton's law

$$\frac{mdv}{dt} = \frac{-B^2l^2v}{R}$$

or $$\frac{dv}{v} = \frac{-B^2l^2}{mR}dt = -\alpha\,dt;\ \alpha = \frac{B^2l^2}{mR}$$

On integration, we have $\log v = -\alpha t + c$, where c is the constant of integration.

Now at $t = 0$, $v = v_0$

$\therefore \quad c = \log v_0$

or $\log v - \log v_0 = -\alpha t$

or $v = v_0\, e^{-\alpha t}$

14.23. Two rods *AB* and *CD* are placed at a distance of y_1 and y_2 respectively from a long straight wire carrying current *I* (see Fig. 14.17). The ends *A* & *C* of the rods are connected to a resistor *R*. Upon the two rods a connector slides freely with a constant velocity v_0. Neglecting the resistance of the two rods & connector find (i) the current in the connector (ii) the force required to keep the connector's velocity constant.

Solution: (i) Consider an element of length *dy* on the connector located at a distance *y* from the wire carrying a current *I*.

$$\text{Field at the element} = \frac{\mu_0}{2\pi}\frac{I}{y}$$

Let the connector moves for a time *dt*. The area described by the element $= dy \cdot v_0 \cdot dt$

$$\text{Flux through the element} = \frac{\mu_0}{2\pi} \cdot \frac{I}{y} \cdot dy \cdot v_0 dt$$

$$\text{Induced emf } de = \frac{\mu_0 I v_0}{2\pi}\frac{dy}{y}$$

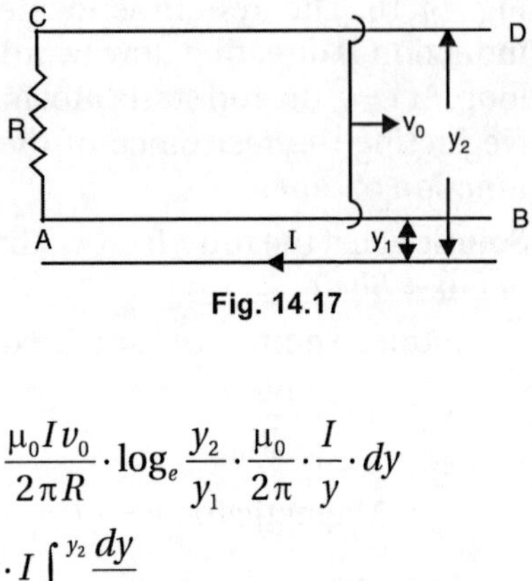

Fig. 14.17

$$\text{Net induced emf} = \frac{\mu_0 I v_0}{2\pi}\int_{y_1}^{y_2}\frac{dy}{y}$$

$$= \frac{\mu_0 I v_0}{2\pi} \cdot \log_e \frac{y_2}{y_1}$$

$$\text{Induced current} = \frac{\mu_0 I v_0}{2\pi R}\log_e \frac{y_2}{y_1}$$

(ii) $$\text{Force acting on the element} = \frac{\mu_0 I v_0}{2\pi R} \cdot \log_e \frac{y_2}{y_1} \cdot \frac{\mu_0}{2\pi} \cdot \frac{I}{y} \cdot dy$$

$$\text{Net force} = \frac{\mu_0 I v_0}{2\pi R} \cdot \log_e \frac{y_2}{y_1} \cdot \frac{\mu_0}{2\pi} \cdot I \int_{y_1}^{y_2}\frac{dy}{y}$$

$$= \frac{v_0}{R}\left[\frac{\mu_0 I}{2\pi}\log_e \frac{y_2}{y_1}\right]^2$$

14.24. Two conducting bars whose one end is connected to a capacitor of capacitance *C* are placed at an angle θ to the horizontal. Upon these bars a connector of mass *m* and length *l* slides down under the action of gravity (see Fig. 14.18). The complete set up is placed in a magnetic field of induction *B* acting in the vertical direction. Neglecting the resistance of the bars and connector find the acceleration of the connector.

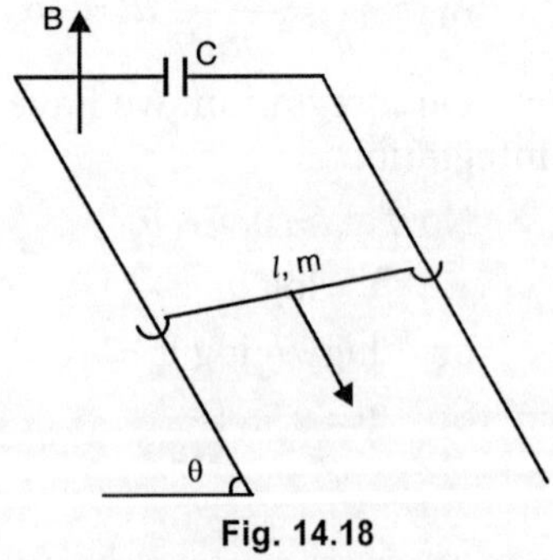

Fig. 14.18

Solution: Let the connector moves through dx in time dt along the plane formed by the two bars. The magnetic flux $d\phi$ is given by

$$d\phi = B \cdot \cos\theta \cdot l \cdot dx$$

$$\text{emf induced} = Bl\cos\theta \cdot v; \quad \frac{dx}{dt} = v$$

Charge in the capacitor $dQ = BlC\cos\theta \cdot v$

$$\text{Current in the closed loop } i = BlC\cos\theta \cdot \frac{dv}{dt}$$

$$= BlC\cos\theta \cdot a; \quad a = \frac{dv}{dt}$$

Magnetic force acting on the connector is given by

$$B \cdot \cos\theta \cdot i \cdot l = B\cos\theta \cdot B\cos\theta \cdot lC \cdot a \cdot l$$

$$= B^2 l^2 \cos^2\theta \cdot C \cdot a$$

Equation of motion of the connector is given by

$$mg\sin\theta - B^2 l^2 \cos^2\theta\, C \cdot a = ma$$

$$\therefore\ a = \frac{g\sin\theta}{1 + \dfrac{B^2 l^2 \cos^2\theta \cdot C}{m}}$$

14.25. A metallic rod OD of mass m moves on the conducting arc AA' of radius r. A capacitor of capacity C is connected between the point of suspension O and the middle of the arc. A uniform magnetic field B is applied perpendicular to the plane of the Fig. 14.19. The centre of mass of the rod is at O' and the moment of inertia of the rod about O' is I_0. The rod is released with an initial angle ϕ_0 without any initial speed. Discuss the nature of its motion. Assume that the total resistance of the circuit is zero.

Solution: Let the rod moves through an angle α from its initial position then flux ϕ is given by

$$\phi = r \cdot \frac{\alpha r}{2} \cdot B$$

$$\text{Induced emf } e = \frac{d\phi}{dt} = \frac{Br^2}{2} \cdot \frac{d\alpha}{dt}$$

Charge on the capacitor

$$= Q = \frac{Br^2}{2} \cdot C \cdot \frac{d\alpha}{dt}$$

$$\text{Current } I = \frac{Br^2 C}{2} \cdot \frac{d^2\alpha}{dt^2}$$

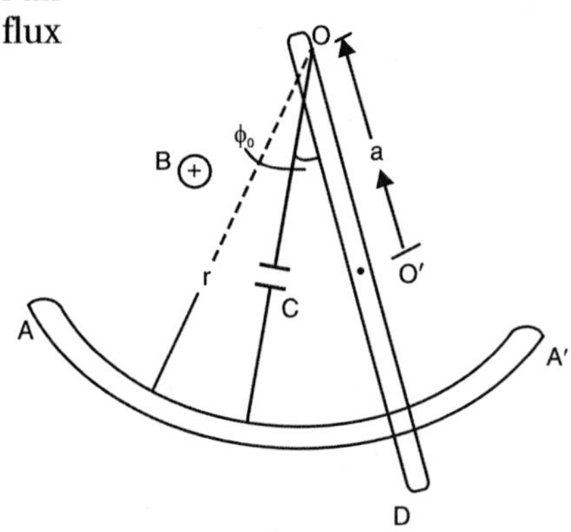

Fig. 14.19

The equation of motion of the rod is given by

$$-B\cdot\frac{Br^2C}{2}\cdot\frac{d^2\alpha}{dt^2}\cdot r\cdot\frac{r}{2}-mg\,a\sin\alpha=\left(I_0+ma^2\right)\frac{d^2\alpha}{dt^2}$$

or $$\frac{d^2\alpha}{dt^2}\left[4\left(I_0+ma^2\right)+B^2r^4C\right]=-\,4mga\sin\alpha$$

For small angles

$$\frac{d^2\alpha}{dt^2}\left[4\left(I_0+ma^2\right)+B^2r^4C\right]=-\,4mga\,\alpha$$

This is a case of simple harmonic motion in which ω is given by

$$\omega=\sqrt{\frac{4mga}{4\left(I_0+ma^2\right)+B^2r^4C}}$$

Let the equation of motion be

$$\phi=x_0\cos\omega t$$

at $t=0,\phi=\phi_0\ \therefore\ x_0=\phi_0$

$\therefore$ $\phi=\phi_0\cos\omega t$ is the equation of motion.

14.26. A metal rod *OA* of mass *m* rotates about *O* by keeping a contact with the circular conductor of radius *r*. The entire set up is located in the magnetic field of induction *B* perpendicular to the plane of the ring (see Fig. 14.20a). The axis and the rim of the ring are connected to the source of emf. Assuming the resistance of the circuit equal to *R*, find the emf of the source when the rod rotates with a constant angular velocity ω.

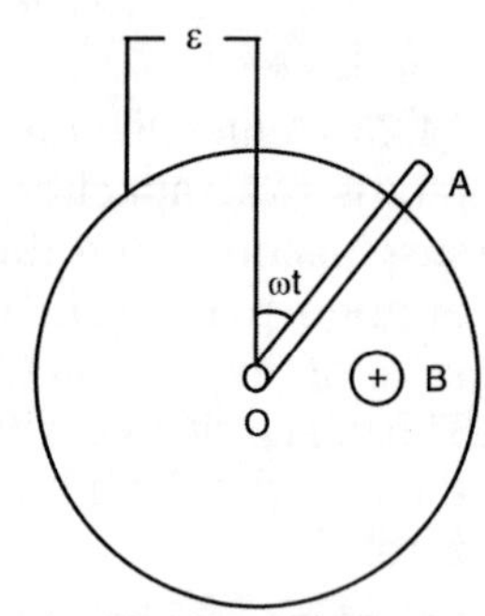

Fig. 14.20a

Solution: Consider an element of length dx located at a distance x from *O*. Linear velocity of the element $v=x\omega$.

Induced emf $=Bx\,\omega\cdot dx$

Total induced emf $e=B\omega\int_0^r xdx=\dfrac{B\omega r^2}{2}$

Let E_0 be the emf of the source. The net emf in the circuit $=E_0-e$.

Net current $I=\dfrac{E_0-\frac{1}{2}B\omega r^2}{R}$

Taking moment about *O* we have (see Fig. 14.20b)

$$F_{mag}\times\frac{r}{2}=mg\sin\omega t\times\frac{r}{2}$$

$$\therefore \quad F_{mag} = BIr = mg \sin \omega t$$

$$\frac{B\left(E_0 - \frac{1}{2}B\omega r^2\right)}{R} \cdot r = mg \sin \omega t$$

$$E_0 = \frac{mgR}{rB} \cdot \sin \omega t + \frac{1}{2}B\omega r^2$$

$$E_0 = \frac{1}{2rB}\left[r^3 \omega B^2 + 2mgR \sin \omega t\right]$$

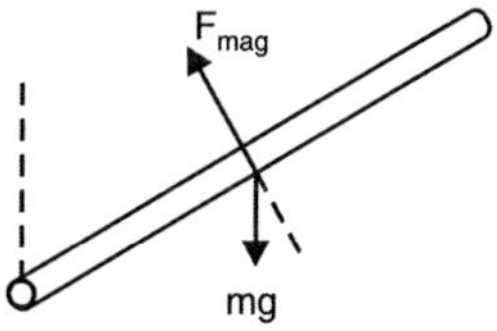

Fig. 14.20b

14.27. A solenoid of inductance L is connected at the left end of two horizontally placed rails over which a conducting rod of length l and mass m slides without friction. The system is placed in a magnetic field of induction B directed downwards (see Fig. 14.21). At $t = 0$, the rod starts moving along the x-axis with an initial velocity v_0. Neglecting the resistance of the rails, conducting rod and the solenoid, find an expression for the displacement of the rod.

Solution: Let the rod move through a distance dx in time dt. The change of flux = $Bl\,dx$

The induced emf = $Bl\dfrac{dx}{dt}$ and, according to electromagnetic induction,

$$Bl\frac{dx}{dt} = -\frac{L\,di}{dt}$$

or $$-\frac{Blx}{L} = i$$

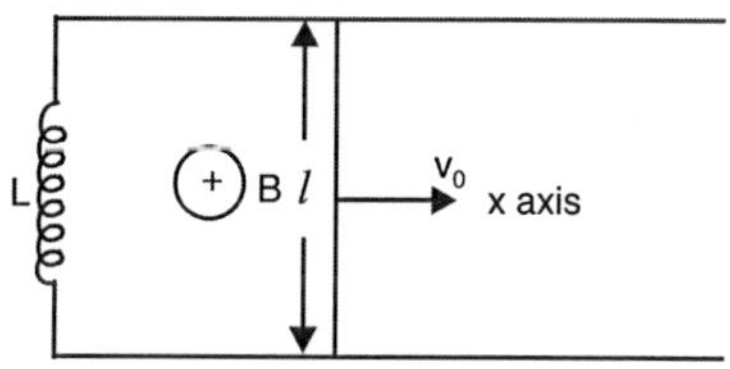

Fig. 14.21

Force on the conductor = $Bil = -\dfrac{B^2l^2x}{L}$

This must be equal to $m \times$ acceleration. Hence,

acceleration $a = -\dfrac{B^2l^2}{mL} \cdot$ displacement.

It is a case of simple harmonic motion in which $\omega = \dfrac{2\pi}{T} = \sqrt{\dfrac{B^2l^2}{mL}}$

The equation of motion of the rod is given by

$$x = x_0 \sin \omega t$$

where x_0 is the amplitude whose value is to be determined using the boundary conditions.

Now $\dfrac{dx}{dt} = \omega x_0 \cos \omega t$

at $t = 0$, $v = v_0$

$$\therefore \quad v_0 = \omega x_0 \text{ or } x_0 = \frac{v_0}{\omega}$$

Hence the displacement is given by

$$x = \frac{v_0}{\omega} \sin \omega t$$

14.28. A solenoid of inductance L is connected at the top end of two vertical rails over which a rod of length l and mass m falls freely without losing contact, as shown in Fig. 14.22. The system is located in a homogenous magnetic field of induction B, perpendicular to the plane of the two vertical rails. At $t = 0$, the rod starts falling from its resting condition. Find the law of motion of the rod.

Solution: Let the rod move vertically downwards through a distance dy in time dt.

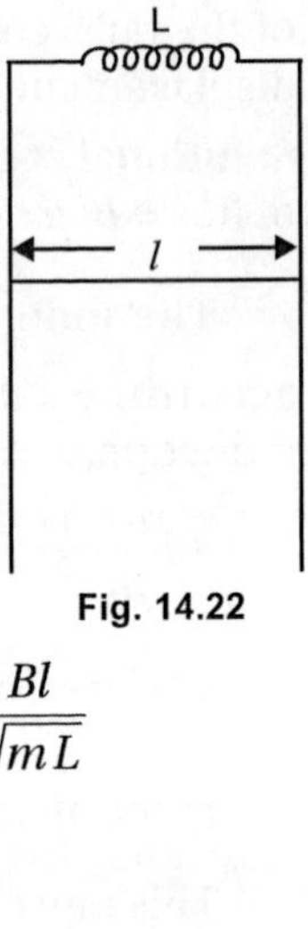

Fig. 14.22

Change of flux $d\phi = Bldy$

$$\text{Induced emf} = \frac{d\phi}{dt} = Bl\frac{dy}{dt} = -\frac{Ldi}{dt}$$

or $\quad Bldy = -Ldi$

or $\quad i = -Bly/L$

$$\text{Force} = m \times \text{acceleration} = -\frac{B^2 l^2 y}{L}$$

$$\therefore \quad \text{Acceleration} = -\frac{B^2 l^2}{mL} \cdot \text{displacement}$$

It is a case of simple harmonic motion in which $\omega = \dfrac{Bl}{\sqrt{mL}}$

Let $y = y_0 \cos \omega t$ be the equation of motion of the rod.

$$\text{Then } \frac{d^2 y}{dt^2} = -\omega^2 y_0 \cos \omega t$$

at $t = 0$, $\omega^2 y_0 = g$, i.e., $y_0 = \dfrac{g}{\omega^2}$

Now at $t = 0$, y should be zero which is possible when y takes the form as

$$\therefore \quad y = g / \omega^2 (1 - \cos \omega t)$$

14.29. A long wire is turned into a flat spiral of $N = 50$ tightly wound turns. The spiral has an external radius $a = 2$ cm and is located in a magnetic field of induction $B = 10^{-3} \sin 50\pi t$ acting perpendicular to the plane of the spiral. Find the maximum value of emf induced in the spiral.

Solution: Consider one turn of the coil of radius r. The induced emf due to this is given by

$$e=-\frac{d\phi}{dt}=-\pi^2 r^2\times 10^{-3}\times 50\cos(50\,\pi t)$$

Total induced emf $E=\int e dN$

Where dN is the number of turns lying between the radii r and $r+dr$

i.e., $\dfrac{dN}{N}=\dfrac{dr}{a}$

$$\therefore \quad E=\frac{N}{a}\pi^2\times 50\times 10^{-3}\cos(50\pi t)\int_0^a r^2 dr$$

$$=\frac{1}{3}\pi^2 a^2 N\times 50\times 10^{-3}\cos(50\pi t)$$

The maximum value of E will be obtained when $\cos(50\,\pi\, t)=1$

Hence $E=\dfrac{1}{3}\pi^2 a^2 N\times 50\times 10^{-3}$

On substituting the values of a and N, we have

$$E=\frac{1}{3}\times 3.14^2\times 4\times 10^{-4}\times 50\times 50\times 10^{-3}$$

$$=3.29 \text{ mV}$$

14.30. A metal rod OA of mass m and length l is kept rotating with a constant angular speed ω in a vertical plane about a horizontal axis at the end O. The free end A is arranged to slide without friction along a fixed conducting circular ring in the same plane as that of rotation. A uniform and constant magnetic field induction B is applied perpendicular and into the plane of rotation as shown in Fig. 14.23a. An inductor and an external resistance R are connected through a switch S between the point O and a point C on the ring to form an electrical circuit. Neglect the resistance of the ring and the rod. Initially, the switch is open. (a) What is the induced emf across the terminals of the switch? (b) The switch S is closed at time $t=0$. (i) Obtain an expression for the current as a function of time. (ii) In the steady state, obtain the time dependence of the torque required to maintain the constant angular speed, given the rod OA was along the positive x axis at $t=0$. (IIT 1995)

Solution: (a) Let the rod moves through an angle θ

$$\text{Flux }\phi=\frac{1}{2}\cdot r\cdot r\theta\cdot B=\frac{1}{2}Br^2\theta$$

$$\text{Induced emf } e=-\frac{d\phi}{dt}=-\frac{1}{2}Br^2\cdot\frac{d\theta}{dt}$$

$$=-\frac{1}{2}Br^2\omega$$

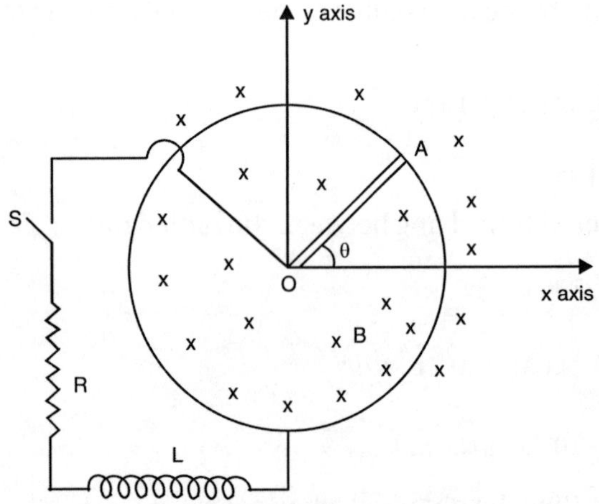

Fig. 14.23a

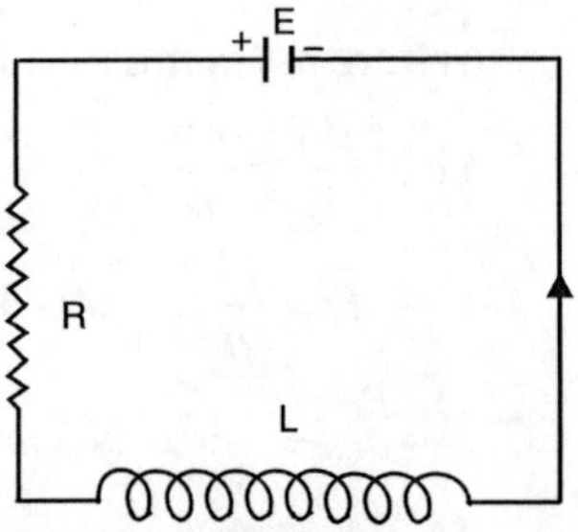

Fig. 14.23b

(b) (i) The system can now be observed as a LR circuit connected to an emf $E = \frac{1}{2}Br^2\omega$ (see Fig. 14.23b).

Applying Kirchhoff's Law we have

$$E - IR - L\frac{dI}{dt} = 0$$

or
$$\frac{dI}{E - IR} = \frac{dt}{L}$$

or
$$\int_0^I -\frac{dI}{E - IR} = -\frac{1}{L}\int_0^t dt$$

$$\left|\log(E - IR)\right|_0^I = -\frac{R}{L}\cdot t$$

$$\log\left(\frac{E - IR}{E}\right) = -\frac{R}{L}t$$

$$I = \frac{E}{R}\left(1 - e^{-Rt/L}\right)$$

$$I = \frac{Br^2\omega}{2R}\left(1 - e^{-Rt/L}\right)$$

(ii) Force acting on the rod is given by

$$F = \frac{Br^2\omega}{2R}\left(1 - e^{-Rt/L}\right)rB$$

$$F = \frac{B^2r^3\omega}{2R}\left(1 - e^{-Rt/L}\right)$$

$$\text{Torque } \tau = F \cdot \frac{r}{2} = \frac{B^2 r^4 \omega}{4R}\left(1 - e^{-Rt/L}\right)$$

An external torque τ has to be applied in anticlock-wise direction.

14.31. A copper rod of length 0.19 m is moving with a uniform velocity 10 m/s parallel to a long straight wire carrying a current of 5.0 amp. The rod itself is perpendicular to the wire with its ends are at distance 0.01 m and 0.2 m from it. Calculate the emf induced in the rod.

Solution: Consider an element of length dx on the rod at distance x from the wire carrying current as shown in Fig. 14.24. The induced emf in the element is given by $dE = B\,v\,dx$; B is magnetic field at a distance x

$$dE = \frac{\mu_0}{4\pi} 2Iv \frac{dx}{x}; \; I \text{ is the current flowing in the wire.}$$

Total induced emf is given by

$$E = \int dE = \frac{\mu_0}{4\pi} 2Iv \int_{0.01}^{0.2} \frac{dx}{x}$$

$$= \frac{\mu_0}{4\pi} 2Iv \log_e 20$$

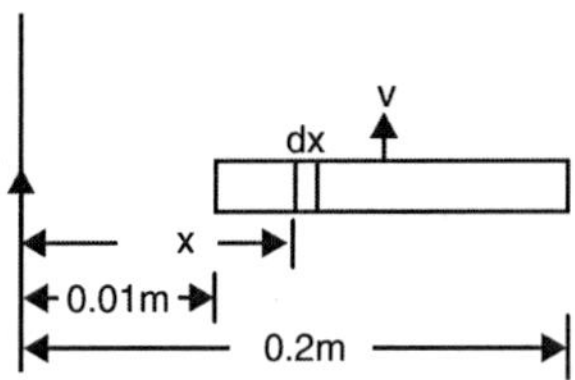

Fig. 14.24

On substituting the values of the various terms, we have

$$E = 10^{-7} \times 2 \times 5 \times 10 \times 2.303 \times 1.3010$$

$$= 29.96\ \mu v$$

$$= 30\ \mu v$$

14.32. A wire is bent in the form of two square loops of sides a = 10 cm and b = 5 cm. The two loops lies in the same plane. A time varying magnetic field of induction $B = B_0 \cos \omega t$ with the peak value B_0 = 10 mT is applied perpendicular to the plane of loops. Neglecting the inductance of the loops and assuming n = 50 c/s and resistance per unit length of the wire ρ = 50 mΩ/m. Find the amplitude of the current induced in the two loops.

Solution: The emf induced in the two loops is given by

$$E_1 = B_0\, b^2\, \omega \sin \omega t \quad \text{(see Fig. 14.25)}$$

and $E_2 = B_0\, a^2\, \omega \sin \omega t$

In view of the fact $E_2 > E_1$ the direction of current is shown in the Fig. 14.25. Applying Kirchhoff's Law we have

$$B_0 (a^2 - b^2)\, \omega \sin \omega t - 4\,\rho\, Ia - 4\,\rho\, Ib = 0$$

$$\therefore \quad I = \frac{B_0\, \omega (a - b)}{4\rho} \cdot \sin \omega t$$

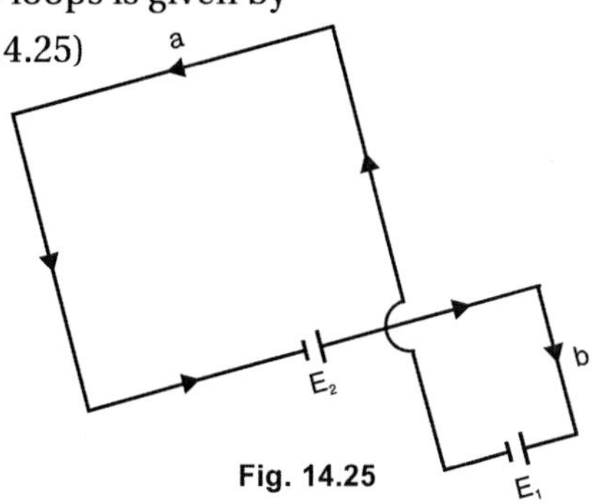

Fig. 14.25

The peak value of the current $I_0 = \dfrac{B_0\,\omega(a-b)}{4\rho}$

On substituting the various values we have

$$I_0 = \frac{10\times10^{-3}\times 2\pi\times 50\times 5\times 10^{-2}}{50\times 10^{-3}}$$

$$= \pi \text{ amp}$$

14.33. A long straight wire is placed along the axis of the toroid of dimensions shown in the Fig. 14.26. The number of turns of coil is N. Find the induced emf in the coil when an alternating $I = I_0 \cos \omega t$ flows through the wire.

Solution: Consider a shell of radius r and $r + dr$

Magnetic field due to the current in the straight wire at a distance r is given by

$$B = \frac{\mu_0}{4\pi}\cdot\frac{2I}{r}$$

Magnetic flux through the rectangular cross-section of the coil

$$= \frac{\mu_0}{2\pi}\cdot Il\cdot\int_{r_1}^{r_2}\frac{dr}{r}$$

$$= \frac{\mu_0}{2\pi}\cdot Il\cdot\log_e\frac{r_2}{r_1}$$

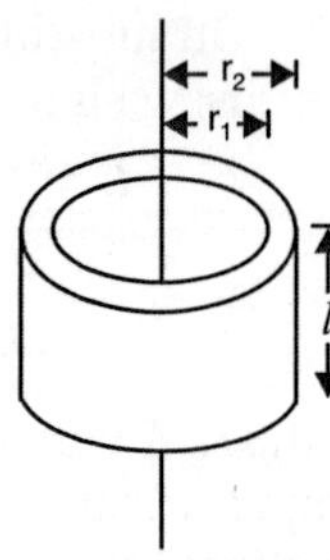

Fig. 14.26

Net magnetic flux $= \dfrac{\mu_0}{2\pi}\cdot NIl\cdot\log_e\dfrac{r_2}{r_1}$

$$= \frac{\mu_0}{2\pi}\cdot NI_0 l\cdot\log_e\frac{r_2}{r_1}\cdot\cos\omega t$$

Induced emf $= E = -\dfrac{d\phi}{dt} = \dfrac{\mu_0}{2\pi}\cdot Nl\omega I_0 \log_e\dfrac{r_2}{r_1}\sin\omega t$

14.34. A thin rod of length a and a long wire carrying a steady current I are placed b apart. Both being in the same plane. The rod moves with uniform veocity v in such a way that it is always perpendicular to the wire carrying current. Find the induced emf across the two ends of the rod.

Solution: Consider an element of length dx situated at a distance x from the wire carrying current (see Fig. 14.27).

Magnetic field at the element $= \dfrac{\mu_0}{2\pi}\cdot\dfrac{I}{x}$

Let the conductor be allowed to move through time dt. The area described by it is $dx\cdot vdt$.

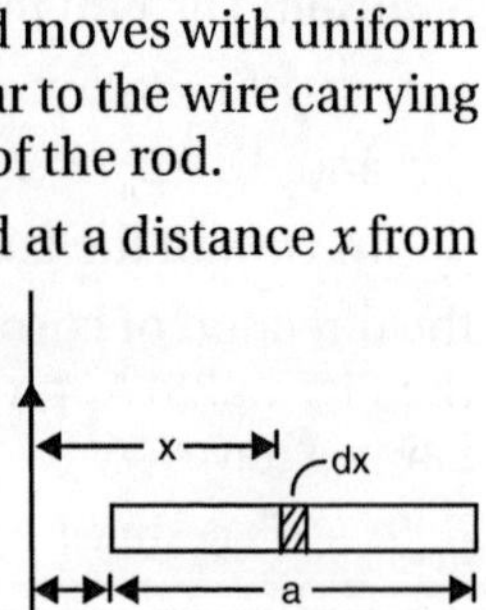

Fig. 14.27

Magnetic flux $d\phi$ through the element

$$d\phi = \frac{\mu_0}{2\pi} \cdot \frac{I}{x} \cdot dx \cdot v \cdot dt$$

Induced emf across the end of the rod = $E = \frac{\mu_0}{2\pi} \cdot I \cdot v \int_a^{a+b} \frac{dx}{x}$

$$= \frac{\mu_0 I v}{2\pi} \cdot \log_e\left(1 + \frac{a}{b}\right)$$

14.35. A metallic rod of length l rotates about a vertical axis passing at a distance of l/n from its one end at an angular velocity ω in a uniform magnetic field of induction B. Calculate the potential difference across the ends of the rod.

Solution: In view of Q. No. 14.1 the potential difference between O and $A = \frac{1}{2} B\omega \frac{l^2}{n^2}$

Similarly the potential difference between O & $B = \frac{1}{2} B\omega l^2 \left(1 - \frac{1}{n}\right)^2$

$\therefore$ Potential difference between the ends A & B of the rod

$$= \frac{1}{2} B\omega l^2 \left[1 - \frac{2}{n} + \frac{1}{n^2} - \frac{1}{n^2}\right]$$

$$= \frac{1}{2} \frac{B\omega l^2}{n} (n - 2)$$

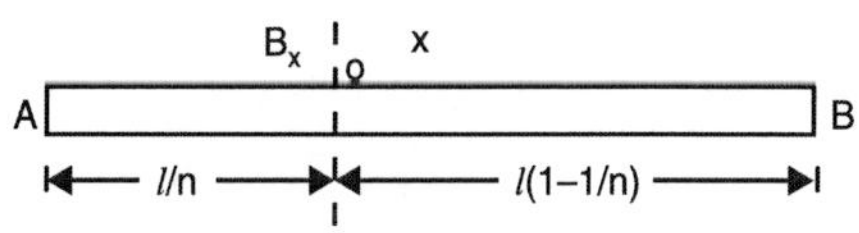

Fig. 14.28

14.36. An iron sphere of radius r moves parallel to a long wire carrying a steady current I with a speed v. During motion the centre of the sphere is always at a distance a from the wire. Determine the induced emf between the extreme points (one closests and the other farthest) on the sphere.

Solution: Magnetic field at a distance x along the perpendicular bisector of the wire $= \frac{\mu_0}{4\pi} \cdot \frac{2I}{x}$

Consider an element of length dx at a distance x from the wire. The area described in time dt due to the motion of the sphere = $v \cdot dt \cdot dx$

$$\text{Change in flux } d\phi = \frac{\mu_0}{4\pi} \cdot \frac{2I}{x} \cdot v \cdot dt \cdot dx$$

$$de = \frac{\mu_0 I}{2\pi} v \cdot \frac{dx}{x}$$

$$\text{Net induced emf} = \int_0^e de = \frac{\mu_0 I v}{2\pi} \cdot \int_{d-r}^{d+r} \frac{dx}{x}$$

$$e = \frac{\mu_0 I v}{2\pi} \log \frac{d+r}{d-r}$$

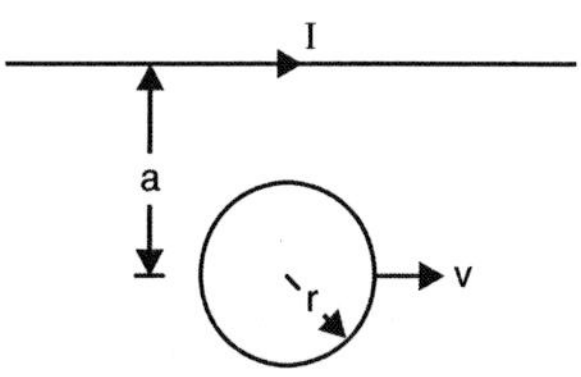

Fig. 14.29

14.37. A wire of length 1.5 m ⊥ to x-z plane is moved with velocity $v = (2i + 6j + k)$ m/sec through a region of uniform magnetic field of induction $B = (i + 2j)$ w/m². Calculate the electric field induced and the potential difference across the wire.

Solution:

The electric field $E = -(v \times B) = -(2i + 6j + k) \times (i + 2j)$

$$= (2i - j + 2k) \text{ V/m}$$

Induced emf $e = -(v \times B) \cdot l$

$$= (2i - j + 2k) \cdot 1.5\,\text{k} = +3 \text{ volt}$$

14.38. A circular loop of radius b as shown in the Fig. 14.30 mounted rigidly on an axle midway, between the two cords carries a current I. The tension in each of the two cords is T_0. Find the tension in the two cords when the system is located in a magnetic field.

Solution: Magnetic force acting on the loop is πIBb.

On taking the moment about O we have

$$T_1 \cdot \frac{L}{2} - T_2 \cdot \frac{L}{2} = \pi BIb \cdot b$$

or $$T_1 - T_2 = \frac{2\pi I b^2 B}{L}$$

$$T_1 + T_2 = 2\,T_0$$

$$T_1 = T_0 + \frac{\pi I b^2 B}{L}$$

$$T_2 = T_0 - \frac{\pi I b^2 B}{L}$$

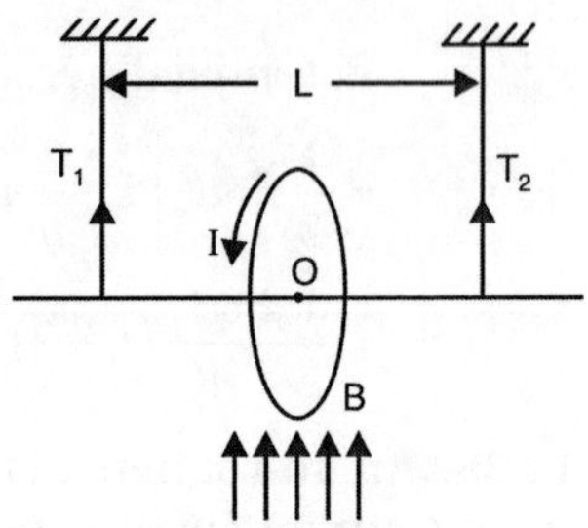

Fig. 14.30

14.39. A simple magnetic hydrodynamic generator consists of a parallel plate capacitor with a plate area S and a separation d which is put into a flowing conducting liquid with a specific conductivity of λ, the liquid moving parallel to the plates with constant velocity v. The capacitor is placed in a magnetic field with a flux density B and directed perpendicular to the velocity of the liquid (Fig. 14.31). How much power will be dissipated in an external circuit containing a resistance R.

Solution: The induced emf generated is given by $\varepsilon = Bvd$

Resistance of the liquid $r = \dfrac{1}{\lambda}\dfrac{d}{S}$

Current $I = \dfrac{\varepsilon}{R + \dfrac{d}{\lambda S}}$

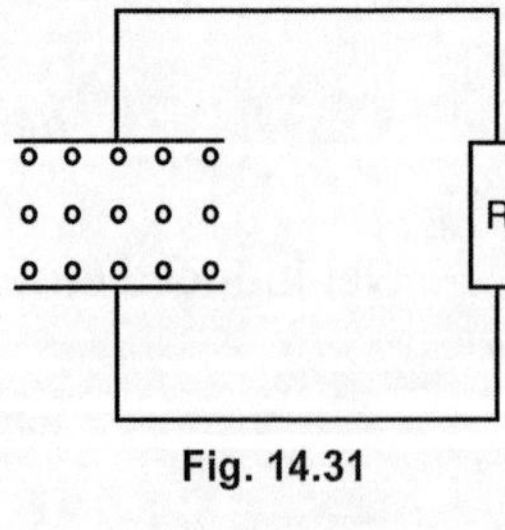

Fig. 14.31

$$\text{Power } P = I^2 R = \frac{\varepsilon^2 R}{\left(R + \frac{d}{\lambda S}\right)^2} = \frac{B^2 v^2 d^2 R}{\left(R + \frac{d}{\lambda S}\right)^2}$$

14.40. A square loop of side 'a' and resistance R moves with a uniform velocity v away from a long wire that carries current I as shown in the Fig. 14.32. The loop is moved away from the wire with side AB always parallel to the wire. Initially, distance between the side AB of the loop and wire is 'a'. Find the work done when the loop is moved through a distance 'a' from the initial position.

Solution: Total magnetic field acting on the loop is given by

$$\frac{\mu_0}{2\pi}\frac{I}{x} - \frac{\mu_0 I}{2\pi(x+a)} = \frac{\mu_0 I a}{2\pi x(x+a)}$$

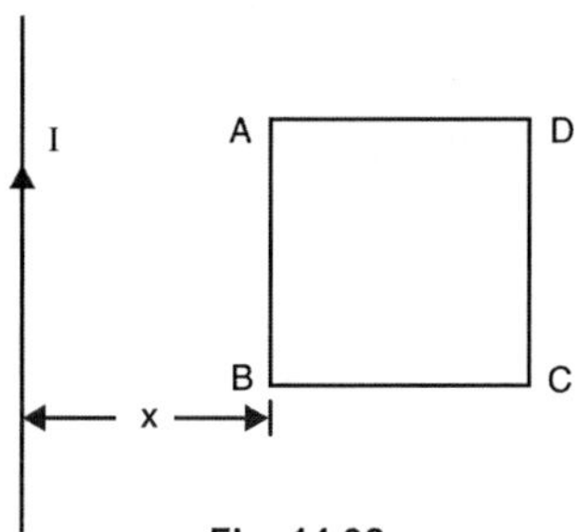

Fig. 14.32

Let the loop moves with velocity v for small time dt then

$$d\phi = \frac{\mu_0 I a v dt \cdot a}{2\pi x(x+a)}$$

$$\text{Induced current } I = \frac{e}{R} = \frac{\mu_0 I a^2 v}{2\pi x(x+a)R}$$

Net work done

$$= \int_a^{2a} \text{Magnetic force} \cdot dx$$

$$= \int_a^{2a} \left[\frac{\mu_0 I}{2\pi x} \cdot \frac{\mu_0 I a^2 v}{2\pi x(x+a)R} \cdot a \cdot dx - \frac{\mu_0 I}{2\pi(x+a)} \cdot \frac{\mu_0 I a^2 v a dx}{2\pi x(x+a)R}\right]$$

$$= \frac{\mu_0^2 I^2 a^3 v}{4\pi^2 R} \cdot \int_a^{2a} \frac{a\,dx}{x^2(x+a)^2} = \frac{\mu_0^2 I^2 a^3 v}{4\pi^2 R} \int_a^{2a} \left[\frac{A}{x} + \frac{B}{x^2} + \frac{C}{(x+a)} + \frac{D}{(x+a)^2}\right]$$

On solving $A = -\frac{2}{a^2}$, $B = \frac{1}{a}$, $C = \frac{2}{a^2}$ and $D = \frac{1}{a}$

$$\text{Work done} = \frac{\mu_0^2 I^2 a^3 v}{4\pi^2 R}\left[-\frac{2}{a^2}\int_a^{2a}\frac{dx}{x} + \frac{1}{a}\int_a^{2a}\frac{dx}{x^2} + \frac{2}{a^2}\int_a^{2a}\frac{dx}{x+a} + \frac{1}{a}\int_a^{2a}\frac{dx}{(x+a)^2}\right]$$

$$= \frac{\mu_0^2 I^2 a^2 v}{4\pi^2 R}\left[\frac{2}{3a} + \frac{2}{a}\log_e \frac{3}{4}\right]$$

14.41. Space is divided by the line AD into two regions. Region I is field free and the region II has a uniform magnetic field B directed into the plane of paper. ACD is semi circular conducting loop of radius r with centre at O (Fig. 14.33a), the plane of the loop being in the plane of the paper. The loop is now made to rotate with a constant angular velocity ω about an axis passing through O and perpendicular to the plane of the paper. The effective resistance of the loop is R.

(i) Obtain an expression for the magnitude of the induced current in the loop

(ii) Show the direction of the current when the loop is entering into the region II

(iii) Plot a graph between the induced emf and time of rotation for two periods of rotation. (IIT 1985)

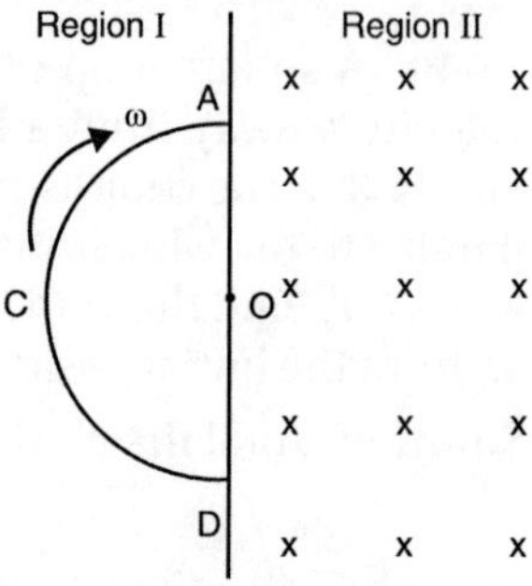

Fig. 14.33a

Solution: (i) Let $d\theta$ be the angle through which the loop is rotated in time dt. Induced emf is given by (see Fig. 14.33b)

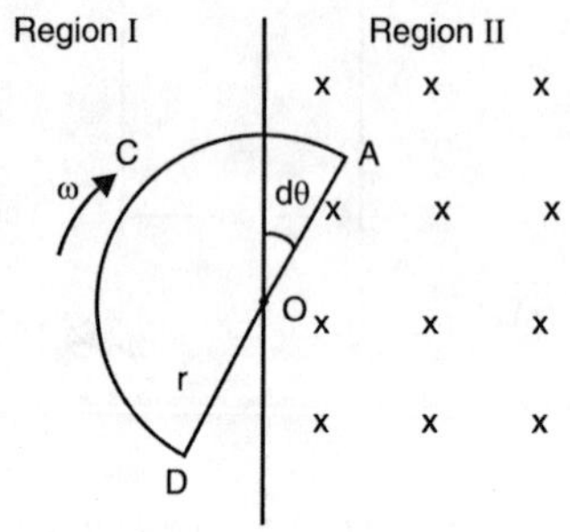

Fig. 14.33b

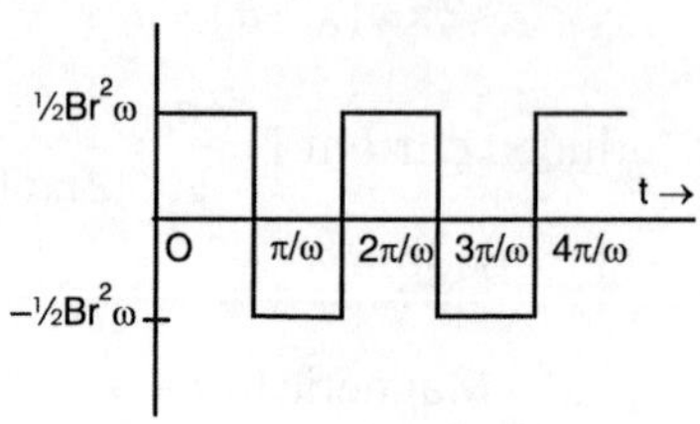

Fig. 14.33c

$$e = B \cdot \frac{1}{2} \cdot r \cdot \frac{r\,d\theta}{dt} = \frac{1}{2} B r^2 \omega$$

$$\text{Induced current} = \frac{1}{2} \frac{Br^2\omega}{R}$$

(ii) The direction of the current is anticlock-wise.

(iii) The variation of induced emf with time is shown in Fig. 14.33c.

14.42. Two long parallel horizontal rails, a distance d apart and each having a resistance λ per unit length, are joined at one end by a resistance R. A perfectly conducting rod MN of mass m is free to slide along the rails without friction (Fig. 14.34). There is a uniform magnetic field of induction B normal to the plane of the paper and directed into the paper. A variable force F is applied to the rod MN such that, as the rod moves, a constant current flows through R.

(i) Find the velocity of the rod and the applied force F as a function of the distance x of the rod from R.

(ii) What fraction of the work done per sec. by F is converted into heat.

(IIT 1988)

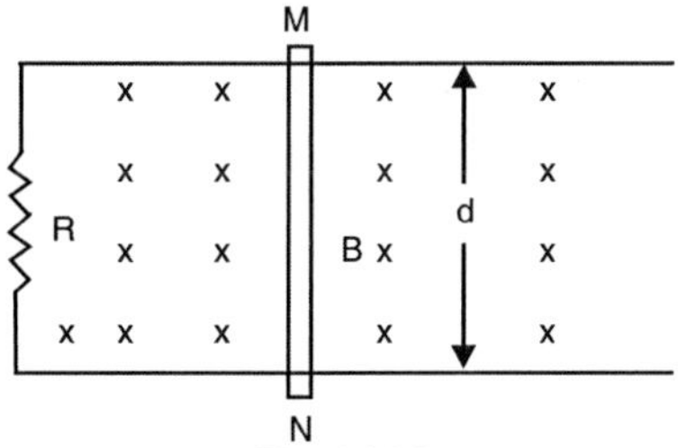

Fig. 14.34

Solution: Let the rod be at a distance x at an instant t

The induced emf $e = Bd \cdot \frac{dx}{dt}$

Total resistance at that instant $= R + 2\lambda x$

Current i through the loop $= \frac{B \cdot d \cdot v}{R + 2\lambda x}$

$\therefore$ Velocity $v = \frac{i(R + 2\lambda x)}{B \cdot d}$

$$\text{Acceleration } a = \frac{dv}{dt} = \frac{i \cdot 2\lambda}{B \cdot d} \cdot \frac{dx}{dt} = \frac{i \cdot 2\lambda}{B \cdot d} \cdot \frac{i(R + 2\lambda x)}{B \cdot d}$$

$$= \frac{2\lambda \cdot i^2 (R + 2\lambda x)}{B^2 \cdot d^2}$$

$$\text{Force } F = m \times \frac{2\lambda i^2 (R + 2\lambda x)}{B^2 \cdot d^2}$$

$$\text{Heat produced per sec} = i^2 (R + 2\lambda x) = \frac{F \cdot B^2 d^2}{2m\lambda}$$

$$W = \text{work done /sec} = F \cdot v = F \cdot \frac{i(R + 2\lambda x)}{B \cdot d}$$

$$\therefore \quad \frac{H}{W} = \frac{F \cdot B^2 d^2}{2m\lambda} \cdot \frac{B \cdot d}{F \cdot i(R + 2\lambda x)} = \frac{B^3 d^3}{2mi\lambda (R + 2\lambda x)}$$

14.43. A very small circular loop of area $5 \times 10^{-4} \text{m}^2$, resistance 2 ohm and negligible inductance is initially coplanar and concentric with a much larger fixed circular loop of radius 0.1 m. A constant current of 1 amp is passed in the bigger loop and the smaller loop is rotated with angular velocity ω rad/sec about a diameter. Calculate (i) the flux linked with the smaller loop, (ii) induced emf and (iii) induced current in the smaller loop as a function of time. (Roorkee 1992)

Solution:

Magnetic field produced by the bigger loop at the centre $= \frac{\mu_0 ni}{2r}$

$$\therefore \quad B = \frac{4\pi \times 10^{-7} \times 1 \times 1}{2 \times 0.1} = 2\pi \times 10^{-6} \text{ w/m}^2$$

Change of flux $\phi = 2BA$

$$\text{Induced emf } e = \frac{\phi}{\text{time for one rotation}}$$

$$= \frac{2BA}{2\pi/\omega}$$

$$e = \frac{\omega BA}{\pi}$$

On putting the values

$$e = \frac{\omega \cdot 2\pi \times 10^{-6} \times 5 \times 10^{-4}}{\pi}$$

$$= 10^{-9}\,\omega$$

$$\text{Induced current} = \frac{10^{-9}\,\omega}{2}$$

Current at any time is given by

$$i = \frac{10^{-9}\,\omega}{2} \sin \omega t$$

14.44. A long solenoid of diameter 0.1 m has 2×10^4 turns/m. At the centre of the solenoid a 100 turn coil of radius 0.01 m is placed with its axis coinciding with the solenoid axis. The current in the solenoid is decreased at a constant rate of $+ 2\,A$ to $- 2\,A$ in 0.05 sec. Find the emf induced in the coil. Also find the total charge flowing through the coil during this time when the resistance of the coil is $10\,\pi^2\Omega$. (Roorkee 1993)

Solution: Magnetic field at the centre of solenoid

$$B = \mu_0 n i \text{ w/m}^2$$

Induced emf in N turn of coil is given by

$$e = N \cdot \frac{d}{dt}(\mu_0 n i A) = N \cdot \mu_0 n A \cdot \frac{di}{dt}$$

On substituting the values we have

$$e = 100 \times 4\,\pi \times 10^{-7} \times 2 \times 10^{-4} \times \pi \times 10^{-4} \times \frac{4}{0.05}$$

$$= 6.4\,\pi^2 \times 10^{-3} \text{ volt}$$

$$q = \frac{e.t}{R} = \frac{6.4\pi^2 \times 10^{-3} \times 0.05}{10\pi^2} = 3.2 \times 10^{-5}\,\text{C}$$

$$= 32\,\mu\text{C}$$

14.45. A wire frame of area 3.92×10^{-4} m^2 and resistance 20 Ω is suspended freely from a 0.392 m long thread. There is a uniform magnetic field of 0.784 tesla and the plane of wire frame is perpendicular to the magnetic field. The frame is made to oscillate under gravity by displacing it through

2×10^{-2}m from its initial position along the direction of magnetic field. The plane of the frame is always along the direction of thread and does not rotate about it. What is the induced emf in wire frame as a function of time? Also find the maximum current in the frame. (Roorkee 1993)

Solution: Let the frame be displaced through an angle θ.

Instantaneous flux $\phi = BA\cos\theta$

emf induced is given by

$$e = -\frac{d\phi}{dt} = BA\sin\theta\cdot\frac{d\theta}{dt}$$

$$= BA\,\theta\cdot\frac{d\theta}{dt};\ \text{for small angles } \sin\theta = \theta$$

The equation of motion of the frame is given by (see Fig. 14.35)

$$\frac{md^2x}{dt^2} = -mg\sin\theta$$

$$= -mg\theta$$

Now $x = l\theta$

$$\therefore \quad \frac{d^2x}{dt^2} = \frac{ld^2\theta}{dt^2}$$

or $$\frac{ld^2\theta}{dt^2} = -g\theta$$

or $$\frac{d^2\theta}{dt^2} = -\frac{g}{l}\cdot\theta$$

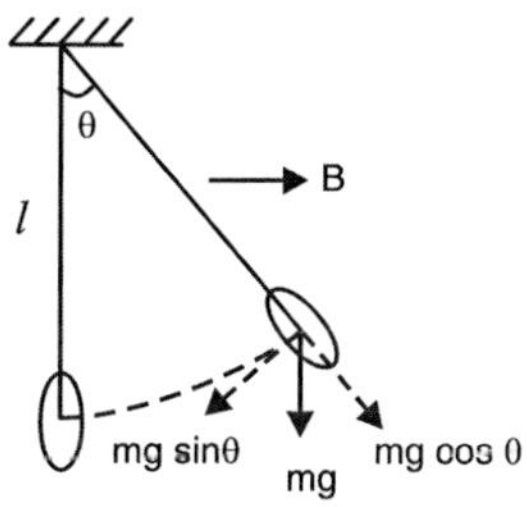

Fig. 14.35

This is the case of simple harmonic motion in which

$$\omega = \sqrt{\frac{g}{l}}$$

Let the solution of the above equation be

$$\theta = \theta_0 \sin\omega t$$

here $\theta_0 = \dfrac{x_0}{l}$

$$\text{Induced emf } e = BA(\theta_0\sin\omega t)\cdot\frac{d}{dt}(\theta_0\sin\omega t)$$

$$= \frac{1}{2}BA\cdot\omega\cdot\theta_0^2\cdot\sin 2\omega t$$

$$\omega = \sqrt{\frac{g}{l}} = \sqrt{\frac{9.8}{0.392}} = 5/\text{sec}$$

$$\theta_0 = \frac{x_0}{l} = \frac{2\times10^{-2}}{0.392}$$

$$e = \frac{1}{2} \times 0.784 \times 3.92 \times 10^{-4} \times 5 \times \left(\frac{2 \times 10^{-2}}{0.392}\right)^2 \sin 10t$$

$$= 2 \times 10^{-6} \sin 10t$$

$$e_{maximum} = 2 \times 10^{-6} \text{ volt}$$

$$i_{max} = \frac{2 \times 10^{-6}}{20} = 10^{-7} \text{ amp.}$$

14.46. A π shaped conductor carrying a connector of length l as shown in the Fig. 14.36 is located in a time varying magnetic field of induction B. At the initial moment the magnetic field is zero and the connector coincides with the side ab of the conductor. The connector starts moving with an acceleration ω parallel to the length of the two bars. Find the induced emf as a function of time.

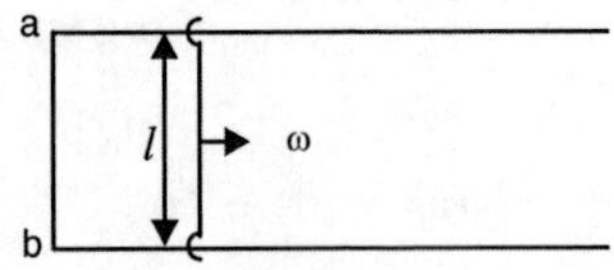

Fig. 14.36

Solution:

$$\frac{d^2x}{dt^2} = \omega$$

$$\frac{dx}{dt} = \omega t + C$$

at $t = 0, \frac{dx}{dt} = 0 \therefore C = 0$

hence $\frac{dx}{dt} = \omega t$

After integration we have

$$x = \frac{\omega t^2}{2} + C'$$

at $t = 0, x = 0 \quad \therefore C' = 0$

$$\therefore \quad x = \frac{\omega t^2}{2}$$

$$\frac{dB}{dt} = \dot{B} \therefore B = \dot{B}t$$

Induced emf $e = \frac{d\phi}{dt} = \frac{d}{dt}(B \cdot A) = \frac{AdB}{dt} + \frac{B \cdot dA}{dt}$

$$= l \times \frac{1}{2}\omega t^2 \cdot \dot{B} + \dot{B}t\frac{d}{dt}\left(\frac{1}{2}\omega t^2 l\right)$$

$$= \frac{1}{2}\omega \dot{B} t^2 l + \dot{B}t \times \omega t l$$

$$= \frac{3}{2}\omega l \dot{B} t^2$$

14.47. A long straight solenoid with cross-sectional radius a having n turns per unit length carries current I. If the current varies with time find the magnitude of the eddy current field strength as a function of the distance r from the axis of the solenoid.

Solution: (i) When $r < a$

$$\text{Field inside the solenoid} = \frac{\mu_0}{4\pi}\cdot\frac{4\pi NI}{L}; \quad N = \text{total number of turns}$$

$$L = \text{Length of the solenoid}$$

$$\text{Flux } \phi = \frac{\mu_0 NI}{L}\cdot \pi r^2 N$$

$$\text{Induced emf } e = \mu_0 n\dot{I}\pi r^2 N$$

$$\text{Strength} = \frac{e}{2\pi r N} = \frac{\mu_0 n\dot{I}\pi r^2 N}{2\pi r N}$$

$$= \frac{1}{2}\mu_0 n\dot{I}r$$

(ii) When $r > a$

$$\text{Magnetic flux } \phi = \frac{\mu_0 NI\pi a^2 N}{L}$$

$$\text{Induced emf } e = \mu_0 n\dot{I}\pi a^2 N$$

$$\text{Strength} = \frac{e}{2\pi r N} = \frac{\mu_0}{2} n\dot{I}\frac{a^2}{r}$$

14.48. A material of resistivity ρ has been used to form a conducting ring of thickness h and having inner and outer radii r_1 and r_2 respectively (Fig. 14.37). The ring is coaxially fitted at the middle of the solenoid. A current through the solenoid produces a time varying field whose induction is $B = \alpha t$ where α is a positive constant. Neglecting the inductance of the ring find the current induced in it.

Solution: Let us divide the ring into n small rings each of width p such that

$$n = \frac{r_2 - r_1}{p}$$

Consider a ring of height h and having internal radius x and external radius $x + p$. The resistance R of such a ring is given by

$$R = \rho\cdot\frac{2\pi x}{ph}$$

Flux through the ring $= \alpha t \cdot \pi x^2$

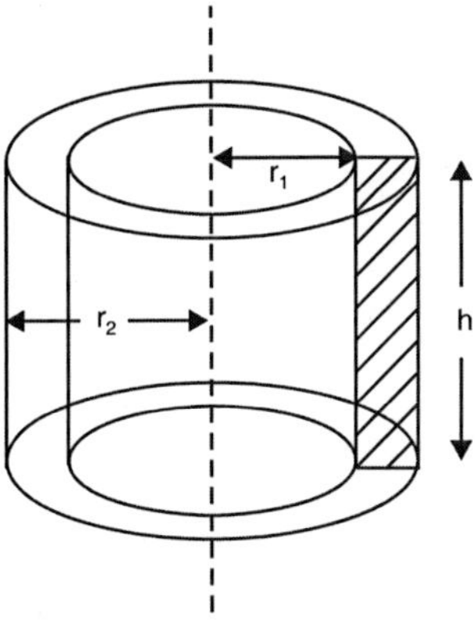

Fig. 14.37

Induced emf through it $= \alpha \cdot \pi x^2$

$$\text{Current through it} = \frac{\alpha \pi x^2 \cdot ph}{\rho \cdot 2\pi x}$$

$$= \frac{\alpha phx}{2\rho}$$

The total current through the entire ring

$$= \frac{\alpha ph}{2\rho}\left[r_1 + (r_1 + p) + (r_1 + 2p) \ldots\ldots r_1 + (n-1)p\right]$$

$$= \frac{\alpha h}{2\rho}\left[\frac{r_1 + r_2 - p}{2}\right][r_2 - r_1]$$

$$= \frac{\alpha h}{4\rho}\left(r_2^2 - r_1^2\right) \text{ here } p \text{ is very small quantity}$$

14.49. A circular coil of radius R and resistance R_0 and having n turn of wire is placed in an external time varying magnetic field. The field is perpendicular to the plane of the coil and its value at a distance r from the centre of the coil is $b_0\left(1 - \frac{r}{3R}\right)\cos \omega t$. Determine the *rms* value of the induced current in the coil.

Solution: Consider a shell of radius r and $r + dr$

Area of the shell $= 2\pi\, rdr$

$$\text{Magnetic flux passing through the shell} = b_0\left(1 - \frac{r}{3R}\right)\cos \omega t \cdot 2\pi rdr$$

$$\text{Total flux through the coil } \phi = 2\pi b_0 \cos\omega t\left[\int_0^R rdr - \frac{1}{3R}\int_0^R r^2 dr\right]$$

$$= \frac{7 b_0 \pi R^2 \cos \omega t}{9}$$

$$\text{Induced emf due to } n \text{ turns} = e = -\frac{d\phi}{dt} = \frac{n \cdot 7 b_0 \pi R^2 \omega \sin \omega t}{9}$$

$$\text{Induced current} = \frac{n \cdot 7 b_0 \pi R^2 \omega \sin \omega t}{9 R_0}$$

$$I_{\text{rms}} = \frac{7 b_0 \pi R^2 \omega n}{9 R_0 \sqrt{2}}$$

14.50. Two infinite long straight parallel wires A and B are separated by 0.1 m distance and carry equal current in opposite directions. A square loop of wire C of side 0.1 m lies in the plane of A and B. The loop of wire C

is kept parallel to both A and B at a distance of 0.1 m from the nearest wire. Calculate the emf induced in loop C while the current in A and B is increasing at the same rate of 10^3 A/sec. Also indicate the direction of current in loop C. (Roorkee 1996)

Solution: Consider an element of width dx at a distance x as shown in Fig. 14.38.

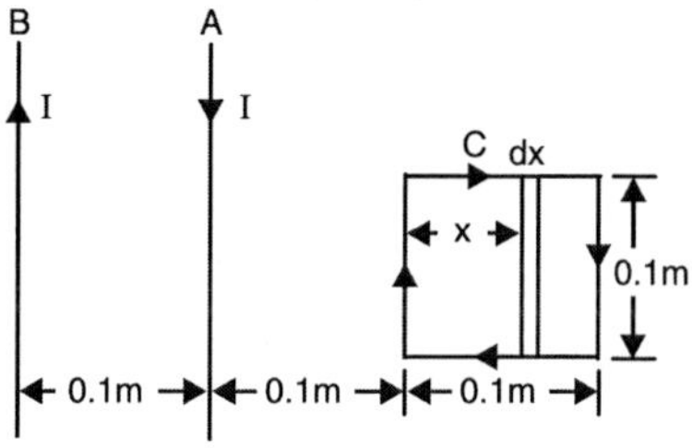

Fig. 14.38

Field at the element B

$$= \frac{\mu_0}{4\pi}\cdot 2I\left[\frac{1}{0.1+x}-\frac{1}{0.2+x}\right]$$

Flux through the element

$$= \frac{\mu_0 I}{2\pi}\left[\frac{1}{0.1+x}-\frac{1}{0.2+x}\right]\times 0.1\times dx$$

$$\text{Net flux} = 0.2\times 10^{-7} I\left[\int_0^{0.1}\frac{dx}{(0.1+x)}-\int_0^{0.1}\frac{dx}{0.2+x}\right]$$

$$= 0.2\times 10^{-7} I\left[\log\left(\frac{0.1+x}{0.2+x}\right)_0^{0.1}\right]$$

$$= 0.2\times 10^{-7} I\log_e \frac{4}{3}$$

$$e = 0.2\times 10^{-7}\times 2.303\log_{10}\frac{4}{3}\times\frac{dI}{dt}$$

$$= 0.2\times 10^{-7}\times 2.303\times 0.125\times 10^3$$

$$= 5.75\times 10^{-6}\text{V}$$

14.51. An infinitely small bar magnet of dipole moment $\vec{M}$ is pointing and moving with the speed v in the x-direction. A small closed circular conducting loop of radius a and negligible self inductance lies in the y-z plane with its centre at $x = 0$ and its axis coinciding with the x-axis. Find the force opposing the motion of the magnet, if the resistance of the loop is R. Assume that the distance x of the magnet from the centre of the loop is much greater than a. (IIT 1997)

Solution:

$$\text{Field due to magnet} = \frac{\mu_0}{4\pi}\cdot\frac{2M}{x^3}$$

$$\text{Flux linked with the loop} = \pi a^2\cdot\frac{\mu_0}{4\pi}\cdot\frac{2M}{x^3}$$

$$e = -\frac{d\phi}{dt} = \frac{\mu_0}{4\pi}\cdot\frac{6\pi\cdot Ma^2}{x^4}\cdot v$$

Induced current $i = \frac{e}{R} = \frac{\mu_0}{4\pi} \cdot \frac{6\pi Ma^2 v}{R \cdot x^4}$

M' = induced dipole moment = $i\pi a^2 = \frac{\mu_0}{4\pi} \cdot \frac{6\pi^2 Ma^4 v}{Rx^4}$

Force between the two dipoles of moments M and M' is given by

$$F = \frac{\mu_0}{4\pi} \cdot \frac{6 \cdot MM'}{x^4} = \frac{\mu_0}{4\pi} \cdot \frac{6 \cdot M}{x^4} \cdot \frac{\mu_0}{4\pi} \cdot \frac{6\pi^2 Ma^4 v}{Rx^4}$$

$$= \frac{9}{4} \frac{\mu_0^2 M^2 a^4 v}{Rx^8}$$

14.52. An airplane, with a 20 m wingspread, is flying at 250 m/sec straight south parallel to the earth's surface. The earth's magnetic field has a horizontal component of 2×10^5 w/m² and the dip angle is 60°. Calculate the induced emf between the plane tips. (Roorkee 1997)

Solution:

Magnetic field $B = H \tan \phi = 2 \times 10^{-5} \sqrt{3}$

Induced emf $e = Bvl = 2 \times 10^{-5} \cdot \sqrt{3} \times 250 \times 20 = 0.173$ volt

14.53. A thin non conducting ring of mass m carrying charge q is capable of rotating freely about its own axis. Initially the ring is at rest. A time varying magnetic field is applied perpendicular to the plane of the ring. Find the angular velocity of the ring as a function of magnetic induction when the field grows with time.

Solution:

Induced emf = $\pi a^2 \frac{d}{dt}[B\langle t\rangle]$; a = radius of the ring

The electric field E is given by the equation

$$2\pi aE = \pi a^2 \frac{d}{dt}[B\langle t\rangle]$$

$$E = \frac{a}{2} \cdot \frac{d}{dt}[B\langle t\rangle]$$

Force due to this field = $Eq = \frac{aq}{2} \cdot \frac{d}{dt}[B\langle t\rangle]$

$$\text{Torque} = a \cdot \frac{aq}{2} \cdot \frac{d}{dt}[B\langle t\rangle]$$

$$I\frac{d}{dt}\omega = ma^2 \frac{d}{dt}(\omega) = \frac{a^2 q}{2} \cdot \frac{d}{dt}[B\langle t\rangle]$$

$$\omega = \frac{q}{2m} B\langle t\rangle$$

14.54. Two current carrying identical circular wires of radius r are located as shown in the Fig. 14.39. The axis of the two wires is d apart. Find out the self-inductance per unit length of this pair of wires.

Solution: Consider an element of unit length and thickness dx in the space between the two wires at a distance x below the upper wire.

Net magnetic field at the element $B = \frac{\mu_0}{4\pi}\left[\frac{2I}{x} + \frac{2I}{d-x}\right]$

Magnetic flux $d\phi$ through the element is given by

$$d\phi = B \cdot dx$$

$$= \frac{\mu_0 I}{2\pi}\left[\frac{dx}{x} + \frac{dx}{d-x}\right]$$

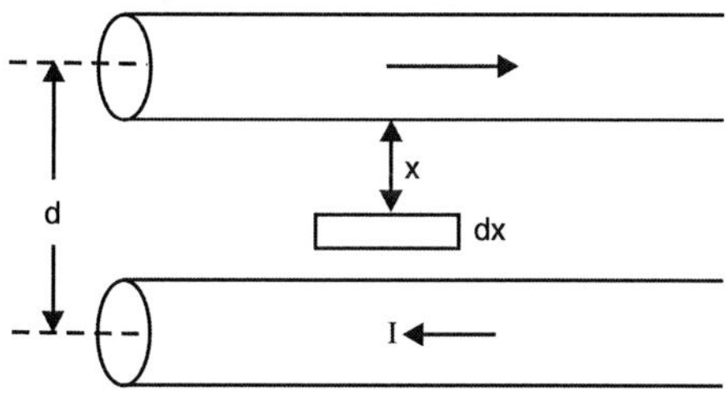

Fig. 14.39

$$\text{Net flux } \phi = \frac{\mu_0 I}{2\pi}\left[\int_r^{d-r} \frac{dx}{x} + \int_r^{d-r} \frac{dx}{d-x}\right]$$

$$= \frac{\mu_0 I}{\pi} \log_e \frac{d-r}{r}$$

$$\text{Self-inductance per unit length} = \frac{\mu_0}{\pi} \log_e \frac{d-r}{r}$$

14.55. A rectangular frame of wire with sides a & b and a long straight wire are located in the same plane by keeping them l apart (see Fig. 14.40). Find the electromotive force induced in the frame if the current in the wire varies as $I - I_0 e^{-\alpha t}$.

Solution: Consider an element of thickness dx at a distance x from the wire.

Magnetic flux through the element

$$d\phi = \frac{\mu_0}{2\pi} \cdot \frac{I_0 e^{-\alpha t}}{x} \cdot b\,dx$$

$$\text{Induced emf } de = -\frac{d\phi}{dt} = \frac{\mu_0}{2\pi} \cdot \alpha I \frac{b\,dx}{x}$$

$$\text{Net induced emf} = \frac{\mu_0 \alpha I b}{2\pi} \int_l^{a+l} \frac{dx}{x}$$

$$= \frac{\mu_0 \alpha I b}{2\pi} \log\left(1 + \frac{a}{l}\right)$$

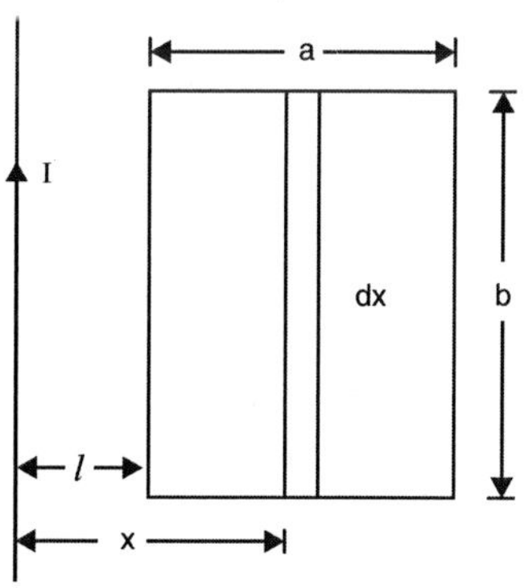

Fig. 14.40

14.56. Calculate the self-inductance per unit lenth of a cable consisting of two thin walled coaxial metallic cylinders of inner and outer radii as r_1 and r_2 respectively. The space between the two cylinders is filled with air.

Solution: Let I be the current through the inner cylinder. The magnetic field at a distance r from the centre of the two cylinders is given by

$$B = \frac{\mu_0}{4\pi} \cdot \frac{2I}{r}$$

Net flux $\phi = \int_{r_1}^{r_2} B \cdot dA = \int_{r_1}^{r_2} \frac{\mu_0}{2\pi} \cdot \frac{I}{r} \cdot l \cdot dr$;

l = length of the cylinder (see Fig. 14.41)

$$\phi = \frac{\mu_0}{2\pi} I l \log_e \frac{r_2}{r_1}$$

$$L = \frac{\mu_0}{2\pi} \cdot l \cdot \log_e \frac{r_2}{r_1}$$

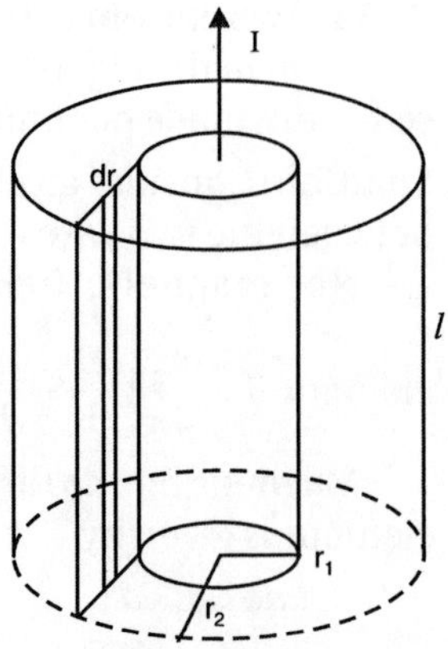

Fig. 14.41

Self-inductance per unit length $= \frac{\mu_0}{2\pi} \cdot \log_e \frac{r_2}{r_1}$

14.57. Find an expression for the self-inductance of a toroid of rectangular cross-section as shown in Fig. 14.42.

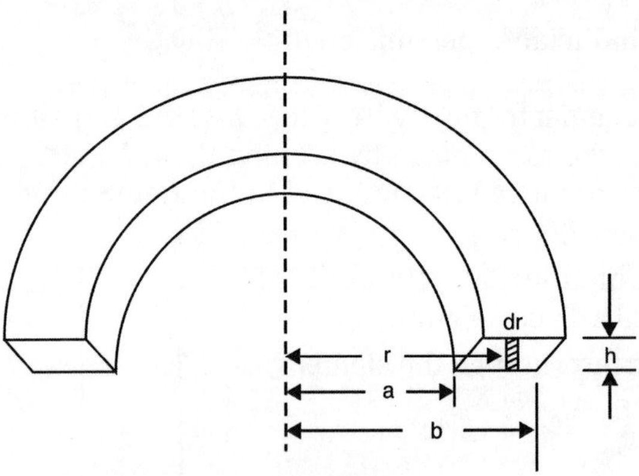

Fig. 14.42

Solution: Magnetic field at a distance r is given by

$$B = \frac{\mu_0 I N}{2\pi r}$$

Where N = number of turns

Consider an element of length dr. Area of the element $= h \cdot dr$

Flux through the element $= \frac{\mu_0 I N}{2\pi r} \cdot h \cdot dr$

$$\text{Total flux } \phi = \frac{\mu_0 I N^2 \cdot h}{2\pi} \int_a^b \frac{dr}{r}$$

$$= \frac{\mu_0 I N^2 h}{2\pi} \cdot \log_e \frac{b}{a}$$

$$\text{Self-inductance} = \frac{\mu_0 N^2 h}{2\pi} \cdot \log_e \frac{b}{a}$$

14.58. A small coil is placed between the poles of an electromagnet. The axis of the coil coincides with the pole shoes of the magnet. The cross-sectional area of the coil is $S = 3$ mm², its number of turns is $N = 60$. When the coil is turned through 180° a charge of $q = 4.5$ μC flows through the ballistic galvanometer which is connected to it. Assuming the total resistance of the connecting wires, coil and galvanometer equal to $R = 40\ \Omega$. Determine the strength of the magnetic field.

Solution:

$$qR = 2\,SN \cdot B$$

$$\therefore \quad B = \frac{qR}{2SN}$$

On substituting the values we have

$$B = \frac{4.5 \times 10^{-6} \times 40}{2 \times 3 \times 10^{-6} \times 60}$$

$$= 0.5 \text{ w/m}^2$$

14.59. Find out the inductance of two inductors connected (i) in series (ii) in parallel combination.

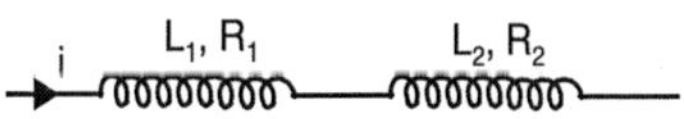

Fig. 14.43a

Solution: (i) Series combination

$$\text{Voltage drop} = R_1 i + L_1 \frac{di}{dt} + M\frac{di}{dt} + L_2 \frac{di}{dt} + M\frac{di}{dt} + R_2 i$$

$$= (R_1 + R_2) i + (L_1 + L_2 + 2M)\frac{di}{dt}$$

Where M is the mutual inductance $= K\sqrt{L_1 L_2}$

The circuit thus contains a resistor $(R_1 + R_2)$ in series with an inductance $L = L_1 + L_2 + 2M = L_1 + L_2 + 2K\sqrt{L_1 L_2}$

When the fluxes in L_1 and L_2 are in the same direction we come across positive coupling and then $L = L_1 + L_2 + 2K\sqrt{L_1 L_2}$. However for negative coupling $L = L_1 + L_2 - 2K\sqrt{L_1 L_2}$.

(ii) In parallel combination

$$i = i_1 + i_2$$

$$V = L_1\frac{di_1}{dt} \pm M\frac{di_2}{dt} + R_1 i_1 = L_2\frac{di_2}{dt} \pm M\frac{di_1}{dt} + R_2 i_2$$

On neglecting the resistance

$$V = L_1\frac{di_1}{dt} \pm M\frac{di_2}{dt}$$

$$V = L_2\frac{di_2}{dt} \pm M\frac{di_1}{dt}$$

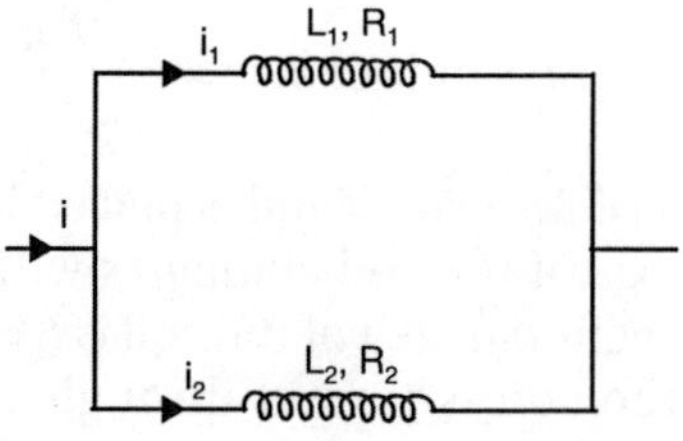

Fig. 14.43b

On solving $V = \frac{L_1 L_2 - M^2}{L_1 + L_2 \pm 2M}\cdot\frac{di}{dt}$

or $$L = \frac{L_1 L_2 - M^2}{L_1 + L_2 \pm 2M}$$

14.60. When the current I in the portion of the circuit shown in Fig. 14.44 is 2 A and increasing at a rate of 1 A/s the potential difference V_{ab} = 8 V. However, when the current I = 2 A and is decreasing at a rate of 1 A/s the potential difference V_{ab} = 4 V. Calculate the value of L and R.

Solution:

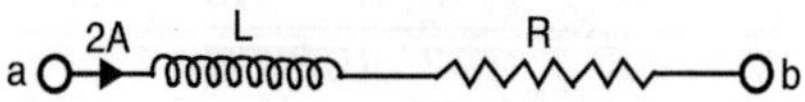

Fig. 14.44

$$8 = RI + L\frac{dI}{dt}$$

$$4 = RI - L\frac{dI}{dt}$$

$$8 = 2R + L$$

$$4 = 2R - L$$

On adding and subtracting the two $R = 3\ \Omega$ and $L = 2\ H$.

14.61. The net work shown in the Fig. 14.45 is a part of a complete circuit. What is the potential difference $V_B - V_A$ when the current I is 5 amp and is decreasing at a rate of 10^3 A/s. (IIT 1997)

Solution:

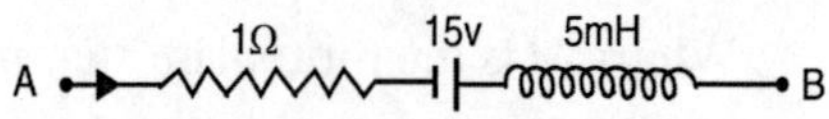

Fig. 14.45

$$V_{BA} = V_B - V_A = -IR + E - L\frac{dI}{dt}$$

$$= -5 + 15 - 5\times 10^{-3} \times -10^{+3}$$

$$= 15\text{ V}$$

14.62. An electric lamp whose resistance in a hot state is 10 Ω is connected via a choke to a 12 V storage battery. The inductance of the choke is 2 H and its resistance is 1 Ω. In how much time after being switched on will the lamp light if it begins to glow noticeably at a voltage of 6 V.

Solution:

$$V = V_0\left[1-e^{-Rt/L}\right]$$

$$6 = 12\left[1-e^{-Rt/L}\right]$$

$$\log_e 2 = \frac{R}{L}\cdot t$$

or $$t = \frac{L\cdot \log_e 2}{R}$$

On substituting the values we have

$$t = \frac{2\times 2.303\times 0.3010}{11}$$
$$= 0.126 \text{ sec.}$$

14.63. A uniformly wound solenoidal coil of self-inductance 1.8×10^{-4} henry and resistance 6 Ω is broken up into two identical coils. These coils are then connected in parallel across a 12 volt battery of negligible resistance. The time constant for the current in the circuit is .... second and the steady current through the battery is .... amp. (IIT 1989)

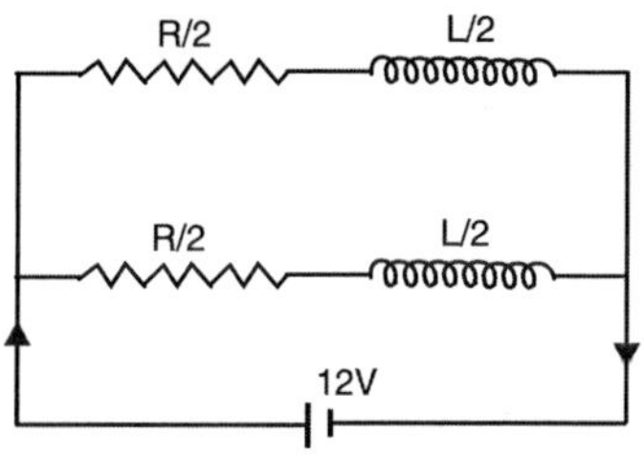

Fig. 14.46

Solution:

The equivalent inductance $L_{eq} = \frac{L}{4}$ and equivalent resistance $R_{eq} = \frac{R}{4}$

Time constant $\tau = \frac{L_{eq}}{R_{eq}} = \frac{L}{R} = \frac{1.8\times 10^{-4}}{6} = 30\,\mu\text{s}$

The steady current $I = \frac{E}{R_{eq}} = \frac{4E}{R} = \frac{4\times 12}{6} = 8$ amp

14.64. The current in a coil of self-inductance 2.0 henry is increasing according to $I = 2\sin t^2$ A. Find the amount of energy spent during the period when the current changes from 0 - 2 A. (Roorkee 1991)

Solution:

Work in time $dt = dw = eIdt = L\frac{dI}{dt}\cdot I\cdot dt$

$$dw = LIdI$$

$$W = L\int_0^2 IdI = \frac{LI^2}{2} = \frac{2\times 4}{2} = 4\text{ J}$$

14.65. Two different coil have self-inductances $L_1 = 8$ mH and $L_2 = 2$ mH. At a certain instant the current in the two coils is increasing at the same constant rate and the power supplied to the two coils is the same. Find the ratio of (a) induced voltage (b) current and (c) energy stored in the two coils at that instant. (IIT 1994)

Solution:

(a) $$\frac{e_1}{e_2} = \frac{L_1 \dfrac{dI}{dt}}{L_2 \dfrac{dI}{dt}} = \frac{L_1}{L_2} = \frac{8\,\text{mH}}{2\,\text{mH}} = 4$$

(b) $P = eI =$ constant hence

$$\frac{I_1}{I_2} = \frac{e_2}{e_1} = \frac{1}{4}$$

(c) $$\frac{E_1}{E_2} = \frac{\frac{1}{2}L_1 I_1^2}{\frac{1}{2}L_2 I_2^2} = \frac{L_1}{L_2}\left(\frac{I_1}{I_2}\right)^2$$

$$= 4 \times \frac{1}{16} = \frac{1}{4}$$

14.66. A solenoid of resistance 50 Ω and inductance 80 H is connected to a 200V battery. How long will it take for the current to reach 50% of its final equilibrium value? Calculate the maximum energy stored. (Roorkee 1995)

Solution:

$$I = I_0\left(1 - e^{-t/\tau}\right); \ \tau = \frac{L}{R}$$

$$\frac{I_0}{2} = I_0\left(1 - e^{-t/\tau}\right)$$

$$t = \tau \log_e 2 = \frac{L}{R} \times 0.693 = \frac{80}{50} \times 0.693 = 1.073 \text{ sec}$$

$$E_{max} = \frac{1}{2} L I_{max}^2 = \frac{1}{2} L \cdot \frac{E^2}{R^2}$$

$$= \frac{1}{2} \times 80 \times \frac{200 \times 200}{50 \times 50}$$

$$= 640 \text{ J}$$

14.67. Two identical resistor of resistance R and an inductance L are connected to a cell of emf E through a switch S as shown in the Fig. 14.47. After shorting the switch find the current through the inductance as a function of time. Neglect the internal resistance of the cell.

Solution: We have

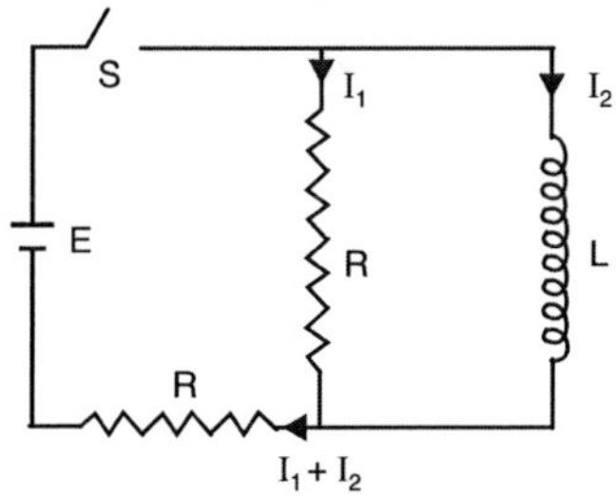

Fig. 14.47

$$E - L\frac{dI_2}{dt} - R(I_1 + I_2) = 0 \quad \text{... (1)}$$

$$E - RI_1 - R(I_1 + I_2) = 0 \quad \text{... (2)}$$

$$E = (2I_1 + I_2)R$$

$$\therefore \quad I_1 = \frac{E - I_2R}{2R}$$

$$I = I_1 + I_2 = \frac{E - RI_2}{2R} + I_2 = \frac{E + RI_2}{2R}$$

On substituting the value of $I_1 + I_2$ in equation 1 we have

$$E - L\frac{dI_2}{dt} - \frac{E}{2} - \frac{RI_2}{2} = 0$$

$$\frac{E}{2} - L\frac{dI_2}{dt} - \frac{RI_2}{2} = 0$$

$$2L\frac{dI_2}{dt} = E - RI_2$$

$$\int \frac{2L\,dI_2}{E - RI_2} = \int dt$$

$$2L\log(E - RI_2) = -Rt + C$$

at $t = 0, I_2 = 0 \therefore C = 2L\log E$

$$2L\log\left(\frac{E - RI_2}{E}\right) = -Rt$$

$$\frac{E \quad RI_2}{E_2} = e^{-Rt/2L}$$

$$I_2 = \frac{E}{R}\left[1 - e^{-\frac{Rt}{2L}}\right]$$

14.68. A coil of inductance *L* and resistance *R* is connected in parallel to a resistance *r*. This combination is connected to a cell of emf *E* through the switch *S*. Neglecting the internal resistance of the cell find the heat produced in the coil after opening the switch.

Solution:

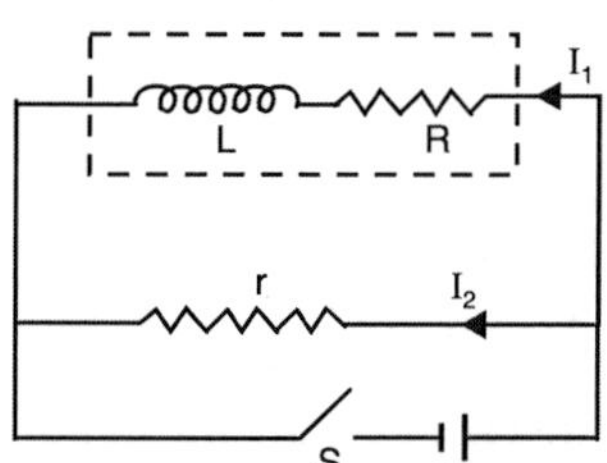

Fig. 14.48

$RI_1 = E$ (see Fig. 14.48)

$$I_1 = \frac{E}{R}$$

At the break of the circuit we have

$$L\frac{dI_1}{dt} + (R + r)I_1 = 0$$

or $$\int \frac{dI_1}{I_1} = -\frac{R+r}{L}\int dt$$

$$\log I_1 = -\left(\frac{R+r}{L}\right)t + C$$

$$\text{at } t=0;\ I_1 = \frac{E}{R}$$

$$\therefore \quad \log\frac{RI_1}{E} = -\left(\frac{R+r}{L}\right)t$$

or $$I_1 = \frac{E}{R}e^{-\frac{R+r}{L}t}$$

Amount of heat produced in the coil is given by

$$Q = \int_0^\infty I_1^2 R\,dt = \frac{E^2}{R^2}\cdot R\int_0^\infty e^{-\frac{2(R+r)t}{L}}\cdot dt$$

$$= \frac{E^2 L}{2R^2\left[1+\frac{r}{R}\right]}$$

14.69. Two coils of inductances L_1 and L_2 respectively are connected to cell of emf E through a switch and a resistance R as shown in the Fig. 14.49. The internal resistance of the cell and the resistance of the two coils is negligible. On closing the switch S find the steady state current in the two coils respectively.

Solution:

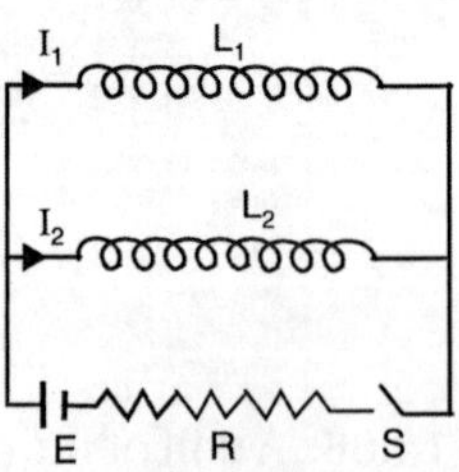

Fig. 14.49

$$I = \frac{E}{R} = I_1 + I_2$$

$$L_1\frac{dI_1}{dt} = L_2\frac{dI_2}{dt}$$

or $$L_1 I_1 = L_2 I_2$$

or $$\frac{E}{R} = \left(\frac{L_2}{L_1}+1\right)I_2$$

or $$I_2 = \frac{EL_1}{R(L_1+L_2)} \text{ and } I_1 = \frac{EL_2}{R(L_1+L_2)}$$

14.70. A coil of inductance L and a resistance R is connected in series to a source of emf E. The internal resistance of the source is negligible. At an instant $t = 0$ the inductance of the coil is suddenly reduced to $\frac{L}{\eta}$. Find the current in the coil as a function of time.

Solution: The current in a LR circuit is given by

$$I = \frac{E}{R}\left[1 - e^{-Rt/L}\right]$$

$$\frac{dI}{dt} = \frac{E}{R}\cdot\frac{R}{L}e^{-Rt/L} = \frac{E}{L}e^{-Rt/L}$$

The value of $\frac{dI}{dt}$ when the value of inductance changes to $\frac{L}{\eta}$

$$\frac{dI}{dt} = \frac{\eta E}{L}\cdot e^{-\frac{\eta R}{L}\cdot t}$$

$$e = L\frac{dI}{dt} = \eta E e^{-\frac{\eta R}{L}t}$$

at $t = 0$; $e = \eta E$ and therefore current $i = \frac{\eta E}{R}$

On abruptly changing the value of the inductance we have

$$\frac{L}{\eta}\cdot\frac{dI'}{dt} + RI' = E$$

or $\int\frac{dI'}{E - RI'} = \frac{\eta}{L}\int dt$

or $\log[E - RI'] = -\frac{\eta R}{L}\cdot t + C$

at $t = 0$; $I = I' = \frac{\eta E}{R}$

$\therefore$ $C = \log(E - \eta E)$

or $\log\left[\frac{E - RI'}{E(1-\eta)}\right] = -\frac{\eta R}{L}\cdot t$

$$E - RI' = E(1-\eta)e^{-\frac{\eta Rt}{L}}$$

$$RI' = E + E(\eta - 1)e^{-\frac{\eta Rt}{L}}$$

or $I' = \frac{E}{R}\left[1 + (\eta - 1)e^{-\frac{\eta Rt}{L}}\right]$

14.71. A long solenoid of length l has been constructed by using a thin wire. The total mass of the wire wound is m and resistance is R. Find the inductance of the solenoid.

Solution:

Magnetic field inside a solenoid $= \frac{\mu_0}{4\pi}\cdot\frac{4\pi ni}{l}$; n = number of turns

Magnetic flux $\phi = \frac{\mu_0}{4\pi} \cdot \frac{4\pi ni}{l} \cdot n \cdot \pi r^2$; r = radius of the solenoid

$2\pi rn = l_1$ = length of the wire wraped

Self-inductance $L = \frac{\mu_0}{4\pi} \cdot \frac{4\pi^2 n^2 r^2}{l}$

$= \frac{\mu_0}{4\pi} \cdot \frac{l_1^2}{l}$

$Al_1\rho_0 = m$; A = area of cross-section of the wire

ρ_0 = density

ρ = resistivity of the wire

$$R = \frac{\rho l_1 \cdot l_1 \rho_0}{m}$$

or $$l_1^2 = \frac{mR}{\rho\rho_0}$$

$$\therefore \quad L = \frac{\mu_0}{4\pi l} \frac{mR}{\rho\rho_0}$$

14.72. A rectangular frame of wire of sides a & b and a long straight conductor carrying current I are located in the same plane. Both are separated by a distance l. Find their mutual inductance.

Solution: Consider an element of thickness dx at a distance x from the wire (see Fig. 14.50).

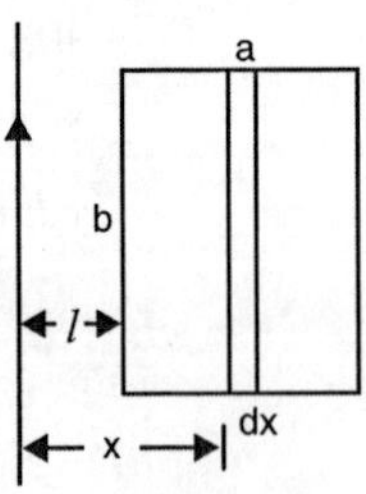

Fig. 14.50

Magnetic field at the element $= \frac{\mu_0}{2\pi} \cdot \frac{I}{x}$

Magnetic flux $d\phi$ through the element $= \frac{\mu_0}{2\pi} \cdot \frac{I}{x} \cdot bdx$

Net flux $\phi = \frac{\mu_0 Ib}{2\pi} \int_l^{a+l} \frac{dx}{x} = \frac{\mu_0 Ib}{2\pi} \log_e \left(\frac{a+l}{l}\right)$

$\therefore$ Mutual inductance $M = \frac{\mu_0 b}{2\pi} \log_e \left(1 + \frac{a}{l}\right)$

14.73. Two circular coils of same radius r and having respectively n_1 and n_2 number of turns are placed coaxially keeping their centres l apart. Find their mutual inductance when $l >> r$.

Solution: Let a current I be set up in the first coil. The magnetic field at the centre of the second coil is given by

$$B = \frac{\mu_0}{4\pi} \cdot \frac{2\pi I n_1 r^2}{(l^2 + r^2)^{3/2}}$$

Magnetic flux through the second coil

$$\phi = \frac{\mu_0}{4\pi} \cdot \frac{2\pi I n_1 r^2}{\left(l^2 + r^2\right)^{3/2}} \cdot n_2 \pi r^2$$

$\therefore$ Mutual inductance $M = \dfrac{\mu_0}{2} \cdot \dfrac{\pi n_1 n_2 r^4}{\left(l^2 + r^2\right)^{3/2}}$

Under the approximation $l >> r$

$$M = \frac{\mu_0}{2} \frac{\pi n_1 n_2 r^4}{l^3}$$

14.74. A very long straight conductor and a triangular conductor lies in a plane and separated from each other as shown in Fig. 14.51. Find their coefficient of mutual inductance.

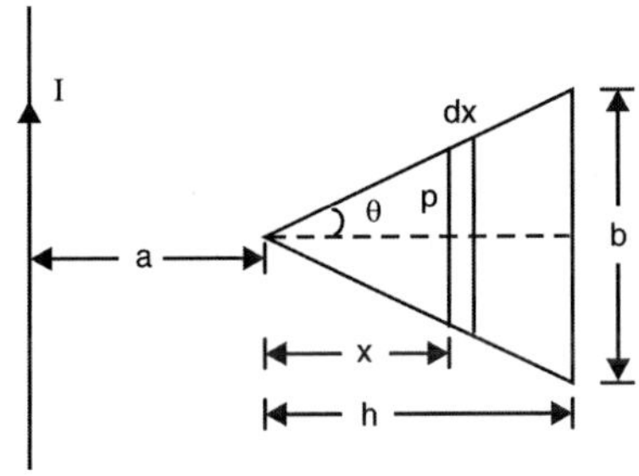

Fig. 14.51

Solution:

$$\tan\theta = \frac{b}{2h}$$

$$p = x \tan\theta$$

Magnetic field at the element of length dx due to current in the straight conductor $= \dfrac{\mu_0}{4\pi} \cdot \dfrac{2I}{a+x}$

$$\text{Flux through the element} = \frac{\mu_0 I \cdot 2p\,dx}{2\pi(a+x)}$$

$$= \frac{\mu_0 I}{\pi} \tan 0 \frac{x\,dx}{(a+x)}$$

$$\text{Net flux through the triangle} = \frac{\mu_0 I \tan\theta}{\pi} \int_0^h \left[1 - \frac{a}{a+x}\right] dx$$

$$= \frac{\mu_0 b I}{2\pi h}\left[h - a\log\left(\frac{a+h}{a}\right)\right]$$

$$M = \text{mutual inductance} = \frac{\mu_0 b}{2\pi h}\left[h - a\log\left(\frac{a+h}{a}\right)\right]$$

14.75. Two long solenoid of equal length l and nearly of equal radius r are fully inserted into the other. The number of turns of wire wound over the two solenoid are n_1 and n_2 respectively. Find their mutual inductance if their self-inductances are L_1 and L_2 respectively.

Solution: The self-inductance of the two solenoids is given by

$$L_1 = \frac{\mu_0 \pi n_1^2 r^2}{l} \text{ and } L_2 = \frac{\mu_0 \pi n_2^2 r^2}{l}$$

On establishing a current I in the inner solenoid having n_1 number of turns of wire the magnetic field is given by

$$B = \frac{\mu_0 n_1 I}{l}$$

Magnetic flux through the other solenoid is equal to

$$\frac{\mu_0 n_1 I}{l} \cdot \pi n_2 r^2$$

$$\therefore \quad \text{Mutual inductance } M = \frac{\mu_0 \pi n_1 n_2 r^2}{l}$$

$$= \sqrt{\frac{\mu_0 \pi n_1^2 r^2}{l}} \cdot \sqrt{\frac{\mu_0 \pi n_2^2 r^2}{l}}$$

$$= \sqrt{L_1 L_2}$$

14.76. With an aim of determining the magnetic moment of a permanent magnet a simple experimental set up has been used in which a small cylindrical magnet of magnetic moment M is placed at the centre of a circular coil of radius r and having n turn of wires. The length of the magnet is perpendicular to the plane of the coil. Further the coil is connected to a ballistic galvanometer. The total resistance of the coil, connecting wires and the galvanometer is R. On sudden removal of the magnet from its position a charge q flows through the galvanometer. Find the magnetic moment of the magnet.

Solution: The magnetic field due to the magnet at a distance x in the broad side on position is given by

$$B = -\frac{\mu_0}{4\pi} \cdot \frac{M}{x^3}; \text{ here } M \text{ is opposite to the field (see Fig. 14.52)}$$

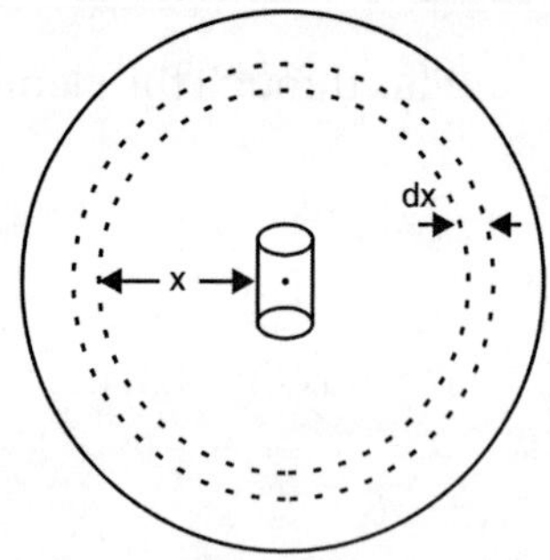

Fig. 14.52

Consider a shell of radius x and $x + dx$

Area of the shell = $2\,\pi\, x dx$

$$\text{Magnetic flux through the shell } \Delta\phi = -\frac{\mu_0}{4\pi} \cdot \frac{M}{x^3} \cdot 2\pi x dx$$

$$\text{Net flux through the coil} = -\frac{\mu_0}{2} M \cdot n \int_0^r \frac{dx}{x^2} = \frac{\mu_0}{2} \frac{Mn}{r}$$

On removing the magnet the final flux is zero

$$\therefore \quad \text{Change in flux } \phi = \frac{\mu_0}{2} \cdot \frac{Mn}{r}$$

$$q = \frac{\phi}{R} = \frac{\mu_0}{2} \cdot \frac{Mn}{rR}$$

or $M = \dfrac{2qrR}{\mu_0 n}$

14.77. Find the interaction energy of two loops carrying currents I_1 and I_2 if both loops are shaped as circles of radii a & b with $a << b$. The loop's centre are located at the same point and their planes form an angle θ between them.

Solution:

Field due to coil of radius $b = \dfrac{\mu_0}{4\pi} \cdot \dfrac{2\pi I_2}{b}$

Magnetic field that will link with the area of the coil of radius a

$$= \frac{\mu_0}{2} \cdot \frac{I_2}{b} \cdot \cos\theta$$

Flux in the second coil $= \dfrac{\mu_0}{2} \cdot \dfrac{I_2}{b} \cdot \cos\theta \cdot \pi a^2$

$\therefore$ Mutual inductance $M = \dfrac{\pi\mu_0 a^2 \cos\theta}{2b}$

Interaction energy $= MI_1 I_2 = \dfrac{\mu_0 \pi a^2 \cos\theta\, I_1 I_2}{2b}$

14.78. Determine the mutual inductance of a toroid and of an infinitely long straight wire located along its axis. The toroid has a rectangular cross-section of width a. The inner radius of the toroid is r_1 and the outer one is r_2. The number of turns of the toroid is N.

Solution:

Magnetic field at a distance r due to current I in the long straight conductor $r_1 < r < r_2$.

$$\frac{\mu_0}{4\pi} \cdot \frac{2I}{r}$$

Consider an element within the rectangular cross-section of the toroid of length dr. Area of the element $= adr$.

Flux through the element $= \dfrac{\mu_0}{4\pi} \cdot \dfrac{2I}{r} \cdot a\,dr$

Net flux through the toroid $\phi = \dfrac{\mu_0 INa}{2\pi} \displaystyle\int_{r_1}^{r_2} \dfrac{dr}{r}$

$$= \frac{\mu_0 NaI}{2\pi} \log_e \frac{r_2}{r_1}$$

Mutual inductance $M = \dfrac{\mu_0 Na}{2\pi} \log_e \dfrac{r_2}{r_1}$

14.79. A small square loop of wire of side l is placed inside a large square loop of wire of side L (>> l). The loops are coplanar and their centres coincide. What is the mutual inductance of the system?

Solution: Let I be the current in the square loop of side L.

Magnetic field at the centre of the loop is given by

$$B = \frac{\mu_0}{4\pi} \cdot \frac{8\sqrt{2}I}{L}$$

Flux linked with te smaller loop of side l.

$$\phi = \frac{\mu_0}{4\pi} \cdot \frac{8\sqrt{2}}{L} . I . l^2$$

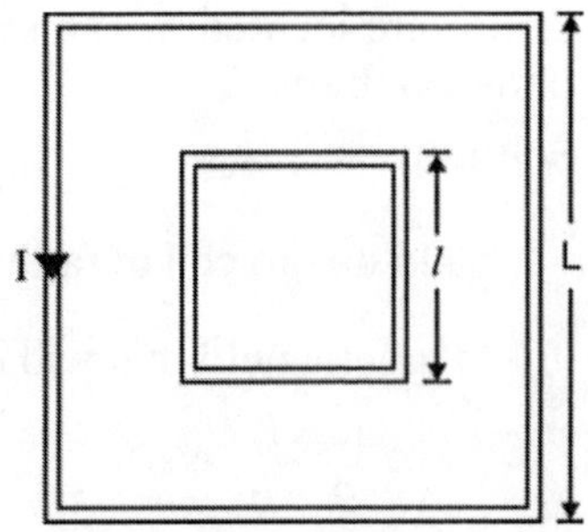

Fig. 14.53

$\therefore$ Mutual inductacne $M = \dfrac{\mu_0 2\sqrt{2}l^2}{\pi L}$

14.80. A coil wound on a transformer of rectangular cross-section has self-inductance L. What will be the value of the self-inductance if all the linear dimensions of the transformer are increased by a factor 2 keeping the number of turns per unit length same?

Solution: Let the rectangular cross-section of the coil of dimension $l \times d$ contains N turns. A current I through the coil will produce a magnetic field at the centre as

$$B = \frac{\mu_0}{4\pi} \cdot 8I \frac{\left(d^2 + l^2\right)^{1/2} N}{d.l}$$

$$\text{Flux } \phi_0 = \frac{\mu_0}{\pi} \cdot 2I \frac{\left(d^2 + l^2\right)^{1/2}}{d.l} \cdot d \cdot l \cdot N^2$$

$$\therefore \quad L = \frac{2\mu_0}{\pi} \cdot \left(d^2 + l^2\right)^{1/2} \cdot N^2$$

Flux ϕ when the linear dimensions are doubled

$$\phi = \frac{\mu_0}{\pi} \cdot 2I \cdot \frac{\left(4d^2 + 4l^2\right)^{1/2}}{2d \cdot 2l} \cdot 2d \cdot 2l \cdot (2N)^2$$

$$= \frac{2\mu_0 I}{\pi} \cdot \left(d^2 + l^2\right)^{1/2} \cdot 2 \cdot 4N^2$$

$$= 8L$$

14.81. Two circular loop with same radii lies in the same plane. Their mutual inductance is M. The current in the first loop varies linearly with time by relation $I_1 = at$. Find the current in the second loop which has a finite inductance L and resistance R as a function of time.

Solution: Flux in the second loop $\phi = MI_1$

Induced emf in the second loop $= \frac{d\phi}{dt} = \frac{M dI_1}{dt} = aM$

Now $aM = \frac{L dI_2}{dt} + RI_2$

$$aM - RI_2 = \frac{L dI_2}{dt}$$

$$\int \frac{L dI_2}{aM - RI_2} = \int dt$$

$$-\frac{L}{R}\log(aM - RI_2) = t + C$$

at $t = 0, I_2 = 0 \therefore C = -\frac{L}{2}\log aM$

or $-\frac{L}{R}\log\left[\frac{aM - RI_2}{aM}\right] = t$

$$1 - \frac{RI_2}{aM} = e^{-\frac{Rt}{L}}$$

$$I_2 = \frac{aM}{R}\left(1 - e^{-\frac{Rt}{L}}\right)$$

14.82. When an electric motor is run at 120 volt, 10 amperes of current flows through it. The back induced emf is 115 volt. Calculate the current flowing in the coil at the time when the motor is switched off.

Solution:

$$I = \frac{E - e}{R}$$

$$\therefore \quad R = \frac{E - e}{I} = \frac{120 - 115}{10} = 0.5\Omega$$

Now $e = I'R; I'$ is the required current

$$I' = \frac{e}{R} = \frac{115}{0.5} = 230A$$

14.83. A dynamo dissipates 20 watt when it supplies a current of 4 amp through it. The terminal potential difference is 220 volt. Calculate the emf produced by it.

Solution:

$$V \times i = P$$

$$\therefore \quad V = \frac{P}{i} = \frac{20}{4} = 5V$$

This is the potential difference developed across the coil of the dynamo.

The emf produced by it = 220 + 5 = 225 volt

14.84. A capacitor of capacity C has q_0 charges in the fully charged conditions. The capacitor is now connected to a coil of inductance L. Calculate the ratio of energy of the magnetic field and the energy of the electric field at the moment of time $T/8$ sec.

Solution:

The total energy $= \frac{1}{2}\frac{q_0^2}{C}$

Fot the LC circuit we have the equation

$$L\frac{dI}{dt} + \frac{q}{C} = 0$$

or $$L\frac{d^2q}{dt^2} = -\frac{q}{C}$$

The solution of the above equation is

$$q = q_0 \cos\omega t$$

At an instant $t = \frac{T}{8}$

$$q = q_0.\cos\frac{2\pi}{T}.\frac{T}{8} = q_0 \cos\frac{\pi}{4}$$

$$= \frac{q_0}{\sqrt{2}}$$

At this instant the electrostatic energy $= \frac{1}{2}.\frac{q_0^2}{2C} = 50\%$ of the total energy

$\therefore$ Magnetic energy = 50% of the total energy

$\therefore$ Ratio is equal to 1

14.85. An oscillatory circuit contains L, C & R in series. Find the ratio of energies of magnetic and electric field when the current assumes a maximum value.

Solution: For an oscillatory circuit

$$L\frac{di}{dt} + Ri + \frac{q}{C} = 0$$

$$\frac{di}{dt} = d \therefore i = -\frac{q}{CR}$$

$$\frac{\text{Magnetic energy}}{\text{Electrostatic energy}} = \frac{\frac{1}{2}Li^2}{\frac{1}{2}\frac{q^2}{C}} = \frac{L\cdot\frac{q^2}{C^2R^2}}{\frac{q^2}{C}} = \frac{L}{CR^2}$$

14.86. An oscillatory circuit consists of an inductance of 10^{-2}H, a capacitance of 0.405 μF and a resistance of 2 Ω. How many times will the difference of potentials across the capacitor plates diminish during one period?

Solution: The difference of potentials across the capacitor plates changes with time according to the law

$$V = V_0 e^{-\frac{R}{2L}.t} \cos\omega t$$

At $t = 0, V_1 = V_0$

At $t = 2\pi\sqrt{LC}$

$$V_2 = V_0 e^{-\frac{R}{2L}2\pi\sqrt{LC}} = V_0 e^{-\pi R\sqrt{\frac{C}{L}}}$$

$$= V_0 e^{-3.14\times 2\sqrt{\frac{0.405\times 10^{-6}}{10^{-2}}}}$$

$$= 0.9606\, V_0$$

$$\frac{V_1}{V_2} = \frac{1}{0.9606} = 1.04$$

14.87. The winding of a coil consist of 500 turns of copper wire with a cross-sectional area of 1 mm². The coil is 50 cm long and 5 cm in diameter. At what frequency of alternating current will the impedance of the coil be twice of its resistance. Resistivity of copper = 1.7×10^{-8} Ω.m.

Solution: The inductance of the solenoid

$$L = \frac{\mu_0 n}{\sqrt{l^2 + D^2}}$$

where n = no. of turns

D = diameter of solenoid

l = length of solenoid

$$\left(\frac{\omega\mu_0 n}{\sqrt{l^2 + D^2}}\right)^2 + R^2 = 4R^2$$

$$\frac{\omega\mu_0 n}{\sqrt{l^2 + D^2}} = \sqrt{3}\cdot R$$

$$\omega = 2\pi\upsilon = \frac{\sqrt{3}R.\sqrt{l^2 + D^2}}{\mu_0 . n}$$

$$v = \frac{\sqrt{3}\cdot\sqrt{l^2+D^2}}{2\pi\mu_0\cdot n}\times\frac{\rho\cdot\pi D\times n}{A}$$

$$= \frac{\sqrt{3}\cdot\sqrt{l^2+D^2}}{2\mu_0}\frac{\rho D}{A}$$

$$= \frac{\sqrt{3}\sqrt{0.25+0.0025}}{2\times 4\pi\times 10^{-7}}\times\frac{1.7\times 10^{-8}\times 5\times 10^{-2}}{10^{-6}}$$

$$= \frac{1.732\times 0.50249\times 1.7\times 5}{8\pi}\times 10^{3}$$

$$= 294.5\text{ Hz}$$

14.88. A capacitor of capacity 2 μF is charged to a potential difference of 12 V. It is then connected across an inductor of inductance 0.6 mH. What is the current in the circuit at a time when the potential difference across the capacitor is 6.0 V?

Solution:

$$q = q_0\cos\omega t; \omega = \frac{1}{\sqrt{LC}}$$

$$\cos\omega t = \frac{q}{q_0} = \frac{V}{V_0} = \frac{6}{12} = \frac{1}{12} \therefore \omega t = \frac{\pi}{3}\text{rad}$$

$$\omega = \frac{1}{\sqrt{LC}} = \frac{1}{\sqrt{2\times 10^{-6}\times 6\times 10^{-4}}} = \frac{10^5}{\sqrt{12}}$$

$$I = \frac{dq}{dt} = -q_0\,\omega\sin\omega t$$

$$|I| = CV_0\omega\sin\omega t$$

$$\text{I} = 2\times 10^{-6}\times 12\times\frac{10^5}{\sqrt{12}}\times\frac{\sqrt{3}}{2} = 0.6A$$

14.89. An inductor of inductance 2.0 mH is connected across a charged capacitor of capacitance 5.0 μF and the resulting $L - C$ circuit is set oscillating at its natural frequency. Let q denote the instantaneous charge on the capacitor, and I the current in the circuit. It is found that the maximum value of q is 200 μC. (a) When q = 100 μC what is the value of $\left(\frac{dI}{dt}\right)$? (b) When q = 200 μC, what is the value of I? Find the maximum value of I? (d) When I is equal to one half its maximum value what is the value of $|q|$? (IIT 1998)

Solution:

(a) $q = q_0 \cos\omega t; \omega = \dfrac{1}{\sqrt{LC}}$

$$I = \frac{dq}{dt} = -q_0\,\omega\,\sin\omega t$$

$$\frac{dI}{dt} = -q_0\omega^2 \cos\omega t$$

$$\therefore \frac{dI}{dt} = \frac{q}{LC} = \frac{100 \times 10^{-6}}{5 \times 10^{-6} \times 2 \times 10^{-3}} = 10^4\, A/\sec$$

(b) When $q = 200\ \mu C = q_0$

$200 \times 10^{-6} = 200 \times 10^{-6} \cos \omega t; = 0$

$I = -200 \times 10^{-6} \times \omega \times \sin 0 = 0$

(c) $I_{max} = q_0\omega$

$$q_0 = 200\mu C; \omega = \frac{1}{\sqrt{LC}} = \frac{1}{\sqrt{2 \times 10^{-3} \times 5 \times 10^{-6}}} = 10^4 \text{ rad/sec}$$

$$I_{max} = 200 \times 10^{-6} \times 10^4 = 2A$$

(d) $I = \dfrac{I_{max}}{2} = \dfrac{q_0\omega}{2}$

$$\frac{q_0\omega}{2} = -q_0\omega \sin\omega t;\ \omega t = \pi + \frac{\pi}{6}$$

$$\therefore q = q_0 \cos \omega t = 200 \times 10^{-6} \cos(\pi + \pi/6)$$

$$|q| = 200 \times 10^{-6} \times \frac{\sqrt{3}}{2} = 100\sqrt{3}\mu C = 17\mu C$$

14.90. An oscillatory circuit consist of a capacitor and a long coil wound of copper wire of resistivity $\rho = 1.7 \times 10^{-8}\,\Omega$.m with a cross-sectional area $A = 0.1$ mm^2. The length of the coil is $l = 40$ cm. What is the capacitance C of the capacitor if the error in calculating the period of oscillation of the circuit by means of the approximate formula $T_1 = 2\pi\sqrt{LC}$ is 1%.

Solution:

True formula period of oscillation $T_2 = \dfrac{2\pi}{\sqrt{\dfrac{1}{LC} - \left(\dfrac{R}{2L}\right)^2}}$

Now $\dfrac{T_2 - T_1}{T_2} = \dfrac{1}{100}$

$$\frac{\dfrac{2\pi}{\sqrt{\dfrac{1}{LC}-\left(\dfrac{R}{2L}\right)^2}}-2\pi\sqrt{LC}}{\dfrac{2\pi}{\sqrt{\dfrac{1}{LC}-\left(\dfrac{R}{2L}\right)^2}}}=\frac{1}{100} \text{ or } 1-\sqrt{LC}\left[\frac{1}{\sqrt{LC}}-\left(\frac{R}{2L}\right)^2\right]^{1/2}=\frac{1}{100}$$

$$C=\frac{4\cdot L\times 199}{R^2\times 100^2};\; L=\frac{\mu_0\pi r^2N^2}{l} \text{ and } R\,\frac{\rho.2\pi rN}{A}$$

where r = radius of the tube of solenoid

N = total number of turns

$$C=\frac{199A^2\mu_0}{l\rho^2\pi\times 100^2}$$

on putting the values we have

$$C=\frac{199\times 10^{-14}\times 4\pi\times 10^{-7}}{0.4\times 1.7^2\times 10^{-16}\times\pi\times 10^4}$$
$$=0.688\mu F$$

14.91. A coil with a length of l = 50 cm and a cross-sectional area of A = 10 cm2 is connected to an alternating current circuit with a frequency of υ = 50 Hz. The number of turns of the coil is N= 3000. Find the resistance of the coil if the phase shift between the voltage and the current is 60°.

Solution:

$$\tan\theta=\frac{\omega L}{R}$$

$$\tan 60=\sqrt{3}=\frac{2\pi\upsilon}{R}\times\frac{\mu_0 AN^2}{l}$$

$$R=\frac{100\pi\times 4\pi\times 10^{-7}\times 10\times 10^{-4}\times 9\times 10^6}{0.5\times 1.732}$$
$$=4.09\ \Omega$$

14.92. A voltage V is supplied to a dc motor. The armature winding resistance is equal to R. At what value of current flowing through the winding will the useful power of the motor be the highest? What is it equal to? What is the motor efficiency?

Solution: The current I in the motor is given by

$$I=\frac{V-e}{R}$$

where e is the back emf produced.

Useful power of the motor $P = eI = \dfrac{eV - e^2}{R}$

This power is a function of *e*, hence find the value of $\dfrac{dP}{de}$ and equate it to zero.

$$\frac{dP}{de} = \frac{V - 2e}{R} = 0 \text{ or } e = \frac{V}{2}$$

$$\frac{d^2P}{de^2} = \frac{-2}{R}$$

which is negative, therefore the power will be maximum when $e = V/2$

$$I = \frac{V}{2R}, P\frac{V^2}{4R} \text{ and efficiency } \eta = \frac{e}{V} = \frac{1}{2} = 50\%$$

14.93. Determine the inductance of a coil in which the emf induced by a change in the current from 5 A to 10 A in 0.1 second is 10 V. What is the change in the energy of the magnetic field of the coil in this case.

Solution:

$$E = -\frac{Ldi}{dt} \text{ or } L = \frac{Edt}{di} = \frac{10 \times 0.1}{10 - 5} = 0.2\text{H}$$

$$\Delta W = \frac{1}{2}Li_2^2 - \frac{1}{2}Li_1^2 = \frac{0.2}{2}\left[10^2 - 5^2\right] = 7.5\text{J}$$

14.94. An ideal inductance *L* and a resistance $R = 30\ \Omega$ are connected in series with an alternating source. If the supply voltage is 120 volt and the current in the circuit be 2 amp, find the potential difference across the inductance and resistance.

Solution: The potential difference across the resistance $V_R = 2 \times 30 =$ 60 volt.

Now $V_S^2 = V_L^2 + V_R^2$

where V_S and V_L are the supply voltage and the potential difference across the inductance, respectively

$$120^2 - 160^2 = V_L^2$$

or

$$V_L = \sqrt{60 \times 80}$$

$$= 104 \text{ V}$$

14.95. When a coil with inductance 0.3 H is connected to a dc circuit at a voltage of 24 V, the current is 0.2 A. Determine the impedance of the coil when it is connected to an ac circuit of frequency 50 Hz.

Solution: According to Ohm's law, $V = IR$

The resistance of the inductive coil $R = \dfrac{24}{0.2} = 120\Omega$

$$\text{Impedance} = \sqrt{R^2 + \omega^2 L^2}$$

$$= \sqrt{120^2 + (2 \times 50\pi \times 0.3)^2}$$

$$= 152.5\ \Omega$$

14.96. Find the *rms* value of emf given by

$$e = 4 \sin \omega t + 6 \sin 2\,\omega t$$

Solution:

$$e^2 = \frac{16}{2\pi}\int_0^{2\pi} \frac{(1\cos 2\,\omega t)}{2} \cdot dt + \frac{36}{2\pi}\int_0^{2\pi} \frac{(1 - \cos 4\omega t)}{2} dt + \frac{48}{2\pi}\int_0^{2\pi} \sin \omega t \sin 2\omega t\, dt$$

$$= 8 + 16 = 26$$

$$e_{\text{rms}} = \sqrt{26}$$

14.97. Find the average power in a $R = 10\ \Omega$ resistor if the current is given by

$$i = 7 \sin \omega t + 5 \sin 3\,\omega t + 3 \sin 5\,\omega t$$

Solution:

$$I_{eff}^2 = \frac{49}{2} + \frac{25}{2} + \frac{9}{2} = \frac{83}{2} = 41.5$$

$$i_{eff} = \sqrt{41.5}$$

$$\text{Power} = I_{eff}^2 \times R = 415 \text{ watt}$$

14.98. Find the average and *rms* value of current for a square wave.

Solution:

$$I_{av} = \frac{1}{T}\left[\int_0^{T/2} I_0 dt + \int_{T/2}^{T} I_0 dt\right] \text{ (see Fig.14.54)}$$

$$= \frac{I_0}{2}$$

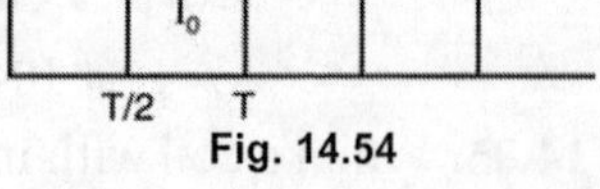

Fig. 14.54

$$\langle I^2 \rangle = \frac{1}{T}\left[\int_0^{T/2} I_0^2 dt + \int_{T/2}^{T} I_0^2 dt\right]$$

$$= \frac{I_0^2}{2}$$

$$I_{rms} = \frac{I_0}{\sqrt{2}}$$

14.99. Find the average value of the current for a triangular wave.

Solution: Divide the wave shape in the two triangles. In triangle *AOB* we have

$$\frac{I_0}{\frac{T}{2}} = \frac{I}{t} \text{ (see Fig. 14.55)}$$

$$\therefore \quad I = \frac{2I_0 t}{T} \qquad ...(1)$$

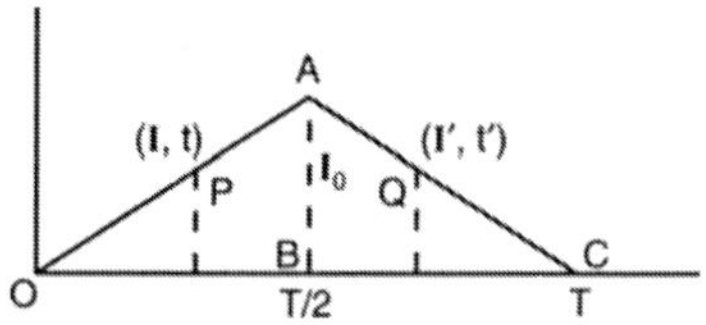

Fig. 14.55

In view of triangle *ACB* we have

$$\frac{I_0}{\frac{T}{2}} = \frac{I'}{T - t'}$$

$$\therefore \quad I' = 2I_0 \frac{(T - t')}{T} \qquad ...(2)$$

The current in the first and second half cycle will vary according to the equation 1 & 2

$$I_{av} = \frac{1}{T}\left[\int_0^{T/2} \frac{2I_0 t \cdot dt}{T} + \int_{T/2}^{T} \frac{2I_0 (T - t')dt'}{T}\right]$$

$$= \left[\frac{I_0}{4} + I_0 - \frac{3I_0}{4}\right] = \frac{I_0}{2}$$

14.100. The circuit shown in the Fig. 14.56 presents a series combination of an inductance, a capacitance and a resistance connected to an alternating source of frequency ω. Find the ratio of output voltage across the resistance to the input voltage.

Solution:

$V_{\text{input}} = I_0 Z$; I_0 = peak value of the current and Z is the total impedance

$$V_{\text{input}} = I_0\left[\left(\frac{1}{C\omega} - L\omega\right)^2 + R^2\right]^{1/2}$$

$$V_{\text{input}} = I_0 R$$

$$\frac{V_{\text{output}}}{V_{\text{input}}} = \frac{1}{\left[1 + \left(\frac{1 - LC\omega^2}{RC\omega}\right)^2\right]^{1/2}}$$

$$= \frac{1}{\sqrt{1 + \left[\frac{1 - (\omega / \omega_0)^2}{RC\omega}\right]^2}}$$

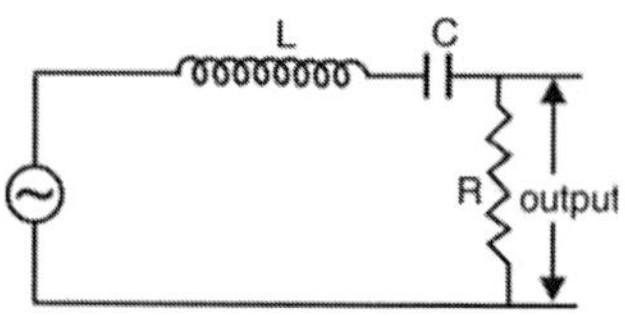

Fig. 14.56

where ω_0 is the resonance frequency.

14.101. In the circuit shown in the Fig. 14.57 find the ratio of output voltage across the resistance R to the input voltage.

Solution: In the first mesh the output voltage across a & b is equal to

$$V_{ab} = \frac{V_{\text{input}} \cdot R}{R\left[1+\left(\frac{1}{R\omega C}\right)^2\right]^{1/2}}$$

This serves the purpose of input voltage for the R, C combination of the second

$$V_{\text{output}} = \frac{V_{ab} \cdot R}{R\left[1+\left(\frac{1}{R\omega C}\right)^2\right]^{1/2}} = \frac{V_{ab}}{\left[1+\left(\frac{1}{R\omega C}\right)^2\right]^{1/2}}$$

$$= \frac{V_{\text{input}}}{\left[1+\left(\frac{1}{R\omega C}\right)^2\right]}$$

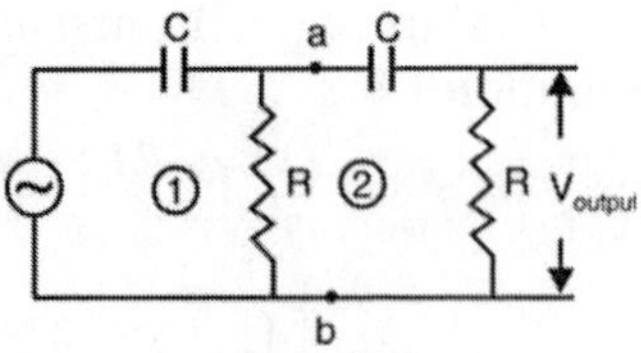

Fig. 14.57

$$\therefore \frac{V_{\text{output}}}{V_{\text{input}}} = \frac{1}{\left[1+\left(\frac{1}{R\omega C}\right)^2\right]}$$

14.102. A pair of capacitor of capacitance C and resistor of resistance R are arranged to form a circuit as shown in the Fig. 14.58. Find the ratio of output voltage across the resistor R to the input voltage.

Solution: In the first mesh

$$V_{ab} = \frac{V_{\text{input}} \times \frac{1}{\omega C}}{\sqrt{R^2 + \frac{1}{\omega^2 C^2}}}$$

The output voltage V_{ab} serves the purpose of input voltage for the second mesh

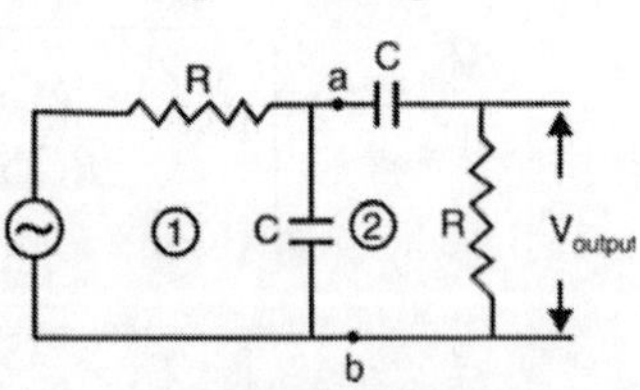

Fig. 14. 58

$$V_{\text{output}} = \frac{V_{ab} \cdot R}{\sqrt{R^2 + \frac{1}{\omega^2 C^2}}}$$

$$= \frac{V_{\text{input}} \times \frac{R}{\omega C}}{\left[R^2 + \frac{1}{\omega^2 C^2}\right]}$$

$$\frac{V_{\text{output}}}{V_{\text{input}}} = \frac{\dfrac{R}{\omega C}}{\left[R^2 + \dfrac{1}{\omega^2 C^2}\right]}$$

14.103. The circuit element shown in the Fig. 14.59 has a current $I = 2.5 \cos (1000t - 60)$ and a voltage $V = 5 \sin (1000t - 60)$. What is the element?

Solution: The current is leading the voltage by 90° hence the element is a capacitor.

Fig. 14.59

$$2.5 = 5 \times 1000 \times C$$

$$C = \frac{2.5}{5 \times 1000} = 5000 \mu F$$

14.104. An ac supply of 90 V and at a frequency of 25 Hz is connected to an inductance coil and a condenser of capacity 700 μF connected in parallel (see Fig. 14.60). The resistance of the coil is 1 Ω and the inductance is 0.1 H. Calculate (i) the current through the capacitor, (ii) the current through the coil, (iii) the total current and (iv) the impedance at resonance under the condition $\omega^2 L^2 >> R^2$.

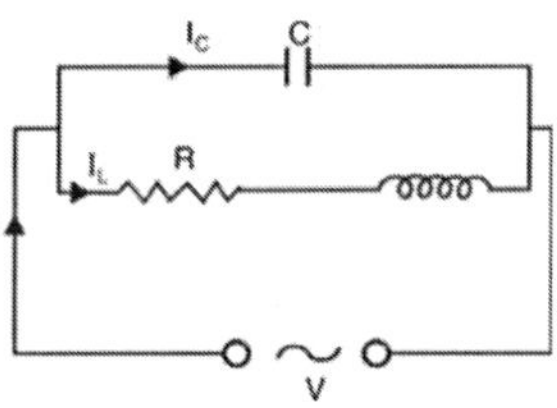

Fig. 14.60

Solution:

$$I_L = \frac{V}{\sqrt{R^2 + \omega^2 L^2}} = \frac{90}{\sqrt{1 + 4\pi^2 \times 625 \times 0.01}}$$

$$= \frac{90}{\sqrt{247.5}} = 5.72 \, A$$

This current will lag behind the voltage by an angle ϕ.

$$I_C = V\omega C = 90 \times 2 \times \pi \times 25 \times 700 \times 10^{-6} = 9.891 \text{ A}$$

This current will lead the voltage by 90°.

$$\text{Net lead of current } \alpha = \frac{\pi}{2} - \phi$$

$$\cos\alpha = \cos\left(\frac{\pi}{2} - \phi\right) = \sin\phi = \frac{\omega L}{\sqrt{R^2 + \omega^2 L^2}}$$

$$= \frac{2\pi \times 25 \times 0.1}{\sqrt{1 + 4\pi^2 \times 625 \times 0.01}}$$

$$= 0.998$$

$$\text{Current } I = \sqrt{I_L^2 + I_C^2 - 2 I_L I_C \cos\alpha}$$

$$= \sqrt{(5.72)^2 + (9.891)^2 - 2 \times 5.72 \times 9.891 \times 0.998}$$

$$= 4.20 \text{ A}$$

$$\text{Impedance } Z = \frac{L}{RC} = \frac{0.1}{1 \times 7 \times 10^{-4}} = 142.85\Omega$$

14.105. A coil with an inductance 0.1 H, a condenser and a resistance R = 100 Ω are connected in series with an ac supply of 220V and a frequency of 50 Hz (see Fig. 14.61). Determine (i) the capacity of the condenser if the active power liberated in the circuit is equal to half the total power, (ii) the current and voltage across the resistor inductor and condenser and (iii) the active, reactive and total powers and the power factor.

Solution: Inductive reactance = $X_L = 2\pi$

$$nL = 2 \times 3.14 \times 50 \times 0.1$$

$$= 3.14\ \Omega$$

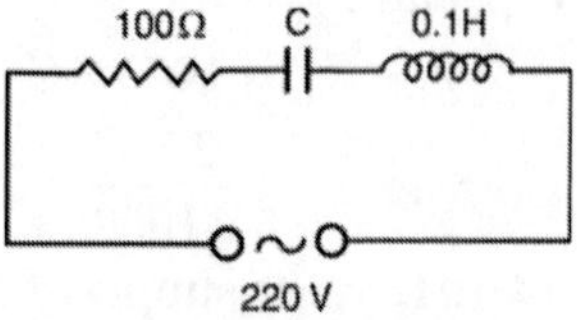

Fig. 14.61

$$\text{Power factor } \cos\phi = \frac{\text{active power}}{\text{total power}} = 0.5$$

Further, $\cos\phi = \dfrac{R}{Z}$ where Z is impedance

$$\therefore \quad R = 0.5Z = 0.5\sqrt{R^2 + (X_C - X_L)^2}$$

$$\therefore \quad 4R^2 = R^2 + (X_C - X_L)^2$$

$$\text{or} \quad X_C = X_L + \sqrt{3}R = 31.4 + \sqrt{3} \times 100 = 31.4 + 173 = 204.4\Omega$$

$$\text{Now } X_C = \frac{1}{\omega C}$$

$$\therefore \quad C = \frac{1}{X_C\omega} = \frac{1}{2\pi n X_C} = \frac{1}{2 \times 3.14 \times 50 \times 204.4} = 15.58\ \mu\text{F}$$

$$\text{Impedence } Z = \sqrt{R^2 + (X_C - X_L)^2} = \sqrt{100^2 + (204.4 - 314)^2} = 199.82\ \Omega$$

Current I in the circuit = supply voltage/impedance = $\dfrac{220}{199.82} = 1.1A$

Voltage across $R = IR = 1.1 \times 100 = 110$V

Voltage across $L = I\omega L = 1.1 \times 31.4 = 34.54$V

$$\text{Voltage across } C = \frac{I}{\omega C} = 1.1 \times 204.4 = 224.84\ V$$

Total power = 1.1 × 220

= 242 watt

$$\text{Active power} = \frac{1}{2} \times \text{total power} = 121 \text{ watt}$$

$$\text{Reactive power} = \text{Total power} \times \sin\phi$$

$$= \text{Total power} \times \sqrt{1 - \cos^2\phi} = 242 \times \frac{\sqrt{3}}{2}$$

$$= 209.3 \text{ watt.}$$

14.106. An ac generator of voltage 30 V and frequency of 50 Hz is connected in parallel to a resistance $R = 3\ \Omega$ and an inductance having reactance $X_L = 5\ \Omega$ as shown in Fig. 14.62. Determine the current in the branches, the total current and $\cos\phi$.

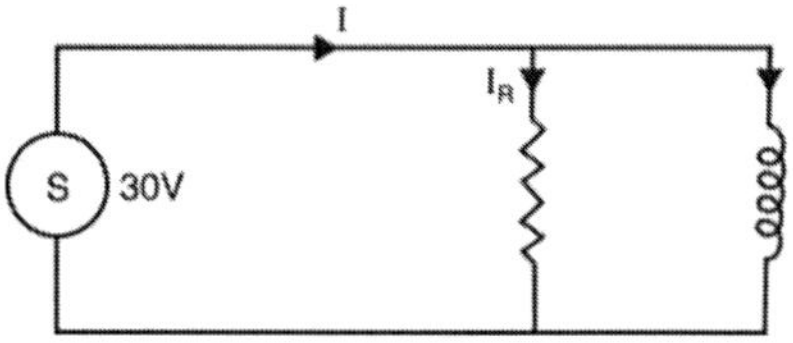

Fig. 14.62

Solution:

$$30 = 3 I_R$$

$$\therefore \quad I_R = 10\text{ A}$$

$$30 = 5 I_L$$

$$\therefore \quad I_L = 6\text{ A}$$

The two currents will be at right angles to each other. Therefore, the total current

$$\text{I} = \sqrt{I_R^2 + I_L^2} = \sqrt{10^2 + 6^2} = 11.66A$$

$$\cos\phi = \frac{\frac{1}{R}}{\sqrt{\frac{1}{R^2} + \frac{1}{X_L^2}}} = \frac{\frac{1}{3}}{\sqrt{\frac{1}{9} + \frac{1}{25}}} = 0.857$$

14.107. An inductive coil, a capacitor and a resistor are connected as shown in Fig. 14.63. The voltage of the alternating source is 50 volt. Calculate the power factor and the total active and reactive powers if (a) $X_L = X_C = 10\ \Omega$ and $R = 5\ \Omega$ (b) $X_C = 15\ \Omega$, $X_L = 10\ \Omega$ and $R = 5\ \Omega$.

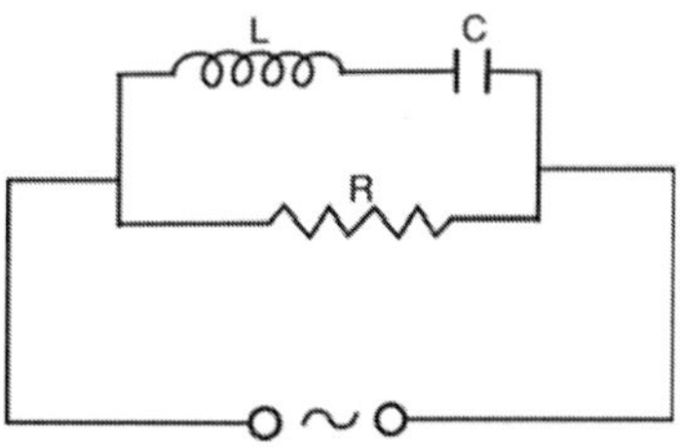

Fig. 14.63

Solution:

(a) Since $X_L = X_C$, the power factor = 1

$$\text{Active power } P = \frac{V^2}{R} = \frac{50 \times 50}{5} = 500 \text{ watt}$$

Reactive power $Q = P \sin\phi = 0$, since $\cos\phi = 1$

(b) Power factor = $\cos\phi$

$$= \frac{\frac{1}{R}}{\sqrt{\frac{1}{R^2} + \frac{1}{(X_C - X_L)^2}}} = \frac{\frac{1}{5}}{\sqrt{\frac{1}{25} + \frac{1}{25}}} = \frac{5}{5\sqrt{2}} = \frac{1}{\sqrt{2}} = 0.707$$

Impedance Z is given by

$$\frac{1}{Z} = \sqrt{\frac{1}{R^2} + \frac{1}{(X_C - X_L)^2}} = \sqrt{\frac{1}{25} + \frac{1}{25}} = \frac{\sqrt{2}}{5}$$

$$\therefore \quad Z = \frac{5}{\sqrt{2}} = 3.535$$

$$\text{Total Power} = \frac{V^2}{Z} = 707 \text{ watt}$$

Active power = Total power × cos ϕ = 707 × 0.707 = 499.85 watt

Reactive power = Total power × sin ϕ = 707 × 0.707 = 499.85 watt

14.108. A potentiometer, an inductive coil and an alternating source of frequency 50 Hz are connected in series as shown in Fig. 14.64. The phase difference between the voltage and the current is 60°. If the resistance of the circuit is increased by 100 Ω, calculate the increase in inductance of the coil in order to get the same phase difference. What is the voltage of the supply if the voltage across the inductance is 103.9 volt?

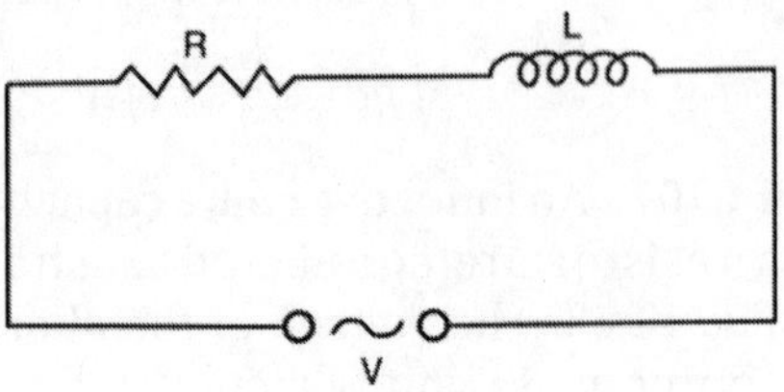

Fig. 14.64

Solution: The problem is essentially related to an LR circuit

$$\text{Now } \tan\phi = \frac{\omega L}{R} = \frac{\omega(L + dL)}{R + dR}$$

where dL and dR represent the increment in the values of inductance and resistance, respectively.

$$\text{Now } \frac{L + dL}{L} = \frac{R + dR}{R} \text{ or } \frac{dL}{L} = \frac{dR}{R}$$

$$\therefore \quad dL = \frac{L}{R} dR = \frac{\tan\phi}{\omega} dR = \frac{\sqrt{3} \times 100}{2 \times 3.14 \times 50} = \frac{1.73}{3.14} = 0.551\text{H}$$

$$\text{Current in the circuit } I = \frac{V}{\sqrt{R^2 + \omega^2 L^2}}; V = \text{supply voltage}$$

$$\text{Voltage across the inductance} = \frac{V\omega L}{\sqrt{R^2 + \omega^2 L^2}} = 103.9$$

$$\text{or } \quad V\sin\phi = 10.9; \frac{\omega L}{\sqrt{R^2 + \omega^2 L^2}} = \sin\phi$$

$$\therefore \quad V\frac{\sqrt{3}}{2} = 103.9$$

or $V = 120\text{V}$

14.109. A variable capacitor and an inductive coil with negligible resistance are connected in series to an alternating source of voltage 100 V. The current in the circuit is 5 A. On halving the capacitance of the capacitor, the current in the circuit becomes 10 A. Calculate the voltage across the circuit elements.

Solution: Let X_L and X_C be the inductive and capacitance reactances. Using these we can have a most general equation for the flow of current as

$$X_L - X_C = \frac{V}{I} \text{ where V is the supply voltage}$$

Now $$X_L - X_C = \frac{100}{5} = 20$$

and $$X_L - 2X_C = \frac{100}{10} = 10$$

Therefore $$X_L - X_C = 2\,(X_L - 2X_C)$$

$$X_L = 3X_C$$

Now $$3X_C - X_C = 20$$

$$2X_C = 20$$

$$X_C = 10\ \Omega$$

$$X_L = 30\ \Omega$$

The voltage across the capacitor $V_C = 5X_C = 50\,\text{V}$

The voltage across the inductance $V_L = 5 \times 30 = 150\,\text{V}$

14.110. A source of ac voltage $V = 180 \sin \omega t$ is connected in series to a resistance and two reactances $X_L = 10\ \Omega$ and $X_C = 15\ \Omega$. Calculate the power factor and the active power if the current in the circuit is 3 A.

Solution:

$$I = \frac{V}{\sqrt{2}\sqrt{R^2 + (X_C - X_L)^2}}$$

$$3\sqrt{2}\sqrt{R^2 + 25} = 180$$

$$R^2 + 25 = 1800$$

$$R = \sqrt{1775}$$

$$\cos\phi = \text{power factor} = \frac{R}{Z} = \sqrt{\frac{1775}{1800}} = 0.993$$

$$\text{Active power} = V_{\text{rms}} \times I_{\text{rms}} \times \cos\phi$$

$$= \frac{180}{\sqrt{2}} \times 3 \times 0.993 = 380.3 \text{ watt.}$$

14.111. With the aim of determining the resonant frequency, an inductance, a capacitance and a resistance are connected to an ac source of variable frequency. In the course of the experiment it was noticed that at frequencies ω_1 and ω_2, the current amplitudes in the circuit assumes the same value. Find the resonance frequency.

Solution: In the *LCR* circuit, the current amplitude is given by

$$I_0 = \frac{E_0}{\sqrt{R^2 + \left(\omega L - \frac{1}{\omega C}\right)^2}}$$

At the two frequencies ω_1 and ω_2, we have

$$R^2 + \left(\omega_1 L - \frac{1}{\omega_1 C}\right)^2 = R^2 + \left(\omega_2 L - \frac{1}{\omega_2 C}\right)^2$$

or $$-\left(\omega_1 L - \frac{1}{\omega_1 C}\right) = \left(\omega_2 L - \frac{1}{\omega_2 C}\right)$$

on dividing it by L, we have

or $$-\omega_1 + \frac{1}{\omega_1 LC} = \omega_2 - \frac{1}{\omega_2 LC}$$

or $$-\omega_1 + \frac{\omega_0^2}{\omega_2} = \omega_2 - \frac{\omega_0^2}{\omega_2}, \text{ where } \omega_0 = \frac{1}{\sqrt{LC}}$$

or $$\omega_0^2\left[\frac{1}{\omega_1} + \frac{1}{\omega_2}\right] = \omega_1 + \omega_2$$

or $$\omega_0 = \sqrt{\omega_1 \omega_2}$$

14.112. In the circuit shown in Fig. 14.65 find (i) resonance frequency (ii) the current flowing in the circuit at resonance.

Solution: The total admittance is given by

$$Y = \frac{1}{Z} = \frac{1}{R_1 + j\omega L} + \frac{1}{R_2 - j/\omega C}$$

$$= \frac{R_1 - j\omega L}{R_1^2 + \omega^2 L^2} + \frac{R_2 + j/\omega C}{R_2^2 + \left(1/\omega C\right)^2}$$

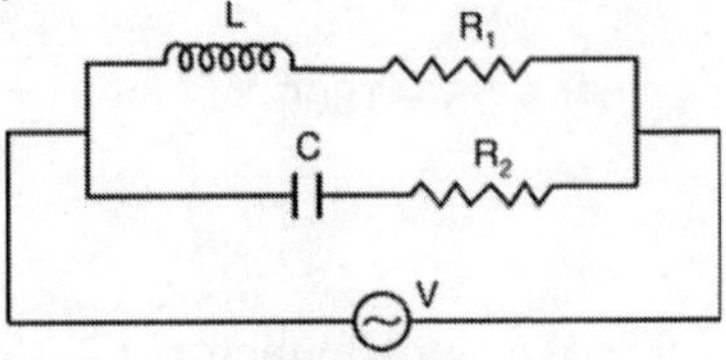

Fig. 14.65

At resonance the admittance is real hence

$$\frac{\omega L}{R_1^2 + \omega^2 L^2} = \frac{1}{\omega C\left(R_2^2 + 1/\omega^2 C^2\right)}$$

$$R_1^2 + \omega^2 L^2 = \omega^2 LC R_2^2 + \frac{L}{C}$$

$$\omega^2\left[L^2 - LC.R_2^2\right] = \frac{L}{C} - R_1^2$$

$$\omega = \frac{1}{\sqrt{LC}}\left[\frac{\frac{L}{C} - R_1^2}{\frac{L}{C} - R_2^2}\right]^{1/2}$$

The admittance at resonance is equal to

$$Y = \frac{R_1}{R_1^2 + \omega^2 L^2} + \frac{R_2}{R_2^2 + 1/\omega^2 C^2}$$

$$\therefore \text{ Current at resonance} = YV = \left(\frac{R_1}{R_1^2 + \omega^2 L^2} + \frac{R_2}{R_2^2 + 1/\omega^2 C^2}\right)V$$

14.113. A coil of inductance L and resistance R_L and a condenser of capacity C are connected in parallel as shown in the Fig. 14.66. The assembly is connected to an ac source through a resistance R. Find the resonance frequency of the circuit.

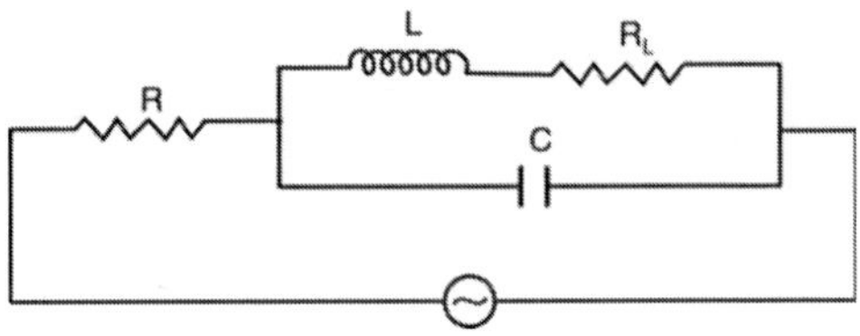

Fig. 14.66

Solution:

The total impedance $Z = R + \dfrac{1}{\dfrac{1}{R_L + j\omega L} + j\omega C}$

$$= R + \frac{R_L + j\omega L}{1 + j\omega C(R_L + j\omega L)}$$

$$= R + \frac{(R_L + j\omega L)(1 - \omega^2 LC - j\omega CR_L)}{(1 - \omega^2 LC)^2 + R_L^2\omega^2 C^2}$$

$$= R + \frac{R_L(1 - \omega^2 LC) + \omega^2 LCR_L}{(1 - \omega^2 LC)^2 + R_L^2\omega^2 C^2} + \frac{j\left[\omega L(1 - \omega^2 LC) - \omega CR_L^2\right]}{(1 - \omega^2 LC)^2 + R_L^2\omega^2 C^2}$$

For impedance to be real we have

$$\omega L(1 - \omega^2 LC) = \omega CR_L^2$$

or $$1 - \omega^2 LC = \frac{CR_L^2}{L}$$

or $$\omega = \sqrt{\frac{1}{LC} - \frac{R_L^2}{L^2}}$$

14.114. A coil of resistance R and inductance L is connected in parallel to a capacitor of capacitance C. Find the total impedance of the circuit.

Solution: In parallel combination admittance

$$Y = \frac{1}{Z} = \frac{1}{R + j\omega L} + j\omega C$$

$$= \frac{R - j\omega L}{R^2 + \omega^2 L^2} + j\omega C$$

$$= \frac{R}{R^2 + \omega^2 L^2} + j\left(\omega C - \frac{\omega L}{R^2 + \omega^2 L^2}\right)$$

Magnitude of the admittance

$$= \sqrt{\left(\frac{R}{R^2 + \omega^2 L^2}\right)^2 + \left(\omega C - \frac{\omega L}{R^2 + \omega^2 L^2}\right)^2}$$

$$= \sqrt{\frac{R^2 + \omega^2 C^2 \left(R^2 + \omega^2 L^2\right)^2 + \omega^2 L^2 - 2\omega^2 LC\left(R^2 + \omega^2 L^2\right)}{\left(R^2 + \omega^2 L^2\right)^2}}$$

$$= \sqrt{\frac{1 + \omega^2 C^2 \left(R^2 + \omega^2 L^2\right) - 2\omega^2 LC}{R^2 + \omega^2 L^2}}$$

$$= \sqrt{\frac{\omega^2 C^2 R^2 + \left(1 - \omega^2 LC\right)^2}{R^2 + \omega^2 L^2}}$$

$$\therefore \quad Z = \sqrt{\frac{R^2 + \omega^2 L^2}{\omega^2 C^2 R^2 + \left(1 - \omega^2 LC\right)^2}}$$

14.115. What must be the relation between R_L and R_C for the circuit shown in the Fig. 14.67 to be resonant at all frequencies?

Solution:

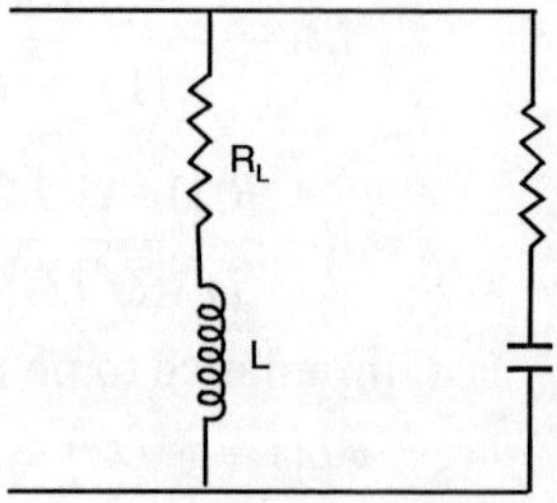

Fig. 14.67

$$Y_L = \text{admittance} = \frac{1}{R_L + j\omega L}$$

$$Y_C = \frac{1}{R_C - \dfrac{j}{\omega C}}$$

$$Y_{\text{Total}} = Y_L + Y_C = \frac{1}{R_L + j\omega L} + \frac{1}{R_C - \dfrac{j}{\omega C}}$$

$$= \frac{R_L - j\omega L}{R_L^2 + \omega^2 L^2} + \frac{R_C + \dfrac{j}{\omega C}}{R_C^2 + \dfrac{1}{\omega^2 C^2}}$$

$$= \frac{R_L\left(R_C^2 + \frac{1}{\omega^2 C^2}\right) - j\omega L\left(R_C^2 + \frac{1}{\omega^2 C^2}\right) + R_C\left(R_L^2 + \omega^2 L^2\right) + \frac{j}{\omega C}\left(R_L^2 + \omega^2 L^2\right)}{\left(R_L^2 + \omega^2 L^2\right)\left(R_C^2 + \frac{1}{\omega^2 C^2}\right)}$$

For admittance to be real

$$\omega L\left(R_C^2 + \frac{1}{\omega^2 C^2}\right) = \frac{1}{\omega C}\left(R_L^2 + \omega^2 L^2\right)$$

$$R_C^2 \omega L + \frac{L}{\omega C^2} = \frac{R_L^2}{\omega C} + \frac{\omega L^2}{C}$$

$$R_C^2 \omega L - \frac{R_L^2}{\omega C} = \frac{\omega L^2}{C} - \frac{L}{\omega C^2}$$

If $R_L = R_C$ then

$$R_L^2\left[\omega L - \frac{1}{\omega C}\right] = \frac{L}{C}\left[\omega L - \frac{1}{\omega C}\right]$$

or $$R_L = R_C = \sqrt{\frac{L}{C}}$$

14.116. In the circuit shown in the Fig. 14.68 find out the necessary condition in which the phase difference between the current and voltage is zero at all frequencies.

Solution:

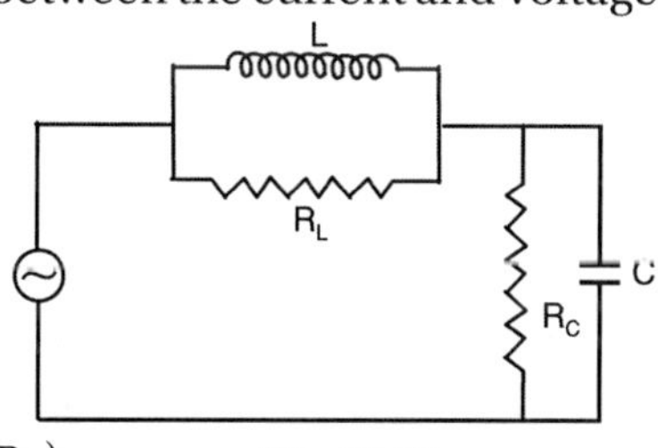

Fig. 14.68

$$Z = \frac{R_L \cdot j\omega L}{R_L + j\omega L} + \frac{R_C \cdot \frac{1}{j\omega C}}{R_C + \frac{1}{j\omega C}}$$

$$= \frac{jR_L\omega L(R_L - j\omega L)}{R_L^2 + \omega^2 L^2} + \frac{R_C(1 - j\omega C R_C)}{1 + \omega^2 C^2 R_C^2}$$

$$= \frac{j R_L^2 \omega L + R_L \omega^2 L^2}{R_L^2 + \omega^2 L^2} + \frac{R_C - j\omega C R_C^2}{1 + \omega^2 C^2 R_C^2}$$

$$= \frac{R_L \cdot \omega^2 L^2}{R_L^2 + \omega^2 L^2} + \frac{R_C}{1 + \omega^2 C^2 R_C^2}$$

$$+ \frac{j\left[R_L^2 \omega L\left(1 + \omega^2 C^2 R_C^2\right) - \omega C R_C^2\left(R_L^2 + \omega^2 L^2\right)\right]}{\left(R_L^2 + \omega^2 L^2\right)\left(1 + \omega^2 C^2 R_C^2\right)}$$

For impedence to be real

$$R_L^2 \cdot L\left(1 + \omega^2 C^2 R_C^2\right) = C \cdot R_C^2\left(R_L^2 + \omega^2 L^2\right)$$

$$\omega^2\left[C^2R_C^2\cdot R_L^2 - R_C^2\cdot LC\right]=\left[\frac{C}{L}R_C^2\cdot R_L^2 - R_L^2\right]$$

$$\omega^2 R_C^2\cdot R_L^2\frac{C}{L}\left[LC-\frac{L^2}{R_L^2}\right]=\frac{C}{L}\cdot R_C^2\cdot R_L^2\left[1-\frac{L}{C}\cdot\frac{1}{R_C^2}\right]$$

$$\omega=\sqrt{\frac{1-\frac{L}{C}\cdot\frac{1}{R_C^2}}{LC-\frac{L^2}{R_L^2}}}$$

14.117. A parallel resonant circuit is shown in the Fig. 14.69. Find the resonance frequency.

Solution: In parallel combination the admittance $Y=\frac{1}{Z}$ is real

$$Y=\frac{1}{Z}=\frac{1}{R+j\omega L}+j\omega C$$

$$=\frac{R-j\omega L}{R^2+\omega^2L^2}+j\omega C$$

$$=\frac{R}{R^2+\omega^2L^2}+j\omega\left(\frac{-L}{R^2+\omega^2L^2}+C\right)$$

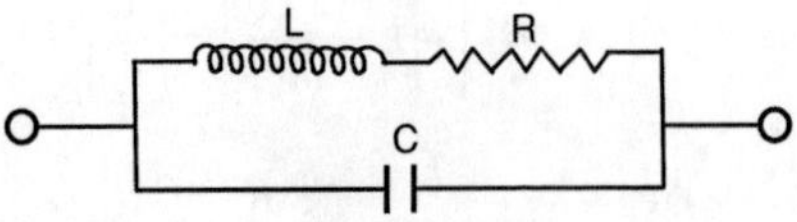

Fig. 14.69

For the real values

$$\frac{L}{R^2+\omega^2L^2}=C$$

or $$\frac{L}{C}=R^2+\omega^2L^2$$

or $$\omega=\sqrt{\frac{1}{LC}-\frac{R^2}{L^2}}$$

14.118. A three element series & parallel circuit contains $L_1=L_2=5$ mH and $C_1=40$ μF. Find the series and parallel resonance frequencies.

Solution:

The impedance $Z=jX=j\omega L_2+\dfrac{j\omega L_1\cdot\frac{1}{j\omega C_1}}{j\omega L_1+\frac{1}{j\omega C_1}}$ (see Fig. 14.70)

$$=j\omega L_2+\frac{j\omega L_1}{1-\omega^2L_1C_1}$$

$$=\frac{j\omega L_2\left(1-\omega^2L_1C_1\right)+j\omega L_1}{1-\omega^2L_1C_1}$$

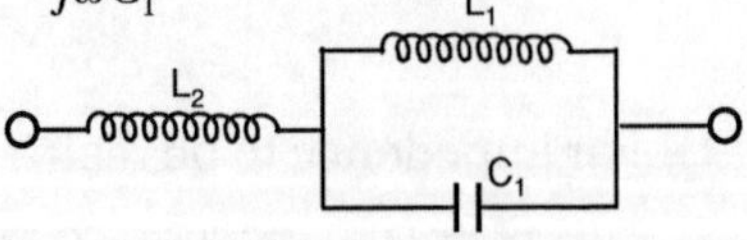

Fig. 14.70

At series resonance $X = 0$ hence

$$L_2 (1 - \omega^2 L_1 C_1) + L_1 = 0$$

$$L_1 + L_2 = \omega^2 L_1 L_2 C_1$$

$$\omega = \left(\frac{L_1 + L_2}{L_1 L_2 C_1} \right)^{1/2}$$

On substituting the values we have

$$\omega = \left(\frac{10 \times 10^{-3}}{25 \times 10^{-6} \times 40 \times 10^{-6}} \right)^{1/2}$$

$$= \sqrt{10} \times 10^3 \text{ rad/sec} = 3.16 \times 10^3 \text{ rad/sec}$$

A parallel resonance $X = \infty$ hence

$$1 - \omega^2 L_1 C_1 = 0$$

or $$\omega = \frac{1}{\sqrt{L_1 C_1}}$$

On putting the values we have

$$\omega = \frac{1}{\sqrt{5 \times 10^{-3} \times 40 \times 10^{-6}}} = \sqrt{5} \times 10^3 \text{ rad/sec} = 2.236 \times 10^3 \text{ rad/sec}$$

14.119. A three element series and parallel circuit contains two capacitors of capacities C_1 and C_2 and an inductance L_1 as shown in the Fig. 14.71. Find the series and parallel resonance frequencies.

Solution: The impedance is given by

$$Z = jX = \frac{1}{j\omega C_2} + \frac{j\omega L_1 \times \frac{1}{j\omega C_1}}{j\omega L_1 + \frac{1}{j\omega C_1}}$$ see Fig. 14.71

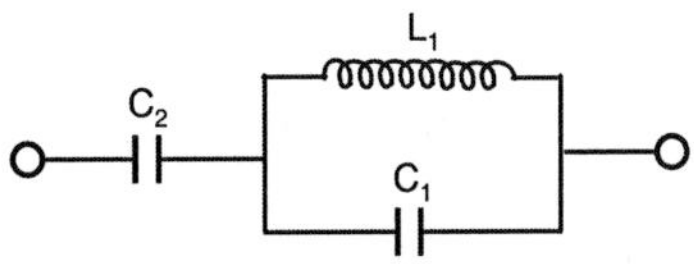

Fig. 14.71

$$= \frac{1}{j\omega C_2} + \frac{j\omega L_1}{1 - \omega^2 L_1 C_1}$$

$$= -\frac{j}{\omega C_2} + \frac{j\omega L_1}{1 - \omega^2 L_1 C_1}$$

$$= -\frac{j\left(1 - \omega^2 L_1 C_1\right) + j\omega^2 L_1 C_2}{\omega C_2 \left[1 - \omega^2 L_1 C_1\right]}$$

At series resonance $X = 0$

$$\therefore \quad 1 - \omega^2 L_1 C_1 = \omega^2 L_1 C_2$$

or $$\omega = \sqrt{\frac{1}{L_1 (C_1 + C_2)}}$$

At parallel resonance $X = \infty$

$$\therefore \quad \omega = \frac{1}{\sqrt{L_1 C_1}}$$

14.120. An oscillatory circuit is designed by using a coil of self-inductance L and a capacitor of capacity C. The resistance of the coil is zero. The coil is placed in a uniform and constant magnetic field. The total magnetic flux linking with the coil is ϕ. At time $t = 0$ the magnetic field is switch off. Calculate the current in the circuit as a function of time.

Solution:

$$L\frac{di}{dt} + \frac{q}{C} = 0$$

$$L\frac{d^2q}{dt^2} + \frac{q}{C} = 0 \qquad \ldots (1)$$

$$\omega^2 = \frac{1}{LC}$$

The solution of equation 1 is

$$q = q_0 \sin \omega t$$

$$I = \frac{dq}{dt} = q_0 \omega \cos \omega t = I_0 \cos \omega t$$

at $t = 0, I = I_0 = q_0 \omega$

The flux ϕ through the coil is given by

$$\phi = LI_0$$

$$\therefore \quad I_0 = \frac{\phi}{L}$$

$$\therefore \quad I = \frac{\phi}{L} \cdot \cos\left(\frac{t}{\sqrt{LC}}\right)$$

14.121. An oscillatory circuit is designed by using a coil of self-inductance L resistance R_L and a capacitor of capacity C connected in series. Due to the resistance of the coil the system represents a damped harmonic oscillator. Assuming the voltage amplitude across the capacitor as V_0. Calculate the mean power to be supplied to the circuit in order to maintain the undamped oscillations.

Solution: Let the supplied alternating source be $e = e_0 \sin \omega t$

At resonance the current is $\frac{e_0}{R_L}$

$$\therefore \quad V_0 = \frac{e_0}{R_L \omega C}$$

$$\text{Mean power} = \frac{e_0}{R_L} \cdot \frac{e_0}{2} = \frac{1}{2R_L} \cdot V_0^2 \omega^2 C^2 R_L^2 = \frac{V_0^2 C \cdot R_L}{2L}$$

14.122. A coil of inductance L and resistance R_L and a capacitor of capacitance C is connected in series to an alternating source $e = e_0 \sin \omega t$. For what value of the capacity C the voltage amplitude across the coil is maximum? What is the voltage amplitude across the coil and capacitor?

Solution: Current flowing in the circuit is given by

$$I = \frac{e_0}{\sqrt{R_L^2 + \left(\frac{1}{C\omega} - L\omega\right)^2}} \sin(\omega t - \phi); \tan \phi = \frac{\frac{1}{C\omega} - L\omega}{R_L}$$

For voltage to be maximum the current I should be maximum and that will be obtained at resonance i.e. when $C = \frac{1}{\omega^2 L}$

$$I_{\text{max}} = \frac{e_0}{R_L} \sin \omega t$$

Z = impedance of the coil $= \sqrt{R_L^2 + \omega^2 L^2}$

Max. voltage across the coil $= \sqrt{R_L^2 + \omega^2 L^2} \cdot \frac{e_0}{R_L} \sin \omega t$

Voltage amplitude across the coil $= e_0 \sqrt{1 + \left(\frac{\omega L}{R}\right)^2}$

Max. voltage across the capacitor $= \frac{e_0}{\omega C R_L} \sin \omega t$

Voltage amplitude across the capacitor $= \frac{e_0}{R_L} \cdot \frac{1}{\omega C}$

14.123. A thin wire ring of resistance R and inductance L rotates about its vertical diameter with constant angular velocity in a uniform magnetic field. The axis of rotation is perpendicular to the magnetic field. In this process the magnetic flux varies with time according to the relation $\phi = \phi_0 \cos \omega t$. Find the current in the ring.

Solution:

Induced emf $e = -\frac{d\phi}{dt} = \phi_0 \omega \sin \omega t$

The system is essentially a LR circuit connected to an alternating source $e = \phi_0 \omega \sin \omega t$.

Now $L\frac{di}{dt} + Ri = \phi_0 \omega \sin \omega t$

Let $i = A \sin \omega t + B \cos \omega t$ be the solution of the above differential equation.

$$LA\omega \cos \omega t - LB\omega \sin \omega t + RA \sin \omega t + RB \cos \omega t = \phi_0 \omega \sin \omega t$$

On equating the coefficient of $\sin \omega t$ and $\cos \omega t$ we have

$$LA\omega = -RB$$

$$RA - LB\omega = \phi_0 \omega$$

On solving

$$A = \frac{\phi_0 \omega R}{R^2 + \omega^2 L^2}, \quad B = \frac{-L\omega\phi_0\omega}{R^2 + L^2\omega^2}$$

$$i = \frac{\phi_0 \omega}{\sqrt{R^2 + \omega^2 L^2}} \left[\sin \omega t \cdot \frac{R}{\sqrt{R^2 + L^2\omega^2}} - \cos \omega t \cdot \frac{L\omega}{\sqrt{R^2 + \omega^2 L^2}} \right]$$

$$= \frac{\phi_0 \omega \sin(\omega t - \theta)}{\sqrt{R^2 + L^2\omega^2}}; \quad \tan \theta = \frac{\omega L}{R}$$

14.124. The circuit shown in the Fig. 14.72 represents a series combination of an ideal inductance L, a capacitor with capacitance C and a resistance R. The source of alternating voltage is an oscillator whose freqency can be varied without changing the voltage amplitude. At what frequency the voltage amplitude across the capacitor is maximum?

Solution:

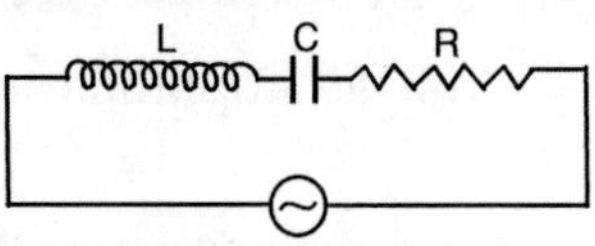

Fig. 14.72

$$I = \frac{V_0}{\sqrt{R^2 + \left(\frac{1}{C\omega} - L\omega\right)^2}} \sin(\omega t - \phi)$$

Where V_0 is the voltage amplitude and ϕ is the phase.

Peak value of the voltage across the capacitor

$$V_C = \frac{V_0}{\omega C \sqrt{R^2 + \left(\frac{1}{C\omega} - L\omega\right)^2}} = \frac{V_0 \cdot \omega_0^2}{\omega \sqrt{\frac{R^2}{L^2} + \left(\frac{\omega_0^2}{\omega} - \omega\right)^2}}; \omega_0^2 = \frac{1}{LC}$$

$$\frac{dV_C}{d\omega} = 0 = V_0 \omega_0^2 \left[\frac{-\omega^2 \left[\frac{R^2}{L^2} + \frac{\left(\omega_0^2 - \omega^2\right)^2}{\omega^2} \right] + \omega_0^4 - \omega^4}{\omega^4 \left[\frac{R^2}{L^2} + \left(\frac{\omega_0^2}{\omega} - \omega\right)^2 \right]^{3/2}} \right]$$

Putting $N_r = 0$

$$\frac{\omega^2 R^2}{L^2} + \left(\omega_0^2 - \omega^2\right)^2 = \omega_0^4 - \omega^4$$

$$2\omega^4 - 2\omega_0^2\omega^2 + \omega^2 \frac{R^2}{L^2} = 0$$

$$\omega^2 = \omega_0^2 - \frac{R^2}{2L^2}$$

$$\omega = \sqrt{\omega_0^2 - \frac{R^2}{2L^2}}$$

15

OPTICS

15.1. Two plane mirrors are placed at angle θ. A ray of light is incident on the first mirror parallel to the second and after reflection from the second mirror it becomes parallel to the first one. Calculate the angle between the two mirrors.

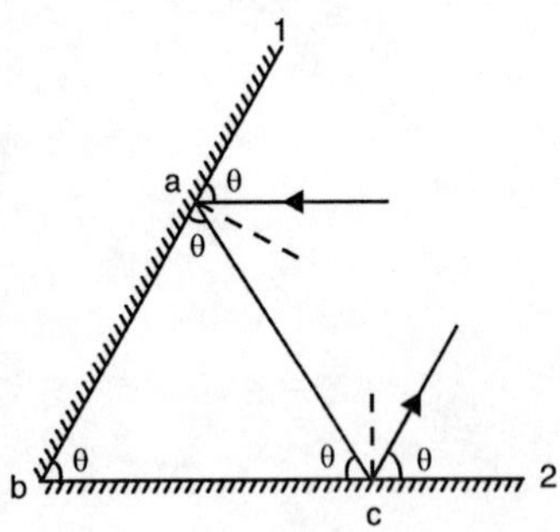

Fig. 15.1

Solution: In view of the geometry of the Fig. 15.1 each of the three angles *a*, *b* & *c* will be equal to θ.

In triangle *abc* $3\theta = 180°$

$\therefore \quad \theta = 60°$

15.2. Two mirros are placed at an angle θ. A ray of light strikes the first mirror at angle of 45°. After reflection from the second mirror it becomes parallel to the first mirror. Find the angle between the mirrors.

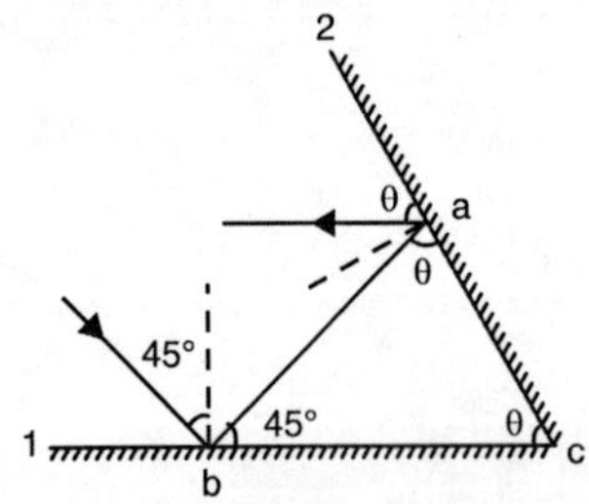

Fig. 15.2

Solution: In view of the geometry of the Fig. 15.2 sum of the angle *a*, *b* & *c* in triangle *abc* is equal to 180°.

$$\therefore \quad 2\theta + 45 = 180°$$

$$2\theta = 135°$$

$$\theta = 67.5°$$

15.3. A ray of light makes an angle of 10° with the horizontal and strikes a plane mirror which is inclined at an angle θ to the horizontal. For what value of θ the reflected ray will become vertical?

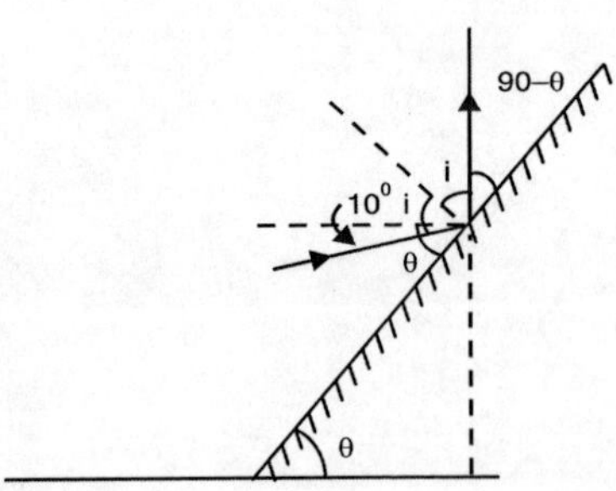

Fig. 15.3a

Solution: Case Ist when the ray strikes the mirror below the horizontal (Fig. 15.3a)

Angle of incidence $i = 90 - \theta + 10$

And this must be equal to θ

$\therefore \quad 100 - \theta = \theta$

or $\quad \theta = 50°$

Case IInd when the ray strikes the mirror above the horizontal (Fig. 15.3b)

Angle of incidence $i = 90 - \theta - 10$

And this must be equal to θ

$\therefore \quad 80 - \theta = \theta$

or $\quad \theta = 40°$

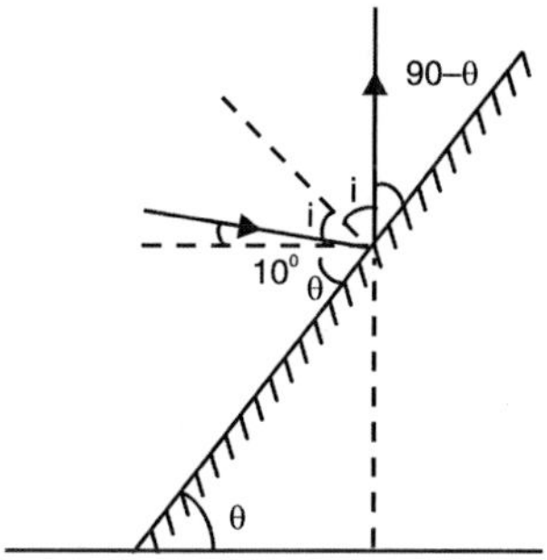

Fig. 15.3b

15.4. A pinhole camera is used to photograph the sun which subtends an angle of 30′ at the surface of earth. The screen is placed at a distance of 100 cm from the pinhole. Calculate the diameter of the image of the sun.

Solution:

$$\theta = \frac{d}{L}$$

where θ is the angle in radians

d = diameter of image

L = distance of the screen from the pin hole

$$\therefore \quad d = \theta \cdot L$$

$$= \frac{30 \times 3.14}{180 \times 60} \times 100$$

$$= 0.87 \text{ cm}$$

15.5. For a particular day the timings of sun rise and sun set as given in almanac is 5.30 am and 6.30 pm respectively, but sun can be seen before 5.30 am and after 6.30 pm. Explain the reason.

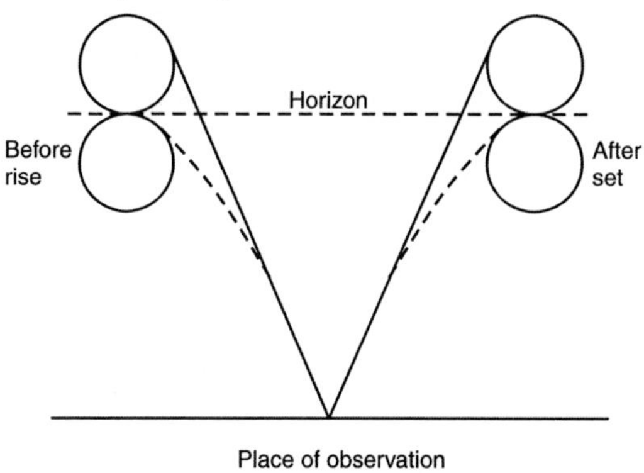

Fig. 15.4

Solution: In view of Fig. 15.4 the sun is below the horizon before rise and after set and therefore one could not see it. In the morning and evening there is variation of refractive index of air. Due to this the rays originating from sun reaches the observer after refraction and the path of ray of light is curved. From the place of observation the observer will see that sun before rise and after the set along the straight line of vision.

15.6. A coin is at a depth d under water of $\mu = \frac{4}{3}$. Find the apparent depth of the coil when it is viewed at an angle ϕ. Assume that the coin is not far off from the place of observation.

Solution: In view of Fig. 15.5

$$\frac{\tan\theta}{\tan\phi} = \frac{\frac{x}{d}}{\frac{x}{y}}$$

$$\therefore \quad y = \text{apparent depth} = d\frac{\tan\theta}{\tan\phi}$$

$$= \frac{d \cdot \cos\phi}{\sqrt{\mu^2 - \sin^2\phi}}$$

$$= \frac{3d \cdot \cos\phi}{\sqrt{7 + 9\cos^2\phi}}$$

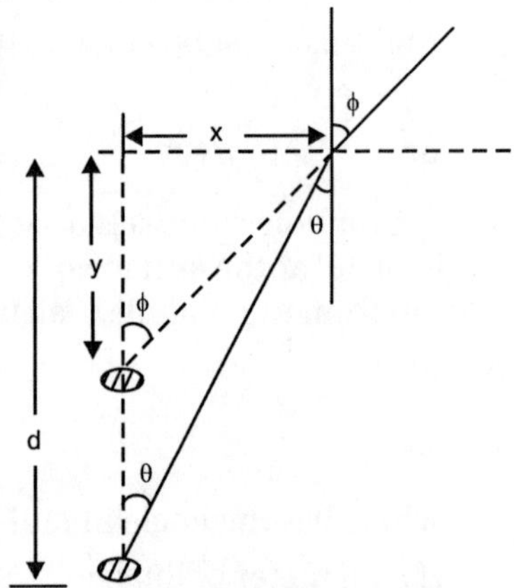

Fig. 15.5

15.7. A glass hemisphere is used as a paper weight with its flat face resting on the paper. The radius of circular cross-section is 4 cm. The refractive index of glass = 1.6. The hemisphere is placed over a paper on which a circle of radius 3 mm is drawn. The centre of the hemisphere is directly above the centre of the circle. Find the radius of the image of the circle when observation is made along a vertical radius.

Solution:

$$\frac{1}{\mu} = \frac{\text{real depth}}{\text{apparent depth}}$$

$$\therefore \quad \text{radius of the image of the circle} = \mu \times 3 \text{ mm}$$

$$= 1.6 \times 3$$

$$= 4.8 \text{ mm}$$

15.8. A swimmer observes from under the water a luminous object above his head at A (Fig. 15.6) at a distance of 30 m above the surface of water.

Calculate the visible distance of the object above the surface of water. The refractive index of water ${}^a\mu_w = \frac{4}{3}$.

Solution: The luminous object will appear to be at *C*.

$$\tan i = \frac{OB}{30}, \tan r = \frac{OB}{OC}$$

$$\therefore \quad \frac{\tan i}{\tan r} = \frac{OC}{30} \text{ or } OC = 30\frac{\tan i}{\tan r}$$

If the angles *i* and *r* are small, then

$$\frac{\tan i}{\tan r} = \frac{\sin i}{\sin r} = {}^a\mu_w \quad \therefore OC = 30 \times {}^a\mu_w$$

$$= 30 \times \frac{4}{3} = 40 \text{ m}$$

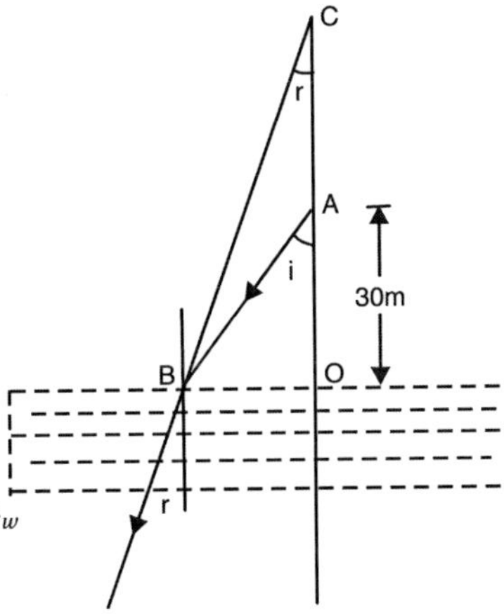

Fig. 15.6

15.9. A pole is standing in clear water. Pole stands 1 m above the water surface. The pond is 2 m deep. What are the lengths of the shadow thrown by the pole on the surface and bottom of the pond if the sun is 30° over the horizon? ${}^a\mu_w = 1.33$, $\cot 30° = \sqrt{3} = 1.732$, $\sin 40.6° = 0.651$ and $\tan 40.6° = 0.8571$. (Roorkee 1971)

Solution: In view of the Fig. 15.7, the length of the shadow thrown on the surface of water is equal to *BF*.

Now $BF = \cot 30 = 1.732$ m

The shadow of the upper end *A* of the pole is formed at *D* after refraction in water. Now

$${}^a\mu_w = \frac{\sin 60}{\sin r}$$

$$\therefore \quad \sin r = \frac{\sqrt{3}/2}{1.33} = 0.651$$

$$\therefore \quad r = 40.6°$$

$$\tan r = \frac{ED}{2}$$

$$\therefore \quad ED = 2 \tan 40.6° = 2 \times 0.8571 = 1.7142$$

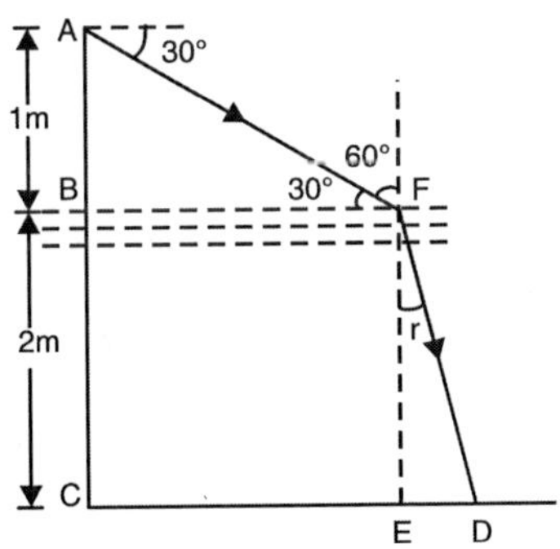

Fig. 15.7

Length of the shadow at the bottom of the pond $= CD = BF + ED$

$= 1.732 + 1.7142$

$= 3.446$ m.

15.10. A bottom weighted verticle pole touching the bottom of a swimming pool extends 2 m above the bottom and 0.5 m above the water surface. The sun is incident at 45°. What is the length of the shadow of the pole on the bottom of the pool? ${}^a\mu_w = \frac{4}{3}$.

Solution: In this problem, we have a similar figure as in Fig. 15.7.

Here $AB = 0.5$ m and $\angle AFB = 45°$

$$\frac{0.5}{BF} = \tan 45°$$

$$\therefore \quad BF = 0.5 \text{ m}$$

$${}^a\mu_w = \frac{\sin 45°}{\sin r}$$

$$\therefore \quad \sin r = \frac{3}{4}\sin 45° = \frac{3}{4\sqrt{2}}$$

$$\tan r = \frac{3}{\sqrt{23}} = 0.625$$

$$\therefore \quad CD = 0.5 + 2\tan r = 0.5 + 2 \times 0.625 = 0.5 + 1.250 = 1.75 \text{ m}$$

15.11. A man standing on the edge of a 10 m deep swimming pool observes a stone lying at its bottom. At what depth below the surface of water will the image of the stone be formed if the line of vision makes an angle of 30° will the normal? ${}^a\mu_w = \frac{4}{3}$.

Solution: The stone is at C. To the observer, it appears to come from C' as shown in Fig. 15.8.

$$\sin dr = \frac{AD}{AC} \text{ and } \sin d\theta = \frac{AD'}{AC'}$$

If the angles dr and $d\theta$ are very small, then we have

$$AD = AC\, dr = \frac{h}{\cos r} dr;\ BB' = h = \text{depth of the swimming pool}$$

$$AD' = AC' \cdot d\theta = \frac{h'}{\cos\theta} d\theta$$

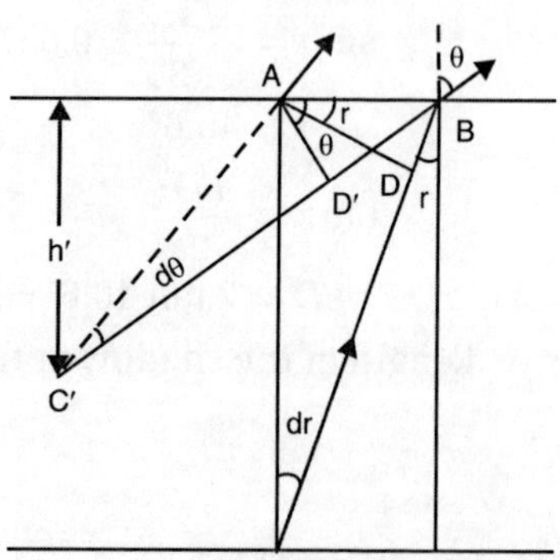

Fig. 15.8

In triangles ABD and ABD', we have

$$\cos r = \frac{AD}{AB} \text{ and } \cos\theta = \frac{AD'}{AB}$$

$$\therefore \quad AB = \frac{h}{\cos^2 r} dr = \frac{h'}{\cos^2\theta} d\theta$$

According to Snell's law

$$\frac{\sin\theta}{\sin r} = {}^a\mu_w = \frac{\sin(\theta + d\theta)}{\sin(r + dr)}$$

Now $\sin\theta\cos d\theta + \cos\theta\sin d\theta = {}^a\mu_w \sin r\cos dr + {}^a\mu_w \cos r\sin dr$

or $\quad \sin\theta + \cos\theta \cdot d\theta = {}^a\mu_w \sin r + {}^a\mu_w \cos r \cdot dr$

$$\therefore \quad \frac{d\theta}{dr} = {}^a\mu_w \frac{\cos r}{\cos\theta}$$

From the above, we have $\dfrac{d\theta}{dr} = \dfrac{h\cos^2\theta}{h'\cos^2 r}$

Hence $^a\mu_w \dfrac{\cos r}{\cos\theta} = \dfrac{h\cos^2\theta}{h'\cos^2 r}$

$$\text{or} \quad h' = \frac{h\cos^3\theta}{{}^a\mu_w \cdot \cos^3 r}$$

$$\sin r = \frac{\sin\theta}{{}^a\mu_w}; \cos r = \frac{\sqrt{({}^a\mu_w)^2 - \sin^2\theta}}{{}^a\mu_w}$$

$$\therefore \quad h' = \frac{h({}^a\mu_w)^2\cos^3\theta}{\left[({}^a\mu_w)^2 - \sin^2\theta\right]^{3/2}}$$

On substituting the values of various terms, we have

$$h' = \frac{10 \cdot \frac{16}{9} \cdot \frac{3\sqrt{3}}{8}}{\frac{55}{36} \cdot \frac{\sqrt{55}}{6}}$$

$$= \frac{20 \times 36 \times 6}{55\sqrt{3}\sqrt{55}} = 6.1 \text{ m}$$

15.12. A vessel containing water up to a height of 60 cm has one-rupee coins at its bottom. At what height above the surface of water should a lighted electric bulb be placed so that its image formed by reflection coincides with the image of the coin? $^a\mu_w = \dfrac{4}{3}$.

Solution:

$$\frac{\text{Real depth}}{\text{Apparent depth}} = \frac{4}{3}$$

$$\therefore \quad \text{apparent depth} = \frac{4}{3} \times 60 = 45 \text{ cm}$$

Thus, the image of the coin will be formed at a depth of 45 cm below the surface of water. If the lighted electric bulb is placed at a height of 45 cm above the surface of water then by reflection it will have its image at a distance of 45 cm below the surface of water and will coincide with the image of the coin.

15.13. In Fig. 15.9, AB is an incident ray and BC is the reflected ray from a plane reflecting surface OO'. If B' be any point on the reflecting plane, show that $AB + BC < AB' + B'C$.

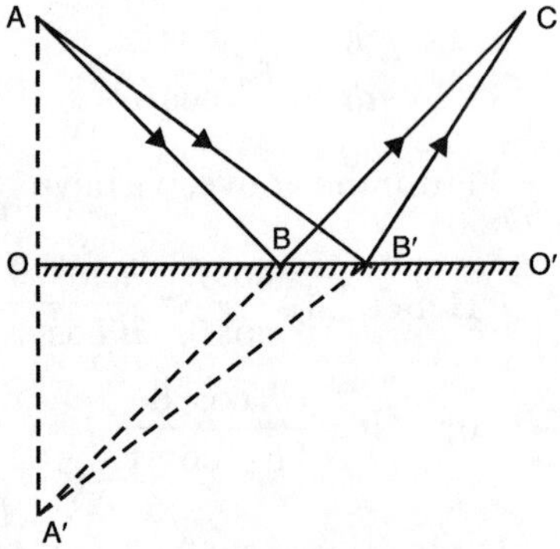

Fig. 15.9

Solution: From the point A drop perpendicular AO on the reflecting surface and extend it so that it should meet the ray BC at A'. Join $A'B'$. From geometry, we have $AO = OA'$. In triangles ABO and $A'BO$, we have, $AB = A'B$. Similarly, we can prove $AB' = A'B'$. Now $A'B' + B'C > A'C$

or $\quad A'B' + B'C > AB + BC$

15.14. Calculate the shift of an image due to refraction when a parallel glass plate of thickness t and refractive index μ is interposed between the eye and a point source.

Solution: A plane glass plate $BCDE$ is interposed between the eye and the object A (Fig. 15.10).

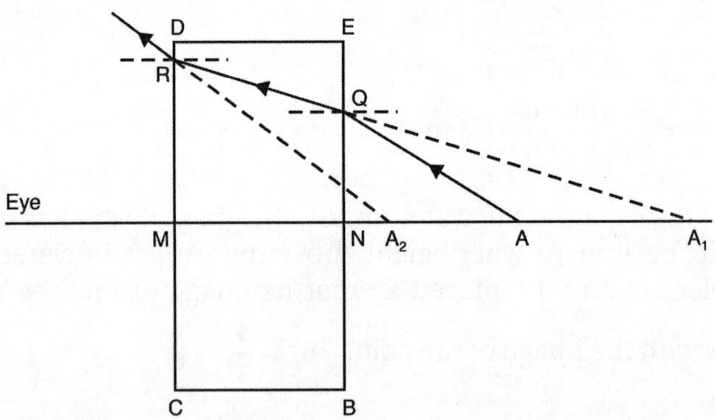

Fig. 15.10

A ray of light AQ is refracting along QR and is appearing to come from A_1.

$$\text{Therefore } \mu = \frac{NA_1}{NA}$$

Now there is a ray A_1R coming from an object A_1 placed in the denser medium and its image is formed at A_2.

Therefore $\mu = \dfrac{MA_1}{MA_2}$

or $\quad MA_2 = \dfrac{MA_1}{\mu} = \dfrac{MN + NA_1}{\mu} = \dfrac{t + \mu NA}{\mu}$; MN is the thickness of the glass plate

$$= \frac{t}{\mu} + NA$$

Shift AA_2 towards the observer $= MA - MA_2 = MA - \dfrac{t}{\mu} - NA$

$$= MN + NA - \frac{t}{u} - NA = t - \frac{t}{\mu} = t\left(1 - \frac{1}{\mu}\right)$$

15.15. A tank contains a slab of glass of 8 cm thickness and of refractive index 1.6. Above this is a depth of 4.5 cm of a liquid of refractive index 1.5 and upon this floats 6 cm of water of refractive index $= \dfrac{4}{3}$. To an observer looking from above, what is the apparent position of a mark on the bottom of the tank.

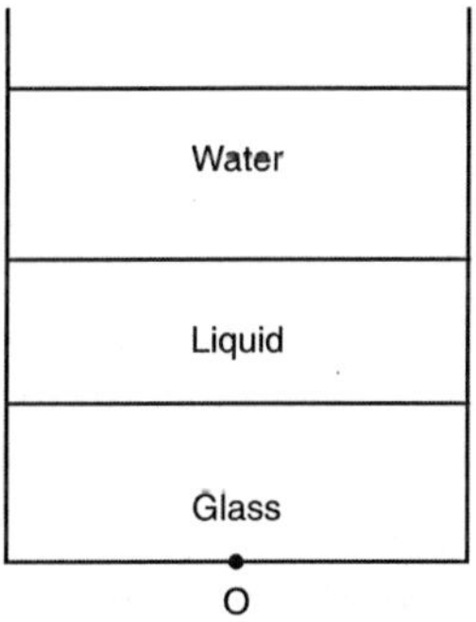

Fig. 15.11

Solution: O is a mark made at the bottom of the tank (Fig. 15.11). The boundaries of the mediums are parallel. The total shift of O is equal to

$$\sum t\left(1 - \frac{1}{\mu}\right)$$

Total apparent shift

$$= 8\left(1 - \frac{1}{1.6}\right) + 4.5\left(1 - \frac{1}{1.5}\right) + 6\left(1 - \frac{1}{4/3}\right)$$

$= 3 + 1.5 + 1.5 = 6$ cm

15.16. There is a test tube of sufficient length. Its bottom is silvered and has a radius of curvature of 40 cm. It contains a liquid of refractive index 1.5 up to a length of 16.5 cm. An object is placed at a distance of 19 cm above the surface of the liquid. Find the position of the image above the surface of the liquid (Fig. 15.12).

Solution:

$$\text{Apparent shift} = t\left(1 - \frac{1}{\mu}\right) = 16.5\left(1 - \frac{1}{1.5}\right) = 5.5 \text{ cm}$$

Position of the object from the lower surface of the liquid = 19 + 16.5 = 35.5 cm. On accounting the shift, the correct position of the object = 35.5 – 5.5 = 30 cm. As the bottom of the test tube is silvered, it will behave as a concave mirror of focal length given by the equation.

$$\frac{1}{f}=\frac{1}{f_{\text{liq}}}+\frac{1}{f_{\text{liq}}}+\frac{1}{f_m}$$

where f_{liq} is the focal length of the plano-convex lens formed by the liquid

$$\frac{1}{f_{\text{liq}}}=\frac{(\mu-1)}{R},\ \frac{1}{f_m}=\frac{2}{R}$$

$$\therefore\quad \frac{1}{f}=\frac{2(\mu-1)}{R}+\frac{2}{R}=\frac{2\mu}{R}=\frac{2\times1.5}{40}=\frac{3}{40}$$

In case of a mirror, we have

$$\frac{1}{v}+\frac{1}{u}=\frac{1}{f}$$

$$\therefore\quad \frac{1}{v}=\frac{1}{f}-\frac{1}{u}=\frac{3}{40}-\frac{1}{30}=\frac{5}{120}$$

$\therefore\quad v = 24$ cm

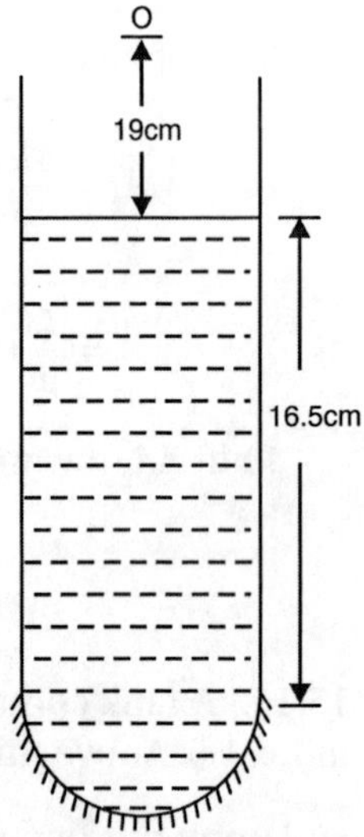

Fig. 15.12

Correct position of the image = 25 + 5.5 = 29.5 cm

The position of the image above the surface of liquid = 29.5 – 16.5 = 13 cm

15.17. An object is placed at 21 cm in front of a concave mirror of radius curvature 10 cm. A glass slab of thickness 3 cm and refractive index 1.5 is placed close to the mirror in the space between the object and the mirror. Find the position of the final image formed. The distance of the nearer surface of the slab from the mirror is 1.0 cm. (IIT 1980)

Solution: The ray of light emerging from O after refraction through the slab becomes parallel to the incident beam but appears to have come from O' (Fig. 15.13).

The shift of the image

$$= t\left(1-\frac{1}{\mu}\right)$$

$$= 3\left(1-\frac{2}{3}\right)=1\,\text{cm}$$

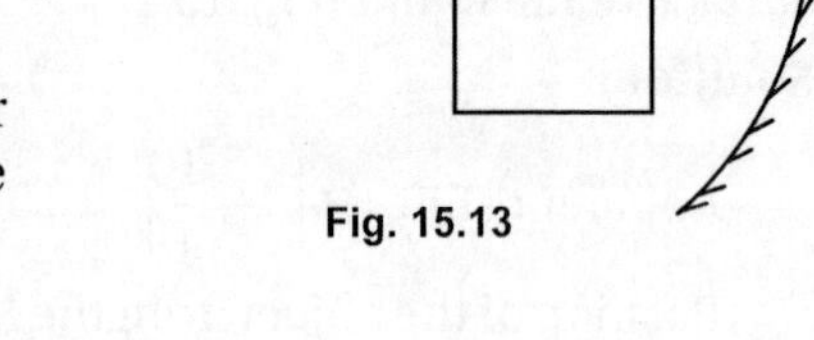

Fig. 15.13

The correct position of object for the purpose of forming the final image = 20 cm

$$\text{Now}\quad \frac{1}{v}=\frac{1}{f}-\frac{1}{u}=-\frac{1}{5}+\frac{1}{20}=-\frac{3}{20}$$

$$\text{or}\quad v=-\frac{20}{3}\text{ cm}$$

Since the ray refracted from the mirror passes through the slab again, the correct position at which the image will be formed is equal to

$$\frac{20}{3}+1=\frac{23}{3}=7.67 \text{ cm}$$

15.18. A ray of light is incident at an angle θ_1 on the upper surface of a transparent glass plate of thickness t as shown in Fig. 15.14. The two surfaces of the plate are plane and parallel to each other. Show that the lateral displacement d of the emergent beam is given by

$$d=t\,\frac{\sin(\theta_1-\theta_1')}{\cos\theta_1'}$$

Solution: In triangle *BCF*

$$\sin CBF=\sin(\theta_1-\theta_1')=\frac{d}{BC}$$

$$\therefore \quad d=BC\sin(\theta_1-\theta_1')$$

In triangle *CBE*

$$\cos CBE=\cos\theta_1'=\frac{t}{BC}$$

$$\therefore \quad BC=\frac{t}{\cos\theta_1'}$$

On substituting the value of *BC* in the above equation, we have

$$d=t\,\frac{\sin(\theta_1-\theta_1')}{\cos\theta_1'}$$

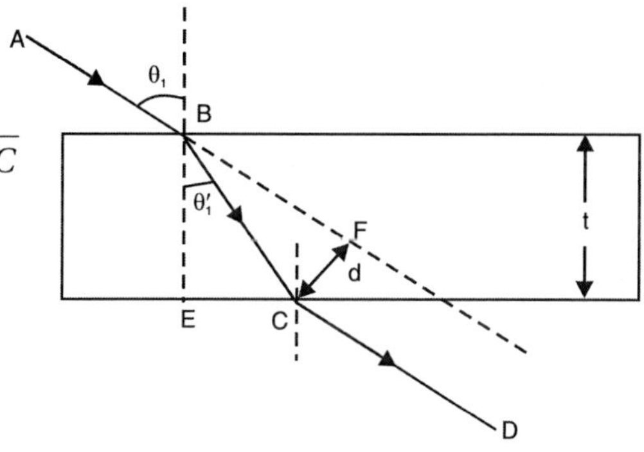

Fig. 15.14

15.19. A ray of light is incident at an angle θ_1 as in Fig. 15.14 on the upper surface of a transparent glass plate of thickness t. The two surfaces of the plate are plane and parallel to each other. Show that the lateral displacement of the emergent beam is given by

$$d=t\sin\theta_1\left[1-\frac{\cos\theta_1}{\sqrt{(\mu^2-\sin^2\theta_1)}}\right]$$

where μ is the refractive index of the glass plate.

Solution: In view of the above problem, we have

$$d=t\,\frac{\sin(\theta_1-\theta_1')}{\cos\theta_1'}$$

$$=\frac{t}{\cos\theta_1'}[\sin\theta_1\cos\theta_1'-\cos\theta_1\sin\theta_1']$$

$$= t\left[\sin\theta_1 - \cos\theta_1 \frac{\sin\theta_1'}{\cos\theta_1'}\right]$$

Now $\mu = \dfrac{\sin\theta_1}{\sin\theta_1'}$ or $\sin\theta_1' = \dfrac{\sin\theta_1}{\mu}$

$$\cos\theta_1' = \sqrt{1 - \frac{\sin^2\theta_1}{\mu^2}}$$

$$\therefore \quad d = t\left[\sin\theta_1 - \cos\theta_1 \frac{\sin\theta_1}{\mu} \times \frac{\mu}{\sqrt{\mu^2 - \sin^2\theta_1}}\right]$$

$$= t\sin\theta_1\left[1 - \frac{\cos\theta_1}{\sqrt{\mu^2 - \sin^2\theta_1}}\right]$$

15.20. A ray of light *AB* is incident on the plane surface of a glass slab of thickness *t* at an angle ϕ. The two surfaces of the plate are plane and parallel to each other. The ray after refraction emerges from the opposite face, parallel to its initial direction but displaced sideways as shown in Fig. 15.15. Show that the lateral displacement *d* of the emergent beam is given by

$$d = t\phi\frac{(\mu - 1)}{\mu}$$

when ϕ is small, where μ is the refractive index of the glass slab.

Solution: In view of the geometry of the figure, we have $\tan\phi = \dfrac{CE}{t}$; *t* is the thickness of the glass slab.

$\therefore \quad CE = t\phi$; $\tan\phi = \phi$ when ϕ is small

$$\tan\theta = \frac{CD}{t}$$

$\therefore \quad CD = t\theta$

Now the lateral displacement

$$d \cong DE = CE - CD = t(\phi - \theta)$$

$$= t\phi\left[1 - \frac{\theta}{\phi}\right]$$

$$t\phi\left[1 - \frac{1}{\mu}\right]; \mu = \frac{\sin\phi}{\sin\theta} = \frac{\phi}{\theta}$$

$$= \frac{t\phi}{\mu}[\mu - 1]$$

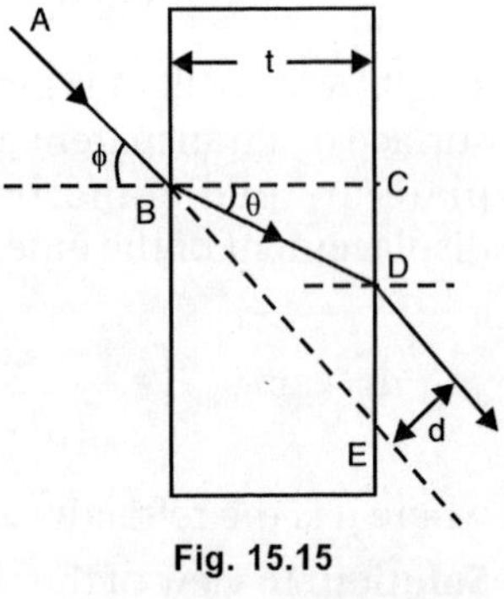

Fig. 15.15

15.21. A cubical vessel with opaque walls, is so placed that the eye of an observer cannot see its bottom but can see the entire wall *CD* (see Fig. 15.16a). A small object is placed at *O* at a distance $b = 10$ cm from the corner *D*. What minimum depth of water $\left(\mu = \frac{4}{3}\right)$ should be poured into the vessel which will enable the observer to see the object?

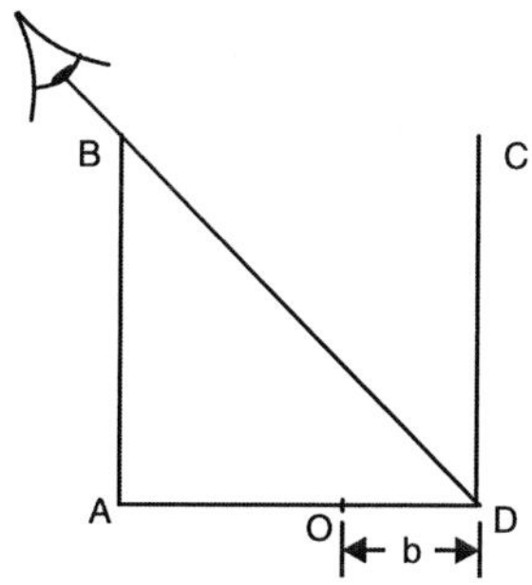

Fig. 15.16a

Solution:

From Snell's law $\mu = \dfrac{\sin i}{\sin r} = \dfrac{1}{\sqrt{2}\sin r}$

since $i = 45°$

In view of Fig. 15.16b, we have

$x \tan r = OQ = x - b$; where x is the depth of water

$$\therefore \quad x = \frac{b}{1 - \tan r} = \frac{b\sqrt{2\mu^2 - 1}}{\sqrt{2\mu^2 - 1} - 1}$$

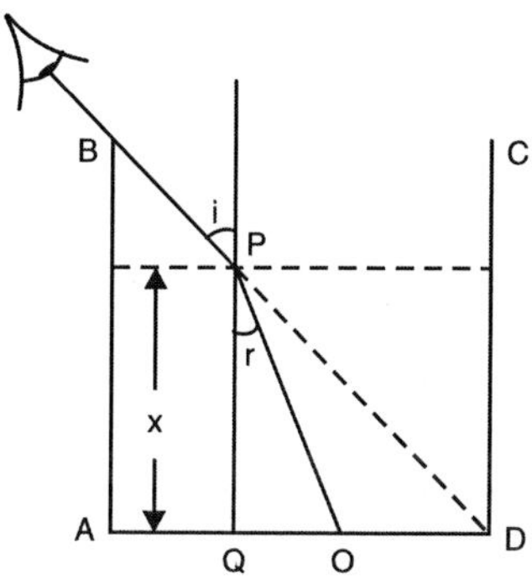

Fig. 15.16b

On substituting the values of b and μ we have

$$x = \frac{10\sqrt{23}}{\sqrt{23} - 3} = 26.7 \text{ cm}$$

15.22. A system consists of *N* transparent slabs of thicknesses d_1, d_2 ..... d_N and refractive indices μ_1, μ_2 ..... μ_N arranged as shown in the Fig. 15.17. Find the equivalent refractive index of the system.

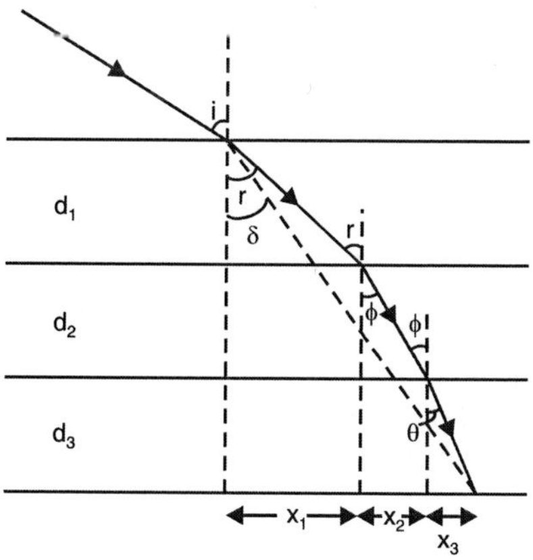

Fig. 15.17

Solution: In view of Fig. 15.17 we have

$$\mu_1 = \frac{\sin i}{\sin r}, \frac{\mu_2}{\mu_1} = \frac{\sin r}{\sin \phi}, \frac{\mu_3}{\mu_2} = \frac{\sin \phi}{\sin \theta} \text{ etc.}$$

$$\mu_2 = \frac{\sin r}{\sin \phi} \times \frac{\sin i}{\sin r} = \frac{\sin i}{\sin \phi}$$

$$\mu_3 = \frac{\sin \phi}{\sin \theta} \times \frac{\sin i}{\sin \phi} = \frac{\sin i}{\sin \theta}$$

$$\text{Now } \frac{d_1}{\mu_1} + \frac{d_2}{\mu_2} + \ldots\ldots \frac{d_N}{\mu_N} = \frac{d_1 \sin r + d_2 \sin \phi + d_3 \sin \theta + \ldots\ldots}{\sin i}$$

$$\cong \frac{x_1 + x_2 + \ldots\ldots}{\sin i}$$

$$= \frac{x}{\sin i}$$

$$\text{Now } \mu_{eq} = \frac{\sin i}{\sin \delta} = \frac{\sin i}{x}(d_1 + d_2 + \ldots\ldots d_N)$$

$$= \frac{d_1 + d_2 + \ldots\ldots d_N}{\dfrac{d_1}{\mu_1} + \dfrac{d_2}{\mu_2} + \dfrac{d_3}{\mu_3} + \ldots\ldots \dfrac{d_N}{\mu_N}}$$

$$= \frac{\sum_{i=1}^{N} d_i}{\sum_{i=1}^{N} \dfrac{d_i}{\mu_i}}$$

15.23. A beam of light is incident on a spherical drop of water at an angle *i*. Find the angle θ through which the beam is deflected from the initial direction after a single reflection from the internal surface of the drop.

Solution: A beam of light is incident at *A* at an angle *i* (see Fig. 15.18). Due to refraction at *A*, it deviates from its path through an angle $i - r$. At *B*, the ray suffers a reflection and travels along *BC*. At *B* it deviates through an angle $\pi - 2r$ with reference to *AB*. At *C*, the ray refracts and emerges out of the drop at an angle *i*. The deviation at *C* with respect to *BC* is $i - r$. Therefore, the net deviation is $\theta = i - r + \pi - 2r + i - r = \pi + 2i - 4r$

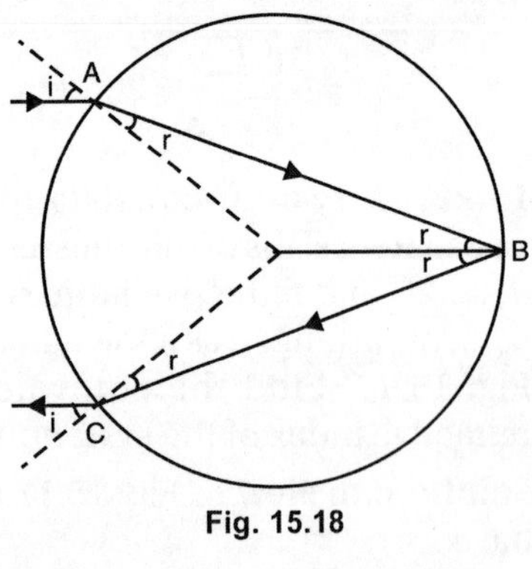

Fig. 15.18

15.24. A ray of sunlight is falling on a rain drop and after a refraction, a reflection from the internal surface of the drop and a refraction it is emerging out of it. Show that $\cos i = \sqrt{\dfrac{\mu^2 - 1}{3}}$ where μ is the refractive index of water and *i* is the angle of incidence.

Solution: In solving this problem, we shall be using the ray diagram of problem 15.23. The total deviation θ is given by $\theta = \pi + 2i - 4r$.

For the angle of deviation to be maximum or minimum, $\frac{d\theta}{di}$ must be equal to zero.

Thus $\frac{d\theta}{di} = 2 - 4\frac{dr}{di} = 0$ or $\frac{dr}{di} = \frac{1}{2}$

Now $\mu = \sin i/\sin r$ or $\sin i = \mu \sin r$

On differentiating it with respect to i, we have

$$\cos i = \mu \cos r \frac{dr}{di} = \frac{\mu \cos r}{2}; \frac{dr}{di} = \frac{1}{2}$$

or $$\cos^2 i = \frac{\mu^2}{4}\cos^2 r = \frac{\mu^2}{4}\left(1 - \sin^2 r\right) = \frac{\mu^2}{4}\left(1 - \frac{\sin^2 i}{\mu^2}\right) = \frac{\mu^2 - \sin^2 i}{4}$$

or $4\cos^2 i = \mu^2 - (1 - \cos^2 i)$

or $3\cos^2 i = \mu^2 - 1$

or $$\cos i = \sqrt{\frac{\mu^2 - 1}{3}}$$

15.25. A ray of sunlight is falling on a rain drop at an angle i. After a refraction, two reflections from the inner surface of the drop and refraction, it is emerging out of it (see Fig. 15.19). Show that

$$\cos i = \sqrt{\frac{\mu^2 - 1}{8}}$$

where μ = refractive index of water.

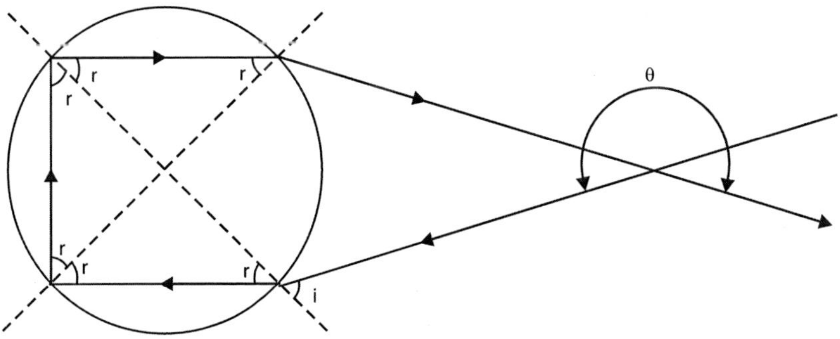

Fig. 15.19

Solution: In this case, the total deviation θ is given by

$$\theta = i - r + \pi - 2r + \pi - 2r + i - r$$
$$= 2\pi + 2i - 6r$$

For the angle of deviation θ to be maximum or minimum $\frac{d\theta}{di} = 0$

Thus $\frac{d\theta}{di} = 2 - 6\frac{dr}{di} = 0$ or $\frac{dr}{di} = \frac{1}{3}$

Now $\mu = \sin i/\sin r$ or $\sin i = \mu \sin r$

On differentiating it with respect to i we have

$$\cos i = \mu \cos r \frac{dr}{di} = \frac{\mu}{3} \cos r$$

$$9\cos^2 i = \mu^2 \cos^2 r = \mu^2 (1 - \sin^2 r)$$

$$= \mu^2 \left(1 - \frac{\sin^2 i}{\mu^2}\right) = \mu^2 - \sin^2 i$$

$$9\cos^2 i = \mu^2 - (1 - \cos^2 i) \text{ or } 8\cos^2 i = (\mu^2 - 1)$$

or $$\cos i = \sqrt{\frac{\mu^2 - 1}{8}}$$

15.26. A ray of light is refracted through a sphere of refractive index μ so as to pass through extremities of second radii, making an angle α. Show that if γ is the deviation of the ray (Fig. 15.20).

$$\cos\left(\frac{\alpha - \gamma}{2}\right) = \mu \cos \alpha/2$$

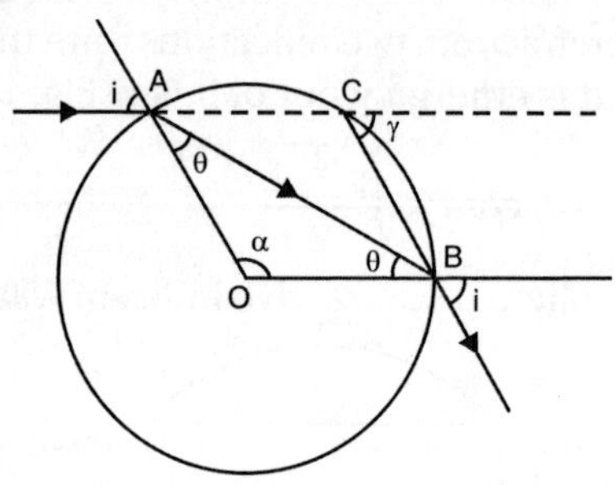

Fig. 15.20

Solution: In triangle ABO,

$$2\theta + \alpha = 180$$

$\therefore$ $$\theta = 90 - \alpha/2$$

Angle of deviation

$$\gamma = \angle CAB + \angle CBA = 2(i - \theta)$$

$\therefore$ $$\frac{\gamma}{2} + \theta = i$$

or $$\frac{\gamma}{2} + 90 - \frac{\alpha}{2} = i$$

$$90 - \left(\frac{\alpha - \gamma}{2}\right) = i$$

According to Snell's law $$\mu = \frac{\sin i}{\sin\theta} = \frac{\sin\left[90 - \left(\frac{\alpha - \gamma}{2}\right)\right]}{\sin\left(90 - \frac{\alpha}{2}\right)}$$

or $$\cos\left(\frac{\alpha - \gamma}{2}\right) = \mu \cos\frac{\alpha}{2}$$

15.27. A ray of light passes through a transparent sphere of refractive index n and radius R. If b is the distance between the incident ray and a parallel diameter of the sphere, show that the angle of deviation is given by the expression. (Roorkee 1987)

$$\sin\frac{\theta}{2}=\frac{b}{nR^2}\left[\left(n^2R^2-b^2\right)^{1/2}-\left(R^2-b^2\right)^{1/2}\right]$$

Solution: A ray of light *AB* is incident at *B* at an angle *i* (Fig. 15.21). After refraction, it deviates through an angle $i-r$. Ray *BC* after refraction at *C* emerges along *CD* and shows a deviation $i-r$. Total deviation

$$\theta = 2i - 2r$$

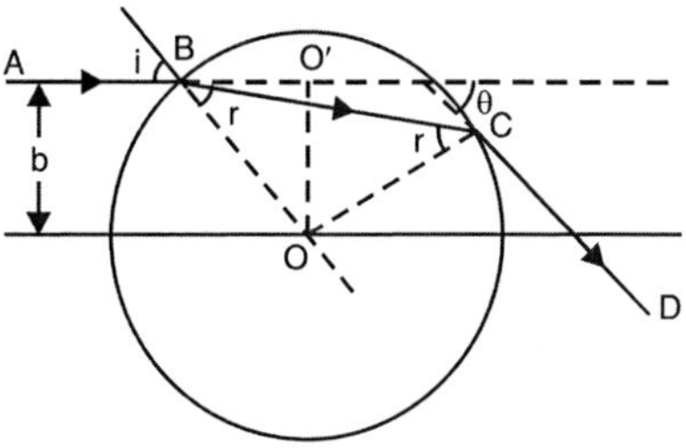

Fig. 15.21

Now $\sin i=\dfrac{b}{R}$ and

$$\cos i=\left(1-\frac{b^2}{R^2}\right)^{1/2}$$

According to Snell's law,

$$\frac{\sin i}{\sin r}=n$$

or $$\sin r=\frac{\sin i}{n}=\frac{b}{nR};\cos r=\left(1-\frac{b^2}{n^2R^2}\right)^{1/2}$$

$$\sin\frac{\theta}{2}=\sin(i-r)=\sin i\cos r-\cos i\sin r$$

$$=\frac{b}{R}\cdot\left(1-\frac{b^2}{n^2R^2}\right)^{1/2}-\left(1-\frac{b^2}{R^2}\right)^{1/2}\frac{b}{nR}$$

$$=\frac{b}{nR^2}\left[\left(n^2R^2-b^2\right)^{1/2}-\left(R^2-b^2\right)^{1/2}\right]$$

15.28. A beam of light passes symmetrically through a prism having the apex angle *A* as shown in the Fig. 15.22. Show that the angle of deviation is given by $\sin\dfrac{A+\theta}{2}=n\sin\dfrac{A}{2}$ where *n* is the refractive index of the material of the prism.

Solution: In the conditions of minimum deviation, the ray of light *B'C* will travel parallel to the base of the prism.

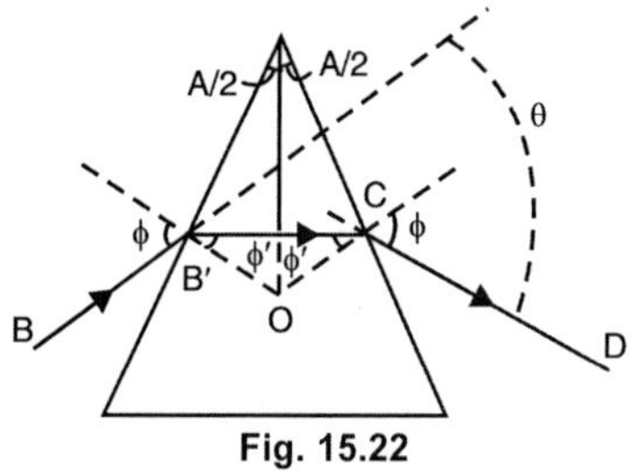

Fig. 15.22

Now $n=\dfrac{\sin\phi}{\sin\phi'}$

The angle of deviation $\theta = 2(\phi-\phi')$

$$\angle B'OC+\angle A=180° \text{ and } \angle 2\phi'+\angle B'OC=180°$$

$$\therefore\quad A=2\phi' \text{ or } \phi'=\frac{A}{2}$$

$$\therefore \quad \phi = \frac{\theta + A}{2}$$

$$\therefore \quad n = \frac{\sin\left(\frac{A+\theta}{2}\right)}{\sin\frac{A}{2}} \quad \text{or} \quad \sin\left(\frac{A+\theta}{2}\right) = n\sin\frac{A}{2}$$

15.29. A beam of light passes through a prism having the apex angle A as shown in Fig. 15.23. Show that for any angle of incidence on a prism,

$$\frac{\sin\left(\frac{A+\theta}{2}\right)}{\sin\frac{A}{2}} = n\frac{\cos(\phi_1' - \phi_2')/2}{\cos(\phi_1 - \phi_2)/2}$$

Solution: For any angle of incidence, the ray of light $B'C$ will not travel parallel to the base of the prism (see Fig. 15.23).

Angle of deviation $\theta = \phi_1 - \phi'_1 + \phi_2 - \phi'_2$ and $A = \phi'_1 + \phi'_2$

Therefore, $\phi_1 + \phi_2 = A + \theta$

According to Snell's law,

$$n = \frac{\sin\phi_1}{\sin\phi_1'} = \frac{\sin\phi_2}{\sin\phi_2'}$$

or $$\frac{\sin\phi_2}{\sin\phi_1} = \frac{\sin\phi_2'}{\sin\phi_1'}$$

or $$1 + \frac{\sin\phi_2}{\sin\phi_1} = 1 + \frac{\sin\phi_2'}{\sin\phi_1'}$$

or $$\frac{\sin\phi_1 + \sin\phi_2}{\sin\phi_1' + \sin\phi_2'} = \frac{\sin\phi_1}{\sin\phi_1'}$$

or $$\frac{2\sin\left(\frac{\phi_1+\phi_2}{2}\right)\cos\left(\frac{\phi_1-\phi_2}{2}\right)}{2\sin\left(\frac{\phi_1'+\phi_2'}{2}\right)\cos\left(\frac{\phi_1'-\phi_2'}{2}\right)} = n$$

or $$\frac{\sin\left(\frac{A+\theta}{2}\right)}{\sin\frac{A}{2}} = n\frac{\cos\left(\frac{\phi_1'-\phi_2'}{2}\right)}{\cos\left(\frac{\phi_1-\phi_2}{2}\right)}$$

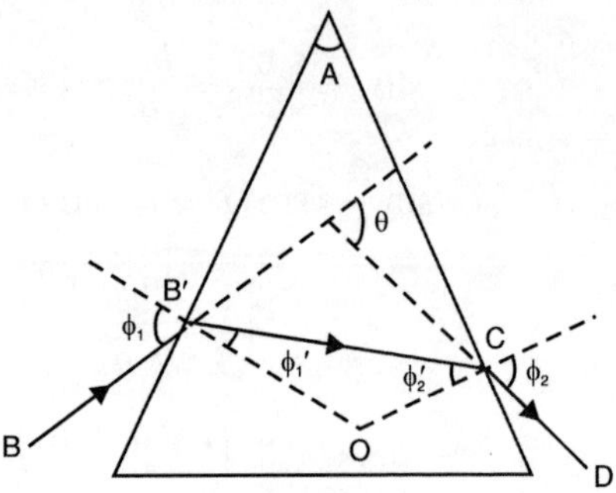

Fig. 15.23

15.30. A ray of light falls at grazing incidence on the face of a prism of refractive index μ and refracting angle A. Show that the angle of emergence is given by

$$\sin e = (\mu^2 - 1)^{1/2}\sin A - \cos A$$

Solution: The path of the ray is shown in Fig. 15.24.

$$\mu = \frac{\sin 90}{\sin \phi}, \frac{1}{\mu} = \frac{\sin \phi'}{\sin e} \text{ and } \phi + \phi' = A$$

$$\sin e = \mu \sin \phi' = \mu \sin (A - \phi)$$

$$= \mu \sin A \cos \phi - \mu \cos A \sin \phi$$

$$= \frac{\mu \left(\mu^2 - 1\right)^{1/2}}{\mu} \sin A - \cos A \text{ since } \mu \sin \phi = 1$$

$$\therefore \quad \sin e = (\mu^2 - 1)^{1/2} \sin A - \cos A$$

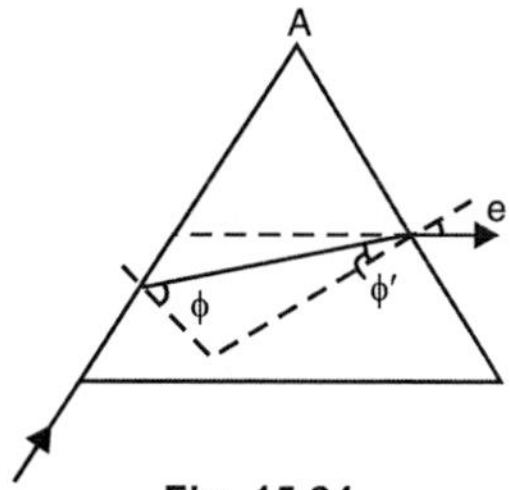

Fig. 15.24

15.31. The sides of an isosceles right prism are silvered. A ray of light falls on the hypotenuse at an angle ϕ_0. Show that the ray leaving the prism is parallel to the incident ray.

Solution: In triangle *DEF* of Fig. 15.25, we have

$$180 = 135 + \phi_1 + 90 - \phi_2$$

or $\quad \phi_2 = 45 + \phi_1$

In triangle *FGH*, we have

$$\phi_2 + \phi_3 = 90$$

In triangle *CGI*, we have

$$180 = 45 + 90 - \phi_3 + 90 - \phi_4$$

or $\quad \phi_3 + \phi_4 = 45$

or $\quad 90 - \phi_2 + \phi_4 = 45$

or $\quad 45 + \phi_4 = \phi_2 = 45 + \phi_1$

or $\quad \phi_1 = \phi_4$

Therefore $\phi_0 = \phi_5$

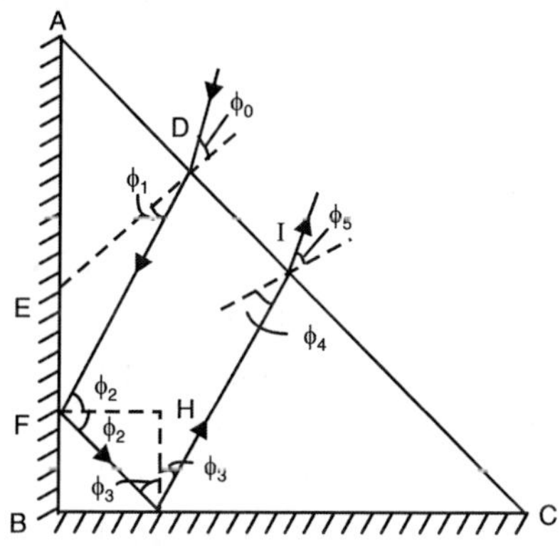

Fig. 15.25

15.32. Calculate the angle of emergence *i*, and the angle of deviation when light is incident at 90° on the face of a 60° prism of refractive index 1.5 (see Fig. 15.26).

Solution:

$$\frac{\sin 90}{\sin \theta} = 1.5$$

or $\quad \sin \theta = \dfrac{1}{1.5} = 0.66$

$$\therefore \quad \theta = 41.8°$$

$$\theta + r = 60°$$

$$\therefore \quad r = 60° - \theta = 60° - 41.8° = 18.2°$$

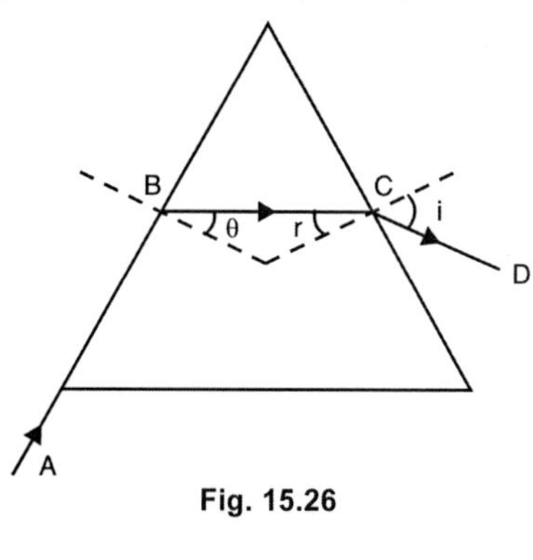

Fig. 15.26

$$\frac{\sin i}{\sin r} = 1.5$$

$\therefore$ $\sin i = 1.5 \sin r = 1.5 \sin 18.2° = 0.4685$

$\therefore$ $i = 27.9°$

Total deviation $= 90° - \theta + i - r$

$= 90° - 41.8° + 27.9° - 18.2° = 57.9°$

15.33. In Fig. 15.27, C and F are two prisms of crown and flint glass of refracting angles α and α' respectively. Assuming the dispersive power of the material of the two prisms as ω and ω' find (i) the condition in which the combination will show only deviation (ii) the total deviation produced by the combination.

Solution: (i) Let μ_v, μ_y and μ_r and μ'_v, μ'_y & μ'_r be the refractive indices of the two materials for violet, yellow and red rays then the deviations of these rays for the two glasses is given by

$$\left.\begin{aligned} \delta_v &= (\mu_v - 1)\,\alpha \\ \delta_y &= (\mu_y - 1)\,\alpha \\ \delta_r &= (\mu_r - 1)\,\alpha \end{aligned}\right\} \text{ for crown glass}$$

and $$\left.\begin{aligned} \delta'_v &= (\mu'_v - 1)\,\alpha' \\ \delta'_y &= (\mu'_y - 1)\,\alpha' \\ \delta'_r &= (\mu'_r - 1)\,\alpha' \end{aligned}\right\} \text{ for flint glass}$$

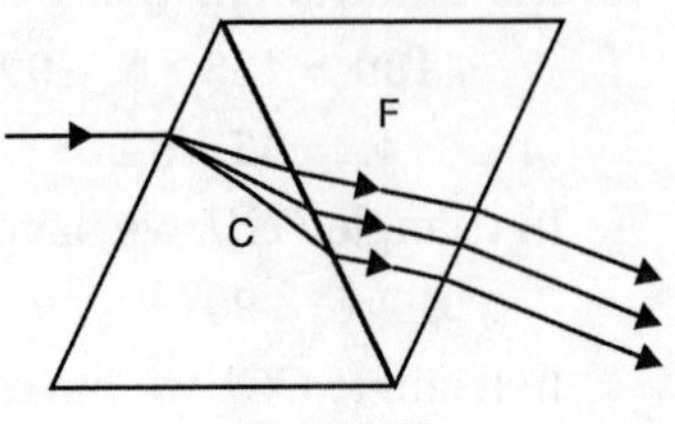

Fig. 15.27

For no dispersion the net angular separation for violet and red rays must be equal to zero i.e.

$$(\delta_v - \delta_r) + (\delta'_v - \delta'_r) = 0$$

or $$(\mu_v - \mu_r)\,\alpha + (\mu'_v - \mu'_r)\,\alpha' = 0$$

or $$\frac{\alpha}{\alpha'} = -\left(\frac{\mu'_v - \mu'_r}{\mu_v - \mu_r}\right)$$

The negative sign shows that the refracting angles of the two prism are opposite.

(ii) Net deviation $\delta = \delta_y + \delta'_y$

$$= (\mu_y - 1)\,\alpha + (\mu'_y - 1)\,\alpha'$$

$$= (\mu_y - 1)\,\alpha + (\mu'_y - 1) \times \frac{-(\mu_v - \mu_r)}{(\mu'_v - \mu'_r)}\,\alpha$$

$$= (\mu_v - \mu_r)\left[\frac{(\mu_y - 1)}{(\mu_v - \mu_r)} - \frac{(\mu'_y - 1)}{(\mu'_v - \mu'_r)}\right]\alpha$$

$$= (\mu_v - \mu_r)\left[\frac{1}{\omega} - \frac{1}{\omega'}\right]\alpha$$

15.34. In Fig. 15.28 C and F are two prisms of crown and flint glass of refracting angles α and α' respectively. Assuming the dispersive power of the material of the two prisms as ω and ω' find (i) the condition in which the combination will show only dispersion (ii) the total dispersion produced by the combination.

Solution: (i) In view of Q. No. 15.33 the deviations δ_y and δ'_y for the two prisms is given by

$$\delta_y = (\mu_y - 1)\,\alpha$$

and $$\delta'_y = (\mu'_y - 1)\,\alpha'$$

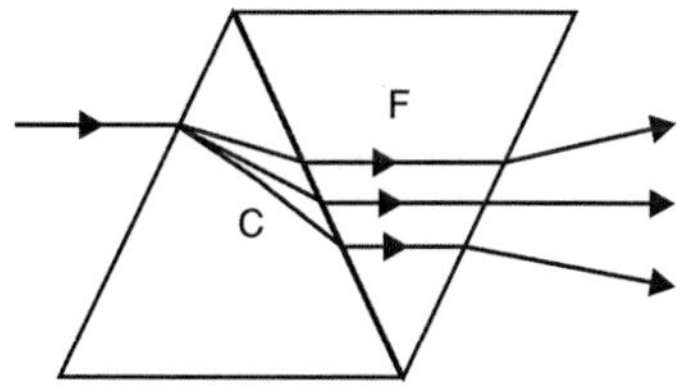

Fig. 15.28

For dispersion only the net deviation must be equal to zero i.e.

$$\delta_y + \delta'_y = (\mu_y - 1)\,\alpha + (\mu'_y - 1)\,\alpha' = 0$$

or $$\alpha' = -\frac{(\mu_y - 1)\,\alpha}{(\mu'_y - 1)}$$

(ii) The dispersion produced by the two prisms is given by

$$\delta_v - \delta_r = (\mu_v - \mu_r)\,\alpha$$

and $$(\delta'_v - \delta'_r) = (\mu'_v - \mu'_r)\,\alpha'$$

Net dispersion $$D = (\delta'_v - \delta'_r) + (\delta_v - \delta_r)$$

$$= (\mu'_v - \mu'_r)\,\alpha' + (\mu_v - \mu_r)\,\alpha$$

$$(\mu_v - \mu_r)\,\alpha - \frac{(\mu'_v - \mu'_r)(\mu_y - 1)\,\alpha}{(\mu'_y - 1)}$$

$$= (\mu_y - 1)\,\alpha\left[\frac{(\mu_v - \mu_r)}{(\mu_y - 1)} - \frac{(\mu'_v - \mu'_r)}{(\mu'_y - 1)}\right]$$

$$D = \alpha\,(\omega - \omega')(\mu_y - 1)$$

15.35. A ray of light is incident at an angle of 60° on one face of a prism which has an angle of 30°. The ray emerging out of the prism makes an angle of 30° with the incident ray. Show that the emergent ray is perpendicular to the face through which it emerges and calculate the refractive index of the material of the prism. (IIT 1978)

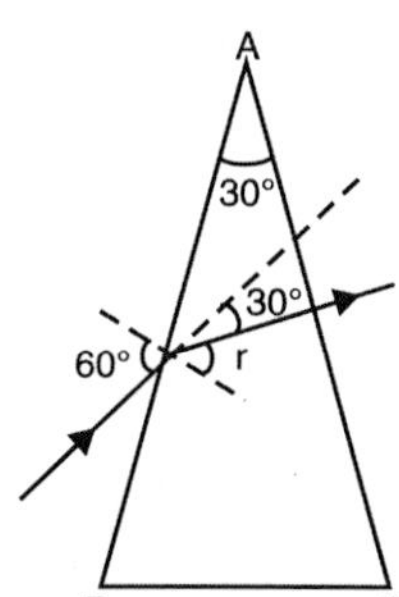

Fig. 15.29

Solution:

$\delta = 60 + i - 30$; i = angle with which the ray will go out of the prism

$30 = 60 + i - 30$

$\therefore \quad i = 0$

The ray will emerge perpendicular to the face *AC* of the prism.

$< r = 30$

$$\mu = \frac{\sin 60}{\sin 30} = \sqrt{3}$$

15.36. A ray of light undergoes a deviation of 30° when incident on equilateral prism of refractive index $\sqrt{2}$. What is the angle subtended by the ray inside the prism with the base of the prism. (IIT 1992)

Solution:

$$\mu = \frac{\sin\frac{A+\delta}{2}}{\sin\frac{A}{2}} = \frac{\sin\frac{60+30}{2}}{\sin 30} = \sqrt{2}$$

This indicates that the value of δ is minimum and under this condition the ray of light will travel parallel to the base of the prism. Therefore the angle subtended by the ray inside the prism with the base of the prism is zero.

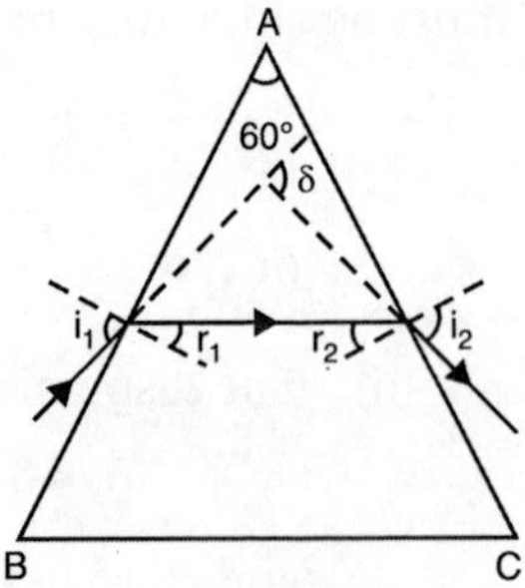

Fig. 15.30

15.37. A prism of refracting angle 45° is made from a glass of refractive index 1.55. Find the maximum value of the angle of incidence of a ray at which no total internal reflection should occur when the ray leaves the prism.

Solution:

$$\frac{1}{1.55} = \sin r_2 = 0.6451$$

$r_2 = 40°\ 10'$

$r_1 = 45 - 40°\ 10' = 4°\ 50'$

$$1.55 = \frac{\sin i}{\sin r_1}$$

$\sin i = 1.55 \times \sin 4°\ 50'$

$= 1.55 \times 0.0943$

$= 0.1462$

$i = 8°\ 24'$

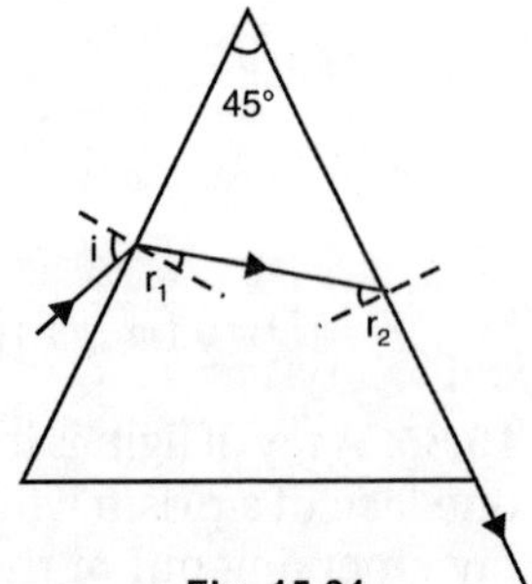

Fig. 15.31

15.38. Two identical isosceles prism of prism angle *A* and refractive index μ are placed with their bases touching each other. This system can act as a crude converging lens. Draw a neat diagram showing the path of rays which are parallel at incidence. Obtain the 'focal length' of this system if *A* = 10°,

$\mu = 1.5$ and the ray of light is incident on a prism at a height of 1 cm from the base. (Roorkee 1977)

Solution: The path of light is shown in Fig. 15.32.

$$\tan\delta = \frac{h}{f}$$

For small angles

$$\tan\delta = \delta = (\mu - 1)A$$

$$\therefore \quad \frac{h}{(\mu-1)A} = f$$

$$10° = \frac{\pi}{18} \text{ radian}$$

$$\therefore \quad f = \frac{1 \times 18}{0.5 \times 3.14} = 11.5 \text{ cm}$$

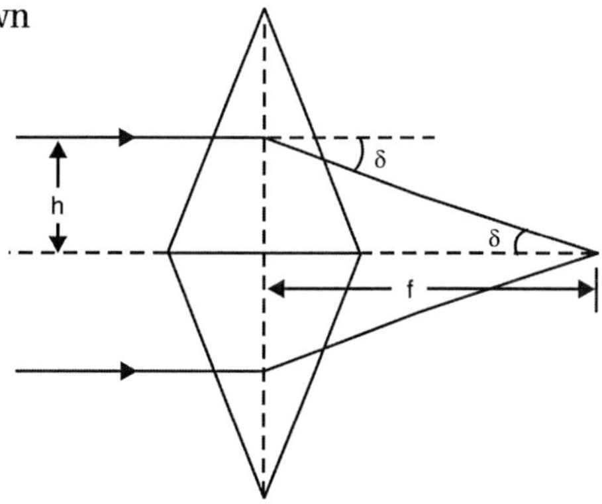

Fig. 15.32

15.39. A ray incident on one face of a prism escapes from the other face after refraction (Fig. 15.33). Calculate the maximum value of the angle of prism if the refractive index of the material of the prism is 1.5.

Solution: Let A be the angle of the prism. In order to get rid of total internal reflection, we must have $\sin\theta \le \frac{1}{\mu}$ where μ is the refractive index of the material of the prism.

$$\text{or} \quad \theta = \sin^{-1}\frac{2}{3}$$

Now $A = \theta + r$

If r is large, A is large, since θ is constant. The maximum value of r can be obtained from the equation.

$$\sin r = \frac{1}{\mu}$$

$$\text{or} \quad r = \sin^{-1}\frac{2}{3}$$

Thus $A_{max} = 2\sin^{-1}\frac{2}{3} = 83°40'$

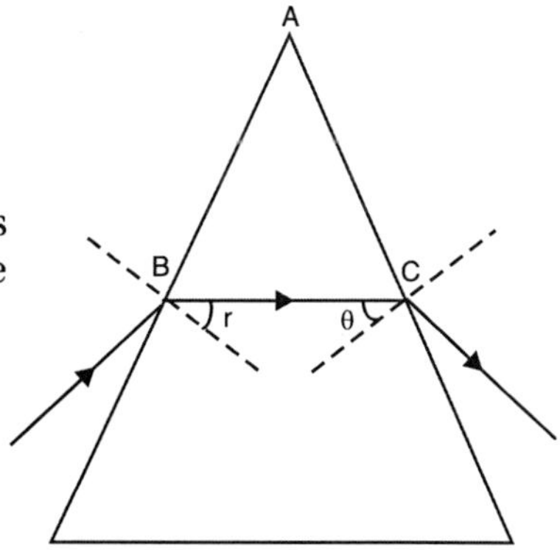

Fig. 15.33

15.40. A beam of light enters one face of a prism at an angle α and emerges from the other face at an angle β (Fig. 15.34). Due to refraction, the total deviation of the ray is γ. Determine the refractive index of the material of the prism.

Solution: The angle of deviation $= \gamma = \alpha - r + \beta - r_1$; $r + r_1 = A$

$\therefore \quad \gamma = \alpha + \beta - A$

or $\quad A = \alpha + \beta - \gamma$

$$n = \frac{\sin\left(\frac{\alpha+\beta}{2}\right)}{\sin\left(\frac{\alpha+\beta-\gamma}{2}\right)}$$

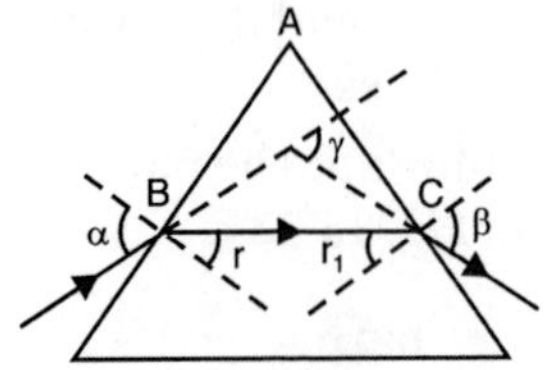

Fig. 15.34

15.41. A ray of light is incident at D on an equilateral prism. After refraction, the ray travels along DE. On the face BC of the prism, the ray is incident at the critical angle for total internal reflection to just occur. The total angle of deviation after the refraction at the face AC is 120°. Find the refractive index of the material of the prism.

Solution: The total deviation $\theta = 120° = i - r + 180 - 2\phi + i' - r'$

In triangle CEF, we have

$90 - \phi + 90 - r' + 60 = 180°$

or $\quad r' + \phi = 60°$

Further $r + \phi = 60°$

$\therefore \quad r = r'$ and therefore $i = i'$

Now $120° = i + 180 - 2\phi + i - 2r$

$120° = 2i + 180 - 120$

$i = 30°$

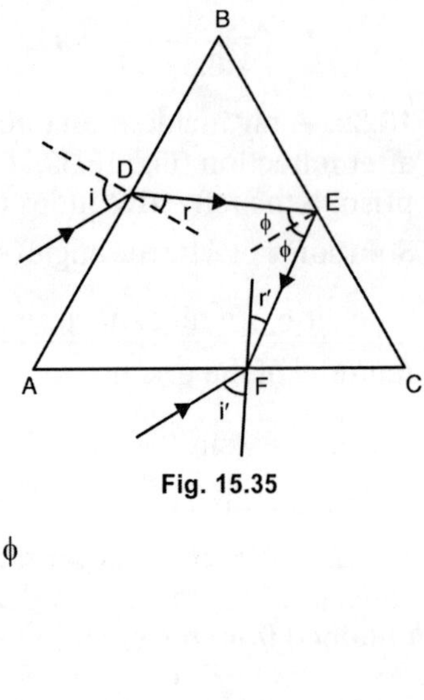

Fig. 15.35

Now $\dfrac{\sin 30}{\sin r} = n = \dfrac{1}{\sin\phi}$

or $\quad 0.5 \sin\phi = \sin r = \sin(60 - \phi)$

or $\quad 0.5 \sin\phi = \sin 60 \cos\phi - \cos 60 \sin\phi$

or $\quad 0.5 \sin\phi = \dfrac{\sqrt{3}}{2}\cos\phi - 0.5\sin\phi$

or $\quad \sin\phi = \dfrac{\sqrt{3}}{2}\sqrt{(1-\sin^2\phi)}$

or $\quad 4\sin^2\phi = 3(1 - \sin^2\phi)$ or $\sin\phi = \sqrt{\dfrac{3}{7}} = 0.6546$

$\therefore \quad \phi = 40°54'$

$r = 60 - \phi = 60 - 40°54' = 19°6'$

$$n = \frac{\sin i}{\sin r} = \frac{0.5}{\sin 19°6'} = \frac{0.5}{0.3272} = 1.528$$

15.42. An observer is at a depth $h = 3$ m below the surface of calm water $\left(\mu = \dfrac{4}{3}\right)$. Calculate the radius of a hole through which objects situated outside the water can be seen.

Solution: The observer can see objects situated outside water only when angle $BAD = \phi$ = critical angle (see Fig. 15.36)

Now $\sin\phi = \dfrac{1}{\mu}$; $\tan\phi = \dfrac{1}{\sqrt{\mu^2 - 1}}$

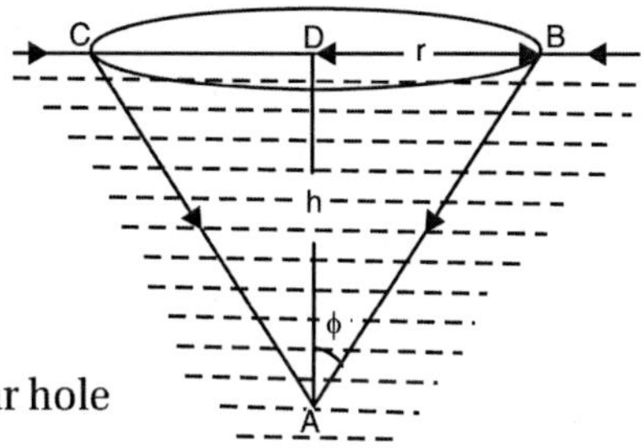

Fig. 15.36

In view of the figure, we have

$$\tan\phi = \frac{1}{\sqrt{\mu^2 - 1}} = \frac{r}{h}$$

where $r = BD$ = radius of the circular hole

$$\therefore \quad r = \frac{h}{\sqrt{\mu^2 - 1}}$$

On substituting the values of h and μ we have

$$r = \frac{3}{\sqrt{(4/3)^2 - 1}} = \frac{3}{0.8819} = 3.40 \text{ m}$$

15.43. A cubical block of glass of refractive index $n_1 = 1.5$ is in contact with the surface of water of refractive index $n_2 = 4/3$. A beam of light is incident on the vertical face of the block (see Fig. 15.37). After refraction, a total internal reflection at the base and a refraction at the opposite vertical face, the ray emerges out at an angle θ. Calculate the value of θ.

Solution:

$$\frac{\sin i}{\sin\theta} = \frac{1}{n_1}$$

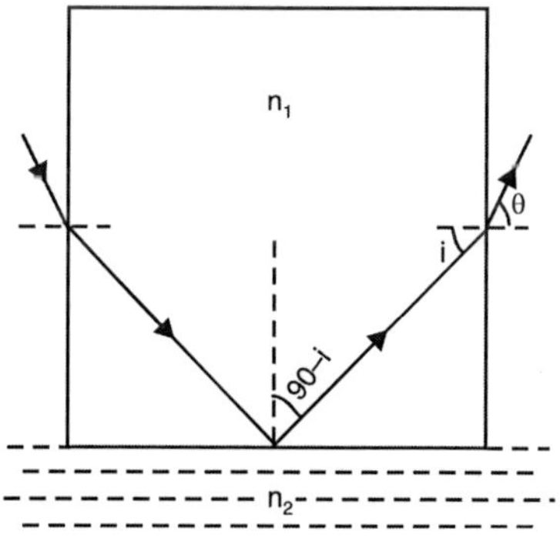

Fig. 15.37

and $$\frac{\sin(90 - i)}{\sin 90} = \frac{n_2}{n_1}$$

$$\therefore \quad \cos i = \frac{n_2}{n_1} = \frac{\sqrt{n_1^2 - \sin^2\theta}}{n_1}$$

or $$n_2 = \sqrt{n_1^2 - \sin^2\theta}$$

or $$n_1^2 - n_2^2 = \sin^2\theta$$

or $$\sin\theta = \sqrt{n_1^2 - n_2^2}$$

$$= \sqrt{(1.5)^2 - (4/3)^2} = 0.6871$$

$$\therefore \quad \theta = 43°24'$$

15.44. A plane boundary separates the two media of refractive indices n_1 and n_2 respectively such that $n_1 > n_2$. A beam of light coming from the denser medium strikes the boundary at an angle θ (see Fig. 15.38), such that the reflected and refracted rays are perpendicular to each other. Find the relative refractive index of these media if $\frac{\sin\phi}{\sin\theta} = \eta = 1.28$ where φ is the critical angle.

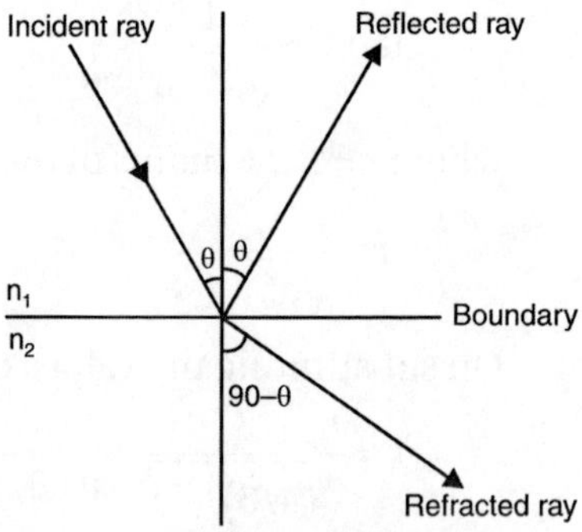

Fig. 15.38

Solution:

$$\sin\phi = \frac{n_2}{n_1} \text{ and } \frac{\sin\theta}{\sin(90-\theta)} = \frac{n_2}{n_1}$$

or $$\frac{n_1}{n_2}\sin\theta = \cos\theta = \sqrt{1-\sin^2\theta}$$

$$\frac{\sin\phi}{\sin\theta} = \frac{n_2}{n_1\sin\theta} = \eta$$

∴ $$\sin\theta = \frac{n_2}{n_1\eta}$$

On substituting the value of sin θ in the above equation, we have

$$\frac{n_1}{n_2}\frac{n_2}{n_1\eta} = \sqrt{1-\frac{n_2^2}{n_1^2\eta^2}}$$

or $$\frac{1}{\eta} = \frac{\sqrt{n_1^2\eta^2 - n_2^2}}{n_1\eta}$$

or $$n_1 = n_2\sqrt{\left(\frac{n_1}{n_2}\eta\right)^2 - 1}$$

or $$1 = \left(\frac{n_1}{n_2}\right)^2(\eta^2-1)$$

or $$\frac{n_1}{n_2} = \frac{1}{\sqrt{\eta^2-1}}$$

On substituting the value of η, we have

$$\frac{n_1}{n_2} = \frac{1}{\sqrt{(1.28)^2-1}} = 1.25$$

15.45. A glass wedge of refractive index $n_1 = 1.5$ is lowered into ethyl alcohol of refractive index $n_2 = 1.36$. At what angle θ will the beam of light, normally incident on *AC* will reach *AB* entirely (see Fig. 15.39)?

Solution: The angle of incident on the face *BC* is equal to θ. When

$\sin\theta = \dfrac{n_2}{n_1}$ then the ray will travel parallel to CB. When $\sin\theta > \dfrac{n_2}{n_1}$ then total internal reflection will take place and the ray will reach the face AB entirely.

$$\sin\theta > \frac{1.36}{1.5} = 0.9066 = \sin 65°$$

$$\therefore \quad \theta > 65°$$

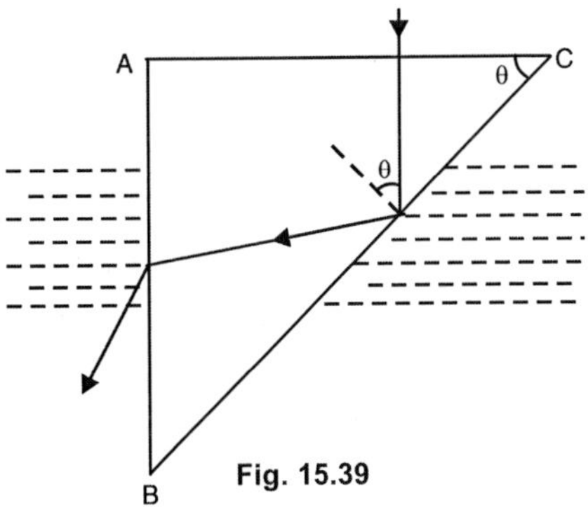

Fig. 15.39

15.46. Show that (i) no ray can pass through a prism whose angle N is greater than twice the critical angle ϕ, (ii) a ray incident at an angle θ can pass through the prism when $\theta > \sin^{-1}[\sin A \cot\phi - \cos A]$.

Solution: (i) A ray of light is incident on face MN at grazing incidence (see Fig. 15.40a). It will refract along AB at the critical angle ϕ. In order to have a grazing emergence at the face NO, the ray will be incident at the critical angle ϕ. Within the triangle ABP, the sum of the angles $PAB + PBA = 2\phi$ and will be equal to the angle of the prism. Thus in order to have refraction at both the faces, the maximum value of the angle of prism should be equal to 2ϕ.

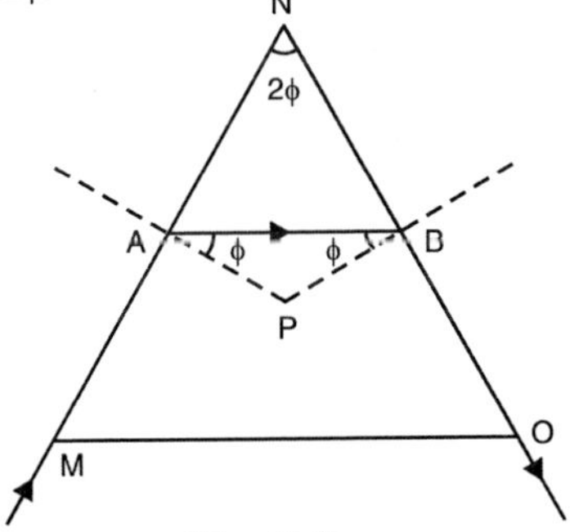

Fig. 15.40a

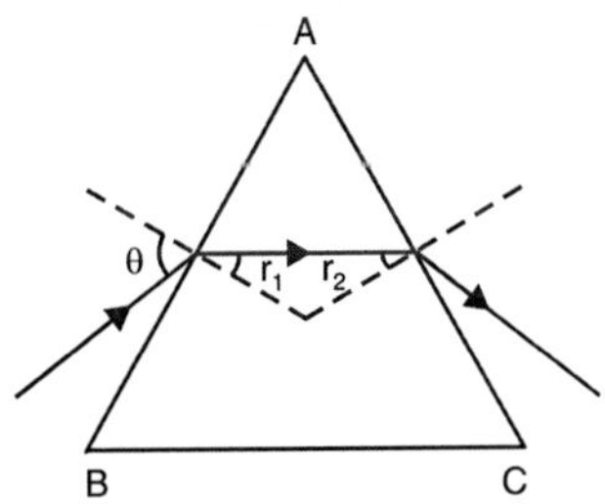

Fig. 15.40b

(ii) According to Snell's law

$$\mu = \frac{\sin\theta}{\sin r_1}; r_1 = r_2 = A \text{ (see Fig. 15.40b)}$$

$$\sin r_1 = \frac{\sin\theta}{\mu}$$

$$\therefore \; r_1 = \sin^{-1}\left(\frac{\sin\theta}{\mu}\right)$$

If the ray follows the path shown in Fig. 15.40b after refraction at face AC, we must have $r_2 < \phi$.

$$\text{Hence } r_2 = A - r_1 = A - \sin^{-1}\left(\frac{\sin\theta}{\mu}\right) < \phi$$

or $\quad \sin(A-\phi) < \sin\theta \sin\phi$ since $\sin\phi = \dfrac{1}{\mu}$

$\quad \sin A \cot\phi - \cos A < \sin\theta$

or $\quad \theta > \sin^{-1}[\sin A \cot\phi - \cos A]$

15.47. A rectangular glass block of thickness 10 cm and refractive index 1.5 placed over a small coin (see Fig. 15.41). A beaker is filled with water of refractive index 4/3 to a height of 10 cm and is placed over the glass block.

(a) Find the apparent position of the object when it is viewed at near normal incidence.

(b) If the eye is slowly moved away from the normal at a certain position, the object is found to disappear, due to total internal reflection. At what surface does this happen and why?

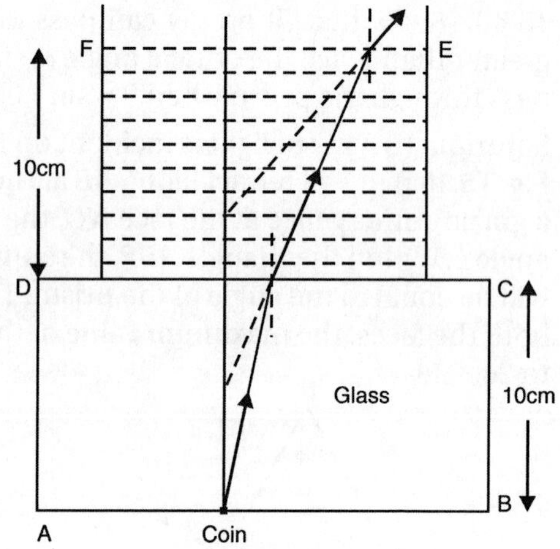

Fig. 15.41

Solution: (a) The two mediums will produce a shift given by

$$x = 10\left(1-\frac{2}{3}\right)+10\left(1-\frac{3}{4}\right)$$

$$= 3.3 + 2.5 = 5.8 \text{ cm}$$

Thus the coin will appear to be raised by 5.8 cm.

The apparent depth of the coin = 20 – 5.8 = 14.2 cm

(b) At CD interface the ray of light will pass from a denser to rare medium.

$$\sin C_1 = \frac{1}{{}^{w}\mu_g} = \frac{{}^{a}\mu_w}{{}^{a}\mu_g} = \frac{\frac{4}{3}}{\frac{3}{2}} = \frac{8}{9}$$

$$\therefore\ C_1 = \sin^{-1}\frac{8}{9}$$

or $\quad C_1 = 62°45'$

At EF interface the ray of light passes from water to air.

$$\therefore \quad \sin C_2 = \frac{1}{{}^a\mu_w} = \frac{3}{4}$$

$$C_2 = \sin^{-1}\frac{3}{4}$$

$$C_2 = 49°$$

The critical angle in the second case is small due to this total internal reflection occurs at the upper surface of water. As the eye is moved away from the normal the object will disappear.

15.48. A right angle prism (45° – 90° – 45°) of refractive index n has a plate of refractive index ($n_1 < n$) cemented to its diagonal face. The assembly is in air. A ray is incident on AB (Fig. 15.42a) (i) Calculate the angle of incidence at AB for which the ray strikes the diagonal face at the critical angle. (ii) Assuming $n = 1.352$, calculate the angle of incidence at AB for which the refracted ray passes through the diagonal face undeviated. (IIT 1996)

Solution: (i) $n \sin \phi = n_1 \sin 90$

or $\quad \sin \phi = \dfrac{n_1}{n}$

$$\cos \phi = \frac{1}{n}\sqrt{n^2 - n_1^2}$$

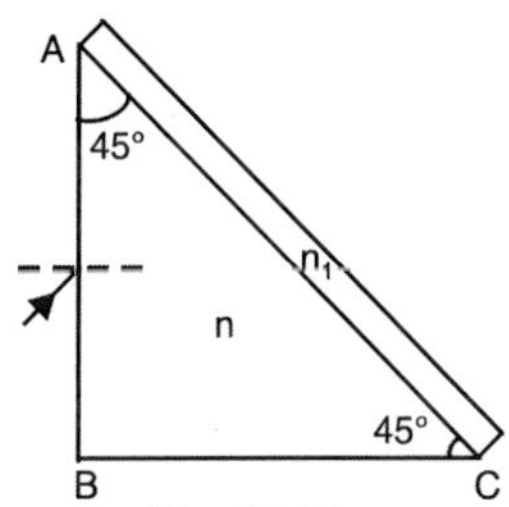

Fig. 15.42a

From Fig. 15.42b

$$90 - r + 90 - \phi + 45 = 180°$$

or $\quad r = 45 - \phi$

$$\sin i = n \sin r = n \sin [45 - \phi]$$

$$= n [\sin 45 \cos \phi - \cos 45 \sin \phi]$$

$$= n\left[\frac{1}{\sqrt{2}}\cdot\frac{1}{n}\cdot\sqrt{n^2 - n_1^2} - \frac{1}{\sqrt{2}}\cdot\frac{n_1}{n}\right]$$

$$\sin i = \frac{1}{\sqrt{2}}\left[\sqrt{n^2 - n_1^2} - n_1\right]$$

or $$i = \sin^{-1}\left[\frac{1}{\sqrt{2}}\left(\sqrt{n^2 - n_1^2} - n_1\right)\right]$$

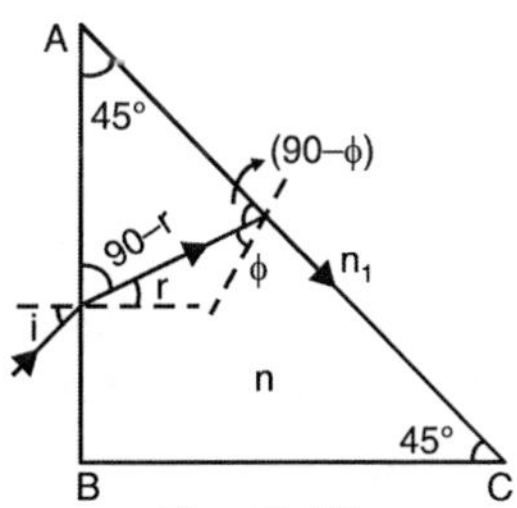

Fig. 15.42b

(ii) In view of Fig. 15.42c

$$r = 45°$$

$$\therefore \quad \sin i = 1.352 \sin 45$$

$$= \frac{1.352}{1.414} = 0.9588$$

$$\therefore \quad i = 73.5°$$

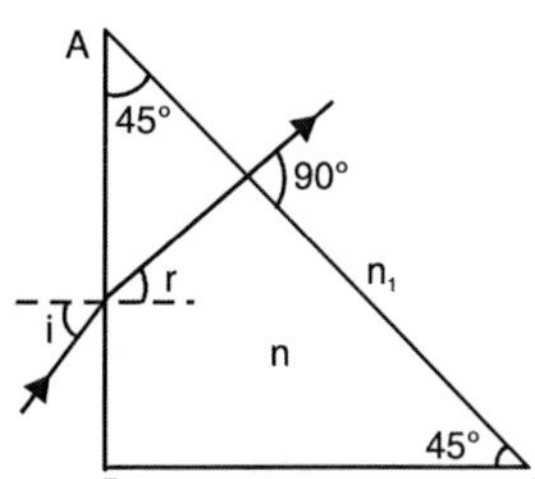

Fig. 15.42c

15.49. A rectangular block of glass is placed on a printed page lying on a horizontal surface. Find the minimum value of the refractive index of glass for which the letters on the page are not visible from any of the vertical faces of the block. (IIT 1979)

Solution: The light coming out will not emerge from the face BC of the block if $i >$ critical angle C.

$$i > C$$

$$\sin i > \sin C$$

or $$\sin i > \frac{1}{\mu}$$

$$\mu = \frac{\sin\theta}{\sin r}$$

Further $r + i = 90 \;\therefore\; r = 90 - i$

$$\mu = \frac{\sin\theta}{\cos i} = \frac{\sin\theta}{\sqrt{1-\sin^2 i}}$$

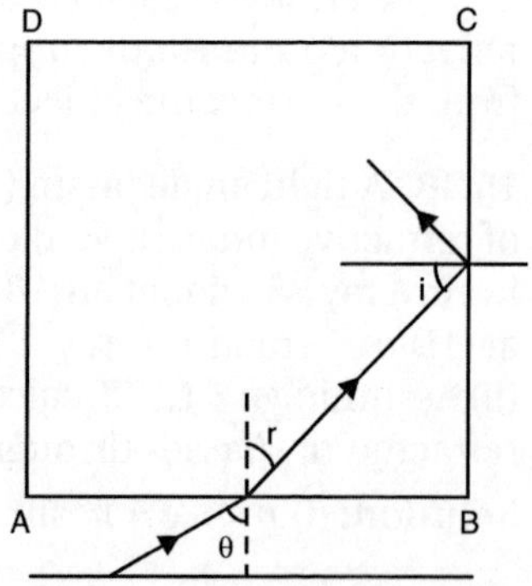

Fig. 15.43

or $$\sqrt{1-\frac{\sin^2\theta}{\mu^2}} > \frac{1}{\mu}$$

or $$\mu^2 > 1 + \sin^2\theta$$

The maximum value of $\sin\theta = 1$

$\therefore$ $$\mu^2 > 2$$

or $$\mu \geq \sqrt{2}$$

15.50. A ray of light travelling in glass $\left(\mu = \frac{3}{2}\right)$ is incident on a horizontal glass air surface at critical angle θ_C. If a thin layer of water $\left(\mu = \frac{4}{3}\right)$ is now poured on the glass air surface, at what angle will the ray of light emerge into air at the water air surface. (MNR 1991)

Solution:

$${}^a\mu_g \sin\theta_C = {}^a\mu_w \sin r$$

$${}^a\mu_g \times \frac{1}{{}^a\mu_g} = {}^a\mu_w \sin r$$

$$\sin r = \frac{1}{{}^a\mu_w}$$

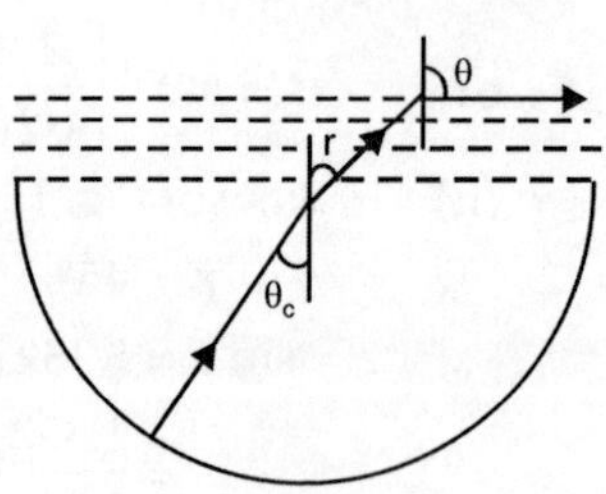

Fig. 15.44

At the water air boundary

$$\frac{\sin r}{\sin\theta} = \frac{1}{{}^a\mu_w}$$

$$\therefore \quad \sin\theta = 1$$

$$\text{or} \quad \theta = 90°$$

15.51. A bulb is placed at a depth of $2\sqrt{7}$ m in water and a floating opaque disc is placed over the bulb so that the bulb is not visible from the surface. What is the minimum diameter of the disc ? ${}^a\mu_w = \frac{4}{3}$. (IIT 1970)

Solution: At the water air surface

$$\sin\theta = \frac{1}{{}^a\mu_w}$$

From the geometry of the Fig. 15.45

$$\tan\theta = \frac{r}{h} = \frac{q}{\sqrt{{}^a\mu_w^2 - 1}}$$

$$\text{or} \quad r = \frac{h}{\sqrt{{}^a\mu_w^2 - 1}} = \frac{2\sqrt{7}}{\sqrt{\frac{16}{9} - 1}} = 6 \text{ m}$$

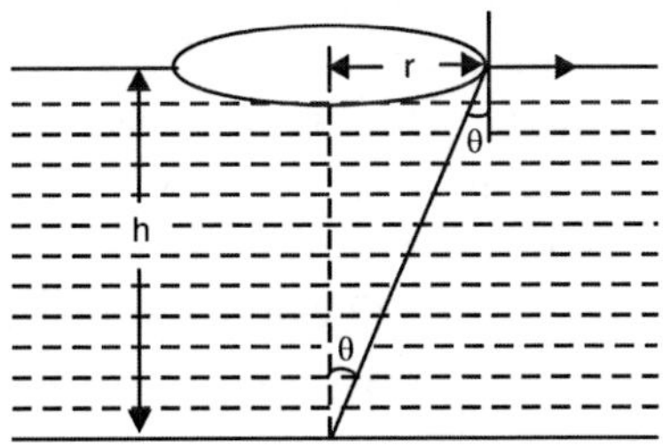

Fig. 15.45

$\therefore$ Diameter of the disc = 12 m

15.52. Light falls on one end face of a glass rod at angle α (see Fig. 15.46). What is the smallest value of refractive index that the glass may have so that light after entering the rod cannot leave it through a lateral face, irrespective of the value of α?

Solution: According to Snell's law

$\frac{\sin\alpha}{\sin\beta} = n$ where n is the refractive index of the glass.

In view of the geometry of the figure, the ray of light will be incident on the face PQ at an angle $(90 - \beta)$. If $(90 - \beta)$ is equal to the critical angle, then the ray will pass along PQ. In order to have total internal reflection, $(90 - \beta)$ should be greater than the critical angle. In other words,

$$\sin(90 - \beta) > \frac{1}{n}$$

$$\text{or} \quad \cos\beta > \frac{1}{n}$$

The maximum value of β will be obtained by taking $\alpha = 90$, i.e.

$$\frac{1}{\sin\beta} = n \text{ or } \sin\beta = \frac{1}{n}$$

$$\sin^2\beta + \cos^2\beta = 1 > \frac{2}{n^2}$$

$$\text{or} \quad n > \sqrt{2}$$

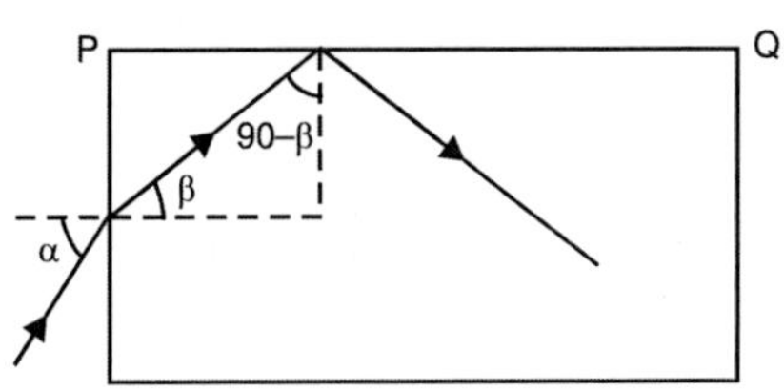

Fig. 15.46

15.53. A typical optical fibre wave guide consists of a thin cylindrical rod of refractive index n_1 clad by glass of slightly lower refractive index n_2. At what angle θ the light guidance takes place through the phenomenon of total internal reflection at the core-cladding interface?

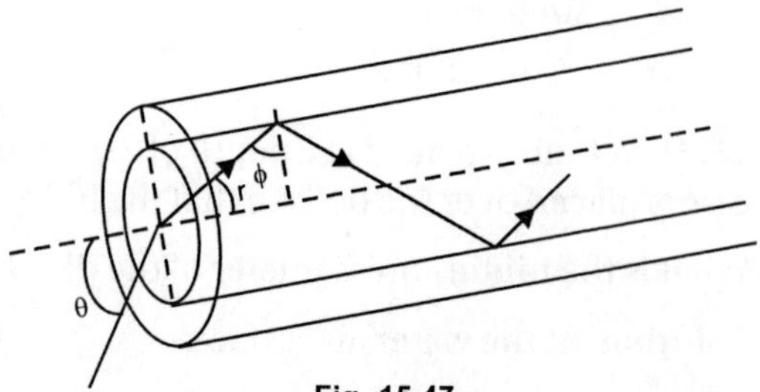

Fig. 15.47

Solution: In the light of Fig. 15.47 total internal reflection will take place when

$$\phi > C; \text{ where } C \text{ is the critical angle}$$

or $\sin\phi > \sin C$

or $\sin\phi > \dfrac{n_2}{n_1}$; $\sin C = \dfrac{1}{\mu} = \dfrac{n_2}{n_1}$

Further $\dfrac{\sin\theta}{\sin r} = n_1$

$$\sin\theta = n_1 \sin(90 - \phi);\ r + \phi = 90°$$

$$= n_1 \cos\phi$$

$$\sin^2\theta = n_1^2\left(1 - \sin^2\phi\right) = n_1^2\left(1 - \frac{n_2^2}{n_1^2}\right)$$

$$\theta = \sin^{-1}\sqrt{n_1^2 - n_2^2}$$

15.54. A point source of light is placed inside a deep lake at a distance h below the surface. Calculate the fraction of the light energy that escapes directly from the water surface.

Solution: Only that part of light will go out which is confined in the cone whose vertex angle = $2C$, where C is the critical angle.

$$\text{Fraction } f = \frac{\text{area of the cap}}{\text{area of the sphere}} = \frac{2\pi r(r-h)}{4\pi r^2};\ r = \text{radius of the sphere}$$

$$= \frac{1}{2}\left[1 - \frac{h}{r}\right]$$

$$= \frac{1}{2}[1 - \cos C]$$

$$= \frac{1}{2}\left[1 - \sqrt{1 - \sin^2 C}\right]$$

$$= \frac{1}{2}\left[1 - \sqrt{1 - \left(\frac{1}{n}\right)^2}\right];\ \sin C = \frac{1}{n}$$

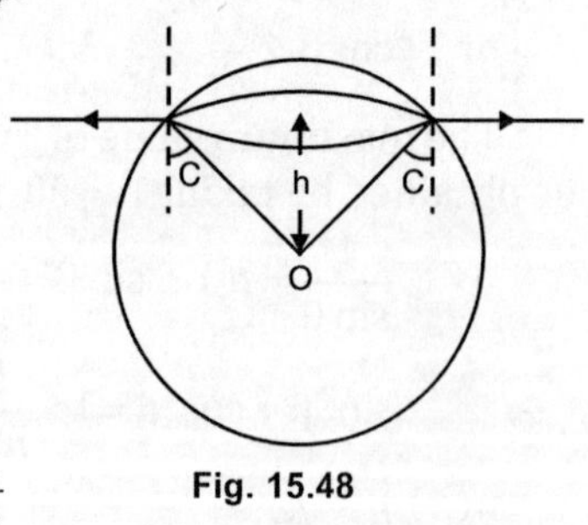

Fig. 15.48

15.55. A glass rod having a square cross-section is bent in the form of horse shoe magnet as shown in Fig. 15.49. The refractive index of the material of the rod is 1.5. The radius of the inner semi circle is R and the width of the rod is d. For what minimum value of $\frac{d}{R}$ the light that enters at A will emerge at B?

Solution: For bending of light we must have the phenomenon of total internal reflection at O.

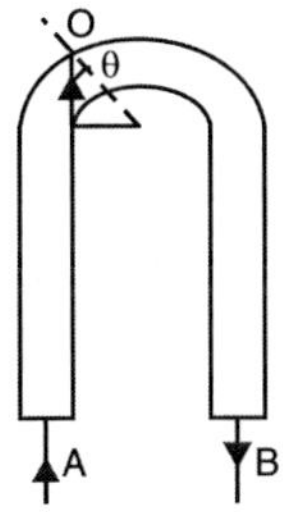

Fig. 15.49

$$\sin\theta = \frac{1}{\mu} = \frac{2}{3} = \frac{R}{R+d}$$

$$3R = 2R + 2d$$

$$R = 2d$$

or $$\frac{d}{R} = 0.5$$

15.56. A parallel beam of light is falling on a transparent spherical shell of inner and outer radii r_1 and r_2 respectively. Find the position of the image from the centre of the sphere after accounting the refraction from the whole shell.

Solution: Consider the refraction at the first surface of radius r_2 (see Fig. 15.50).

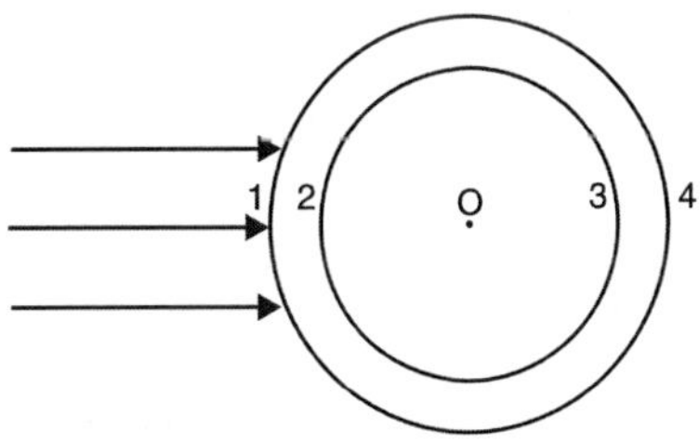

Fig. 15.50

$$\frac{\mu}{v} - \frac{1}{\infty} = \frac{(\mu - 1)}{r_2}$$

$$\therefore \quad v = \frac{\mu r_2}{\mu - 1}$$

The image formed will be in the transparent medium of refractive index μ. For refraction at the second surface

$$u = \text{position of object} = \frac{\mu r_2}{\mu - 1} - (r_2 - r_1) = \frac{\mu r_2 - (r_2 - r_1)(\mu - 1)}{\mu - 1}$$

$$\frac{1}{v_1} - \frac{\mu(\mu - 1)}{\mu r_2 - (r_2 - r_1)(\mu - 1)} = \frac{1 - \mu}{r_1}$$

or $$v_1 = -r_1 \frac{[\mu r_2 - (r_2 - r_1)(\mu - 1)]}{(\mu - 1)(r_2 - r_1)}$$

–ve sign shows that the image is in air and on the side of the object. Position of the image from the centre of the shell

$$= r_1 \frac{[\mu r_2 - (r_2 - r_1)(\mu - 1)]}{(\mu - 1)(r_2 - r_1)} + r_1 = \frac{\mu r_1 r_2}{(\mu - 1)(r_2 - r_1)}$$

For refraction at surface 3
$$u = \frac{\mu r_1 r_2}{(\mu - 1)(r_2 - r_1)} + r_1$$
$$= \frac{\mu r_1 r_2 + r_1(\mu - 1)(r_2 - r_1)}{(\mu - 1)(r_2 - r_1)}$$

Now $\frac{\mu}{v_2} + \frac{(\mu - 1)(r_2 - r_1)}{\mu r_1 r_2 + r_1(\mu - 1)(r_2 - r_1)} = \frac{(\mu - 1)}{-r_1}$ here u is negative

$$v_2 = -r_1 \frac{[\mu r_2 + (\mu - 1)(r_2 - r_1)]}{(\mu - 1)(2r_2 - r_1)}$$

For refraction at surface 4
$$u = (r_2 - r_1) + r_1 \frac{[\mu r_2 + (\mu - 1)(r_2 - r_1)]}{(\mu - 1)(2r_2 - r_1)}$$
$$= \frac{(\mu - 1)(r_2 - r_1)2r_2 + \mu r_1 r_2}{(\mu - 1)(2r_2 - r_1)}$$

Now $\frac{1}{v_3} + \frac{\mu(\mu - 1)(2r_2 - r_1)}{(\mu - 1)(r_2 - r_1)2r_2 + \mu r_1 r_2} = \frac{1 - \mu}{-r_2}$ here u is negative

$$v_3 = -r_2 \frac{[2(\mu - 1)(r_2 - r_1) + \mu r_1]}{2(\mu - 1)(r_2 - r_1)}$$

Position of the image from the centre of the shell
$$= \frac{r_2[2(\mu - 1)(r_2 - r_1) + \mu r_1]}{2(\mu - 1)(r_2 - r_1)} - r_2$$
$$= \frac{\mu r_1 r_2}{2(\mu - 1)(r_2 - r_1)}$$

15.57. The slab of a material of refractive index 2 shown in Fig. 15.51 has a curved surface *APB* of radius of curvature 10 cm and a plane surface *CD*. On the left of *APB* is air on the right of *CD* in water with refractive indices as given in the Fig. 15.51. An object *O* is placed at a distance of 15 cm from the pole *P* as shown. The distance of the final image of *O* from *P* as viewed from the left is .... (IIT 1991)

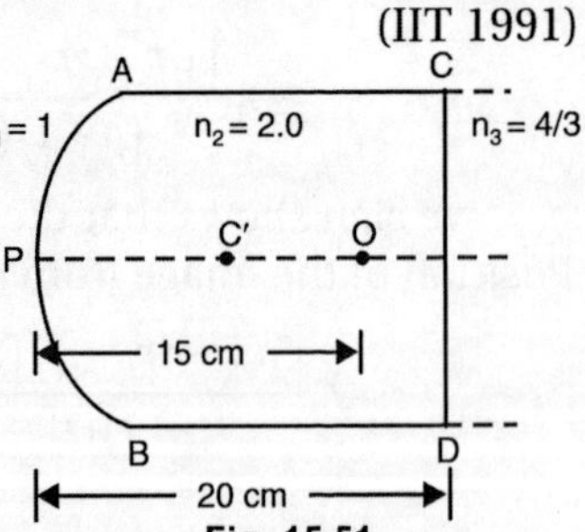

Fig. 15.51

Solution: For refraction we have
$$\frac{n_1}{v} - \frac{n_2}{u} = \frac{n_1 - n_2}{R}$$
$u = -15$ cm, $R = -10$ cm
$$\frac{1}{v} + \frac{2}{15} = \frac{-1}{-10}$$
$v = -30$ cm

$$ {}^{g}n_{w} = \frac{\text{real distance in glass}}{\text{apparent distance}} $$

$$ \frac{4/3}{2} = \frac{30-20}{x} $$

$\therefore \quad x = 15 \text{ cm}$

Final position of image from $P = 20 + 15 = 35$ cm

15.58. A parallel beam of light having radius of cross-section r is incident vertically on a glass hemisphere of radius R lying with its plane surface on the table. The axis of the beam coincides with the vertical diameter of the hemisphere. Find the radius of the image formed on the table.

Solution: Let l be the radius of the image in the glass (see Fig. 15.52).

In view of the geometry of the figure we have

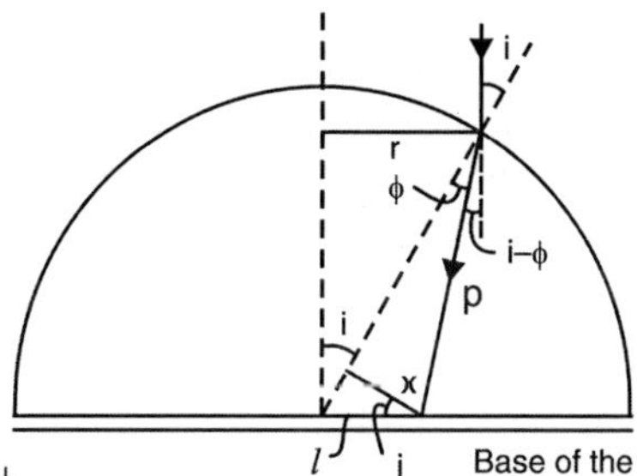

Fig. 15.52

$$ \cos i = \frac{x}{l} $$

$\therefore \quad x = l \cos i$

Further $\sin \phi = \dfrac{x}{p}$

$$ \therefore \quad p = \frac{x}{\sin \phi} = \frac{l \cos i}{\sin \phi} $$

$$ \sin(i-\phi) = \frac{r-l}{p} = \frac{(r-l)\sin \phi}{l \cos i} $$

$$ l\left[\cos i \cdot \sin(i-\phi) + \sin \phi\right] = r \sin \phi $$

$$ \therefore \quad l = \frac{r \sin \phi}{\left[\cos i \cdot \sin(i-\phi) + \sin \phi\right]} $$

$$ = \frac{r\dfrac{\sin \phi}{\sin i}}{\left[\dfrac{\cos i}{\sin i} \cdot \sin(i-\phi) + \dfrac{\sin \phi}{\sin i}\right]} $$

$$ = \frac{r\dfrac{\sin \phi}{\sin i}}{\left[\cos i \cos \phi + \sin i \sin \phi\right]} $$

$$ = \frac{r\dfrac{\sin \phi}{\sin i}}{\left[\cos i\sqrt{1-\sin^2 \phi} + \dfrac{\sin^2 i}{\mu}\right]}; \quad \frac{\sin \phi}{\sin i} = \frac{1}{\mu} $$

$$= \frac{r}{\left[\cos i\sqrt{\mu^2 - \sin^2 i} + \sin^2 i\right]}$$

$$= \frac{r}{\left\{\left[1-\left(\frac{r}{R}\right)^2\right]^{1/2}\left[\mu^2-\left(\frac{r}{R}\right)^2\right]^{1/2}+\left(\frac{r}{R}\right)^2\right\}}$$

Radius of the spot formed on the table

$$r' = \frac{l}{\mu} = \frac{\frac{r}{\mu}}{\left\{\left[1-\left(\frac{r}{R}\right)^2\right]^{1/2}\left[\mu^2-\left(\frac{r}{R}\right)^2\right]^{1/2}+\left(\frac{r}{R}\right)^2\right\}}$$

15.59. A hollow spherical shell of glass of refractive index μ having inner and outer radii as R and $2R$ respectively has a small mark on its inner surface which is observed from point outside the sphere on the side opposite to the centre (see Fig. 15.53). By what distance the mark will appear nearer than its real position.

Solution:

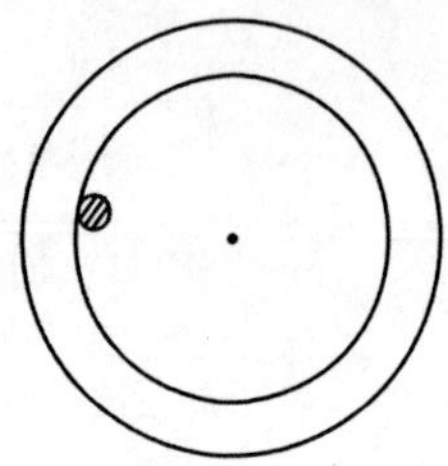

Fig. 15.53

$$\frac{\mu}{v} - \frac{1}{u} = \frac{(\mu-1)}{R}$$

$$\frac{\mu}{v} = \frac{(\mu-1)}{R} + \frac{1}{2R} = \frac{(2\mu-1)}{2R}$$

∴ For refraction at the second surface

$$u = \frac{2\mu R}{2\mu-1} + R = \frac{R(4\mu-1)}{2\mu-1}$$

At the second surface we have

$$\frac{1/\mu}{v} - \frac{1}{u} = \frac{\left(\frac{1}{\mu}-1\right)}{2R}$$

or $$\frac{1}{v} - \frac{\mu}{u} = \frac{1-\mu}{2R}$$

∴ $$\frac{1}{v} = \frac{1-\mu}{2R} + \frac{\mu(2\mu-1)}{R(4\mu-1)} = \frac{3\mu-1}{2R(4\mu-1)}$$

∴ $$v = \frac{2R(4\mu-1)}{3\mu-1}$$

$$\text{Required position} = 3R - \frac{2R(4\mu-1)}{(3\mu-1)} = \frac{(\mu-1)R}{(3\mu-1)}$$

15.60. A glass rod of length $3R$ and refractive index μ has curvatures of radii R and $R/2$ at its ends as shown in the Fig. 15.54. An object O is placed at a distance $2R$ from the surface of radius of curvature R. Find the position of the image formed from the first surface.

Solution: At the first surface

$$\frac{\mu}{v'} - \frac{1}{u'} = \frac{(\mu-1)}{R_1}; u' = -2R, R_1 = +R$$

$$\therefore \quad \frac{\mu}{v'} + \frac{1}{2R} = \frac{(\mu-1)}{R}$$

$$v' = \frac{2\mu R}{2\mu - 3}$$

For the second surface we have $u = -\left[3R - \dfrac{2\mu R}{2\mu-3}\right]$

$$\frac{1}{v} + \frac{\mu}{\left(3R - \dfrac{2\mu R}{2\mu-3}\right)} = \frac{1-\mu}{\dfrac{R}{2}}$$

$$\therefore \quad v = \frac{(9-4\mu)R}{(10\mu-9)(\mu-2)}$$

$$= \frac{(9-4\mu)R}{10\mu^2 - 29\mu + 18}$$

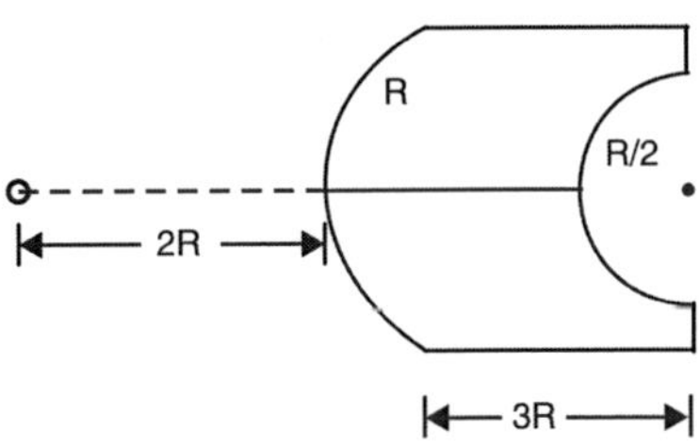

Fig. 15.54

Position of the image from the first surface

$$= \frac{9R - 4\mu R}{10\mu^2 - 29\mu + 18} + 3R$$

$$= \frac{9R - 4\mu R + 30\mu^2 R - 87\mu R + 54R}{(10\mu^2 - 29\mu + 18)}$$

$$= \frac{R\left[30\mu^2 - 91\mu + 63\right]}{(10\mu^2 - 29\mu + 18)}$$

15.61. A point object O is situated on the axis at a distance R/μ from the centre C of a sphere of radius R and refractive index μ. Find the position of the image from the centre of the sphere.

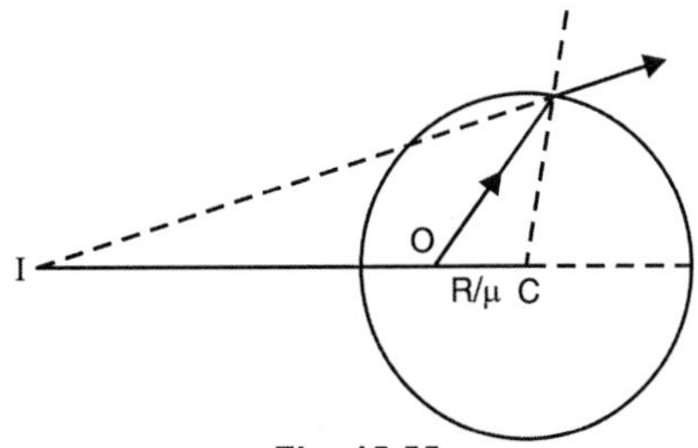

Fig. 15.55

Solution: In view of Fig. 15.55 we have

$$\frac{1}{v}-\frac{\mu}{R\left[1+\frac{1}{\mu}\right]}=\frac{1-\mu}{R} \text{ here } u, v \text{ and } R \text{ all are negative}$$

$$\frac{1}{v}=\frac{\mu^2}{R(\mu+1)}-\frac{(\mu-1)}{R}=\frac{1}{R(\mu+1)}$$

$$\therefore \quad v = R(\mu+1)$$

$$CI = R(\mu+1) - R = \mu R$$

15.62. A object is placed at a distance of 40 cm. from the surface of a glass sphere of refractive index $\mu = 1.5$ and radius $R = 10$ cm along the horizontal diameter (see Fig. 15.56). Calculate the position of final image after accounting the refraction at both the surfaces.

Solution: For refraction at the first surface we have

$$\frac{1.5}{v}+\frac{1}{40}=\frac{0.5}{10}$$

$$\therefore \quad v = 60 \text{ cm}$$

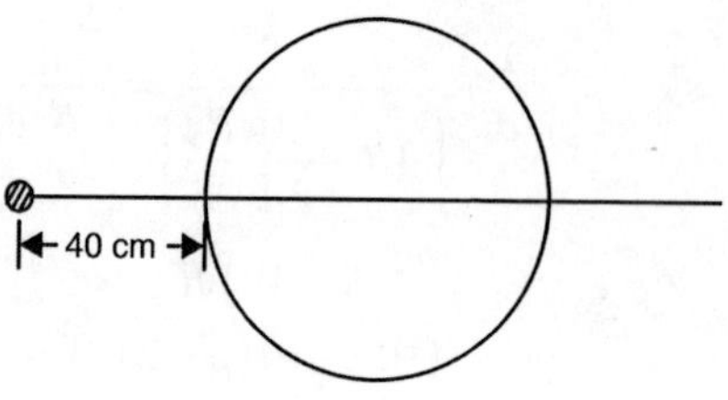

Fig. 15.56

For refraction at second surface

$$u = 40 \text{ cm}$$

$$\therefore \quad \frac{1}{v}-\frac{1.5}{40}=\frac{0.5}{10}$$

$$\frac{1}{v}=\frac{3.5}{40}$$

$$\therefore \quad v=\frac{40}{3.5}$$

Position of image from the centre of the sphere $= 10 + \dfrac{40}{3.5} = 21.4$ cm.

15.63. A parallel beam of light travelling in water (refractive index = 4/3) is refracted by a spherical air bubble of radius 2 mm, situated in water. Assuming the light rays to be parallel (i) find the position of the image due to refraction at the first surface and the position of the final image (ii) draw a ray diagram showing the position of both the images.

(IIT 1988)

Solution: (i) For refraction at the first surface

$$\frac{1}{v}-\frac{4/3}{\infty}=\frac{1-4/3}{2}$$

$$\therefore \quad v = -6 \text{ mm}$$

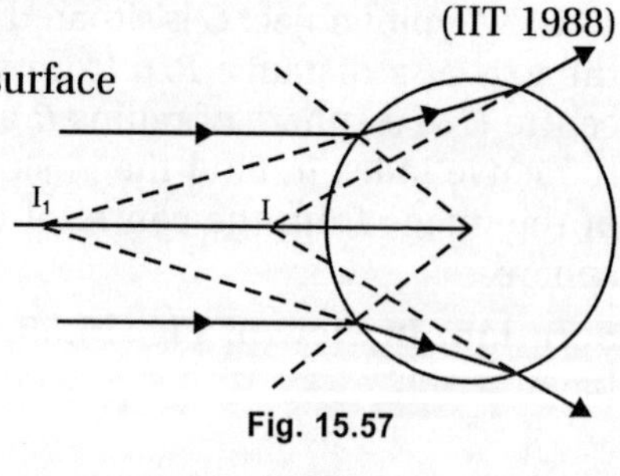

Fig. 15.57

The image I_1 is at a distance of 6 mm from the first surface (see Fig. 15.57). For second surface

$$\frac{4/3}{v}+\frac{1}{10}=\frac{4/3-1}{-2} \text{ here } u=6+4=10 \text{ mm}$$

$$\therefore \qquad v=-5 \text{ mm}$$

The final image I is at a distance of 5 mm from the second surface.

(ii) The ray diagram is shown in the figure.

15.64. A glass sphere of radius R = 30 cm and refractive index μ = 1.5 is silvered as shown in the Fig. 15.58. A point object is placed on the horizontal diameter of the sphere at a distance of 60 cm from the pole of the unsilvered surface. Find the position of the image formed after accounting all the refractions and reflections.

Solution: When the refraction is at A at the unsilvered surface we have

$$\frac{1.5}{v}+\frac{1}{60}=\frac{1.5-1}{30}$$

$$\therefore \qquad v=\infty$$

i.e. after refraction the ray of light will be parallel to the principal axis. It will now suffer a reflection at B and the image will be formed at O' such that $OO' = \frac{R}{2}$ = 15 cm. Now it will again have a refraction at unsilvered surface and after that the ray will be out of sphere.

$$\frac{1}{v}-\frac{1.5}{45}=\frac{1-1.5}{30}$$

$$\frac{1}{v}-\frac{1}{30}=-\frac{1}{60}$$

or $\qquad v = 60$ cm

hence the final image will be formed at the pole of the silvered hemisphere and it will be virtual.

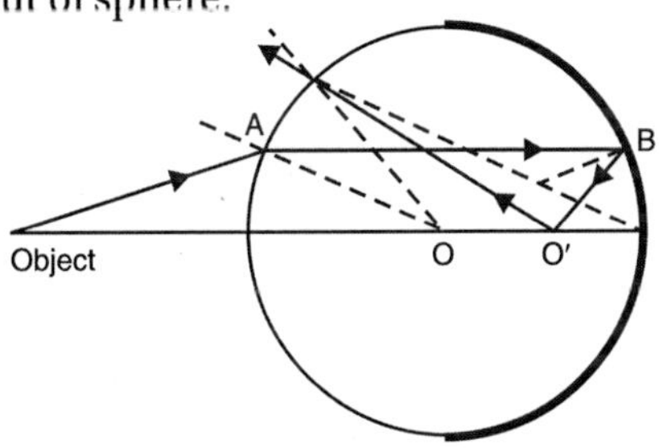

Fig. 15.58

15.65. A vertical ray of light is incident on a transparent sphere of radius R and refractive index μ at a distance $d < R$ from the vertical axis. For what value of μ and d will the point of intersection of the ray and the vertical axis lie inside the sphere.

Solution: In view of Fig. 15.59 we have

$$d \cot (i-r) = x+y$$

$$x = d\,[\cot (i-r) - \cot i]$$

For image inside the sphere $x < R$ hence

$$\frac{d\left[\sin i \cos(i-r)-\cos i \sin(i-r)\right]}{\sin(i-r)\sin i}<R$$

$$\frac{d\sin r}{\sin(i-r)\sin i} < R$$

$$\frac{\sin(i-r)\sin i}{d\sin r} > \frac{1}{R}$$

$$\frac{\mu R}{d}\sin(i-r) > 1$$

$$\frac{\mu R}{d}[\sin i\cos r - \cos i\sin r] > 1$$

$$\frac{\mu R}{d}\left[\sin i\sqrt{1-\sin^2 r} - \sqrt{1-\sin^2 i}\cdot\frac{\sin i}{\mu}\right] > 1$$

$$\frac{\mu R}{d}\left[\frac{\sin i}{\mu}\sqrt{\mu^2-\sin^2 i} - \sqrt{1-\sin^2 i}\cdot\frac{\sin i}{\mu}\right] > 1$$

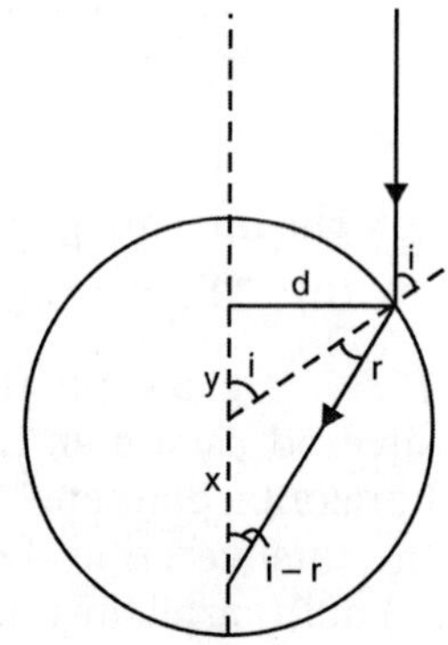

Fig. 15.59

or $$\left[\sqrt{\mu^2-\sin^2 i} - \sqrt{1-\sin^2 i}\right] > 1$$

$$\mu^2 - \sin^2 i + 1 - \sin^2 i - 2\sqrt{\mu^2-\sin^2 i}\cdot\sqrt{1-\sin^2 i} > 1$$

or $$\mu^2 - 2\sin^2 i - 2\sqrt{\mu^2-\sin^2 i}\cdot\sqrt{1-\sin^2 i} > 0$$

$$\left(\mu^2 - \frac{2d^2}{R^2}\right)^2 > 4\left[\mu^2 - \frac{d^2}{R^2}\right]\left[1-\frac{d^2}{R^2}\right]$$

$$\mu^4 - 4\mu^2\frac{d^2}{R^2} + 4\frac{d^2}{R^2} > 4\left[\mu^2 - \frac{\mu^2 d^2}{R^2} - \frac{d^2}{R^2} + \frac{d^4}{R^4}\right]$$

$$\mu^4 > 4\mu^2 - \frac{4d^2}{R^2}$$

$$\frac{4d^2}{R^2} > 4\mu^2 - \mu^4$$

$$\frac{4d^2}{R^2} > 4 - \left(\mu^2 - 2\right)^2$$

$$\frac{d^2}{R^2} > 1 - \frac{\left(\mu^2-2\right)^2}{4}$$

For upper limit

$4 - (\mu^2 - 2)^2 > 0$

$4 > (\mu^2 - 2)^2$

$2 > (\mu^2 - 2)$

$4 > \mu^2$

$2 > \mu$

For lower limit

$$\mu^2 - 2 > 0$$
$$\mu^2 > 2$$
$$\mu > \sqrt{2}$$

Hence we must have $\sqrt{2} \le \mu \le 2$

15.65. A cylindrical glass rod of radius 0.1 m and refractive index $\sqrt{3}$ lies on horizontal plane mirror. A horizontal ray of light moving perpendicular to the axis of the rod is incident on it. At what height from the mirror should the ray be incident so that it leaves the rod at a height of 0.1 m above the plane mirror? At what distance a second similar rod, parallel to the first, be placed on the mirror, such that the emergent ray from the second rod is in line with the incident ray on the first rod.

(Roorkee 1990)

Solution:

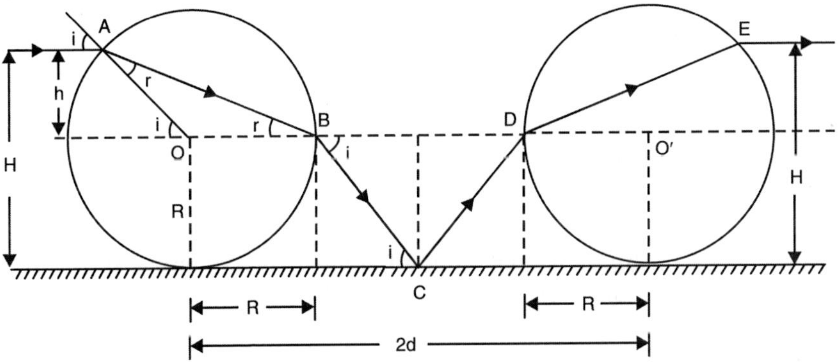

Fig. 15.60

In view of Fig. 15.60 we have

$$h = AO \sin i = R \sin i = R \sin 2r = 2R \sin r \cos r; \; i = 2r$$

$$\mu = \frac{\sin i}{\sin r} = \frac{\sin 2r}{\sin r} = 2 \cos r$$

$$h = 2R \sin r \cos r = \mu R \sin r = \mu R \sqrt{1 - \cos^2 r}$$

$$= \mu R \left[1 - \frac{\mu^2}{4}\right]^{1/2} = \frac{\mu R}{2}\sqrt{4 - \mu^2}$$

On substituting the values we have

$$= \frac{\sqrt{3} \times 0.1}{2}\sqrt{(4-3)} = 0.086 \text{ m}$$

$$H = R + h = 0.1 + 0.086 = 0.186 \text{ metre}$$

From the geometry

$$d = R(1 + \cot i)$$

$$= R(1 + \cot 2r) = R\left[1 + \frac{\cos 2r}{\sin 2r}\right]$$

$$= R\left[1 + \frac{2\cos^2 r - 1}{2 \sin r \cos r}\right]$$

$$= R\left[1 + \frac{2\frac{\mu^2}{4} - 1}{\mu\sqrt{1 - \frac{\mu^2}{4}}}\right]$$

On substituting the values we have

$$d = R\left[1 + \frac{\frac{1}{2}}{\frac{\sqrt{3}}{2}}\right] = 0.1\left[1 + \frac{1}{\sqrt{3}}\right]$$

Distance of the second rod

$$2d = 0.2\left[1 + \frac{1}{\sqrt{3}}\right] = 0.3155 \text{ metre}$$

15.67. A small sphere is located at a height $h = 4.9$ m above the convex surface of radius of curvature $R = 10$ cm of a block of glass of refractive index $\mu = 1.5$ (see Fig. 15.61). At an instant $t = 0$ the ball falls. Find the speed of the image at $t = 0.8$ sec.

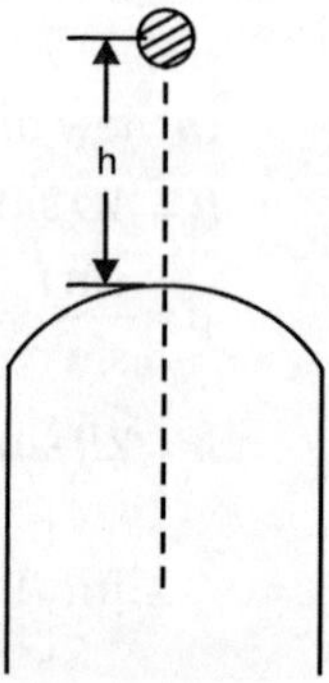

Fig. 15.61

Solution: For refraction we have

$$\frac{\mu_2}{V} + \frac{\mu_1}{u} = \frac{\mu_2 - \mu_1}{R}$$

Let the ball falls through a distance x then

$$u = h - x, V = v + x'; x = \frac{1}{2}gt^2, \frac{dx}{dt} = gt$$

$$\frac{\mu}{v + x'} = \frac{(\mu - 1)}{R} - \frac{1}{h - x} = \frac{(\mu - 1)(h - x) - R}{R(h - x)}$$

$$v + x' = \frac{\mu R(h - x)}{(\mu - 1)(h - x) - R}$$

On differentiating it we have

$$\frac{dx'}{dt} = \text{speed of the image}$$

$$= -\frac{\mu R \cdot \frac{dx}{dt}}{(\mu-1)(h-x)-R} + \frac{\mu R(h-x)(\mu-1)\frac{dx}{dt}}{[(\mu-1)(h-x)-R]^2}$$

$$= \frac{\mu R^2 \frac{dx}{dt}}{[(\mu-1)(h-x)-R]^2} = \frac{\mu R^2 gt}{\left[(\mu-1)\left(h-\frac{1}{2}gt^2\right)-R\right]^2}$$

On substituting the values of various terms we have

$$\frac{dx'}{dt} = \frac{1.5\times10^{-2}\times9.8\times0.8}{\left[0.5\left(4.9-\frac{1}{2}\times9.8\times0.8^2\right)-0.1\right]^2}$$

$$= 19.2 \text{ cm/sec.}$$

15.68. A transparent solid sphere of refractive index ${}^a\mu_g = 1.5$ and radius $r = 10$ cm is fitted at the mouth of a long cylindrical tube containing water of refractive index ${}^a\mu_w = \frac{4}{3}$. The tube is held horizontal as shown in the Fig. 15.62. Parallel beam of light is falling on the sphere. Find the position of the image from the centre of the sphere.

Solution: For refraction at the first surface we have

$$\frac{{}^a\mu_g}{v} - \frac{1}{\infty} = \frac{{}^a\mu_g - 1}{r}$$

or

$$\frac{1.5}{v} = \frac{1.5-1}{10}$$

$\therefore$ $v = 1.5\times20 = 30$ cm.

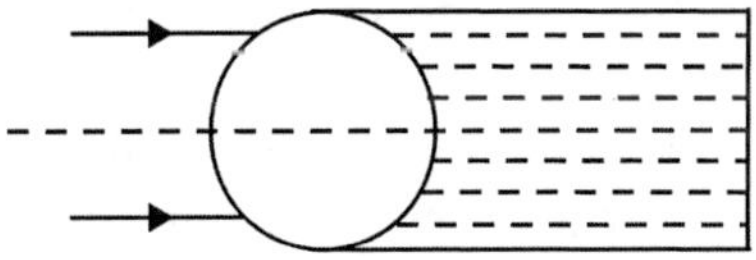

Fig. 15.62

Position of the object from the second surface = 30 – 20 = 10 cm.

$$\frac{\frac{4}{3}}{v'} - \frac{1.5}{10} = \frac{\frac{4}{3}-\frac{3}{2}}{-10}$$

$$\frac{4/3}{v'} = \frac{3}{20} + \frac{1}{60}$$

$$v' = 8 \text{ cm}$$

$\therefore$ Position of the image from the centre of the sphere

$= 10 + 8$

$= 18$ cm.

15.69. A glass sphere of radius 5×10^{-2} m has a small bubble 2×10^{-2} m from its centre. The bubble is viewed along a diameter of the sphere from the side on which it lies. How far from the surface will it appear? Refractive index of glass is 1.5. (Roorkee 1997)

Solution:

$u = -3 \times 10^{-2}$m, $R = -5 \times 10^{-2}$m

From equation

$$\frac{\mu_2}{v} - \frac{\mu_1}{u} = \frac{\mu_2 - \mu_1}{R}$$

We have

$$\frac{1}{v} - \frac{3/2}{-3\times10^{-2}} = \frac{1-3/2}{-5\times10^{-2}}$$

$\therefore \quad v = -2.5 \times 10^{-2}$ m

$= -2.5$ cm

The bubble will appear at a distance of 2.5 cm towards the centre of sphere from the surface through which the observation is made.

15.70. A polished hemisphere of glass, having radius of curvature R = 60 cm contains a thin layer of benzene. When an object is placed at the centre of curvature, two real images (one due to reflection and the other due to refraction) are formed. Of the two, the image formed by reflection coincides with the object and the other formed by the refraction is at a separation of 30 cm from it. (i) Draw a ray-diagram showing the position of the image formed by refraction and (ii) determine the refractive index of benzene.

Solution: (i) A ray of light *OA* originating from the object refracts along *AB*. At *B*, the ray suffers a reflection along *BC*. At *C*, the ray refracts along *CD*. The final image is formed at *D* (see Fig. 15.63).

(ii) Due to liquid in the hemisphere, a plano-convex lens is formed. The total focal length of the system is given by

$$\frac{1}{F} = \frac{1}{f_{\text{liq}}} + \frac{1}{f_{\text{liq}}} + \frac{1}{f_m}$$

where $\frac{1}{f_{\text{liq}}} = \frac{(\mu-1)}{R}$, μ is the refractive index of the liquid and f_m is the focal length of the mirror

$$\frac{1}{F} = \frac{2(\mu-1)}{R} + \frac{2}{R} = \frac{2\mu}{R}$$

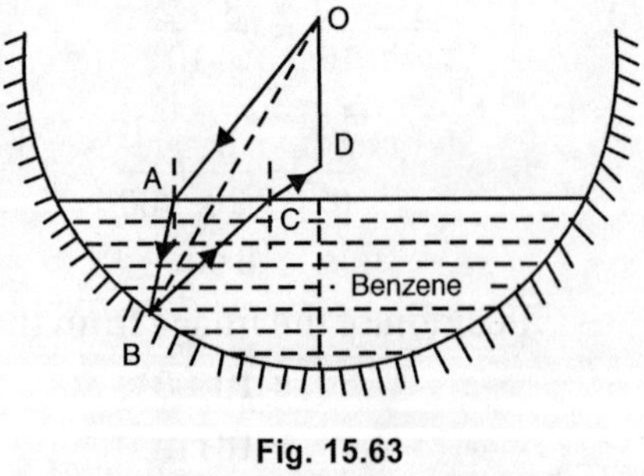

Fig. 15.63

On applying the formula applicable to a mirror, we have

$$\frac{1}{v}+\frac{1}{u}=\frac{1}{F}$$

$$\frac{1}{30}+\frac{1}{60}=\frac{2\mu}{60}$$

$$\frac{3}{60}=\frac{2\mu}{60}$$

$$\therefore \quad \mu=\frac{3}{2}=1.5$$

15.71. A beam of light is incident on a glass sphere of radius $R = 12$ cm and refractive index $\mu = 1.5$ along its diameter (see Fig. 15.64a). (i) Calculate the distance from the centre to the point at which the rays are focused. (ii) Under what condition are the rays focused inside the sphere? Calculate the distance in terms of μ and R at which the image will be formed. Draw a ray-diagram showing the position of the image.

Solution: (i) In view of the geometry of the figure $\angle BOE = \angle 90 - i$

Along the line $O'OD$ we have

$$180 = i + \angle BOC + \theta$$

$$180 = i + 180 - 2r + \theta; \text{ In triangle } BOC,\ 2r + \angle BOC = 180$$

$$\therefore \quad \theta = 2r - i$$

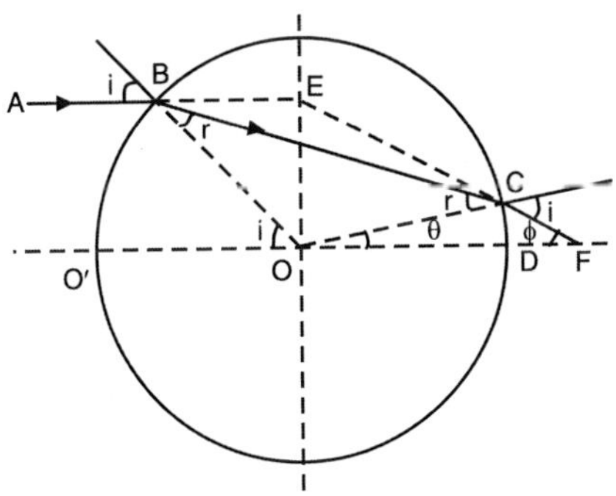

Fig. 15.64a

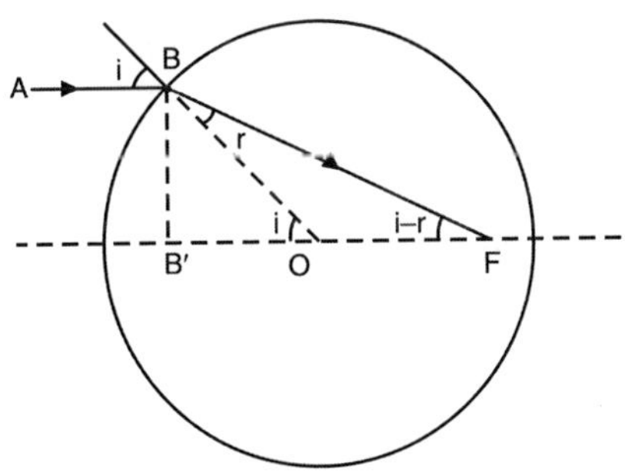

Fig. 15.64b

Now $CD = R\sin\theta = R\sin(2r - i)$

When the angles are small

$$CD = R(2r - i) = Rr\left(2-\frac{i}{r}\right)$$

or $$CD = \frac{Ri}{\mu}(2-\mu)$$

Further $\theta + \phi = i$

$\therefore \quad \phi = i - \theta = i - 2r + i = 2(i - r)$

Now $DF = CD \cot \phi = \dfrac{CD}{\tan \phi} = \dfrac{CD}{\phi} = \dfrac{CD}{2(i-r)} = \dfrac{R(2r-i)}{2(i-r)} = \dfrac{R(2-\mu)}{2(\mu-1)}$

Total distance at which image will be formed is given by

$$OF = OD + DF = R + DF = R + \frac{R(2-\mu)}{2(\mu-1)} = \frac{\mu R}{2(\mu-1)}$$

On substituting the values of μ and R, we have

$$OF = \frac{1.5 \times 12}{2 \times 0.5} = 18 \text{ cm}$$

(ii) We have $CD = \dfrac{Ri}{\mu}(2-\mu)$

Now when $\mu > 2$ the image will be inside the sphere.

The ray diagram is shown in the Fig. 15.64b.

Now $OF = B'F - OB'$

$$= BB' \cot(i - r) - R$$

$$= \frac{R \sin i}{\tan(i-r)} - R = \frac{iR}{i-r} - R$$

$$= \frac{\mu R}{\mu - 1} - R = \frac{R}{\mu - 1}$$

15.72. A partially reflecting plane glass plate and a convex mirror of focal length 40 cm are 10 cm apart (see Fig. 15.65). At what distance from the plane mirror should an object be placed so that the images formed by the two mirrors coincide ?

Solution: The image of the object O as formed by the plane mirror is at a distance x from the mirror. For the concave mirror, the object is at a distance of x + 10 cm and the image is formed at a distance equal to x – 10 cm.

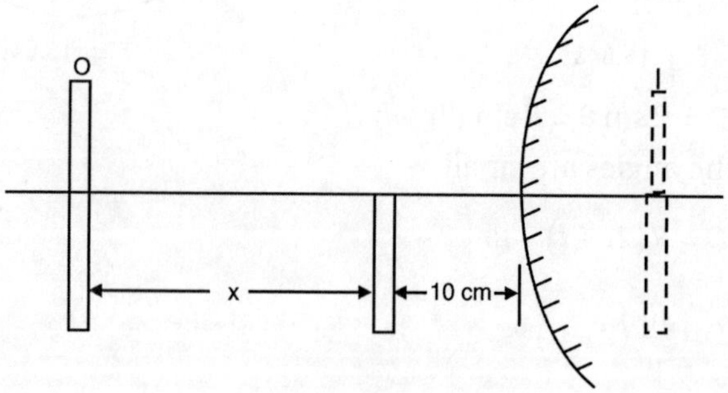

Fig. 15.65

Now for a mirror

$$\frac{1}{u}+\frac{1}{v}=\frac{1}{f};-\frac{1}{x+10}+\frac{1}{x-10}=\frac{1}{40} \text{ or } \frac{-x+10+x+10}{x^2-100}=\frac{1}{40}$$

or $$\frac{20}{x^2-100}=\frac{1}{40} \text{ or } 800+100=x^2$$

or $x=30$ cm

15.73. Determine the radius of curvature and the focal length of a convex spherical mirror if $OO'=2$ cm and $AB=20$ cm.

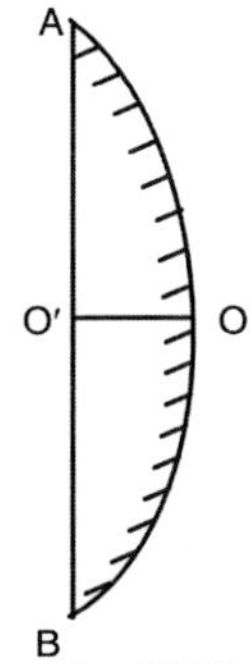

Fig. 15.66

Solution: Segment AOB in Fig. 15.66 is a part of a circle whose cord is AB. Let R be the radius of curvature, then we have

$$AO'\times O'B = OO'\times(2R-OO')$$

or $$10\times 10 = 2\times(2R-2)$$

or $$2R-2 = 50$$

$$R = 26 \text{ cm}$$

$\therefore$ $$F = 13 \text{ cm}$$

15.74. A convex mirror and a concave mirror of radii 10 cm each are placed facing each other and are 15 cm apart. An object is placed mid-way between them. Find the position of the final image if the reflection first takes place in the concave and then in the convex mirror.

Solution: u = position of the object = 15/2 cm, f = 5 cm

Now $$\frac{1}{v}+\frac{1}{u} = \frac{1}{f}$$

or $$\frac{1}{v} = \frac{1}{f}-\frac{1}{u}=\frac{1}{5}-\frac{2}{15}=\frac{1}{15}$$

$\therefore$ $$v = 15 \text{ cm}$$

That is, the final image will be formed at the pole of the convex mirror. It will be virtual and will be of the same size.

15.75. A convex mirror and a concave mirror having an identical focal length of 40 cm are placed 120 cm apart (Fig. 15.67). The two mirrors have the same optical axis. Where should an object be placed between the two mirrors so that its image formed by the two mirrors are identical?

Solution: Let the object be at a distance of x cm from the concave mirror. On applying the formula connecting u, v and f we have

$$\frac{1}{v}+\frac{1}{u} = \frac{1}{f}$$

or $$\frac{u}{v} = \frac{u}{f}-1=\frac{x-40}{40}$$

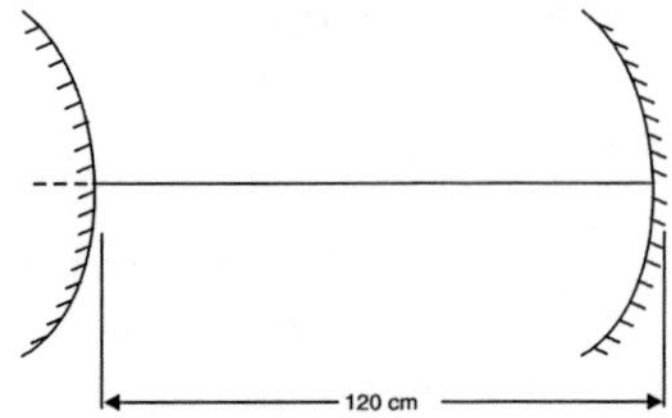

Fig. 15.67

For finding the position of image using convex mirror, we have

$$\frac{1}{v}-\frac{1}{u} = \frac{1}{f}; \; u \text{ is negative.}$$

or $$\frac{u}{v} = \frac{u}{f}+1=\frac{120-x+40}{40}$$

Now $\frac{u}{v}$ in the two cases must be the same, hence

$$x-40 = 120-x+40 \text{ or } 2x=200$$

$$\therefore \quad x = 100 \text{ cm}$$

15.76. The sun subtends an angle of 1/2° at the pole of a concave mirror which has a radius of curvature of 10 m. Find the size of the image of sun formed by the concave mirror.

Solution: For a concave mirror, we have

$$\frac{1}{v}+\frac{1}{u}=\frac{1}{f} \text{ here } u=\infty \;\therefore\; v=f=5 \text{ m} = 500 \text{ cm}$$

$$\frac{\text{Diameter of sun}}{\text{Distance of sun}} = 0.5 \text{ degree} = \frac{\pi}{360} \text{ radian}$$

$$\text{Now } \frac{\text{Diameter of image of sun}}{\text{Diameter of sun}} = \frac{v}{u}$$

$$= \frac{\text{Distance of image}}{\text{Distance of sun}}$$

$$= \frac{500}{\text{Distance of sun}}$$

$$\therefore \quad \text{Diameter of image of sun} = \frac{\text{Diameter of sun}}{\text{Distance of sun}} \times 500$$

$$= \frac{500 \times 3.14}{360} = 4.36 \text{ cm}$$

15.77. A concave mirror of radius of curvature R = 15 cm is placed on a horizontal surface. Water of refractive index ${}^a\mu_w = 4/3$ is poured into it upto a height h = 3 cm where should an object on the axis OO' (see Fig. 15.68) be placed so that its image formed by the system coincides with itself.

Solution: It is a case of plano convex lens polished at the curved surface. It will behave like a concave mirror whose focal length is defined by the equation

$$\frac{1}{F} = \frac{2\,{}^a\mu_w}{R}$$

Let the object be at a distance x from the pole O. The liquid will show a shift $= h\left[1-\frac{1}{{}^a\mu_w}\right]$

The correct object distance $u = x-h\left[1-\frac{1}{{}^a\mu_w}\right]$. This will also be equal to v. The correct image position will be equal to x after adding the shift $h\left[1-\frac{1}{{}^a\mu_w}\right]$

$$\frac{2}{x-h\left[1-\frac{1}{{}^a\mu_w}\right]} = \frac{2\,{}^a\mu_w}{R}$$

$$\therefore \quad \frac{R}{{}^a\mu_w} = x-h+\frac{h}{{}^a\mu_w}$$

$$x = \frac{R}{{}^a\mu_w}+h\left(1-\frac{1}{{}^a\mu_w}\right)$$

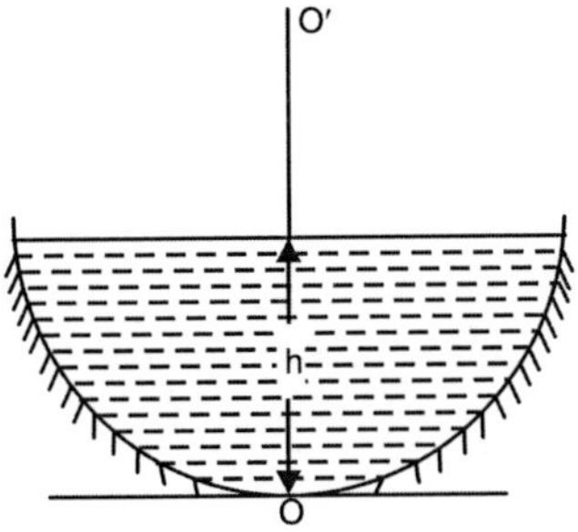

Fig. 15.68

$$x-h = \frac{R-h}{{}^a\mu_w}$$ is the position of the object above the surface of water.

On substituting the values we have

$$x-h = \frac{(15-3)\times 3}{4} = 9\text{ cm}$$

Thus the object should be placed at a distance of 9 cm above the surface of water.

15.78. An object is moving towards a concave mirror of radius of curvature $R = 15$ cm. along its principal axis with a constant speed of $V = 9$ cm/sec. Find the speed of the image formed by the concave mirror when the object is at a distance of $L = 30$ cm from the pole of the mirror.

Solution: Let the particle moves through a distance dx in time dt then $\frac{dx}{dt} = V$

Position of the object $u = L - dx$

Let the image position be $v + dx'$

Using the mirror formula we have

$$\frac{1}{v+dx'} = \frac{1}{f} - \frac{1}{L-dx}$$

$$\frac{1}{v+dx'} = \frac{L-dx-f}{f(L-dx)}$$

$$\therefore \quad v + dx' = \frac{f(L - dx)}{L - f - dx}$$

$$= \frac{fL - fdx}{(L-f)\left(1 - \dfrac{dx}{L-f}\right)}$$

$$= \frac{fL - fdx}{L-f}\left[1 + \frac{dx}{L-f}\right]$$

$$v + dx' = \frac{fL}{L-f} - \frac{fdx}{L-f} + \frac{fL \cdot dx}{(L-f)^2}$$

$$= \frac{fL}{L-f} + fdx\left[\frac{L}{(L-f)^2} - \frac{1}{(L-f)}\right]$$

$$= \frac{fL}{L-f} + fdx\left[\frac{L - L + f}{(L-f)^2}\right]$$

$$= v + \frac{f^2 dx}{(L-f)^2}$$

$$\frac{dx'}{dt} = \text{Speed of image } V' = \frac{f^2 \cdot \dfrac{dx}{dt}}{(L-f)^2}$$

$$= \frac{R^2 \cdot V}{(2L-R)^2}$$

On substituting the values we have

$$V' = \frac{15^2 \times 9}{(2 \times 30 - 15)^2} = 1 \text{ cm/sec.}$$

15.79. A convex and a concave mirror of focal lengths 50 and 40 cm respectively are placed coaxially and facing each other at a distance of 2 m. A circle of radius 6 mm is drawn on the convex mirror with its pole as centre. Find the radius of the image after two reflections first at concave and then at convex mirror.

Solution:

$$\frac{1}{v} + \frac{1}{u} = \frac{1}{f}$$

$$\frac{1}{v} = \frac{1}{f} - \frac{1}{u} = \frac{1}{40} - \frac{1}{200} = \frac{1}{50}$$

$$\therefore \quad v = 50 \text{ cm}$$

$$\frac{v}{u} = \frac{I}{O} \quad \therefore I = \frac{v}{u} \cdot O = \frac{50}{200} \times 6 = 1.5 \text{ mm}$$

For the convex mirror $u = 200 - 50 = 150$ cm

$$\frac{1}{v} - \frac{1}{u} = \frac{1}{f} \text{ here } u \text{ is negative}$$

$$\frac{1}{v} = \frac{1}{f} + \frac{1}{u} = \frac{1}{50} + \frac{1}{150} = \frac{3+1}{150} = \frac{4}{150}$$

$$\therefore \quad v = \frac{150}{4}$$

$$\frac{v}{u} = \frac{150}{4 \times 150} = \frac{I'}{O'}$$

$$\therefore \quad I' = \frac{O'}{4} = \frac{1.5}{4} = 0.375 \text{ mm}$$

15.80. A point mass is placed on the principal axis of a concave mirror of radius of curvature $R = 15$ cm at a distance of 30 cm from the pole of the mirror. The point mass is executing simple harmonic motion along the principal axis with an amplitude of 1.5 cm. Find the amplitude of oscillation of the image of the mass formed by the mirror.

Solution: Let u be equal to $30 + 1.5 \sin \omega t$.

$$\text{Now } \frac{1}{v'} = \frac{1}{f} - \frac{1}{30 + 1.5 \sin \omega t}$$

$$= \frac{30 + 1.5 \sin \omega t - f}{f(30 + 1.5 \sin \omega t)}$$

$$\therefore \quad v' = \frac{30f + f \times 1.5 \sin \omega t}{(30 - f)\left[1 + \frac{1.5}{30 - f} \cdot \sin \omega t\right]}$$

$$= \frac{30f + 1.5f \sin \omega t}{30 - f}\left[1 - \frac{1.5}{30 - f} \sin \omega t\right]$$

$$= \frac{30f}{30 - f} + \frac{1.5 f \sin \omega t}{30 - f} - \frac{30f \times 1.5 \sin \omega t}{(30 - f)^2}$$

$$= v - \frac{1.5f \sin \omega t}{(30 - f)^2}[30 - 30 + f]$$

$$= v - \frac{1.5 \cdot f^2 \cdot \sin \omega t}{(30 - f)^2}$$

$\therefore$ Amplitude of the image is equal to $\dfrac{1.5 f^2}{(30-f)^2}$

On substituting the value of f we have

$$\text{Amplitude of image} = \frac{1.5\times 7.5^2}{22.5^2} = 1.6 \text{ mm}$$

15.81. A parallel ray is incident on a spherical surface at a point P at a height h from the axis (see Fig. 15.69). Calculate the amount of spherical aberration due to a spherical surface of radius R.

Solution: The ray of light meets the axis at F after refraction. This is known as the focal point of h zone. OF is known as focal length of h zone.

$$OF_h = f_h = R + CF$$

In $\Delta\, CPF$

$$\frac{CF}{\sin r} = \frac{R}{\sin(i-r)}$$

$$\therefore \quad CF = \frac{R\sin r}{\sin(i-r)}$$

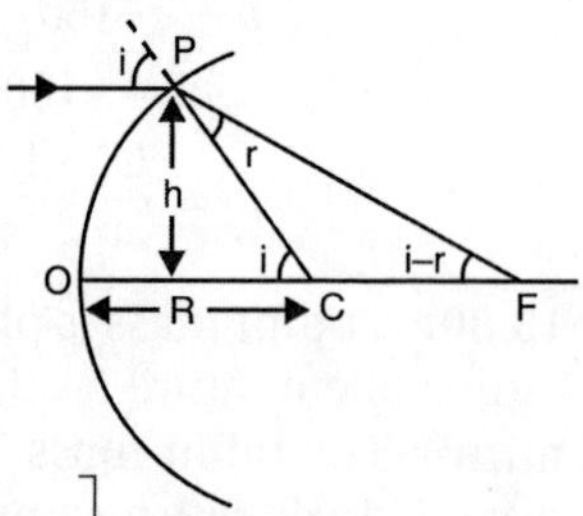

Fig. 15.69

$$f_h = R + \frac{R\sin r}{\sin(i-r)} = R\left[1+\frac{1}{\mu\cos r-\cos i}\right]$$

When $h \to 0$ i.e. for axial ray cos i and cos r each tends to 1.

$$\therefore \quad f_0 = \frac{\mu R}{(\mu-1)}$$

Change in focal length

$$\Delta f = f_0 - f_h = \left[\frac{\mu R}{(\mu-1)} - R - \frac{R}{\mu\cos r-\cos i}\right]$$

$$= R\left[\frac{1}{\mu-1} - \frac{1}{\mu\cos r-\cos i}\right]$$

This is the expression for longitudinal spherical aberration. To an approximation $\sin i = \dfrac{h}{R}$

$$\cos i = \sqrt{1-\sin^2 i} = 1 - \frac{h^2}{2R^2}$$

$$\mu\cos r - \cos i = (\mu-1)\left[1+\frac{h^2}{2\mu R^2}\right]$$

$$\Delta f = \frac{R}{(\mu-1)}\left[1-\frac{1}{1+\frac{h^2}{2\mu R^2}}\right]$$

$$\cong \frac{R}{(\mu-1)}\left[1-\left(1-\frac{h^2}{2\mu R^2}\right)\right]$$

$$= \frac{h^2}{2(\mu-1)\mu R} = \frac{h^2}{2(\mu-1)^2 f_0}$$

15.82. A broad source of light whose width is known in terms of angle θ travelling parallel to the principal axis strikes at the edge of a spherical mirror see Fig. 15.70. After reflection it crosses the optical axis at Q at a distance FQ from the focus. Calculate the distances FQ and FL in terms of the radius of curvature R and angle θ.

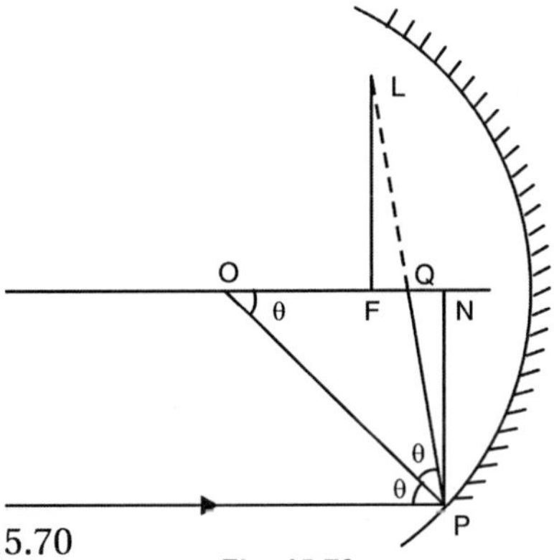

Fig. 15.70

Solution: PN is perpendicular to the principal axis.

In view of the geometry of the Fig. 15.70

$$OQ = PQ$$

$$ON = OQ + QN = R\cos\theta;\ PN = R\sin\theta$$

$$QN^2 + PN^2 = PQ^2 = OQ^2$$

$$\therefore \quad QN = \sqrt{OQ^2 - R^2\sin^2\theta}$$

$$(R\cos\theta - OQ)^2 = OQ^2 - R^2\sin^2\theta$$

$$R^2\cos^2\theta - 2R\cdot OQ\cdot\cos\theta + OQ^2 = OQ^2 - R^2\sin^2\theta$$

$$R^2 = 2R\cdot OQ\cdot\cos\theta$$

$$\therefore \quad OQ = \frac{R}{2\cos\theta}$$

Longitudinal spherical aberration $x = FQ = OQ - OF = OQ - \frac{R}{2}$

$$x = \frac{R}{2}\left[\frac{1}{\cos\theta} - 1\right]$$

From geometry $< OPQ = 180 - 2\theta$

$$\therefore \quad < NQP = 2\theta = < FQL$$

Transverse spherical aberration $FL = y = FQ\tan 2\theta$

$$= \frac{R}{2}\left[\frac{1}{\cos\theta} - 1\right]\tan 2\theta$$

15.83. A concave spherical mirror with an aperture diameter of 40 cm has radius of curvature of 60 cm. Calculate the longitudinal and transverse spherical aberrations of the edge rays moving parallel to the principal axis.

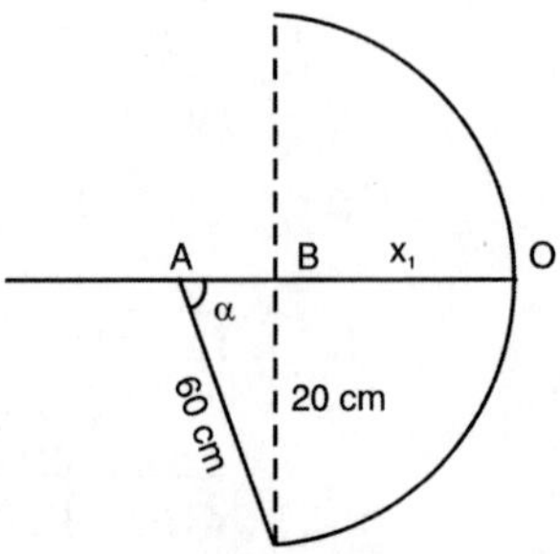

Fig. 15.71

Solution:

$$AB = \sqrt{60^2 - 20^2} = 56.56 \text{ cm}$$

$$OB = x_1 = 60 - 56.56 = 3.44 \text{ cm}$$

$$\cos \alpha = \frac{56.56}{60}$$

$$\alpha = 19°30'$$

$$\text{Longitudinal spherical aberration } x = \frac{60}{2}\left[\frac{1}{\cos \alpha} - 1\right]$$

$$= 30\left[\frac{60}{56.56} - 1\right] = 1.82 \text{ cm}$$

$$\text{Transverse spherical aberration } y = 1.82 \tan 39°$$

$$= 1.82 \times 0.80978$$

$$= 1.47 \text{ cm}$$

15.84. A concave spherical mirror has a focal length of 20 cm. At what minimum height h from the optical axis should an object be placed so that the longitudinal spherical aberration does not exceed 2% of the focal length.

Solution: In view of the Fig. 15.72, the longitudinal spherical aberration

$$FA = \frac{R}{2}\left[\frac{1}{\cos \alpha} - 1\right]$$

$$= f\left[\frac{1}{\cos \alpha} - 1\right]$$

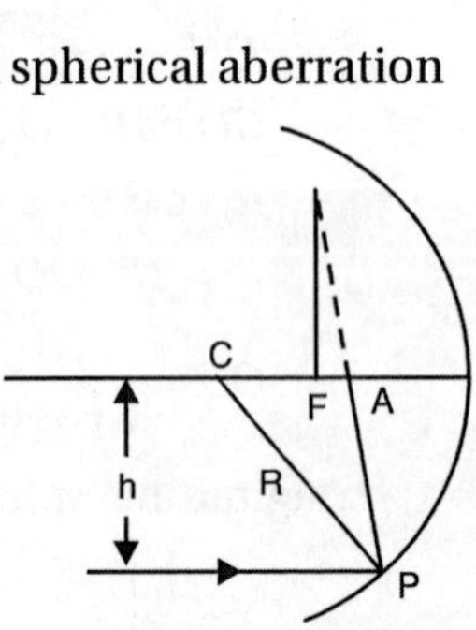

Fig. 15.72

Now according to the problem

$$\frac{2f}{100} = f\left[\frac{1}{\cos \alpha} - 1\right]$$

$$\frac{102}{100} = \frac{1}{\cos \alpha}$$

$$\cos \alpha = \frac{\sqrt{R^2 - h^2}}{R} = \frac{100}{102}$$

$$1-\frac{h^2}{R^2} = \frac{100^2}{102^2}$$

$$h = R\sqrt{1-\frac{100^2}{102^2}}$$

$$= 2f\sqrt{\frac{2\times 202}{102^2}}$$

$$= \frac{40}{102}\sqrt{404}$$

$$= 7.88 \text{ cm.}$$

15.85. An image Y is formed of a point object X by a lens whose optic axis AB as shown in Fig. 15.73a. Draw a ray diagram to locate the lens and its focus. If the image Y of the object X is formed by a concave mirror (having the same optic axis AB) instead of lens, draw another ray diagram to locate the mirror and focus. Write down the steps of construction of the ray diagram. (IIT 1994)

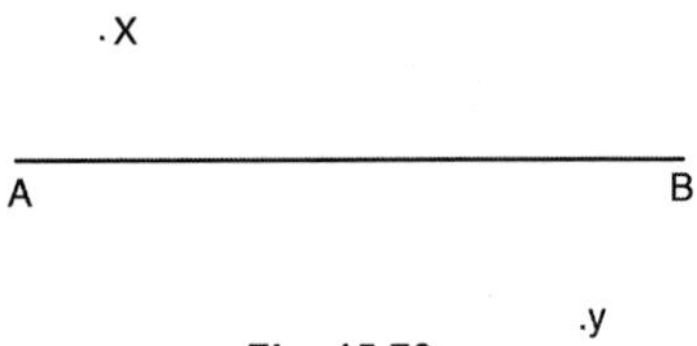

Fig. 15.73a

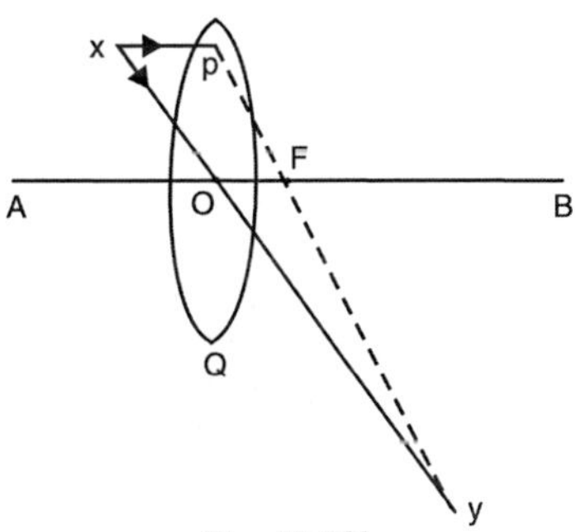

Fig. 15.73b

Solution:

Ist construction (see Fig. 15.73 b)

(i) Join the line XY which intersects the line AB at O.

(ii) A line POQ passing through O and $\perp$ to AB will represent the position of the lens.

(iii) Draw a line from X parallel to AB. Join P to Y. This line intersects AB at F. OF = focal length of the lens.

IInd construction (see Fig. 15.73c)

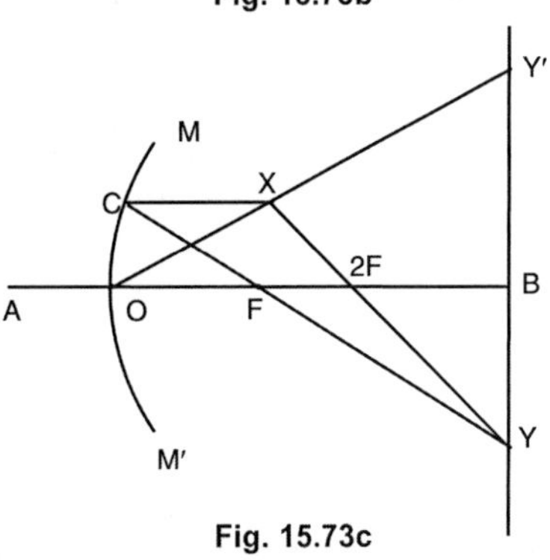

Fig. 15.73c

(i) Draw a line $Y'BY \perp$ to AB. Take $BY = BY'$.

(ii) Join XY' and extend it to O. O is the pole of mirror.

(iii) Join XY which intersects AB at $2F$.

(iv) Take $2F$ as centre and distance O to $2F$ as radius draw an arc MOM'. This represents the position of mirror.

(v) Draw XC parallel to AB. Joint CY. It intersetcs AB at F. Which is the focus of the mirror.

15.86. A equi-convex lens of refractive index ${}^a\mu_g = 3/2$ has radius of each face as R. The region on the left side of the lens known as object space is filled with air. The region on the right side of the lens known as image space is filled with water of refractive index ${}^a\mu_w = 4/3$. A parallel beam of light travels from left of right. Find the distance at which the image will be formed.

Solution: For refraction at the air glass interface we have

$$\frac{\mu_2}{v} - \frac{\mu_1}{u} = \frac{\mu_2 - \mu_1}{R}$$

$$\frac{1.5}{v} + \frac{1}{\infty} = \frac{0.5}{R}$$

or
$$\frac{1.5}{v} = \frac{0.5}{R}$$

For refraction at the second surface we have

$$\frac{4/3}{f} - \frac{1.5}{v} = \frac{4/3 - 3/2}{-R}$$

$$\frac{4}{3f} = \frac{1}{2R} + \frac{1}{6R} = \frac{4}{6R}$$

$\therefore$
$$f = 2R$$

15.87. In the above problem if the object medium is filled with water and image medium contains air, then find the distance at which the image will be formed.

Solution: For refraction at the water glass interface we have

$$\frac{1.5}{v} + \frac{4/3}{\infty} = \frac{3/2 - 4/3}{R} \text{ or } \frac{1.5}{v} = \frac{1}{6R}$$

For refraction at the glass air interface

$$\frac{1}{f} - \frac{1.5}{v} = \frac{1 - 1.5}{-R}$$

$$\frac{1}{f} = \frac{1}{6R} + \frac{1}{2R} = \frac{2}{3R} \text{ or } f = \frac{3R}{2}$$

15.88. The greatest thickness of a plano convex lens appears to be 2 cm when observed normally through the plane surface. On observing through the curved surface the greatest thickness comes out to be 20/9 cm. Find the radius of curvature of the lens. Assume the real thickness of the lens as 3 cm.

Solution:

$$\mu = \frac{\text{real depth}}{\text{apparent depth}} = \frac{3}{2} = 1.5$$

For refraction at the curved surface we have

$$\frac{\mu_2}{v} - \frac{\mu_1}{u} = \frac{\mu_2 - \mu_1}{R}$$

$$\frac{1}{20/9} - \frac{1.5}{3} = \frac{1-1.5}{R} = -\frac{1}{2R}$$

$$\frac{9}{20} - \frac{1}{2} = \frac{-1}{2R}$$

$$\frac{1}{20} = \frac{1}{2R}$$

$$R = 10\,\text{cm}$$

15.89. A plano convex lens when silvered in the plane side behaves like a concave mirror of focal length 30 cm. However, when silvered on the convex side it behaves like a concave mirror of focal length 10 cm. Find the radius of curvature of the lens.

Solution: When the plane side is silvered then the focal length of the concave mirror formed is given by

$$\frac{1}{f_m} = \frac{2(\mu - 1)}{R}$$

When the curved surface is silvered then the focal length of the concave mirror formed is given by

$$\frac{1}{f'_m} = \frac{2\mu}{R}$$

On dividing the two we have

$$\frac{f'_m}{f_m} = \frac{(\mu - 1)}{\mu}$$

$$\frac{10}{30} = \frac{\mu - 1}{\mu}$$

$$\mu = 3\mu - 3$$

$$2\mu = 3$$

or $$\mu = \frac{3}{2}$$

$$\frac{1}{30} = \frac{2 \times (1.5 - 1)}{R}$$

$\therefore$ $$R = 30\,\text{cm}.$$

15.90. We have an ordinary camera in which the distance between the lens and the film is 10 cm. The focal length of the lens $f = 9$ cm. Calculate the depth at which objects lying inside water of refractive index 4/3 can be photographed with the position of the camera lens close to the surface of water.

Solution:

$$\text{Apparent depth} = \frac{d}{\mu}\ ;\ d = \text{real depth}$$

$$\frac{1}{v} + \frac{1}{u} = \frac{1}{f}$$

$$\frac{1}{v} + \frac{\mu}{d} = \frac{1}{f}$$

or $$v = \frac{d \cdot f}{d - \mu f}$$

$$10 = \frac{9d}{d - \frac{4}{3} \times 9}$$

$$10d - 120 = 9d$$

or $$d = 120 \text{ cm}$$

15.91. A convex lens of focal length $f_1 = 12$ cm and a concave lens of focal length $f_2 = 10$ cm are placed co-axially at a distance of 10 cm from each other (see Fig. 15.74). An object which is nearer to the convex lens on the common axis of the two lenses, is placed 48 cm away from the convex lens. Find the position, magnification and nature of the final image formed.

Solution: For convex lens, $u_1 = 48$ cm, $f_1 = 12$ cm, $v_1 = ?$

$$\frac{1}{v_1} = \frac{1}{f_1} - \frac{1}{u_1} = \frac{1}{12} - \frac{1}{48} = \frac{1}{16}\ \therefore v_1 = 16 \text{ cm}$$

That is the image will be formed behind the concave lens at a distance of 6 cm from the lens.

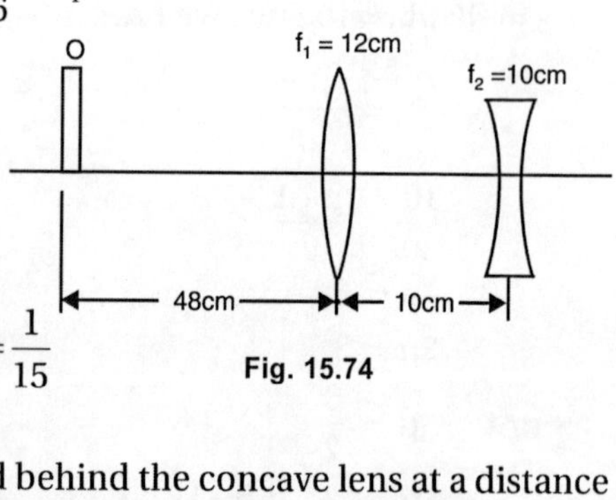

Fig. 15.74

For concave lens,

$$u_2 = 6 \text{ cm}, f_2 = -10 \text{ cm}$$

$$\frac{1}{v_2} = \frac{1}{f_2} + \frac{1}{u_2} = -\frac{1}{10} + \frac{1}{6} = \frac{1}{15}$$

$$\therefore \quad v_2 = 15 \text{ cm}$$

The final image will be formed behind the concave lens at a distance 15 cm. The magnification is 5/6. The image is real and inverted.

15.92. A equi-convex lens of refractive index 1.5 and focal length 10 cm is floating on the surface of water. Find (i) the distance at which parallel rays falling on it are focused and (ii) the focal length when the lens is fully submerge ${}^a\mu_w = \frac{4}{3}$.

Solution:

(i) $$\frac{1}{f} = (\mu - 1) \times \frac{2}{R}$$

$$\frac{1}{10} = \frac{0.5 \times 2}{R}$$

$$\therefore \quad R = 10 \text{ cm}$$

The rays are falling on the first surface of radius of curvature + 10 cm. The rays are focused at v, given by

$$\frac{1.5}{v} - \frac{1}{\infty} = \frac{1.5 - 1}{10} \quad \therefore v = 30 \text{ cm}$$

This image will serve the purpose of the object and the final image will be formed at v' given by

$$\frac{\frac{4/3}{3/2}}{v'} - \frac{1}{30} = \frac{\frac{4/3}{3/2} - 1}{-10} \quad \text{or} \quad \frac{8}{9v'} - \frac{1}{30} = \frac{1}{90}$$

$$\frac{8}{9v'} = \frac{1}{30} + \frac{1}{90} = \frac{4}{90}$$

$$\therefore \quad v' = \frac{8 \times 10}{4} = 20 \text{ cm}$$

(ii) $$\frac{1}{f'} = \left(\frac{3/2}{4/3} - 1\right) \times \frac{2}{R} = \left(\frac{9}{8} - 1\right) \times \frac{2}{10} = \frac{1}{8} \times \frac{2}{10}$$

$$\therefore \quad f' = 40 \text{ cm}$$

15.93. A convex lens of focal length 10 cm and a convex mirror of unknown curvature are coaxially placed as shown in Fig. 15.75. The lens and the mirror are 5 cm apart. An object placed at a distance of 15 cm from the lens has its image at the object. Find the radius of curvature of the mirror.

Solution: The rays are reflected back normally because the image coincides with the object. Therefore, $v = R + 5$, where R is the radius of curvature of the mirror.

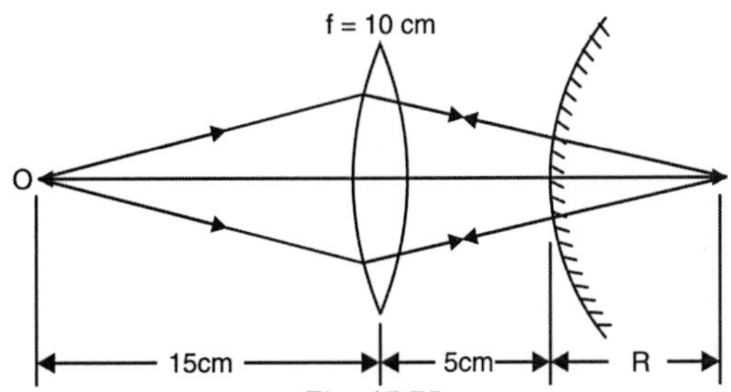

Fig. 15.75

Now $\frac{1}{v}+\frac{1}{u} = \frac{1}{f}$

$$\frac{1}{R+5} = \frac{1}{f}-\frac{1}{u}=\frac{1}{10}-\frac{1}{15}$$

$$= \frac{1}{30}$$

$$R+5 = 30$$

$\therefore$ $R = 25\text{ cm}$

15.94. A converging lens of refractive index $\mu_1 = 1.69$ and a diverging lens of refractive index $\mu_2 = 1.53$, both having the same curvature of radius $R = 12$ cm are placed in contact and are submerged in water of refractive index $\mu_0 = \frac{4}{3}$. Find the focal length of this system in water.

Solution:

The focal length of the convex lens in air is given by $\frac{1}{f_1}=\frac{2(\mu_1-1)}{R}$

The focal length of concave lens in air is given by $\frac{1}{f_2}=\frac{-2(\mu_2-1)}{R}$

The focal length of convex lens in water is given by $\frac{1}{f_1'}=\frac{2}{R}\left(\frac{\mu_1}{\mu_0}-1\right)$

The focal length of concave lens in water is given by $\frac{1}{f_2'}=\frac{-2}{R}\left(\frac{\mu_2}{\mu_0}-1\right)$

where μ_0 is the refractive index of water.

Let F be the focal length of the combination in water, then

$$\frac{1}{F}=\frac{1}{f_1'}+\frac{1}{f_2'}=\frac{2}{R}\left[\frac{\mu_1}{\mu_0}-1-\frac{\mu_2}{\mu_0}+1\right]=\frac{2}{R\mu_0}[\mu_1-\mu_2]$$

or $$F=\frac{R\mu_0}{2(\mu_1-\mu_2)}$$

On substituting the values of various terms, we have

$$F=\frac{12\times 4}{2\times 3\,(1.69-1.53)}$$

$$=\frac{12\times 4}{2\times 3\times 0.16}=50\text{ cm}$$

15.95. A convex lens of focal length 25 cm forms a sharp image of an object on a screen placed at a distance 450 cm from the lens. Now the screen has been displaced 20 cm towards the lens. By what distance should the object be shifted in order to obtain a sharp image on the screen?

Solution: $\frac{1}{u_1}+\frac{1}{v_1}=\frac{1}{f}$, where u_1 and v_1 represent the initial position of the object and screen, respectively.

$$\frac{1}{u_1}=\frac{1}{f}-\frac{1}{v_1}=\frac{v_1-f}{fv_1}$$

or $$u_1=\frac{fv_1}{v_1-f}=\frac{25\times 450}{425}=26.47\text{ cm}$$

On displacing the screen, let u_2 and v_2 be the final positions of the object and the screen, respectively.

Then, we have

$$u_2=\frac{fv_2}{v_2-f}=\frac{25\times 430}{405}=26.54\text{ cm}$$

Shift of the object = 26.54 – 26.47 = 0.7 mm

15.96. A convex lens of focal length 20 cm and a concave mirror of radius of curvature 16 cm are placed at a distance of 25 cm from each other. A luminous object is placed at a distance of 25 cm from the lens. Where will the image be formed and what would be its magnification?

(Roorkee 1986)

Solution: For a convex lens, we have

$$\frac{1}{v_1}-\frac{1}{u_1}=\frac{1}{f}$$

$$\therefore \quad \frac{1}{v_1}=\frac{1}{f}+\frac{1}{u_1},\ f-20\text{ cm},\ u_1=-25\text{ cm}$$

$$\frac{1}{v_1}=\frac{1}{20}-\frac{1}{25}$$

$$v_1=100\text{ cm}$$

Magnification of the lens

$$m_1=-\frac{v_1}{u_1}=-\frac{100}{25}=-4$$

Now this image will act as a source for the concave mirror. Now, for the mirror

$$u_2=100-25=75\text{ cm},\ f_m=-8\text{ cm}$$

$$\frac{1}{v_2}+\frac{1}{u_2}=\frac{1}{f_m}$$

$$\therefore \quad \frac{1}{v_2}=\frac{1}{f_m}-\frac{1}{u_2}=-\frac{1}{8}-\frac{1}{75}$$

$$\therefore \quad v_2 = -\frac{600}{83}\text{ cm}$$

$$\text{Magnification } m_2 = -\frac{v_2}{u_2} = \frac{600}{83\times 75} = 0.096$$

The image thus formed will act as an object for the lens.

Now the position of the object $u_3 = 25 - \frac{600}{83} = \frac{1475}{83}$ cm, $f = -20$ cm

$$\frac{1}{v_3} - \frac{83}{1475} = \frac{-1}{20}$$

$$\therefore \quad \frac{1}{v_3} = \frac{83}{1475} - \frac{1}{20}$$

$$\therefore \quad v_3 = \frac{1475\times 20}{185} = 159.46\text{ cm}$$

Magnification produced by the lens, $m_3 = \frac{v_3}{u_3} = \frac{159.46\times 83}{1475} = 8.973$

Final magnification $= m_1 \times m_2 \times m_3 = -4 \times 0.096 \times 8.973 = -3.44$

15.97. An object is placed at a distance of 45 cm from an equi-convex lens of focal length 45 cm. With the aim of getting the image in a denser medium in the direction perpendicular to the optical axis, a plane mirror inclined at an angle of 45° with respect to the optical axis is placed at a distance of 3 m behind the lens. At what depth below the optical axis should a tray containing 16 cm thick water layer be placed in order to get a sharp image of the object at its bottom? ${}^a\mu_w = \frac{4}{3}$.

Solution: In the absence of the plane mirror the image will be at a distance v given by the relation

$$\frac{1}{v} = \frac{1}{f} - \frac{1}{u} = \frac{1}{40} - \frac{1}{45};\ v = 360\text{ cm}$$

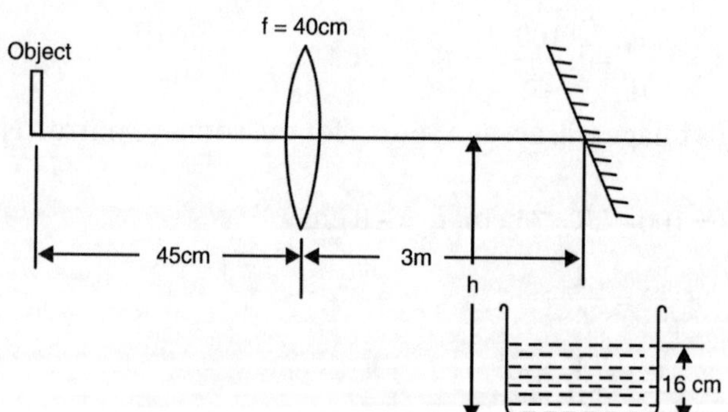

Fig. 15.76

i.e. the image will be formed at a distance of 360 cm from the lens. We have a plane mirror in the path. It will reflect the image into the tray containing water (Fig. 15.76). This image will occupy a position at a depth of 60 cm from the principal axis. The water layer of thickness 16 cm will shift the image through a distance = 16 [1 – (3/4)] = 4 cm. The correct depth $h = 60 + 4 = 64$ cm.

15.98. A convex lens of focal length 20 cm and a plane mirror placed at a distance of 10 cm behind the lens in a direction perpendicular to the principal axis form an optical system. An object is placed at a distance of 10 cm from the lens. Find the position of the image formed by the optical system.

Solution: In Fig. 15.77 an object is placed at a distance less than the focal length of the lens. It will give rise to virtual image I_1 at a distance v given by the relation

$$\frac{1}{v} = \frac{1}{f} - \frac{1}{u} = \frac{1}{20} - \frac{1}{10} = -\frac{1}{20}$$

$$\therefore \quad v = -20 \text{ cm}$$

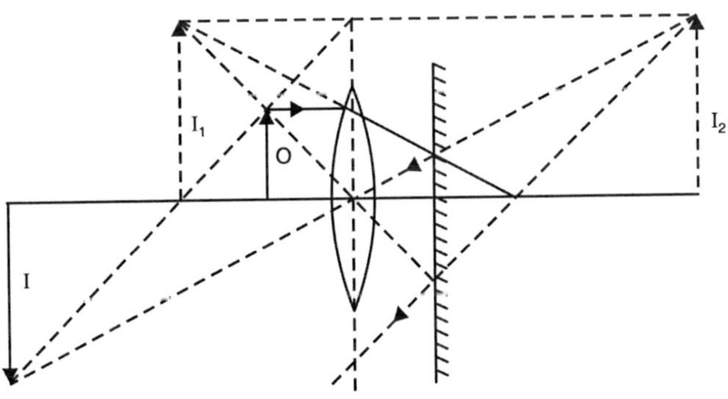

Fig. 15.77

This virtual image is at a distance of 30 cm from the mirror. This image will serve the purpose of object for the mirror and will form a image I_2 at a distance 30 cm behind the mirror. Image I_2 lying at a distance of 40 cm, from the convex lens will serve the purpose of object for forming the final image I. The distance of image I is given by the following relation:

$$\frac{1}{v'} = \frac{1}{f} - \frac{1}{u}$$

Here $u = 40$ cm

$$= \frac{1}{20} - \frac{1}{40}$$

$$\therefore \quad v' = 40$$

The final image I is at a distance of 40 cm from the lens.

15.99. A planoconvex lens has a thickness of 4 cm. When placed on a horizontal table with the curved surface in contact with it, the apparent depth of the bottommost point of the lens is 3 cm. If the lens is inverted so that the plane face is in contact with the table, the apparent depth of the centre of the plane face of the lens is found to be 25/8 cm. Find the focal length of the lens. (CEE 1990, IIT 1984)

Solution: When the lens is in the first position the refraction takes place at the plane surface. Now we have

$$^{a}\mu_{g} = \frac{\text{real depth}}{\text{apparent depth}} = \frac{4}{3}$$

When the lens is in the second position then refraction takes place at the curved surface. Now we have

$$\frac{^{g}\mu_{a}}{v} - \frac{1}{v} = \frac{^{g}\mu_{a} - 1}{R}$$

On substituting the values we have

$$\frac{3/4}{-25/8} + \frac{1}{4} = \frac{(3/4) - 1}{R}; R = -25 \text{ cm}$$

Now $$\frac{1}{f} = \left(\frac{4}{3} - 1\right)\left(\frac{1}{\infty} + \frac{1}{25}\right)$$

$\therefore$ $$f = 75 \text{ cm}$$

15.100. The distance between two point sources of light is 24 cm. Find out where would you place a converging lens of focal length 9 cm, so that the image of both sources are formed at the same point? (Roorkee 1988)

Solution: *A* and *B* are the objects whose images are to be formed at one point. For the object *A*, we have

$$\frac{1}{v} + \frac{1}{x} = \frac{1}{9} \text{(see Fig. 15.78)}$$

or $$\frac{1}{v} = \frac{1}{9} - \frac{1}{x}$$

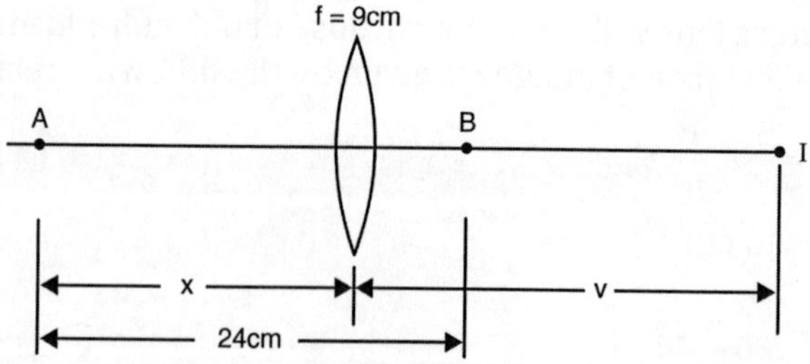

Fig. 15.78

For the object B, we have

$$\frac{1}{v} - \frac{1}{(24-x)} = -\frac{1}{9}$$

or $$\frac{1}{v} = \frac{1}{(24-x)} - \frac{1}{9}$$

On equating the values of v, we have

$$\frac{1}{9} - \frac{1}{x} = \frac{1}{24-x} - \frac{1}{9} \text{ or } \frac{2}{9} = \frac{1}{x} + \frac{1}{(24-x)} \text{ or } \frac{2}{9} = \frac{24}{x(24-x)}$$

$$24x - x^2 = 108; \; x^2 - 24x + 108 = 0$$

or $$(x-18)(x-6) = 0; \; x = 18 \text{ cm or } 6 \text{ cm}$$

15.101. Fig. 15.79 illustrates an aligned system consisting of three lenses located in air. An object is placed at a distance of 10 cm from the first lens. Find the position of the image formed by the system.

Solution: The position of the image formed by the first lens is given by

$$\frac{1}{v_1} = \frac{1}{f_1} - \frac{1}{u_1}$$

Here $u_1 = 10$ cm

$$= \frac{1}{5} - \frac{1}{10}$$

$$= \frac{1}{10}$$

$\therefore$ $v_1 = 10$ cm

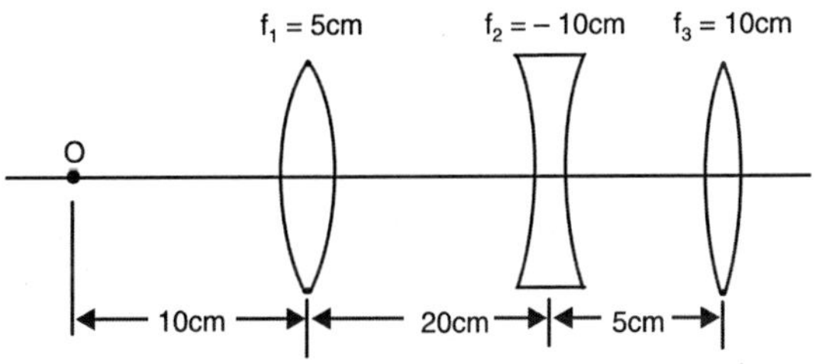

Fig. 15.79

Thus the image is between first and the second lenses and will be at a distance of 10 cm from the second lens. This image serves the purpose of the object for the second lens. The position of the image is given by

$$\frac{1}{v_2} = -\frac{1}{f_2} - \frac{1}{u_2} \quad \text{Here } u_2 = 10 \text{ cm}$$

$$= -\frac{1}{10} - \frac{1}{10}$$

$$v_2 = -5 \text{ cm}$$

The image is between the first and the second lenses and will be at a distance of 10 cm from the third lens. For the third lens we have

$$\frac{1}{v_3} = \frac{1}{f_3} - \frac{1}{u_3}; \; u_3 = 10 \text{ cm}$$

$$\frac{1}{v_3} = \frac{1}{10} - \frac{1}{10} = 0$$

$\therefore \quad v_3 = \infty$

Thus a parallel beam of light will emerge from the third lens and will be focused at infinity.

15.102. A concave lens of focal length 20 cm is placed at a distance of 15 cm in front a concave mirror of radius of curvature 26 cm, and an object is placed further 10 cm away from the lens. The principal axis of the lens and the mirror are coincident and the object is on the axis. Find the position and the nature of the image.

Solution: The position of the image formed by the concave lens is given by

$$\frac{1}{v} - \frac{1}{u} = \frac{1}{f} \quad \text{Here } u, v \text{ and } f \text{ are all negative.}$$

$$\frac{1}{v} = \frac{1}{f} + \frac{1}{u}$$

or $$\frac{1}{v} = \frac{1}{20} + \frac{1}{10}$$

or $$v = 20/3 \text{ cm.}$$

This image will serve the purpose of an object and will be at a distance $20/3 + 15 = 65/3$ cm from the mirror.

For the mirror we have

$$\frac{1}{v} = \frac{1}{f} - \frac{1}{u}$$

$$= \frac{1}{13} - \frac{3}{65}$$

$\therefore$ $$v = 32.5 \text{ cm}$$

The final image is at a distance of 32.5 cm from the mirror and is real one.

15.103. An object is placed at the principal axis of a equi-convex lens of focal length f at a distance L from the optical centre of the lens. The object is executing simple harmonic motion along the principal axis and its displacement is given by $x = x_0 \sin \omega t$. Calculate the amplitude of oscillation of the image formed by the lens.

Solution: Let the position of the object be $u = L + x_0 \sin \omega t$

Using the lens equation we have

$$\frac{1}{v'} = \frac{1}{f} - \frac{1}{L + x_0 \sin \omega t}$$

$$\frac{1}{v'} = \frac{L + x_0 \sin \omega t - f}{f(L + x_0 \sin \omega t)}$$

$$v' = \frac{f(L + x_0 \sin \omega t)}{(L-f)\left[1 + \frac{x_0 \sin \omega t}{(L-f)}\right]}$$

$$= \frac{f(L + x_0 \sin \omega t)}{(L-f)}\left[1 - \frac{x_0}{(L-f)} \sin \omega t\right]$$

$$= \frac{fL}{L-f} + \frac{f x_0 \sin \omega t}{L-f} - \frac{f L x_0 \sin \omega t}{(L-f)^2}$$

$$= v - \frac{f x_0}{(L-f)^2}[L - L + f] \sin \omega t$$

$$= v - \frac{x_0 f^2 \sin \omega t}{(L-f)^2}$$

Amplitude of image executing simple harmonic motion $= \dfrac{x_0 f^2}{(L-f)^2}$.

15.104. A thin convex lens of focal length f and diameter D is used to photograph a distant object of Luminance L. Find the illuminance of the image if the loss of light in the lens is α.

Solution: Light incident on the lens has Luminosity $\dfrac{L\pi D^2}{4}$

Loss of light in the lens $= \dfrac{\alpha L\pi D^2}{4}$

Light emerging out of the lens $= (1-\alpha)\dfrac{L\pi D^2}{4}$

Illuminance of the image $= (1-\alpha)\dfrac{L\pi D^2}{4f^2}$

Since the final image is formed at the focus of the lens.

15.105. A $\dfrac{f}{2.5}$ camera lens has focal length of 7.5 cm. What will be its effective diameter. If the exposure time of a $\dfrac{f}{2.5}$ lens is $\dfrac{1}{100}$ sec, find the exposure time for a $\dfrac{f}{5}$ lens.

Solution: If the effective diameter of a lens of focal length f be d then $\dfrac{f}{d}$ is known as f number.

A $\frac{f}{2.5}$ lens has effective diameter as $\frac{f}{2.5}$, therefore f number is 2.5.

d = effective diameter = $\frac{f}{2.5} = \frac{7.5}{2.5} = 3$ cm

The f number for second lens is 2 times that of the first one.

Now exposure time $\propto$ square of f number.

or exposure time = K. square of f number

$$\frac{1}{100} = K\,(2.5)^2$$

$$t = K \times 5^2$$

$$\therefore \quad t = \frac{1}{25} \text{ sec.}$$

15.106. A $\frac{f}{4.5}$ lens has an exposure time of $\frac{1}{25}$ sec. Find the effective diameter of the lens whose exposure time is $\frac{1}{100}$ sec.

Solution: Let x be the f number of the second lens, then we have

$$\frac{1}{25} = K \cdot (4.5)^2$$

and $$\frac{1}{100} = K \cdot x^2$$

On dividing the two we have

$$4 = \frac{(4.5)^2}{x^2}$$

or $$2 = \frac{4.5}{x}$$

or $$x = 2.25$$

$\therefore$ Effective diameter of the second lens is $\frac{f}{2.25}$.

15.107. A small fish 0.4 m below the surface of a lake is viewed through a simple converging lens of focal length 3m. The lens is kept at 0.2 m above the water surface such that the fish lies on the optical axis of the lens. Find the image of the fish seen by the observer. The refractive index of water is 4/3. (Roorkee 1996)

Solution: From air the fish will appear at a depth x given by

$$\frac{0.4}{x} = \frac{4}{3} \quad \text{(see Fig. 15.80)}$$

$$\therefore \quad x = 30 \text{ cm.}$$

For the formation of image the position of object $u = 20 + 30 = 50$ cm.

$$f = 300 \text{ cm}$$

$$\frac{1}{v} + \frac{1}{50} = \frac{1}{300}$$

here u and v both will be negative.

$$\frac{1}{v} = \frac{1}{300} - \frac{1}{50} = -\frac{1}{60}$$

$$\therefore \quad v = -60 \text{ cm}$$

The image of fish will be at a distance of 60 cm from the lens and will be on the side of the fish.

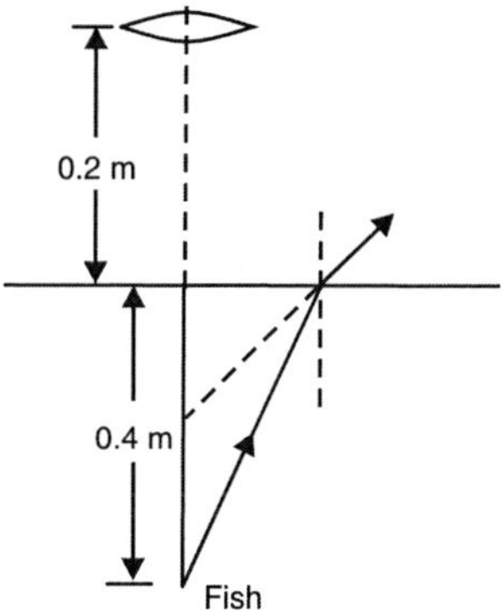

Fig. 15.80

15.108. A thin equi-convex lens of glass of refractive index $\mu = \frac{3}{2}$ and of focal length 0.3 m in air is sealed into an opening at one end of a tank filled with water $\left(\mu = \frac{4}{3}\right)$. On the opposite side of the lens, a mirror is placed inside the tank on the tank wall perpendicular to the lens as shown in Fig. 15.81a. The separation between the lens and the mirror is 0.8 m. A small object is placed outside the tank in front of the lens at a distance of 0.9 m from the lens along its axis. Find the position (relative to the lens) of the image of the object formed by the system. (IIT 1997)

Solution: Let us consider the refraction at the first surface of the lens

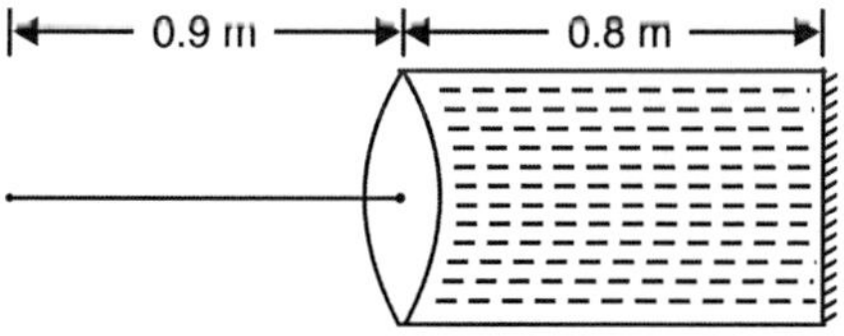

Fig. 15.81a

$$\frac{3/2}{v} - \frac{1}{u} = \frac{3/2 - 1}{R}$$

$$\frac{3/2}{v} + \frac{1}{90} = \frac{0.5}{30}; \; R = 30 \text{ cm}$$

$$v = 270 \text{ cm}$$

The image thus formed will act as an object for the second surface.

At second surface

$$\frac{4/3}{v_1} - \frac{3/2}{270} = \frac{4/3 - 3/2}{-30}$$

$$\frac{4}{3v_1} = \frac{1}{180} + \frac{1}{180} = \frac{1}{90}$$

$$v_1 = 120 \text{ cm}$$

Thus the lens will form the image I_1 at a distance of 120 cm from it.

Shift produced by water is given by

$$x = \left(1 - \frac{1}{\mu}\right)t = 80\left(1 - \frac{3}{4}\right) = 20 \text{ cm}$$

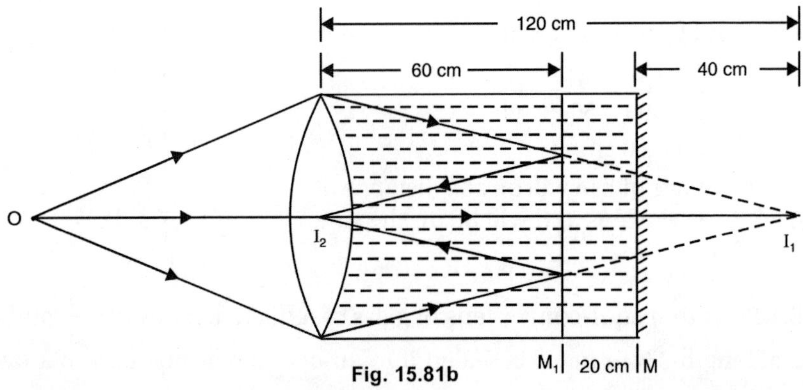

Fig. 15.81b

Thus the position of mirror M is now at M_1. The image I_1 is at a distance of 60 cm from M_1 (see Fig. 15.81b).

The mirror M_1 will form the image I_2 for the object I_1 at the optical centre of the lens.

15.109. Two thin similar convex glass pieces are joined together, front to front, with its rear portion silvered such that a sharp image is formed at 0.2 m. When the air between the glass pieces is replaced by water $\left(\mu = \frac{4}{3}\right)$. Find the position of the image. (Roorkee 1995)

Solution: On introducing air between the glasses the system will act as a concave mirror (see Fig. 15.82).

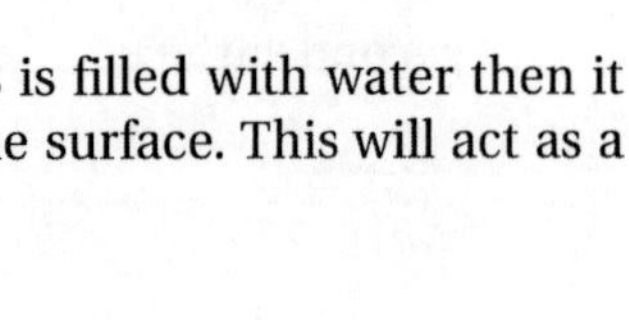

Fig. 15.82

$$\frac{1}{v} + \frac{1}{u} = \frac{2}{R}$$

$$\frac{2}{20} = \frac{2}{R}$$

$$\therefore \quad R = 20 \text{ cm.}$$

When the space between the two pieces is filled with water then it will be a convex lens of water polished at one surface. This will act as a concave mirror of focal length given by

$$\frac{1}{F} = 2 \times 2\frac{(\mu - 1)}{R} + \frac{2}{R}$$

$$= \frac{4\mu - 2}{R} = \frac{4 \times 4/3 - 2}{20} = \frac{1}{6}$$

$$\text{Now } \frac{1}{u}+\frac{1}{v} = \frac{1}{F}$$

$$\frac{2}{u} = \frac{1}{6}$$

$$\therefore \quad u = 12 \text{ cm.}$$

15.110. A thin equi-convex lens of refractive index μ and optical power ϕ_0 is immersed in a liquid of refractive index μ_0. Find the optical power of the lens.

Solution:

$$\frac{1}{f_{\text{air}}} = \phi_0 = \frac{2(\mu-1)}{R}$$

$$\frac{1}{f_{\text{liq}}} = \frac{2(\mu-\mu_0)}{\mu_0 R} = \frac{(\mu-\mu_0)\phi_0}{\mu_0(\mu-1)}$$

$$\therefore \quad \text{Optical power in liquid} = \frac{\mu_0}{f_{\text{liq}}} = \frac{(\mu-\mu_0)\phi_0}{(\mu-1)}$$

15.111. An optical system consists of two thin equi-convex lens of focal lengths f_1 and f_2 respectively separated by a distance d. What should be the equivalent focal length of a lens and where it should be placed so that it may provide the same transverse magnification as that through the system of the two lenses?

Solution: The system of lenses L_1 and L_2 can be replaced by an equivalent lens of focal length H_2F_2. The placement of the lens must be along the plane M_2H_2. Point H_2 is known as principal point and the plane passing through H_2 and perpendicular to the principal axis is known as principal plane (see Fig. 15.83). Similarly the first focal point F_1 and its respective principal point H_1 corresponds to a ray which passes through F_1 and becomes parallel to the principal axis after refraction. H_1F_1 is known as first principal focal length.

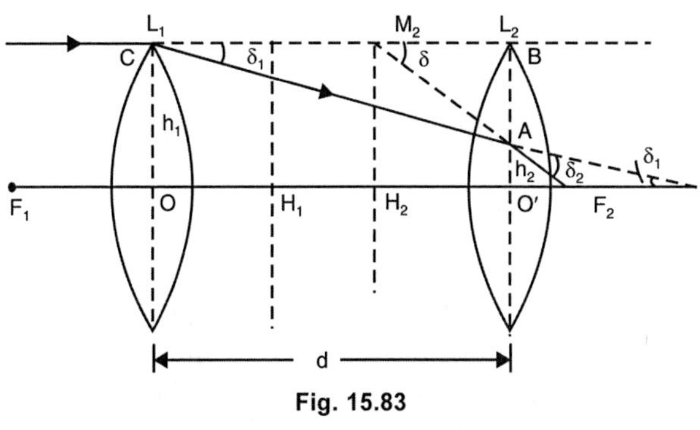

Fig. 15.83

$$\delta_1 = \frac{h_1}{f_1},\ \delta_2 = \frac{h_2}{f_2} \text{ and } \delta = \frac{h_1}{F}$$

$$\frac{h_1}{F} = \frac{h_1}{f_1} + \frac{h_2}{f_2}$$

$$h_2 = h_1 - AB$$

$$= h_1 - BC \cdot \delta_1$$

$$= h_1 - d\,\frac{h_1}{f_1}$$

$$h_2 = h_1\left(1 - d/f_1\right)$$

Now $$\frac{h_1}{F} = \frac{h_1}{f_1} + \frac{h_1}{f_2}\left(1 - d/f_1\right)$$

or $$\frac{1}{F} = \frac{1}{f_1} + \frac{1}{f_2} - \frac{d}{f_1 f_2}$$

or $$F = \frac{f_1 f_2}{f_1 + f_2 - d} = \frac{-f_1 f_2}{\Delta};\ \Delta = d - f_1 - f_2$$

Consider triangles M_2H_2F and $AO'F_2$

$$\frac{O'F_2}{H_2F_2} = \frac{h_2}{h_1}$$

$$O'F_2 = H_2F_2 \cdot \frac{h_2}{h_1} = F \cdot \frac{h_1}{h_1}\left(1 - \frac{d}{f_1}\right) = F\left(1 - \frac{d}{f_1}\right)$$

$$O'H_2 = H_2F_2 - O'F_2 = F - F\left(1 - \frac{d}{f_1}\right) = \frac{F \cdot d}{f_1}$$

$$= \frac{-f_1 f_2}{\Delta} \cdot \frac{d}{f_1} = \frac{-df_2}{\Delta}$$

This is the distance of the second principal point from second lens. Similarly the distance of first principal point with respect to first lens

$$= -\frac{f_1 d}{\Delta}$$

So the single lens should be placed at a distance $\dfrac{f_1 d}{\Delta}$ from the first lens or $\dfrac{f_2 d}{\Delta}$ from the second lens.

15.112. A equi-convex lens of refractive index μ and radius of curvature R is fitted at one end of a cylindrical tube of length l. At the other end of the tube a plane mirror is placed as shown in the Fig. 15.84. The space between the mirror and the lens is filled with a liquid of refractive index μ_0. Find the optical power of this system.

Solution: Consider the refraction of a parallel beam of light passing through this system.

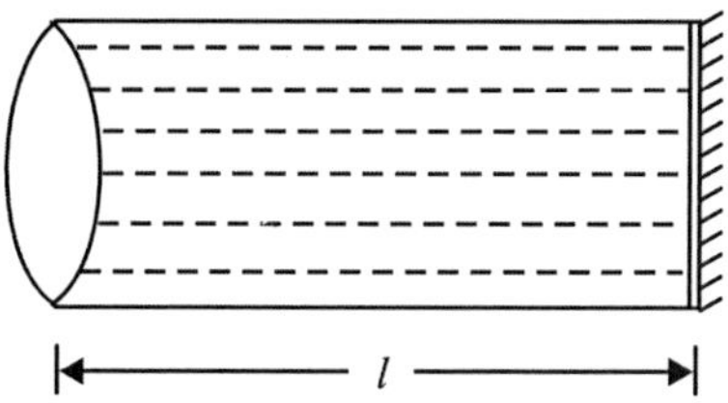

Fig. 15.84

At the first surface we have

$$\frac{\mu}{v} - \frac{1}{\infty} = \frac{\mu - 1}{R}$$

For the refraction at second surface we have

$$\frac{\mu_0}{f} - \frac{\mu}{v} = \frac{\mu_0 - \mu}{-R}$$

or $$\frac{\mu_0}{f} = \frac{\mu - \mu_0}{R} + \frac{(\mu - 1)}{R} = \frac{2\mu - \mu_0 - 1}{R}$$

$\therefore$ $$f = \frac{\mu_0 R}{2\mu - \mu_0 - 1} = \frac{\mu_0}{\phi'}; \phi' = \frac{2\mu - \mu_0 - 1}{R}$$

The plane mirror behind the lens will form the image of the lens at a distance of l. In this way we have two convex lens each of focal length $f = \frac{\mu_0}{\phi'}$ separated by a distance $2l$ and the space between them is filled with a liquid of refractive index μ_0. The focal length of this system is given by

$$\frac{1}{F} = \frac{\phi'}{\mu_0} + \frac{\phi'}{\mu_0} - \frac{2l \cdot \phi'^2}{\mu_0^2}$$

Optical power of the system is given by

$$\frac{\mu_0}{F} = 2\phi' - \frac{2l\phi'^2}{\mu_0}$$

15.113. A boy with his spectacles consisting of concave lens of focal length 1 m can see clearly the objects lying at a distance of 25 cm. How far he should keep a newspaper from his eyes so that he can read it without glasses.

Solution: Using glasses we have

$$\frac{1}{f} - \frac{1}{100} = \frac{1}{x} + \frac{1}{25}$$; x is the position of the image and f is the focal length of the eye.

Without glasses we have

$$\frac{1}{f} = \frac{1}{x} + \frac{1}{d}$$; d is the distance between the eye and newspaper

Now $$\frac{1}{d} = \frac{1}{25} + \frac{1}{100} = \frac{1}{20}$$

$\therefore$ $d = 20$ cm.

15.114. For a myopic eye the far point is 80 cm. Calculate the power of the lens required to correct the defect.

Solution:

$$\frac{1}{v}+\frac{1}{\infty}=\frac{1}{f_e}+\frac{1}{f};$$ f_e = focal length of the eye

f = focal length of the lens required

Now $\frac{1}{v}+\frac{1}{80}=\frac{1}{f_e}$

Using the two equation

$$\frac{1}{f}=-\frac{1}{80}$$

$$P=\text{power}=-\frac{1}{0.8}=-1.25\text{ D}$$

15.115. A man suffering from short sight is unable to see objects distinctly at a distance greater than 2 m. Find the power required to correct the defect.

Solution:

$\frac{1}{v}+\frac{1}{200}=\frac{1}{f_e}$; f_e = focal length of the eye

$\frac{1}{v}+\frac{1}{\infty}=\frac{1}{f_e}+\frac{1}{f}$; f = focal length of the lens to be added

On substituting the two we have

$$\frac{1}{f}=-\frac{1}{200}$$

$$P=-0.5\text{ diopter}$$

15.116. Derive an expression for the focal length of a (a) planoconvex lens silvered at its plane surface, (b) a planoconvex lens silvered at its convex surface, and (c) a double convex lens with a convex surface of radius R_2 is silvered.

Solution:

(a) $\frac{1}{F}=\frac{1}{f_l}+\frac{1}{f_m}+\frac{1}{f_l};\frac{1}{f_l}=\frac{(\mu-1)}{R},\frac{1}{f_m}=0$ (see Fig. 15.85a)

Fig. 15.85a

$\therefore \quad \frac{1}{F}=\frac{2(\mu-1)}{R}$

(b) $\frac{1}{F}=\frac{1}{f_l}+\frac{1}{f_m}+\frac{1}{f_l};\frac{1}{f_l}=\frac{(\mu-1)}{R};\frac{1}{f_m}=\frac{2}{R}$ (see Fig. 15.85b)

$$\therefore \quad \frac{1}{F} = \frac{2(\mu-1)}{R} + \frac{2}{R} = \frac{2\mu}{R}$$

(c) $\frac{1}{F} = \frac{1}{f_l} + \frac{1}{f_m} + \frac{1}{f_l}; \frac{1}{f_l} = (\mu-1)\left(\frac{1}{R_1} + \frac{1}{R_2}\right); \frac{1}{f_m} = \frac{2}{R_2}$

$$\therefore \quad \frac{1}{F} = \frac{2(\mu-1)}{R_1} + \frac{2\mu}{R_2} \text{ (see Fig. 15.85c).}$$

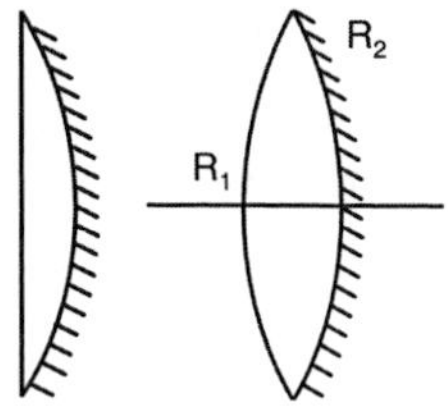

Fig. 15.85b Fig. 15.85c

15.117. The radius of curvature of the convex face of a planoconvex lens is 12 cm and its refractive index is 1.5.

(a) Find the focal length of the lens.

The plane surface of the lens is now silvered.

(b) At what distance from the lens will the parallel rays incident on the convex surface coverage?

(c) Sketch the ray diagram to locate the image, when a point object is placed on the axis 20 cm from the lens.

(d) Calculate the image distance when the object is placed as in (c).

(IIT 1979)

Solution:

(a) $\frac{1}{f_l} = \frac{(\mu-1)}{R} = \frac{0.5}{12}$

$\therefore \ f_l = 24$ cm

(b) On silvering the plane surface it will behave like a concave mirror whose focal length is given by the equation.

$$\frac{1}{F'} = \frac{2}{f_l} = \frac{2}{24}$$

$\therefore \ F' = 12$ cm

(c) The ray diagram is shown in Fig. 15.86.

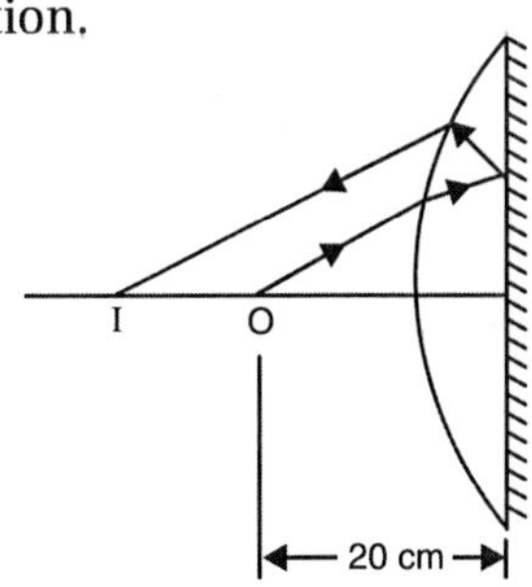

Fig. 15.86

(d) On applying the formula applicable in case of mirror

$$\frac{1}{v} + \frac{1}{u} = \frac{1}{F'}$$

$$\frac{1}{v} = \frac{1}{F'} - \frac{1}{u} \quad \text{(here } u, v, F' \text{ are all negative)}$$

$$= \frac{1}{12} - \frac{1}{20}$$

$$= \frac{1}{30}$$

$\therefore \ v = 30$ cm.

15.118. The convex surface of a thin concavoconvex lens of glass of refractive index 1.5 has a radius of curvature 20 cm. The concave surface has a radius of curvature 60 cm. The convex side is silvered and placed on a horizontal surface (Fig. 15.87). (a) Where should a pin be placed on the axis such that its image is formed at the same place? (b) If the concave part is filled with water of refractive index 4/3, find the distance through which the pin should be moved so that the image of the pin again coincides with the pin. (IIT 1981)

Solution:

(a) Focal length of the concavoconvex lens is given by

$$\frac{1}{f_g}=(\mu-1)\left(\frac{1}{R_1}-\frac{1}{R_2}\right) \text{ (here } R_1 \text{ and } R_2 \text{ are both negative)}$$

$$=(1.5-1)\left(-\frac{1}{60}+\frac{1}{20}\right)=0.5\times\frac{2}{60}=\frac{1}{60}$$

As the convex surface is silvered it will behave like a concave mirror of focal length given by

$$\frac{1}{F}=\frac{2}{f_g}+\frac{2}{R_2}=\frac{1}{30}+\frac{1}{10}=\frac{4}{30}$$

$$\therefore \quad F=\frac{30}{4}=7.5 \text{ cm}$$

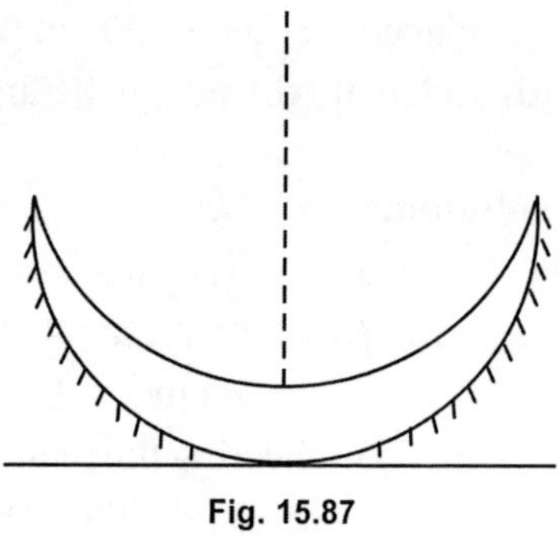

Fig. 15.87

On applying the formula applicable to mirrors, we have

$$\frac{1}{v}+\frac{1}{u}=\frac{4}{30}; \text{ here } v \text{ and } u \text{ are same.}$$

$$\therefore \quad \frac{2}{u}=\frac{4}{30}$$

$$u=15 \text{ cm}$$

(b) On filling the concave part with water, we have a planoconvex lens formed by the liquid and a concavoconvex lens made of glass.

$$\frac{1}{f_{liq}}=\left(\frac{4}{3}-1\right)\times\frac{1}{60}=\frac{1}{180}$$

The focal length of the system is given by

$$\frac{1}{F'}=\frac{2}{f_{liq}}+\frac{2}{f_g}+\frac{2}{R_2}$$

$$=\frac{1}{90}+\frac{1}{30}+\frac{1}{10}=\frac{13}{90}$$

$$\therefore \quad F' = \frac{90}{13}\text{ cm}$$

On applying the formula for concave mirror, we have

$$\frac{1}{u'} + \frac{1}{u'} = \frac{1}{F'}$$

$$\therefore \quad \frac{2}{u'} = \frac{13}{90}$$

$$u' = \frac{180}{13} = 13.85\text{ cm}$$

The pin has to be shifted downwards towards the lens by a distance = 1.15 cm.

15.119. An equi-convex lens of crown glass having radii $R_1 = 10$ cm forms an achromatic lens in contact with a double concave lens of flint glass having radii $R_1 = 10$ cm and $R_2 = 11$ cm. Calculate the ratio of the change in refractive indices of the extreme rays of the visible region in the two glasses.

Solution: For an achromatic lens $\frac{\omega}{f} + \frac{\omega'}{f'} = 0$

where ω and ω' correspond to the dispersive power of crown and flint glass respectively.

Now $\frac{\omega}{f} = -\frac{\omega'}{f'}$

$$\frac{\mu_v - \mu_r}{\mu_y - 1} \times (\mu_y - 1)\frac{2}{R_1} = \frac{\mu'_v - \mu'_r}{\mu'_y - 1}(\mu'_y - 1)\left(\frac{1}{R_1} + \frac{1}{R_2}\right)$$

or
$$\frac{\mu_v - \mu_r}{\mu'_v - \mu'_r} = \frac{R_1 + R_2}{2R_2}$$

On substituting the values of R_1 and R_2, we have

$$\frac{\mu_v - \mu_r}{\mu'_v - \mu'_r} = \frac{10 + 11}{22} = 0.9545$$

15.120. A thin convex lens of crown glass forms an achromatic lens of focal length 30 cm in contact with a thin concave lens of flint glass. The radius of the crown surface in contact is 20 cm. Find the radius of curvature of the second surface of each lens. The dispersive power and mean refractive index are respectively 0.017 and 1.5 for crown glass and 0.034 and 1.7 for flint glass.

Solution: For an achromatic lens, we have

$$\frac{\omega_1}{f_1} + \frac{\omega_2}{f_2} = 0$$

Further the focal length of the combination is given by

$$\frac{1}{F} = \frac{1}{f_1} + \frac{1}{f_2}$$

On solving for f_1 and f_2, we have

$$f_2 = \frac{\omega_1 - \omega_2}{\omega_1} F, \; f_1 = \frac{\omega_2 - \omega_1}{\omega_2} F$$

$$\text{Now } f_1 = \frac{0.034 - 0.017}{0.034} \times 30 = 15 \text{ cm (convex lens)}$$

$$f_2 = \frac{0.017 - 0.034}{0.017} \times 30 = -30 \text{ cm (concave lens)}$$

$$\text{Now } \frac{1}{f_1} = (\mu - 1)\left(\frac{1}{R} + \frac{1}{20}\right)$$

$$\frac{1}{15} = 0.5\left(\frac{1}{R} + \frac{1}{20}\right)$$

$$\frac{2}{15} - \frac{1}{20} = \frac{1}{R}$$

$$\therefore \quad R = 12 \text{ cm}$$

Similarly

$$-\frac{1}{f_2} = -(\mu - 1)\left[\frac{1}{20} + \frac{1}{R'}\right]$$

$$\frac{1}{30} = \frac{0.7}{20} + \frac{0.7}{R'}$$

$$\text{or} \quad \frac{1}{30} - \frac{0.7}{20} = \frac{0.7}{R'}$$

$$-\frac{0.1}{60} = \frac{0.7}{R'}$$

$$\therefore \quad R' = -420 \text{ cm.}$$

15.121. A convex lens of crown glass having focal length of 30 cm and a concave lens of flint glass forms an achromatic lens when placed in contact. The mean refractive indices for the two glasses are 1.56 and 1.63 respectively and the difference in refractive indices for these two glasses is respectively 0.014 and 0.021 for two spectral lines. Find the focal length of the concave lens.

Solution: The dispersive power is defined by the relation.

$$\omega = \frac{\mu_v - \mu_r}{\mu_y - 1}$$

The above equation can be expressed as

$$\omega = \frac{d\mu}{\mu_y - 1}$$

where $d\mu$ and μ_y corresponds to the change in refractive indices and mean refractive indices respectively.

$$\text{Now } \omega_1 = \frac{0.014}{1.56-1} = \frac{0.014}{0.56} = \frac{1}{40}$$

$$\text{and } \omega_2 = \frac{0.021}{1.63-1} = \frac{0.021}{0.63} = \frac{1}{30}$$

In view of the conditions of achromatism, we have

$$\frac{f_1}{f_2} = -\frac{\omega_1}{\omega_2} = -\frac{3}{4}$$

$$\therefore \quad f_2 = -\frac{4}{3} f_1 = -\frac{4}{3} \times 30 = -40 \text{ cm}$$

15.122. A compound microscope has an objective of focal length 4 cm and an eye-piece of focal length 10 cm. The two lenses are separated by a distance of 15 cm. Find the magnification of the image when it is formed at the distance of most distinct vision ($D = 25$ cm).

Solution:

$v_e = 25$ cm and $f_e = 10$ cm

$$\frac{1}{v_e} - \frac{1}{u_e} = \frac{1}{f_e} \qquad \text{(here } u_e \text{ and } v_e \text{ both are negative)}$$

$$\therefore \quad \frac{1}{u_e} = \frac{1}{f_e} + \frac{1}{v_e} = \frac{1}{10} + \frac{1}{25} = \frac{7}{50}$$

$$\therefore \quad u_e = \frac{50}{7} = 7.1 \text{ cm}$$

Now $v_0 = 15 - 7.1 = 7.9$ cm

$$\frac{1}{v_0} - \frac{1}{u_0} = \frac{1}{f_0} \qquad \text{(here is } u_0 \text{ negative)}$$

$$\therefore \quad \frac{1}{u_0} = \frac{1}{f_0} - \frac{1}{v_0}$$

$$= \frac{1}{4} - \frac{1}{7.9} = \frac{3.9}{4 \times 7.9}$$

$$u_0 = \frac{4 \times 7.9}{3.9} = 8.10 \text{ cm}$$

$$\text{Magnification } m = -\frac{v_0}{u_0}\left(1+\frac{D}{f_e}\right)$$

$$= -\frac{7.9}{8.1}\left(1+\frac{25}{10}\right)$$

$$= -\frac{7.9\times 3.5}{8.1}$$

$$= -3.41$$

15.123. What should the radius of curvature of the surfaces limiting a magnifying glass ($R_1 = R_2$) be for it to give a magnification of 10 for a normal eye? The refractive index of glass = 1.5.

Solution:

$$\frac{1}{f} = (\mu - 1)\times\frac{2}{R}$$

$$\frac{1}{f} = (1.5-1)\times\frac{2}{R} = \frac{1}{R}$$

$$\text{Magnification} = 10 = \frac{25}{f} = \frac{25}{R}$$

$\therefore \quad R = 2.5 \text{ cm} = 25 \text{ mm}$

15.124. A telescope has an objective of focal length 150 cm and an eye-piece of focal length 10 cm. At what viewing angle can the full moon be observed through this telescope, if it can be seen at an angle of 31′ through the naked eye.

Solution:

$$31' = \frac{31\times 3.14}{60\times 180}\text{ rad}$$

$$\frac{31\times 3.14}{60\times 180} = \frac{\text{Diameter of image}}{f_0}$$

$$\therefore \quad \text{Diameter of image} = \frac{31\times 3.14\times 150}{60\times 180}$$

$$\text{Required angle } \phi = \frac{\text{Diameter of image}}{f_e} = \frac{31\times 3.14\times 150}{60\times 180\times 10}$$

$$= 7^\circ\, 45'$$

15.125. A telescope with a focal length of 50 cm is adjusted to infinity. When the eye-piece of the telescope is moved over a certain distance, all objects at a distance of 50 m from the objective becomes clearly visible. Calculate the distance through which the eye-piece is moved.

Solution:

$$\frac{1}{v}=\frac{1}{f_0}-\frac{1}{u}=\frac{1}{50}-\frac{1}{50\times 100}$$

$$\therefore \quad v=\frac{50\times 100}{99}=50.505$$

Distance through which the eye piece has to be displaced

$= 50.505 - 50 = 0.505$ cm

$= 5$ mm

15.126. A lens with a focal length $f = 50$ cm and diameter $D = 5$ cm is illuminated by a parallel monochromatic beam of light of wavelength $\lambda = 6300\ A$. How many times is the light intensity at the lens's focus than the intensity of the wave incident on the lens. Calculate the size of the spot in the focal plane.

Solution: This is the case of Fraunhofer diffraction at a circular aperture. It consists of a central bright disc (Ary's disc) surrounded by alternate bright and dark rings.

The angular radius of the Ary's disc is given by

$$\theta=\frac{r}{f}=\frac{\lambda}{D};$$ r = radius of the disc, f = focal length of the lens

D = diameter of the lens

$$\therefore \quad r=\frac{\lambda f}{D}$$

Energy at the lens = $I_0\pi\frac{D^2}{4}$

Energy at the image (Ary's disc) = $I\pi\frac{\lambda^2 f^2}{D^2}$; I = intensity of the light in the disc.

If we assume that all the energy is confined in the Ary's disc then

$$I_o\pi\frac{D^2}{4}=I\pi\frac{\lambda^2 f^2}{D^2}$$

or $$\frac{I}{I_0}=\frac{D^4}{4\lambda^2 f^2}$$

On putting the values of various terms we have

$$\frac{I}{I_0}=\frac{25^2\times 10^{-8}}{4\times 63^2\times 10^{-16}\times 0.25}$$

$$= 1.57 \times 10^7$$

$$r=\frac{63\times 10^{-8}\times 0.5}{5\times 10^{-2}}=63\times 10^{-4}\text{ cm.}$$

15.127. The focal lengths of objective and the eye-piece of a telescope are 50 and 5 cm respectively. Least distance of distinct vision is 25 cm. The telescope is focused for distinct vision on a scale placed at a distance of 200 cm away from the objective. Calculate : (i) separation between the objective and the eye-piece, and (ii) magnification. (IIT 1980)

Solution:

(i) $u_0 = 200$ cm, $f_0 = 50$ cm, $v_0 = ?$

$$\frac{1}{v_0} - \frac{1}{u_0} = \frac{1}{f_0} \quad \text{(here } u_0 \text{ is negative)}$$

$$\frac{1}{v_0} = \frac{1}{50} - \frac{1}{200} = \frac{3}{200}$$

$$\therefore \quad v_0 = \frac{200}{3} \text{ cm}$$

$v_e = 25$ cm, $f_e = 5$ cm, $u_e = ?$

$$\frac{1}{v_e} - \frac{1}{u_e} = \frac{1}{f_e} \quad \text{(here } u_e \text{ and } v_e \text{ both are negative)}$$

$$\frac{1}{u_e} = \frac{1}{f_e} + \frac{1}{v_e} = \frac{1}{5} + \frac{1}{25} = \frac{6}{25}$$

$$\therefore \quad u_e = \frac{25}{6} \text{ cm}$$

Separation between the objective and eye-piece

$$= v_0 + u_e = \frac{200}{3} + \frac{25}{6} = 70.8 \text{ cm}$$

(i) Magnification $m = m_0 \times m_e$

$$= \frac{-200}{3 \times 200} \times \frac{25 \times 6}{25}$$

$$= -2$$

15.128. The focal lengths of the objective and the eye-piece of a microscope are 1 and 6 cm, respectively. The distance between them is 10.84 cm. Calculate the position of the object from the objective when the final image is seen at the distance of most distinct vision ($D = 25$ cm). Also find the magnifying power.

Solution: $v_e = -25$ cm, $f_e = 6$ cm, $u_e = ?$

$$\frac{1}{v_e} - \frac{1}{u_e} = \frac{1}{f_e}$$

$$\frac{1}{u_e} = \frac{1}{f_e} + \frac{1}{v_e} = \frac{1}{6} + \frac{1}{25} = \frac{31}{150} \quad \text{(here } u_e \text{ is negative)}$$

$\therefore \; u_e = 4.84$ cm

$v_0 = 10.84 - 4.84 = 6$ cm

Now $\dfrac{1}{v_0} - \dfrac{1}{u_0} = \dfrac{1}{f_0}$ (here u_0 is negative)

$$\frac{1}{u_0} = 1 - \frac{1}{6} = \frac{5}{6}$$

$$\therefore \quad u_0 = \frac{6}{5} = 1.2 \text{ cm}$$

$$\text{Magnifying power} = -\frac{v_0}{u_0}\left(1 + \frac{D}{f_e}\right)$$

$$= -\frac{6}{1.2}\left(1 + \frac{25}{6}\right)$$

$$= -\frac{31}{1.2}$$

$$= -25.8$$

15.129. A microscope has a magnification of 125. The focal length of the objectives is 0.5 cm. An object is placed at a distance of 0.52 cm. Calculate (i) the magnification and the focal length of eye-piece, and (2) the length of the tube when the final image is at the distance of most distinct vision ($D = 25$ cm).

Solution: $u_0 = 0.52$ cm, $f_0 = 0.5$ cm, $v_0 = ?$

$\dfrac{1}{v_0} - \dfrac{1}{u_0} = \dfrac{1}{f_0}$ (here u_0 is negative)

$$\frac{1}{v_0} = \frac{1}{f_0} - \frac{1}{u_0} = \frac{1}{0.5} - \frac{1}{0.52}$$

$$v_0 = \frac{0.52 \times 0.5}{0.52 - 0.5} = 13 \text{ cm}$$

$$m_0 = \frac{v_0}{u_0} = \frac{13}{0.52} = 25$$

$$m_e = \frac{m}{m_0} = \frac{125}{25} = 5$$

Now $\dfrac{v_e}{u_e} = 5 \; \therefore u_e = \dfrac{25}{5} = 5$ cm

$\dfrac{1}{v_e} - \dfrac{1}{u_e} = \dfrac{1}{f_e}$ (here u_e and v_e are both negative)

$$\frac{1}{f_e} = \frac{1}{u_e} - \frac{1}{v_e} = \frac{1}{5} - \frac{1}{25} = \frac{4}{25}$$

$$f_e = \frac{25}{4} = 6.25 \text{ cm}$$

Length of the tube $v_0 + u_e = 13 + 5 = 18$ cm.

15.130. An astronomical telescope has an objective of focal length = 125 cm and an eye-piece of focal length = 25 cm. The diameter of the objective is 4 cm. Find (i) the magnifying power (ii) the separation between the lenses, and (iii) the diameter of the image.

Solution:

(i) Magnifying power $= \frac{f_0}{f_e} = \frac{125}{25} = 5$

(ii) Separation between the lenses = 125 + 25 = 150 cm

(iii) Diameter of the image $= \frac{\text{diameter of the objective}}{\text{magnifying power}}$

$$= \frac{4}{5} = 0.8 \text{ cm.}$$

15.131. If the focal length of the objective and eye-piece of a microscope are 2 & 5 cm respectively and the distance between them is 20 cm. What is the distance of the object from the objective when the image seen by the eye is 25 cm from eye-piece? Also find the magnifying power.

(Roorkee 1987)

Solution:

$$\frac{1}{v_e} - \frac{1}{u_e} = \frac{1}{f_e}; v_e = -25, f_e = 5$$

$$\frac{1}{u_e} = \frac{1}{f_e} + \frac{1}{v_e} = \frac{1}{5} + \frac{1}{25} = \frac{6}{25}$$

$$\therefore \quad u_e = \frac{25}{6} \text{ cm}$$

$$v_0 = 20 - \frac{25}{6} = \frac{95}{6} \text{ cm}$$

$$f_0 = 2 \text{ cm}$$

$$\frac{1}{u_0} = \frac{1}{f_0} - \frac{1}{v_0} = \frac{1}{2} - \frac{6}{25}$$

$$u_0 = \frac{190}{83} \cong 2.3 \text{ cm}$$

$$\text{Magnifying power} = \frac{v_0}{u_0}\left(1+\frac{D}{f_e}\right) = \frac{95}{6}\times\frac{83}{190}\left(1+\frac{25}{5}\right) = 41.5$$

15.132. In a compound microscope the objective and eye-piece have focal length of 0.95 cm and 5 cm respectively, and are kept at a distance of 20 cm. The last image is formed at a distance of 25 cm from eye-piece. Calculate the position of the object and the total magnification. (Roorkee 1988)

Solution:

$$\frac{1}{v_e} - \frac{1}{u_e} = \frac{1}{f_e}; v_e = -25\text{ cm}, f_e = 5\text{ cm}$$

$$\frac{1}{u_e} = \frac{1}{v_e} - \frac{1}{f_e} = -\frac{1}{25} - \frac{1}{5} = \frac{-6}{25}$$

$$u_e = \frac{-25}{6}\text{ cm}$$

$$v_0 = 20 - \frac{25}{6} = \frac{95}{6}\text{ cm}$$

$$f_0 = 0.95\text{ cm}$$

$$\frac{1}{v_0} - \frac{1}{u_0} = \frac{1}{f_0}$$

$$\frac{1}{u_0} = \frac{1}{v_0} - \frac{1}{f_0} = \frac{6}{5} - \frac{1}{0.95}$$

$$u_0 = -\frac{95}{94}\text{ cm}$$

$$M = \frac{95/6}{-95/94}\left(1+\frac{25}{5}\right) = -94$$

15.133. The eye-piece and objective of a microscope of focal length 0.03 m and 0.04 m respectively, are separated by a distance of 0.2 m. The eye-piece and the objective are to be interchanged such that the angular magnification of the instrument remains same. What is the new separation between the lenses? (Roorkee 1991)

Solution:

$$M_1 = \frac{v_0}{u_0}\left(1+\frac{D}{f_e}\right)$$

$$f_e = 0.03\text{ m}, D = 25\text{ cm}$$

$$M_1 = \frac{v_0}{u_0} \times \frac{28}{3}$$

In the second case

$$M_2 = \frac{V_0}{u_0}\left(1+\frac{D}{f_e}\right) = \frac{V_0}{u_0}\left(1+\frac{25}{4}\right) = \frac{V_0}{u_0} \times \frac{29}{4}$$

$$V_0 = \frac{112}{87} v_0$$

In the first case let x be the distance between objective & eye-piece

$$-\frac{1}{25} + \frac{1}{x} = \frac{1}{3} \text{ or } x = \frac{75}{28} \text{ cm}$$

$$v_0 = 20 - \frac{75}{28} = \frac{485}{28} \text{ cm}$$

$$V_0 = \frac{112}{87} \times \frac{485}{28} = \frac{1940}{87} \text{ cm}$$

In the second case let y be the distance between the object and eye-piece

$$-\frac{1}{25} + \frac{1}{y} = \frac{1}{4} \therefore y = \frac{100}{29}$$

$$\text{Distance between lenses} = \frac{1940}{87} + \frac{100}{29} = 25.75 \text{ cm.}$$

15.134. When a flux of light passes through a telescope it has been noticed that the intensity increases $\eta = 5.76 \times 10^4$ times. Find the angle subtended by an object whose image formed by a telescope has an angular dimension $\phi' = 3°$.

Solution: In view of Fig. 15.88 we have

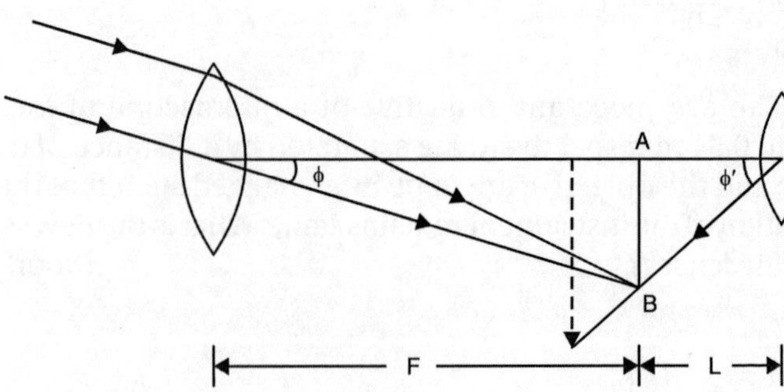

Fig. 15.88

$$\phi' = \frac{AB}{L}, \ \phi = \frac{AB}{F}$$

$$\frac{\phi}{\phi'} = \frac{L}{F}$$

L and F are the distances and the intensity depends upon the square of the distances

$$\therefore \quad \frac{L}{F} = \frac{1}{\sqrt{\eta}} \quad \therefore \phi = \frac{\phi'}{\sqrt{\eta}}$$

On substituting the values of ϕ' and η we have

$$\phi = \frac{3\times 60}{\sqrt{5.76\times 10^4}} = \frac{3\times 60}{240} = 0.75'$$

15.135. Two equi-convex lenses of focal length F & f and diameters D & d respectively forms a Keplerian telescope. Find the magnification when the telescope is adjusted to infinity.

Solution: It is essentially an astronomical telescope. From the geometry of Fig. 15.89 we have

$$\frac{F}{D} = \frac{f}{d}$$

Where D and d are the diameters of objective & eye-piece respectively.

$$\therefore \quad \text{Magnification } M = \frac{F}{f} = \frac{D}{d}$$

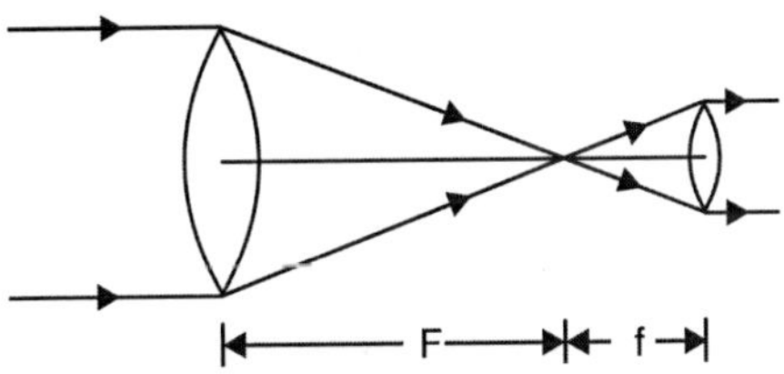

Fig. 15.89

15.136. A Galilian telescope having a 10 fold magnification is adjusted to infinity. Under this condition the tube length is 54 cm (see Fig. 15.90). Find (i) the focal length of objective and eye-piece (ii) the distance through which the eye-piece is to be displaced when the telescope is adjusted to a distance of 60 m.

Solution:

$$\frac{F}{f} = 10$$

$$F - f = 54$$

$$f = 6 \text{ cm}, F = 60 \text{ cm}$$

$$\frac{1}{v} = \frac{1}{F} - \frac{1}{u} = \frac{1}{60} - \frac{1}{6000} = \frac{99}{6000}$$

$$\therefore \quad v = \frac{6000}{99}$$

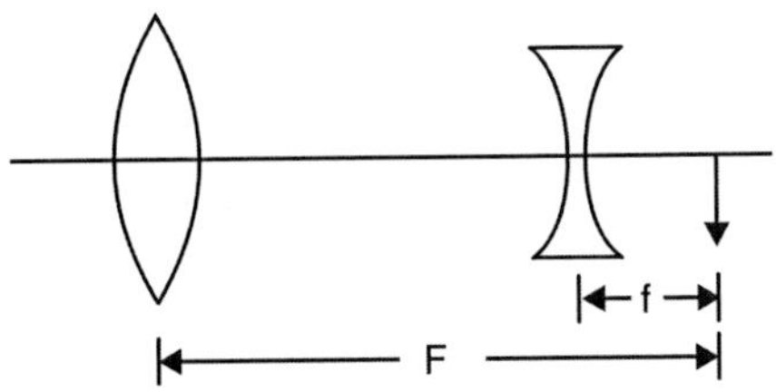

Fig. 15.90

The distance through which the eye-piece has to be displaced

$$= \frac{6000}{99} - 60 = 0.6 \text{ cm.}$$

15.137. For what thickness a thick convexo concave lens of refractive index 1.6 behave like a telescope in which the radius of curvature of the convex surface is $\Delta R = 1.8$ cm larger than that of the concave surface?

Solution: Let a parallel beam of light falls on the convex surface (see Fig. 15.91) then the image will be formed in the denser medium.

$$\frac{\mu}{f'} - \frac{1}{\infty} = \frac{(\mu-1)}{R+\Delta R}$$

$$\therefore \quad f' = \frac{\mu(R+\Delta R)}{(\mu-1)}$$

Let x be the thickness of the lens then u for the concave surface is equal to $x - \frac{\mu(R+\Delta R)}{(\mu-1)}$

$$u = \frac{x(\mu-1) - \mu(R+\Delta R)}{(\mu-1)}$$

Now for the refraction at the second surface we have

$$\frac{1}{\mu/v} - \frac{1}{u} = \frac{1/\mu - 1}{R}$$

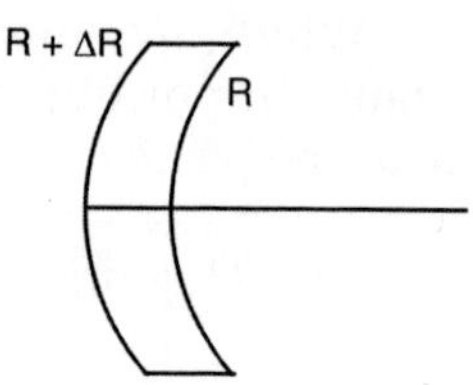

Fig. 15.91

here u & v both are negative and R is positive.

$$\frac{1}{u} = \frac{1-\mu}{\mu R}; \; v = \infty$$

$$\frac{-(\mu-1)}{\mu R} = \frac{(\mu-1)}{x(\mu-1) - \mu(R+\Delta R)}$$

or $$x = \frac{\mu \Delta R}{(\mu-1)}$$

On substituting the values of μ and ΔR we have

$$x = \frac{1.6 \times 1.8}{1.6 - 1} = 4.8 \text{ cm.}$$

15.138. A telescope has been designed by using two transparent spheres of radii R_1 and R_2 such that $R_1 > R_2$. The refractive index of the material of the two spheres is $\mu < 2$. Using the bigger sphere as an objective find (i) the distance between the centres of the two spheres (ii) the magnification of the system.

Solution:

(i) Here μ is less than 2 hence the image will be formed outside the sphere. The parallel beam of light falling on the sphere of radius R_1 will be focused at B (see Fig. 15.92) such that $OB = \frac{\mu R_1}{2(\mu - 1)}$.

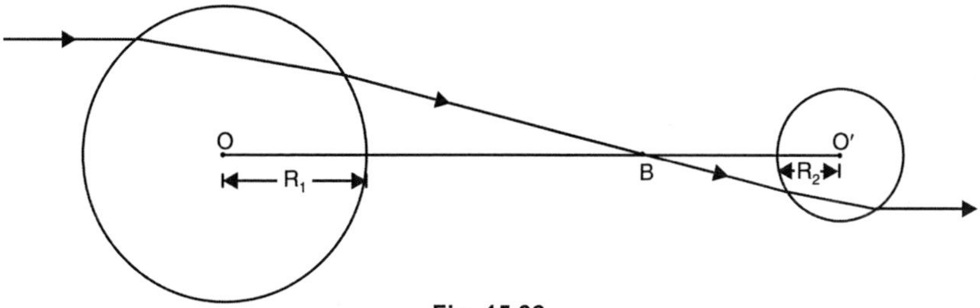

Fig. 15.92

Now if $O'B = \frac{\mu R_2}{2(\mu - 1)}$ then the final image will be formed at infinity. For such a telescope the distance between the centres is given by

$$OO' = \frac{\mu(R_1 + R_2)}{2(\mu - 1)}$$

(ii) Magnification $= \frac{f_0}{f_e} = \frac{\mu R_1}{2(\mu-1)} \Big/ \frac{\mu R_2}{2(\mu-1)} = \frac{R_1}{R_2}$

15.139. There is a telescope whose objective has a diameter $D = 5$ cm. Find its resolving power. Find also the minimum separation between two points lying at a distance of $l = 3$km from the telescope which can be fully resolved. Wavelength of light $\lambda = 5500$ A.

Solution:

$$\text{Resolving power} = \frac{D}{1.22\lambda} = \frac{5 \times 10^{-2}}{1.22 \times 55 \times 10^{-8}}$$
$$= 7.45 \times 10^4$$

$$\frac{\Delta x}{l} = \frac{1}{\text{resolving power}}$$

$$\therefore \quad \Delta x = \frac{l}{\text{resolving power}} = \frac{3 \times 10^3}{7.45 \times 10^4} = 4.026 \text{ cm.}$$

15.140. There is a microscope whose objective's numerical aperture is $\sin\alpha = 0.24$ where α is the half angle subtended by the rim of the objective. Find the minimum separation resolved by this microscope when the object is illuminated by light of wavelength $\lambda = 5500$ A.

Solution:

$$d_{\min} = \frac{0.61\lambda}{\sin\alpha} = \frac{0.61 \times 55 \times 10^{-8}}{0.24} = 1.4\ \mu\text{m}$$

15.141. On the surface of earth the sun subtends an angle of 32′. Sun light is falling normally on a double slit arrangement. A filter corresponding to 5500 A is placed between the double slit and sun. Calculate the separation between the two slits in order to have the fringes of good contrast.

Solution:

$$\text{Lateral coherent length } l_{\omega} = \frac{\lambda}{\theta}$$

$$l_{\omega} = \frac{55\times10^{-6}}{0.01};\ \theta = 32' = \frac{32\pi}{180\times60} = 0.01 \text{ rad}$$

$$= 0.0055 \text{ cm}$$

In order to have fringes of good contrast the two slits must be placed at a separation of 0.0055 cm.

15.142. For the sodium D_1 line calculate the monochromaticity using the data $\lambda = 5890$ A, $\tau_c = 10^{-10}$ sec and $\Delta\upsilon = 10^{10}$ Hz.

Solution:

$$\text{Monochromaticity} = \frac{\Delta\upsilon}{\upsilon} = \frac{\Delta\upsilon\cdot\lambda}{C}$$

$$= \frac{10^{10}\times5890\times10^{-10}}{3\times10^{8}}$$

$$= \frac{589}{3}\times10^{-7}$$

$$= 1.96\times10^{-5}$$

15.143. A source of light with a diameter of $d = 25$ cm is at a distance of $l = 200$ m from the place of observation. The radiation coming out of the source contains wavelength ranging from 480 to 500 mm. Calculate (i) coherent time (ii) coherent length (iii) coherent radius (iv) coherent volume.

Solution:

(i) Coherent time $\tau_c = \dfrac{\lambda^2}{C\cdot\Delta\lambda} = \dfrac{480\times480\times10^{-18}}{3\times10^{8}\times200\times10^{-10}}$

$$= 3.84\times10^{-14} \text{ sec}$$

(ii) Coherent length $L = \dfrac{\lambda^2}{\Delta\lambda} = \dfrac{480\times480\times10^{-18}}{200\times10^{-10}}$

$$= 0.0115 \text{ mm}$$

(iii) Coherent radius $r = \dfrac{\lambda}{\theta} = \dfrac{\lambda}{d/l}$

$$= \frac{480 \times 10^{-9} \times 200}{0.25}$$
$$= 0.384 \text{ mm}$$

(iv) Coherent volume $= L\pi r^2 = 0.0115 \times 3.14 \times 0.384^2$
$$= 5.3 \times 10^{-3} \text{ mm}^3$$

15.144. Two light waves producing oscillations of the same direction at a point in space are represented by $A \cos \omega t$ and $A \cos (\omega + \Delta\omega) t$. Assuming $\Delta\omega = 0.628 \text{ sec}^{-1}$ discuss the intensity at that point.

Solution:

$$\text{Resultant } R = A \cos \omega t + A \cos (\omega + \Delta\omega) t$$
$$= 2A \cos\left(\omega + \frac{\Delta\omega}{2}\right) t \cos \frac{\Delta\omega}{2} \cdot t$$
$$= 2A \cos\left(\frac{2\pi}{20} t\right) \cos\left(\omega + \frac{\Delta\omega}{2}\right) t$$

The intensity it max. at $t = 0$ and $t = 10$ sec. Therefore the intensity will pulsate with a period of 10 sec.

15.145. How much power P must a laser beam of diameter D will have in order to initiate an electric break down in a gas? Ionisation potential = 10 V. Rest energy of electron $m_0c^2 = 10^{-13}$ J.

Solution: Energy 10 eV $= 1.6 \times 10^{-18}$ J is much less than the rest energy of the electron. Consider the equation of motion of the electron.

$$\frac{md^2x}{dt^2} = -eE_0 \cos \omega t$$

On integration we have

$$v = \frac{-eE_0}{m\omega} \sin \omega t$$

$$\text{Max kinetic energy} = \frac{1}{2} m \left(\frac{eE_0}{m\omega}\right)^2 = eU$$

or $$E_0^2 = \frac{2Um\omega^2}{e}$$

$$\text{Radiation power flux density} = \frac{P}{\pi D^2/4} = c\epsilon_0 E_{av}^2 = c\epsilon_0 \frac{E_0^2}{2}$$

$$\therefore \quad P = \frac{\pi D^2}{4} \cdot c\epsilon_0 \cdot \frac{Um\omega^2}{e}$$

15.146. In a ruby laser the rod is 4 mm in diameter and 35 mm long. The laser emits radiations of wavelength 6943 A and 6929 A. Find the lowest angular divergence.

Solution:

$$2\theta = \frac{2\lambda}{D}\text{ rad}$$

$$= \frac{2\lambda}{D} \times \frac{180 \times 60}{3.14}\text{ min}$$

$$= \frac{2 \times 6929 \times 10^{-10} \times 180 \times 60}{4 \times 10^{-3} \times 3.14}$$

$$= 1.2'$$

15.147. Two coherent plane light waves propogating with very small angle of divergence ψ, fall almost normally on a screen. Assuming the wavelength of the wave as λ calculate the fringe width.

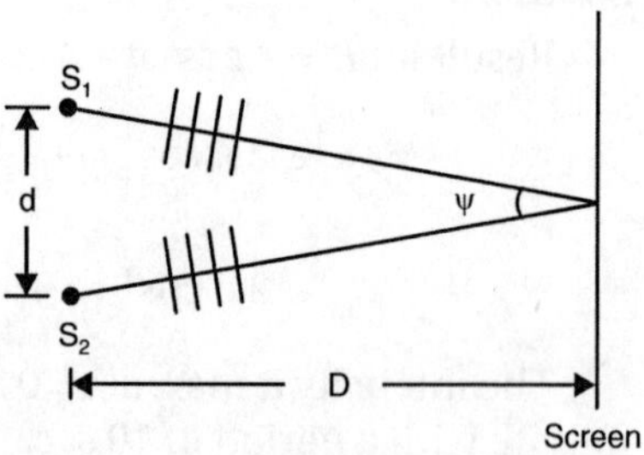

Fig. 15.93

Solution:

$$\psi = \frac{d}{D}$$

$$\text{Fringe width } \bar{X} = \frac{D\lambda}{d} = \frac{\lambda}{\psi}$$

15.148. A system consists of a chain of sources separated by a distance d as shown in Fig. 15.94a. The phase varies linearly along the chain. Find the time dependence of the phase difference $\Delta\phi$ between the neighbouring sources at which the principal maxima of the system will be scanning the surrounding with angular velocity ω.

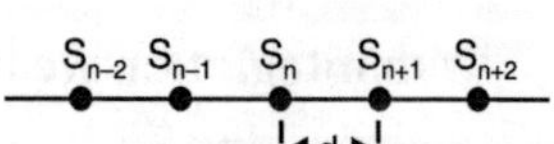

Fig. 15.94a

Solution: Let the maxima be observed at an angle θ see Fig. 15.94b. In view of the requirement $\theta = \omega t + \phi_0$

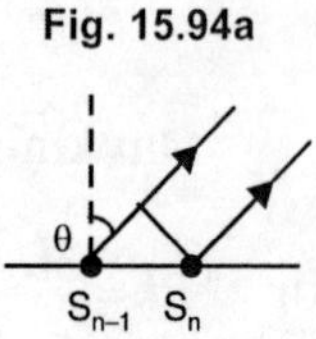

Fig. 15.94b

$$\text{Phase diff} = \frac{2\pi}{\lambda} \cdot d \sin\theta$$

$$\text{Total phase diff.} = \Delta\phi + \frac{2\pi}{\lambda} d \sin\theta = 2N\pi$$

$$\therefore \quad \Delta\phi = 2\pi\left[N - \frac{d}{\lambda}\sin(\omega t + \phi_0)\right]$$

15.149. Determine the depth of water that contains the same number of waves as 16 cm of glass when each is traversed by the same monochromatic light. The refractive indices of water and glass are 4/3 and 1.5 respectively.

Solution: Number of waves is 16 cm of glass $= n_g = \dfrac{16}{\lambda_g}$

Number of waves in x cm of water $= n_w = \dfrac{x}{\lambda_w}$

where λ_g and λ_w are the wavelengths of the same light wave in glass and water.

Now $n_g = n_w$ (given)

$$\therefore \quad \frac{16}{\lambda_g} = \frac{x}{\lambda_w}$$

$$\text{or} \quad x = \frac{\lambda_w}{\lambda_g} \times 16$$

$$= \frac{9}{8} \times 16; \quad \frac{\lambda_w}{\lambda_g} = \frac{{}^a\mu_g}{{}^a\mu_w}$$

$$= 18 \text{ cm.}$$

15.150. Two coherent sources of intensity ratio β interfere. Show that in the interference pattern

$$\frac{I_{max} - I_{min}}{I_{max} + I_{min}} = \frac{2\sqrt{\beta}}{1+\beta}$$

Solution: Let the two waves of amplitudes a_1 and a_2 be superimposed on each other and produce interference pattern. Thc maximum and minimum intensities are given by

$I_{max} = K(a_1 + a_2)^2$ (here K is a constant)

$I_{min} = K(a_1 - a_2)^2$

Now $I_{max} - I_{min} = K \cdot 4a_1 a_2$

$$I_{max} + I_{min} = K \cdot 2\left(a_1^2 + a_2^2\right)$$

$$\therefore \quad \frac{I_{max} - I_{min}}{I_{max} + I_{min}} = \frac{2a_1 a_2}{a_1^2 + a_2^2} = \frac{2a_1/a_2}{\left[(a_1/a_2)^2 + 1\right]}$$

$$\frac{a_1^2}{a_2^2} = \beta \qquad \text{hence } \frac{a_1}{a_2} = \sqrt{\beta}$$

$$\therefore \quad \frac{I_{max} - I_{min}}{I_{max} + I_{min}} = \frac{2\sqrt{\beta}}{1+\beta}$$

15.151. Two straight and narrow parallel slits 2 mm apart are illuminated by monochromatic light of wavelength 5.9×10^{-5} cm. Fringes are obtained on a screen 60 cm distant from the slit. Determine the fringe width.

Solution: The fringe width $\overline{X}$ is given by the relation

$$\overline{X} = \frac{D\lambda}{2d}$$

where D is the distance between the slit and screen

λ = wavelength of monochromatic light used

$2d$ = separation between the slits.

$$\overline{X} = \frac{60 \times 5.9 \times 10^{-5}}{0.2} = 17.7 \times 10^{-3} \text{ cm}$$

15.152. A monochromatic radiation of wavelength 6000 Å falls on a double slit of separation 0.3 mm. Calculate the width of the fringes observed in the focal plane of a convex lens of focal length 1.5 m placed near the double slit (see Fig. 15.95).

Solution:

$$\text{Fringe width } \overline{X} = \frac{D\lambda}{2d}$$

Here $D = f$ = focal length of the lens

$$\therefore \quad \overline{X} = \frac{f\lambda}{2d}$$

On substituting the values of the various terms, we have

$$\overline{X} = \frac{150 \times 6 \times 10^{-5}}{0.3 \times 10^{-1}}$$

$$= 0.3 \text{ cm}$$

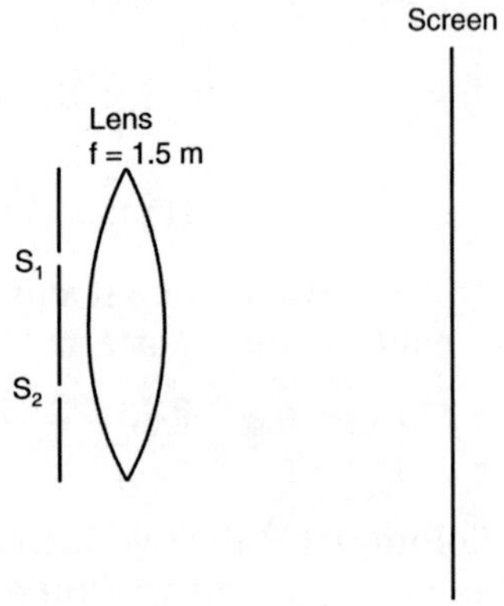

Fig. 15.95

15.153. Interference fringes produced by a double slit arrangement using a mercury radiation of wavelength 5,461 Å, have an angular fringe width 0.24°. What percentage decrease in the angular fringe width will be observed when the entire arrangement is immersed in water? ${}^{a}\mu_{w} = \frac{4}{3}$.

Solution:

$$\text{Angular fringe width } \theta = \frac{\overline{X}}{D} = \frac{\lambda}{2d}$$

$$0.24° = \frac{\lambda_{\text{in air}}}{2d}$$

$$\theta = \frac{\lambda_{\text{in water}}}{2d}$$

$$\therefore \quad \frac{0.24°}{\theta} = \frac{\lambda_{\text{in air}}}{\lambda_{\text{in water}}} = \frac{4}{3}$$

$$\therefore \quad \theta = \frac{3}{4} \times 0.24° = 0.18°$$

$$\% \text{ decrease} = \frac{100 \times 0.06°}{0.24°} = 25\%$$

15.154. In Fig. 15.96, S_1 and S_2 are the sources emitting electromagnetic radiations of wavelength 2 m. A detector D moves only along the positive side of the x-axis in order to record the gain. How many points will be obtained at which the gain is maximum and minimum? Also, at what distance will these points be from the origin?

Solution: Let us assume that $S_1D = x$. Then $S_2D = \sqrt{x^2 + 25}$

Path difference between the two waves originating from S_2 and S_1 and reaching the detector D is given by

$$S_2D - S_1D = \sqrt{x^2 + 25} - x$$

If $S_2D - S_1D = m\lambda$ then we have the condition for maximum gain and if $S_2D - S_1D = (2m + 1)\lambda/2$ then we have the condition for minimum gain. For maximum gain,

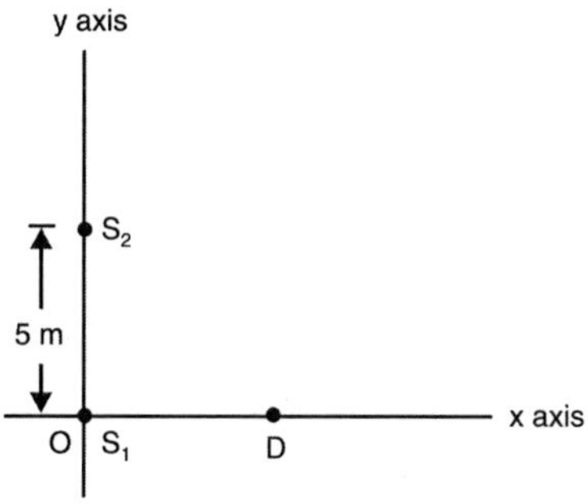

Fig. 15.96

$$\sqrt{x^2 + 25} - x = 1 \times 2; \text{ for first maxima}$$

or $\quad x^2 + 25 = (x + 2)^2 = x^2 + 4x + 4$

or $\quad x = \dfrac{21}{4} = 5.25 \text{ m}$

$$\sqrt{x^2 + 25} - x = 4 \text{ for second maxima}$$

or $\quad x^2 + 25 = (x + 4)^2$

or $\quad x^2 + 25 = x^2 + 8x + 16$

or $\quad x = \dfrac{9}{8} = 1.125 \text{ m}$

$$\sqrt{x^2 + 25} - x = 6 \text{ for third maxima or } x^2 + 25 = (x + 6)^2$$

From this equation, x turn out to be negative. Hence along the positive side of the x-axis there are only two points at distances 1.125 m and 5.25 m at which the gain will be maximum.

For minimum gain, we have

$$\sqrt{x^2 + 25} = (x + 1)$$

or $\quad x^2 + 25 = x^2 + 2x + 1$

or $\quad x = 12 \text{ m} \qquad$ which is the first minima

$$\sqrt{x^2 + 25} = (x + 3)$$

or $\quad x^2 + 25 = x^2 + 6x + 9$

$$x = \frac{16}{6} = 2.67 \text{ m which is the second minima}$$

$$\sqrt{x^2+25}=(x+5)$$

or $$x^2 + 25 = x^2 + 10x + 25$$

$x = 0$ i.e. at the origin O lies the third minima.

Thus there will be three points at distances 0, 2.67 m and 12 m at which the gain is minimum.

15.155. A beam of light consisting of two wavelengths 6,500 Å and 5,200 Å is used to obtain interference fringes in a double-slit experiment.

(i) Find the distance of the third bright fringe on the screen from the central maximum for wavelength 6,500 Å.

(ii) What is the least distance from the central maximum where the bright fringes due to both the wavelength coincide?

The distance between the slits is 2 mm and the distance between the plane of the slit and the screen is 120 cm. (IIT 1985)

Solution:

(i) The position of the nth bright fringe on the screen from the central maxima is given by $x=\dfrac{nD\lambda}{2d}$

where n is the order of the bright fringe. D is the distance between the slit and screen, λ is the wavelength of the wave and $2d$ is the separation between the slits.

$$x=\frac{3\times 120\times 6{,}500\times 10^{-8}}{0.2}=0.117\text{ cm}$$

(ii) Let y be the least distance at which nth maxima due to 6,500 Å wavelength coincides with the mth maxima due to 5,200 Å wavelength.

$$y=\frac{n\times 6{,}500\times D}{2d}=\frac{m\times 5{,}200\times D}{2d}$$

or $$\frac{m}{n}=\frac{6{,}500}{5{,}200}=\frac{5}{4}$$

i.e. the fourth bright fringe of 6,500 Å coincides with the fifth bright fringe of 5,200 Å hence

$$y=\frac{4\times 6{,}500\times 10^{-8}\times 120}{0.2}=0.156\text{ cm.}$$

15.156. In Young's double slit experiment white light fringes are formed. When one of the slits is covered with a thin sheet of glass or refractive index $\mu = 1.5$, it is noticed that the central bright fringe shifts by four fringes. Taking the effective wavelength as 5,500 Å, calculate (i) the increase in the optical path due to the introduction of the glass sheet, and (ii) the thickness of the glass sheet.

Solution:

(i) When a thin transparent film of refractive index μ and thickness t is introduced in the path of one of the slits, it introduces an optical path $(\mu - 1)\,t$ which is equal to $n\lambda$.

$\therefore \quad (\mu - 1)\,t = n\lambda = 4 \times 55 \times 10^{-6} = 2.2 \times 10^{-4}$ cm

(ii) $(\mu - 1)\,t = 2.2 \times 10^{-4}$

$0.5 \times t = 2.2 \times 10^{-4}$

$t = 4.4 \times 10^{-4}$ cm.

15.157. In a double-slit experiment when a transparent thin sheet of thickness 4.72×10^{-2} cm is introduced in the path of one of the slits the central bright fringe shifts to a position originally occupied by the fourth bright fringe. If the pattern is formed with sodium light of wavelength $\lambda = 5{,}900$ Å find the refractive index of the sheet.

Solution:

$(\mu - 1)\,t = n\lambda$

$$\therefore \quad \mu = \frac{n\lambda}{t} + 1; \frac{4 \times 59 \times 10^{-6}}{4.72 \times 10^{-4}} + 1; = 1.5$$

15.158. A convex lens of focal length 5 cm is cut into two halves. The two portions are then separated by 0.5 mm. A source of monochromatic radiation of wavelength 5,000 Å is placed at a distance of 10 cm from this lens. Calculate the number of interference fringes which will be observed on a screen placed at a distance of 30 cm from the lens.

Solution: The two portions of the lens will form images s_1 and s_2 which will be in the same plane. Thus s_1 and s_2 will serve the purpose of coherent sources (see Fig. 15.97).

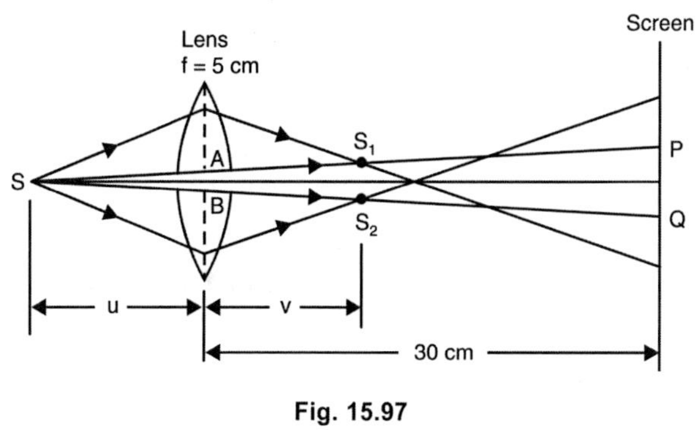

Fig. 15.97

Now $\frac{1}{v} - \frac{1}{u} = \frac{1}{f}$ (u is negative)

$$\frac{1}{v}=\frac{1}{f}-\frac{1}{u}=\frac{1}{5}-\frac{1}{10}=\frac{1}{10}$$

$\therefore \quad v = 10$ cm

In view of the geometry of the figure, we have

$$\frac{s_1 s_2}{AB}=\frac{u+v}{u}=\frac{20}{10}=2$$

$s_1 s_2 = 2d =$ separation between the coherent sources

$= 2 \times AB = 2 \times 5 \times 10^{-2}$ cm $= 0.1$ cm

$D =$ distance between the screen and coherent sources

$= 30 - 10 = 20$ cm

$$\text{Fringe width} = \frac{D\lambda}{2d}=\frac{20\times 5\times 10^{-5}}{0.1}=10^{-2}\text{ cm}$$

$$\text{Now } \frac{PQ}{AB}=\frac{30+u}{u}=\frac{40}{10}=4$$

$\therefore \quad PQ = 4 \times AB = 4 \times 5 \times 10^{-2} = 0.2$ cm; PQ is the region of overlap in which the fringes will appear.

$\therefore$ Number of interference fringes observed on the screen

$$=\frac{PQ}{\text{fringe width}}=\frac{0.2}{10^{-2}}=20$$

15.159. A convex lens of focal length $f = 20$ cm is cut into two halves. The two portions are fixed by separating them 1.3 mm apart. A monochromatic source of wavelength $\lambda = 5000$ Å is placed at a distance of 10 cm from this lens (a) At what distance from the lens a screen be fixed in order to obtain 5 interferences fringes on it? (b) what is the maximum number of interference fringes that can be obtained with this optical system?

Solution:

$u = 10$ cm, $f = 20$ cm

$$\frac{1}{v}=\frac{1}{u}-\frac{1}{f}=\frac{1}{20}$$

$\therefore \quad v = 20$ cm

$$\frac{OO'}{S_1S_2}=\frac{u}{v-u}=1 \therefore OO' = S_1S_2 = 1.3\text{ mm (see Fig. 15.98)}$$

$$\text{Fringe width } \overline{X}=\frac{(D+20)\lambda}{S_1S_2};$$ D is the distance between the lens and screen

$$PQ=\frac{5(D+20)\lambda}{S_1S_2}$$

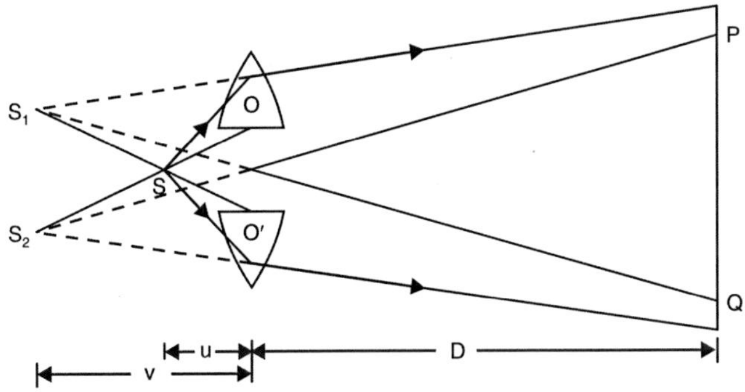

Fig. 15.98

$$\frac{PQ}{S_1S_2} = \frac{D}{v}$$

$$\frac{5(D+20)\lambda}{S_1S_2^2} = \frac{D}{20}$$

On substituting the values of the various terms we have

$$D = \frac{5(D+20)\times 5\times 10^{-5}\times 20}{169\times 10^{-4}}$$

$$50\ (D+20) = 169\ D$$

$$119\ D = 1000$$

$$\therefore \qquad D = \frac{1000}{119} = 8.4\ \text{cm}$$

(b) $$v = \frac{uf}{f-u}; v-u = \frac{u^2}{f-u}$$

$$S_1S_2 = \frac{d\cdot u^2}{(f-u)u} = \frac{du}{f-u};\ d = \text{separation between the lenses}$$

$$\overline{X} = \frac{\left(D + \frac{fu}{f-u}\right)\lambda}{\frac{d\cdot u}{f-u}} = \frac{[D(f-u)+uf]}{d\cdot u}\cdot\lambda$$

$$\text{Number of fringes} = N = \frac{PQ}{\overline{X}} = \frac{D\cdot S_1S_2}{v\cdot\overline{X}} = \frac{D\cdot d\cdot u(f-u)}{(f-u)\overline{X}\cdot u\cdot f} = \frac{D\cdot d}{f\cdot\overline{X}}$$

$$N = \frac{D\cdot d^2\cdot u}{f[D(f-u)+uf]\lambda}$$

$$N\lambda f^2 D - Nf Du\lambda + Nuf^2\lambda = Dd^2u$$

or $$D = \frac{Nuf^2\lambda}{d^2 \cdot u - Nf^2\lambda + Nfu\lambda}$$

For having maximum number of fringes $D = \infty$

$\therefore$ $d^2u = N(\lambda f^2 - fu\lambda)$

or $$N = \frac{d^2 \cdot u}{\lambda f^2 - f \cdot u \cdot \lambda}$$

On substituting the values we have

$$N = \frac{1.3^2 \times 10^{-2} \times 10}{5 \times 10^{-5} \times 400 - 20 \times 10 \times 5 \times 10^{-5}}$$

$$= \frac{169 \times 10^{-4}}{10^{-3}} = 16.9 = 17$$

15.160. L_1 and L_2 are the two halves of a equi-convex lens separated by a distance *y*. The two lenses forms the images of the mono chromatic source *S* at *A* & *B* respectively. Discuss the nature of the interference fringes produced on the screen placed normal to *A* & *B* and situated in between *A* and *B*.

Solution: Here the images formed at *A* & *B* are coherent (see Fig. 15.99). Consider a point *P* in the interfering region. Light reaching at *P* through the lens L_1 covers a path $= p_1 + AP$ where p_1 is the optical path between source *S* and its image *A*. In the similar manner the light reaching at *P* through the lens L_2 covers a path $p_2 - BP$ where p_2 is the optical path between the source *S* and its image *B*.

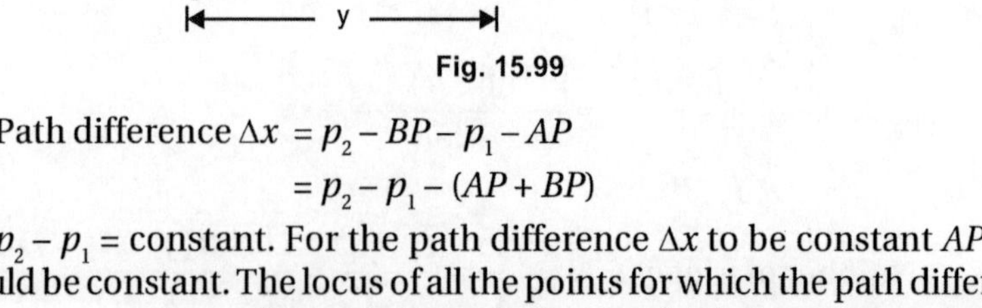

Fig. 15.99

$$\text{Path difference } \Delta x = p_2 - BP - p_1 - AP$$
$$= p_2 - p_1 - (AP + BP)$$

$p_2 - p_1$ = constant. For the path difference Δx to be constant $AP + BP$ should be constant. The locus of all the points for which the path difference Δx is constant is an ellipsoid of revolution in which *A* and *B* are the focii. The cross-section of this surface by a plane normal to *AB* will be a circle. Therefore alternate bright and dark circular rings will be observed on the screen. As the beam interface only on one side the rings will be semi circular.

15.161. In a modified Young's double-slit experiment, a monochromatic uniform and parallel beam of light of wavelength 6000 Å and intensity $10/\pi$ W/m^2 is incident normally on two circular apertures A and B of radii 0.001 m and 0.002 m respectively. A perfectly transparent film of thickness 2,000 Å and refractive index 1.5 for wavelength of 6,000 Å is placed in front of aperture A (see Fig. 15.100). Calculate the power (in watts) received at the focal spot F of the lens. The lens is symmetrically placed with respect to the apertures. Assume that 10% of the power received by each aperture goes in the original direction and is brought to the focal spot. (IIT 1989)

Solution: Intensity of parallel beam travelling from left to right

$$= \frac{10}{\pi} \times \frac{1}{10} = \frac{1}{\pi} \text{ w/m}^2$$

Intensity I_1 of the beam emerging from aperture $A = \frac{1}{\pi} \pi \times (0.001)^2$

$$= 10^{-6} \text{W}$$

Intensity I_2 of the beam emerging from aperture

$$B = \frac{1}{\pi} \pi \times (0.002)^2 = 4 \times 10^{-6} \text{ W}$$

The path difference introduced between the two beams of light

$$= (\mu - 1)\, t$$
$$= (1.5 - 1) \times 2 \times 10^{-7} \text{ m}$$
$$= 10^{-7} \text{ m}$$

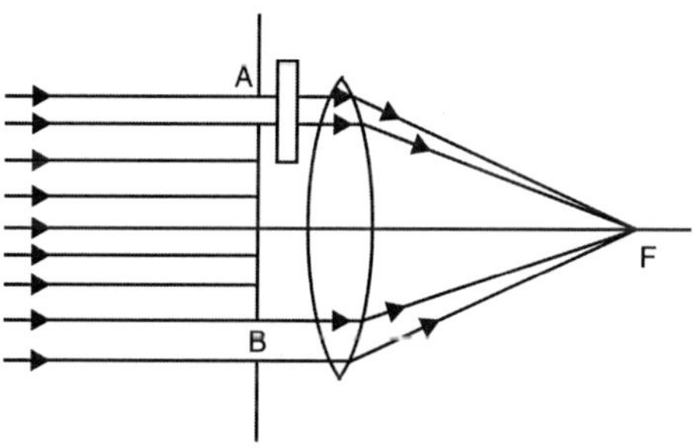

Fig. 15.100

Corresponding to above path difference, the phase difference δ is given by the following formula :

$$\delta = \frac{2\pi}{\lambda} \times \text{path difference}$$

$$\therefore \quad \delta = \frac{2\pi}{6 \times 10^{-7}} \times 10^{-7} = \frac{\pi}{3} \text{ radian}$$

Net intensity at $F = I = I_1 + I_2 + 2\sqrt{I_1 I_2} \cos \delta$

$$= 10^{-6} + 4 \times 10^{-6} + 2 \times 10^{-6} \times 2 \cos 60$$
$$= 7 \times 10^{-6} \text{W}$$

15.162. Interference fringes are obtained with Young's double-slit experiment using a monochromatic source of amplitude a. On the screen the separation between a bright and the adjacent dark fringe is found to be 0.48 mm. Find the intensity at a point lying on the screen at a distance of

5.52 mm from the central fringe in a direction perpendicular to the incident beam.

Solution: Distance 0.48 mm is half the fringe width. Thus it involves a path difference $\lambda/2$. The total path difference corresponding to the distance 5.52 mm $= 11.5 \times \lambda/2$.

The phase difference corresponding to this path difference is given by

$$\Delta\phi = \frac{2\pi}{\lambda} \times 11.5 \times \frac{\lambda}{2} = 11.5\pi = 270°$$

Resultant intensity $= I = a^2 + a^2 + 2a^2 \cos 270 = 2a^2$

15.163. A thin vertical slit is illuminated with light of wavelength 5,000 Å. In front of the slit and parallel to it a thin wire of diameter 0.1 cm is placed at a distance of 20 cm (see Fig. 15.101). The shadow of the wire falls on a screen placed at a distance of 180 cm from the wire. Calculate the number of fringes in the geometrical shadow region.

Solution: $A'B'$ is the geometrical shadow region in which interference fringes with equal space are observed. The fringe width $\overline{X}$ is given by $\overline{X} = \frac{D\lambda}{2d}$

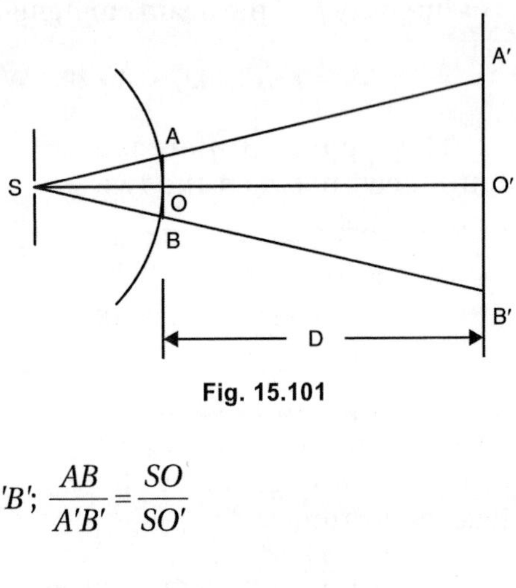

Fig. 15.101

$D = 180$ cm, $2d = 0.1$ cm; here separation between the coherent sources is equal to the diameter of the wire.

$$\overline{X} = \frac{180 \times 5 \times 10^{-5}}{0.1} = 0.09 \text{ cm}$$

Consider triangles SAB and $SA'B'$; $\frac{AB}{A'B'} = \frac{SO}{SO'}$

$$\therefore \quad A'B' = \frac{SO'}{SO} \times AB$$

$$= \frac{200 \times 0.1}{20} = 1 \text{ cm}$$

$$\text{Number of fringes} = \frac{A'B'}{\overline{X}} = \frac{1}{0.09} = \frac{100}{9} = 11$$

15.164. Two point sources S_1 and S_2 as shown in Fig. 15.102 are placed along y axis at a separation d. Discuss the shape of interference fringes produced on a screen PP' placed normal to the line joining S_1 and S_2.

Solution:

$$S_1A = \sqrt{x^2 + \left(D + d/2\right)^2 + z^2} \text{ and } S_2A = \sqrt{x^2 + \left(D - d/2\right)^2 + z^2}$$

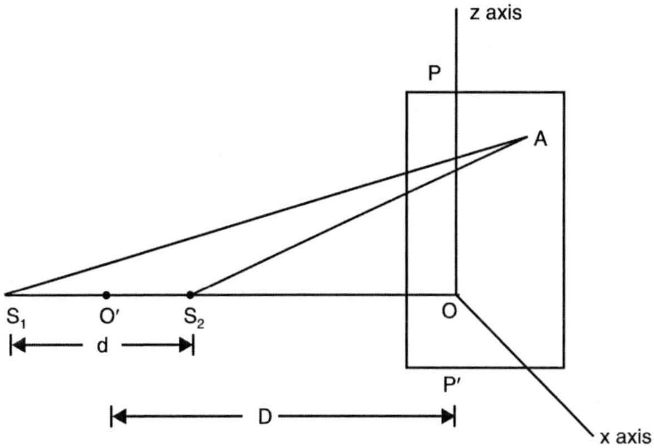

Fig. 15.102

$$S_1A - S_2A = n\lambda = \sqrt{x^2 + \left(D + d/2\right)^2 + z^2} - \sqrt{x^2 + \left(D - d/2\right)^2 + z^2}$$

or $$n\lambda + \sqrt{x^2 + \left(D - d/2\right)^2 + z^2} = \sqrt{x^2 + \left(D + d/2\right)^2 + z^2}$$

On squaring the two sides we have

$$n^2\lambda^2 + x^2 + z^2 + D^2 - Dd + d^2/4 + 2n\lambda\sqrt{x^2 + \left(D - d/2\right)^2 + z^2}$$
$$= x^2 + z^2 + D^2 + d^2/4 + Dd$$

or $$2n\lambda\sqrt{x^2 + \left(D - d/2\right)^2 + z^2} = 2Dd - n^2\lambda^2$$

or $$x^2 + \left(D - d/2\right)^2 + z^2 = \left(\frac{Dd}{n\lambda} - \frac{n\lambda}{2}\right)^2$$

or $$x^2 + z^2 = \frac{D^2d^2}{n^2\lambda^2} + \frac{n^2\lambda^2}{4} - Dd - D^2 - \frac{d^2}{4} + Dd$$
$$= D^2\left[\frac{d^2}{n^2\lambda^2} + \frac{n^2\lambda^2}{4D^2} - 1 - \frac{d^2}{4D^2}\right]$$

Assuming D to be large

$$x^2 + z^2 = D^2\left(\frac{d^2}{n^2\lambda^2} - 1\right) = \frac{D^2d^2}{n^2\lambda^2}\left(1 - \frac{n^2\lambda^2}{d^2}\right)$$

Now $d \cong n\lambda$

$$x^2 + z^2 = 2D^2\left(1 - \frac{n\lambda}{d}\right)$$

Thus the interference fringes will be concentric circles and the radius of *n*th bright fringe

$$R_n = D\sqrt{2}\left(1-\frac{n\lambda}{d}\right)^{1/2}$$

15.165. In the Young's double slit arrangement as shown in the Fig. 15.103 the amplitudes of light reaching to the central point P_0 of the screen from individual slits S_2 and S_1 are in the ratio of 3:2. Find the resultant intensity on the screen at direction θ.

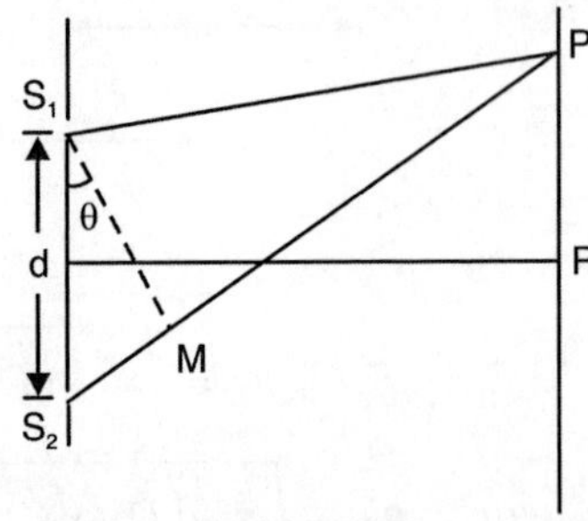

Fig. 15.103

Solution: Resultant amplitude at $P_0 = 5a$

Intensity $I_0 = 25\,a^2$

Intensity of source $S_1 = \dfrac{4I_0}{25} = 4a^2$

Intensity of source $S_2 = \dfrac{9I_0}{25} = 9a^2$

Path difference $S_2P - S_1P = d\sin\theta$

Phase difference $\delta = \dfrac{2\pi}{\lambda}\cdot d\sin\theta$

$$\text{Intensity at } P = \frac{4I_0}{25}+\frac{9I_0}{25}+2\sqrt{\frac{4I_0}{25}\cdot\frac{9I_0}{25}}\cdot\cos\frac{2\pi d\sin\theta}{\lambda}$$

$$= \frac{13I_0}{25}+\frac{12I_0}{25}\cdot\frac{\cos 2\pi d\sin\theta}{\lambda}$$

$$= \frac{13I_0}{25}+\frac{12I_0}{25}\left[2\cos^2\frac{\pi d\sin\theta}{\lambda}-1\right]$$

$$= \frac{I_0}{25}+\frac{24I_0}{25}\,\frac{\cos^2\pi d\sin\theta}{\lambda}$$

$$= \frac{I_0}{25}\left[1+24\cos^2\frac{\pi d\sin\theta}{\lambda}\right]$$

15.166. Two slits each of width b are arranged to observe interference pattern on the screen as shown in the Fig. 15.104. The centres of the two slits are l apart and the distance of the slit from the screen is D. (i) Determine the total optical path diff. of the waves originating from the

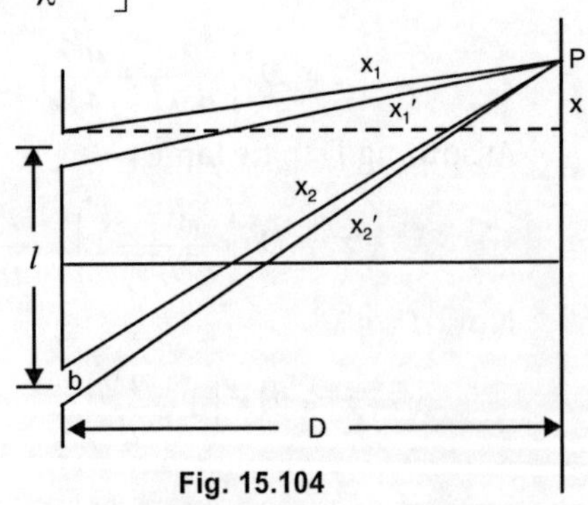

Fig. 15.104

two slits and reaching at P (ii) calculate the maximum width of the slit at which the interference fringes on the screen are distinguished clearly.

Solution: The position of the edges of the two slits are marked in the figure.

$$x_1 = \sqrt{\left[x - \left(\frac{l+b}{2}\right)\right]^2 + D^2},\ x_2 = \sqrt{\left[x + \left(\frac{l-b}{2}\right)\right]^2 + D^2}$$

$$x_1' = \sqrt{\left[x - \left(\frac{l-b}{2}\right)\right]^2 + D^2}\ \&\ x_2' = \sqrt{\left[x + \left(\frac{l+b}{2}\right)\right]^2 + D^2}$$

$$\text{Net path diff.} = (x_2' - x_1') - (x_2 - x_1)$$

$$= x_2' - x_2 - (x_1' - x_1)$$

$$= \sqrt{\left[x + \left(\frac{l+b}{2}\right)\right]^2 + D^2} - \sqrt{\left[x + \left(\frac{l-b}{2}\right)\right]^2 + D^2}$$

$$- \sqrt{\left[x - \left(\frac{l-b}{2}\right)\right]^2 + D^2} + \sqrt{\left[x - \left(\frac{l+b}{2}\right)\right]^2 + D^2}$$

$$= D\left\{1 + \frac{1}{2}\left[\frac{x + \left(\frac{l+b}{2}\right)}{D}\right]^2 - 1 - \frac{1}{2}\left[\frac{x + \left(\frac{l-b}{2}\right)}{D}\right]^2\right.$$

$$\left. - 1 - \frac{1}{2}\left[\frac{x - \left(\frac{l-b}{2}\right)}{D}\right]^2 + 1 + \frac{1}{2}\left[\frac{x - \left(\frac{l+b}{2}\right)}{D}\right]^2\right\}$$

$$= \frac{1}{4D}\left[(l+b)^2 - (l-b)^2\right] = \frac{4lb}{4D} = \frac{lb}{D}$$

(ii) $\dfrac{2\pi}{\lambda} \cdot \dfrac{lb}{D}$ = Phase difference

$$b_{\text{max}} = \text{Phase difference}\ \frac{\lambda D}{2\pi l} = \frac{\pi}{2} \cdot \frac{\lambda D}{2\pi l} = \frac{\lambda D}{4l}$$

15.167. Using the arrangement shown in the Fig. 15.105 interference fringes are observed on the screen. For what width of the slit SS' the interference fringes on the screen will disappear?

Solution: Let the source S produces the interference fringes of good contrast on the screen. Point O will be the position of central maxima. Let S' be another source such that $SS' = b$. If the width b is slowly increased then the contrast of the fringes becomes poorer.

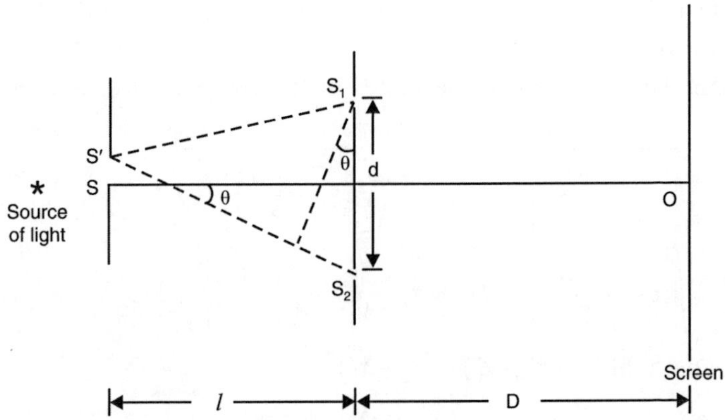

Fig. 15.105

If $S'S_2 - S'S_1 = \dfrac{\lambda}{2}$

i.e. the minima of pattern produced by S falls on the maxima produced by S' no fringes are observed on the screen.

From geometrical considerations

$$S'S_2 = \left[l^2 + \left(\frac{d}{2}+b\right)^2\right]^{1/2} = l\left[1+\frac{1}{2l^2}\left(\frac{d}{2}+b\right)^2\right]$$

$$= l + \frac{1}{2l}\left(\frac{d}{2}+b\right)^2$$

Similarly $S'S_1 = \left[l^2 + \left(\dfrac{d}{2}-b\right)^2\right]^{1/2} = l + \dfrac{1}{2l}\left(\dfrac{d}{2}-b\right)^2$

$$S'S_2 - S'S_1 = \frac{1}{2l}\cdot 4\cdot\frac{d}{2}\cdot b = \frac{d\cdot b}{l}$$

The fringes will disappear when $\dfrac{d\cdot b}{l} = \dfrac{\lambda}{2}$

or $\quad b = \dfrac{\lambda l}{2d}$

15.168. A double slit apparatus is immersed in a liquid of refractive index 1.33. It has slit separation of 1 mm, and distance between the plane of slits and screen in 1.33 m. The slits are illuminated by a parallel beam of light whose wavelength in air is 6300 Å. (i) Calculate the fringe width (ii) one of the slits of the apparatus is covered by a thin glass sheet of refractive index 1.53. Find the smallest thickness of the sheet to bring the adjacent minimum on the axis. (IIT 1996)

Solution:

(i) Fringe width in water is given by

$$\overline{X} = \frac{D\lambda'}{2d} \quad \text{where } \lambda' = \frac{\lambda}{{}^a\mu_l}$$

$$= \frac{D\lambda}{2d \cdot {}^a\mu_l}$$

On putting the values of various terms we have

$$\overline{X} = \frac{1.33 \times 6300 \times 10^{-10}}{10^{-3} \times 1.33} = 0.63 \text{ mm}$$

(ii) Shift $= x = \frac{D}{2d}\left({}^l\mu_g - 1\right)t$ and as distance of minima from maxima

$$= \frac{D}{2d} \cdot \frac{\lambda}{2} = \frac{\overline{X}}{2}$$

or $$\frac{1.33}{10^{-3}}\left[\frac{1.53}{1.33} - 1\right] \times t = \frac{0.63}{2}$$

or $$t = 1.575 \text{ µm}$$

15.169. In Young's experiment the source is red light of wavelength 7×10^{-7} m. When a thin glass plate of refractive index 1.5 at this wavelength is put in the path of one of the interfering beams, the central bright fringe shifts by 10^{-3} m to the position previously occupied by the 5th bright fringe. Find the thickness of the plate. When the source is now changed to green light of wavelength 5×10^{-7} m, the central fringe shifts to a position initially occupied by the 6th bright fringe due to red light. Find the refractive index of glass for the green light. Also estimate the change in fringe width due to the change in wavelength.

Solution:

$$(\mu_r - 1)\, t = 5\lambda_r$$

$$\therefore \quad t = \frac{5\lambda_r}{(\mu_r - 1)}$$

On putting the values of λ_r and μ_r we have

$$t = \frac{5 \times 7 \times 10^{-7}}{(1.5 - 1)} = 7 \text{ µm}$$

For green light we have

$$(\mu_g - 1)\, t = 6\, \lambda_g$$

or $$\frac{(\mu_r - 1)}{\mu_g - 1} = \frac{5 \times 7 \times 10^{-7}}{6 \times 7 \times 10^{-7}} = \frac{5}{6}$$

$$\frac{0.5 \times 6}{5} = \mu_g - 1$$

or $\mu_g = 1.6$

Further $5\overline{X}_r = 10^{-3}$ $\therefore \overline{X}_r = 2 \times 10^{-4}$

$$\frac{\overline{X}_g}{\overline{X}_r} = \frac{\lambda_g}{\lambda_r} = \frac{5}{7}$$

$$\frac{\overline{X}_g - \overline{X}_r}{\overline{X}_r} = -\frac{2}{7}$$

$$\therefore \quad \Delta\overline{X} = \overline{X}_g - \overline{X}_r = -\frac{2}{7}\overline{X}_r = -\frac{2}{7} \times 2 \times 10^{-4}$$

$$= -0.57 \times 10^{-4}$$

Therefore fringe width will decrease by an amount of 0.57×10^{-4} when the red light is replaced by green light.

15.170. Interference fringes are produced by Fresnel's biprism of refractive index $\mu = 1.5$ using monochromatic source of wavelength $\lambda = 5000$ Å. The distance of the source and biprism is $a = 27$ cm and the distance between the biprism and screen is $b = 73$ cm. Find (i) the prism angle α assuming the fringe width $\overline{X} = 0.3$ mm (ii) the maximum number of fringes observed on the screen.

Solution: The deviation produced is given by (see Fig. 15.106).

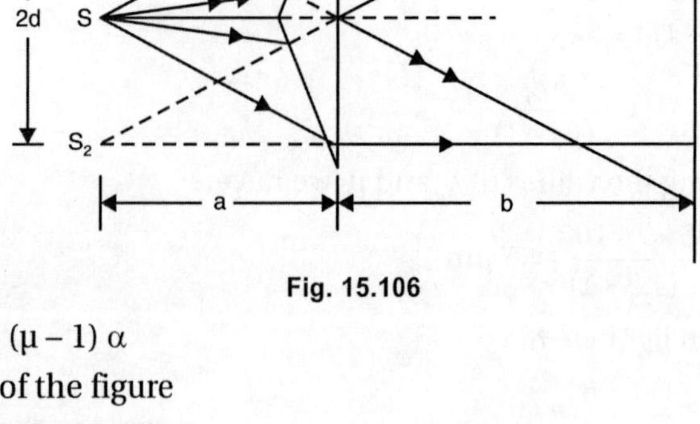

Fig. 15.106

$$\delta = (\mu - 1)\,\alpha$$

In view of the figure

$$2d = 2a\,(\mu - 1)\,\alpha$$

$$\text{Fringe width } \overline{X} = \frac{(a+b)\lambda}{2a(\mu-1)\alpha}$$

$$\therefore \qquad \alpha = \frac{(a+b)\lambda}{2a(\mu-1)\overline{X}}$$

On substituting the values we have

$$\alpha = \frac{5\times10^{-7}}{0.54\times0.5\times0.3\times10^{-3}} = \frac{1}{162}\text{ rad.}$$

$$= 21.2'$$

(ii) Let x be the length on the screen then $\delta = \dfrac{x}{b}$

$x = \delta \cdot b = (\mu - 1)\,\alpha \cdot b$

Total length on the screen $= 2x = 2\,(\mu - 1)\,\alpha \cdot b$

Max no. of fringes $N = \dfrac{2x}{\overline{X}} = \dfrac{2(\mu-1)\alpha\cdot b}{\overline{X}}$

On putting the values we have

$$N = \frac{2\times0.5}{0.3\times10^{-3}}\times\frac{1}{162}\times0.73 = \frac{730}{162\times0.3} = 15$$

15.171. Fresnel's double mirror arrangement shown in the Fig. 15.107 has been used to produce interference fringes with monochromatic source of wavelength $\lambda = 5000$ Å. Using $r = 0.2$ m and $b = 0.85$ m. Find (i) the angle θ assuming the width $\overline{X}$ of the interference fringes = 0.5 mm (ii) the maximum number of fringes observed on the screen.

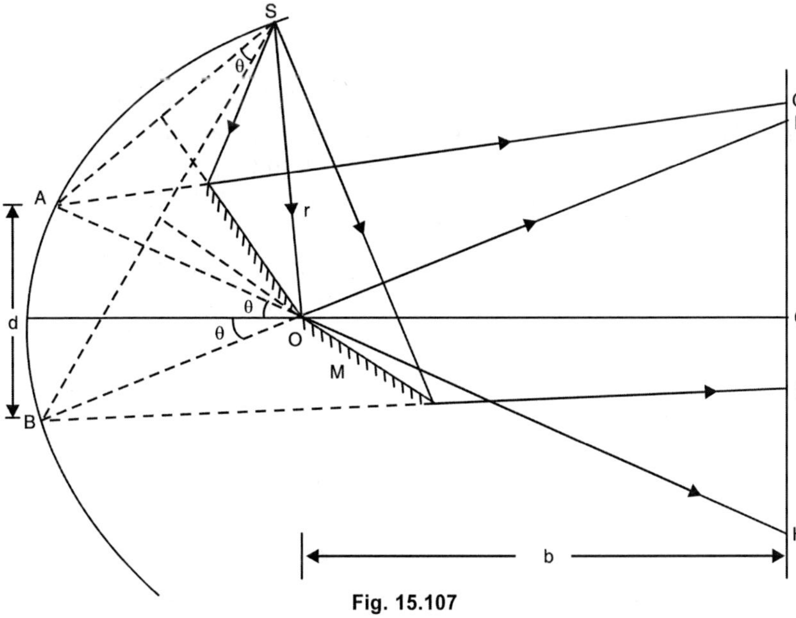

Fig. 15.107

Solution:

(i) In view of the geometry D = distance between the coherent sources and screen $= r + b$.

$$\overline{X} = \frac{(r+b)\lambda}{d} = \frac{\lambda(r+b)}{2r\theta}$$

$$\therefore \quad \theta = \frac{\lambda(r+b)}{2r \cdot \overline{X}}$$

On substituting the values of terms occurring

$$\theta = \frac{5 \times 10^{-7} \times 1.05}{2 \times 0.2 \times 5 \times 10^{-4}} = \frac{1.05}{4} \times 10^{-2} = 9'$$

(ii) $$\frac{EC}{b} = \frac{d/2}{r}$$

$$\therefore \quad EC = \frac{b}{r} \cdot \frac{d}{2} = \frac{b}{r} \cdot r\theta = b\theta$$

$$\text{Max no. of fringes} = \frac{2EC}{\overline{X}} = \frac{b \cdot 2\theta}{\overline{X}}$$

$$= \frac{0.85 \times 1.05 \times 10^{-2} \times 2}{5 \times 10^{-4} \times 4}$$

$$= \frac{85 \times 1.05}{10}$$

$$= 8.5 \times 1.05 = 8.9 = 9$$

15.172. A plane monochromatic wave of wavelength λ falls normally on the base of a biprism of refractive index μ and refracting angle ϕ (see Fig. 5.108a). Behind the biprism there is a thin parallel glass plate and the space in between is filled with transparent liquid of refractive index μ'. Calculate the width of the interference fringes observed on the screen.

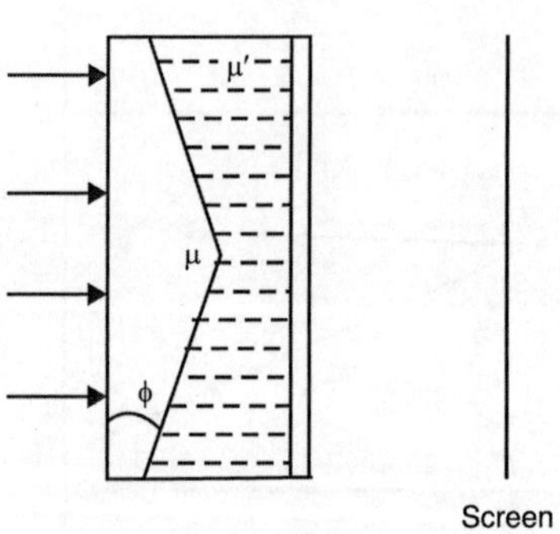

Fig. 15.108a

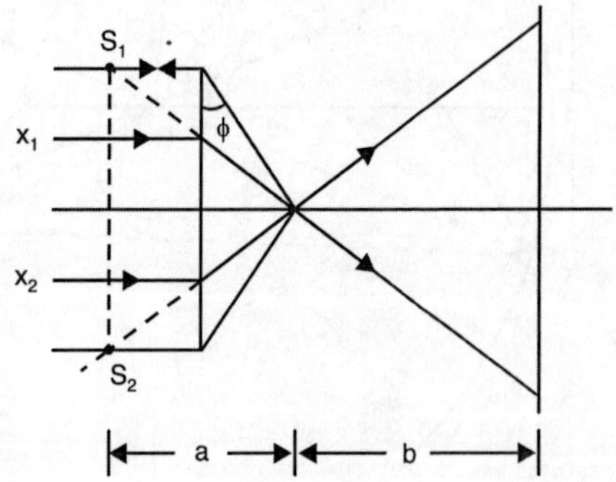

Fig. 15.108b

Solution: The deviation produced δ is given by

$$\delta = \phi\,(\mu - \mu') = (\mu - 1)\,\theta$$

$$\therefore \qquad \theta = \frac{\phi\,(\mu - \mu')}{(\mu - 1)}$$

This system is equivalent to a biprism of refracting angle θ.

Now $2\delta = \frac{x_1}{a} + \frac{x_2}{a} = \frac{S_1 S_2}{a}$ (see Fig. 15.88b)

or $S_1 S_2 = 2\delta a = 2\phi\,(\mu - \mu')\,a$

$$\text{Fringe width } \overline{X} = \frac{(a+b)\lambda}{2\phi \cdot a(\mu - \mu')}$$

$$= \frac{\lambda}{2\phi(\mu - \mu')}\left(1 + \frac{b}{a}\right)$$

here a is very large hence

$$\overline{X} = \frac{\lambda}{2\phi(\mu - \mu')}$$

15.173. The arrangement shown in Fig. 15.109 is known as Lloyed single mirror arrangement for producing interference fringes on the screen. Using $h = 1.5$ mm, $l = 1.5$ m, $\lambda = 5000$ Å. Find the fringe width.

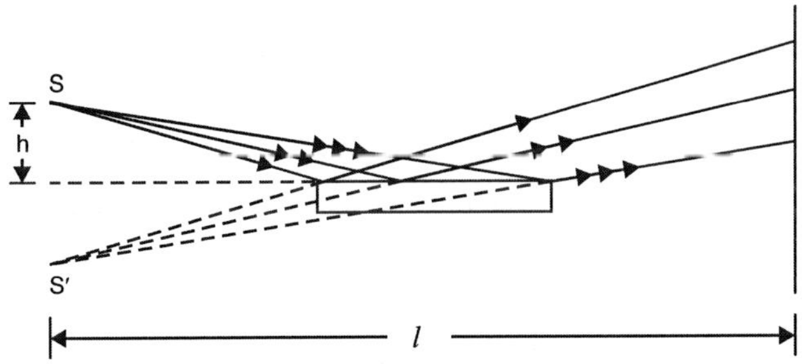

Fig. 15.109

Solution:

$$\text{Fringe width } \overline{X} = \frac{l\lambda}{2h}$$

On putting the values

$$\overline{X} = \frac{1.5 \times 5 \times 10^{-7}}{2 \times 1.5 \times 10^{-3}} = 2.5 \times 10^{-4} \text{ m}$$

$$= 0.25 \text{ mm}$$

15.174. Consider three equispaced slits S_1, S_2 & S_3 as shown in Fig. 15.110a. These slits are placed at a larger distance from the screen. The assembly is illuminated by a monochromatic source of wavelength λ. Discuss the distribution of intensity of the pattern observed on the screen.

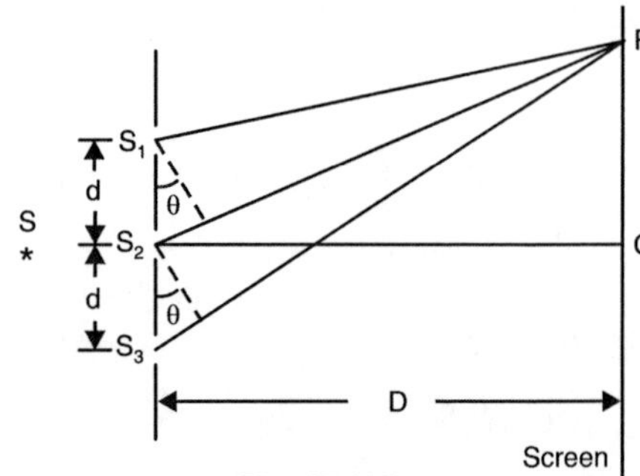

Fig. 15.110a

Solution: The screen is placed at a larger distance and due to this the rays reaching at the point P on the screen are approximately parallel. The path diff. between the rays originating from S_1 & S_2 is equal to $d \sin\theta$ and between S_1 & S_3 equal to $2d\sin\theta$. The waves emerging from these slits and interfering at point P will differ from each other in phase by an amount $\phi = \frac{2\pi}{\lambda}\cdot d\sin\theta$. The displacement produced at P by the waves originating from these slits may be taken as

$$y_1 = a\sin\omega t$$

$$y_2 = a\sin(\omega t + \phi)$$

$$y_3 = a\sin(\omega t + 2\phi)$$

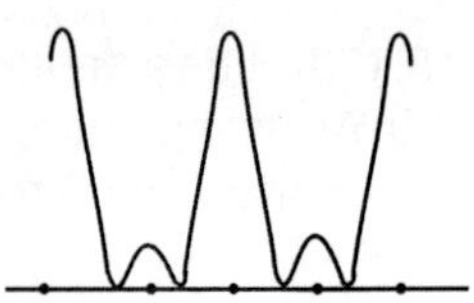
Fig. 15.110b

where a = amplitude

Resultant displacement at P will be

$$\begin{aligned} y = y_1 + y_2 + y_3 &= a\,[\sin\omega t + \sin(\omega t + \phi) + \sin(\omega t + 2\phi)] \\ &= a\,[\sin\omega t\,(1 + \cos\phi + \cos 2\phi) + \sin\omega t\,(\sin\phi + \sin 2\phi)] \\ &= A\sin(\omega t + \delta) \end{aligned}$$

where $A = a(2\cos\phi + 1)$ and $\tan\delta = \dfrac{\sin\phi + \sin 2\phi}{1 + \cos\phi + \cos 2\phi}$

When $\phi = 2n\pi$ where $n = 0, 1, 2$ etc. the resultant amplitude $= 3a$ and the intensity is equal to $9a^2$. We call this as central maxima. As we move further from the central maxima on either side θ increases which results in the increase of ϕ. The amplitude and intensity decreases and becomes zero at $\phi = \frac{2\pi}{3}$. On further increasing ϕ the intensity increases and attains a maximum value a^2 at $\phi = 180°$. We call it as secondary maxima. Its intensity is $\frac{1}{9}$th of the central maxima. At $\phi = 240°$ we again get a minima and at $\phi = 360°$ we again get a maximum intensity of $9a^2$. Thus between two maxima we have a secondary maxima and two minima as shown in Fig. 15.110b.

15.175. Consider a system consisting of four equispaced slits S_1, S_2, S_3 & S_4 as shown in Fig. 15.111a. The plane of the slit is far off from the screen. All the slits are illuminated by a monochromatic source of light. Discuss the distribution of intensity of the pattern obtained on the screen as a result of superposition.

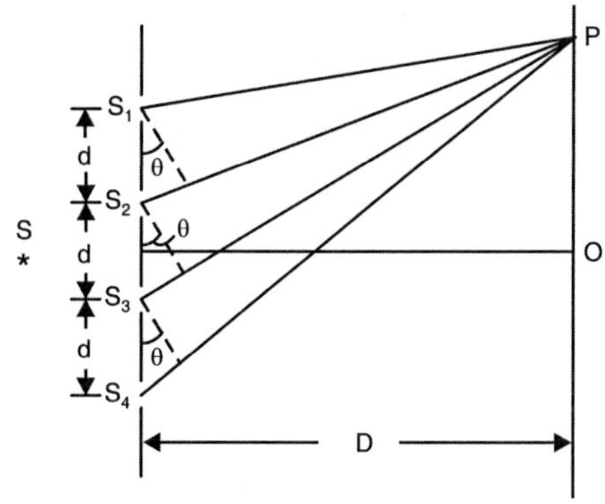

Fig. 15.111a

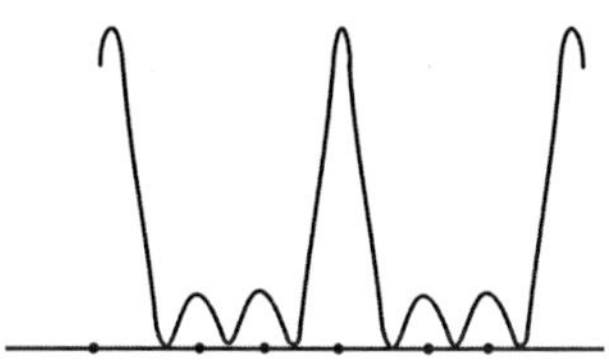

Fig. 15.111b

Solution: The rays originating from the four slits and reaching at the point P on the screen will be approximately parallel. In the light of Q. 15.174 the displacement at P due to the waves emnating from these slits are represented as

$$y_1 = a \sin \omega t$$
$$y_2 = a \sin (\omega t + \phi)$$
$$y_3 = a \sin (\omega t + 2\phi)$$
$$y_4 = a \sin (\omega t + 3\phi)$$

where a = amplitude and $\phi = \frac{2\pi}{\lambda} \cdot d \sin \theta$

The resultant displacement at the point P is given by

$$\begin{aligned} y &= y_1 + y_2 + y_3 + y_4 \\ &= a [\sin \omega t + \sin (\omega t + \phi) + \sin (\omega t + 2\phi) + \sin (\omega t + 3\phi)] \\ &= a [\sin \omega t (1 + \cos \phi + \cos 2\phi + \cos 3\phi) \\ &\qquad + \cos \omega t (\sin \phi + \sin 2\phi + \sin 3\phi)] \\ &= A \sin (\omega t + \delta) \end{aligned}$$

Where $A = a\sqrt{8(1 + \cos \phi)} \cdot \cos \phi$

and $\tan \delta = \dfrac{\sin \phi + \sin 2\phi + \sin 3\phi}{1 + \cos\phi + \cos 2\phi + \cos 3\phi}$

The resultant intensity = $16a^2$ when $\phi = 2n\pi$; $n = 0, 1, 2.....$ We define it as a central maxima. On increasing θ and hence ϕ we get a min. intensity at $\phi = \frac{\pi}{2}$. At $\phi = \frac{3\pi}{4}$ and $\frac{5\pi}{4}$ we obtain a maximum amplitude $A = 1.08\, a$ and the corresponding intensity is nearly $\frac{1}{16}$ th of that of central maxima. These maximas are known a secondary maxima. In between the secondary maxima we get a minima at $\phi = \pi$. Again at $\phi = 2\pi$ we get a maximum intensity of $16a^2$. In this set up we observe two secondary maxima and two minima as shown in Fig. 15.111b.

15.176. A plano convex lens of radius of curvature R is placed on a transparent glass plate as shown in the Fig. 15.112a. Light from an extended monochromatic source of wavelength λ falls on the transparent plate P. The plate reflects a part of the incident light towards an air film enclosed between the plate and lens. When the reflected beam from the air film is view normally through microscope M circular bright and dark fringes are observed. Find the diameter of the nth dark ring.

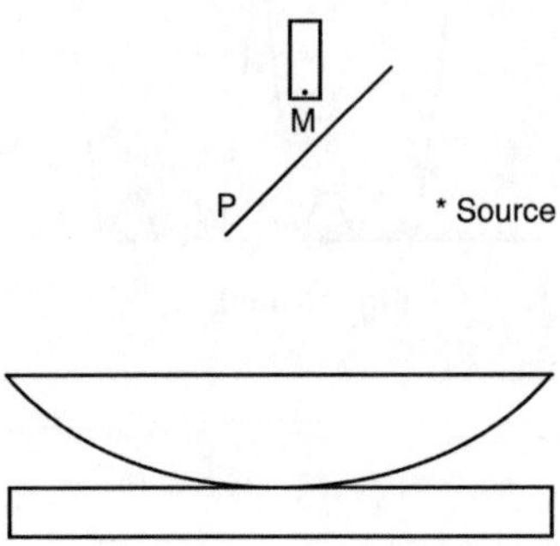

Fig. 15.112a

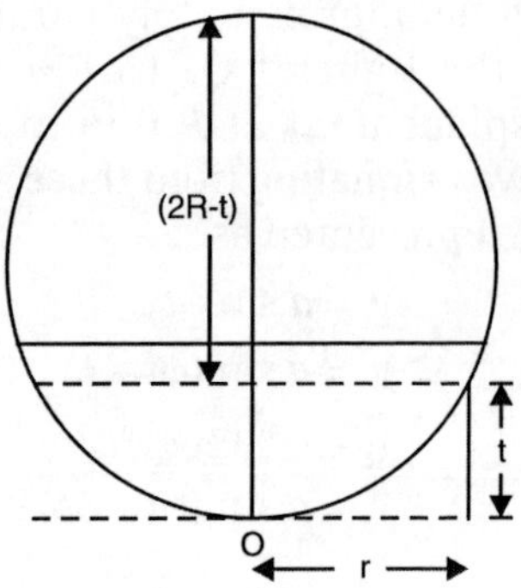

Fig. 15.112b

Solution: For dark rings we must have

$$2\mu t \cos\theta = n\lambda$$

where μ = refractive index of the material of the film, t = thickness of the film and θ = angle of refraction.

In the present case $\mu = 1$ (air film) and $\theta = 0$

$$\therefore \quad 2t = n\lambda$$

Using the geometry of Fig. 15.112b we have

$$r^2 = (2R - t)\,t \cong 2Rt; \; r = \text{radius of the ring}$$

or $\quad D_n^2 = 8\,Rt = 4n\lambda\,R$

15.177. The curved surface of a plano convex lens is placed in contact with a glass plate. The radius of curvature of the lens is R. An extended monochromatic source of wavelength λ has been used to form Newton's ring by reflected light. Find the width Δr of the rings as a function of r, the radius of the rings lying in the region where $\Delta r << r$.

Solution: The radius of the nth bright and nth dark ring is given by

$$r_b = \sqrt{(2n-1)\frac{\lambda R}{2}}$$

$$r_d = \sqrt{n\lambda R}$$

Let r be the radius of the pth bright ring then

$$r + \Delta r = \sqrt{p\lambda R}\,;\ \Delta r = \text{width of the ring}$$

$$(r+\Delta r)^2 = p\lambda R$$

$$r = \sqrt{(2p-1)\frac{\lambda R}{2}} = \sqrt{p\lambda R - \frac{\lambda R}{2}}$$

$$(r+\Delta r)^2 = r^2 + \frac{\lambda R}{2}$$

or $$r^2 + 2r\Delta r = r^2 + \frac{\lambda R}{2}$$

or $$\Delta r = \frac{\lambda R}{4r}$$

15.178. A plano convex lens of radius of curvature R_1 is placed on the concave surface of a glass block as shown in the Fig. 15.113a. The radius of curvature of the concave surface is R_2. The two surfaces are in contact at the point O. The lens is illuminated from the above in order to observe interference fringes. What shape do the interference fringes have?

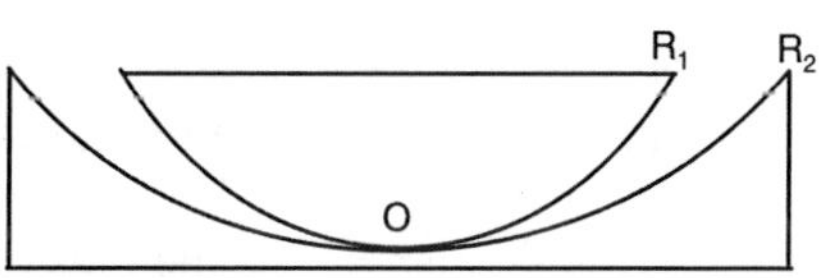

Fig. 15.113a

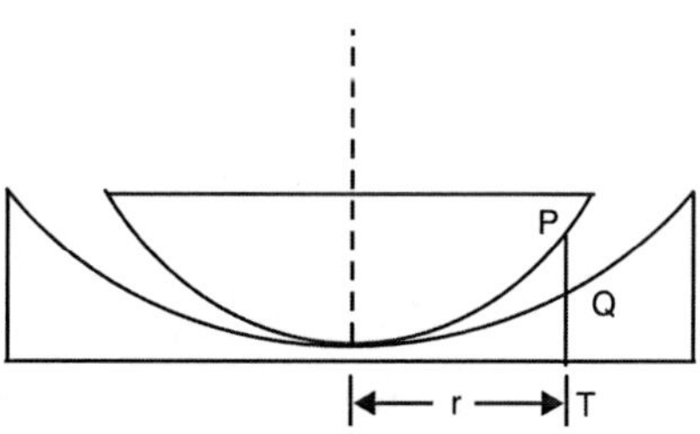

Fig. 15.113b

Solution:

In view of Fig. 15.113b the thickness of air film at $P = PQ = PT - QT$

From geometrical considerations $PT = \dfrac{r^2}{2R_1}$ and $QT = \dfrac{r^2}{2R_2}$

$\therefore$ Thickness of air film $= t = \dfrac{r^2}{2}\left[\dfrac{1}{R_1} - \dfrac{1}{R_2}\right]$

Since R_1 and R_2 are constant the shape of the fringes is therefore circle.

For reflected light

$$2t = n\lambda \text{ for dark rings}$$

$$2\left[\frac{1}{R_1} - \frac{1}{R_2}\right]\frac{r^2}{2} = n\lambda$$

or $$r = \sqrt{\frac{n\lambda R_1 R_2}{(R_2 - R_1)}}; \ r = \text{radius of } n\text{th dark ring}$$

For bright rings

$$2t = (2n+1)\frac{\lambda}{2}$$

or $$r = \sqrt{\frac{(2n+1)\lambda R_1 R_2}{2(R_2 - R_1)}}; \ r = \text{radius of the } n\text{th bright ring.}$$

14.179. A plano convex lens of radius of curvature R_1 is placed on a cylindrical glass block of radius R_2. The system thus formed is illuminated from the above. What shape do the interference fringes have?

Solution: In order to define the shape of the interference fringes let the x axis be along the axis of the cylinder and passing through the point of contact O and the y axis be perpendicular to it.

The thickness of the air column at point P is equal to $t = t_1 - t_2$

$$t = \frac{r^2}{2R_1} - \frac{y^2}{2R_2} \quad \text{(see Fig. 15.114)}$$

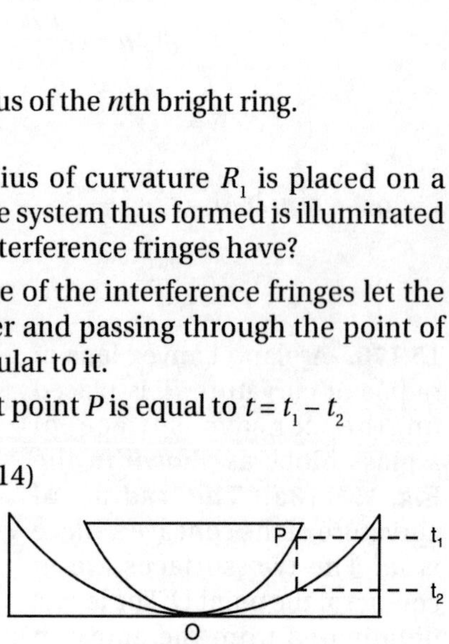

Fig. 15.114

Substituting $r^2 = x^2 + y^2$ we have

$$t = \frac{x^2}{2R_1} + \frac{y^2}{2}\left[\frac{1}{R_1} - \frac{1}{R_2}\right]$$

or $$1 = \frac{x^2}{2R_1 t} + \frac{y^2}{2t}\left[\frac{1}{R_1} - \frac{1}{R_2}\right]$$

Thus the shape of interference fringes is an elliptical one. In reflected light the thickness of air column t is given by the relation.

$$2t = n\lambda \text{ for dark fringes}$$

and $$2t = (2n+1)\frac{\lambda}{2} \text{ for bright fringes}$$

where $n = 0, 1, 2\ldots$

15.180. While performing an interference experiment with a monochromatic source of wavelength $\lambda = 6000$ A in reflected light using a plane glass plate and a plano convex lens of radius of curvature $R = 100$ cm a small dust particle has got into the lens and plate. Determine the diameter of the particle if the diameter of the 10th dark ring is 4 mm.

Solution: Let d be the diameter of the particle. The thickness of air between the lens and the plate is equal to

$$t = \frac{r^2}{2R} + d$$

where r is the radius of a particular dark or bright ring

In reflected light the thickness of air column t is given by the relation

$2t = n\lambda$ for dark rings

$2t = (2n+1)\dfrac{\lambda}{2}$ for bright rings

In case of dark rings

$$\frac{n\lambda}{2} = \frac{r^2}{2R} + d = \frac{D^2}{8R} + d$$

$$\therefore \quad d = \frac{n\lambda}{2} - \frac{D^2}{8R}$$

On putting the values

$$d = \frac{10}{2} \times 6 \times 10^{-7} - \frac{6 \times 10^{-6}}{8}$$

$$= 10^{-6} \text{ m}$$

$$= 10^{-3} \text{ mm}$$

15.181. On account of the repeated use a ground off spot of radius R_0 has formed at the crest of the curved surface of a plano convex lens of radius of curvature R. This lens forms a contact with a glass plate through this spot. On illuminating this lens plate system from the above by a monochromatic light of wavelength λ interference frings are produced. Calculate the diameter of the nth bright fringe when observed in reflected light.

Solution: The thickness of the air film enclosed is given by

$$t = \frac{r_n^2}{2R} - \frac{R_0^2}{2R} = \frac{D_n^2}{8R} - \frac{R_0^2}{2R}$$

For bright fringes we have

$$2t = (2n-1)\frac{\lambda}{2} = \frac{D_n^2}{4R} - \frac{R_0^2}{R}$$

$$\therefore \quad D_n = 2\sqrt{R_0^2 + (2n-1)\frac{\lambda R}{2}}$$

15.182. The two thin plano convex lenses of refractive index μ and unknown radius of curvature are placed as shown in the Fig. 15.115. The system is illuminated from the above by a monochromatic source of light of wavelength λ to produce interference fringes. If D_n be the diameter of the nth bright ring find the power of the system of lenses.

Solution: The thickness of the air film enclosed is given by

$$t = \frac{r_n^2}{2R_1} + \frac{r_n^2}{2R_2} = \frac{D_n^2}{8}\left(\frac{1}{R_1} + \frac{1}{R_2}\right)$$

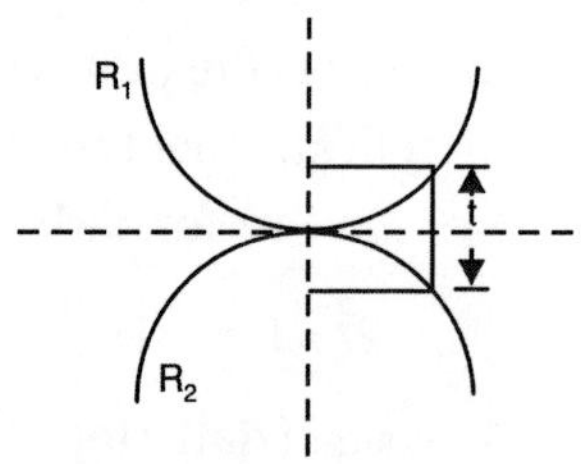

Fig. 15.115

where R_1 and R_2 represents the radius of curvature of the two lenses

For bright fringes

$$2t = (2n-1)\frac{\lambda}{2} = \frac{D_n^2}{4}\left(\frac{1}{R_1} + \frac{1}{R_2}\right)$$

or $$P = (\mu - 1)\left(\frac{1}{R_1} + \frac{1}{R_2}\right) = \frac{2(2n-1)\lambda}{D_n^2}(\mu - 1)$$

15.183. A soap film of refractive index 1.33 is illuminated with light of different wavelengths at an angle of 45°. There is complete destructive interference for $\lambda = 5890$ Å. Find the thickness of the film. (IAS 1991)

Solution:

$$2\mu t \cos r = n\lambda$$

$$t = \frac{n\lambda}{2\mu \cos r}$$

On putting the values

$$t = \frac{1 \times 5890 \times 10^{-10}}{2 \times 1.33 \cos 45}$$

$$= 3.1 \times 10^{-4} \text{ mm}$$

15.184. A thin film of refractive index $n_2 = 1.2$ is applied on the surface of a lens whose refractive index is $n_1 = 1.5$. At what minimum thickness of this film will the reflected light be attenuated most in the middle of the visible spectrum ($\lambda = 5500$ A)?

Solution: The required thickness d is given by

$$d = \frac{\lambda}{4n_2}$$

$$= \frac{5.5 \times 10^{-7}}{4 \times 1.2}$$

$$= 1.15 \times 10^{-7} \text{ m}$$

15.185. A vertical soap film forms a wedge due to the liquid trickling down. On observing the interference fringes in the reflected light of wavelength 5461A it has been found that five fringes are confined in a space of 2 cm. Find the wedge angle; Refractive index of soap solution = 1.33.

Solution:

$$\text{Fringe width } \beta = \frac{\lambda}{2\theta\mu}$$

$$\text{or} \qquad \theta = \frac{\lambda}{2\beta\mu}$$

On putting the values we have

$$\theta = \frac{5461\times10^{-10}}{2\times0.4\times10^{-2}\times1.33} = \frac{5461\times10^{-7}}{10.64}\text{ rad.}$$

$$= \frac{5461\times10^{-7}\times180\times60\times60}{10.64\times3.14} = 10.5''$$

15.186. A source of light S is placed at a distance $l = 1$ m from a thin mica sheet of thickness $h = 0.1$ mm and with refractive index $\mu = 1.4$. At the same distance from the plate a small screen is placed at right angles to the reflected ray to observe the interference fringes. Taking the angle of incidence $i = 60°$. Determine (i) the order of the fringes located at the screen's centre (ii) the admissible nonmonochromaticity $\Delta\lambda$. Wavelength of light used $\lambda = 5600$ A.

Solution:

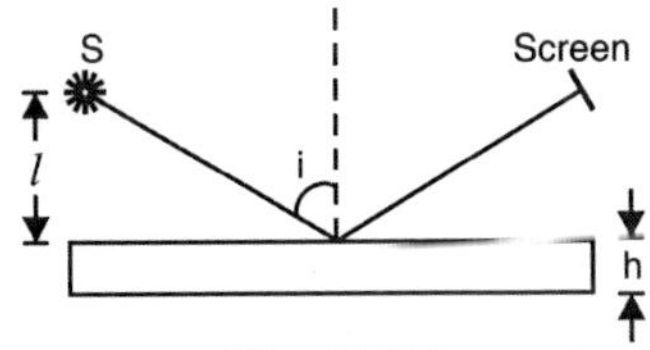

Fig. 15.116

(i) $$\Delta = 2\mu\, h\cos r = \frac{2\mu h}{\mu}\sqrt{\mu^2-\sin^2 i}$$

$$= 2h\sqrt{\mu^2-\sin^2 i}$$

$$= 2\times10^{-4}\sqrt{1.4^2-0.75}$$

$$= 2.2\times10^{-4}$$

$$n = \text{order of the fringes} = \frac{2h\sqrt{\mu^2-\sin^2 i}}{\lambda}$$

$$= \frac{2\times10^{-4}\sqrt{1.4^2-0.75}}{56\times10^{-8}}$$

$$= 393 \text{ fringes}$$

$$\text{nonmonochromaticity } \Delta\lambda = \frac{\lambda^2}{\Delta} = \frac{56^2\times10^{-16}}{2.2\times10^{-4}}$$

$$= 14.25\text{ A}$$

15.187. Using a telescope set at infinity, interference fringes are observed in a thin glass sheet of thickness $h = 0.2$ mm and with refractive index $\mu = 1.41$. The angle of observation i being from 0 to 90°. Find (i) the maximum and minimum orders of the fringes observed (ii) the admissible non- monochromaticity of the source for which all the interference fringes are fairly distinct. Wavelength of the source $\lambda = 5600$ A.

Solution: The angle of incidence = angle of observation.

Path difference $2\mu h \cos r = n\lambda$

For maximum orders $r = 0$

$$\therefore \quad n_{max} = \frac{2\mu h}{\lambda} = \frac{2 \times 1.41 \times 0.2 \times 10^{-3}}{56 \times 10^{-8}} = 1007$$

For minimum orders $\cos r$ has to be calculated in view of total internal reflection.

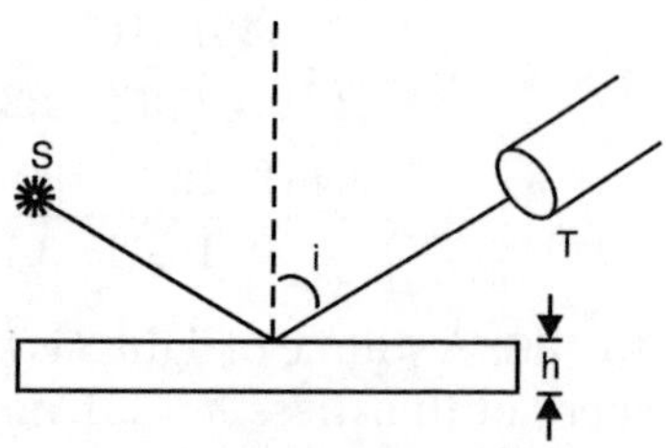

Fig. 15.117

$$\sin r = \frac{1}{\mu};\ \cos r = \frac{\sqrt{\mu^2 - 1}}{\mu}$$

$$n_{min} = \frac{2\mu h\sqrt{\mu^2 - 1}}{\mu\lambda} = \frac{2h}{\lambda}\sqrt{\mu^2 - 1}$$

$$= \frac{2 \times 0.2 \times 10^{-3}\sqrt{1.41^2 - 1}}{56 \times 10^{-8}}$$

$$= 710$$

$$\text{Nonmonochromaticity } \Delta\lambda = \frac{\lambda^2}{\Delta};\ \Delta = \text{path diff. for maximum orders}$$

$$= \frac{\lambda^2}{2\mu h} = 5.56 \text{ A}$$

15.188. A beam of monochromatic light of wavelength 5.82×10^{-7} m falls normally on a glass wedge with the wedge angle of 20 sec of an arc. If the refractive index of glass is 1.5, find the number of dark interference fringes per cm of the wedge length. (IAS 1987)

Solution:

$$\theta = \frac{20\pi}{60 \times 60 \times 60} \text{ rad.}$$

$$\beta = \frac{\lambda}{2\theta\mu} = \frac{5.82 \times 10^{-7} \times 60 \times 60 \times 60}{2 \times 20\pi \times 1.5}$$

$$= 0.2 \text{ cm}$$

$$\text{No. of fringes per cm} = \frac{1}{0.2} = 5$$

15.189. When one of the tubes of Rayleigh refractometer is fully evacuated, the other tube remaining filled with air at atmospheric pressure, 150 fringes cross the field of view. The wavelength of light used is 5000 A. The tube is 25 cm long. Calculate the refractive index of air at atmospheric pressure.

(IAS 1983)

Solution:

$$(\mu - 1)\, l = n\lambda$$

$$\therefore \quad \mu = \frac{n\lambda}{l} + 1$$

On substituting the values we have

$$\mu = 1 + \frac{150 \times 5 \times 10^{-7}}{0.25}$$

$$= 1.0003$$

15.190. The arrangement shown in the Fig. 15.118 is an interferrometer used to determine the refractive index of gases. In this set S is a monochromatic source of wavelength 5890Å. A & B are two identical cells of length l = 15 cm. Initially both the cells contains air and interference pattern has been observed on the screen. When the air of the cell A has been replaced by the experimental gas it has been found that the entire pattern shifts upward by N = 20 fringes. Assuming the refractive index of air equal to μ = 1.000277. Determine the refractive index of the gas under investigation.

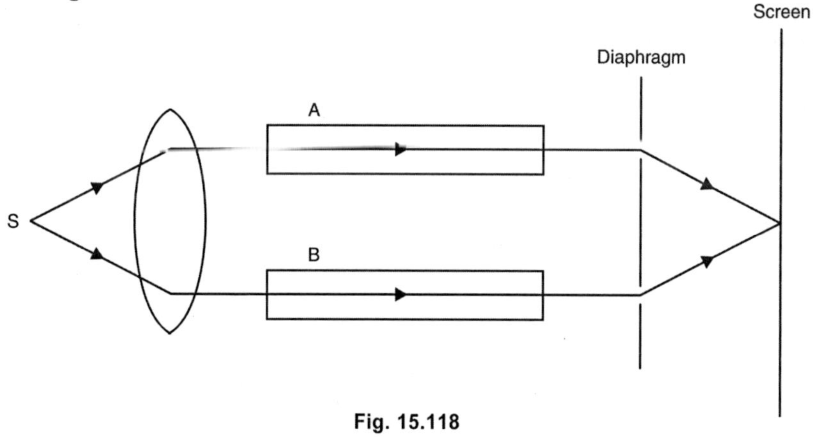

Fig. 15.118

Solution:

For air $(\mu - 1)\, l = n_1\, \lambda$

For gas $(\mu' - 1)\, l = n_2\, \lambda$

$(\mu' - \mu)\, l = (n_2 - n_1)\, \lambda = N\lambda$

or $\quad \mu' = \dfrac{N\lambda}{l} + u$

On substituting the values we have

$$\mu' = \frac{20 \times 589 \times 10^{-9}}{15 \times 10^{-2}} + 1.000277$$

$$= 1.000355$$

15.191. A slit of width d is placed in front of a lens of focal length 0.5 m and is illuminated normally with light of wavelength 5.89×10^{-7} m. The diffraction minima on either side of the central diffraction maximum are separated by 2×10^{-3} m. The width d of the slit is....*m*. (IIT 1997)

Solution:

$$d \sin\theta = n\lambda$$

$$d \cdot \theta = \lambda \quad \text{for small values of } \theta$$

$$\theta = \frac{y}{f}$$

$$\frac{d \cdot y}{f} = \lambda \text{ or } d = \frac{\lambda f}{y}$$

On substituting the values

$$d = \frac{5.89 \times 10^{-7} \times 0.5}{10^{-3}} = 2.95 \times 10^{-4}\,\text{m}$$

15.192. A lamp of intensity 800 cd is suspended 10 m above the ground. Over what area of the ground will the illuminance be at least 1 lx?

Solution: The illuminance at the edge A (see Fig. 15.119) is given by

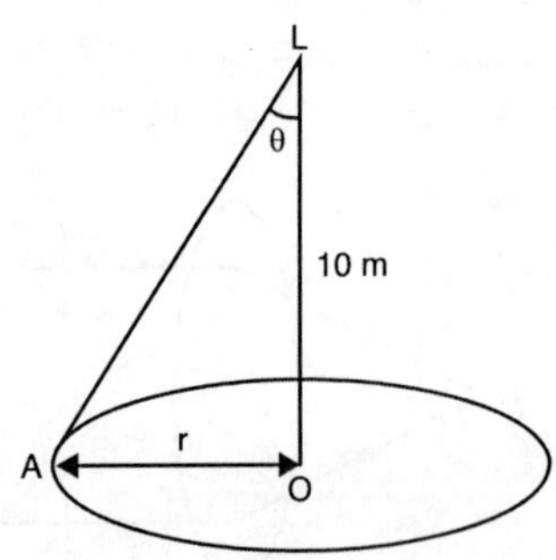

Fig. 15.119

$$E = \frac{I\cos\theta}{LA^2} = \frac{10\,I}{LA^3} = \frac{8000}{(r^2+100)^{3/2}}$$

$$1 = \frac{8000}{(r^2+100)^{3/2}}$$

$$\therefore \quad r^2 + 100 = 400$$

$$\therefore \quad r^2 = 300$$

Area of the ground $\pi r^2 = 3.14 \times 300 = 942\ \text{m}^2$

15.193. Two lamps A and B with intensities $I_1 = 36$ cd and $I_2 = 81$ cd are placed 1 m apart. Where should a screen be placed between them for equalising the intensities?

Solution: Let the intensities be equal at a distance x from lamp A (Fig. 15.120). Then

$$E_1 = \frac{I_1}{x^2} \text{ and } E_2 = \frac{I_2}{(1-x)^2}$$

Since $E_1 = E_2$,

$$\frac{\sqrt{I_1}}{x} = \frac{\sqrt{I_2}}{1-x}$$

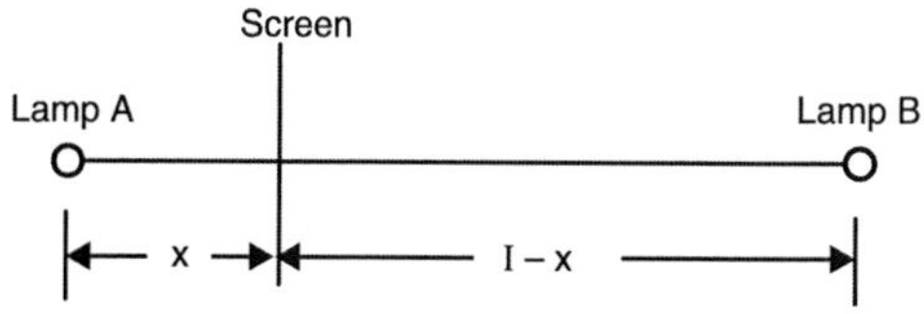

Fig. 15.120

or $$\frac{1-x}{x} = \frac{\sqrt{I_2}}{\sqrt{I_1}}$$

or $$\frac{1}{x} = \frac{\sqrt{I_2}+\sqrt{I_1}}{\sqrt{I_1}}$$

or $$x = \frac{\sqrt{I_1}}{\sqrt{I_1}+\sqrt{I_2}} = \frac{\sqrt{36}}{\sqrt{36}+\sqrt{81}} = \frac{6}{6+9} = \frac{6}{15} = 0.4 \text{ m} = 40 \text{ cm.}$$

15.194. Two lamps A and B of intensities $I_1 = 300$ cd and $I_2 = 400$ cd are suspended from a 4 m high pole (see Fig. 15.121). The distance between the lamps is 3 m. Find out the illuminance at points P, Q and R.

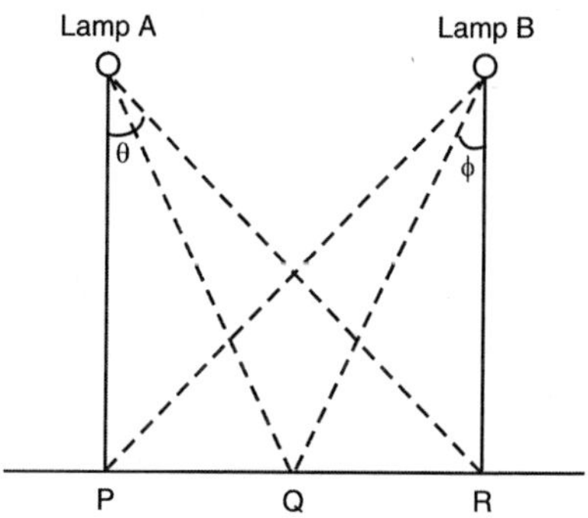

Fig. 15.121

Solution: Illuminance at P due to lamp $A = \frac{300}{16} = 18.75$ lx

Illuminance at P due to lamp B

$$= \frac{400 \cos\theta}{5^2} = \frac{400 \times 4}{125} = 12.8 \text{ lx}$$

Net illuminance at $P = 31.55$ lx

Illuminance at R due to lamp $A = \frac{300\cos\theta}{5^2} = \frac{300 \times 4}{125} = 9.6$ lx

Illuminance at R due to lamp $B = \frac{400}{16} = 25$ lx

Total illuminance at $R = 34.6$ lx

Total illuminance at $Q = \frac{300\cos\phi}{18.25} + \frac{400\cos\phi}{18.25}$

$$= \frac{700}{18.25} \times \frac{4}{\sqrt{18.25}} = 35.91 \text{ lx}$$

15.195. At what height should an electric bulb be hung above the centre of a circular table in order to get a maximum illuminance at its edges?

Solution: In view of Fig. 15.122, the illuminance at A is given by $= I\cos\theta/l^2$

or $$E = \frac{I\cos\theta \times \sin^2\theta}{r^2}$$

In the above equation, I and r are constant. Hence E is a function of θ.

Now $$\frac{dE}{d\theta} = \frac{I}{r^2}\left(\cos\theta \cdot 2\sin\theta\cos\theta - \sin^3\theta\right) = 0$$

or $$2\cos^2\theta = \sin^2\theta$$

or $$2 - 2\sin^2\theta = \sin^2\theta$$

or $$\sin\theta = \sqrt{\frac{2}{3}}$$

$$\theta = \sin^{-1}\sqrt{\frac{2}{3}}$$

Now $$h = r\cot\theta = \frac{r}{\sqrt{2}} = 0.71\,r$$

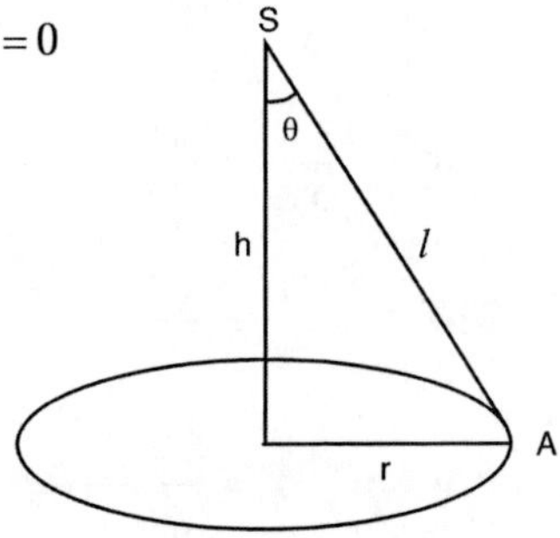

Fig. 15.122

15.196. A screen S is illuminated by two point sources A and B. Another source C sends a parallel beam of light towards the point P on the screen (Fig. 15.123). The line AP is normal to the screen and the line AP, BP and CP are in one plane. The distances AP, BP and CP are 3, 1.5 and 1.5 m respectively. The radiant powers of sources A and B are 90 W and 180 W respectively. The beam coming from C is of intensity 20 W/m². Calculate the intensity at P on the screen. (IIT 1982)

Solution: Intensity at P due to source

$$A = \frac{90}{4\pi \times 9} = \frac{10}{4\pi}\ \text{W/m}^2$$

Intensity at P due to source

$$B = \frac{180\cos 60}{4\pi(1.5)^2} = \frac{10}{\pi}\ \text{W/m}^2$$

Intensity due to source C along CP = $20\cos 60 = 10\ \text{W/m}^2$

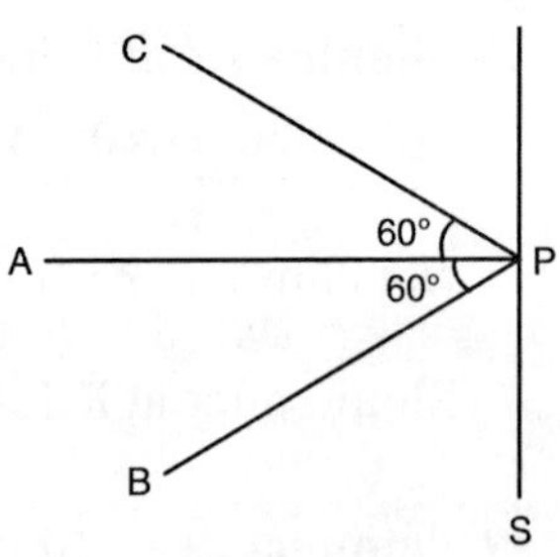

Fig. 15.123

$$\text{Net intensity} = \frac{10}{4\pi} + \frac{10}{\pi} + 10 = 13.981\ \text{W/m}^2$$

15.197. Find the pressure due to radiation of energy density uJ/m³ falling normally on a perfectly black surface placed in its path.

Solution: Let S be the area of the given surface (see Fig. 15.124). The energy passing through this surface per sec will be contained in the volume cSm³ where c is the velocity of light.

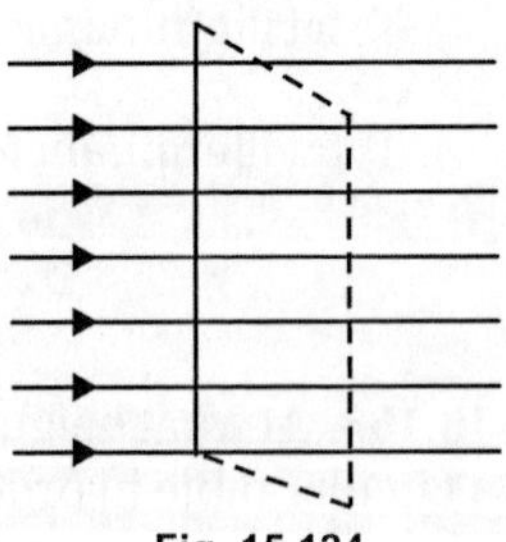
Fig. 15.124

Energy passing through this area per sec = cSu

Momentum associated with it per sec = Su

Change in momentum per sec = force = Su

$$\therefore \quad \text{Pressure} = \frac{Su}{S} = u$$

15.198. If on a surface the radiation is incident from various possible directions it is called a diffused radiation. Calculate the pressure exerted by the diffused radiation of energy density uJ/m^3 on the surface of a perfectly black body.

Solution: Let a parallel beam of radiation of energy density uJ/m^3 be incident on the surface PQ (Fig. 15.125a) of surface area A at an angle of incidence θ.

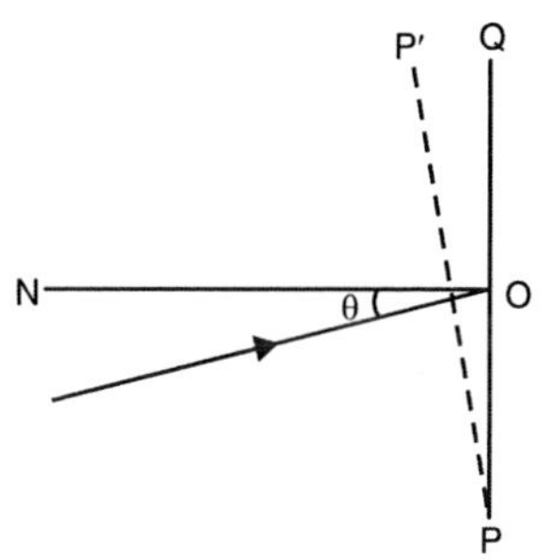

Fig. 15.125a

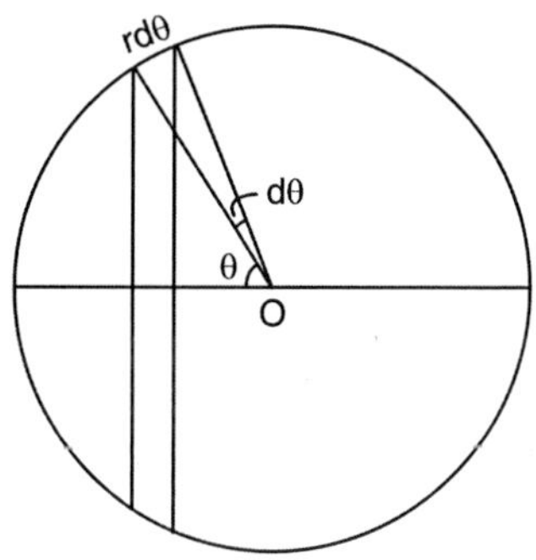

Fig. 15.125b

Energy incident on area PP' per sec = uc

Total energy on surface PP' per sec = $ucA \cos \theta$

Energy incident on PQ per unit area per sec = $uc \cos \theta$

Momentum per sec per unit area = $u \cos \theta$

Change of momentum of plane PQ per sec per unit area = $u \cos \theta$

This momentum is in the direction of the beam which makes an angle θ with the normal to the plane PQ. For calculating pressure we have to calculate the component of this momentum perpendicular to the surface PQ.

$\therefore$ Rate of change of momentum of the plane PQ, normal to its surface = $u \cos \theta \cdot \cos \theta = u \cos^2 \theta$

$\therefore$ Pressure exerted by the radiation on the surface = $u \cos^2 \theta$

In diffused radiation beam is incident from all possible directions hence we will have to consider the average value of $\cos^2 \theta$. Take a sphere of radius r (Fig. 15.125b) and consider all the rays lying between θ and $\theta + d\theta$.

The surface area of the portion of sphere containing rays

$$= 2\pi r \sin \theta \cdot r d\theta; \quad r = \text{radius of the sphere.}$$

Solid angle at centre O by all these rays between θ and $\theta + d\theta$

$$= \frac{2\pi r^2 \sin\theta\, d\theta}{r^2} = 2\pi \sin\theta \cdot d\theta$$

$$\text{Average of } \cos^2\theta = \overline{\cos^2\theta} = \frac{\int_0^\pi \cos^2\theta \cdot 2\pi \sin\theta\, d\theta}{4\pi}$$

$$= \frac{1}{3}$$

Hence pressure due to diffused radiation on surface $PQ = \frac{u}{3}$.

15.199. A plane light wave of intensity $I = 0.25\,\text{W/cm}^2$ strikes a plane mirror of reflectivity $\rho = 0.6$ at an angle $\phi = 30°$ with the mirror. Calculate the normal pressure exerted by the light.

Solution: Let S be the area of the surface AB on which the wave is incident see Fig. 15.126.

Area of the surface $AB' = AB \sin\phi = S \sin\phi$

Energy falling on the surface AB' per sec $= IS \sin\phi$

Momentum per sec associated with the radiation on the surface AB'

$$= \frac{I}{c} S \sin\phi$$

Rate of change of momentum on the plane AB = force $= \frac{I}{c} S \sin^2\phi$

Force after reflection $= \rho \frac{I}{c} S \sin^2\phi$

Net force $\Delta F = (1+\rho)\frac{IS}{c}\sin^2\phi$

Net pressure $p = (1+\rho)\frac{I}{c}\sin^2\phi$

On substituting the values of various terms we have

$$p = \frac{(1+0.6)}{3\times 10^8} \times \frac{0.25}{10^{-4}} \times \frac{1}{4}$$

$$= \frac{0.4 \times 0.25}{3\times 10^4} = 33 \times 10^{-7}\ \text{N/m}^2$$

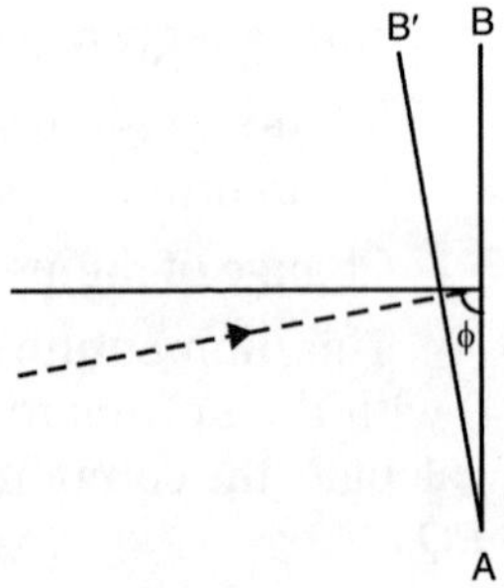

Fig. 15.126

15.200. A pulse laser emitting visible radiation of energy $E = 12$ J at a duration of $t = 0.1$ ms falls normally on a surface of reflectivity $\rho = 0.6$. It is focused as a circular spot of diameter $d = 12$ μm. Calculate the average pressure exerted by the light pulse on the surface.

Solution:

$$\text{Incident momentum} = \frac{E}{c}$$

$$\text{Reflected momentum} = \frac{\rho E}{c}$$

$$\text{Change of momentum} = \frac{E}{c}(1+\rho)$$

$$\text{Force on the surface} = \frac{E}{ct}(1+\rho)$$

$$\text{Pressure } p = \frac{4E(1+\rho)}{ct \cdot \pi d^2}$$

On substituting the values we have

$$p = \frac{4 \times 12 \times 1.6}{3 \times 10^8 \times 10^{-4} \times 3.14 \times 144 \times 10^{-12}}$$

$$= 0.0566 \times 10^8 \text{ N/m}^2$$

$$= 56.6 \text{ atmosphere}$$

15.201. A narrow and parallel beam of short light pulse of energy $E = 10$ J falls on a horizontally placed mirror of reflectivity $\rho = 0.5$ at an angle of $\phi = 60°$ to the mirror. Calculate the momentum imparted to the mirror.

Solution: In view of Fig. 15.127 change in momentum along x axis

$$= \rho \frac{E}{c} \cos\phi - \frac{E}{c} \cos\phi$$

$$= \frac{E}{c} \cos\phi\,(\rho - 1)$$

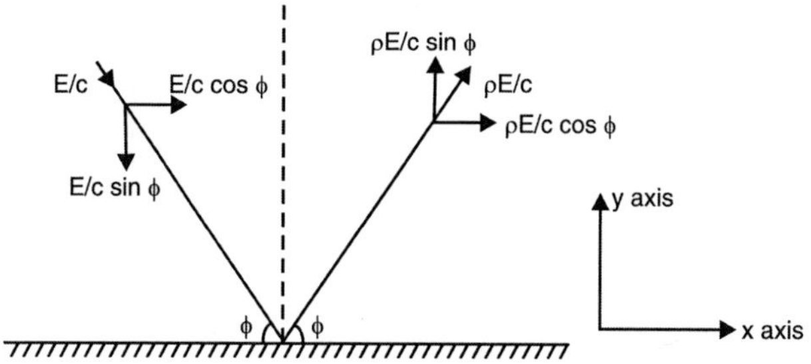

Fig. 15.127

$$\text{Change in momentum along } Y \text{ axis } = \rho \frac{E}{c} \sin\phi + \frac{E}{c} \sin\phi$$

$$= \frac{E}{c} \sin\phi\,(\rho + 1)$$

Net change in momentum = momentum imparted to the mirror

$$\Delta p = \frac{E}{c}\sqrt{(\rho-1)^2\cos^2\phi + (\rho+1)^2\sin^2\phi}$$

or $$\Delta p = \frac{E}{c}\sqrt{\rho^2 - 2\,\rho\cos 2\phi + 1}$$

On substituting the values of the various terms we have

$$\Delta p = \frac{10}{3\times 10^8}\sqrt{0.25 - 2\times 0.5\cos 120 + 1}$$

$$= \frac{10}{3}\sqrt{1.75}\times 10^{-8}$$

$$= 4.409\times 10^{-8}\ \text{Nsec.}$$

15.202. A small mirror with a mass m is suspended by a weighless thin quartz fibre of length l. A short laser flash of energy E travelling horizontally strikes the mirror at right angles to it (see Fig. 15.128a). Calculate the angle θ through which the quartz fibre will be deflected.

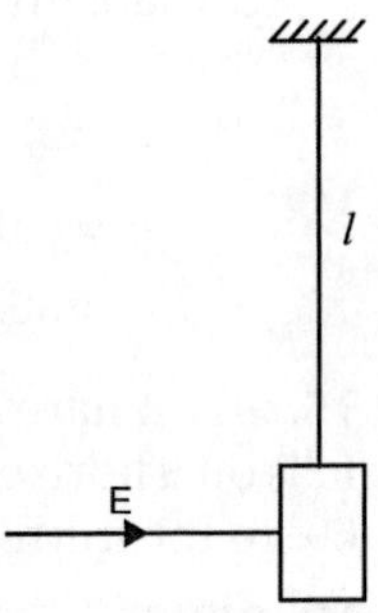

Fig. 15.128a

Solution: Let E' be the energy of the flash after reflection, then according to the conservation of momentum we have (see Fig. 15.128b)

$$\frac{E}{c} = \frac{-E'}{c} + mv \quad \text{... (1)} \quad m = \text{mass of mirror}$$

v = velocity of mirror

According to conservation of energy we have

$$E = E' + \frac{1}{2}mv^2 \quad \text{... (2)}$$

Using these two relations we have

$$\frac{2E}{c} = mv + \frac{1}{2}\frac{mv^2}{c} \quad \text{... (3)}$$

The contribution by the second term in equation 3 is negligible hence

$$2E = mvc$$

If h be the height through which the mirror rises then

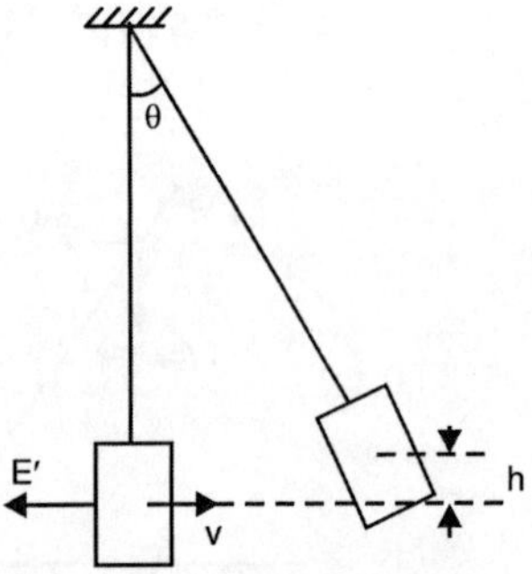

Fig. 15.128b

$$\frac{1}{2}mv^2 = mgh = mg\,(l - l\cos\theta) = 2\,mgl\sin^2\frac{\theta}{2}$$

or $$v = 2\sqrt{gl}\,\sin\frac{\theta}{2}$$

$$\text{or} \quad \frac{2E}{mc} = 2\sqrt{gl}\,\sin\frac{\theta}{2}$$

$$\text{or} \quad \sin\frac{\theta}{2} = \frac{E}{mc\sqrt{gl}}$$

$$\text{or} \quad \theta = 2\sin^{-1}\left[\frac{E}{mc\sqrt{gl}}\right]$$

15.203. A sphere of radius r = 4 cm is placed in the path of a plane light wave of intensity I = 0.8 W/cm². Calculate the force that light exerts on the sphere when the reflection coefficient is equal to one.

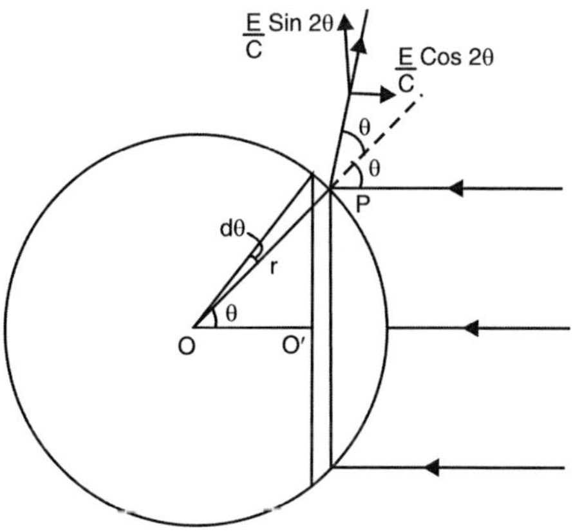

Fig. 15.129

Solution: Consider a small strip of area dA. This strip is a ring of radius $r \sin\theta$ and width $rd\theta$ (Fig. 15.129).

$$dA = 2\pi r \sin\theta \cdot r d\theta$$

Energy of light falling on this strip in time dt is given by

$$E = I dt \cdot dA \cos\theta$$

$$\text{Incident momentum} = \frac{E}{c} = \frac{I}{c} dt\, dA \cos\theta$$

In view of the figure the reflected momentum will have two components $\frac{E}{c}\cos 2\theta$ along the horizontal and $\frac{E}{c}\sin 2\theta$ along the vertical.

$$\therefore \quad \text{Change in momentum} = dp = \sqrt{\frac{E^2}{c^2}(1+\cos 2\theta)^2 + \frac{E^2}{c^2}\sin^2 2\theta}$$

$$= \frac{2E}{c}\cos\theta$$

$$= \frac{2I}{c} \cdot dt \cdot dA \cdot \cos^2\theta$$

$$\therefore \quad \frac{dp}{dt} = \text{Force} = \frac{2I}{c} dA \cos^2\theta$$

This force is along PO. The resultant force should be along OO'

$$\text{Required force on the strip} = \frac{2I}{c} \cdot dA \cos^2\theta \cdot \cos\theta$$

$$= \frac{2I}{c} \cdot 2\pi r^2 \cos^3\theta \sin\theta d\theta$$

$$\text{Resultant force } F = \frac{4\pi r^2}{c} I \int_0^{\pi/2} \cos^3\theta \cdot \sin\theta\, d\theta$$

$$= \frac{\pi r^2 I}{c}$$

On substituting the values of various terms we have

$$F = \frac{3.14 \times 16 \times 10^{-4} \times 0.8}{3 \times 10^8 \times 10^{-4}} = 13.397 \times 10^{-8}\,\text{N}$$

15.204. A perfectly black particle P of density $\rho = 2 \times 10^3$ kg/m³ is illuminated by a plane wave of intensity $I = 0.98$ W/cm² by putting a source of light S at the focus of a parabolic mirror (see Fig. 15.130). Calculate the dimensions of the particle when it is in equilibrium with earth gravitational force.

Solution:

Force due to radiation = $\frac{\pi r^2 I}{c}$; r = radius of the particle (treating it as a sphere)

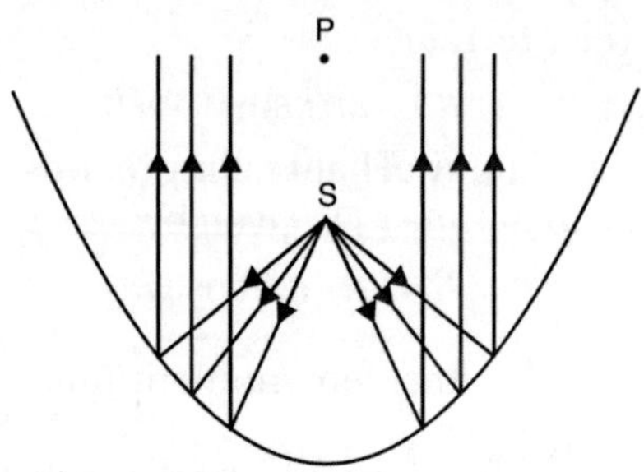

Fig. 15.130

$$\text{Now} \quad \frac{\pi r^2 I}{c} = \frac{4}{3}\pi r^3 \rho \cdot g$$

$$\text{or} \quad r = \frac{3I}{4\rho c g}$$

On putting the values of various terms we have

$$r = \frac{3}{4 \times 2 \times 10^3} \times \frac{0.98}{10^{-4}} \times \frac{1}{3 \times 10^8 \times 9.8}$$

$$= 12.5 \times 10^{-10}\,\text{m}$$

15.205. A perfectly black particle in between earth and sun is illuminated by sun light. The force on the particle due to the light is balanced by the sun's gravitational force. Assuming the density of the particle to be $\rho = 1.92 \times 10^3$ kg/m³ find the dimensions of the particle. Solar constant S = 1.36 kw/m². Mass of sun $m_s = 1.98 \times 10^{30}$ kg and distance between earth and sun $R_0 = 149.46 \times 10^9$ m.

Solution: Let the particle be at a distance R from sun the gravitational force on the particle is given by

$$F = \frac{G \cdot m \cdot m_s}{R^2}\ ;\ m = \text{mass of particle and } m_s = \text{mass of sun}$$

In equilibrium

$$\frac{G \cdot m \cdot m_s}{R^2} = \frac{\pi r^2 I}{c}\ ;\ r = \text{radius of the particle (treating it as a sphere)}$$

Further $IR^2 = SR_0^{\,2}$ where R_0 is the distance between earth and sun.

$$\frac{G \cdot m \cdot m_s}{R^2} = \frac{\pi r^2 \cdot SR_0^2}{cR^2}$$

or $$G \cdot \frac{4}{3} \pi r^3 \rho m_s = \frac{\pi r^2 SR_0^2}{c}$$

or $$r = \frac{3SR_0^2}{4Gc\, m_s \rho}$$

On substituting the values of various quantities we have

$$r = \frac{3 \times 1.36 \times 10^3 \times (149.46)^2 \times 10^{18}}{4 \times 6.672 \times 10^{-11} \times 3 \times 10^8 \times 1.98 \times 10^{30} \times 1.92 \times 10^3}$$

$$= \frac{1.36 \times (149.46)^2 \times 10^{-9}}{4 \times 6.672 \times 1.98 \times 1.92}$$

$$= 0.299 \times 10^{-6} \text{ m} = 0.3\ \mu\text{m}$$

15.206. In darkness the human eye is highly sensitive and at a wavelength of 5600 A it can deduct to a minimum of 70 photons per sec. Estimate the power of a light source placed at a distance of $R = 8$ km from such an eye having pupil diameter of $d = 6$ mm.

Solution: Energy at wavelength λ is equal to

$$\frac{hc}{\lambda}$$

Total energy received per sec by an eye = $\dfrac{nhc}{\lambda}$

Energy per unit area per sec = $\dfrac{4nhc}{\lambda \cdot \pi d^2}$

Let P be the power then

$$\frac{P}{4\pi R^2} = \frac{4nhc}{\lambda \cdot \pi d^2}$$

$$\therefore \quad P = \frac{16nhc \cdot R^2}{\lambda \cdot d^2}$$

On substituting the values of various terms we have

$$P = \frac{16 \times 70 \times 6.62 \times 10^{-34} \times 3 \times 10^8 \times 64 \times 10^6}{56 \times 10^{-8} \times 36 \times 10^{-6}}$$

$$= 706.1 \times 10^{-6} \text{ watt.}$$

16

Atomic Physics

16.1. A beam of cathode rays is passed through an electric field as shown in Fig. 16.1. Under the influence of the electric field the cathode rays deviate from their path and after emerging from the field the particles moves along a straight line and strikes the screen at O'. The final deflection of the beam is as though the deviation had taken place at the point O. Find the distance OB.

Solution: Let m be the mass and v be the velocity of the particles constituting cathode rays. Initially they are moving parallel to the length of the parallel-plate capacitor being used for applying the electric field. After entering into the field of strength E the particles will experience an acceleration Ee/m. The velocity of the particles in the direction BB' is given by

$v_1 = at = \frac{Ee}{m}t = \frac{Ee}{m}\frac{l}{v}$; where a is acceleration and l is the length of the plate.

The displacement y in the direction

$$BB' = \frac{1}{2}at^2$$

$$\therefore \quad y = \frac{1}{2}\frac{Ee}{m}\left(\frac{l}{v}\right)^2; t = \frac{l}{v}$$

$$\text{Now } \frac{y}{OB} = \frac{v_1}{v} = \tan\theta$$

$$\therefore \quad OB = y\frac{v}{v_1} = \frac{1}{2}\frac{Ee}{m}\frac{l^2}{v^2}\frac{v}{Eel}mv = \frac{l}{2}$$

OB = half of the lenth of the plate of the capacitor.

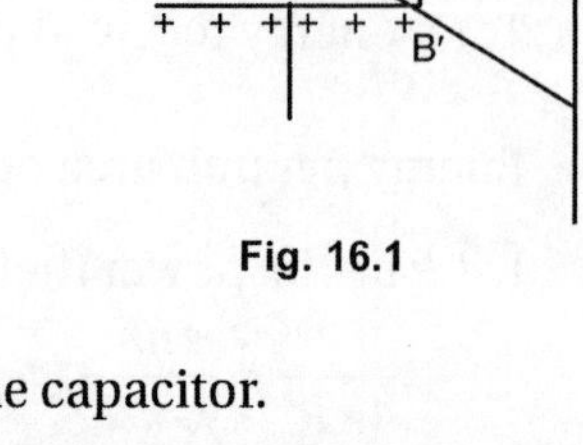

Fig. 16.1

16.2. An electric field of 2×10^4 N has been produced between the 1 cm long plates of parallel plate capacitor. An electron enters this field in a direction perpendicular to it with a kinetic energy 3.2×10^{-16} J. Calculate the displacement of the electron.

Solution: In the electric field of strength E, the electron experiences a force $= \frac{Ee}{m}$.

The displacement of the electron in a direction perpendicular to its initial direction of motion is given by

$$y = \frac{1}{2}at^2 = \frac{1}{2}\frac{Ee}{m}t^2$$

$$= \frac{1}{2}\frac{Ee}{m}\left(\frac{l}{v}\right)^2 \quad \text{where } l \text{ is the length of the plates of the capacitor}$$

$$= \frac{Eel^2}{4 \times \text{kinetic energy of the electron}}$$

$$= \frac{2\times 10^4 \times 1.6\times 10^{-19}\times 10^{-4}}{4\times 3.2\times 10^{-16}} = 0.25 \text{ mm}$$

16.3. An electron of mass 9×10^{-31} kg moving with a velocity of 4.8×10^6 m/s enters into a region in which there is an electric field of 900 N/C in a direction parallel to the field. In this state of motion the electron experiences a retarding force. Calculate (i) the distance travelled by the electron in the field before coming to rest for a moment (ii) the time that will elapse and (iii) the percentage loss in its initial energy after traversing a distance of 1 cm in the field.

Solution: The retardation acting on the electron is given by

$$a = \frac{Ee}{m} = \frac{900\times 1.6\times 10^{-19}}{9\times 10^{-31}} = 1.6\times 10^{14}\,\text{m/s}^2$$

$v^2 = 0 = u^2 - 2\,as$ where v is the final velocity and u is the initial velocity.

$$s = \frac{u^2}{2a} = \frac{4.8\times 4.8\times 10^{12}}{2\times 1.6\times 10^{14}} = 7.2 \text{ cm}$$

(ii) $v = 0 = u - at$

$$\therefore \quad t = \frac{u}{a} = \frac{4.8\times 10^6}{1.6\times 10^{14}} = 3\times 10^{-8}\,\text{s}$$

(iii) $v^2 = u^2 - 2\,as$; v is the velocity of the electron after traversing a distance of 1 cm.

$$= (4.8 \times 10^6)^2 - 2 \times 1.6 \times 10^{14} \times 10^{-2}$$

$$= (4.8^2 - 3.2) \times 10^{12} = 19.84 \times 10^{12}$$

$$\therefore \quad v = 4.454 \times 10^6 \text{ m/s}$$

$$\text{Fractional loss in kinetic energy} = \frac{1/2\,mu^2 - 1/2\,mv^2}{1/2\,mu^2}$$

$$= \left[1 - \left(\frac{v}{u}\right)^2\right]$$

$$= 1 - 0.861 = 0.139$$

Percentage loss in energy = 13.9%

16.4. Two plates each of length l = 8 cm separated by d = 1.5 cm of a parallel plate capacitor produce an electric field $E = 4.5 \times 10^3$ N/C. In this field an electron of mass $m = 9 \times 10^{-31}$ kg is projected with a velocity of 8×10^6 m/s at an angle of $\theta = 45°$ with the horizontal (see Fig. 16.2). Where does the electron strikes either of the plates?

Solution: The acceleration acting on the electron is given by

$$a = \frac{Ee}{m} = \frac{4.5 \times 10^3 \times 1.6 \times 10^{-19}}{9 \times 10^{-31}} = 8 \times 10^{14} \text{ m/s}^2 = 8 \times 10^{16} \text{ cm/s}^2$$

Under the influence of the electric field the electron describes a parabolic path. The equation of motion is given by

$$y = x \tan\theta - \frac{ax^2}{2v^2 \cos^2\theta}$$

Where y is vertical and x is the horizontal displacement (the distance after traversing which the electron strikes the upper plate)

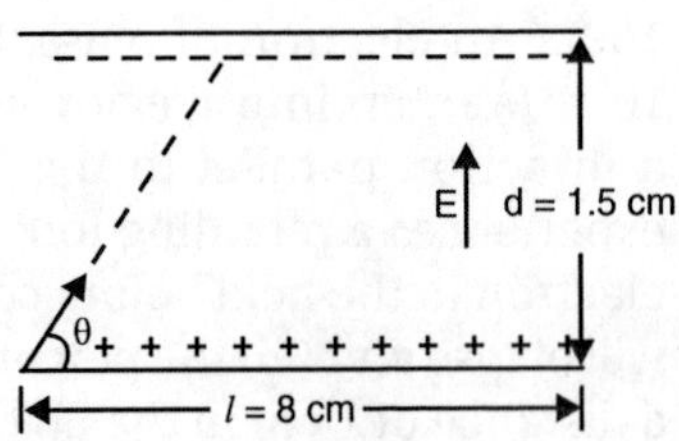

Fig. 16.2

$$1.5 = x - \frac{8 \times 10^{16} x^2}{2 \times 64 \times 10^{16} \times (1/2)}$$

or $\quad x^2 - 8x + 12 = 0$

$$(x - 2)(x - 6) = 0$$

$$x = 2, \text{ and } 6 \text{ cm}$$

16.5. An electron flies freely with a velocity 4×10^7 m/s. At an instant $t = 0$ an electric field $E = 4.5 \times 10^3$ N/C making an angle of 30° with the direction of the velocity is applied on it. Find: (i) the radius of curvature of the path at the point where the electron has minimum velocity, and (ii) the increment in the momentum of the electron in time 0.5 s. Mass of electron 9×10^{-31} kg.

Solution:

(i) At the point where the velocity is minimum, the centripetal force is balanced by the electric force, i.e.,

$$\frac{mv^2 \sin^2\theta}{r} = eE; \quad r = \text{radius of curvature}$$

$$\therefore \quad r = \frac{mv^2 \sin^2\theta}{eE}$$

On substituting the values, we have

$$r = \frac{9 \times 10^{-31} \times 16 \times 10^{14}}{4 \times 1.6 \times 10^{-19} \times 4.5 \times 10^3}$$

$$= 0.5 \text{ m}$$

(ii) $\frac{\Delta p}{t} = (-eE - 0) = -eE$; Δp = increment in momentum

$\therefore \quad \Delta p = -eEt = -1.6 \times 10^{-19} \times 4.5 \times 10^{3} \times 0.5$

$= -3.6 \times 10^{-16}$ Ns

16.6. A proton is travelling along the x-axis with a velocity of 4×10^6 m/s. Calculate (i) force experienced in a magnetic field $B = 0.2\,\hat{j} + 0.4\,\hat{k}$ W/m² and (ii) acceleration in an electric field $E = 2\,\hat{i} + \hat{j}$ V/cm. Mass of proton = 1.67×10^{-27} kg.

Solution:

Force $q\,(V \times B) = 1.6 \times 10^{-19}\,[4 \times 10^6\,\hat{i} \times (0.2\,\hat{j} + 0.4\,\hat{k})]$

$= 1.6 \times 10^{-19}\,[0.8 \times 10^6\,\hat{k} - 1.6 \times 10^6\,\hat{j}\,]$

$= 1.6 \times 0.8 \times 10^{-13}\,(\hat{k} - 2\,\hat{j})$

Magnitude of the force $= 1.28 \times 10^{-13} \times \sqrt{5} = 2.863 \times 10^{-13}$ N

(ii) In an electric field, we have $ma = qE$

$$\therefore \quad a = \frac{qE}{m} = \frac{1.6 \times 10^{-19}\left(2\hat{i} + \hat{j}\right)}{1.67 \times 10^{-27} \times 10^{-2}}$$

$$= \frac{1.6}{1.67} 10^{10} \left(2\hat{i} + \hat{j}\right)$$

$$= \frac{1.6}{1.67} 10^{10} \sqrt{5}$$

$= 2.14 \times 10^{10}$ m/s²

16.7. At an instant $t = 0$ an electron starts moving from one of the plates of a parallel capacitor without any initial velocity under the action of time varying voltage $V = \alpha t$ where $\alpha = 400$ V/s. The separation between the two plates is $l = 2.5$ cm. What velocity the electron will acquire when it reaches the other plate. Mass of electron = 9×10^{-31} kg.

Solution:

$$V = \alpha t$$

$$\frac{dV}{dx} \cdot \frac{dx}{dt} = \alpha$$

$$\frac{dV}{dx} = \frac{\alpha}{v}$$

Force acting on electron $= \frac{edV}{dx} = \frac{e\alpha}{v}$

Acceleration acting on the electron $= \dfrac{e\alpha}{mv}$

Let the time taken by the electron to reach the other plate be t. The equation of motion of the electron is given by

$$l = 0 + \frac{1}{2}\frac{e\alpha}{mv}\cdot t^2$$

or $$t = \left(\frac{2l\,mv}{e\alpha}\right)^{1/2}$$

Now $$V = \alpha\left(\frac{2lmv}{e\alpha}\right)^{1/2} = \left(\frac{2\alpha lmv}{e}\right)^{1/2}$$

$$\frac{dV}{dx} = \frac{\alpha}{v} = \left(\frac{2\alpha lm}{e}\right)^{1/2} \times \frac{1}{2}v^{-1/2}\cdot\frac{dv}{dx}$$

$$2\sqrt{\alpha}\int_0^l dx = \left(\frac{2lm}{e}\right)^{1/2}\int_0^v v^{1/2}\cdot dv$$

$$\frac{3}{2}\cdot 2\sqrt{\alpha}\cdot l = \left(\frac{2lm}{e}\right)^{1/2}\cdot v^{3/2}$$

$$3\sqrt{\frac{\alpha le}{2m}} = v^{3/2}$$

or $$\left(\frac{9e\alpha l}{2m}\right) = v^3$$

or $$\left(\frac{9e\alpha l}{2m}\right)^{1/3} = v$$

On substituting the values of various terms we have

$$v = \left(\frac{9\times 1.6\times 10^{-19}\times 400\times 2.5\times 10^{-2}}{2\times 9\times 10^{-31}}\right)^{1/3}$$

$$= 20 \text{ km/sec.}$$

16.8. At the moment $t = 0$ a proton accelerated through a potential difference V enters into the space between the parallel plate capacitor parallel to the direction of its length l. In the region between the plates there exists an electric field which varies linearly with time via equation $E = \alpha t$ where α is a constant. At what angle with respect to initial direction of motion the proton will emerge out of the capacitor?

Solution: The velocity of the proton before entering into the electric field is obtained by

$$\frac{1}{2}mv_0^2 = eV$$

$$\therefore \quad v_0 = \sqrt{\frac{2eV}{m}}$$

Time taken by the proton to travel through length l is given by

$$t = \frac{l}{v_0} = \frac{l\sqrt{m}}{\sqrt{2eV}}$$

Force acting on proton = $eE = e\,\alpha\, t = e\alpha l \cdot \sqrt{\frac{m}{2eV}}$

Acceleration acting on the proton = $\frac{e\alpha l}{\sqrt{2meV}}$

The displacement of the proton in a direction perpendicular to initial motion is given by

$$y = \frac{1}{2} \cdot \frac{e\alpha l}{\sqrt{2meV}} \times \frac{l^2 m}{2eV}$$

$$\tan\theta = \frac{y}{l} = \frac{\alpha l^2}{4} \cdot \left(\frac{m}{2eV^3}\right)^{1/2}$$

16.9. A particle of charge q and mass m lying at the origin moves along x axis under the action of a variable electric field given by $E = E_0 - \alpha x$ where α is a positive constant. Calculate (i) The distance travelled by the particle till the moment it comes to a standstill (ii) The acceleration of the particle at that moment.

Solution:

(a) $\quad E = E_0 - \alpha x$

$$\int dV = -\int (E_0 - \alpha x)\, dx$$

$$V = -E_0 x + \frac{\alpha x^2}{2}$$

In the conditions of standstill $V = 0$ hence

$$x = \frac{2E_0}{\alpha}$$

(b) $$K \cdot E = \frac{1}{2} m v^2 = \left(-E_0 x + \frac{\alpha x^2}{2}\right) q$$

$$mv \cdot \frac{dv}{dt} = -E_0 q \cdot v + \alpha x v q$$

$$m\frac{dv}{dt} = -E_0 q + \alpha x q$$

$$= -E_0 q + \alpha q \cdot \frac{2E_0}{\alpha} = E_0 q$$

$$\therefore \quad \frac{dv}{dt} = \text{acceleration} = E_0 q/m$$

16.10. Two square plates each of length l and separated by a distance h forms a parallel plate capacitor. An electric field is established between them (see Fig. 16.3). From the lower plate an electron enters into this field with an initial speed v at an angle θ. At what angle and with what field intensity the electron will pass between the two plates without striking either of them?

Solution: Let E be the electric field between the two plates. For the parabolic path into the field we have

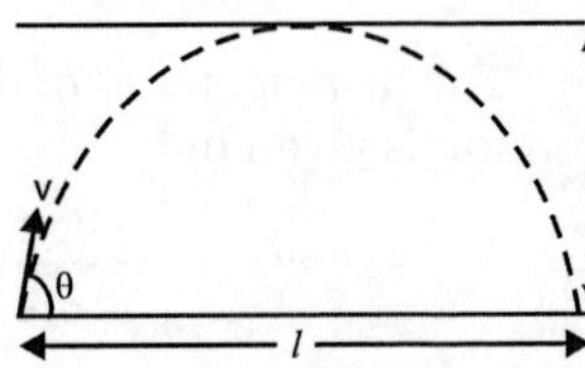

Fig. 16.3

$$v \sin\theta - \frac{Ee}{m} \cdot t = 0$$

$$\therefore \quad t = \frac{mv \sin\theta}{Ee}$$

$$O = v^2 \sin^2\theta - \frac{2Ee}{m} \cdot h$$

$$\therefore \quad E = \frac{mv^2 \sin^2\theta}{2eh}$$

Further

$$v \cos\theta \cdot \frac{2\,mv \sin\theta}{Ee} = l$$

$$\text{or} \quad E = \frac{mv^2 \sin 2\theta}{el}$$

$$\frac{mv^2 \sin^2\theta}{Ee} \bigg/ \frac{2mv^2 \sin\theta \cos\theta}{eE} = \frac{2h}{l}$$

$$\tan\theta = \frac{4h}{l}$$

Therefore we must have the following intensity

$$O < \tan\theta < \frac{4h}{l}$$

$$\text{and} \quad \frac{mv^2 \sin^2\theta}{2eh} < E < \frac{mv^2 \sin 2\theta}{el}$$

16.11. A particle of charge q and mass m moving under the influence of a uniform electric field $E\hat{i}$ and a uniform magnetic field $B\hat{k}$ follows a trajectory from P to Q as shown in the Fig. 16.4. The velocities at P and Q are $v\hat{i}$ and $-2v\hat{j}$ which of the following statement(s) is/are correct?

(a) $E = \dfrac{3}{4}\dfrac{mv^2}{qa}$

(b) Rate of work done by the electric field at P is $\frac{3}{4}\frac{mv^3}{a}$

(c) Rate of work done by the electric field at P is zero.

(d) Rate of work done by both the fields at Q is zero. (IIT 1991)

Solution:

(a) In view of conservation of energy

$$\frac{1}{2}m(2v)^2 - \frac{1}{2}mv^2 = qE\cdot 2a$$

$$\frac{3}{2}mv^2 = 2qaE$$

$$\therefore\ E = \frac{3}{4}\frac{mv^2}{q\cdot a}$$

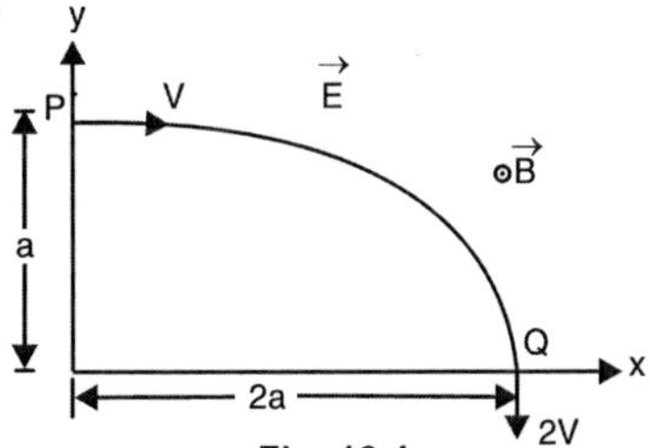

Fig. 16.4

(b) Rate of work done $= F\cdot v$

(c) Rate of work done by electric force at $P = qE\cdot v = qv\cdot\frac{3}{4}\frac{mv^2}{qa}$

$$= \frac{3}{4}\frac{mv^3}{a}$$

(d) Rate of work done at Q by electric field $= qE\hat{i}\cdot\left(-2v\hat{j}\right) = 0$

Rate of work done by magnetic field at Q

$$= \left\{q\left(-2v\hat{j}\right)\times B\hat{k}\right\}\cdot 2v\hat{j}$$

$$= -4qv^2B\left(\hat{j}\times\hat{k}\right)\cdot\hat{j}$$

$$= -4qv^2B\,\hat{i}\cdot\hat{j} = 0$$

16.12. An electron gun G emits electron of energy 2 kev travelling in the positive x direction. The electrons are required to hit the spot S where $GS = 0.1$ m and the line GS makes an angle of 60° with the x-axis as shown in Fig. 16.5. A uniform magnetic field B parallel to GS exists in the region outside the electron gun. Find the minimum value of B needed to make the electron hit S. (IIT 1993)

Solution:

Kinetic energy $\frac{1}{2}mv^2 = 2\times 10^3\times 1.6\times 10^{-19}$

or $mv = \left[2\times 9.1\times 10^{-31}\times 2\times 10^3\times 1.6\times 10^{-19}\right]^{1/2}$ ... (1)

In the magnetic field we have

$$Bev\sin 60 = \frac{mv^2}{R}$$

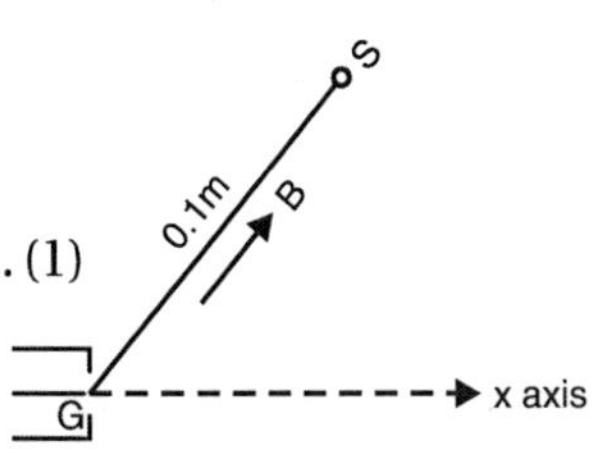

Fig. 16.5

$\therefore \quad mv = BeR \sin 60$

$$= B \times 1.6 \times 10^{-19} \times 0.05 \times \frac{\sqrt{3}}{2} \qquad \text{... (2)}$$

Using the two equation we have

$$\left(2 \times 9.1 \times 10^{-31} \times 2 \times 1.6 \times 10^{-16}\right)^{1/2} = B \times 0.8 \times 10^{-19} \times 0.05 \times \sqrt{3}$$

$$B = 3.48 \times 10^{-3} \text{ Tesla.}$$

16.13. A potential of $V = 300$ volt is applied across a parallel plate capacitor of length $l = 10$ cm and plate separation $d = 2$ cm. An electron enters into the electric field of the capacitor at a height of 1 cm above the lower plate and parallel to the plates with an initial velocity v_0 as shown in the Fig. 16.6. Find the maximum value of v_0 at which the electron will not fly out of the capacitor.

Solution: The vertical displacement of the electron is given by

$$y = \frac{1}{2}at^2 = \frac{1}{2}\frac{eV}{md}\left(\frac{l}{v_0}\right)^2$$

The electron will not fly out of the capacitor if

$$y \geq \frac{d}{2}$$

In the light of this

$$\frac{d}{2} = \frac{eV}{2md}\left(\frac{l}{v_0}\right)^2$$

or $\quad v_0 \leq l\sqrt{\frac{eV}{md^2}}$

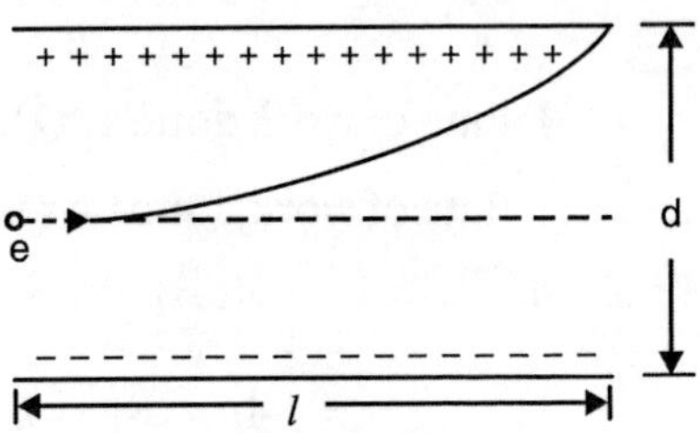

Fig. 16.6

On putting the values of the various terms we have

$$v_0 = 0.1\sqrt{\frac{1.6 \times 10^{-19} \times 300}{9 \times 10^{-31} \times 4 \times 10^{-4}}}$$

$$= 3.65 \times 10^7 \text{ m/s.}$$

16.14. An electron beam enters into a region of uniform electric field of 15 KV/m created between the plates of a parallel plate capacitor of length $l = 10$ cm at an angle of $\theta = 45°$ to the plates and leaves it at an angle $\phi = 30°$ as in Fig. 16.7. Neglecting the effect of the earth's gravitational field find the initial velocity of the electrons.

Solution: Time taken by the electron is given by

$$t = \frac{l}{v_x}; \; v_x = \text{horizontal component of velocity}$$

Acceleration along y axis $= \dfrac{-eE}{m}$

For velocity along y axis we have

$$v_x \tan\phi = v_x \tan\theta - \frac{eE}{m}\cdot\frac{l}{v_x}$$

or $$v_x^2(\tan\theta - \tan\phi) = \frac{eEl}{m}$$

or $$v_0^2\cos^2\theta(\tan\theta - \tan\phi) = \frac{eEl}{m}$$

or $$v_0 = \sqrt{\frac{eEl}{m\cos^2\theta(\tan\theta - \tan\phi)}}$$

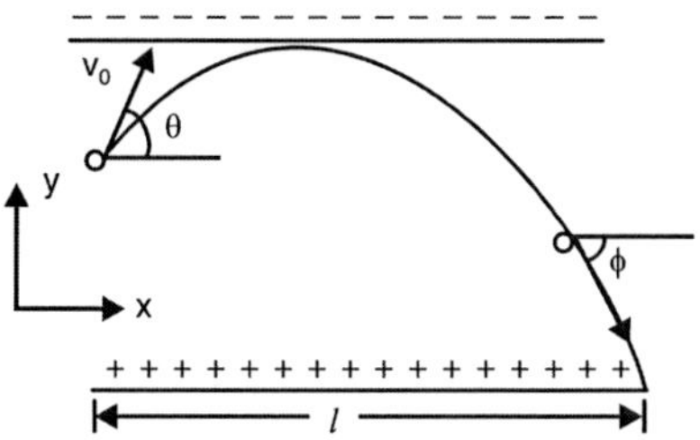

Fig. 16.7

On substituting the values of various terms we have

$$v_0 = \sqrt{\frac{1.6\times10^{-19}\times15\times10^3\times0.1}{9\times10^{-31}\cos^2 45(1-\tan 30)}}$$

$$= \sqrt{\frac{2.4\times10^{-16}}{4.5\times10^{-31}(0.4226)}}$$

$$= \sqrt{\frac{24}{4.5\times0.4226}}\times10^7$$

$$= 3.55\times10^7 \text{ m/s}$$

16.15. An electron flies into a horizontally placed parallel plate capacitor at an equal distance from them. The plates are 20 cm long and the separation between them is $d = 2$ cm. An electric field of $E = 1$ V/cm is established between the plates. Calculate (i) the time in which the electron will reach on one of the plates after it has entered into the field (ii) the distance from the beginning at which the electron will strike the plate if it has been accelerated by a potential of 20 volt.

Solution: For the displacement along the vertical direction we have

$$10^{-2} = \frac{1}{2}\frac{eE}{m}t^2$$

or $$t = \sqrt{\frac{2\times10^{-2}\times m}{eE}}$$

$$= \sqrt{\frac{2\times10^{-2}\times9\times10^{-31}}{1.6\times10^{-19}\times100}}$$

$$= 3.35\times10^{-8}\text{ sec}$$

$$\frac{1}{2}mv^2 = eV$$

$\therefore$ $$v = \sqrt{\frac{2eV}{m}}$$

$$= \sqrt{\frac{2 \times 1.6 \times 10^{-19} \times 20}{9 \times 10^{-31}}}$$

$$= \frac{8}{3} \times 10^{6} \text{ m/sec}$$

$$l = t \cdot v = 3.35 \times 10^{-8} \times \frac{8}{3} \times 10^{6}$$

$$= 8.9 \text{ cm.}$$

16.16. Two similar slits S_1 and S_2 of width d (Fig. 16.8) have been placed in an evacuated vessel and separate out a sheet electron beam with energy W. How far from the slit S_2 will the beam's width have doubled due to the Coulomb repulsion? The electron current/length is J.

Solution: Electric field at the boundary of the beam is given by

$$E = \frac{ne}{2\epsilon_0} = \frac{J}{2\epsilon_0 v}$$

where n = number of electrons/area of the sheet and v is the velocity

Time required to double the width

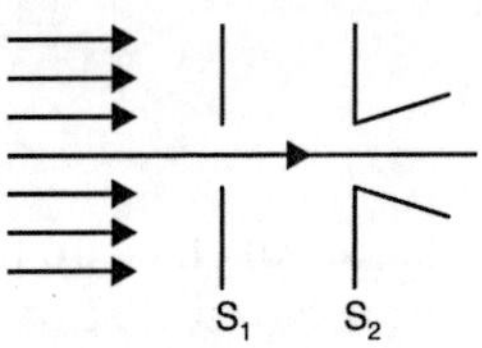

Fig. 16.8

$$t = \sqrt{\frac{2d}{a}}; \; a = \text{acceleration}$$

$$= \sqrt{\frac{2d \times 2m\epsilon_0 v}{eJ}}$$

$$= \sqrt{\frac{4\epsilon_0 d \cdot v}{J} \cdot \frac{m}{e}}$$

$$x = v \cdot t = \sqrt{\frac{4\epsilon_0 d}{J} \cdot \frac{m}{e}} \cdot v^{3/2}$$

$$\therefore \quad x = \sqrt{\frac{4\epsilon_0 d}{J} \cdot \frac{m}{e}} \left(\frac{2W}{m}\right)^{3/4}$$

16.17. A particle of mass m and charge e enters a homogenous and stationary electric field E with a velocity v_0 perpendicular to the direction of the field. Discuss the motion of the particle under the assumption that it is acted upon by a resistive force proportional to the first power of the velocity.

Solution:

$$\frac{mdv}{dt} = eE - \gamma v$$

$$\frac{-\gamma dv}{eE - \gamma v} = \frac{-\gamma}{m} dt$$

On integration we have

$$\log(eE - \gamma v) = \frac{-\gamma}{m}t + k_1$$

At $t = 0$, $v = v_0$ $\therefore$ $k_1 = \log(eE - \gamma v_0)$

$$\therefore \quad eE - \gamma v = (eE - \gamma v_0)\, e^{\frac{-\gamma}{m}t}$$

$$= C_0 e^{\frac{-\gamma}{m}t};\; C_0 = (eE - \gamma v_0)$$

Along x axis

$$eE - \gamma\frac{dx}{dt} = C_0 e^{\frac{-\gamma}{m}t}$$

$$\gamma\, dx = eE\, dt - C_0 e^{\frac{-\gamma}{m}t} \cdot dt$$

On integration we have

$$x = \frac{eEt}{\gamma} + \frac{C_0}{\gamma} \times \frac{m}{\gamma} e^{\frac{-\gamma}{m}t} + C_1$$

$$x = C_1 + C_2\, e^{\frac{-\gamma}{m}t} + \frac{eE}{\gamma} \cdot t;\; C_2 = \frac{mC_0}{\gamma_2}$$

Along y axis

$$\frac{mdv_y}{dt} = -\gamma\, v_y$$

$$\frac{dv_y}{v_y} = -\frac{\gamma}{m} \cdot dt$$

On integration we have

$$v_y = e^{\frac{-\gamma}{m}t}$$

$$\frac{dy}{dt} = e^{\frac{-\gamma}{m}t}$$

$$y = -\frac{m}{\gamma} e^{\frac{-\gamma}{m}t} + C_3$$

$$= C_3 + C_4\, e^{\frac{-\gamma}{m}t};\, C_4 = -\frac{m}{\gamma}$$

Similarly we have along z axis

$$z = C_5 + C_6\, e^{\frac{-\gamma}{m}t}$$

16.18. A beam of electrons falls on the middle of a vane of the Crookes radiometer (see Fig. 16.9). Determine the law that governs the motion of the vane wheel, which consists of six vanes, if the electrons are accelerated by a potential V_0 and the electron current is I_0. The radius of the vane wheel is R, its width is l, the thickness of each vane is d and the density of the vane material is ρ. Neglect the secondary electron emission, reflection of the electrons and the friction at the bearing.

Solution: The velocity of the electrons is given by

$$\frac{1}{2}mv^2 = eV_0$$

$$\therefore \quad v = \sqrt{\frac{2eV_0}{m}}$$

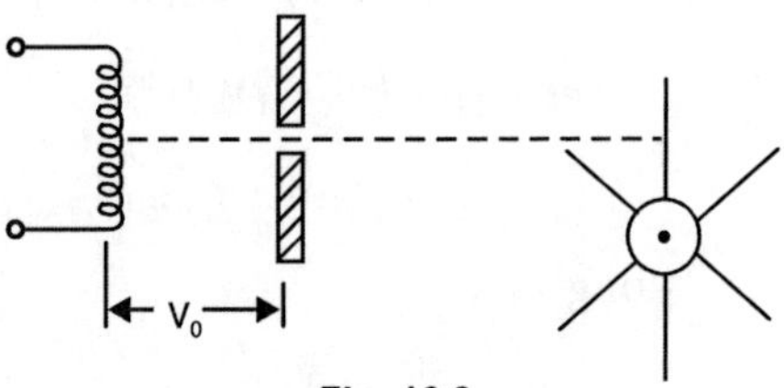

Fig. 16.9

$$\frac{dv}{dt} = \sqrt{\frac{2V_0}{m}} \times \frac{1}{2} \times \frac{1}{\sqrt{e}} \cdot I_0$$

$$\text{Force } = m\sqrt{\frac{2V_0}{m}} \times \frac{I_0}{2\sqrt{e}}$$

$$= \sqrt{\frac{mV_0}{2e}} \cdot I_0$$

$$\text{Torque} = \sqrt{\frac{mV_0}{2e}}\, I_0 \cdot \frac{R}{2} = I\frac{d^2\phi}{dt^2};$$ I = Total moment of inertia of the vane wheel about an axis passing through its centre.

$$I = 2ld\,\rho\, R^3$$

On integration we have

$$\frac{d\phi}{dt} = \frac{RI_0}{2I}\sqrt{\frac{mV_0}{2e}} \cdot t$$

On further integration we have

$$\phi = \frac{Rt^2 I_0}{4I}\sqrt{\frac{mV_0}{2e}}$$

16.19. In Fig. 16.10 plane K is a source of electrons. The electrons are accelerated through a potential difference V between apart plane K and grid G and then after coming out of the grid they fly into space. On accounting the space charge calculate (i) the reaction force applied to the system (ii) the power required for accelerating the electrons.

Solution:

(i) Force of reaction is given by

$$F = \frac{dM}{dt} \cdot v; \; M = \frac{mQ}{e}$$

where M and Q are the total mass and total charge of the electrons.

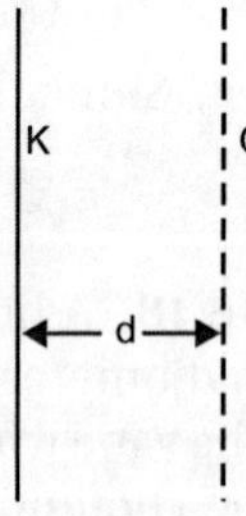

Fig. 16.10

$$\text{Further } \frac{1}{2}mv^2 = eV \text{ or } v = \sqrt{\frac{2eV}{m}}$$

$$F = \frac{m}{e}\sqrt{\frac{2eV}{m}}\cdot\frac{dQ}{dt} = \frac{m}{e}\sqrt{\frac{2eV}{m}}\,J\cdot S$$ where J is current density and

is equal to $\frac{8}{9}\epsilon_0\left(\frac{e}{2m}\right)^{1/2}\frac{V^{3/2}}{d^2}$ and S is the surface area of plane K.

$$F = \frac{m}{e}\sqrt{\frac{2eV}{m}}\cdot\frac{8}{9}\cdot\epsilon_0\left(\frac{e}{2m}\right)^{1/2}\cdot\frac{V^{3/2}}{d^2}\cdot S = \frac{8}{9}\epsilon_0\cdot\frac{V^2 S}{d^2}$$

(ii) Power $P = V\times I = JSV = \frac{8}{9}\cdot\epsilon_0\left(\frac{e}{2m}\right)^{1/2}\frac{V^{5/2}}{d^2}S$

16.20. Consider a homogenous electric field that changes according to the law $E = E_0 \cos \omega t$. A particle of mass m and charge e enters the field with a velocity v_0 perpendicular to the direction of the field at an instant t_0. Find the path of the particle.

Solution: At time t we have the displacement as

$$v_0\,(t - t_0) = x$$

or $$t = \left(\frac{x}{v_0} + t_0\right)$$

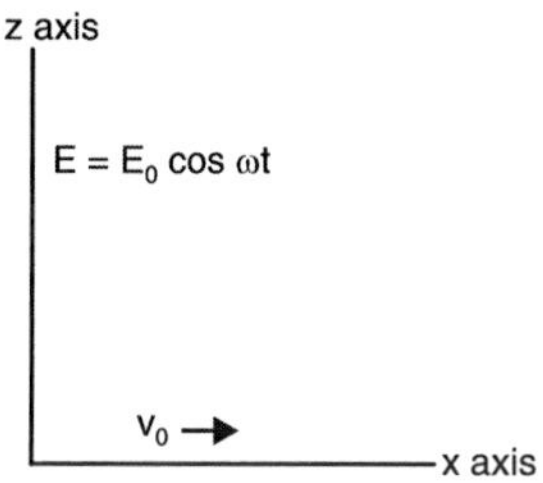

Fig. 16.11

Acceleration along z axis is given by

$$\frac{dv_z}{dt} = \frac{eE_0}{m}\cos\left(\frac{\omega x}{v_0} + \omega t_0\right)$$

$$\frac{dv_z}{dt} = \frac{dv_z}{dx}\cdot\frac{dx}{dt} = v_0\frac{dv_z}{dx} = \frac{eE_0}{m}\cdot\cos\left(\frac{\omega x}{v_0} + \omega t_0\right)$$

On integration we have

$$v_0\cdot v_z = \frac{v_0 eE_0}{m\omega}\sin\left(\frac{x\omega}{v_0} + \omega t_0\right) + C;\ C = \text{constant of integration}$$

At $x = 0$, $v_z = 0$ $\therefore$ $C = \frac{-eE_0}{m\omega}\sin\omega t_0$

$$\therefore\quad v_z = \frac{eE_0}{m\omega}\sin\left(\frac{x\omega}{v_0} + \omega t_0\right) - \frac{eE_0}{m\omega}\cdot\sin\omega t_0$$

$$v_z = \frac{dz}{dt} = \frac{dz}{dx}\cdot\frac{dx}{dt} = v_0\frac{dz}{dx} = \frac{eE_0}{m\omega}\sin\left(\frac{x\omega}{v_0} + \omega t_0\right) - \frac{eE_0}{m\omega}\cdot\sin\omega t_0$$

On integration we have

$$z = \frac{-eE_0}{m\omega^2}\cos\left(\frac{x\omega}{v_0} + \omega t_0\right) - \frac{eE_0 x}{m\omega v_0}\sin\omega t_0 + C'$$

At $x = 0, z = 0 \quad \therefore\ C' = \frac{eE_0}{m\omega^2}\cos\omega t_0$

$$\therefore \quad z = \frac{-eE_0}{m\omega^2}\cos\left(\frac{x\omega}{v_0} + \omega t_0\right)\frac{-eE_0 x}{m\omega v_0}\sin\omega t_0 + \frac{eE_0}{m\omega^2}\cos\omega t_0$$

$$= \frac{-eE_0}{m\omega^2}\left[\cos\frac{\omega x}{v_0}\cos\omega t_0 - \sin\frac{\omega x}{v_0}\sin\omega t_0\right]\frac{-eE_0}{m\omega^2}\cdot\frac{\omega x}{v_0}\sin\omega t_0 + \frac{eE_0}{m\omega^2}\cos\omega t_0$$

$$= \frac{eE_0}{m\omega^2}\left[\left(1 - \cos\frac{\omega x}{v_0}\right)\cos\omega t_0 - \left(\frac{\omega x}{v_0} - \sin\frac{\omega x}{v_0}\right)\sin\omega t_0\right]$$

16.21. A beam of electrons enters the space between two pairs of deflecting plates which have the following voltages $V_x = V_1 \sin\omega t$ on the vertical plates and $V_y = V_2\cos\omega t$ on the horizontal ones. Find the path of the beam on the screen if all the electrons before entering have the same initial velocity v_0 parallel to the plates. The length of the plates is l and their distance from the screen is also l.

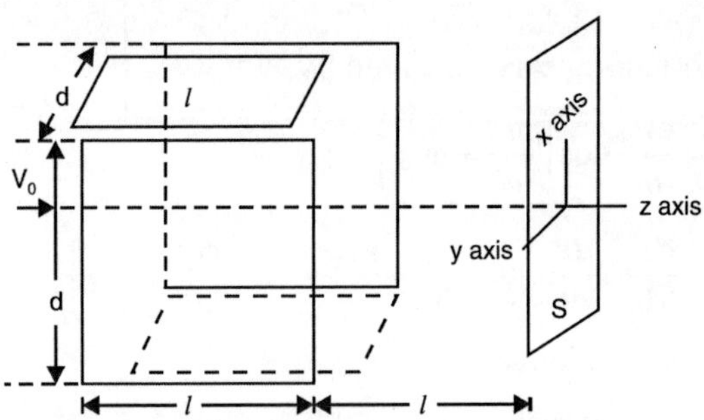

Fig. 16.12

Solution: The acceleration along x axis is given by

$$\frac{dv_x}{dt} = \frac{eV_1}{md}\sin\omega t$$

On integration we have

$$v_x = -\frac{a_1}{\omega}\cos\omega t + C_0;\ a_1 = \frac{eV_1}{md} \qquad \text{... (1)}$$

Let us assume that the electron enters into the space between the plates at time t_0 then

$$O = -\frac{a_1}{\omega}\cos\omega t_0 + C_0 \text{ or } C_0 = \frac{a_1}{\omega}\cos\omega t_0$$

$$\frac{dx_1}{dt} = v_x = -\frac{a_1}{\omega}(\cos\omega t - \cos\omega t_0)$$

On integration we have

$$x_1 = -\frac{a_1}{\omega^2}\sin\omega t + \frac{a_1}{\omega}t\cos\omega t_0 + C'$$

At $t = t_0$, $x = 0$ $\therefore$ $C' = \frac{a_1}{\omega^2}\sin\omega t_0 - \frac{a_1 t_0}{\omega}\cos\omega t_0$

$$\therefore \quad x_1 = \frac{-a_1}{\omega^2}[\sin\omega t - \sin\omega t_0] + \frac{a_1}{\omega}[t - t_0]\cos\omega t_0$$

$$= \frac{-a_1}{\omega^2}[\sin\omega t - \sin\omega t_0] + \frac{a_1 l}{\omega v_0}\cos\omega t_0;\ t - t_0 = \frac{l}{v_0} \qquad \text{... (2)}$$

This is the displacement within the field. The electron moves outside the field for time l/v_0 before striking the screen. The displacement along x axis in the absence of the field = velocity at the upper plate × l/v_0.

$$\therefore \quad x_2 = -\frac{a_1 l}{\omega v_0}(\cos\omega t - \cos\omega t_0) \qquad \text{... (3)}$$

Net displacement along x axis = $x = x_1 + x_2$

$$x = \frac{-a_1 l}{\omega v_0}\left[\cos\omega\left(t_0 + \frac{l}{v_0}\right) - 2\cos\omega t_0\right] - \frac{a_1}{\omega^2}\left[\sin\omega\left(t_0 + \frac{l}{v_0}\right) - \sin\omega t_0\right]$$

$$= A\cos\omega t_0 + B\sin\omega t_0 \qquad \text{... (4)}$$

The acceleration along y axis is represented by

$$\frac{dv_y}{dt} = \frac{eV_2}{md}\cos\omega t$$

On integration we have $v_y = \frac{a_2}{\omega}\sin\omega t + C$

At $t = t_0$, $v_y = 0$ $\therefore$ $C = \frac{-a_2}{\omega}\sin\omega t_0$

$$\therefore \quad v_y = \frac{a_2}{\omega}(\sin\omega t - \sin\omega t_0) \qquad \text{... (5)}$$

On integration we have

$$y_1 = \frac{a_2}{\omega}\left[-\frac{\cos\omega t}{\omega} - t\sin\omega t_0\right] + C_1$$

At $t = t_0$, $y_1 = 0$ $\therefore$ $C_1 = \frac{a_2}{\omega^2}\cos\omega t_0 + \frac{a_2 t_0}{\omega}\sin\omega t_0$

$$y_1 = -\frac{a_2}{\omega^2}[\cos\omega t - \cos\omega t_0] - \frac{a_2}{\omega}(t - t_0)\sin\omega t_0$$

$$= -\frac{a_2}{\omega^2}\left[\cos \omega t - \cos \omega t_0\right] - \frac{a_2 l}{\omega v_0} \sin \omega t_0 \qquad \text{... (6)}$$

The displacement along y axis in the absence of the field

$$y_2 = \frac{a_2}{\omega} \frac{l}{v_0} (\sin \omega t - \sin \omega t_0)$$

$$\text{Net displacement } y = \frac{a_2 l}{\omega v_0}\left[\sin \omega\left(t_0 + \frac{l}{v_0}\right) - 2 \sin \omega t_0\right]$$

$$- \frac{a_2}{\omega^2}\left[\cos \omega\left(t_0 + \frac{l}{v_0}\right) - \cos \omega t_0\right]$$

$$= C \sin \omega t_0 + D \cos \omega t_0 \qquad \text{... (7)}$$

The path on the screen is given by

$$(Ay - Dx)^2 + (By - Cx)^2 = (AC - BD)^2$$

which is an ellipse.

16.22. The potential difference between two flat electrodes is V and the distance between them is d. The field induced emission of electrons from one of the electrodes continues until a space charge forms between the electrodes, which opposes the external field. Find the current density as a function of potential difference.

Solution: The potential difference satisfies the Poisson's equations.

$$\nabla\phi = \frac{-\rho}{\epsilon_0}$$ here ρ is to be represented in terms of current density J and velocity v as $\rho = \frac{J}{v}$

Further $\frac{1}{2}mv^2 = e\phi$

or $\quad v = \sqrt{\frac{2e\phi}{m}}$

Let the motion be along x axis then

$$\frac{d^2\phi}{dx^2} = \frac{-J}{\epsilon_0}\sqrt{\frac{m}{2e\phi}}$$

On integration we have

$$\frac{d\phi}{dx} = \frac{-J}{\epsilon_0}\sqrt{\frac{m}{2e\phi}} \cdot x + C_1; \; C_1 = \text{constant of integration}$$

On substituting the condition that at $x = 0$, $\frac{d\phi}{dx} = 0 \; \therefore C_1 = 0$

$$\therefore \quad \phi^{1/2} d\phi = \frac{-J}{\epsilon_0}\sqrt{\frac{m}{2e}} \cdot x dx$$

On further integration

$$\phi^{3/2} = \frac{-J}{\epsilon_0} \cdot \sqrt{\frac{m}{2e}} \times \frac{3}{2} \cdot \frac{x^2}{2} + C_2$$

At $x = 0, \phi = 0 \;\therefore\; C_2 = 0$. Further at $x = d, \phi = V$ we have

$$V^{3/2} = \frac{-J}{\epsilon_0}\sqrt{\frac{m}{2e}} \cdot \frac{3d^2}{4}$$

or $$J = -\frac{4\epsilon_0}{3d^2}\sqrt{\frac{2e}{m}} \cdot V^{3/2}$$

16.23. Find the force acting on an electron at the time when it intersects the axis of a long current carrying solenoid at right angles in direct proximity to its end with a velocity of 3×10^7 m/sec. The solenoid has 4000 number of turns per meter and carries a current $I = 3$ amp.

Solution:

Magnetic field at the end $B = \frac{\mu_0}{4\pi} 2\pi ni$

Magnetic force $= Bev = \frac{\mu_0}{4\pi} 2\pi ni \times ev$

On substituting the values of various terms we have

$$\text{Magnetic force} = 10^{-7} \times 2 \times 3.14 \times 4000 \times 3 \times 1.6 \times 10^{-19} \times 3 \times 10^7$$
$$= 3.62 \times 10^{-14}\ \text{N}$$

16.24. A magnetic field with an intensity of 8×10^3 A/m and an electric field with an intensity of $E = 10$V/cm are acting in the same direction. An electron flies into the two fields with velocity $v = 10^5$ m/sec. Find the normal, tangential and total acceleration of the electron when (i) the velocity of the electron is parallel to the lines of force (ii) the velocity of the electron is perpendicular to the lines of force.

Solution:

(i) In this case the normal acceleration = 0 and the total acceleration is equal to the tangential acceleration.

$$a = a_t = \frac{eE}{m} = \frac{1.6 \times 10^{-19} \times 10^3}{9 \times 10^{-31}} = 1.78 \times 10^{14}\ \text{m/sec}^2$$

(ii) In this case the tangential acceleration is zero and the total acceleration = normal acceleration.

$$a = a_n = \sqrt{\left(\frac{Bev}{m}\right)^2 + \left(\frac{eE}{m}\right)^2}$$

$$= \sqrt{\left(\frac{8\times10^3 \times 4\pi\times10^{-7} \times 1.6\times10^{-19} \times 10^5}{9\times10^{-31}}\right)^2 + \left(\frac{1.6\times10^{-19} \times 10^3}{9\times10^{-31}}\right)^2}$$

$$= 2.5 \times 10^{14} \text{ m/sec}^2$$

16.25. An α - particle of mass $m_\alpha = 6.65 \times 10^{-27}$ kg moving with a velocity $v = 0.5 \times 10^7$ m/sec. enters at right angles to a magnetic field of induction B = 1 Tesla. Find (i) the radius of the circular path (ii) the magnitude of the magnetic moment (iii) the ratio of the magnetic moment to the angular momentum.

Solution:

(i) $Be'v = \dfrac{m_\alpha v^2}{r}$; $e' = 2e$ and r = radius of the circular path

$$r = \frac{m_\alpha v}{e'B}$$

$$= \frac{6.65\times10^{-27} \times 0.5\times10^7}{2\times1.6\times10^{-19} \times 1}$$

$$= \frac{6.65\times0.5}{3.2}\times10^{-1}$$

$$= 10.4 \text{ cm}$$

(ii) Magnetic moment $p = iA = \dfrac{2e\omega}{2\pi}\cdot\pi r^2$

$$= e\frac{v}{r}\cdot r^2 = evr$$

$$= 1.6 \times 10^{-19} \times 0.5 \times 10^7 \times 10.4 \times 10^{-2}$$

$$= 8.32 \times 10^{-14} \text{ amp. m}^2$$

(iii) ratio $= \dfrac{evr}{m_\alpha vr} = \dfrac{e}{m_\alpha} = \dfrac{1.6\times10^{-19}}{6.65\times10^{-27}}$

$$= 2.4 \times 10^7$$

16.26. An electron accelerated by a potential difference 400 volt moves parallel to a long rectilinear conductor at a distance of 3 mm from it. Calculate the magnetic force experienced by the electron if a current of 5 amp flows through the conductor.

Solution: The velocity of the electron is given by

$$v = \sqrt{\frac{2eV}{m}}$$

Magnetic field acting on the electron is given by

$$B = \frac{\mu_0}{4\pi} \cdot \frac{2I}{r}$$

Magnetic force experienced by the electron is equal to

$$F = Bev = \frac{\mu_0}{4\pi} \cdot \frac{2I}{r} \cdot e\sqrt{\frac{2eV}{m}}$$

On substituting the values

$$F = \frac{10^{-7} \times 2 \times 5 \times 1.6 \times 10^{-19}}{3 \times 10^{-3}} \sqrt{\frac{2 \times 1.6 \times 10^{-19} \times 400}{9 \times 10^{-31}}}$$

$$= \frac{16}{3} \times 10^{-23} \times 1.19 \times 10^{+7}$$

$$= 6.36 \times 10^{-16} \text{ N}$$

16.27. A proton and an electron move with the same velocity and enters into a uniform magnetic field at right angles to it. How many times the radius of curvature of the path of proton is greater than that of the electron.

Solution:

$$Bev = \frac{m_p v^2}{r_p} \text{ for proton}$$

$$Bev = \frac{m_e v^2}{r_e} \text{ for electron}$$

$$\frac{m_p}{r_p} = \frac{m_e}{r_e}$$

$$\therefore \quad \frac{r_p}{r_e} = \frac{m_p}{m_e} = \frac{1.67 \times 10^{-27}}{9 \times 10^{-31}} = 1855$$

16.28. A proton and an electron accelerated by the same potential difference acquires different velocities. Now they enter into the uniform magnetic field perpendicular to the lines of force with their acquired velocities. Find the ratio of the radius of curvature of proton to that of electron.

Solution:

$$\frac{1}{2} m_p v_p^2 = eV = \frac{1}{2} m_e v_e^2$$

$$\therefore \quad v_p = \sqrt{\frac{2eV}{m_p}} \text{ and } v_e = \sqrt{\frac{2eV}{m_e}}$$

Under the action of the magnetic field

$$Bev_p = \frac{m_p v_p^2}{r_p} \text{ and } B_e v_e = \frac{m_e v_e^2}{r_e}$$

$$\frac{r_p}{r_e} = \frac{m_p \cdot v_p}{m_e v_e} = \frac{m_p}{m_e} \times \sqrt{\frac{m_e}{m_p}} = \sqrt{\frac{m_p}{m_e}} = \sqrt{\frac{1.67 \times 10^{-27}}{9 \times 10^{-31}}}$$

$$= 43.06$$

16.29. A proton and an α - particle fly into a homogenous magnetic field perpendicular to the lines of force. How many times the period of revolution of proton is greater than that of α - particle.

Solution:

$$Bev_p = m_p r_p \omega_p^2$$

or $$Be\, r_p \omega_p = m_p r_p \omega_p^2$$

$$\therefore \quad \omega_p = \frac{Be}{m_p} = \frac{2\pi}{T_p}$$

$$\therefore \quad T_p = \frac{2\pi m_p}{Be}$$

Similarly $$T_\alpha = \frac{2\pi m_\alpha}{B \cdot 2e}$$

$$\frac{T_p}{T_\alpha} = \frac{2m_p}{m_\alpha} = \frac{2m_p}{4m_p} = \frac{1}{2}$$

16.30. A cyclotron is operated by a variable oscillator of 10 KV, 12 MHz. If the radius of the dee is 50 cm find (i) the magnetic field required to accelerate protons (ii) kinetic energy acquired by the protons (iii) the number of revolutions made and time taken by the protons.

Solution:

(i) $$\omega = 2\pi n = \frac{qB}{m_p}$$

$$\therefore \quad B = \frac{2\pi m_p \cdot n}{q}$$

$$= \frac{2 \times 3.14 \times 1.67 \times 10^{-27} \times 12 \times 10^{6}}{1.6 \times 10^{-19}}$$

$$= 0.78 \text{ Tesla}$$

(ii) $$K.E. = \frac{1}{2}\frac{(q B \cdot r_D)^2}{m_p}; \; r_D = \text{radius of the dee}$$

$$= \frac{1}{2}\frac{(1.6 \times 10^{-19} \times 0.78 \times 0.5)^2}{1.67 \times 10^{-27}} = 7.3 \text{ Mev}$$

(iii) $$N = \frac{K.E \text{ in } eV}{2 \times \text{applied potential}}$$

$$= \frac{7.3 \times 10^6}{2 \times 10^4}$$

$$= 365$$

$$t = \frac{N}{n} = \frac{365}{12 \times 10^6} = 3 \times 10^{-5} \text{ sec.}$$

16.31. How many fixed singly ionised +ve ions must be within a parallel homogenous electron beam of a circular cross-section so that the radius of the beam remains constant if the beam moves with velocity v?

Solution: Let n_0 be the concentration of the electrons in the beam

Electric field at a distance $r = \dfrac{ern_0}{2\epsilon_0}$

The magnetic field can be obtained by the relation

$$B \times 2\pi r = \mu_0 \pi r^2 v e n_0$$

or $$B = \frac{\mu_0 revn_0}{2}$$

$$\text{Electrostatic force} = \frac{ern_0}{2\epsilon_0} \cdot e = \frac{e^2rn_0}{2\epsilon_0}$$

$$\text{Magnetic force} = \frac{\mu_0 re^2v^2n_0}{2}$$

$$\text{Net force} = \frac{e^2n_0r}{2\epsilon_0}\left[1 - \frac{v^2}{c^2}\right]; \; \mu_0\epsilon_0 = \frac{1}{c^2}$$

Let there be n ions then

$$\frac{ne^2r}{2\epsilon_0} = \frac{e^2rn_0}{2\epsilon_0}\left[1 - \frac{v^2}{c^2}\right]$$

or $$n = n_0\left[1 - \frac{v^2}{c^2}\right]$$

16.32. An electron enters into a magnetic field of induction $B = 4.00$ mT at an angle. It describes a helical path. The helical line has a diameter $d = 40$ mm and a pitch of $l = 150$ mm. Find the speed of the electron. Mass of electron = 9×10^{-31} kg.

Solution: The perpendicular component of the velocity will be responsible for the radius of the helix and the parallel component for the pitch of the helix.

The radius of the helix r is given by

$$r = \frac{mv_{\perp}}{eB} = 20 \times 10^{-3}\,\text{m}$$

The pitch l of the helix is given by

$$l = \frac{2\pi m v_{11}}{eB} = 150 \times 10^{-3}\,\text{m}$$

$$v_{\perp} = \frac{eB}{m} \times 20 \times 10^{-3};\; v_{11} = \frac{eB}{m} \times \frac{150 \times 10^{-3}}{2\pi}$$

$$v = \frac{eB}{m} \cdot 10^{-3} \cdot \sqrt{20^2 + \left(\frac{75}{\pi}\right)^2}$$

$$= \frac{1.6 \times 10^{-19} \times 4 \times 10^{-3} \times 10^{-3}}{9 \times 10^{-31}} \sqrt{20^2 + \left(\frac{75}{\pi}\right)^2}$$

$$= \frac{1.6 \times 4 \times 10^{6}}{9} \sqrt{20^2 + \left(\frac{75}{\pi}\right)^2}$$

$$= 2.215 \times 10^7\,\text{m/s}$$

16.33. An electron of velocity $v = \left(\hat{i} + 2\hat{j}\right) \times 10^6$ m/s enters a region occupied by a magnetic field $B = 0.05\ \hat{i}$ ω/m². Show that it will describe a helical path. Calculate the radius and the pitch of the helix. Mass of the electron = 9×10^{-31} kg.

Solution: The magnetic field is applied along the x-axis. The velocity of the electron has two components, one along the x-axis and the other along the y-axis. The magnetic field has no effect on the x-component of the velocity. This component velocity will provide a linear motion and will be responsible for the pitch of the helix. The y-component of velocity will give rise to circular motion and thus responsible for the radius of the helix.

The radius r of the helix is given by

$$r = \frac{mv_y}{eB} = \frac{9 \times 10^{-31} \times 2 \times 10^{6}}{1.6 \times 10^{-19} \times 0.05} = 2.25 \times 10^{-4}\,\text{m}$$

$$\text{Pitch of helix} = \frac{2\pi m v_x}{eB} = \frac{2 \times 3.14 \times 9 \times 10^{-31} \times 10^{6}}{1.6 \times 10^{-19} \times 0.05} = 7.065 \times 10^{-4}\,\text{m}$$

16.34. A diverging beam of electron having speed $v = 6 \times 10^6$ m/s emerges from a point A along the x-axis and enters in the region where there is a uniform magnetic field $B = 10 \times 10^{-3}$ T along the x axis. At what distance from A the beam will be focused (see Fig. 16.13)?

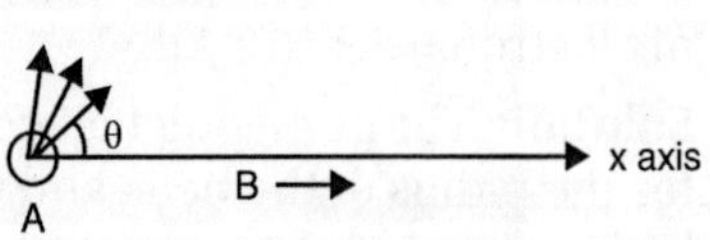

Fig. 16.13

Solution: Consider an electron in the diverging beam at A inside a uniform magnetic field of induction B, moving with a velocity v in a direction making an angle θ with the field. The electron will describe a helical path. The $v \sin\theta$ component will given rise to a circular path whose radius is given by $r = \frac{mv\sin\theta}{eB}$ and the $v\cos\theta$ component will give rise to the pitch of the helix l given by

$$l = \frac{2\pi\, mv\cos\theta}{eB}$$

The magnetic field acts on the electron beam in the same manner as a convex lens acts on a diverging beam of light because of this, this is known as a magnetic lens whose focal length is equal to l.

$$l = \frac{2\pi\, mv}{eB} \quad \text{taking } \theta \text{ to be very small}$$

On substituting the values of the various terms, we have

$$l = \frac{2\times 3.14\times 9\times 10^{-31}\times 6\times 10^{6}}{1.6\times 10^{-19}\times 10\times 10^{-3}}$$

$$= \frac{2\times 3.14\times 9\times 6}{1.6}\times 10^{-4} = 21.2 \text{ mm}$$

16.35. An electron accelerated through a potential of 100 V enters a space occupied by a magnetic field of induction $B = 2.83 \times 10^{-4}$ T in a direction perpendicular to the field. Calculate the radius of the circular path described by the electron.

Given $e/m = 1.73 \times 10^{11}$ C/kg

Solution:

$$eV = \frac{1}{2}mv^2; \quad V \text{ is the accelerating potential.}$$

$$Bev = \frac{mv^2}{r}$$

or $$v = \frac{Ber}{m}$$

$\therefore$ $$eV = \frac{1}{2}m\frac{B^2e^2r^2}{m^2}$$

$\therefore$ $$2V = B^2\frac{e}{m}r^2$$

$\therefore$ $$r = \sqrt{\frac{2V}{B^2\times e/m}}$$

On substituting the values of various terms, we have

$$r = \sqrt{\frac{2\times100\times10^8}{(2.83)^2\times1.73\times10^{11}}}$$

$$= \sqrt{\frac{2}{(2.83)^2\times17.3}}$$

$$= 0.120\text{ m}$$

$$= 12\text{ cm}$$

16.36. In Thompson's setup for determining the e/m of electron, the deflecting plates are 5.0 cm long and 2.0 cm apart. The plates are maintained at a potential difference of 240 V. Electrons accelerated to an energy of 1.0 keV enter from one edge of the plates in a direction parallel to the plates (see Fig. 16.14). Find (i) the deflection at the other edge of the plates, and (ii) the position of the beam on the screen with reference to its undeflected position if the screen lies 32.5 cm away from the other edge of the plates.

Solution:

$$E = \frac{V}{d} = \frac{240}{2\times10^{-2}} = 12\times10^3\text{ V/m}$$

Acceleration acting on the electron,

$$a = \frac{Ee}{m}$$

Now $y = \frac{1}{2}at^2 = \frac{1}{2}\frac{Ee}{m}\times\left(\frac{l}{v}\right)^2 = \frac{Eel^2}{4\times(1/2)mv^2} = \frac{Eel^2}{4\times10^3 eV}$

On substituting the values of various terms, we have

$$y = \frac{12\times10^3\times1.6\times10^{-19}\times25\times10^{-4}}{4\times10^3\times1.6\times10^{-19}}$$

$$= 0.75\text{ cm}$$

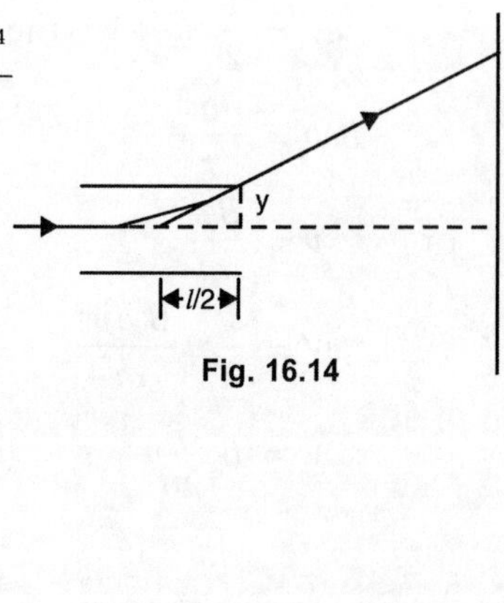

Fig. 16.14

$\frac{y'}{y} = \frac{L+l/2}{l/2}$; L is the distance between the other edge of the plate and the screen

$$= \frac{32.5+2.5}{2.5}$$

$$\therefore \quad y' = 0.75\times\frac{35}{2.5}$$

$$= 10.5\text{ cm}$$

16.37. In the positive ray analysis by Thomson's parabola method, similar parabolas were photographed for singly and dubly ionized particles. Find the ratio of the masses of the two particles when

(i) $\frac{B_1}{B_2}=\frac{1}{2}$ and $\frac{E_1}{E_2}=1$ (ii) $\frac{B_1}{B_2}=1$ and $\frac{E_1}{E_2}=\frac{1}{2}$

where B_1 and B_2 represent the magnetic fields and E_1 and E_2 the electric fields.

Solution:

$y^2 = K\frac{B^2}{E}\frac{q}{M}x$ is the equation of parabola ontained by Thomson's method.

For singly ionised particles,

$$y^2 = K\frac{B_1^2}{E_1}\frac{q}{M_1}x$$

For doubly ionized particles,

$$y^2 = K\frac{B_2^2}{E_2}\frac{2q}{M_2}x$$

Since the two parabolas are similar, we have

$$\frac{B_1^2}{E_1}\times\frac{1}{M_1}=\frac{2B_2^2}{E_2M_2}$$

or $$\frac{M_1}{M_2}=\frac{E_2}{2E_1}\times\left(\frac{B_1}{B_2}\right)^2$$

(i) When $\frac{B_1}{B_2}=\frac{1}{2}$ and $\frac{E_1}{E_2}=1$, we have

$$\frac{M_1}{M_2}=\frac{1}{2}\times\frac{1}{4}=\frac{1}{8}$$

(ii) When $\frac{B_1}{B_2}=1$ and $\frac{E_1}{E_2}=\frac{1}{2}$

$$\frac{M_1}{M_2}=\frac{2}{2}=1$$

16.38. A stream of protons and deuterons, both accelerated by the same potential, enters a region occupied by a magnetic field in a direction perpendicular to it. Both the particles traverse circular paths. Calculate the radius of the circular path described by the deuterons if the radius of the circular path described by proton = 10 cm; $m_d = 2\,m_p$.

Solution: The equation of motion of the two particles is given by

$Bev_p = \frac{m_p v_p^2}{r_p}$; r_p is the radius of the circular path described by proton.

and $Bev_d = \frac{m_d v_d^2}{r_d}$; r_d is the radius of the circular path described by deuteron

$$\frac{m_p v_p}{r_p} = \frac{m_d v_d}{r_d}$$

$$\frac{1}{2}\left(\frac{m_p v_p}{r_p}\right)^2 = \frac{1}{2}\left(\frac{m_d v_d}{r_d}\right)^2$$

or $$\frac{m_p}{r_p^2} = \frac{m_d}{r_d^2} \text{ since } \frac{1}{2}m_p v_p^2 = \frac{1}{2}m_d v_d^2$$

or $$r_d = \sqrt{\frac{m_d}{m_p}}\, r_p$$

$$r_d = \sqrt{2} \times 10$$
$$= 1.41 \times 10$$
$$= 14.1 \text{ cm}$$

16.39. A uniform magnetic field with a slit system as shown in Fig. 16.15 has been used as a momentum filter for charged particles. At a certain field B it has been found that α - particles with energy of 4.5 Mev are transmitted. On increasing the magnetic field 2 times deuterons are passed into the filter. Calculate the kinetic energy of the deuteron.

Solution: In a magnetic field the radius of the charge particle is given by

$$r = \frac{mv}{qB}$$

In the light of this equation the K.E. of α particle is given by

$$K_\alpha = \frac{1}{2}\frac{r^2 \cdot 4e^2 B^2}{4m_p}$$

Similarly the K.E of deuteron is given by

$$K_d = \frac{1}{2}\frac{r^2 \cdot e^2 (2B)^2}{2m_p}$$

$$\therefore \quad \frac{K_d}{K_\alpha} = 2$$

or $$K_d = 2K_\alpha = 9 \text{ Mev}$$

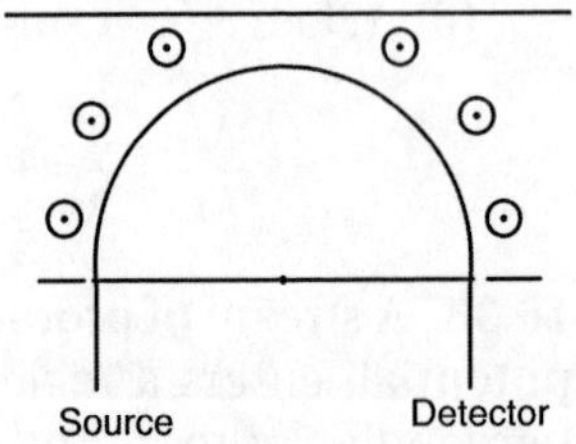

Fig. 16.15

16.40. In the set up shown in Fig. 16.16 the velocity filter consists of mutually perpendicular electric field of strength E and magnetic field of induction B_0. A beam of ions of isotopes of masses m_1 and m_2 after emerging out of the velocity filter enters into a region of uniform magnetic field of induction B perpendicular to the lines of force. The two ions describes circular path and accumulates at S_1 and S_2. Find the separation between S_1 and S_2.

Solution: The velocity of the two ions is given by

$$v = \frac{E}{B_0}$$

The equation of motion of the two ions is given by

$$qvB = \frac{m_1 v^2}{r_1} = \frac{m_2 v^2}{r_2}$$

$$\therefore \quad \frac{r_2}{r_1} = \frac{m_2}{m_1}$$

$$\frac{r_2 - r_1}{r_1} = \frac{m_2 - m_1}{m_1}$$

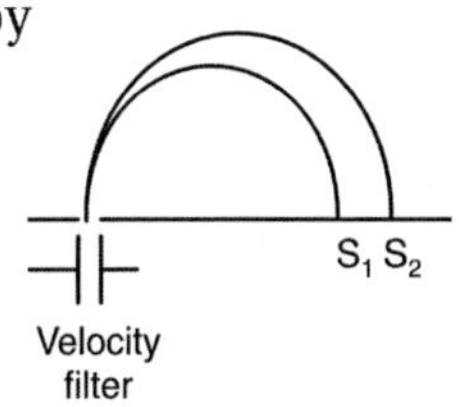

Fig. 16.16

Δr = separation between S_1 and S_2 is given by

$$\frac{\Delta r}{2r_1} = \frac{m_2 - m_1}{m_1}; \ \Delta r = 2(r_2 - r_1)$$

$$\text{or} \quad \Delta r = \frac{2r_1}{m_1}(m_2 - m_1)$$

$$= \frac{2v}{qB}(m_2 - m_1)$$

$$= \frac{2E}{qBB_0}(m_2 - m_1)$$

16.41. A beam of ions of singly ionised U^{235} and U^{238} having energy W = 5 KeV enters through the slit S into a uniform magnetic field of induction B. The direction of motion of the ions is perpendicular to the field. Under the action of the field the two ions deposits at position S_1 and S_2 respectively. At what value of B, Δ the separation between S_1 and S_2 will be equal to 5 mm?

Solution: Let m_1 and m_2 be the mass of the two ions

$$\frac{1}{2}m_1 v_1^2 = \frac{1}{2}m_2 v_2^2 = W$$

$$v_1 = \sqrt{\frac{2W}{m_1}} \text{ and } v_2 = \sqrt{\frac{2W}{m_2}}$$

Under the action of magnetic field of induction B

$$Bev_1 = \frac{m_1 v_1^2}{r_1} \text{ and } Bev_2 = \frac{m_2 v_2^2}{r_2}$$

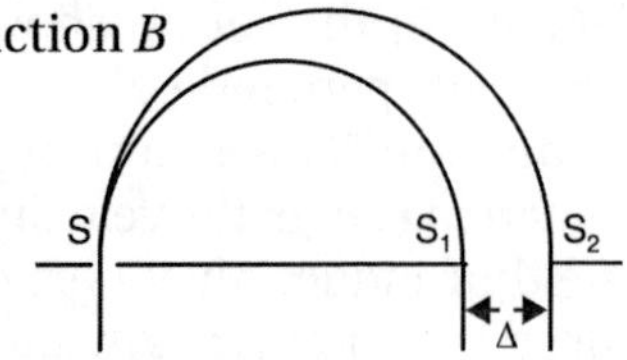

Fig. 16.17

Now $$\frac{Ber_2}{Ber_1} = \sqrt{\frac{m_2}{m_1}}$$

$$\frac{r_1 + \Delta/2}{r_1} = \sqrt{\frac{m_2}{m_1}}$$

$$\frac{\Delta}{2r_1} = \frac{\sqrt{m_2} - \sqrt{m_1}}{\sqrt{m_1}}$$

or $$r_1 = \frac{\Delta \cdot \sqrt{m_1}}{2\left[\sqrt{m_2} - \sqrt{m_1}\right]}$$

$$r_1 = \frac{m_1 v_1}{Be} = \frac{m_1\sqrt{\frac{2W}{m_1}}}{Be} = \frac{\Delta\sqrt{m_1}}{2\left[\sqrt{m_2} - \sqrt{m_1}\right]}$$

or $$B = \frac{2\left[\sqrt{m_2} - \sqrt{m_1}\right]\sqrt{2W}}{e \cdot \Delta}$$

$$= \frac{2\left[\sqrt{A_2} - \sqrt{A_1}\right]m}{e \cdot \Delta}\sqrt{\frac{2W}{m}};$$ m = mass of proton

A_1 and A_2 are atomic wt.

On substituting the values we have

$$B = \frac{2\left[\sqrt{0.238} - \sqrt{0.235}\right] \times 1.67 \times 10^{-27}}{1.6 \times 10^{-19} \times 5 \times 10^{-3}}\sqrt{\frac{2 \times 5 \times 10^3 \times 1.6 \times 10^{-19}}{1.67 \times 10^{-27}}}$$

$$= \frac{2 \times 0.003 \times 1.67}{8} \times 10^{-5}\sqrt{\frac{1.6}{1.67}} \times 10^6$$

$= 1.22 \times 10^{-3}$ Tesla.

16.42. A proton beam with energy W travels through a distance l in vacuum before striking a target. The distribution of proton in the beam is symmetric about the axis. Calculate the current density of the beam if the radius of the beam expands by 10% of its initial value due to Coulombian repulsion of the protons.

Solution:

Electric field at the boundary of the beam $= \frac{enr}{2\epsilon_0}$; r = radius of the beam

Electric force acting on proton $= \dfrac{e^2 nr}{2\epsilon_0}$

Acceleration acting on proton $= \dfrac{e^2 nr}{2\epsilon_0 m}$

$$\frac{10r}{100} = \frac{1}{2}\frac{e^2 nr}{2\epsilon_0\, m}\cdot\frac{l^2}{v^2}$$

$$0.1 = \frac{e^2 nl^2}{4\epsilon_0\, mv^2}$$

Current density $J = nev = \dfrac{0.1\times 4\,\epsilon_0 m\cdot v^2}{e^2 l^2}\, ev$

$$= \frac{4\epsilon_0\times 0.1 m}{el^2}\cdot\left(\frac{2W}{m}\right)^{3/2}$$

16.43. An α - particle is accelerated by a potential difference of 10^4 V. Find the change in its direction of motion, if it enters normally in a region of thickness 0.1 m having transverse magnetic induction of 0.1 Tesla (Given : mass of α - particle 6.4×10^{-27} kg) (Roorkee 1994)

Solution: The kinetic energy of the particle is given by

$$\frac{1}{2}mv^2 = qV$$

$$\therefore \qquad v = \sqrt{\frac{2qV}{m}}$$

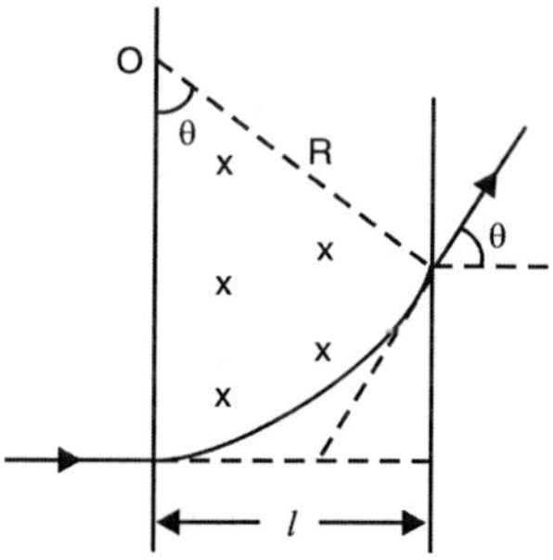

Fig. 16.18

Under the action of the magnetic field the equation of motion is given by

$$qvB = \frac{mv^2}{R}$$

$$\therefore \quad R = \frac{mv}{qB} = \frac{m}{qB}\sqrt{\frac{2qV}{m}} = \frac{1}{B}\sqrt{\frac{2mV}{q}}$$

In view of the figure 16.18; $\sin\theta = \dfrac{l}{R} = l\cdot B\sqrt{\dfrac{q}{2mV}}$

On substituting the values of the various terms we have

$$\sin\theta = 0.1\times 0.1\sqrt{\frac{3.2\times 10^{-19}}{2\times 6.4\times 10^{-27}\times 10^4}}$$

$$= \frac{1}{2}$$

$$\theta = 30°$$

16.44. A particle of charge e, mass m and having kinetic energy E passes through a uniform transverse magnetic field of width l and induction B as shown in the Fig. 16.19. Find the angle at which the particle will be deviated with respect to its initial direction of motion.

Solution:

$$\frac{1}{2}mv^2 = E$$

or $$v = \sqrt{\frac{2E}{m}}$$

Under the action of magnetic field the particle will move in circular path. The equation of motion is given by

$$evB = \frac{mv^2}{r}$$

or $$r = \frac{mv}{eB}$$

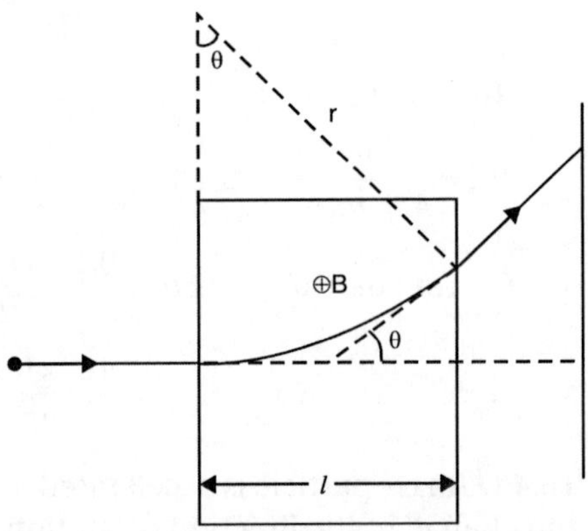

Fig. 16.19

From the geometry of the Fig. 16.19 we have

$$\sin\theta = \frac{l}{r} = \frac{leB}{mv} = \frac{leB}{m}\sqrt{\frac{m}{2E}} = \frac{leB}{\sqrt{2mE}}$$

16.45. A beam of singly ionised Ne atom of energy E = 50 KeV consisting of two isotopes of atomic weight 20 and 22 propogates into a region occupied by a magnetic field of width d = 1 cm and strength B = 0.5 T. The ions are collected on a screen 99 cm away from the field region. Calculate the separation between the points P & Q where the two isotopes strikes the screen. Mass of ion = $A \times 1.6 \times 10^{-27}$ kg where A = atomic weight.

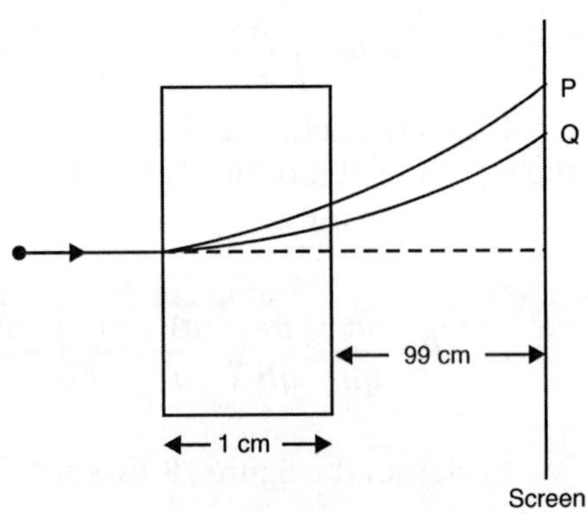

Fig. 16.20

Solution:

$$\frac{1}{2}mv^2 = E$$

$$\therefore \quad mv = \sqrt{2mE}$$

In view of Q. No. 16.44 we have

$$\sin\theta = \frac{d\,Bq}{mv} = \frac{d\,Bq}{\sqrt{2mE}}$$

For small values of θ we have

$$\sin\theta = \tan\theta = \frac{d\,Bq}{\sqrt{2mE}}$$

In view of the Fig. 16.20 we have two equations for the two ions

$$\frac{dBq}{\left(2\times 20\times 1.6\times 10^{-27}\cdot E\right)^{1/2}} = \frac{L_1}{(1+99)\times 10^{-2}}$$

Similarly

$$\frac{dBq}{\sqrt{2\times 22\times 1.6\times 10^{-27}\cdot E}} = \frac{L_2}{(1+99)\times 10^{-2}}$$

Separation between two points is given by

$$L_1 - L_2 = \frac{dBq}{\sqrt{2\times 20\times 1.6\times 10^{-27}E}} - \frac{dBq}{\sqrt{2\times 22\times 1.6\times 10^{-27}\times E}}$$

On substituting the values of various terms we have

$$\frac{10^{-2}\times 0.5\times 1.6\times 10^{-19}}{\sqrt{2\times 20\times 1.6\times 10^{-27}\times 50\times 10^{3}\times 1.6\times 10^{-19}}}$$

$$-\frac{10^{-2}\times 0.5\times 1.6\times 10^{-19}}{\sqrt{2\times 22\times 1.6\times 10^{-27}\times 50\times 10^{3}\times 1.6\times 10^{-19}}}$$

$$8\times 10^{-22}\left[\frac{1}{1.6\times 10^{-20}\sqrt{2}} - \frac{1}{1.6\times 10^{-20}\sqrt{2.2}}\right] = \frac{8\times 10^{-2}}{1.6}[0.707 - 0.674]$$

$$= 0.165 \text{ cm.}$$

16.46. Electrons obtained from an electron gun are accelerated through a potential difference of V volt and then are allowed to move from a point O lying on the axis of a long solenoid having n turns per unit length in the form of a slightly divergent beam. The electrons are focused at a distance l from the point O at two successive values of solenoid current I_1 and I_2. Calculate the e/m of electron.

Solution: Let us assume that the electron is moving at an angle θ with the axis of solenoid.

The equation of motion is given by

$$\frac{mv^2\sin^2\theta}{r} = B_1 ev\sin\theta \text{ where } B_1 = \mu_0 n I_1$$

Period of revolution $t = \frac{2\pi m}{eB_1} = \frac{2\pi m}{\mu_0 e n I_1}$

$$l = \frac{2\pi m}{\mu_0 e n I_1} \cdot v\cos\theta = \frac{2\pi m v}{\mu_0 e n I_1} \quad \text{since } \theta \text{ is very small}$$

Similarly for the solenoid current I_2 the electron beam will be focused under the condition.

$$\frac{l}{2} = \frac{2\pi m v}{\mu_0 e n I_2}$$

$$I_1 = \frac{2\pi m v}{\mu_0 e n l} \text{ and } I_2 = \frac{4\pi m v}{\mu_0 e n l}$$

$$I_2 - I_1 = \frac{2\pi m v}{\mu_0 e n l}$$

$$\frac{1}{2}mv^2 = eV$$

$$mv = \sqrt{2eVm}$$

$$(I_2 - I_1)^2 = \frac{4\pi^2 (mv)^2}{\mu_0^2 e^2 n^2 l^2} = \frac{4\pi^2 \cdot 2eVm}{\mu_0^2 e^2 n^2 l^2}$$

$$\frac{e}{m} = \frac{8\pi^2 V}{\mu_0^2 n^2 l^2 (I_2 - I_1)^2}$$

16.47. A particle of charge q and mass m moving under the influence of a magnetic field describes a circular path. Now if the particle moving in a magnetic field penetrates into a medium where it looses 50% of its kinetic energy. Calculate the % change in its radius of curvature.

Solution:

$$\frac{mv^2}{r_0} = Bqv$$

or $mv = Bq r_0$

Kinetic energy $E = \frac{1}{2}mv^2 = \frac{1}{2}m \cdot \frac{B^2 q^2 r_0^2}{m^2} = \frac{1}{2} Bq^2 \frac{r_0^2}{m}$

Now $0.5\,E = \frac{1}{2}\frac{Bq^2 r^2}{m}$

$\therefore \quad r^2 = 0.5\, r_0^2$

or $r = \sqrt{0.5}\, r_0$

% change in radius $= \frac{r_0 - \sqrt{0.5}\, r_0}{r_0} \times 100 = \left(1 - \sqrt{0.5}\right) \times 100$

$= 29.3\%$

16.48. A particle of mass 9×10^{-31} kg and negative charge of 1.6×10^{-19} C is projected horizontally with a velocity of 10^6 m/sec into a region between two infinite horizontal parallel plates of metal. The distance between the plates is 0.3 cm and the particle enters 0.1 cm below the top plate. The top and bottom plates are connected respectively to the positive and negative terminals of a 30 volt battery. Find the components of the velocity just before it hits one of the plates.

Solution:

Electric field between the plates $E = \dfrac{30}{0.3 \times 10^{-2}} = 10^4$ V/m

Acceleration acting on the electron $a = \dfrac{eE}{m} = \dfrac{1.6 \times 10^{-19} \times 10^4}{9 \times 10^{-31}}$

$$= \frac{1.6}{9} \times 10^{16} \text{ m/sec}^2$$

For vertical component of velocity we have

$$v^2 = 0 + 2a \times 0.1 \times 10^{-2}$$

$$= \frac{2 \times 1.6 \times 10^{-3} \times 10^{16}}{9}$$

$\therefore \quad v = 1.88 \times 10^6$ m/sec

The horizontal component of velocity remains unchanged.

16.49. Under the proper pressure of the surrounding a wire of radius r is heated by direct current I. The emitted electrons escapes from the surface in a direction perpendicular to the wire with a velocity v_0. Calculate the maximum distance upto which the electron will move before it turns back under the action of the magnetic force produced by the current flowing in the wire.

Solution: In view of Fig. 16.21 let the initial direction of motion of the electron be x axis.

Under the action of magnetic field the equation of motion is given by

$$Bev_0 = \frac{mv_0^2}{R}; \; R = \text{radius of curvature of the circular path.}$$

$$\frac{\mu_0}{2\pi}\frac{I}{x}e = \frac{mv_0}{R}$$

or $\quad R = \dfrac{mv_0 \cdot 2\pi x}{\mu_0 I e}$

Let ds be an element on the circular path then

$$R\,d\phi = ds$$

$\therefore \quad \dfrac{1}{R} = \dfrac{d\phi}{ds} = \dfrac{d\phi}{dx} \cdot \dfrac{dx}{ds} = \dfrac{d\phi}{dx} \cdot \cos\phi; \; ds\cos\phi = dx$

$$\therefore \quad \frac{\mu_0 Ie}{mv_0 2\pi x} = \frac{d\phi}{dx} \cdot \cos\phi$$

$$\frac{\mu_0 Ie}{mv_0 2\pi} \int \frac{dx}{x} = \int \cos\phi \cdot d\phi$$

$$\frac{\mu_0 Ie}{mv_0 \cdot 2\pi} \cdot \log x = \sin\phi + C$$

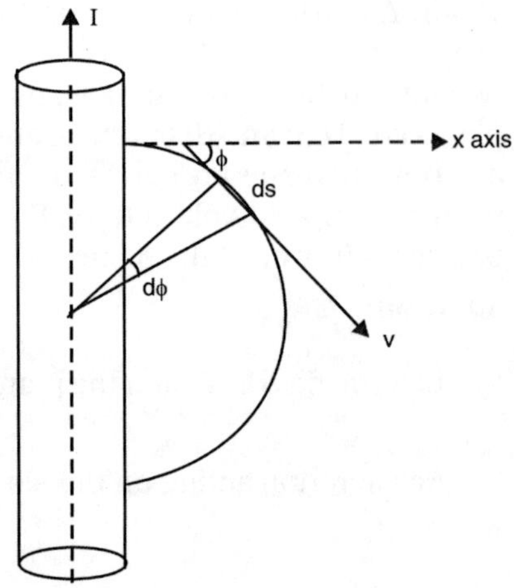

Fig. 16.21

The value of C has to be calculated by using the boundary conditions.

$$x = r, \phi = 0$$

$$\therefore \quad C = \frac{\mu_0 Ie \log r}{mv_0 \cdot 2\pi}$$

$$\sin\phi = \frac{\mu_0 Ie}{mv_0 2\pi} \cdot \log\left(\frac{x}{r}\right)$$

when $x = l$ (max value) ϕ attains $\frac{\pi}{2}$

$$\frac{mv_0 \cdot 2\pi}{\mu_0 Ie} = \log\left(\frac{l}{r}\right) = a \text{ (constt.)}$$

$$\therefore \quad l = re^a$$

16.50. An electron starting from the point O moves with a velocity of $v = 10^6$ m/sec. at an angle of $\theta = 30°$ with the axis of the solenoid (Fig. 16.22a). A current through the solenoid produces an uniform magnetic field of induction $B = 0.15T$. Find the distance r between the point O and the point on the fluorescent screen at which electron strikes. The screen is at right angles to the axis of the solenoid and is at a distance of $l = 10$ cm from the point O. Mass of electron = 9×10^{-31} kg.

Solution: The equation of motion of the electron is given by

$$Bev \sin\theta = \frac{mv^2 \sin^2\theta}{R}$$

$$R = \frac{mv \sin\theta}{Be}$$

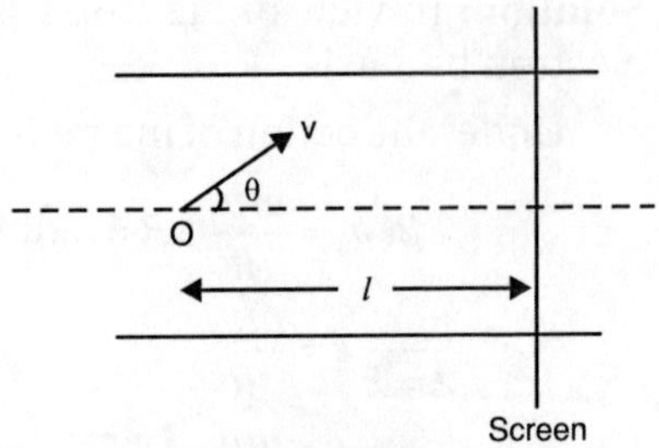

Fig. 16.22a

The period of revolution $T = \frac{2\pi m}{Be}$

Pitch of the helix $p = \frac{2\pi m}{Be} \cdot v\cos\theta$

Number of revolutions within the distance $l = \frac{l}{p} = \frac{Bel}{2\pi mv \cos\theta}$

Let ϕ be the angle through which the electron rotates then

$$\phi = \frac{Bel}{2\pi mv\cos\theta}\cdot 2\pi = \frac{Bel}{mv\cos\theta}$$

In Fig. 16.22b P is the point at which electron strikes. Then $OP = r$

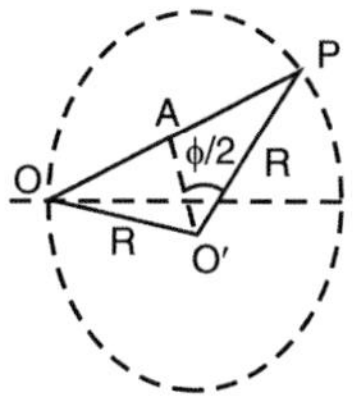

Fig. 16.22b

$$\phi = < OO'P;\ O' \text{is the centre of the circle}$$

$$r = OP = 2OA = 2R\sin\left(\frac{\phi}{2}\right)$$

$$\therefore \quad r = 2\cdot\frac{mv\sin\theta}{Be}\times\sin\left(\frac{Bel}{2\,mv\cos 30}\right)$$

On substituting the values of various terms we have

$$r = \frac{2\times 9\times 10^{-31}\times 10^{6}\sin 30}{0.15\times 1.6\times 10^{-19}}\sin\left(\frac{0.15\times 1.6\times 10^{-19}\times 10^{-1}}{2\times 9\times 10^{-31}\times 10^{6}\sqrt{3/2}}\right)$$

$$= \frac{9}{15\times 16}\times 10^{-3}\sin\left(\frac{15\times 16}{9\times 1.732}\times 10^{2}\right)$$

$$= \frac{9\times 0.9858}{15\times 16}\times 10^{-3}$$

$$= 3.7\times 10^{-5}\text{ m}$$

16.51. A positively charged particle of change e and mass m is located at the origin O of the frame of reference. It moves without any initial velocity in the region where there is uniform electric and magnetic fields of strengths E and B acting along y and z axes (Fig. 16.23a). At an instant t find (a) the trajectory of the particle (b) the length of the segment of the path between two nearest point at which velocity of the particle turns out to be zero.

Solution: The magnetic force will have two values Bev_x along y axis and Bev_y along x axis.

The equation of motion is given by

$$m\frac{dv_x}{dt} = Bev_y$$

$$m\frac{dv_y}{dt} = eE - Bev_x$$

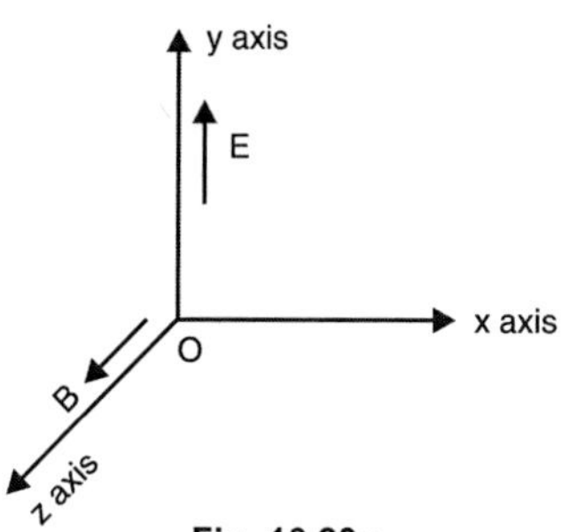

Fig. 16.23a

Let us substitute $\omega = \dfrac{eB}{m}$ and $K = \dfrac{E}{B}$

The two equations will now be written as

$$\frac{dv_x}{dt} = \omega v_y$$

and $\dfrac{dv_y}{dt} = K\omega - \omega v_x$

On differentiation we have

$$\frac{d^2 v_y}{dt^2} = -\omega \cdot \frac{dv_x}{dt} = -\omega^2 v_y$$

Let the solution of the above equation be

$$v_y = A \sin(\omega t + \theta)$$

At $t = 0$, $v_y = 0$

$O = A \sin\theta$. Since A is not zero hence $\sin\theta = 0$ or $\theta = 0$

$\therefore \quad v_y = A \sin \omega t$

$$m \frac{d}{dt}(A \sin \omega t) = eE - Bev_x$$

$$mA\omega \cos \omega t = eE - Be\, v_x$$

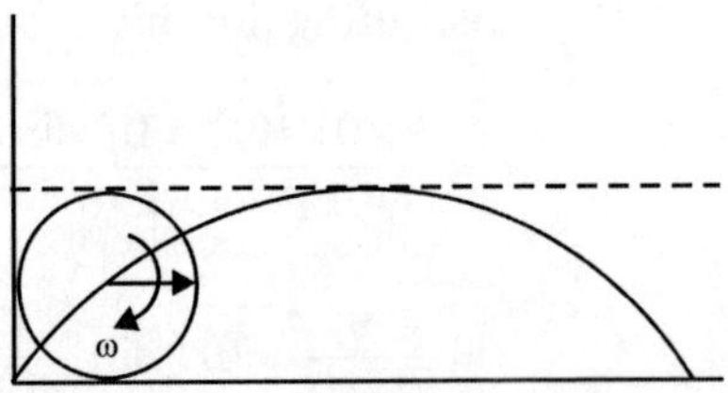

Fig. 16.23b

At $t = 0$, $v_x = 0$ hence

$$A = \frac{eE}{m\omega}$$

$$\therefore \quad v_y = \frac{eE}{m\omega} \sin \omega t$$

On integration we have

$$y = -\frac{eE}{m\omega^2} \cos \omega t + C$$

At $t = 0, y = 0 \;\therefore\; C = \dfrac{eE}{m\omega^2}$

$$\therefore \quad y = \frac{eE}{m\omega^2}(1 - \cos \omega t)$$

$$m \frac{dv_x}{dt} = Be \cdot \frac{eE}{m\omega} \cdot \sin \omega t$$

$$v_x = -\frac{Be}{m} \cdot \frac{eE}{m\omega^2} \cos \omega t + C'; \quad \frac{Be}{m} \cdot \frac{Ee}{m \cdot \omega^2} = \frac{eE}{m\omega} = \frac{E}{B}$$

$$v_x = -\frac{E}{B} \cdot \cos \omega t + C'$$

At $t = 0$, $v_x = 0 \;\therefore\; C' = \dfrac{E}{B}$

$$\therefore \quad v_x = \frac{E}{B}(1 - \cos \omega t)$$

On integration

$$x = \frac{E}{B}\left(t - \frac{\sin \omega t}{\omega}\right) + C''$$

At $t=0, x=0 \therefore C''=0$

$$\therefore \quad x=\frac{E}{B\omega}(\omega t-\sin \omega t)$$

This is the equation of a cycloid. It is the trajectory of a particle on the circumference of a circle. (See Fig. 16.23b)

$$v^2 = v_x^2+v_y^2$$

$$v = \sqrt{\frac{E^2}{B^2}(1-\cos \omega t)^2+\frac{e^2E^2}{m^2\omega^2}\sin^2 \omega t}$$

$$= \frac{E}{B}\sqrt{(1-\cos \omega t)^2+\sin^2 \omega t}$$

$$= \frac{2E}{B}\sin\frac{\omega t}{2}$$

v will be equal to zero when $\frac{\omega t}{2}=0, \pi, 2\pi$ etc.

Time interval at which velocity turns out to be zero is $\frac{2\pi}{\omega}$

$$S=\int_0^{2\pi/\omega} v\,dt = \frac{2E}{B}\int_0^{2\pi/\omega}\sin\frac{\omega t}{2}\cdot dt=-\frac{4E}{B\omega}\left|\cos\frac{\omega t}{2}\right|_0^{2\pi/\omega}$$

$$=\frac{8E}{B\omega}=\frac{8E}{BeB}\cdot m=\frac{8mE}{eB^2}$$

16.52. The two plates of a parallel plate capacitor having plate separation l are placed as shown in the Fig. 16.24. A potential difference V across the two plates produces an electric field along y axis. A filament placed at the negative plate produces electrons in various directions with negligible speed. A magnetic field of induction B is applied along the z axis. For what value of the magnetic field none of the electrons will reach the positive plate of the capacitor?

Solution: The problem is identical to Q. No. 16.51. The path of the electron is cycloid.

The x and y coordinates as derived earlier are

$$x=\frac{E}{B\omega}(\omega t-\sin \omega t)$$

$$y=\frac{eE}{m\omega^2}(1-\cos \omega t)$$

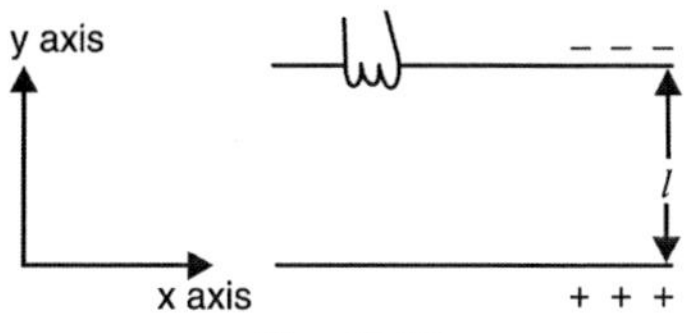

Fig. 16.24

The maximum displacement of electron along the y axis will be equal to the diameter of the rolling circle

$$r=\frac{eE}{m\omega^2}=\frac{eE}{m\cdot e^2}\frac{m^2}{B^2}=\frac{mE}{eB^2}=\frac{l}{2};\ r=\text{radius of the circle}$$

$$\frac{m \cdot V}{eB^2 l} = \frac{l}{2}; \; \frac{V}{l} = E$$

$$\therefore \quad B = \sqrt{\frac{2mV}{el^2}}$$

When $B > \sqrt{\dfrac{2mV}{el^2}}$ none of the electrons will reach the positive plate.

16.53. A proton of charge e, mass m located at the origin starts moving with an initial velocity v_0 along x axis. An electric field of strength E and magnetic field of induction B is applied along y axis as shown in the Fig. 16.25. Find (i) the coordinate of the proton when it crosses y axis for the nth time (ii) the angle θ between velocity vector of the proton and the y axis.

Solution: The forces acting on the proton along the three mutually perpendicular axis is given by

$$m\frac{dv_x}{dt} = -eBv_z, m\frac{dv_y}{dt} = eE \text{ and } m\frac{dv_z}{dt} = Bev_x$$

$$\frac{dv_y}{dt} = \frac{eE}{m} = \text{Constt.}$$

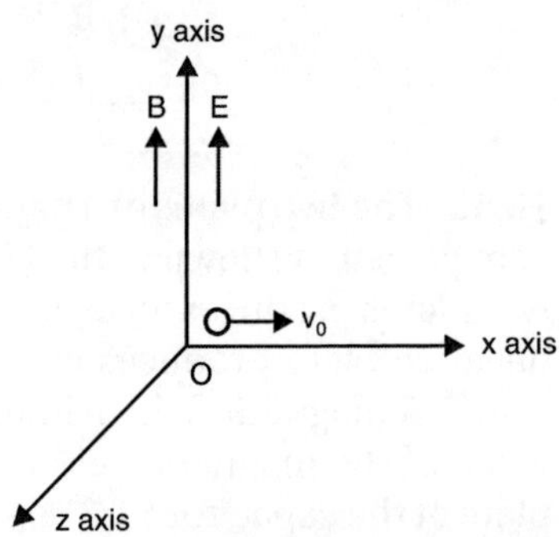

Fig. 16.25

Displacement along y axis given by

$$y = \frac{1}{2}\left(\frac{eE}{m}\right) \cdot t^2$$

$$v_y = \frac{eE}{m} \cdot t$$

$$m\frac{d^2 v_x}{dt^2} = -eB\frac{dv_z}{dt} = -\frac{e^2B^2}{m} \cdot v_x$$

This is equation of simple harmonic motion in which angular velocity ω is given by

$$\omega = \frac{eB}{m}$$

Let $\quad v_x = A \sin(\omega t + \phi)$

$$\frac{dv_x}{dt} = A\omega \cos(\omega t + \phi)$$

At $t = 0$, $v_x = v_0 = A \sin \phi$

At $t = 0$, $\dfrac{dv_x}{dt} = 0 = A\omega \cos \phi$

This is possible when $\phi = \dfrac{\pi}{2}$. Under this condition $A = v_0$

$$\therefore \quad v_x = v_0 \sin\left(\omega t + \frac{\pi}{2}\right) = v_0 \cos \omega t$$

On integration

$$x = \frac{v_0}{\omega} \sin \omega t + C$$

At $t = 0, x = 0 \ \therefore \ C = 0$

Thus $x = \frac{v_0}{\omega} \cdot \sin \omega t$

$$m \frac{dv_z}{dt} = Be \cdot v_x = Be \cdot v_0 \cos \omega t$$

$$\frac{dv_z}{dt} = \omega \cdot v_0 \cos \omega t$$

$$v_z = v_0 \sin \omega t + C'$$

At $t = 0, v_z = 0 \ \therefore \ C' = 0$

$$\therefore \quad \frac{dz}{dt} = v_0 \sin \omega t$$

$$z = -\frac{v_0}{\omega} \cos \omega t + C''$$

At $t = 0, z = 0 \ \therefore \ C'' = \frac{v_0}{\omega}$

$$\therefore \quad z = \frac{v_0}{\omega}(1 - \cos \omega t)$$

$$x^2 + \left(z - \frac{v_0}{\omega}\right)^2 = \frac{v_0^2}{\omega^2}$$

This is a equation of circle with centre at $x = 0$ and $z = \frac{v_0}{\omega}$ in the $x\,z$ plane.

Total time of revolution $t' = n \cdot T = \frac{n \cdot 2\pi}{\omega}$

$$= \frac{n \cdot 2\pi}{Be} m$$

$$\therefore \quad y = \frac{1}{2}\left(\frac{eE}{m}\right) t^2 = \frac{1}{2} \frac{eE}{m} \cdot \frac{n^2 \cdot 4\pi^2 m^2}{B^2 e^2} = \frac{2\pi^2 m \cdot n^2 E}{eB^2}$$

$$v = i \cdot v_0 \cos \omega t + j \frac{eE}{m} \cdot t + k v_0 \sin \omega t$$

$$\cos\theta = \frac{\frac{eE}{m}t}{\sqrt{v_0^2\cos^2\omega t + \frac{e^2E^2t^2}{m^2} + v_0^2\sin^2\omega t}}$$

$$= \frac{\frac{eE}{m}\cdot\frac{2\pi mn}{BE}}{\sqrt{v_0^2 + \frac{e^2E^2}{m^2}\cdot\frac{n^2\cdot 4\pi^2m^2}{B^2e^2}}} = \frac{2\pi\frac{nE}{B}}{\sqrt{v_0^2 + 4\pi^2n^2\frac{E^2}{B^2}}}$$

16.54. Consider a cylindrical capacitor with a homogenous and stationary magnetic field H directed along its axis. A particle of mass m and charge e flies into the capacitor. At the entrance slit the kinetic energy of the particle is W. Calculate the potential difference between the two electrodes of the capacitor if the particle has to fly along the mid-line of the capacitor.

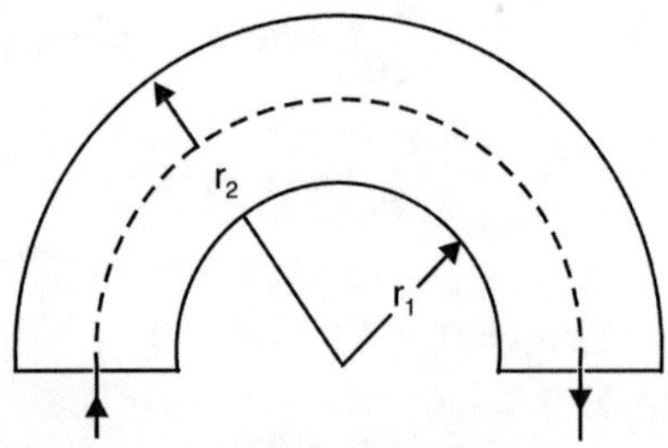

Fig. 16.26

Solution: The electric field inside the capacitor at a distance r is given by

$$E = \frac{V}{r\log_e\frac{r_2}{r_1}};\ V = \text{potential difference between the two cylinders}$$

Field along the mid line $= \dfrac{2V}{(r_1+r_2)\log_e\frac{r_2}{r_1}}$

Electrostatic force $= \dfrac{2eV}{(r_1+r_2)\log_e\frac{r_2}{r_1}}$

$$\frac{1}{2}mv^2 = W\,(\text{given})$$

$$\therefore \quad v = \sqrt{\frac{2W}{m}}$$

Magnetic force on the particle $= e\mu_0 H v = e\mu_0 H\cdot\sqrt{\dfrac{2W}{m}}$

Equation of motion of the particle is given by

$$\frac{mv^2}{r} = \text{electric force} + \text{magnetic force}$$

$$\frac{4W}{r_1+r_2}=\frac{2eV}{(r_1+r_2)\log_e\frac{r_2}{r_1}}+e\mu_0 H\cdot\sqrt{\frac{2W}{m}}$$

$$\therefore \quad V=\left[\frac{4W}{r_1+r_2}-e\mu_0 H\cdot\sqrt{\frac{2W}{m}}\right]\frac{(r_1+r_2)}{2e}\log_e\frac{r_2}{r_1}$$

16.55. In Fig. 16.27 we have a homogenous and stationary electric and magnetic fields E & B in a certain region of space. The angle between the two fields is α. A particle of mass m and charge e enters this region with a velocity v_0. Find the path of the particle.

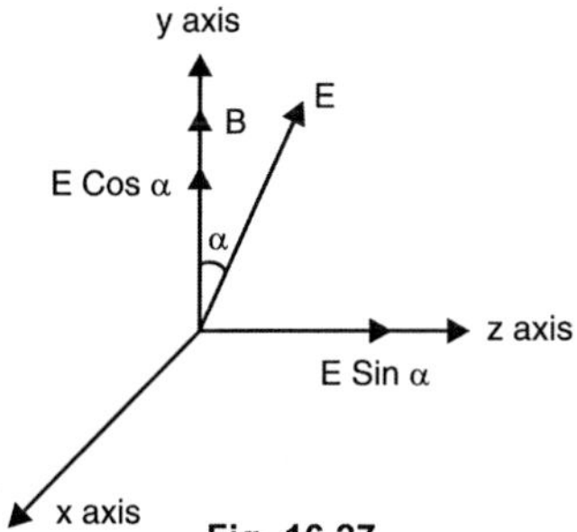

Fig. 16.27

Solution: Equation of motion of the particle

$$-\frac{mdv_x}{dt}=ev_z B \text{ or } \frac{dv_x}{dt}=-\omega v_z;\ \omega=\frac{eB}{m} \quad \text{... (1)}$$

$$\frac{mdv_y}{dt}=eE\cos\alpha \quad \text{... (2)}$$

$$\frac{dv_z}{dt}=\frac{ev_x B}{m}+\frac{Ee}{m}\cdot\sin\alpha$$

$$=\omega v_x+\frac{eE}{m}\sin\alpha \quad \text{... (3)}$$

Let us assume that $v_x = A_3 \sin \omega t + A_4 \cos \omega t + C$ and $v_z = A_1 \sin \omega t + A_2 \cos \omega t$. On substituting the values of v_z and v_x in equation 3 we have

$$A_1\omega\cos\omega t - A_2\omega\sin\omega t=\omega A_3\sin\omega t+\omega A_4\cos\omega t+\omega C+\frac{eE}{m}\cdot\sin\alpha$$

On comparing the coefficients of $\sin \omega t$ and $\cos \omega t$ we have

$$A_1 = A_4,\ A_3=-A_2 \text{ and } C=-\frac{eE}{m\omega}\sin\alpha$$

$$\therefore \quad v_x=-A_2\sin\omega t+A_1\cos\omega t-\frac{eE}{m\omega}\sin\alpha$$

At $t = 0$, $v_x = v_{0x}$ $\therefore A_1 = v_{0x}+\frac{eE}{m\omega}\sin\alpha$

$$\therefore \quad v_z=\left(v_{0x}+\frac{eE}{m\omega}\sin\alpha\right)\sin\omega t+A_2\cos\omega t$$

At $t = 0$, $v_z = v_{0z}$ $\therefore A_2 = v_{0z}$

$$\therefore \quad v_z = \left(v_{0x} + \frac{eE}{m\omega}\sin\alpha\right)\sin\omega t + v_{0z}\cos\omega t$$

On integration we have

$$z = -\left(\frac{v_{0x}}{\omega} + \frac{eE}{m\omega^2}\sin\alpha\right)\cos\omega t + \frac{v_{0z}}{\omega}\sin\omega t + C_1$$

At $t = 0, z = 0 \;\therefore\; C_1 = \frac{v_{0x}}{\omega} + \frac{eE}{m\omega^2}\sin\alpha$

$$z = \left(\frac{v_{0x}}{\omega} + \frac{eE}{m\omega^2}\sin\alpha\right)(1 - \cos\omega t) + \frac{v_{0z}}{\omega}\sin\omega t \quad \text{... (4)}$$

$$v_x = \frac{dx}{dt} = -v_{0z}\sin\omega t + \left(v_{0x} + \frac{eE}{m\omega}\sin\alpha\right)\cos\omega t - \frac{eE}{m\omega}\sin\alpha$$

On integration we have

$$x = \frac{v_{0z}}{\omega}\cos\omega t + \left(\frac{v_{0x}}{\omega} + \frac{eE}{m\omega^2}\sin\alpha\right)\sin\omega t - \frac{eE}{m\omega}\sin\alpha\cdot t + C_2$$

At $t = 0, x = 0 \;\therefore\; C_2 = \frac{-v_{0z}}{\omega}$

$$\text{or} \quad x = \frac{-v_{0z}}{\omega}(1 - \cos\omega t) + \left(\frac{v_{0x}}{\omega} + \frac{eE}{m\omega^2}\sin\alpha\right)\sin\omega t - \frac{eE}{m\omega}\sin\alpha\cdot t \quad \text{... (5)}$$

$$v_y = \frac{eE}{m}\cdot\cos\alpha\cdot t + C_0$$

At $t = 0, v_y = v_{0y} \;\therefore\; C = v_{0y}$

$$v_y = \frac{dy}{dt} = \frac{eE}{m}\cos\alpha\cdot t + v_{0y}$$

$$y = \frac{eE}{2m}\cos\alpha\cdot t^2 + v_{0y}\cdot t + C_3$$

At $t = 0, y = 0 \;\therefore\; C_3 = 0$

$$\text{or} \quad y = \frac{eE}{2m}\cos\alpha\cdot t^2 + v_{0y}\cdot t \quad \text{... (6)}$$

16.56. A long straight wire carries a current i (see Fig. 16.28). A particle of mass m and having positive charge q kept at a distance x_0 from the wire is projected towards it with a speed V. Find the distance of the closed approach of the particle.

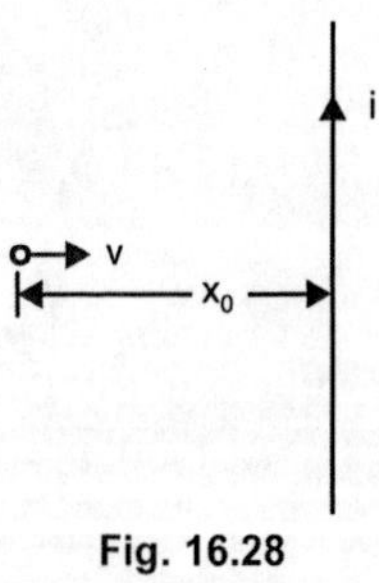

Fig. 16.28

Solution:

$$\frac{mdv}{dt} = \frac{mdv}{dx}\cdot v = \frac{\mu_0}{4\pi}\cdot\frac{2i}{x}qv$$

$$2\pi m \int_V^0 dv = \mu_0 i q \int_{x_0}^{x} \frac{dx}{x}$$

$$-2\pi m V = \mu_0 i q \log_e \frac{x}{x_0}$$

or $$x = x_0 e^{\frac{-2\pi m V}{\mu_0 q i}}$$

16.57. A particle of charge e, mass m is moving with a velocity v_0 along x axis. The particle is acted upon by a magnetic field B along z axis (see Fig. 16.29). In the state of motion the particle experiences a resistive force $-\gamma v$. Find the path of the particle.

Solution: The equation of motion of the particle is given by

$$\frac{m dv_x}{dt} = e v_y B - \gamma v_x$$

or $$\frac{dv_x}{dt} = \omega v_y - \frac{\gamma}{m} v_x; \frac{eB}{m} = \omega \qquad \text{... (1)}$$

$$\frac{m dv_y}{dt} = -e v_x B - \gamma v_y$$

or $$\frac{dv_y}{dt} = -\omega v_x - \frac{\gamma}{m} v_y \qquad \text{... (2)}$$

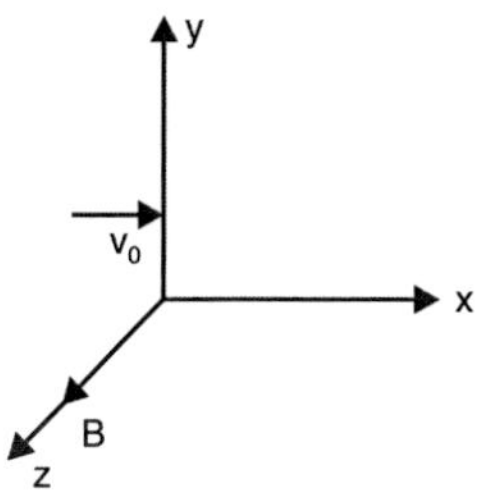

Fig. 16.29

$$\frac{m dv_z}{dt} = -\gamma v_z$$

or $$\frac{dv_z}{dt} = -\frac{\gamma}{m} v_z \qquad \text{... (3)}$$

Let the solution of the differential equation be

$$v_x = A_1 e^{\alpha t} \sin \omega t + A_2 e^{\alpha t} \cos \omega t \text{ and } v_y = A_3 e^{\alpha t} \sin \omega t + A_4 e^{\alpha t} \cos \omega t$$

Equation 1 takes the form

$$A_1 \alpha e^{\alpha t} \sin \omega t + A_1 \omega e^{\alpha t} \cos \omega t + A_2 \alpha e^{\alpha t} \cos \omega t - A_2 \omega e^{\alpha t} \sin \omega t$$

$$= \omega A_3 e^{\alpha t} \sin \omega t + \omega A_4 e^{\alpha t} \cos \omega t - \frac{\gamma}{m} A_1 e^{\alpha t} \sin \omega t - \frac{\gamma}{m} A_2 e^{\alpha t} \cos \omega t$$

On comparing the coefficients of sine and cosine terms we have

$$A_1 \alpha - A_2 \omega = \omega A_3 - \frac{\gamma A_1}{m} \text{ and } A_1 \omega + A_2 \alpha = \omega A_4 - \frac{\gamma}{m} A_2$$

and from these we have $\alpha = -\frac{\gamma}{m}$, $A_3 = -A_2$ and $A_2 = A_4$

From equation 2 we have

$$A_3 \alpha e^{\alpha t} \sin \omega t + A_3 \omega e^{\alpha t} \cos \omega t + A_4 \alpha e^{\alpha t} \cos \omega t - A_4 \omega e^{\alpha t} \sin \omega t$$
$$= -\omega A_1 e^{\alpha t} \sin \omega t - \omega A_2 e^{\alpha t} \cos \omega t$$

$$\frac{-\gamma}{m} A_3 e^{\alpha t} \sin \omega t - \frac{\gamma}{m} A_4 e^{\alpha t} \cos \omega t$$

Proceeding on similar lines we have

$$\alpha = \frac{-\gamma}{m}, A_1 = A_4 \;\&\; A_2 = -A_3$$

$$v_x = A_1 e^{\frac{-\gamma}{m}t} \sin \omega t + A_2 e^{\frac{-\gamma}{m}t} \cos \omega t$$

$$v_y = -A_2 e^{\frac{-\gamma}{m}t} \sin \omega t + A_1 e^{\frac{-\gamma}{m}t} \cos \omega t$$

At $t = 0$, $v_x = v_0$ & $v_y = 0$. On the basis of this $A_2 = v_0$ and $A_1 = 0$

$$\therefore \quad v_x = v_0 e^{\frac{-\gamma}{m}t} \cos \omega t \text{ and } v_y = -v_0 e^{\frac{-\gamma}{m}t} \sin \omega t \qquad \text{... (4)}$$

$$dx = v_0 e^{\frac{-\gamma}{m}t} \cos \omega t \cdot dt$$

On integration we have

$$x = C_1 - \frac{v_0 e^{\frac{-\gamma}{m}t}}{\omega \sqrt{1 + \frac{\gamma^2}{m^2\omega^2}}} \left[\cos \omega t \cdot \frac{\frac{\gamma}{m\omega}}{\sqrt{1 + \frac{\gamma^2}{m^2\omega^2}}} - \sin \omega t \cdot \frac{1}{\sqrt{1 + \frac{\gamma^2}{m^2\omega^2}}} \right]$$

$$x = C_1 - C_2 e^{\frac{-\gamma}{m}t} \cos(\omega t + \alpha); \; C_2 = \frac{v_0}{\omega \sqrt{1 + \frac{\gamma^2}{m^2\omega^2}}} \text{ and } \tan \alpha = \frac{1}{\frac{\gamma}{m\omega}} \qquad \text{... (5)}$$

Now $\dfrac{dy}{dt} = -v_0 e^{\frac{-\gamma}{m}t} \sin \omega t$

On integration we have

$$y = \frac{v_0 e^{\frac{-\gamma}{m}t}}{\omega \sqrt{1 + \frac{\gamma^2}{m^2\omega^2}}} [\sin(\omega t + \alpha)] + C_4$$

$$= C_4 + C_3 e^{\frac{-\gamma}{m}t} \sin(\omega t + \alpha) \qquad \text{... (6)}$$

For z axis

$$\log v_z = -\frac{\gamma}{m} t + C$$

At $t = 0$, $v_z = 0 \;\therefore\; C = 0$

Now $\dfrac{dz}{dt} = e^{\frac{-\gamma}{m}t}$

$$z = \frac{e^{\frac{-\gamma}{m}t}}{\frac{-\gamma}{m}} + C_5$$

$$= C_5 - \frac{m}{\gamma} e^{\frac{-\gamma}{m}t} \qquad \text{... (7)}$$

16.58. A particle of mass $m = 1 \times 10^{-26}$ kg and charge $q = 1.6 \times 10^{-19}$ C travelling with a velocity $v = 1.28 \times 10^6$ m/s in the $+x$ direction enters a region in which a uniform electric field E and a uniform magnetic induction B are present such that $E_x = E_y = 0$, $E_z = -$ 102.4 KV/m and $B_x = B_z = 0$, $B_y = 8 \times 10^{-2}$ weber/m^2. The particle enters this region at time $t = 0$. Determine the location (x, y and z coordiates) of the particle at $t = 5 \times 10^{-6}$ s. If the electric field is switched off at this instant (with the magnetic field still present); what will be the position of the particle at $t = 7.45 \times 10^{-6}$ sec.

(IIT 1982)

Solution: Fig. 16.30 shows the directions of the two fields.

Net force F on the particle is given by

$$F = qE + q\,(\vec{v} \times B)$$

$$= -\,i\,q\,v_z\,B + (q\,v_x\,B - q\,E)\,k$$

or $$\frac{m dv_x}{dt} = -qv_zB, \frac{m dv_y}{dt} = 0 \text{ and } \frac{m dv_z}{dt} = qv_xB - qE$$

$$\frac{dy}{dt} = v_y = 0 \;\therefore\; y = C$$

At $t = 0$ particle is at origin $\therefore\; C - 0$

$$\frac{m\,dv_z}{dt} = q\,v_xB - qE$$

$$\frac{d^2v_z}{dt^2} = \frac{qB}{m}\frac{d}{dt}v_x = -\frac{q^2B^2}{m^2}\cdot v_z$$

This is equation of simple harmonic motion

Let $v_z = A \sin(\omega t + \theta)$ be the solution

where $\omega = \dfrac{qB}{m}$

At $t = 0$, $v_z = 0 \;\therefore\; A \sin\theta = 0$

$$\frac{dv_z}{dt} = A\omega\cos(\omega t + \theta)$$

At $t = 0$, $$\frac{dv_z}{dt} = A\omega\cos\theta = \frac{q}{m}[v_x\,B - E]$$

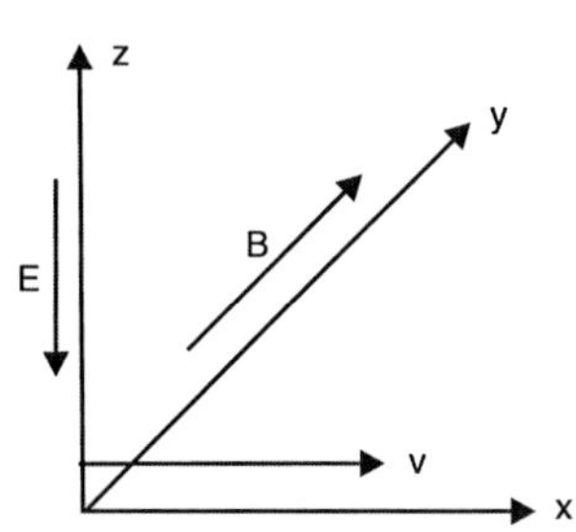

Fig. 16.30

$$A\omega\cos\theta = \frac{q}{m}\left[8\times10^{-2}\times1.28\times10^{6} - 102.4\times10^{3}\right] = 0$$

$\therefore \quad A\cos\theta = 0$ and $A\sin\theta = 0$

The two identities are simultaneously satisfied when $A = 0$.

$\therefore \quad v_z = 0$ or $z =$ constt.

At $t = 0, z = 0$

Thus the y and z coordinate of the particle is zero.

$$x = vt = 1.28\times10^{6}\times5\times10^{-6} = 6.4 \text{ m}$$

When the electric field is switched off then the particle will describe a circular path. The equation of motion is given by

$$qvB = \frac{mv^2}{R}$$

$$\therefore \quad R = \frac{mv}{qB}$$

$$= \frac{10^{-26}\times1.28\times10^{6}}{1.6\times10^{-19}\times8\times10^{-2}} = 1 \text{ m}$$

$$\text{Time } T = \frac{2\pi R}{v} = \frac{2\pi\times1}{1.28\times10^{6}} = 4.9\times10^{-6} \text{ sec}$$

Time in the circular path $= 7.45\times10^{-6} - 5\times10^{-6} = 2.45\times10^{-6}$ s

In this way the particle will complete half the revolution.

$\therefore$ Coordinate of the particle at $t = 7.45\times10^{-6}$ sec are (6.4, 0, 2) meter.

16.59. A kenotron tube consists of a wire of radius r_1 as a filament and a coaxial hollow cylindrical tube of radius r_2 as anode. The tube is covered with a short solenoid by keeping its axis parallel to the filament. A potential difference V is applied across the two electrodes of the tube. A current through the solenoid produces a magnetic field parallel to the filament. At what value of the magnetic field the electrons starting from the filament without any initial velocity will reach the anode?

Solution: Let us assume that the electron is moving along x axis and the magnetic field is applied along z axis. Due to the potential difference V applied between filament and anode the electron experiences an electric field along x axis.

$$\vec{F}_{el} = -e\vec{E} \quad \text{(see Fig. 16.31)}$$

The electron therefore moves in xy plane.

Magnetic force acting on the electron $F = -ev\times B$

In polar coordinates

$$F_r = -eB\,v_\theta \text{ and } F_\theta = eB\,v_r$$

Torque $\vec{\tau} = \vec{r} \times \vec{F}$

$$\tau_\theta = -rF_\theta,\ \tau_r = 0,\ \tau_z = 0$$

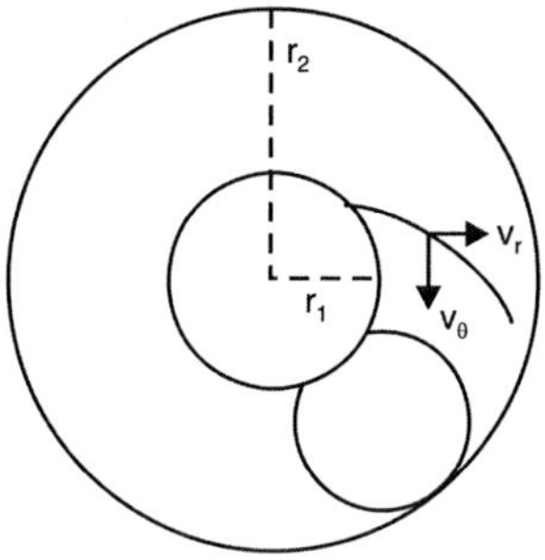

Fig. 16.31

We know that

$$\tau = \frac{dL}{dt};\ L = \text{angular momentum}$$

$$L_r = 0,\ L_\theta = -mrv_\theta \text{ and } L_z = 0$$

$$\tau_\theta = \frac{dL_\theta}{dt} = \frac{d}{dt}(-mrv_\theta)$$

$$rF_\theta = \frac{d}{dt}(mrv_\theta);\ v_r = \frac{dr}{dt} \text{ and } v_\theta = \frac{r\,d\theta}{dt}$$

$$F_\theta = eB\frac{dr}{dt}$$

$$reB\frac{dr}{dt} = \frac{d}{dt}\left(mr^2\frac{d\theta}{dt}\right)$$

$$r\frac{dr}{dt} = \frac{m}{eB}\cdot\frac{d}{dt}\left(r^2\frac{d\theta}{dt}\right)$$

On integration we have

$$\frac{r^2}{2} = \frac{m}{eB}r^2\frac{d\theta}{dt} + C$$

At $r = r_1$, $v_\theta = r\frac{d\theta}{dt} = 0$

$$\therefore \quad C - \frac{r_1^2}{2}$$

$$\therefore \quad r^2 = r_1^2 + \frac{2m}{eB}r^2\omega$$

$$v^2 = v_r^2 + v_\theta^2$$

Under the condition of just touching the anode we have

$$v_r = 0 \text{ at } r = r_2$$

$$\therefore \quad v^2 = v_\theta^2 \text{ at } r = r_2$$

$$v_\theta^2 \text{ at } r = r_2 = \frac{2eV}{m}$$

and $\omega = \frac{v_\theta}{r_2} = \frac{1}{r_2}\sqrt{\frac{2eV}{m}} = \frac{eB}{2m}\left[1 - \left(\frac{r_1}{r_2}\right)^2\right]$

$$= \frac{eB}{2m}\left[\frac{r_2^2 - r_1^2}{r_2^2}\right]$$

$$\therefore \quad B = \frac{2r_2}{r_2^2 - r_1^2} \cdot \sqrt{\frac{2mV}{e}}$$

16.60. A potential difference V is maintained across the two electrodes of inner and outer radii r_1 and r_2 respectively of a cylindrical capacitor (see Fig. 16.32). A proton of charge e and mass m enters into the uniform transverse magnetic field of induction B through the electric field of the capacitor and describes a circular path of radius r. Find the velocity and e/m of the proton.

Solution: Let λ be the charge per unit length, then the electric field inside the capacitor is given by

$$E = \frac{\lambda}{2\pi\epsilon_0 r}$$

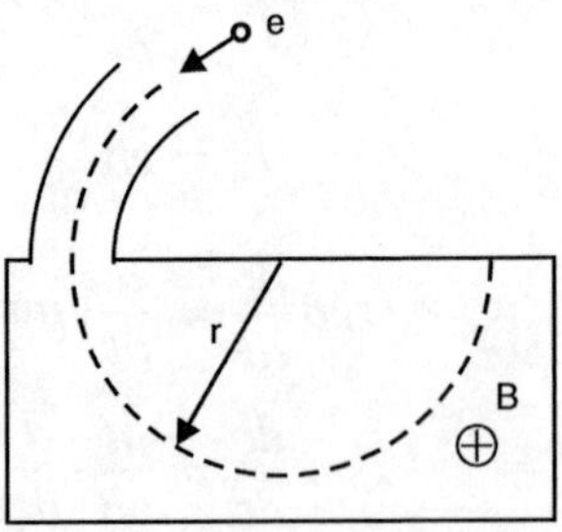

Fig. 16.32

Due to the electric field the proton experiences a force equal to $\frac{\lambda e}{2\pi\epsilon_0 r}$ and moves in a circular path of radius r, such that

$$\frac{mv^2}{r} = \frac{\lambda e}{2\pi\epsilon_0 r}$$

In the magnetic field the equation of motion is given by

$$\frac{mv^2}{r} = Bev$$

Thus $Bev = \frac{\lambda e}{2\pi\epsilon_0 r}$

or $$v = \frac{\lambda}{2\pi\epsilon_0 \cdot Br}$$

$$V = -\int_V^0 dV = \int_{r_1}^{r_2} E dr = \int_{r_1}^{r_2} \frac{\lambda}{2\pi\epsilon_0 r} dr = \frac{\lambda}{2\pi\epsilon_0} \log_e \frac{r_2}{r_1}$$

$$v = \frac{V}{rB\log_e \frac{r_2}{r_1}}$$

In the circular path we have

$$\frac{mv}{r} = Be$$

$$\therefore \quad \frac{e}{m} = \frac{v}{rB} = \frac{V}{r^2 B^2 \log_e \frac{r_2}{r_1}}$$

16.61. A proton beam moving along x axis passes through a region where there is a uniform electric field of strength E and magnetic field of induction B acting along y and z axes respectively. The beam forms a circular spot of radius r on the fluorescent screen placed at right angles to its path without suffering any deviation in the two fields. Assuming the beam current as I, find the pressure on the screen.

Solution: Let e be the charge and m be the mass of the proton

For no deviation from its path we have

$$evB = eE$$

or $$v = \frac{E}{B}$$

Momentum of each proton $= m\dfrac{E}{B}$

Let there be n protons per sec which are causing the current I then

$$ne = I$$

or $$n = \frac{I}{e}$$

$$\text{Total momentum per sec} = m\frac{E}{B}\cdot\frac{I}{e} = \text{Force}$$

$$\therefore \quad \text{Pressure on the screen} = \frac{mEI}{eB\cdot\pi r^2}$$

16.62. A proton moves along a linear path in a region in which there is a uniform electric field of strength E and magnetic field of induction B as shown in the Fig. 16.33. The path is confined in the xz plane and forms an angle θ with the x axis. The electric field is switched off. Determine the pitch of the helical path described by the proton in the magnetic field.

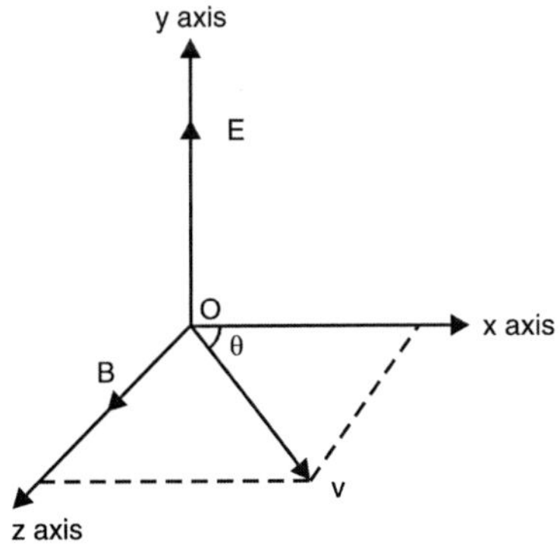

Fig. 16.33

Solution: Under the action of the magnetic field we have

$$\frac{mv^2\cos^2\theta}{r} = Bev\cos\theta$$

or $$r = \frac{mv\cos\theta}{eB}$$

$$\text{The period of revolution } T = \frac{2\pi mv\cos\theta}{eB\cdot v\cos\theta} = \frac{2\pi m}{eB}$$

$$\text{Pitch } l = v\sin\theta\cdot\frac{2\pi m}{eB}$$

Further $Bev\cos\theta = Ee$

$$\therefore \quad v = \frac{E}{B\cos\theta}$$

$$\therefore \quad \text{Pitch } l = \frac{2\pi mE}{eB^2}\cdot\tan\theta$$

16.63. A beam of positive ions of charge *e* and mass *m* and having different velocities moves along *x* axis. The beam enters into a region where an electric and magnetic field of strength *E* and induction *B* acts along *y* axis as shown in the Fig. 16.34a. A fluorescent screen is placed at a distance *l* from origin *O* at right angles to the initial direction of motion of the ions. Find the trajectory of the ions on the screen.

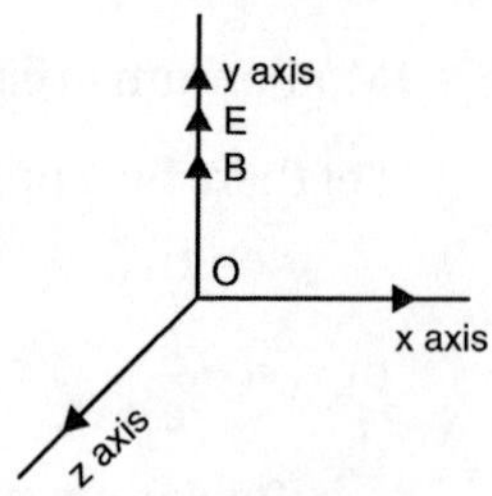

Fig. 16.34a

Solution: In electric field we have

$$\frac{y}{l/2} = \tan\theta = \frac{v_\perp}{v};$$ $v_\perp$ is the velocity perpendicular to the direction of motion

$$\frac{y}{l} = \frac{1}{2}\frac{eE}{m}\cdot\frac{l}{v^2}$$

$$\text{or} \quad y = \frac{1}{2}\frac{eEl^2}{mv^2} \quad \text{(see Fig. 16.34b)}$$

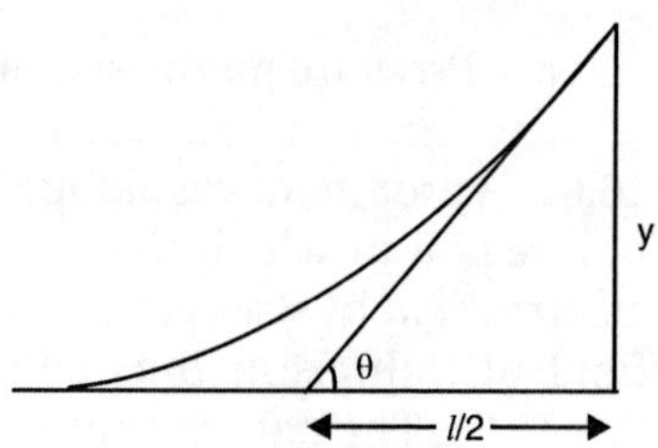

Fig. 16.34b

In magnetic field we have

$$\phi' = \frac{l}{\rho}; \ \rho = \text{radius of curvature}$$

$$\frac{mv^2}{\rho} = Bev$$

$$\therefore \quad \frac{1}{\rho} = \frac{Be}{mv}$$

$$\phi'^2 = \frac{l^2}{\rho^2} = \frac{l^2\cdot B^2e^2}{m^2v^2} = \frac{l^2B^2e^2}{m^2}\cdot\frac{2my}{eEl^2} = \frac{2B^2ey}{mE} \quad \text{(see Fig. 16.34c)}$$

$$\text{or} \quad \phi' = \sqrt{\frac{2B^2ey}{mE}}$$

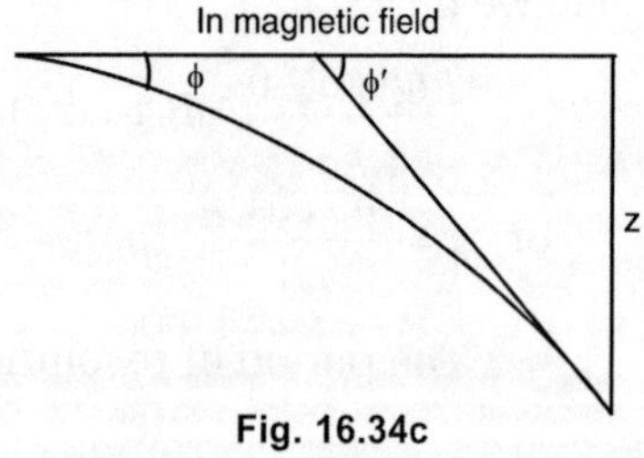

Fig. 16.34c

Further $\tan\phi = \dfrac{z}{l}$

$$\tan\phi' = \frac{2z}{l}$$

For small angles $\dfrac{\phi'}{\phi} = 2$

$$\text{or} \quad \phi = \frac{\phi'}{2} = \sqrt{\frac{B^2 ey}{2mE}}$$

$$z = l \tan\phi = l \tan\sqrt{\frac{B^2 e y}{2mE}}$$

16.64. A beam of electrons moves through a region of space where there is an electric field of strength E and magnetic field of induction B. Both the fields are perpendicular to each other and also perpendicular to the direction of motion of electrons. In the presence of both the fields the beam does not deviates from its rectilinear path and strikes the screen at O (Fig. 16.35) but when the magnetic field is switched off the electrons strikes the screen at O' and produces a shift Δx. Calculate the e/m electron.

Solution: Let y be the displacement produced in the electric field then

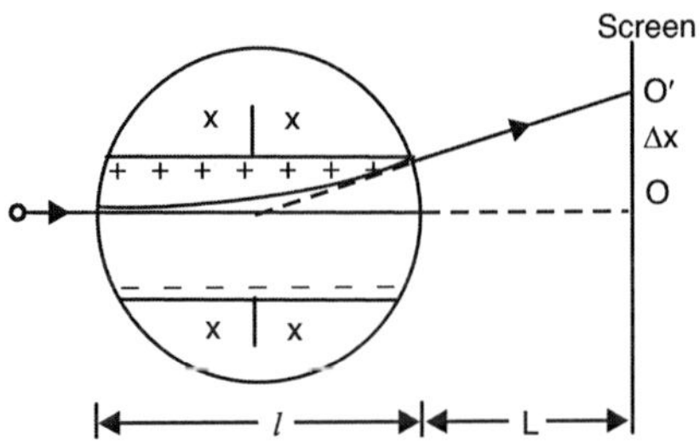

Fig. 16.35

$$\frac{y}{l/2} = \frac{\Delta x}{L + l/2}$$

$$\text{or} \quad y = \frac{l \cdot \Delta x}{(l + 2L)}$$

$$Bev = Ee$$

$$\text{or} \quad v = \frac{E}{B}$$

Time required for the displacement $y = \frac{l}{E/B} = \frac{Bl}{E}$

$$y = \frac{l \cdot \Delta x}{l + 2L} = \frac{1}{2}\frac{Ee}{m} \cdot \frac{B^2 l^2}{E^2}$$

$$\therefore \quad \frac{e}{m} = \frac{2 \cdot \Delta x \cdot E}{l(l + 2L) B^2}$$

16.65. A particle of charge q and mass m initially located at the origin O in which a constant magnetic field of induction B acts along z axis and an oscillating electric field $E = E_0 \cos\omega t$ acts along y axis as shown in the Fig. 16.36. Calculate the displacement of the particle along x and y axis as a function of time.

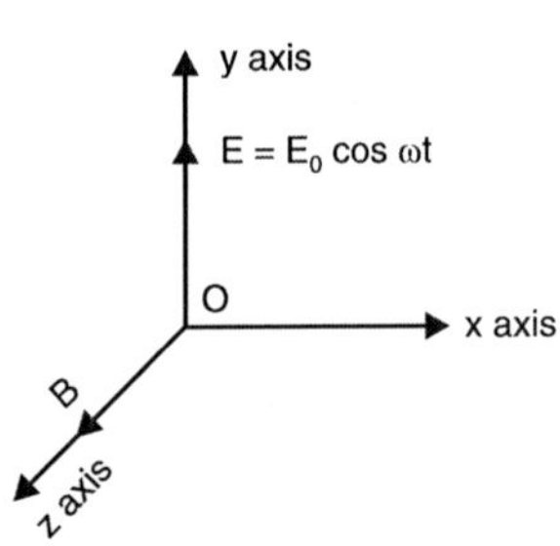

Fig. 16.36

Solution: Under the action of the two fields the force acting on the particle is given by

$$F = q\,(E + v \times B)$$

$$m\frac{dv_x}{dt} = qv_yB$$

$$\frac{dv_x}{dt} = \frac{qB}{m}\cdot v_y = \omega v_y;\ \frac{qB}{m} = \omega$$

$$m\frac{dv_y}{dt} = qE_0\cos\omega t - q\,Bv_x$$

or
$$\frac{dv_y}{dt} = \frac{qE_0}{m}\cos\omega t - \frac{qB}{m}\cdot v_x$$

$$\frac{dv_y}{dt} = a\cos\omega t - \omega\,v_x;\ a = \frac{qE_0}{m}$$

On differentiating it with respect to t we have

$$\frac{d^2v_y}{dt^2} = -\,a\,\omega\sin\omega t - \omega\cdot\frac{dv_x}{dt}$$

$$\frac{d^2v_y}{dt^2} = -\,a\omega\sin\omega t - \omega^2 v_y$$

Let $v_y = A\sin\omega t + A't\cos\omega t$ be the solution of the above equation

$$\frac{dv_y}{dt} = A\omega\cos\omega t + A'\cos\omega t - A't\,\omega\sin\omega t$$

At $t = 0$, $\dfrac{dv_y}{dt} = a = A\omega + A'$

$$\frac{d^2v_y}{dt^2} = -A\omega^2\sin\omega t - A'\omega\sin\omega t - A'\omega^2 t\cos\omega t - A'\omega\sin\omega t$$

$$= -a\,\omega\sin\omega t - A'\omega\sin\omega t - \omega^2 A't\cos\omega t$$

On equating the coefficient of $\sin\omega t$ we have

$$2A' = a$$

$$\therefore \quad A' = \frac{a}{2}$$

and $\quad a = A\omega + \dfrac{a}{2}$

or $\quad A = \dfrac{a}{2\omega}$

Hence $v_y = \dfrac{a}{2\omega}\sin\omega t + \dfrac{a}{2}t\cos\omega t$

$$y = -\frac{a}{2\omega^2}\cos\omega t + \frac{a}{2\omega}t\sin\omega t + \frac{a}{2\omega^2}\cos\omega t + C$$

$$y = \frac{at}{2\omega}\sin\omega t + C$$

At $t = 0, y = 0 \therefore C = 0$

Thus $y = \dfrac{at}{2\omega}\sin\omega t$

$$\frac{dv_x}{dt} = \omega\left[\frac{a}{2\omega}\sin\omega t + \frac{at}{2}\cos\omega t\right]$$

On integration

$$v_x = -\frac{a}{2\omega}\cos\omega t + \frac{a}{2}t\sin\omega t + \frac{a}{2\omega}\cos\omega t + C'$$

$$v_x = \frac{a}{2}t\sin\omega t + C'$$

At $t = 0, v_x = 0 \therefore C' = 0$

$$\therefore \quad v_x = \frac{a}{2}t\sin\omega t$$

$$x = \frac{-a}{2\omega}t\cdot\cos\omega t + \frac{a}{2}\int\frac{\cos\omega t}{\omega}dt + C''$$

$$= \frac{-at}{2\omega}\cos\omega t + \frac{a}{2\omega^2}\sin\omega t + C''$$

At $t = 0, x = 0 \therefore C'' = 0$

$$\therefore \quad x = \frac{a}{2\omega^2}(\sin\omega t - \omega t\cos\omega t)$$

16.66. A helium ion of mass m and charge e accelerated in a cyclotron acquires a radius r. Assuming the oscillator frequency as ν and accelerating voltage across the dees as V find the approximate distance covered by the ion during acceleration.

Solution: Total time of acceleration is given by

$$t = \frac{\pi}{\omega}n$$

$$= \frac{1}{2\nu}\cdot n; \text{ } n \text{ is the number of passages on the Dees.}$$

$$\text{Kinetic energy } T = neV = \frac{1}{2}m\left(\frac{B^2e^2r^2}{m^2}\right)$$

$$= \frac{B^2e^2r^2}{2m}$$

$$\therefore \quad n = \frac{B^2er^2}{2mV}$$

$$t = \frac{\pi}{\frac{eB}{m}} \times \frac{B^2 er^2}{2mV}$$

$$= \frac{\pi B r^2}{2V}$$

$$= \frac{\pi}{2V} \cdot r^2 \cdot \frac{2\pi m \nu}{e};\ B = \frac{2\pi m \nu}{e}$$

$$= \frac{\pi^2 m \nu r^2}{eV}$$

The distance covered $s = \sum v_n \cdot \frac{1}{2\nu}$

Where $v_n = \sqrt{\frac{2eV}{m}} \cdot \sqrt{n}$

$$s = \sqrt{\frac{eV}{2m\nu^2}} \sum \sqrt{n}$$

$$= \sqrt{\frac{eV}{2m\nu^2}} \int \sqrt{n}\, dn$$

$$= \frac{2}{3} n^{3/2} \cdot \sqrt{\frac{eV}{2m\nu^2}}$$

$$n = \frac{B^2 e^2 r^2}{2eVm} = \frac{2\pi^2 m \nu^2 r^2}{eV}$$

$$s = \frac{4}{3} \frac{\pi^3 \nu^2 m r^2}{eV}$$

16.67. A particle of charge e and mass m is located at a distance r from the axis of a solenoid. A current through the solenoid produces a magnetic field of induction B. By assuming that during the growth of the current in the solenoid the particle practically assumes the same position find the velocity and radius of curvature of the path of the particle.

Solution: Due to the magnetic field in the solenoid an electric field will be induced and the particle will accelerated. If $\frac{dB}{dt}$ be the rate of growth of the magnetic field then the electric field is given by the equation.

$$2\pi r E = \pi r^2 \frac{dB}{dt}$$

or $$E = \frac{r}{2} \frac{dB}{dt}$$

$$\text{Force} = eE = m \cdot \frac{dv}{dt} = e\frac{r}{2}\frac{dB}{dt}$$

or $$v = \frac{erB}{2m}$$

In the magnetic field the particle will describe a circular path of radius ρ.

Now $$\frac{mv^2}{\rho} = Bev$$

or $$\frac{mv}{Be} = \rho = \frac{r}{2}$$

16.68. Calculate the number of photons of wavelength 5.9×10^{-7} m which produces a force of 10^{-5} N when it falls on a blackened plate. $h = 6.62 \times 10^{-34}$ Js.

Solution: Momentum of a single photon = $\frac{h}{\lambda}$

If n photons are falling per second then total momentum of the incident photon = $\frac{nh}{\lambda}$

According to Newton's second law of motion momentum per second is equal to force. Therefore, $\frac{nh}{\lambda} = 10^{-5}$

On substituting the various values, we have

$$n = \frac{5.9 \times 10^{-7} \times 10^{-5}}{6.62 \times 10^{-34}} = \frac{5.9}{6.62} \times 10^{22}/\text{s}$$
$$= 8.912 \times 10^{21}/\text{s}$$

16.69. Radiations of wavelength $4{,}358 \times 10^{-10}$ m and of intensity 3×10^{-3} W/m^2 fall on a photocell. Find out the number of photoelectrons produced per square metre per second assuming that each photon creates an electron. $h = 6.62 \times 10^{-34}$ Js, $c = 3 \times 10^8$ m/s.

Solution: Let n be the number of photons per square metre per second. Then,

$$nh\nu = \frac{nhc}{\lambda} = 3 \times 10^{-3}$$

or $$n = \frac{\lambda \times 3 \times 10^{-3}}{3 \times 10^8 \times 6.62 \times 10^{-34}} = \frac{4{,}358 \times 10^{-10} \times 3 \times 10^{-3}}{3 \times 10^8 \times 6.62 \times 10^{-34}}$$
$$= \frac{4{,}358}{6.62} \times 10^{13}/\text{m}^2/\text{s}$$
$$= 6.58 \times 10^{15}/\text{m}^2/\text{s}$$

16.70. A ray of ultraviolet light of wavelength 3,000 A falls on the surface of a material of work function 2.85 *eV*, and ejects an electron. What will be the velocity of the emitted electron?

$h = 6.62 \times 10^{-34}$ Js, $c = 3 \times 10^8$ m/s, $m = 9 \times 10^{-31}$ kg.

Solution:

$$\frac{hc}{\lambda} = \phi + \frac{1}{2}mv^2$$

On substituting the various values, we have

$$\frac{6.62 \times 10^{-34} \times 3 \times 10^8}{3 \times 10^{-7}} - 2.85 \times 1.6 \times 10^{-19} = \frac{1}{2}mv^2$$

or $\quad mv^2 = 2 \times 2.06 \times 10^{-19}$

$$\therefore \quad v = \sqrt{\frac{2 \times 2.06 \times 10^{-19}}{9 \times 10^{-31}}} = 6.766 \times 10^5 \text{ m/s}$$

16.71. Using the law of conservation of energy and momentum show that a free electron cannot absorb a photon completely.

Solution: According to theory of relativity the energy of a particle is given by

$$E = \sqrt{m_0^2c^4 + p^2c^2}\,; \; m_0 \text{ is the rest mass and } p \text{ is momentum}$$

According to conservation of momentum we have

$$\frac{h\nu}{c} = p$$

According to conservation of energy we have

$$h\nu + m_0c^2 = \sqrt{m_0^2c^4 + p^2c^2}$$

$$h^2\nu^2 + m_0^2c^4 + 2h\nu\, m_0c^2 = m_0^2c^4 + p^2c^2$$

$$c^2p^2 + 2h\nu\, m_0c^2 = p^2c^2$$

or $\quad 2m_0pc^3 = 0$

This is absurd identity hence the complete absorption of photon by a free electron is not possible.

16.72. Electromagnetic radiation of wavelength $\lambda = 2500$ A falls on a photocell whose spectral sensitivity for this wavelength is $J = 3.2$ mA/W. Calculate the number of photoelectrons produced.

Solution:

$$ne = J \cdot \frac{hc}{\lambda}$$

$$\therefore \quad n = \frac{Jhc}{e\lambda}$$

On substituting the values we have

$$n = \frac{3.2 \times 10^{-3} \times 6.62 \times 10^{-34} \times 3 \times 10^{8}}{1.6 \times 10^{-19} \times 25 \times 10^{-8}}$$

$$= \frac{2 \times 6.62 \times 30 \times 10^{-3}}{25}$$

$$= 0.0159$$

16.73. Electromagnetic radiation having time varying electric component $E = a\,(1 + \cos \omega t) \cos \omega_0 t$ falls on a metal surface of work function ϕ. Calculate the maximum velocity of photoelectrons. Assume $\omega > \omega_0$.

Solution:

$$E = a \cos \omega_0 t + a \cos \omega_0 t \cdot \cos \omega t$$

$$= a \cos \omega_0 t + \frac{a}{2} \cos(\omega + \omega_0) t + \frac{a}{2} \cos(\omega - \omega_0) t$$

The electromagnetic radiation contains three different frequencies.

Maximum velocity will be obtained with frequency $\frac{(\omega + \omega_0)}{2\pi}$

$$\frac{1}{2} m v_{max}^2 = \frac{h}{2\pi}(\omega + \omega_0) - \phi$$

or $$v_{max} = \sqrt{\frac{2}{m}\left[\mathrm{h}\,(\omega + \omega_0) - \phi\right]}$$

16.74. The electric field of a monochromatic radiation becomes zero 1.3×10^{15} times in one second. Calculate the velocity of the photoelectron emitted when these radiations falls on a metal surface of work function $\phi = 2.5\, eV$. Mass of electron $= 9.1 \times 10^{-31}$ kg.

Solution:

$$\text{Frequency of radiation} = \frac{1.3}{2} \times 10^{15} = 0.65 \times 10^{15}$$

$$\frac{1}{2} m v^2 = h\nu - e\phi$$

On substituting the values we have

$$\frac{1}{2} m v^2 = 6.62 \times 10^{-34} \times 0.65 \times 10^{15} - 2.5 \times 1.6 \times 10^{-19}$$

$$= 4.303 \times 10^{-19} - 4 \times 10^{-19}$$

$$= 0.303 \times 10^{-19}$$

$$v = \sqrt{\frac{2 \times 0.303 \times 10^{-19}}{9.1 \times 10^{-31}}} = \sqrt{0.066} \times 10^{6}$$

$$= 2.57 \times 10^{5} \text{ m/sec.}$$

16.75. When the surface of a certain metal is successively illuminated with electromagnetic radiations of wavelength $\lambda_1 = 3{,}500 \times 10^{-10}$ m and $\lambda_2 = 5{,}400 \times 10^{-10}$ m, the maximum velocities of photoelectrons are related as $v_1 = 2v_2$. Calculate the work function of the metal. $h = 6.62 \times 10^{-34}$ Js, $c = 3 \times 10^8$ m/s.

Solution: According to Einstein's photoelectric equation, we have

$$\frac{hc}{\lambda_1} = \phi + \frac{1}{2}mv_1^2$$

and $$\frac{hc}{\lambda_2} = \phi + \frac{1}{2}mv_2^2$$

where v_1 and v_2 are the velocities of the electrons emitted under the action of light of wavelengths λ_1 and λ_2 respectively.

Now $$hc\left[\frac{1}{\lambda_1} - \frac{1}{\lambda_2}\right] = \frac{1}{2}m\left[v_1^2 - v_2^2\right] = \frac{3}{2}mv_2^2;\ v_1 = 2v_2$$

or $$hc\left[\frac{1}{\lambda_1} - \frac{1}{\lambda_2}\right] = 3\left[\frac{hc}{\lambda_2} - \phi\right]$$

or $$3\phi = hc\left[\frac{3}{\lambda_2} + \frac{1}{\lambda_2} - \frac{1}{\lambda_1}\right]$$

$$3\phi = hc\left[\frac{4}{\lambda_2} - \frac{1}{\lambda_1}\right]$$

or $$\phi = \frac{hc}{3}\left[\frac{4}{\lambda_2} - \frac{1}{\lambda_1}\right]$$

On substituting the values, we have

$$\phi = \frac{6.62 \times 10^{-34} \times 3 \times 10^8}{3 \times 10^{-9}}\left[\frac{4}{540} - \frac{1}{350}\right] = 3.012 \times 10^{-19}\ \text{J}$$

16.76. Up to what potential V can a zinc ball removed from other bodies be charged by irradiating it with ultraviolet radiation of wavelength = 2500 Å. Work function of zinc = 3.74 eV. $h = 6.62 \times 10^{-34}$ Js, $c = 3 \times 10^8$ m/s.

Solution:

$$\frac{hc}{\lambda} = \phi + \frac{1}{2}mv^2$$

$$\frac{1}{2}mv^2 = \frac{hc}{\lambda} - \phi$$

$$= \left[\frac{6.62 \times 10^{-34} \times 3 \times 10^8}{25 \times 10^{-8} \times 1.6 \times 10^{-19}} - 3.74\right] eV$$

$$= (4.965 - 3.74)\ eV$$
$$= 1.225\ eV$$

An energy of 1.225 *eV* corresponds to a potential of 1.225 V. Thus the potential of the zinc ball will be 1.225 V.

16.77. A photocell has one electrode made of cesium and the other of copper. The two electrodes are interconnected outside the cell. Calculate the maximum velocity of electrons reaching the copper electrode when a radiation of wavelength 2,200 Å falls on the cesium electrode. Work function of cesium = 1.89 *eV*. Work function of copper = 4.47 *eV*.

$h = 6.62 \times 10^{-34}$ Js, $c = 3 \times 10^8$ m/s, $m = 9 \times 10^{-31}$ kg.

Solution:

$$\frac{hc}{\lambda} = \phi + \frac{1}{2}mv^2$$

On substituting the values, we have

$$\left(\frac{6.62\times10^{-34}\times3\times10^{8}}{22\times10^{-8}} - 4.47\times1.6\times10^{-19}\right) = \frac{1}{2}mv^2$$

$$\sqrt{\frac{(9.028-7.152)\times10^{-19}\times2}{9\times10^{-31}}} = v$$

$$\sqrt{\frac{1.876\times2}{9}}\times10^{6} = v$$

$$v = 6.457 \times 10^5\ \text{m/s}$$

16.78. A radiation of wavelength 2340 Å can produce photoelectric effect in platinum. Calculate the kinetic energy acquired by the electrons due to radiation of wavelength 2000 Å. $h = 6.62 \times 10^{-34}$ Js, $c = 3 \times 10^8$ m/s.

Solution:

$$\frac{hc}{\lambda} = \frac{hc}{\lambda_0} + \frac{1}{2}mv^2$$

$$\therefore \quad \frac{1}{2}mv^2 = \text{kinetic energy of electron} = hc\left(\frac{1}{\lambda} - \frac{1}{\lambda_0}\right) = hc\left(\frac{\lambda_0 - \lambda}{\lambda\lambda_0}\right)\ \text{J}$$

On substituting the values, we have

$$KE = \frac{6.62\times10^{-34}\times3\times10^{8}\times34\times10^{-9}}{200\times234\times10^{-18}}$$

$$= 1.443 \times 10^{-19}\ \text{J}$$

16.79. If 3% of the energy supplied to an incandescent lamp is radiated as visible light, how many visible quanta are emitted per second by a 100 W bulb? Assume wavelength of all the visible light to be 5500 Å.

$h = 6.62 \times 10^{-34}$ Js, $c = 3 \times 10^8$ m/s.

Solution: Let there be n photons per second then according to conservation of energy we have

$$\frac{nhc}{\lambda} = 100 \times \frac{3}{100}$$

or $$n = \frac{3\lambda}{hc}$$

On substituting the values, we have

$$n = \frac{3 \times 55 \times 10^{-8}}{6.62 \times 10^{-34} \times 3 \times 10^{8}}$$

$$= \frac{55}{6.62} \times 10^{18}$$

$$= 8.31 \times 10^{18} \text{ /s}$$

16.80. What will be the change in the stopping potential for photoelectrons emitted from a surface if the wavelength of the incident light is reduced from 3000 to 2500 Å? $h = 6.62 \times 10^{-34}$ Js, $c = 3 \times 10^{8}$ m/s.

Solution:

$$\frac{hc}{\lambda_1} = \phi + eV_1$$

and $$\frac{hc}{\lambda_2} = \phi + eV_2$$

$$hc\left(\frac{1}{\lambda_1} - \frac{1}{\lambda_2}\right) = e(V_1 - V_2)$$

$$\frac{hc}{10^{-9}}\left(\frac{1}{300} - \frac{1}{250}\right) = -e\Delta V$$

$$\frac{hc \times 1}{15 \times 10^{-7}} = e\Delta V$$

$$\therefore \quad \Delta V = \frac{6.62 \times 10^{-34} \times 3 \times 10^{8}}{15 \times 10^{-7} \times 1.6 \times 10^{-19}}$$

$$= \frac{6.62}{8}$$

$$= 0.828 \text{ V}$$

16.81. The threshold wavelength for photoelectric emission in tungsten is 2300 Å. What wavelength of light must be used for electrons with a maximum energy of 1.4 eV to be ejected? $h = 6.62 \times 10^{-34}$ Js, $c = 3 \times 10^{8}$ m/s.

Solution:

$$\frac{hc}{\lambda} = \frac{hc}{\lambda_0} + \frac{1}{2}mv^2; \quad \frac{1}{2}mv^2 = eV$$

$$\frac{hc}{\lambda} = \left(\frac{6.62 \times 10^{-34} \times 3 \times 10^{8}}{23 \times 10^{-8}} + 1.4 \times 1.6 \times 10^{-19}\right)$$

$$= \left(\frac{6.62 \times 30}{23} + 2.24\right) \times 10^{-19}$$

$$= 10.875 \times 10^{-19}$$

$$\text{Now } \frac{hc}{\lambda} = 10.875 \times 10^{-19}$$

$$\therefore \quad \lambda = \frac{hc}{10.875 \times 10^{-19}}$$

$$= \frac{6.62 \times 10^{-34} \times 3 \times 10^{8}}{10.875 \times 10^{-19}}$$

$$= 1.826 \times 10^{-7} \text{ m}$$

$$= 1826 \text{ Å}$$

16.82. The two electrodes of a vacuum photocell are made of zinc of work function 3.74 *eV*. When an electromagnetic radiation of wavelength 2620 Å falls on the electrode photoelectric current is obtained. If an external decelerating potential 1.5 V is applied the current becomes zero. Find the contact potential difference of the photocell.
$h = 6.62 \times 10^{-34}$ Js, $c = 3 \times 10^{8}$ m/s.

Solution:

Energy corresponding to 2620 Å is $\frac{hc}{\lambda} = \frac{6.62 \times 10^{-34} \times 3 \times 10^{8}}{262 \times 10^{-9}} = 4.74\, eV$

Contact potential difference $= (1.5 + 3.74) - 4.74$

$= 0.5$ V

16.83. A piece of aluminium is placed at a distance of 10 cm from a large charged copper plate. The surface charge density of the copper plate facing the aluminium piece is 1.5×10^{-9} c/m^2. Monochromatic ultraviolet radiation of wavelength $\lambda = 3000$ Å falls on the aluminium. The maximum kinetic energy of the electron emitted from aluminium and reaching copper plate is 17.3535 *eV*. Calculate the work function of aluminium.

Solution:

Energy of photon $= \frac{hc}{e\lambda} eV = \frac{6.62 \times 10^{-34} \times 3 \times 10^{8}}{1.6 \times 10^{-19} \times 3 \times 10^{-7}} = 4.1375\, eV$

K.E of the electron emitted from the Aluminium $= 4.1375 - \phi$;

ϕ is work function in electron volt.

Electric field due to the charged plate $= \frac{\sigma}{\epsilon_0}$

$$\text{Work done} = \frac{e\sigma}{\epsilon_0}\cdot x;\ x \text{ is the distance between the two plates}$$

$= $ K.E. of electrons reaching the copper plate without any initial energy

$$= \frac{4\pi\sigma x}{4\pi\epsilon_0} \text{ in } eV$$

$$= 9\times 10^9 \times 4 \times 3.14 \times 1.5 \times 10^{-9} \times 0.1$$

$$= 3.6 \times 3.14 \times 1.5 = 16.956\ eV$$

$$\text{Max K.E.} = 4.1375 - \phi + 16.956 = 17.3535$$

$$\therefore \quad \phi = 3.74\ eV$$

16.84. Ultraviolet radiations of wavelength 2500 Å falls on a aluminium plate of work function 3.74 eV. An electron absorbs the photon completely and before coming out of the plate it suffers collision and at each collision 10% of the energy is lost. Calculate the number of collision after which the electron will be incapable of coming out of the aluminium plate.

Solution: 2500 Å radiation will have energy = 4.965 eV

Energy left after Ist collision = 4.965 – 0.4965 = 4.468 eV

Energy left after second collision = 4.468 – 0.4468 = 4.021

Energy left after third collision = 4.021 – 0.402 = 3.619 eV

This energy of 3.619 eV is less than work function and due to this electron will be incapable of coming out of the aluminium plate.

Hence the number of collision = 3.

16.85. A small plate of a metal (work function = 1.17 eV) is placed at a distance of 2 m from a monochromatic light source of wavelength 4.8×10^{-7} m and power 1.0 watt. The light falls normally on the plate. Find the number of photons striking the metal plate per square meter/sec. If a constant magnetic field of strength 10^{-4} tesla is applied parallel to the metal surface, find the radius of the largest circular path followed by the emitted photoelectrons. (Roorkee 1991)

Solution:

$$n\frac{hc}{\lambda} = 1;\ n = \text{number of photons per sec.}$$

$$\therefore \quad n = \frac{\lambda}{hc}$$

$$= \frac{4.8\times 10^{-7}}{6.62\times 10^{-34}\times 3\times 10^{8}}$$

$$= \frac{48}{6.62\times 3}\times 10^{18}$$

$$= 2.417\times 10^{18}$$

Number of photons striking per unit area per sec

$$= \frac{2.417\times10^{18}}{4\times3.14\times2^2} = 4.81\times10^{16}$$

Kinetic energy of electrons $= \dfrac{hc}{\lambda} - \phi$; ϕ = work function

$$\frac{1}{2}mv^2 = \left[\frac{6.62\times10^{-34}\times3\times10^8}{4.8\times10^{-7}} - 1.17\times1.6\times10^{-19}\right]$$

$$= [4.1375 - 1.872]\times10^{-19}$$

$$\frac{1}{2}mv^2 = 2.2655\times10^{-19}$$

$$\therefore \quad v = \sqrt{\frac{2\times2.2655\times10^{-19}}{9.1\times10^{-31}}}$$

$$= 7.056\times10^5 \text{ m/sec}$$

Radius of the circular path

$$r = \frac{mv}{eB} = \frac{9.1\times10^{-31}\times7.056\times10^5}{1.6\times10^{-19}\times10^{-4}}$$

$$= 40.13\times10^{-3}$$

$$= 4.013 \text{ cm}$$

16.86. In an experiment on photoelectric emission, following observations are made : (i) wavelength of incident light = 1.98×10^{-7} m (ii) stopping potential = 2.5 volt. Find (a) threshold frequency (b) work function and (c) energy of photoelectrons with maximum speed. $h = 6.6\times10^{-34}$ J.sec.

(Roorkee 1992)

Solution:

$$\frac{1}{2}mv^2 = eV;\ V = \text{stopping potential}$$

$$= 1.6\times10^{-19}\times2.5$$

$$= 4.0\times10^{-19} \text{ Joule}$$

Energy of the incident photon $= \dfrac{hc}{\lambda}$

$$= \frac{6.6\times10^{-34}\times3\times10^8}{1.98\times10^{-7}}$$

$$= 10^{-18} \text{ Joule}$$

Work function $= 10^{-18} - 4\times10^{-19}$

$$= 6\times10^{-19} \text{ Joule}$$

Threshold frequency $= \dfrac{6\times10^{-19}}{6.6\times10^{-34}}$

$$= 9.09\times10^{18} \text{ Hz}$$

16.87. The stopping potential for the photoelectron emitted from a metal surface of work function 1.7 *eV* is 10.4 *V*. Find the wavelength of the radiation used. Also identify the energy levels in hydrogen atom which will emit this wavelength.

Solution: Let V_0 be the stopping potential then

$$eV_0 = \frac{hc}{\lambda} - \phi; \ \phi = \text{work function}$$

$$\therefore \qquad \lambda = \frac{hc}{eV_0 + \phi}$$

On substituting the values of various terms we have

$$\lambda = \frac{6.62 \times 10^{-34} \times 3 \times 10^8}{(1.7 + 10.4) \times 1.6 \times 10^{-19}}$$

$$= 1.026 \times 10^{-7} \text{ m} = 1026 \text{ Å}$$

Using the relation

$$\frac{1}{\lambda} = R\left[\frac{1}{n_1^2} - \frac{1}{n_2^2}\right]$$

We have

$$\frac{1}{1.026 \times 10^{-7}} = 1.1 \times 10^7 \left[\frac{1}{1} - \frac{1}{n_2^2}\right]$$

$$1 - \frac{1}{n_2^2} = \frac{1}{1.1 \times 1.026}$$

$$\frac{1}{n_2^2} = 1 - \frac{1}{1.1 \times 1.026}$$

$$= \frac{0.1286}{1.1 \times 1.026}$$

$$n_2^2 = \frac{1.1 \times 1.026}{0.1286}$$

$$n_2^2 = 8.776$$

$$n = 2.96 = 3$$

16.88. A light source, emitting three wavelengths 5000 Å, 6000 Å and 7000 Å has a total power of 10^{-3} W and a beam diameter 2×10^{-3} m. The power density is distributed equally amongst the three wavelengths. The beam shines normally on a metallic surface of area on 10^{-4} m^2 and having a work function of 1.9 *eV*. Assume that each photon liberates an electron, calculate the charge emitted per unit area in one second. (Roorkee 1995)

Solution:

$$\text{Power per unit area} = \frac{10^{-3} \times 4}{3.14 \times (2 \times 10^{-3})^2} = 318 \text{ W/m}^2$$

Power per unit area with each wavelength $= \frac{318}{3} = 106\ \text{W/m}^2$

Energy corresponding to the three wavelengths in electron volt is equal to

For 5000 Å energy $= \frac{hc}{e\lambda} = \frac{6.62\times10^{-34}\times3\times10^{8}}{5\times10^{-7}\times1.6\times10^{-19}} = 2.48\ eV$

For 6000 Å energy $= \frac{hc}{e\lambda} = \frac{6.62\times10^{-34}\times3\times10^{8}}{6\times10^{-7}\times1.6\times10^{-19}} = 2.07\ eV$

For 7000 Å energy $= \frac{hc}{e\lambda} = \frac{6.62\times10^{-34}\times3\times10^{8}}{7\times10^{-7}\times1.6\times10^{-19}} = 1.77\ eV$

In view of the energy of the photon the electron emission will take place with 5000 and 6000 Å

No. of electrons emitted with 5000 Å/m²/sec

$$= \frac{106\times5\times10^{-7}}{6.62\times10^{-34}\times3\times10^{8}} = 2.67\times10^{20}$$

No. of electrons emitted with 6000 Å/m²/sec

$$= \frac{106\times6\times10^{-7}}{6.62\times10^{-34}\times3\times10^{8}} = 3.20\times10^{20}$$

Charge emitted per unit area per sec $= (2.67 + 3.2)\times10^{20}\times1.6\times10^{-19}$

$= 5.87\times16 = 93.92\ \text{c/m}^2\text{/sec.}$

16.89. Ultraviolet light of wavelength 800 A and 700 A when allowed to fall on hydrogen atom in their ground state is found to liberate electrons with kinetic energies 1.8 eV and 4.0 eV respectively. Find the value of Planck's constant.

Solution:

$$\frac{hc}{e\lambda_1} - 13.6 = E_1$$

and $$\frac{hc}{e\lambda_2} - 13.6 = E_2$$

On substituting the two, we have

$$\frac{hc}{e}\left[\frac{1}{\lambda_2} - \frac{1}{\lambda_1}\right] = E_2 - E_1$$

or $$h = \frac{(E_2 - E_1)\lambda_1\lambda_2 e}{(\lambda_1 - \lambda_2)c}$$

On putting the values of various terms, we have

$$h = \frac{(4-1.8)\times 800\times 10^{-10}\times 700\times 10^{-10}\times 1.6\times 10^{-19}}{100\times 10^{-10}\times 3\times 10^{8}}$$

$= 6.57 \times 10^{-34}$ J.sec.

16.90. A mercury vapour lamp with quartz envelop emits radiation of wavelengths 5790, 5770, 5461, 4358, 4046 and 2536A. Calculate the maximum kinetic energy of the emitted electrons when these radiations fall on a silver plate with work function of 4.28 *eV*.

Solution: The energy of photon in electron volt is given by

$$E = \frac{hc}{e\lambda}$$

On putting the values we have

$$E = \frac{6.62\times 10^{-34}\times 3\times 10^{8}}{1.6\times 10^{-19}\times 2536\times 10^{-10}} = 4.8945\, eV$$

K.E of electrons = 4.8945 – 4.28 = 0.9832×10^{-19} Joule

16.91. A free mercury atom initially at rest emits a photon of wavelength $\lambda = 5461$ Å. What change in the wavelength of the photon will be observed due to the recoil experienced by the atom? $N = 6.02 \times 10^{23}$. Atomic wt of mercury = 200.5 gm.

Solution: According to conservation of energy we have

$$\frac{hc}{\lambda'} = \frac{hc}{\lambda} + \frac{1}{2}mv^2;\ m = \text{mass of mercury atom and } v \text{ is its velocity}$$

Further $mv = \dfrac{h}{\lambda}$

$$\therefore \quad \frac{hc}{\lambda'} = \frac{hc}{\lambda} + \frac{1}{2}m\cdot\frac{h^2}{m^2\lambda^2}$$

or
$$\frac{1}{\lambda'} = \frac{1}{\lambda} + \frac{h}{2mc\lambda^2}$$

$$\frac{\lambda}{\lambda'} = 1 + \frac{h}{2mc\lambda}$$

or
$$1 - \frac{\lambda'}{\lambda} = \frac{h}{2mc\lambda}$$

or
$$\Delta\lambda = \lambda - \lambda' = \frac{h}{2mc}$$

On putting the values we have

$$\Delta\lambda = \frac{6.62\times 10^{-34}\times 6.02\times 10^{23}}{2\times 200.5\times 10^{-3}\times 3\times 10^{8}}$$

$= 0.033 \times 10^{-16}$ m

16.92. An ionization chamber is constructed by using two electrodes of area $A = 120$ cm^2 placed at a distance of $d = 8$ cm apart. In the process of ionisation per sec $n = 10^9$ monovalent ions of both the signs are obtained in 1 cm^3. Calculate the saturation current.

Solution:

$$I = \frac{Adne}{10^{-6}}$$

$$= \frac{120\times10^{-4}\times8\times10^{-2}\times10^{9}\times1.6\times10^{-19}}{10^{-6}} = 0.1536\ \mu\text{A}$$

16.93. In a discharge tube the two electrodes are $d = 5$ cm apart and a potential difference of $V = 10$ volt is applied across it. As a result of this the gas in the tube becomes singly ionised and the number of +ve and –ve ions per m^3 is 10^9. Further $v_+ = 3 \times 10^{-2}$ m^2/v.s and $v_- = 3 \times 10^2$ m^2/v.s. Find (i) the current density (ii) the part of the total current constituted by +ve ions.

Solution:

(i) $$J = en\left(v_+ + v_{-1}\right)\frac{V}{d}$$

$$= 1.6\times10^{-19}\times10^{9}\left[3\times10^{-2}+3\times10^{2}\right]\times\frac{10}{5\times10^{-2}}$$

$$= 1.6\times10^{-10}\times300.03\times\frac{2}{10^{-2}}$$

$$= 9.6\ \mu\text{A/m}^2$$

(ii) $$\frac{0.03}{300}\times100 = 0.01\%$$

16.94. A 100 cm long sealed tube and having area of cross-section 10 mm^2 is filled with air. The air is ionised using radiation. The tube contains 10^7 pairs of monovalent ions in 1 cm^3 volume with mobilities $v_+ = 1.4 \times 10^{-4}$ m^2/v.s. and $v_- = 1.8 \times 10^{-4}$ m^2/v.s. Calculate the resistance of the tube.

Solution:

$$J = \frac{I}{A} = en\left[v_+ + v_-\right]\frac{V}{d}$$

or $$V = \frac{Id}{Aen\left[v_+ + v_-\right]}$$

$\therefore$ $$R = \frac{d}{Aen\left[v_+ + v_{-1}\right]}$$

On putting the values we have

$$R = \frac{10^{-6}}{10\times10^{-6}\times1.6\times10^{-19}\times10^{7}\times3.2\times10^{-4}}$$

$$= \frac{100 \times 10^{14}}{16 \times 3.2}$$

$$= 1.95 \times 10^{14}\ \Omega$$

16.95. Air at N.T.P is irradiated with X-rays with a radiation dose of 0.5 R_0. Calculate the number of ion pairs obtained in 1 cm^3 volume. Mol wt of air = 29 kg/K mol, 1R_0 = 2.58 × 10^{-4} C/kg.

Solution:

$$PV = \frac{m}{M} RT$$

$$\therefore \quad m = \frac{PVM}{RT}$$

$$q = 0.5\, R_0 \cdot m = 0.5 R_0 \cdot \frac{PVM}{RT}$$

$$n = \frac{q}{e} = \frac{0.5\, R_0 \cdot PVM}{eRT}$$

On substituting the values of various terms we have

$$n = \frac{0.5 \times 2.58 \times 10^{-4} \times 0.76 \times 13.6 \times 10^{3} \times 9.8 \times 10^{-6} \times 29}{8.31 \times 10^{3} \times 273 \times 1.6 \times 10^{-19}}$$

$$= \frac{0.5 \times 2.58 \times 0.76 \times 13.6 \times 9.8 \times 29}{8.31 \times 273 \times 1.6} \times 10^{9}$$

$$= 1.04 \times 10^{9}$$

16.96. An ionisation chamber of volume 10 cc contains air at N.T.P. The air in the chamber is ionised by irradiating it with X-rays with a dose of 0.5 mR/hr. Calculate the saturation current. Mol wt = 29 × 10^{-3} kg.

Solution:

$$I = \frac{0.5 \times 10^{-3} \times 2.58 \times 10^{-4}}{3600} \times m$$

$$= \frac{0.5 \times 10^{-3} \times 2.58 \times 10^{-4}}{3600} \times \frac{PVM}{RT}$$

$$= \frac{0.5 \times 10^{-3} \times 2.58 \times 10^{-4} \times 0.76 \times 13.6 \times 10^{3} \times 9.8 \times 10 \times 10^{-6} \times 29 \times 10^{-3}}{3600 \times 8.31 \times 273}$$

$$= \frac{0.5 \times 2.58 \times 0.76 \times 13.6 \times 9.8 \times 29}{36 \times 831 \times 273} \times 10^{-12}$$

$$= 4.64 \times 10^{-16} \text{ amp.}$$

16.97. Air enclosed in a certain volume is irradiated with X-ray of dose 5R. Calculate the part of the atoms which will be ionised. Mol wt of air = 29 × 10^{-3} kg.

Solution:

$$n = \frac{29 \times 10^{-3} \times 5 \times 2.58 \times 10^{-4}}{6.02 \times 10^{23} \times 1.6 \times 10^{-19}}$$

$$= \frac{29 \times 5 \times 2.58 \times 10^{-11}}{6.02 \times 1.6}$$

$$= 3.88 \times 10^{-10}$$

16.98. A potential difference of 60 KV is applied across the electrodes of an X-ray tube. The minimum wavelength of X-ray produced is 0.207 A. Find the value of the Planck's constant.

Solution:

$$eV = \frac{hc}{\lambda_{\min}}$$

$$\therefore \quad h = \frac{eV}{c} \lambda_{\min}$$

On substituting the values of various terms we have

$$h = \frac{1.6 \times 10^{-19} \times 60 \times 10^{3} \times 0.207 \times 10^{-10}}{3 \times 10^{8}}$$

$$= 6.624 \times 10^{-34} \text{ J.sec.}$$

16.99. Find the value of screening constant for a L radiation of wavelength $\lambda = 1.43$ A produced from a tungsten target. Atomic number of tungsten = 74 and $R = 1.097 \times 10^{7}$ /m.

Solution:

$$\sqrt{\upsilon} = \sqrt{Rc\left[\frac{1}{n_1^2} - \frac{1}{n_2^2}\right]}(Z - b)$$

$$\sqrt{\frac{c}{\lambda}} = \sqrt{Rc\left[\frac{1}{2^2} - \frac{1}{3^2}\right]}(Z - b)$$

$$= \sqrt{\frac{5Rc}{36}}(Z - b)$$

$$\sqrt{\frac{36}{5\lambda R}} = Z - b$$

On substituting the values we have

$$\sqrt{\frac{36}{5 \times 1.43 \times 10^{-10} \times 1.097 \times 10^{7}}} = 74 - b$$

$$67.75 = 74 - b$$

or $\quad b = 74 - 67.75$

$$= 6.25$$

16.100. On increasing the potential applied to an X-ray tube by $\eta = 2$ times the λ_{min} of continuous X-ray spectrum shifts by $\Delta\lambda = 20$ pm. Calculate the initial voltage applied across the tube.

Solution: According to Duane & Hunt law

$$eV = \frac{hc}{\lambda_{min}}$$

Now $$eV_1 = \frac{hc}{\lambda_1}$$

and $$e\eta V_1 = \frac{hc}{\lambda_2}$$

$$\therefore \quad \lambda_1 - \lambda_2 = \Delta\lambda = \frac{hc}{eV_1}\left(1 - \frac{1}{\eta}\right)$$

or $$V_1 = \frac{hc}{e \cdot \Delta\lambda}\left(1 - \frac{1}{\eta}\right)$$

On substituting the values of various terms in the above equation we have

$$V_1 = \frac{6.62 \times 10^{-34} \times 3 \times 10^8}{1.6 \times 10^{-19} \times 20 \times 10^{-12}} \times \frac{1}{2}$$
$$= 3.104 \times 10^4 \text{ volt.}$$

16.101. The wavelength of K_α line for tungsten ($Z = 74$) is 0.021381 nm and 0.056378 for silver ($Z = 47$). Find the value of the two constant a and b of the Mosley's relation.

Solution:

$$\sqrt{v} = a(Z - b)$$

or $$\sqrt{\frac{c}{\lambda}} = a(Z - b)$$

On substituting the values we have

$$\sqrt{\frac{c}{0.021381 \times 10^{-9}}} = a(74 - b)$$

and $$\sqrt{\frac{c}{0.056378 \times 10^{-9}}} = a(47 - b)$$

On dividing the two equations we have

$$\sqrt{\frac{56378}{21381}} = \frac{74 - b}{47 - b}$$

$$1.6238 = \frac{74 - b}{47 - b}$$

$$76.3186 - 1.6238b = 74 - b$$

$$2.3186 = 0.6238b$$

$$\therefore \quad b = \frac{2.3186}{0.6238} = 3.71$$

Now $\sqrt{\dfrac{3 \times 10^8}{21381 \times 10^{-15}}} = a\,(74 - 3.71)$

$$= a \times 70.29$$

$$\therefore \quad a = \frac{3.7458 \times 10^9}{70.29}$$

$$= 0.533 \times 10^8$$

16.102. The wavelength of K_α line of iron ($Z = 26$) is equal to 193 pm. Find the wavelength of K_α line of copper ($Z = 29$).

Solution: From Mosley's law, the wavelength of K_α line is given by

$$\frac{1}{\lambda} = R(Z - 1)^2 \left(\frac{1}{1^2} - \frac{1}{2^2}\right)$$

For the two elements we have

$$\frac{1}{\lambda_{cu}} = R(29 - 1)^2 \left(\frac{1}{1^2} - \frac{1}{2^2}\right)$$

and $\dfrac{1}{193 \times 10^{-12}} = R(26 - 1)^2 \left(\dfrac{1}{1^2} - \dfrac{1}{2^2}\right)$

On dividing the two equations we have

$$\frac{193 \times 10^{-12}}{\lambda_{cu}} = \frac{28^2}{25^2}$$

$$\therefore \quad \lambda_{cu} = 193 \times \frac{25^2}{28^2} \times 10^{-12}$$

$$= 153.86 \text{ pm.}$$

16.103. How many elements are there in a row between those whose wavelengths of K_α lines are equal to 250 and 179 pm? $R = 1.097 \times 10^7$ /m.

Solution: According to Mosley's Law we have

$$\frac{1}{\lambda_1} = R(Z_1 - 1)^2 \left(\frac{1}{1^2} - \frac{1}{2^2}\right)$$

and $\dfrac{1}{\lambda_2} = R(Z_2 - 1)^2 \left(\dfrac{1}{1^2} - \dfrac{1}{2^2}\right)$

Now $(Z_1 - 1) = \sqrt{\dfrac{4}{3\lambda_1 R}}$

and $(Z_2 - 1) = \sqrt{\dfrac{4}{3R\lambda_2}}$

$$Z_1 = \sqrt{\frac{4}{3\lambda_1 R}} + 1 = \sqrt{\frac{4}{3 \times 179 \times 10^{-12} \times 1.097 \times 10^7}} + 1 = 27.057$$

$$Z_2 = \sqrt{\frac{4}{3\lambda_2 R}} + 1 = \sqrt{\frac{4}{3 \times 250 \times 10^{-12} \times 1.097 \times 10^7}} + 1 = 23.05$$

The number of elements lying between the two Z values are three with atomic numbers 24, 25 and 26.

16.104. On increasing the voltage across an X-ray tube from $V_1 = 10$ KV to $V_2 = 20$ KV the wavelength separation between the characteristic K_α line and λ_{min} of continuous X-ray becomes three times the initial value. Find the atomic number of the target. $R = 1.097 \times 10^7$ /m.

Solution: We have the following equations for the characteristic K_α line and continuous X-rays.

$$\frac{1}{\lambda} = R(Z-1)^2\left(\frac{1}{1^2} - \frac{1}{2^2}\right) = \frac{3}{4}R(Z-1)^2$$

and $\dfrac{hc}{\lambda_{min}} = eV$

In view of the problem we have

$$3\left[\frac{4}{3R(Z-1)^2} - \frac{hc}{eV_1}\right] = \frac{4}{3R(Z-1)^2} - \frac{hc}{eV_2}$$

or $$\frac{8}{3R(Z-1)^2} = \frac{hc}{e}\left[\frac{3}{V_1} - \frac{1}{V_2}\right]$$

or $$\frac{8}{3R(Z-1)^2} = \frac{hc}{4e \times 10^3}$$

or $$\sqrt{\frac{32\, e \times 10^3}{3R \cdot h \cdot c}} + 1 = Z$$

On substituting the values we have

$$Z = \sqrt{\frac{32 \times 1.6 \times 10^{-19} \times 10^3}{3 \times 1.097 \times 10^7 \times 6.62 \times 10^{-34} \times 3 \times 10^8}} + 1$$

$$= 29$$

16.105. The K_α radiation obtained from an X-ray tube with Zinc ($Z = 30$) as target falls on an atom and liberates photoelectrons from the K shell of the atom whose K absorption edge has wavelength $\lambda = 174$ pm. Find the Kinetic energy of photoelectrons. $R = 1.097 \times 10^7$ /m.

Solution: The wavelength of K_α radiation is given by

$$\frac{1}{\lambda} = R(Z-1)^2 \times \left(\frac{1}{1^2} - \frac{1}{2^2}\right)$$

$$\text{Energy of } K_\alpha \text{ radiation in } eV = \frac{hc}{e\lambda} = \frac{Rhc(Z-1)^2}{3} \times \frac{3}{4} eV$$

On putting the values we have

Energy of K_α radiation in eV

$$= \frac{1.097 \times 10^7 \times 6.62 \times 10^{-34} \times 3 \times 10^8 \times 29^2 \times 3}{1.6 \times 10^{-19} \times 4}$$

$$= 8.5886\, KeV$$

Energy corresponding to K absorption edge in eV

$$= \frac{hc}{e\lambda} = \frac{6.62 \times 10^{-34} \times 3 \times 10^8}{1.6 \times 10^{-19} \times 174 \times 10^{-12}}$$

$$= 7.1336\, KeV$$

$\therefore$ K.E of photoelectrons $= (8.5886 - 7.1336)\, KeV$

$$= 1.455\, KeV$$

16.106. When an electron is transferred in an atom from L shell to K shell the emitted X-rays have a wavelength of 0.788 Å. Taking $\sigma = 1$ identify the element.

Solution:

$$\frac{1}{\lambda} = R(Z-1)^2 \left[\frac{1}{1^2} - \frac{1}{2^2}\right]$$

$$\frac{10^{10}}{0.788} = 1.097 \times 10^7 (Z-1)^2 \times \frac{3}{4}$$

$$Z - 1 = \sqrt{\frac{4 \times 10^3}{0.788 \times 1.097 \times 3}}$$

$$= 39$$

$$Z = 40$$

The element is Zirconium.

16.107. Continuous X-rays are produced when a voltage of 20 KV is applied to an X-ray tube. By what amount the short wavelength limit of continuous X-rays will be shifted when the voltage applied across the tube is increased by 40 KV?

Solution:

$$\lambda_1 = \frac{hc}{e \times 20 \times 10^3}$$

$$\lambda_2 = \frac{hc}{e \times 40 \times 10^3}$$

$$\Delta\lambda = \lambda_1 - \lambda_2 = \frac{hc}{e \times 40 \times 10^3}$$

On substituting the values of various terms we have

$$\Delta\lambda = \frac{6.62 \times 10^{-34} \times 3 \times 10^8}{1.6 \times 10^{-19} \times 40 \times 10^3}$$

$$= 31 \text{ pm}$$

16.108. In the Thomson's model of the atom +ve charge e is assumed to be distributed inside a sphere of radius 10^{-8} cm. How should the +ve charge density depend on the distance from the centre of the sphere? If an electron inside the sphere oscillates harmonically find the frequency of oscillation.

Solution:

$$\int_0^e dq = \int_0^R \rho \cdot 4\pi r^2 dr; \ R = \text{radius of the sphere}$$

$$e = \frac{4\pi}{3} \cdot R^3 \cdot \rho$$

or $$\rho = \frac{3e}{4\pi R^3}$$

Electric field inside a solid sphere of radius $x = \frac{\rho x}{3\epsilon_0}$

$$\text{Force} = m \times \text{acceleration} = \frac{\rho x e}{3\epsilon_0} = \frac{e}{3\epsilon_0} \cdot \frac{3e}{4\pi R^3} \cdot x$$

$$= \frac{e^2 x}{4\pi R^3}$$

$$\text{Acceleration} = \frac{e^2}{4\pi m R^3} \cdot x$$

$$\omega = 2\pi \upsilon = \sqrt{\frac{e^2}{4\pi m R^3}}$$

or $$\upsilon = \frac{1}{2\pi} \cdot \sqrt{\frac{e^2}{4\pi m R^3}}$$

16.109. An electron in the hydrogen atom moves around the nucleus in a circular orbit of radius $r = 10^{-10}$ m.

(a) What will be frequency of the light emitted by the electron.

(b) How the frequency of the light emitted varies with the radius of the orbit?

(c) How long will the electron take to reach the nucleus of radius $R = 10^{-14}$ m? Assume that an electron moving with an acceleration a loses its energy due to radiation as

$$\frac{dE}{dt} = -\frac{2e^2a^2}{3\cdot 4\pi\epsilon_0 c^3}$$

Solution: The equation of motion in a circular orbit of radius r is given by

(a) $$\frac{mv^2}{r} = \frac{e^2}{4\pi\epsilon_0 r^2}$$

$$v = \sqrt{\frac{e^2}{4\pi\epsilon_0 mr}}$$

Frequency of electron around the nucleus

$$n = \frac{v}{2\pi r} = \sqrt{\frac{e^2}{4\pi\epsilon_0 mr}}\cdot\frac{1}{2\pi r}$$

$$= \frac{e}{\sqrt{4\pi\epsilon_0 mr}}\cdot\frac{1}{2\pi r}$$

$$= \frac{1.6\times 10^{-19}\times(9\times 10^9)^{1/2}}{\sqrt{9\times 10^{-31}\times 10^{-10}}}\times\frac{1}{2\times 3.14\times 10^{-10}}$$

$$= 2.547\times 10^{15}$$

(b) The frequency varies as $r^{-3/2}$

(c) Acceleration $a = \dfrac{v^2}{r} = \dfrac{e^2}{4\pi\epsilon_0 mr^2}$

$$\frac{dE}{dt} = -\frac{2e^2}{3\cdot 4\pi\epsilon_0}\frac{1}{c^3}\cdot\frac{e^4}{(4\pi\epsilon_0)^2\cdot m^2r^4}$$

$$= -\frac{2}{3}\cdot\frac{e^6}{(4\pi\epsilon_0)^3}\cdot\frac{1}{m^2r^4\cdot c^3}$$

$$E = \text{total energy} = \frac{1}{2}mv^2 - \frac{e^2}{4\pi\epsilon_0 r}$$

$$\frac{1}{2}mv^2 = \frac{e^2}{8\pi\epsilon_0 r}$$

$$E = -\frac{e^2}{8\pi\epsilon_0 r}$$

$$\frac{dE}{dt} = \frac{e^2}{8\pi\epsilon_0 r^2}\cdot\frac{dr}{dt} = -\frac{2}{3}\frac{e^6}{(4\pi\epsilon_0)^3 m^2r^4c^3}$$

$$r^2\frac{dr}{dt}=-\frac{4}{3}\frac{e^4}{(4\pi\epsilon_0)^2m^2c^3}$$

$$\int_{10^{-10}}^{10^{-14}} r^2dr=-\frac{4}{3}\frac{e^4}{(4\pi\epsilon_0)^2}\frac{1}{m^2}\frac{1}{c^3}\int_0^T dt$$

$$\left(\frac{10^{-30}-10^{-42}}{3}\right)=\frac{4}{3}\cdot\frac{1.6^4\times10^{-76}\times81\times10^{18}T}{81\times10^{-62}\times27\times10^{24}}$$

$$\frac{10^{-30}\times27\times10^{20}}{4\times1.6^4}=T$$

$$T=10^{-10}\text{ sec.}$$

16.110. In hydrogen like ion electron moves in a circular orbit of radius r. Calculate the magnetic field induction at the centre using C.G.S. system.

Solution:

$$\frac{mv^2}{r}=\frac{Ze^2}{r^2}$$

$$\therefore\quad mv^2=\frac{Ze^2}{r}$$

According to quantisation condition we have

$$mvr=\text{h}$$

$$\therefore\quad v=\frac{\text{h}}{mr}$$

$$\therefore\quad \frac{m\text{h}^2}{m^2r^2}=\frac{Ze^2}{r}$$

$$\therefore\quad r=\frac{\text{h}^2}{Zme^2}$$

Further $mr^2\omega=\text{h}$

or $mr^2\cdot2\pi\nu=\text{h}$; ν = frequency

$$e\nu=\frac{e\text{h}}{2\pi mr^2}$$

$$\text{current } i=\frac{e\nu}{c}=\frac{e\text{h}}{2\pi mcr^2}$$

$$B=\frac{2\pi i}{r}=\frac{2\pi\cdot e\text{h}}{2\pi mcr^3}=\frac{e\text{h}}{mcr^3}$$

$$=\frac{e\text{h}}{mc\cdot\text{h}^6}Z^3m^3e^6$$

$$=\frac{m^2e^7\cdot Z^3}{c\text{h}^5}$$

16.111. Calculate the minimum distance upto which an α - particle of kinetic energy E will reach to a stationary free nucleus of mass M in the process of a head on collision. Atomic number of the element = Z.

Solution: Consider the motion along x axis. The X component of centre of mass is given by

$$X = \frac{m_\alpha \cdot O + M \cdot x}{m_\alpha + M};$$ x is the distance of closest approach and m_α is the mass of α - particle.

$$E = \frac{2e \cdot Ze}{X} = \frac{2Ze^2(m_\alpha + M)}{Mx}$$

$$\therefore \quad x = \frac{2Ze^2(m_\alpha + M)}{M \cdot E}$$

$$= \frac{2Ze^2}{E}\left(1 + \frac{m_\alpha}{M}\right)$$

16.112. To what minimum distance an α - particle of kinetic energy $E = 0.5$ MeV will approach to a stationary Pb nucleus in the process of head on collision. Atomic number of $Pb = 82$.

$$E = \frac{Ze \cdot 2e}{4\pi\epsilon_0 \cdot r}$$

$$r = \frac{Ze \cdot 2e}{4\pi\epsilon_0 \cdot E}$$

On substituting the values of the various terms, we have

$$r = \frac{82 \times 2 \times e^2 \times 9 \times 10^9}{0.5 \times 10^6 \times e}$$

$$= \frac{82 \times 2 \times 9 \times 10^9 \times 1.6 \times 10^{-19}}{0.5 \times 10^6}$$

$$= 82 \times 4 \times 9 \times 1.6 \times 10^{-16}$$

$$= 0.47 \text{ pm}$$

16.113. An electron moving with an acceleration a loses its energy given by the relation

$$\frac{dE}{dt} = -\frac{2e^2}{3c^3} \cdot a^2$$

where c is velocity of light. Calculate the time in which the energy of the electron which is executing harmonic oscillation of frequency ω reduces to $1/\eta$ of its value.

Solution: During motion we have

$$\frac{e^2}{r^2} = \frac{mv^2}{r} \text{ (In C.G.S. system)}$$

$$K.E. = \frac{e^2}{2r},\ P.E. = \int F dr = \int_{\infty}^{r} \frac{e^2}{r^2} dr = -\frac{e^2}{r}$$

Total energy $E = -\dfrac{e^2}{2r}$

Further $\dfrac{e^2}{r} = mv^2 = mr^2\omega^2$

$$\therefore \qquad r^3 = \frac{e^2}{m\omega^2}$$

Acceleration $a = \dfrac{e^2}{mr^2}$

$$\frac{dE}{dt} = -\frac{2e^2}{3c^3} \times \frac{e^4}{m^2 r^4}$$

$$\frac{dE}{E} = +\frac{2e^2}{3c^3} \cdot \frac{e^4}{m^2 r^4} \cdot \frac{2r}{e^2} \cdot dt$$

$$= \frac{4e^4}{3c^3 m^2 \cdot r^3} dt = \frac{4e^4 m\omega^2}{3c^3 \cdot m^2 e^2} \cdot dt$$

$$\int_{E_1/\eta}^{E_1} \frac{dE}{E} = \int_0^t \frac{4e^2\omega^2}{3mc^3} dt$$

$$\log \eta = \frac{4e^2\omega^2}{3mc^3} \cdot t$$

$$t = \frac{3mc^3}{4e^2\omega^2} \log \eta$$

16.114. An α - particle of kinetic energy $E = 0.2$ MeV is deflected through an angle of 60° by a tin foil of atomic number 50. Calculate the value of the aiming parameter.

Solution: In the given Fig. 16.37 N is the position of the stationary nucleus. The α-particles are deflected along the hyperbolic path PAP'. Here N is one of the focus and PO and $P'O$ are asymptotes. These asymptotes gives the initial and final position of α-particles. When α-particle is directed straight towards N so that $p = 0$. On account of the repulsive force the α-particles are stopped at a certain distance b and made to retrace its path in which $\phi = 180°$

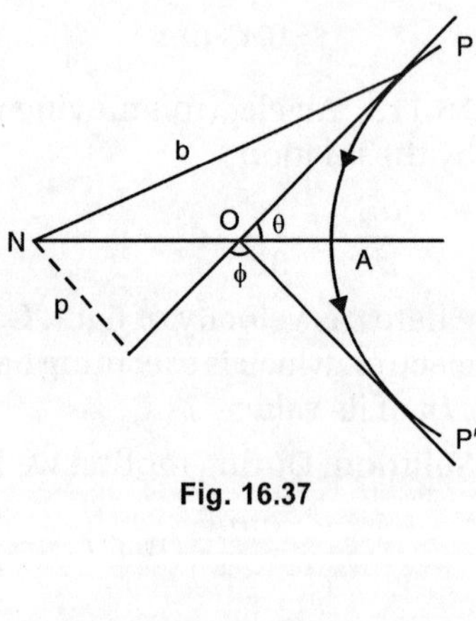

Fig. 16.37

$$E = \frac{Ze \cdot 2e}{4\pi\epsilon_0\, b}$$

$$\therefore \quad b = \frac{Ze \cdot 2e}{4\pi\epsilon_0 \cdot E}$$

Using the properties of hyperbola $\varepsilon = \dfrac{1}{\cos\theta}$

where $\theta = \dfrac{\pi - \phi}{2}$

$b = 2\, p \tan\dfrac{\phi}{2}$ here p = aiming parameter

$$\therefore \quad p = \frac{b}{2\tan\frac{\phi}{2}} = \frac{Ze \cdot 2e}{4\pi\epsilon_0 \cdot 2 \cdot E \cdot tam\frac{\phi}{2}} = \frac{Ze^2}{4\pi\epsilon_0 \cdot E\tan\frac{\phi}{2}}$$

On substituting the values we have

$$p = \frac{50 \times \left(1.6 \times 10^{-19}\right)^2 \times 9 \times 10^9}{0.2 \times 10^6 \times 1.6 \times 10^{-19} \cdot \tan 30}$$

$$= \frac{50 \times 1.6 \times 10^{-19} \times 9 \times 10^9 \cdot \sqrt{3}}{2 \times 10^5}$$

$$= 25 \times 1.6 \times 9 \times 1.732 \times 10^{-15}$$

$$= 0.62 \text{ pm}$$

16.115. An α - particle of energy E is deflected from the Coulombian field of a nucleus through an angle ϕ. Calculate the minimum distance of approach between the α - particle and the nucleus.

Solution: In view of Fig. 16.37 *NA* is the minimum distance of approach and is given by

$$NA = p\cot\frac{\theta}{2}$$

Further $b = 2\, p\tan\dfrac{\phi}{2}$

$$NA = \frac{b}{2}\,\frac{\cos\theta/2}{\tan\phi/2} = \frac{b}{2\tan\phi/2} \cdot \frac{\cos\left(\pi/4 - \phi/4\right)}{\sin\left(\pi/2 - \phi/4\right)}$$

$$= \frac{b}{\tan\phi/2 \cdot \sin\left(\pi/2 - \phi/2\right)} \times \frac{1 + \cos\left(\pi/2 - \phi/2\right)}{2}$$

$$= \frac{b}{2\sin\phi/2} \cdot \left(1 + \sin\phi/2\right) = \frac{b}{2}\left(1 + \operatorname{cosec}\phi/2\right)$$

$$E = \frac{Ze \cdot 2e}{4\pi\epsilon_0\, b}$$

$$\therefore \quad NA = \frac{Ze^2}{4\pi\epsilon_0\, E}\left(1 + \operatorname{cosec} \phi/2\right)$$

16.116. On the basis of Thomson's model, calculate (i) the radius of the hydrogen atom and (ii) the wavelength of the light emitted by the atom. The ionisation energy of hydrogen is equal to E.

Solution: Let r be the radius of the atom

(i) Charge per unit volume = $\dfrac{3e}{4\pi r^3}$

Consider an element of radius x and $x + dx$

Charge confined in the element = $\dfrac{3e}{4\pi r^3} \cdot 4\pi x^2 dx$

Electrostatic potential energy = Ionisation energy

$$E = \int_0^r e \cdot \frac{3e}{4\pi r^3} \cdot \frac{4\pi x^2 dx}{x} \text{ (In C.G.S. system)}$$

$$= \frac{3e^2}{2r}$$

$$\therefore \ r = \frac{3e^2}{2E}$$

(ii) $$\frac{mv^2}{r} = \frac{e^2}{r^2}$$

$$v = \frac{e}{\sqrt{mr}}$$

$$\nu = \text{frequency of revolution} = \frac{v}{2\pi r} = \frac{e}{2\pi r\sqrt{mr}}$$

$$\nu = \frac{c}{\lambda} = \frac{e}{2\pi\sqrt{mr^3}}$$

$$\therefore \quad \lambda = \frac{2\pi c}{e}\sqrt{mr^3}$$

16.117. Using the concept of quantisation of angular momentum calculate the energy of an electron moving in a circular orbit of radius r.

Solution:

$$mvr = n\hbar$$

$$\therefore \quad v = \frac{n\hbar}{mr}$$

$$K.E. = \frac{1}{2}mv^2 = \frac{1}{2}m \cdot \frac{n^2\hbar^2}{m^2r^2} = \frac{1}{2}\frac{n^2\hbar^2}{mr^2}$$

$$\text{In circular path } P.E. = -2K.E. = -\frac{n^2\hbar^2}{mr^2}$$

$$\therefore \quad \text{Total energy } E = K.E + P.E = -\frac{n^2\hbar^2}{2mr^2}$$

16.118. Using Bohr's quantum condition calculate the energy of the electron which is moving in a circular orbit in a central field where its potential energy $U = -\frac{\beta}{r}$ where β is a +ve constant.

Solution:

$$U = -\frac{\beta}{r}$$

$$\text{Force } F = -\frac{dU}{dr} = \frac{\beta}{r^2} = \frac{mv^2}{r}$$

$$\therefore \quad KE = \frac{1}{2}mv^2 = \frac{1}{2}\frac{\beta}{r}$$

$$P.E = -\frac{\beta}{r}$$

$$\therefore \quad \text{Total energy } E = -\frac{\beta}{2r}$$

According to the Bohr's quantum condition

$$mvr = n\hbar$$

$$m \cdot \frac{\sqrt{\beta}}{\sqrt{mr}} \cdot r = n\hbar$$

$$\therefore \quad \sqrt{r} - \frac{n\hbar}{\sqrt{m\beta}}$$

$$\therefore \quad \text{Total energy } E = -\frac{\beta}{2n^2\hbar^2} \cdot m\beta = -\frac{m\beta^2}{2n^2\hbar^2}$$

16.119. In a hydrogen like ion the electron is moving in the nth orbit. Find (i) the magnetic moment (ii) the ratio of the magnetic moment and the angular momentum.

Solution: The motion of the electron in a circular orbit is equivalent to a current in a circular wire. The magnetic moment μ_N is defined as

$$\mu_N = iA = ev \cdot A = \frac{e\omega}{2\pi} \cdot \pi r_n^2; \; r_n = \text{radius of the } n\text{th orbit}$$

$$= \frac{e}{2}\omega r_n^2$$

The equation of motion in the circular orbit is given by

$$\frac{Ze^2}{4\pi\epsilon_0 r_n^2} = mr_n \cdot \omega^2$$

$$\therefore \quad \omega = \sqrt{\frac{Ze^2}{4\pi\epsilon_0 \, m r_n^3}}$$

According to quantisation condition

$$mvr_n = n\hbar$$

$$\frac{Ze^2}{4\pi\epsilon_0 r_n^2} = \frac{m}{r_n} \cdot \frac{n^2\hbar^2}{m^2 r_n^2}$$

$$\therefore \quad r_n = \frac{4\pi\epsilon_0 \cdot n^2\hbar^2}{mZe^2}$$

$$\mu_N = \frac{e}{2} \cdot \sqrt{\frac{Ze^2}{4\pi\epsilon_0 \, m}} \cdot \frac{r_n^2}{r_n^{3/2}} = \frac{\sqrt{Z}\, e^2}{2\sqrt{4\pi\epsilon_0 \, m}} \cdot \sqrt{r_n}$$

$$= \frac{en\hbar}{2m}$$

Angular momentum $= n\hbar$

$$\therefore \quad \frac{\text{Magnetic moment}}{\text{Angular momentum}} = \frac{e}{2m}$$

16.120. On the basis of Bohr's theory, calculate: (i) the velocity, and (ii) the energy of an electron in the first orbit of the hydrogen atom. $h = 6.62 \times 10^{-34}$ Js, $\varepsilon_0 = 8.85 \times 10^{-12}\ C^2/N\text{-}m^2$, $m = 9.1 \times 10^{-31}$ kg.

Solution: Consider an electron moving around the nucleus in a hydrogen like atom in a circular orbit of radius r as shown in Fig. 16.38. The equation of motion is given by

$$\frac{1}{4\pi\varepsilon_0} \frac{Ze^2}{r^2} = \frac{mv^2}{r} \quad \text{... (i)}$$

or $$mv^2 = \frac{1}{4\pi\varepsilon_0} \frac{Ze^2}{r}$$

or $$\frac{1}{2} mv^2 = \frac{Ze^2}{8\pi\varepsilon_0 r} \quad \text{... (ii)}$$

$$\text{Potential energy of electron} = \int_\infty^r F\,dr = \int_\infty^r \frac{Ze^2\,dr}{4\pi\varepsilon_0 r^2} = \frac{-Ze^2}{4\pi\varepsilon_0 r} \quad \text{... (iii)}$$

$$\text{Total energy of the electron} = KE + PE = \frac{Ze^2}{8\pi\varepsilon_0 r} - \frac{Ze^2}{4\pi\varepsilon_0 r} = \frac{-Ze^2}{8\pi\varepsilon_0 r} \quad \text{... (iv)}$$

According to Bohr's quantum condition, we have

$$mvr = \frac{nh}{2\pi} \quad \text{... (v)}$$

On substituting the value of v in Eq. (ii), we have

$$\frac{mn^2h^2}{4\pi^2m^2r^2} = \frac{Ze^2}{4\pi\varepsilon_0 r}$$

or $$r = \frac{\varepsilon_0 h^2}{\pi me^2}\frac{n^2}{Z} = a_0\frac{n^2}{Z} \qquad \text{... (vi)}$$

where a_0 is the radius of the first Bohr orbit.

On substituting the values of r from Eq. (vi) to Eq. (iv), we have

$$E = -\frac{Ze^2\pi me^2 Z}{8\pi\varepsilon_0\cdot\varepsilon_0 n^2h^2} = -\frac{me^4Z^2}{8\varepsilon_0^2h^2n^2} = -\frac{me^4Z^2ch}{8\varepsilon_0^2ch^3n^2} = \frac{-RchZ^2}{n^2}$$

where R is Rydberg's constant $= \dfrac{me^4}{8\varepsilon_0^2ch^3}$

(i) For hydrogen atom the velocity of the electron is given by

$$v^2 = \frac{e^2}{4\pi\varepsilon_0 mr}$$

On substituting the value of r from Eq. (vi), we have

$$v^2 = \frac{e^2\pi me^2}{4\pi\varepsilon_0 m\varepsilon_0 h^2n^2} = \frac{e^4}{4\varepsilon_0^2n^2h^2}$$

or $v = \dfrac{e^2}{2\varepsilon_0 h}$ for first orbit ($n = 1$)

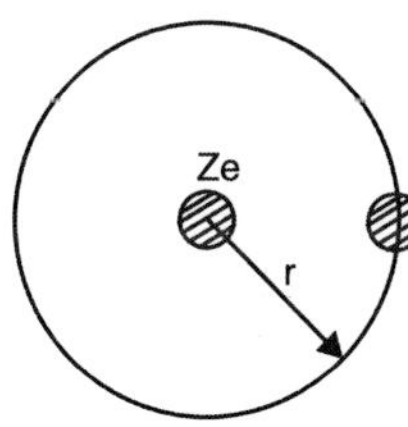

Fig. 16.38

On substituting the various values, we have

$$v = \frac{\left(1.6\times10^{-19}\right)^2}{2\times8.85\times10^{-12}\times6.62\times10^{-34}}$$

$= 2.185 \times 10^6$ m/s

(ii) Energy of the electron in the first orbit

$$= -\frac{me^4}{8\varepsilon_0^2h^2}$$

$$= \frac{-9.1\times10^{-31}\times\left(1.6\times10^{-19}\right)^4}{8\times\left(8.85\times10^{-12}\right)^2\times\left(6.62\times10^{-34}\right)^2}$$

$= -21.71 \times 10^{-19}$ J

16.121. Calculate the ionization energy of hydrogen atom.

Solution: The energy of an electron in an orbit is given by

$$E = -\frac{me^4}{8\varepsilon_0^2n^2h^2}$$

Let E_1 and E_2 be the energies of the electron in two orbits for which $n = n_1$ and $n = n_2$, respectively. Then we have

$$E_2 = -\frac{me^4}{8\varepsilon_0^2 n_2^2 h^2} \text{ and } E_1 = -\frac{me^4}{8\varepsilon_0^2 n_1^2 h^2}$$

Now $E_2 - E_1 = \Delta E = \dfrac{me^4}{8\varepsilon_0^2 h^2}\left(\dfrac{1}{n_1^2} - \dfrac{1}{n_2^2}\right)$

On substituting $n_1 = 1$ and $n_2 = \infty$, ΔE corresponds to the ionization energy.

$$\therefore \quad \Delta E = \frac{me^4}{8\varepsilon_0^2 h^2}$$

In view of Q. 16.120, $\Delta E = 21.71 \times 10^{-19}$ J = 13.58 *eV*

16.122. The ionization potential of hydrogen atom is 13.58 V. Find the second ionization potential of the helium atom.

Solution: In a hydrogen-like atom, the ionization energy is given by

$$\Delta E = \frac{me^4 Z^2}{8\varepsilon_0^2 h^2}$$

The second ionization energy of helium = $\dfrac{4me^4}{8\varepsilon_0^2 h^2}$

But the ionization energy of hydrogen = $\dfrac{me^4}{8\varepsilon_0^2 h^2} = 13.58 \times e$ J

Second ionization energy of helium = $4 \times 13.58 \times e$ J

Second ionization potential of helium atom = $\dfrac{4 \times 13.58 \times e}{e} = 54.32$ V

16.123. A particle of mass *m* moves in a circular orbit in a centrosymmetrical potential field $U(r) = \dfrac{kr^2}{2}$. Calculate the radius of the permissible orbits and the energy levels of the particle in view of Bohr's quantum condition.

Solution: The potential energy of the particle = $\dfrac{kr^2}{2}$. Therefore, the force acting on the particle = $F = kr$. For the purpose of motion in the circular orbit this force must be equal to centripetal force. Hence

$$kr = \frac{mv^2}{r} \qquad \text{... (1)}$$

or $$\omega = \sqrt{\frac{k}{m}} \qquad \text{... (2)}$$

According to Bohr's quantum condition, we have

$$m\omega r^2 = n\hbar$$

or $\quad r = \sqrt{\dfrac{n\text{h}}{m\omega}}$ ... (3)

Total energy of the particle = $PE + KE$

$PE = \dfrac{1}{2}kr^2$ and in view of Eq. (1), kinetic energy = $\dfrac{1}{2}kr^2$. Hence

$$\text{total energy} = kr^2 = \frac{kn\text{h}}{m\omega} = \frac{m\omega^2 \cdot n\text{h}}{m\omega} = n\text{h}\omega$$

16.124. The binding energy of an electron in the ground state of the helium atom is equal to $E_b = 24.6\ eV$. Find the energy required to remove both electrons.

Solution: As pointed out in Q. 16.122 the second ionization potential of helium atom = 54.32 V.

The corresponding energy = 54.32 eV.

The energy required to remove both the electrons from the atom

$$= 24.6 + 54.32\ eV = 78.92\ eV$$

16.125. Calculate the binding energy of an electron in the ground state of hydrogen-like ions in whose spectrum the second line of Balmer series is equal to 1215 Å.

$h = 6.62 \times 10^{-34}$ Js, $c = 3 \times 10^8$ m/s

Solution: The energy difference between two levels of a hydrogen-like ion is given by

$$\Delta E = E_2 - E_1 = \frac{me^4Z^2}{8\varepsilon_0^2h^2}\left(\frac{1}{n_1^2} - \frac{1}{n_2^2}\right)$$

As a consequence of an electron jump if the ion is emitting a spectral line then the wavelength of that line is given by the equation

$$\frac{hc}{\lambda} = \frac{me^4Z^2}{8\varepsilon_0^2h^2}\left(\frac{1}{n_1^2} - \frac{1}{n_2^2}\right)$$

On substituting the values, we have

$$\frac{6.62 \times 10^{-34} \times 3 \times 10^8}{1215 \times 10^{-10}} = \frac{me^4Z^2}{8\varepsilon_0^2h^2}\left(\frac{1}{2^2} - \frac{1}{4^2}\right)$$

$$\therefore \quad \frac{me^4Z^2}{8\varepsilon_0^2h^2} = \text{Binding energy} = \frac{6.62 \times 3 \times 16}{1215 \times 3} \times 10^{-16}\ \text{J}$$

$$\text{Binding energy} = \frac{6.62 \times 16 \times 10^3 \times 10^{-19}}{1215 \times 1.6 \times 10^{-19}}\ eV$$

$$= \frac{6.62 \times 10^4}{1215}\ eV = 54.49\ eV$$

The ion under consideration is He^+.

16.126. Which hydrogen-like ion has a wavelength difference of 59.3 nm between the first line of the Balmer and Lyman series.

$h = 6.62 \times 10^{-34}$ Js, $c = 3 \times 10^8$ m/s, $m = 9.1 \times 10^{-31}$ kg.

Solution: The wavelength of the radiation emitted as a result of an electron jump between the two energy states of the hydrogen like ion is given by the equation.

$$\frac{hc}{\lambda} = \frac{me^4 Z^2}{8\varepsilon_0^2 h^2}\left(\frac{1}{n_1^2} - \frac{1}{n_2^2}\right)$$

$$\frac{hc}{\lambda_1} = \frac{me^4 Z^2}{8\varepsilon_0^2 h^2}\left(\frac{1}{2^2} - \frac{1}{3^2}\right) = \frac{me^4 Z^2}{8\varepsilon_0^2 h^2} \times \frac{5}{36}$$ for the first line of Balmer series.

$$\therefore \quad \lambda_1 = \frac{36}{5} hc \frac{8\varepsilon_0^2 h^2}{me^4 Z^2}$$

Similarly $$\frac{hc}{\lambda_2} = \frac{me^4 Z^2}{8\varepsilon_0^2 h^2}\left(\frac{1}{1^2} - \frac{1}{2^2}\right) = \frac{3}{4}\frac{me^4 Z^2}{8\varepsilon_0^2 h^2}$$ for the first line of Lyman series.

$$\therefore \quad \lambda_2 = \frac{4}{3} hc \frac{8\varepsilon_0^2 h^2}{me^4 Z^2}$$

$$\lambda_1 - \lambda_2 = \frac{8ch^3\varepsilon_0^2}{me^4 Z^2}\left(\frac{36}{5} - \frac{4}{3}\right) = \frac{8ch^3\varepsilon_0^2}{me^4 Z^2} \cdot \frac{88}{15} = 593 \times 10^{-10}$$

$$\therefore \quad Z = \sqrt{\frac{8 \times 88 \times ch^3 \times \varepsilon_0^2}{me^4 \times 15 \times 593 \times 10^{-10}}}$$

On substituting the various values, we have

$$Z = \sqrt{\frac{8 \times 88 \times 3 \times 10^8 \times \left(6.62 \times 10^{-34}\right)^3 \times \left(8.85 \times 10^{-12}\right)^2}{15 \times 9.1 \times 10^{-31} \times \left(1.6 \times 10^{-19}\right)^4 \times 593 \times 10^{-10}}} = 3$$

The ion under consideration is Li^{++}.

16.127. Calculate the Rydberg constant R if He^+ ions are known to have a wavelength difference $\Delta\lambda$ = 958.75 Å between the second lines of the Balmer and Lyman series.

Solution: The wavelength of the radiation as a result of an electron jump is given by the relation

$$\frac{1}{\lambda} = RZ^2\left(\frac{1}{n_1^2} - \frac{1}{n_2^2}\right)$$

$$\frac{1}{\lambda_1} = 4R\left(\frac{1}{2^2} - \frac{1}{4^2}\right) = \frac{3}{4}R$$ for the second line of the Balmer series.

Similarly $\frac{1}{\lambda_2} = 4R\left(\frac{1}{1^2} - \frac{1}{3^2}\right) = \frac{32R}{9}$ for the second line of Lyman series

$$\therefore \quad \lambda_1 - \lambda_2 = \frac{4}{3R} - \frac{9}{32R} = 958.75 \times 10^{-10}$$

$$\frac{101}{96R} = 958.75 \times 10^{-10}$$

$$\therefore \quad R = \frac{101}{96 \times 958.75 \times 10^{-10}}$$

$$= 1.097 \times 10^7 \text{ /m}$$

16.128. Find the velocity of photoelectrons liberated by electromagnetic radiation of wavelength $\lambda = 200 \times 10^{-10}$ m from a stationary He^+ ion in the ground state. $h = 6.62 \times 10^{-34}$ Js, $c = 3 \times 10^8$ m/s, $m = 9.1 \times 10^{-31}$ kg.

Solution: The energy of electromagnetic radiation = $\frac{hc}{200 \times 10^{-10}}$ J. Out of this an energy equal to 54.4 *eV* will be required to make the electron free from the ion and the rest will impart kinetic energy to the photoelectron, i.e.,

$$\frac{hc}{200 \times 10^{-10}} - 54.4 \times 1.6 \times 10^{-19} = \frac{1}{2} mv^2$$

$$\frac{6.62 \times 10^{-34} \times 3 \times 10^8}{2 \times 10^{-8}} - 54.4 \times 1.6 \times 10^{-19} = \frac{1}{2} mv^2$$

On substituting the value of m we have

$$\sqrt{\frac{2 \times (99.3 - 87.04) \times 10^{-19}}{9.1 \times 10^{-31}}} = v$$

$$\sqrt{\frac{2 \times 12.26}{9.1}} \times 10^6 = v$$

$$v = 1.641 \times 10^6 \text{ m/s}$$

16.129. A stationary hydrogen atom emits a photon corresponding to the first line of the Lyman series. What velocity does the atom acquire? Wavelength of the first line of Lyman series = 1216 Å, Atomic weight of hydrogen = 1.008 g, $h = 6.62 \times 10^{-34}$ Js, $N = 6.02 \times 10^{23}$ mol^{-1}.

Solution: Let m be the mass of the hydrogen atom and v be its velocity, then according to the conservation of momentum.

$$mv = \frac{h}{\lambda}$$

$$\therefore \quad v = \frac{h}{\lambda m}$$

$$\text{Mass of hydrogen atom} = \frac{1.008 \times 10^{-26}}{6.02} \text{ kg}$$

Wavelength of first line of Lyman series = 1216 Å

$$\therefore \quad v = \frac{6.62 \times 10^{-34} \times 6.02}{1216 \times 10^{-10} \times 1.008 \times 10^{-26}} = 3.252 \text{ m/s}$$

16.130. Two stationary hydrogen atoms are initially in ground state. With what minimum kinetic energy one of them should move with for its inelastic head on collision with the other atom to make one of them capable of emitting a photon? $R = 1.097 \times 10^7$ /m, $h = 6.62 \times 10^{-34}$ Js, $c = 3 \times 10^8$ m/s.

Ist Solution: In the process of collision there will be an exchange of energy between the two atoms. If one hydrogen atom has an energy $2x$, then after collision each atom will have energy equal to x. By virtue of this, each atom is capable of emitting a photon corresponding to the first line of Lyman series of wavelength 1216 Å.

$$\text{Therefore energy required} = 2x = \frac{2hc}{\lambda} = \frac{2 \times 6.62 \times 10^{-34} \times 3 \times 10^8}{1{,}216 \times 10^{-10}} \text{ J}$$

$$= \frac{2 \times 6.62 \times 3 \times 10^{-16}}{1216 \times 1.6 \times 10^{-19}} eV$$

$$= 20.4 \, eV$$

IInd Solution: Let u be the initial velocity of hydrogen atom. Then, according to the law of conservation of momentum, we have

$mu + 0 = 2mv$; m is mass of hydrogen atom and v is the velocity of hydrogen atom after collision.

$$\therefore \quad v = \frac{u}{2}$$

According to conservation of energy, we have

$$\frac{1}{2} mu^2 = \frac{1}{2} mv^2 + \frac{1}{2} mv^2 + \Delta X; \Delta X = \text{loss in energy in collision}$$

$$\therefore \quad \Delta X = \frac{1}{4} mu^2$$

Now ΔX should be equal to excitation energy, i.e., energy corresponding to the first line of the Lyman series $= \frac{3}{4} Rch$

$$\text{i.e.} \quad \frac{1}{4} mu^2 = \frac{3}{4} Rch$$

$$\text{or} \quad \frac{1}{2} mu^2 = \text{kinetic energy of the hydrogen atom} = \frac{3}{2} Rch$$

$$= \frac{3}{2} \times 1.097 \times 10^7 \times 3 \times 10^8 \times 6.62 \times 10^{-34}$$
$$= 1.5 \times 1.097 \times 3 \times 6.62 \times 10^{-19}$$
$$= 32.68 \times 10^{-19} \text{ J}$$
$$= 20.4 \; eV$$

16.131. A stationary He^+ ion emitted a photon corresponding to the second line of the Lyman series. This photon liberated a photoelectron from a stationary hydrogen atom in the ground state. Find the velocity of the photoelectrons; $R = 1.096 \times 10^7$ /m, $h = 6.62 \times 10^{-34}$ Js, $c = 3 \times 10^8$ m/s, $m = 9.1 \times 10^{-31}$ kg.

Solution: Energy of the photon emitted is given by

$$\frac{hc}{\lambda} = Rch\, Z^2 \left(\frac{1}{1^2} - \frac{1}{3^2}\right)$$
$$= 1.096 \times 10^7 \times 3 \times 10^8 \times 6.62 \times 10^{-34} \times 4 \times \frac{8}{9} \text{ J}$$
$$= 1.096 \times 6.62 \times \frac{32}{3} \times 10^{-19} \text{ J}$$
$$= 77.39 \times 10^{-19} \text{ J}$$

Out of this, $13.6 \times 1.6 \times 10^{-19}$ J is absorbed in making the electron free from the stationary hydrogen atom. The rest is used in imparting the kinetic energy to the electron, i.e.

$$(77.39 - 21.76) \times 10^{-19} = \frac{1}{2} m v^2$$

Where m is the mass and v is the velocity of electron.

On substituting the values of the various terms, we have

$$v = \sqrt{\frac{55.63 \times 10^{-19} \times 2}{9.1 \times 10^{-31}}}$$
$$= \sqrt{\frac{111.26}{9.1}} \times 10^6 = 3.496 \times 10^6 \text{ m/s}$$

16.132. Find the quantum number n corresponding to the exciting state of He^+ ion if on transition to the ground state that ion emits two photons in succession with wavelengths 1026.7 and 304 Å. $R = 1.096 \times 10^7$ /m.

Solution: According to conservation of energy

$$\frac{hc}{\lambda_1} + \frac{hc}{\lambda_2} = Rch\, Z^2 \left(\frac{1}{1^2} - \frac{1}{n^2}\right)$$

or $$\frac{1}{\lambda_1} + \frac{1}{\lambda_2} = RZ^2 \left(1 - \frac{1}{n^2}\right)$$

or $$\frac{1}{n^2} = 1 - \frac{\lambda_1 + \lambda_2}{\lambda_1 \lambda_2} \times \frac{1}{RZ^2}$$

$$= 1 - \frac{1330.7 \times 10^{-10}}{1026.7 \times 304 \times 10^{-20} \times 4 \times 1.096 \times 10^7}$$

$$\frac{1}{n^2} = 1 - \frac{1330.7 \times 10^3}{1026.7 \times 304 \times 4 \times 1.096}$$

$$\frac{1}{n^2} = 0.0275$$

$$n = 6.03$$

Hence the quantum number = 6

16.133. Calculate the speed acquired by a free mercury atom initially at rest when it absorbs energy corresponding to a wavelength of 2,536 Å. Atomic weight of mercury = 200.5 g, $h = 6.62 \times 10^{-34}$ Js, $N = 6.02 \times 10^{23}$ /mol.

Solution: In this case conservation of momentum will take place, i.e.

$$mv = \frac{h}{\lambda}$$

when m and v are the mass and velocity of the mercury atom.

$$v = \frac{h}{m\lambda} = \frac{6.62 \times 10^{-34} \times 6.02 \times 10^{23}}{200.5 \times 10^{-3} \times 2536 \times 10^{-10}}$$

$$= \frac{6.62 \times 6.02}{200.5 \times 2536} \times 10^2$$

$$= 7.84 \times 10^{-3} \text{ m}$$

$$= 0.784 \text{ cm/s}$$

16.134. Spectral lines of what wavelength will appear if a hydrogen atom is transferred to the 3s electron configuration?

$h = 6.62 \times 10^{-34}$ Js, $c = 3 \times 10^8$ m/s.

Solution: The energy level diagram of the hydrogen atom together with the possible transitions is shown in Fig. 16.39. For a transition from n = 3 to n = 2, we have

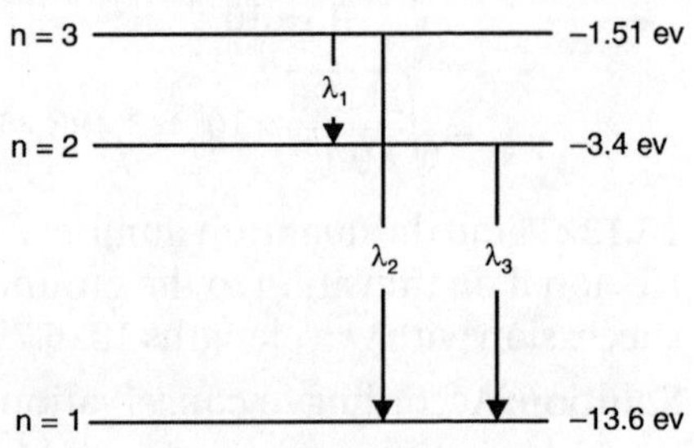

Fig. 16.39

$$\frac{hc}{\lambda_1} = 1.89 \times 1.6 \times 10^{-19} \text{ J}$$

$$\therefore \quad \lambda_1 = \frac{6.62 \times 10^{-34} \times 3 \times 10^8}{1.89 \times 1.6 \times 10^{-19}} = 6567 \times 10^{-10} \text{ m}$$

For a transition from $n = 3$ to $n = 1$, we have

$$\frac{hc}{\lambda_2} = 12.09 \times 1.6 \times 10^{-19} \text{ J}$$

$$\therefore \quad \lambda_2 = \frac{6.62 \times 10^{-34} \times 3 \times 10^8}{12.09 \times 1.6 \times 10^{-19}} = 1026.7 \times 10^{-10} \text{ m}$$

For a transition from $n = 2$ to $n = 1$, we have

$$\frac{hc}{\lambda_3} = 10.2 \times 1.6 \times 10^{-19} \text{ J}$$

$$\therefore \quad \lambda_2 = \frac{6.62 \times 10^{-34} \times 3 \times 10^8}{10.2 \times 1.6 \times 10^{-19}} = 1217 \times 10^{-10} \text{ m}$$

16.135. Doubly ionized lithium atom with atomic number $Z = 3$ is isoelectronic to hydrogen. Find the wavelength of the radiation required to transfer an electron from the first to the third orbit in Li^{++} ion. Ionization energy of hydrogen = 13.6 *eV*. How many spectral lines will be emitted by such an excited atom?

Solution:

$$\frac{1}{\lambda} = RZ^2 \left[\frac{1}{1^2} - \frac{1}{3^2}\right] = \frac{Rch\, Z^2}{ch} \times \frac{8}{9}; \; Rch = 13.6 \times 1.6 \times 10^{-19}$$

$$\therefore \quad \frac{1}{\lambda} = \frac{13.6 \times 1.6 \times 10^{-19} \times 9 \times 8}{3 \times 10^8 \times 6.62 \times 10^{-34} \times 9}$$

$$\lambda = \frac{3 \times 6.62}{13.6 \times 1.6 \times 8} \times 10^{-7} \text{ m} = 114.1 \times 10^{-10} \text{m}$$

In all, three spectral lines will be emitted.

16.136. The wavelength 912 Å corresponds to the last line of the Lyman series. Calculate the ionization potential of the hydrogen atom. $h = 6.62 \times 10^{-34}$ Js, $c = 3 \times 10^8$ m/s.

Solution: Energy corresponding to 912 Å is given by

$$\frac{hc}{\lambda} = \frac{6.62 \times 10^{-34} \times 3 \times 10^8}{912 \times 10^{-10}} \text{ J}$$

$$= \frac{6.62 \times 10^{-34} \times 3 \times 10^8}{912 \times 10^{-10} \times 1.6 \times 10^{-19}} \, eV$$

$$= 13.61 \, eV$$

When an energy of 13.61 *eV* is given to the atom it will be ionized. The ionization potential is therefore equal to 13.6 *V*.

16.137. A photon of energy 5.4852 *eV* liberates an electron from the Li atom initially at rest. The emitted electron moves at right angles to the direction in which photon moves. Find the speed and the direction in which the Li^{++} ion will move. Ionization potential of Li atom = 5.3918 *V*. Atomic weight of Li = 6.94 g, $N = 6.02 \times 10^{23}$ / mol, $m = 9.1 \times 10^{-31}$ kg.

Solution:

$$\text{Energy of the emitted electron} = \frac{1}{2}mv^2$$
$$= (5.4852 - 5.3918) \times 1.6 \times 10^{-19}$$
$$= 0.0934 \times 1.6 \times 10^{-19}\text{ J}$$
$$\therefore \quad v = \sqrt{\frac{2\times 0.0934\times 1.6\times 10^{-19}}{9.1\times 10^{-31}}}$$
$$= 0.18 \times 10^6\text{ m/s}$$

Let the ion move with velocity V at an angle θ with the direction of the photon. The $MV\sin\theta = mv$ where M is the mass of the lithium ion.

$$\therefore \quad V\sin\theta = \frac{mv}{M}$$

On substituting the values, we have

$$V\sin\theta = \frac{0.18\times 10^6 \times 9.1\times 10^{-31}}{6.94\times 10^{-3}}\times 6.02\times 10^{23}$$
$$= \frac{0.18\times 9.1\times 10^{-2}}{6.94\times 10^{-3}}\times 6.02 = 14.2\text{ m/s}$$

Similarly

$$MV\cos\theta = \frac{h}{\lambda} = \frac{5.4852\times 1.6\times 10^{-19}}{3\times 10^8}$$
$$\therefore \quad V\cos\theta = \frac{5.4852\times 1.6\times 10^{-19}\times 6.02\times 10^{23}}{3\times 10^8\times 6.94\times 10^{-3}}$$
$$= \frac{5.4852\times 0.16\times 6.02}{3\times 6.94}$$
$$= 0.2538\text{ m/s}$$
$$V = \sqrt{(V\cos\theta)^2 + (V\sin\theta)^2} = 14.2\text{ m/s}$$
$$\tan\theta = \frac{V\sin\theta}{V\cos\theta} = 55.95$$
$$\theta = 89°$$

16.138. A photon with the energy of 15.0 eV liberates an electron from a hydrogen atom which is at rest in the ground state. At what speed v will the electron travel a distance from the nucleus? Ionization potential of hydrogen atom = 13.6 V, $m = 9 \times 10^{-31}$ kg.

Solution: The amount of energy required to make the electron free from the atom is equal to 13.6 eV. Thus the kinetic energy of the electron is equal to

$$\frac{1}{2}mv^2 = (15.0 - 13.6)\ eV$$
$$= 1.4 \times 1.6 \times 10^{-19}\ \text{J}$$
$$\therefore \quad v = \sqrt{\frac{2 \times 1.4 \times 1.6 \times 10^{-19}}{9 \times 10^{-31}}}$$
$$= 7.05 \times 10^5\ \text{m/s}$$

16.139. The radiations emitted by an excited hydrogen atom are observed at an angle of 45° with respect to its direction of motion. Calculate its velocity if the first line of the Lyman series shows a shift of $\Delta\lambda = 0.2$ nm. Rydberg constant $R = 1.097 \times 10^7$ /m

Solution:

$$\frac{1}{\lambda} = R\left(\frac{1}{1^2} - \frac{1}{2^2}\right) = \frac{3R}{4}$$

In view of Doppler's effect we have

$$\frac{v\cos\theta}{c} = \frac{\Delta\lambda}{\lambda}; \quad v = \text{velocity of the atom}$$
$$\therefore \quad v = \frac{c \cdot \Delta\lambda}{\lambda\cos\theta} = \frac{3c \cdot \Delta\lambda}{4\cos\theta} \cdot R$$

On substituting the values we have

$$v = \frac{3 \times 3 \times 10^8 \times 0.2 \times 10^{-9} \times 1.097 \times 10^7 \cdot \sqrt{2}}{4}$$
$$= 4.5 \times \sqrt{2} \times 1.097 \times 10^5$$
$$= 6.98 \times 10^5\ \text{m/sec.}$$

16.140. A stationary hydrogen atom emits first line of the Lyman series and as a result of that the atom acquires a velocity. How much per cent the energy of the emitted photon differs from the energy of the same line in a hydrogen atom? Rydberg constant $R = 1.097 \times 10^7$ /m, Avogadro number = 6.02×10^{23}.

Solution:

$$\frac{h}{\lambda} = mv; \quad m = \text{mass of hydrogen atom}$$

λ = wavelength of the first line of Lyman series.

Change in energy is $\frac{1}{2}mv^2 = \frac{1}{2}m \cdot \frac{h^2}{m^2\lambda^2} = \frac{h^2}{2m\lambda^2}$

% change in energy $\Delta E = \frac{h^2}{2m\lambda^2} \times \frac{\lambda}{hc} \times 100$

$$= \frac{h \times 100}{2\lambda mc}$$

$$\Delta E = \frac{h \times 100}{2mc} \times \frac{3R}{4}$$

On substituting the values of various terms we have

$$\Delta E = \frac{6.62 \times 10^{-34} \times 3 \times 1.097 \times 10^{7} \times 100 \times 6.02}{8 \times 10^{-26} \times 3 \times 10^{8}}$$

$$= \frac{6.62 \times 1.097 \times 6.02}{8} \times 10^{-7}$$

$$= 0.546 \times 10^{-6}\%$$

16.141. Calculate the wavelength of the first line of the spectral series of He^+ ion whose interval between the extreme lines is 5.18×10^{15} rad/sec. Rydberg constant = 1.097×10^7/m.

Solution:

$$\omega_1 = 2\pi R c Z^2 \left[\frac{1}{(n-1)^2} - \frac{1}{n^2}\right] \quad \text{(see Fig. 16.40)}$$

$$\omega_2 = 2\pi R c Z^2 \left[\frac{1}{(n-1)^2} - \frac{1}{\infty^2}\right]$$

$$\Delta\omega = \frac{2\pi R c Z^2}{n^2} = 5.18 \times 10^{15}$$

$$\therefore \quad n = \sqrt{\frac{2\pi R c Z^2}{5.18 \times 10^{15}}}$$

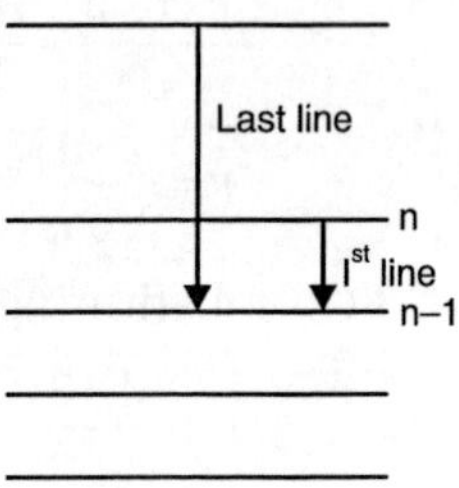

Fig. 16.40

On substituting the values we have

$$n = \sqrt{\frac{2 \times 3.14 \times 1.097 \times 10^{7} \times 3 \times 10^{8} \times 4}{5.18 \times 10^{15}}}$$

$$= 4$$

$$\frac{hc}{\lambda} = RchZ^2 \left[\frac{1}{(n-1)^2} - \frac{1}{n^2}\right]$$

$$\frac{1}{\lambda} = RZ^2 \left(\frac{1}{3^2} - \frac{1}{4^2}\right) = \frac{7\,RZ^2}{16 \times 9}$$

$$\lambda = \frac{16 \times 9}{7RZ^2}$$

On substituting the values we have

$$\lambda = \frac{16 \times 9}{7 \times 1.097 \times 10^{7} \times 4}$$

$$= 4.688 \times 10^{-7} \text{ m}$$

$$= 0.4688\ \mu\text{m}$$

16.142. Hydrogen atom in its ground state is excited by means of monochromatic radiation of wavelength 975 Å. How many lines are possible in the resulting emission spectrum? Calculate the longest wavelength amongst them. You may assume the ionization energy for hydrogen atom as 13.6 eV. Planck's constant $= 6.63 \times 10^{-34}$ Js, $c = 3 \times 10^8$ m/s.

(Roorkee 1989, IIT 1982)

Solution:

$$\text{Energy corresponding to 975 Å} = \frac{hc}{\lambda}$$

$$= \frac{6.63 \times 10^{-34} \times 3 \times 10^8}{975 \times 10^{-10} \times 1.6 \times 10^{-19}} = 12.75\, eV$$

Energy of hydrogen atom in $n = 1$ state $= -13.6\ eV$

Energy of hydrogen atom in $n = 2$ state $= -3.4\ eV$

Energy of hydrogen atom in $n = 3$ state $= -1.51\ eV$

Energy of hydrogen atom in $n = 4$ state $= -0.85\ eV$

Energy of the atom after it absorbs photon of energy 12.75 eV

$= -13.6 + 12.75 = -0.85\ eV$

This shows that the electron of the hydrogen atom will be excited up to the $n = 4$ state. On its return it will give rise to six lines as shown in the Fig. 16.41. Amongst these, the longest wavelength will be due to change of energy

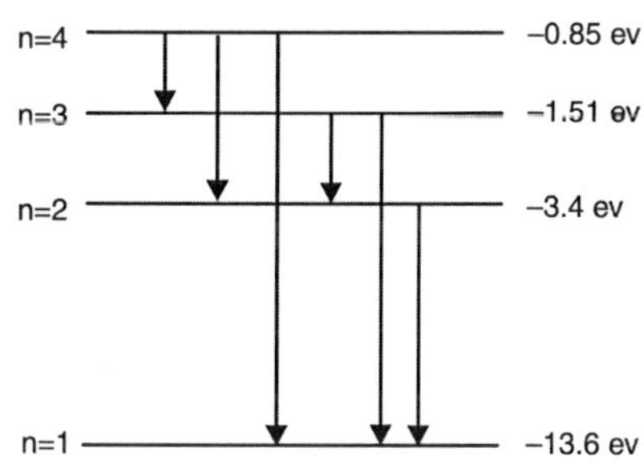

Fig. 16.41

$\Delta E = -0.85 - (-1.51) = 0.66\ eV$

$$\therefore \quad \frac{hc}{\lambda} = 0.66 \times 1.6 \times 10^{-19}\ \text{J}$$

$$\text{or} \quad \lambda = \frac{6.63 \times 10^{-34} \times 3 \times 10^8}{0.66 \times 1.6 \times 10^{-19}} = 18{,}835 \times 10^{-10}\,\text{m}$$

16.143. A gas of identical hydrogen-like atoms has some atoms in the lowest (ground) energy level A and some atoms in a particular upper (excited) energy level B and there are no atoms in any other energy level. The atoms of the gas make transition to a higher energy level by absorbing monochromatic light of photon energy 2.7 eV. Subsequently, the atoms emit radiation of only six different photon energies. Some of the emitted photons have energy 2.7 eV, some have energy more and some have less than 2.7 eV. (i) Find the principal quantum number of the initially excited level B. (ii) Find the ionization energy for the gas atoms. (iii) Find the maximum and minimum energies of the emitted photons. (IIT 1989)

Solution: (i) The gas atoms will emit radiations of six different wavelengths (see Fig. 16.42). By absorbing a photon of energy 2.7 *eV* the atom in the state *A* will not be excited because for going to any higher energy state more than 2.7 *eV* energy is needed. Electrons from $n = 2$ will reach $n = 4$ by absorbing the photon therefore the principal quantum number of the initially excited level *B* is 2. (ii) Let E_1 be the energy in the state $n = 1$, then the energies in the $n = 2$ and $n = 4$ states are given by

$$E_2 = \frac{E_1}{4} \text{ and } E_4 = \frac{E_1}{16}$$

Now $E_2 - E_4 = \frac{3E_1}{16} = 2.7\,eV$

$\therefore \quad E_1 = 14.4\,eV$

$\therefore$ the ionization energy of the gas atoms = 14.4 *eV*

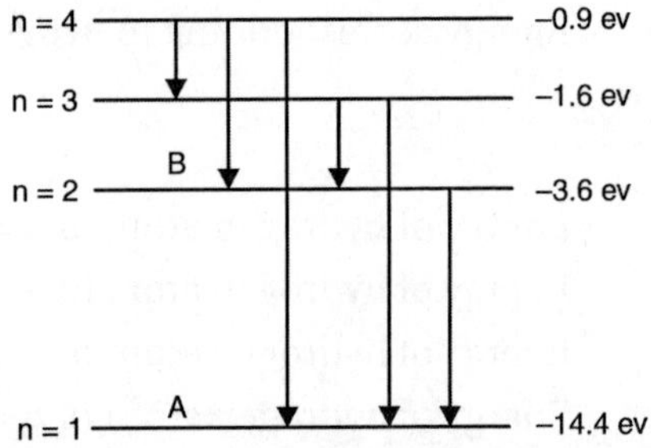

Fig. 16.42

(iii) Maximum energy

$$E_4 - E_1 = -0.9 + 14.4$$
$$= 13.5\,eV$$

Minimum energy $= E_4 - E_3 = -0.9 + 1.6 = 0.7\,eV$

Since $E_3 = -\frac{14.4}{3^2} = -1.6\,eV$

16.144. Electrons in hydrogen-like atoms ($Z = 3$) makes transition from the fifth to the fourth orbit and from the fourth to the third orbit. The resulting radiations are incident normally on a metal plate and eject photoelectrons. The stopping potential for the photoelectrons ejected by the shorter wavelength is 3.95 volt. Calculate the work function of the metal and the stopped potential for the photoelectrons ejected by the longer wavelength. Rydberg constant $R = 1.094 \times 10^7$ /m. (IIT 1990)

Solution:

$$\frac{1}{\lambda_1} = RZ^2 \left[\frac{1}{4^2} - \frac{1}{5^2}\right]$$

On substituting the values we have

$$\frac{1}{\lambda_1} = 1.094 \times 10^7 \times 9 \left[\frac{1}{16} - \frac{1}{25}\right]$$

$$\lambda_1 = 4.514 \times 10^{-7}\text{ m}$$

Similarly

$$\frac{1}{\lambda_2} = RZ^2 \left[\frac{1}{3^2} - \frac{1}{4^2}\right]$$

On substituting the values we have

$$\frac{1}{\lambda_2} = 1.094 \times 10^7 \times 9\left[\frac{1}{9} - \frac{1}{16}\right]$$

$$\lambda_2 = 2.089 \times 10^{-7}\ \text{m}$$

Using photo electric equation

$$\frac{hc}{\lambda_2} = E = W + eV_2;\ W = \text{work function}$$

$$\therefore \quad W = \frac{hc}{\lambda_2} - eV_2$$

On substituting the values of the various terms

$$W = \frac{6.62 \times 10^{-34} \times 3 \times 10^8}{2.089 \times 10^{-7}} - 1.6 \times 10^{-19} \times 3.95$$

$$= 3.2 \times 10^{-19}\ \text{J}$$

$$= 2\ eV$$

For the other wavelength λ_1 we have

$$\frac{hc}{\lambda_1} = W + eV_1$$

$$V_1 = \frac{hc}{e\lambda_1} - W$$

On putting the values we have

$$V_1 = \frac{6.62 \times 10^{-34} \times 3 \times 10^8}{1.6 \times 10^{-19} \times 4.514 \times 10^{-7}} - 2$$

$$= 0.75\ \text{V}$$

16.145. An energy of 68.0 *eV* is required to excite a hydrogen-like atom from its second Bohr orbit to the third. The nuclear charge is *Ze*. Find the value of *Z*, the kinetic energy of the electron in the first Bohr orbit and the wavelength of the electromagnetic radiation required to eject the electron from the first Bohr orbit to infinity. (Roorkee 1992)

Solution: Energy of an electron in the hydrogen atom is given by

$$E = -\frac{RchZ^2}{n^2} = -\frac{13.6Z^2}{n^2}\ eV$$

$$\Delta E = 68\ eV = 13.6Z^2\left[\frac{1}{2^2} - \frac{1}{3^2}\right]$$

$$= 13.6\ Z^2 \times \frac{5}{36}$$

$$\therefore \quad Z = \sqrt{\frac{68 \times 36}{5 \times 13.6}} = 6$$

K.E. of the electron in the first orbit $= 0 - (-13.6 \times 36)$

$= 489.6\ eV$

$$\frac{hc}{\lambda} = 489.6 \times 1.6 \times 10^{-19}$$

$$\therefore \quad \lambda = \frac{6.62 \times 10^{-34} \times 3 \times 10^{8}}{489.6 \times 1.6 \times 10^{-19}}$$

$$= 25.28 \times 10^{-10}\ \text{m}$$

16.146. Light from a discharge tube containing hydrogen atom falls on the surface of a piece of sodium. The kinetic energy of the fastest photoelectrons emitted from sodium is 0.73 *eV*. The work function for sodium is 1.82 *eV*. Find:

(a) the energy of the photons causing the photoelectric emission,

(b) the quantum numbers of the two levels involved in the emission of these photons,

(c) the change in the angular momentum of the electron in the hydrogen atom in the above transition,

(d) the recoil speed of the emitting atom assuming it to be at rest before the transition. (Ionisation potential of hydrogen is 13.6 *eV*) (IIT 1982)

Solution:

(a) Energy of photon E_0 = work function + kinetic energy

$= 1.82 + 0.73 = 2.55\ eV$

(b) Energy of the hydrogen atom is given by $E = -\dfrac{13.6}{n^2}$

Energies corresponding to the values of n = 1, 2, 3, 4 etc. are – 13.6, – 3.4, 1.51, – 0.85 electron volt.

When the electron jump takes place between 4th & 2nd orbit.

$\Delta E = -0.85 + 3.4 = 2.55\ eV$

(c) Change in angular momentum $= 4\hbar - 2\hbar = \dfrac{2h}{2\pi}$

$$= \frac{h}{\pi} = \frac{6.62 \times 10^{-34}}{3.14}$$

$$= 2.11 \times 10^{-34}\ \text{kgm}^2\ \text{sec}^{-1}$$

(d) By conservation of momentum $mv = \dfrac{E_0}{c}$

$$\therefore \quad v = \frac{E_0}{mc} = \frac{2.55 \times 1.6 \times 10^{-19}}{1.67 \times 10^{-27} \times 3 \times 10^{8}} = 0.81\ \text{m/sec.}$$

16.147. Two hydrogen like atoms *A* and *B* are of different masses and each atom contains equal number of protons and neutrons. The energy difference between the first Balmer lines emitted by A and B is 5.667 eV. When the atoms A and B moving with the same velocity, strikes a heavy target they rebound back with the same velocity. In this process the atom B imparts twice the momentum to the target than that A imparts. Identify the atoms A and B. (Roorkee 1993)

Solution: The energies of the first line of the Balmer series in the two atoms are

$$E_A = 13.6\, Z_A^2 \left[\frac{1}{2^2} - \frac{1}{3^2}\right] = \frac{5}{36} \times 13.6\, Z_A^2 \; eV$$

and $$E_B = 13.6\, Z_B^2 \left[\frac{1}{2^2} - \frac{1}{3^2}\right] = \frac{5}{36} \times 13.6\, Z_B^2 \; eV$$

$$\Delta E \quad = E_B - E_A = 5.667 = 13.6 \times \frac{5}{36}\left(Z_B^2 - Z_A^2\right)$$

or $$Z_B^2 - Z_A^2 = 3$$

According to conservation of linear momentum

$$m_A\, v + 0 = MV - m_A\, v_1;\; M = \text{mass of the target}$$

$$m_B\, v + 0 = 2MV - m_B v_1$$

$$m_B\,(v + v_1) = 2m_A\,(v + v_1)$$

$$m_B = 2m_A$$

$$m_A = 2Z_A \text{ and } m_B = 2Z_B$$

$\therefore$ $$Z_B = 2Z_A$$

$$4Z_A^2 - Z_A^2 = 3$$

$$Z_A = 1$$

$$Z_B = 2$$

Hence atom A is $1H^2$ (deuterium) and atom B is $2H_e^4$ (helium).

16.148. A neutron of kinetic energy 65 eV collides inelastically with a singly ionised helium atom at rest. It is scattered at an angle of 90° with respect of its original direction.

(i) Find the allowed values of the energy of the neutron and that of the atom after collision.

(ii) If the atom gets de-excited subsequently be emitting radiation, find the frequencies of the emitted radiation.

Given mass of He atom = 4 × mass of neutron, Ionisation energy of H atom = 13.6 eV. (IIT 1993)

Solution: In view of the conservation of momentum we have

$$mu = 4\, m\, v_2 \cos\theta \text{ (see Fig. 16.43)}$$

$$mv_1 = 4m\, v_2 \sin\theta$$

$$u^2 + v_1^2 = 16v_2^2$$

$$\frac{1}{2m}\cdot mu^2 + \frac{1}{2m}mv_1^2 = \frac{4}{2m}\times 4mv_2^2$$

$$K + K_1 = 4K_2$$

where K = K.E of neutron before impact, K_1 = K.E of neutron after impact and K_2 is the K.E of He atom.

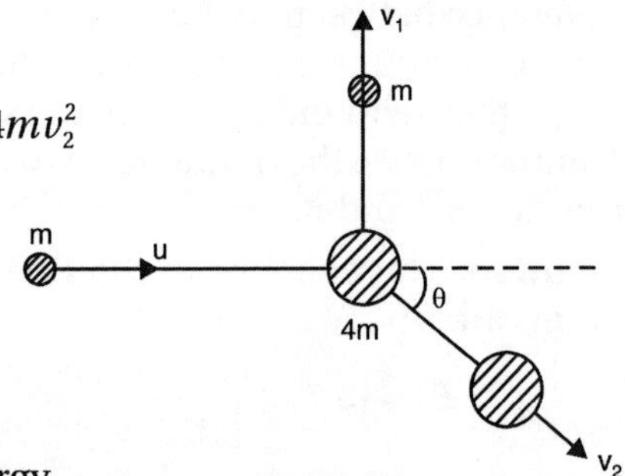

Fig. 16.43

According to conservation of energy

$K = K_1 + K_2 + Q$; Q is the energy stored

$$K = K_1 + \frac{K + K_1}{4} + Q$$

$$4K = 5K_1 + K + 4Q$$

$$K_1 = \frac{3K - 4Q}{5} = \frac{3\times 65 - 4Q}{5} = 39 - 0.8Q$$

Allowed value of $Q = 13.6\left(1 - \frac{1}{n^2}\right)$ where $n = 2, 3, 4$ etc.

$$\therefore \quad K_1 = 39 - 0.8\times 13.6\left(1 - \frac{1}{n^2}\right) = 28.12 + \frac{10.88}{n^2}$$

$$E_{atom} = K - K_1 = 65 - 28.12 - 10.88/n^2$$

$$= 36.88 - 10.88/n^2$$

(ii) When the atom is de-excited then

$$h\nu = 13.6\times 1.6\times 10^{-19}\left(\frac{1}{n_1^2} - \frac{1}{n^2}\right)$$

$$\text{or}\quad \nu = \frac{13.6\times 1.6\times 10^{-19}}{6.62\times 10^{-34}}\left(\frac{1}{n_1^2} - \frac{1}{n^2}\right)$$

$$= 3.287\times 10^{15}\left(1 - \frac{1}{n^2}\right); n = 2, 3, 4.... \text{ Lyman series}$$

$$\nu = 3.287\times 10^{15}\left(\frac{1}{2^2} - \frac{1}{n^2}\right); n = 3, 4, 5.... \text{ Balmer series}$$

$$\nu = 3.287\times 10^{15}\left(\frac{1}{3^2} - \frac{1}{n^2}\right); n = 4, 5.... \text{ Paschen series}$$

16.149. The wavelength of the first line of Lyman series for hydrogen is identical to that of the second line of the Balmer series for some hydrogen-like ion X. Calculate the energies of the four levels of X. Also find its ionisation potential (Given: Ground state binding energy of hydrogen atom = 13.6 eV). (Roorkee 1994)

Solution: For the first line of the Lyman series of hydrogen we have

$$\frac{1}{\lambda}=R\left[\frac{1}{1^2}-\frac{1}{2^2}\right]=\frac{3R}{4}$$

For second line of the Balmer series of hydrogen-like ion we have

$$\frac{1}{\lambda}=RZ^2\left[\frac{1}{2^2}-\frac{1}{4^2}\right]=\frac{3RZ^2}{16}$$

On equating the two we have $Z^2 = 4$ or $Z = 2$

Thus the ion is singly ionised Helium

$$E=\frac{-13.6\times 4}{n^2}$$

The energy of the first four levels are $-54.4\ eV$, $-13.6\ eV$, $-6.04\ eV$ and $-3.4\ eV$.

The second ionisation potential = 54.4 volt.

16.150. A hydrogen-like atom (atomic no Z) is in a higher excited state of quantum number n. This excited atom can make transition to the first excited state by successively emitting two photons of energies 10.2 eV and 17.0 eV respectively. Alternatively, the atom from the same excited state can make transition to the second excited state by successively emitting two photons of energies 4.25 eV and 5.95 eV respectively. Determine the values of n and Z. (Ionisation energy of hydrogen atom = 13.6 eV)

(IIT 1994)

Solution: Energy of photon emitted in electron volt in the first case is given by

$$10.2+17.0=13.6\,Z^2\left[\frac{1}{2^2}-\frac{1}{n^2}\right]$$ (see Fig. 16.44)

$$27.2=13.6\,Z^2\left[\frac{1}{4}-\frac{1}{n^2}\right]$$

In the second case we have

$$4.25+5.95=10.20=13.6\,Z^2\left[\frac{1}{9}-\frac{1}{n^2}\right]$$

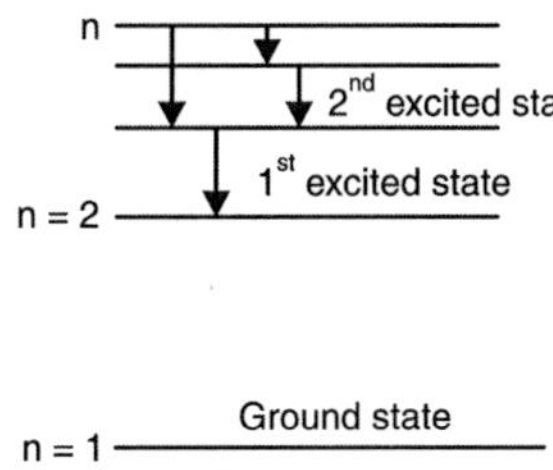

Fig. 16.44

On dividing the two equation we have

$$\frac{27.2}{10.2}=\frac{\frac{1}{4}-\frac{1}{n^2}}{\frac{1}{9}-\frac{1}{n^2}}$$

$$3.02-\frac{27.2}{n^2}=2.55-\frac{10.2}{n^2}$$

$$0.472=\frac{17}{n^2}$$

$$n^2 = \left(\frac{17}{0.472}\right)$$

or $$n = \sqrt{\frac{17}{0.472}} = 6$$

$$10.2 = 13.6\, Z^2 \left[\frac{1}{9} - \frac{1}{36}\right]$$

$$= \frac{13.6\, Z^2}{12}$$

$$\therefore \quad Z = \sqrt{\frac{10.2 \times 12}{13.6}} = 3$$

16.151. A particle of charge equal to that of an electron, $-e$ and mass 208 times the mass of the electron (called a mu-meson) moves in a circular orbit around a nucleus of charge $+3e$. (Take the mass of the nucleus to be infinite). Assuming that the Bohr's model of the atom is applicable to this system: (i) derive an expression for the radius of the nth Bohr orbit, (ii) find the value of n for which the radius of the orbit is approximately the same as that of the first Bohr orbit for the hydrogen atom, and (iii) find the wavelength of radiation emitted when the mu-meson jumps from the third orbit to the first orbit. Rydberg constant = 1.097×10^7 /m.

(IIT 1988)

Solution: The equation of motion of mu-meson is given by

$$\frac{208\, mv^2}{r} = \frac{1}{4\pi\varepsilon_0} \frac{3\, e \times e}{r^2} \quad \text{... (i)}$$

According to Bohr's quantum condition

or $$208\, mvr = \frac{nh}{2\pi} \quad \text{... (ii)}$$

Eliminating v from the two equations

$$208\, m \frac{n^2 h^2}{4\pi^2 m^2\, 208^2 r^2} = \frac{1}{4\pi\varepsilon_0} \frac{3e^2}{r}$$

or $$\frac{n^2 h^2}{208\, m\pi r} = \frac{3e^2}{\varepsilon_0}$$

$$\therefore \quad r = \frac{n^2 h^2 \varepsilon_0}{3\pi(208\, m)\, e^2}$$

Radius of the first Bohr orbit $a_0 = \dfrac{\varepsilon_0 h^2}{\pi m e^2}$ (see Q. 16.120)

On equating the two, we have

$$\frac{n^2 h^2 \varepsilon_0}{3\pi\,(208\, m) e^2} = \frac{\varepsilon_0 h^2}{\pi m e^2}$$

$\therefore \quad n^2 = 3 \times 208 = 624$

$n \cong 25$

(iii) $\frac{1}{\lambda} = R \times 208 \times 9\left(\frac{1}{1^2} - \frac{1}{3^2}\right)$

On substituting the values of R, we have

$\therefore \quad \lambda = 0.5478 \times 10^{-10}$ m

16.152. Derive an expression for the energy of the hydrogen atom assuming the finite mass of the nucleus.

Solution: Let r_e and r_n be the distance between the centre of mass O and the electron and nucleus. Such that $r = r_e + r_n$ (see Fig. 16.45).

In equilibrium $mr_e = Mr_n$; m and M represents the mass of the electron and nucleus respectively.

The equation of motion of electron is given by

$$\frac{mv_e^2}{r_e} = \frac{e^2}{4\pi\varepsilon_0 r^2}\text{ ; } v_e = \text{velocity of the electron}$$

Similarly for nucleus we have

$$\frac{Mv_n^2}{r_n} = \frac{e^2}{4\pi\varepsilon_0 r^2}\text{ ; } v_n = \text{velocity of the nucleus}$$

On combining the two equations we have

$$\frac{mv_e^2}{r_e} = \frac{Mv_n^2}{r_n}$$

or $$\frac{mv_o^2}{r_e} = \frac{mr_o}{r_n} \cdot \frac{v_n^2}{r_n}$$

or $$\frac{v_e^2}{r_e^2} = \frac{v_n^2}{r_n^2}$$

$$\therefore \quad \frac{v_e}{r_e} = \frac{v_n}{r_n} = \omega$$

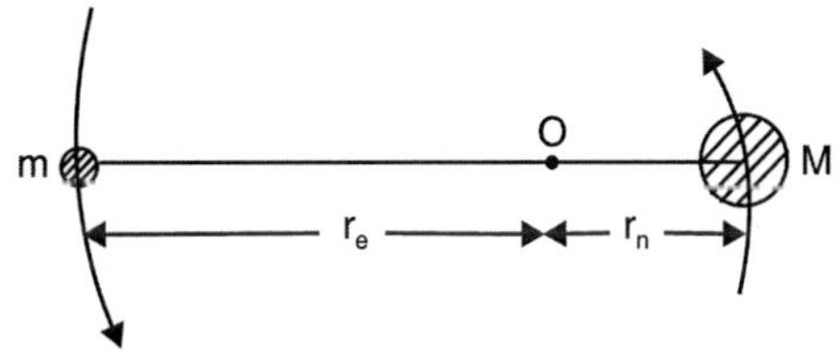

Fig. 16.45

Thus the angular velocity of the electron and the nucleus will have the same value.

$$L = \text{total angular momentum} = \left(mr_e^2 + Mr_n^2\right)\omega$$

$$r = r_e + r_n$$

$$= r_e + \frac{m}{M} \cdot r_e$$

$$\therefore \quad r_e = \frac{Mr}{m+M} \text{ and } r_n = \frac{mr}{M+m}$$

$$\therefore \quad L=\left[m\cdot\frac{M^2r^2}{(M+m)^2}+M\cdot\frac{m^2r^2}{(M+m)^2}\right]\omega=\mu r^2\omega;\quad \mu=\frac{Mm}{M+m}$$

On applying the quantisation conditions we have

$$\mu r^2\omega = n\,\text{h}$$

Total energy of the electron is given by

$$E=\frac{1}{2}\left[mv_e^2+Mv_n^2\right]-\frac{e^2}{4\pi\varepsilon_0 r}$$

$$=\frac{1}{2}\left[mr_e^2+Mr_n^2\right]\omega^2-\frac{e^2}{4\pi\varepsilon_0 r}$$

$$=\frac{1}{2}\left[m\cdot\frac{M^2r^2}{(M+m)^2}+M\cdot\frac{m^2r^2}{(M+m)^2}\right]\omega^2-\frac{e^2}{4\pi\varepsilon_0 r}$$

$$=\frac{1}{2}\mu r^2\omega^2-\frac{e^2}{4\pi\varepsilon_0 r}$$

From the equation of motion of the electron we have

$$mr_e\cdot\omega^2=\frac{e^2}{4\pi\varepsilon_0 r^2}$$

$$\frac{m\cdot Mr}{M+m}\cdot r\omega^2=\frac{e^2}{4\pi\varepsilon_0 r}$$

or $\quad \mu r^2\omega^2=\dfrac{e^2}{4\pi\varepsilon_0 r}$

$$\therefore \quad E=\frac{e^2}{8\pi\varepsilon_0 r}-\frac{e^2}{4\pi\varepsilon_0 r}=\frac{-e^2}{8\pi\varepsilon_0 r}$$

Further $\quad \mu r^2\cdot\omega^2=\dfrac{e^2}{4\pi\varepsilon_0 r}$

$$\frac{\mu r^2\cdot n^2\text{h}^2}{\mu^2\cdot r^4}=\frac{e^2}{4\pi\varepsilon_0 r}$$

$$\frac{n^2h^2}{4\pi^2\mu r}=\frac{e^2}{4\pi\varepsilon_0}$$

$$\therefore \quad r=\frac{\varepsilon_0 n^2h^2}{\mu\pi e^2}$$

With this value of r the total energy E comes out to be

$$E=-\frac{e^2\cdot\mu\pi e^2}{8\pi\varepsilon_0\,\varepsilon_0 n^2h^2}$$

$$= \frac{-\mu e^4 \cdot ch}{8\varepsilon_0^2 ch^3 n^2}$$

$$= -R_\mu \frac{ch}{n^2}$$

$$R_\mu = \frac{\mu}{m} \frac{me^4}{8\varepsilon_0^2 ch^3}$$

$$= \frac{\mu}{m} R$$

$$\text{or} \quad m \cdot R_\mu = \mu R$$

16.153. In a positronium atom the electron moves around the positron in a circular orbit. Using Bohr's model calculate the allowed energy values and ionisation energy.

Solution:

$$E = -\frac{R_\mu ch}{n^2}$$

$$= -\frac{\mu Rch}{m \cdot n^2}; \ \mu = \frac{m}{2} \text{ where } m \text{ is the mass of electron}$$

$$= -\frac{1}{2}\frac{Rch}{n^2}$$

$$= -\frac{6.8}{n^2} eV; Rch = 13.6\, eV$$

Ionisation energy = 6.8 eV

16.154. Calculate the difference in binding energies of the electron in the ground state of hydrogen and heavy hydrogen. $m = 9 \times 10^{-31}$ kg, $m_p = 1.67 \times 10^{-27}$ kg.

Solution:

$$E_D = -R_D ch, E_H = -R_H ch$$

$$E_D = -\frac{\mu_D chR}{m}; E_H = -\frac{\mu_H chR}{m}; m = \text{mass of electron}$$

$$\frac{1}{\mu_D} = \frac{1}{m} + \frac{1}{2m_p}, \frac{1}{\mu_H} = \frac{1}{m} + \frac{1}{m_p}$$

$$E_D = -\frac{2m_p}{m + 2m_p} Rch; \ E_H = -\frac{m_p \cdot Rch}{m + m_p}$$

$$\Delta E = E_H - E_D = Rch\left[\frac{2m_p}{m + 2m_p} - \frac{m_p}{m + m_p}\right]$$

$$= Rch\left[\frac{2m \cdot m_p - m_p m}{(m + m_p)(m + 2m_p)}\right] = \frac{Rch \cdot m \cdot m_p}{(m + m_p)(m + 2m_p)}$$

On substituting the values of the various terms we have

$$\Delta E = \frac{13.6\times 9\times 10^{-31}\times 1.67\times 10^{-27}}{\left(9\times 10^{-31}+1.67\times 10^{-27}\right)\left(9\times 10^{-31}+2\times 1.67\times 10^{-27}\right)}$$

$$= \frac{13.6\times 1.67\times 10^{-27}}{\left(1+\frac{1.67}{9}\times 10^{4}\right)9\times 10^{-31}\left(1+\frac{2\times 1.67}{9}\times 10^{4}\right)}$$

$$= \frac{13.6\times 1.67\times 10^{-4}}{9\times 1856.5\times 3712.1}$$

$$= 3.66\times 10^{-3}\ eV$$

16.155. Calculate the difference in wavelengths of first lines of Lyman series for hydrogen and heavy hydrogen.

$m = 9.1\times 10^{-31}$ kg, $m_p = 1.67\times 10^{-27}$ kg, $R = 1.097\times 10^{7}$/m.

Solution: According to Bohr's theory the wavelength of the first line of Lyman series is given by

$$\frac{1}{\lambda} = R_{\mu}\left[\frac{1}{1^2}-\frac{1}{2^2}\right]$$

$$= \frac{3}{4}R_{\mu}$$

$$\lambda_H = \frac{4}{3R_H}$$

$$\lambda_D = \frac{4}{3R_D}$$

$$R_H = \frac{\mu_H}{m}\cdot R = \frac{m_p\cdot R}{m+m_p}$$

$$R_D = \frac{\mu_D R}{m} = \frac{2m_p\cdot R}{m+2m_p}$$

$$\Delta\lambda = \lambda_H - \lambda_D = \frac{4}{3R}\left[\frac{m+m_p}{m_p}-\frac{m+2m_p}{2m_p}\right]$$

$$= \frac{2m}{3Rm_p}$$

On substituting the values of the various terms we have

$$\Delta\lambda = \frac{2\times 9.1\times 10^{-31}}{3\times 1.097\times 10^{7}\times 1.67\times 10^{-27}}$$

$$= \frac{2\times 9.1}{3\times 1.097\times 1.67}\times 10^{-11}$$

$$= 3.3115\times 10^{-11}\ \text{m}$$

$$= 33.115\ \text{pm}$$

16.156. While calculating the energy of an electron in hydrogen atom using Bohr's model we consider only the Coulombian interaction between the electron and proton. In what way the energy pattern will change if in addition to the electrostatic interaction, the magnetic interaction between the electron and proton is also accounted?

Solution: The total energy E will now be

$$E = E_{\text{electrostatic}} + E_{\text{magnetic}} = \frac{-e^2}{4\pi\varepsilon_0 r} \pm \frac{2\mu_0}{4\pi} \cdot \frac{\mu_p \mu_e}{r^3}$$

where μ_p and μ_e represents the magnetic moments.

The relative error involved in this situation will be

$$\frac{E - E_{\text{electrostatic}}}{E} \cong \frac{E_{\text{mag}}}{E_{\text{electrostatic}}} = \frac{2\mu_p \mu_e}{e^2 c^2 a_0^2}$$

where a_0 = radius of the first Bohr orbit.

In view of the above each level will split up into two sublevels in which the upper one will be $E + E_{\text{magnetic}}$ and the lower one as $E - E_{\text{magnetic}}$ (see Fig. 16.46). If one accounts a transition between these two sub levels then

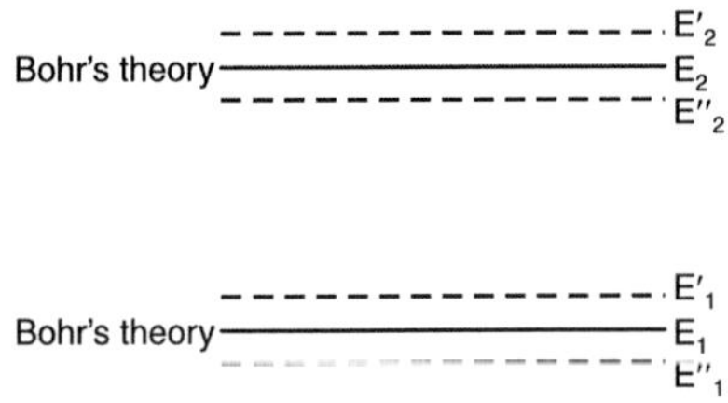

Fig. 16.46

$$\Delta E = \frac{hc}{\lambda} = E_1' - E_1'' = 2\,E_{\text{mag}}$$

$$\therefore \quad \lambda = \frac{hc}{E_{\text{mag}}} = \frac{4\pi hc\, a_0^3}{4\mu_0 \mu_e \mu_p}$$

$$= \frac{4 \times 3.14 \times 6.62 \times 10^{-34} \times 3 \times 10^8 \times 0.529^3 \times 10^{-30}}{4 \times 4 \times 3.14 \times 10^{-7} \times 1.41 \times 10^{-26} \times 9.28 \times 10^{-24}}$$

$$= 56 \text{ cm}$$

However no such wavelength could be detected.

16.157. A free lithium atom at rest absorbs a photon with a frequency of $\omega = 2.81 \times 10^{15}$ /sec. As a result of absorption of energy it passes over to the first excited state and also acquired some speed. Now the atom returns to the normal state after emitting a photon in a direction at right angles to its motion. At what speed does the atom move after this. Atomic wt of $Li = 6.94$ gm.

Solution:

$$mv_1 = \frac{h}{\lambda}; \; mv_2 = \frac{h}{\lambda}$$

$$v_1 = v_2 = \frac{h}{m\lambda}$$

The two velocities are perpendicular to each other hence the final velocity acquired by the atom = $\sqrt{2}\ v_1$ or $\sqrt{2}\ v_2$

$$v = \sqrt{2}\,\frac{h}{m\lambda} = \sqrt{2}\cdot\frac{\frac{h}{2\pi}\cdot\omega}{mc} = \frac{1.414\times 6.62\times 10^{-34}\times 2.81\times 10^{15}}{2\times 3.14\times 6.94\times 10^{-3}\times 3\times 10^{8}}$$

$$= 0.12\ \text{m/sec.}$$

16.158. An excited sodium atom emits a photon of wavelength 5893 Å as a result of an electron jump from $3P - 3S$ transition. Calculate the recoil energy and the speed gained by the atom. Atomic wt of sodium $= 22.98977 \times 10^{-3}$ kg.

Solution:

$$\text{Mass of sodium atom} = \frac{22.98977\times 10^{-3}}{6\times 10^{23}}$$

$$v = \frac{6.62\times 10^{-34}\times 6\times 10^{23}}{5893\times 10^{-10}\times 22.98977\times 10^{-3}} = 2.9\ \text{cm/sec}$$

$$\text{Recoil energy} = \frac{1}{2}mv^2 = \frac{1}{2}\times\frac{22.98977\times 10^{-3}\times (0.029)^2}{6\times 10^{23}}$$

$$= 1.61 \times 10^{-29}\ \text{J}$$

16.159. The wavelength of the yellow doublet of sodium are 589.00 and 589.59 nm. Determine the difference $\Delta\omega$ between the frequencies of the doublets of the sharp series.

Solution:

$$h\omega_1 - h\omega_2 = \frac{h}{2\pi}\Delta\omega = \frac{hc}{\lambda_1} - \frac{hc}{\lambda_2}$$

$$\therefore \quad \Delta\omega = \frac{2\pi c(\lambda_2 - \lambda_1)}{\lambda_1\lambda_2}$$

On substituting the values of various terms we have

$$\Delta\omega = \frac{2\times 3.14\times 3\times 10^{8}\times 0.59\times 10^{-9}}{589\times 10^{-9}\times 589.59\times 10^{-9}}$$

$$= 3.2 \times 10^{12}\ \text{/sec.}$$

16.160. Find the spin orbit interaction energy of $3^2 P_{3/2}$ and $3^2 P_{1/2}$ of sodium atom in eV. The two lines are having wavelengths 589.00 nm and 589.59 nm.

Solution:

$$\Delta E = \frac{hc(\lambda_2 - \lambda_1)}{e\cdot\lambda_1\lambda_2}$$

On putting the values we have

$$\Delta E = \frac{6.62 \times 10^{-34} \times 3 \times 10^{8} \times 0.59 \times 10^{-9}}{1.6 \times 10^{-19} \times 589 \times 10^{-9} \times 589.59 \times 10^{-9}}$$

$$= 2.1 \times 10^{-3}\, eV$$

16.161. What is the maximum possible total angular momentum of sodium atom whose valence electron is in the state with $n = 4$. Find also the atomic state.

Solution: Orbital quantum no $l = 3$, spin $= \frac{1}{2}$ $\therefore J = \frac{7}{2}$

$$\text{Angular momentum} = \text{h}\sqrt{J(J+1)} = \frac{\text{h}}{2}\sqrt{63}$$

$$\text{Atomic state} = {}^{2}F_{7/2}$$

16.162. An atom in a state whose multiplicity is four has an angular momentum $\frac{\text{h}}{2}\sqrt{63}$. What values can L have?

Solution:

$$2 \times \text{spin} + 1 = 4$$

$$\therefore \quad \text{Spin} = \frac{3}{2}$$

$$\text{h}\sqrt{(J+1)J} = \frac{\text{h}}{2}\sqrt{63}$$

$$J^2 + J = \frac{63}{4}$$

$$J = \frac{7}{2}$$

$$L + S = \frac{7}{2}$$

$$L = \frac{7}{2} - S = \frac{7}{2} - \frac{3}{2} = 2$$

$$L + S - 1 = \frac{7}{2}$$

$$L = \frac{7}{2} + 1 - S = 3$$

$$L + S - 2 = \frac{7}{2}$$

$$L = 2 + \frac{7}{2} - S = 4$$

$$L + S - 3 = \frac{7}{2}$$

$$L = 3 + \frac{7}{2} - S = 5$$

Last value is $L - S = \frac{7}{2}$

$$\therefore \qquad L = \frac{7}{2} + \frac{3}{2} = 5$$

Therefore L can have values 2, 3, 4 & 5.

16.163. The isotopes U^{238} and U^{235} occur in nature in the ratio 140 : 1. Assuming that at the time of earth's formation they were present in equal ratio, make an estimation of the age of the earth. The half-lives of U^{238} and U^{235} are 4.5×10^9 years and 7.13×10^8 years respectively. ($\log_{10} 140 = 2.1461$, $\log_{10} 2 = 0.3010$) (Roorkee 1983)

Solution: Let N_1 and N_2 be the number of atoms at time t, then

$$N_1 = N_0 e^{-\lambda_1 t} \text{ and } N_2 = N_0 e^{-\lambda_2 t}$$

or $$\frac{N_1}{N_2} = e^{(\lambda_2 - \lambda_1)t}$$

$$\log_e\left(\frac{N_1}{N_2}\right) = (\lambda_2 - \lambda_1)t$$

or $$t = \frac{\log_e\left(\frac{N_1}{N_2}\right)}{\lambda_2 - \lambda_1} = \frac{\log_e\left(\frac{N_1}{N_2}\right)}{\frac{\log_e 2}{T_2} - \frac{\log_e 2}{T_1}} = \frac{\log_e \frac{N_1}{N_2}}{\log_e 2} \times \frac{T_1 T_2}{T_1 - T_2}$$

$$= \frac{\log_{10} \frac{N_1}{N_2}}{\log_{10} 2} \times \frac{T_1 T_2}{T_1 - T_2}$$

On substituting the values, we have

$$t = \frac{\log_{10} 140}{\log_{10} 2} \times \frac{4.5 \times 10^9 \times 7.13 \times 10^8}{4.5 \times 10^9 - 7.13 \times 10^8}$$

$$= \frac{2.1461}{0.301} \times \frac{4.5 \times 7.13 \times 10^9}{37.87}$$

$$= 6 \times 10^9 \text{ years}$$

16.164. The disintegration rate of a certain radioactive sample at any instant is 4750 disintegrations per minute, 5 minutes after the rate becomes 2700 disintegration per minute. Calculate the (i) decay constant and (ii) half-life of the sample. $\log_{10} 1.76 = 0.2455$.

(CEE 1985, Roorkee 1986)

Solution:

The rate of disintegration is $\frac{dN}{dt}$.

Now $\left(\frac{dN}{dt}\right)_i = \lambda N_i$

After a lapse of t minutes,

$$\left(\frac{dN}{dt}\right) = \lambda N$$

Thus $\frac{N_i}{N} = e^{\lambda t} = \frac{4750}{2700} = 1.76$

$$\lambda\, t = \log_e 1.76 = 2.303 \log_{10} 1.76$$

$$\therefore \quad \lambda = \frac{2.303 \times 0.2455}{5} = 0.11/\text{min}$$

(ii) Half-life $= \frac{\log_e 2}{\lambda} = \frac{0.693}{0.11} = 6.3$ min

16.165. The mean lives of a radioactive substance are 1620 years and 405 years for α-emission and β-emission respectively. Find out the time during which three-fourth of a sample will decay if it is decaying both by α emission and β-emission simultaneously $\log_e 4 = 1.386$.

(Roorkee 1985)

Solution: Let λ_1 and λ_2 be the radioactive constants for α and β emission respectively.

Total decay constant $= \lambda = \lambda_1 + \lambda_2$

Mean life $= \frac{1}{\text{decay constant}}$

$$\therefore \quad \lambda_1 = \frac{1}{1620} \text{ per year and } \lambda_2 = \frac{1}{405} \text{ per year}$$

$$\lambda = \frac{1}{1620} + \frac{1}{405} = \frac{1}{324} \text{ per year}$$

$$\frac{N_0}{4} = N_0 e^{-\lambda t}$$

$$\frac{1}{4} = \frac{1}{e^{\lambda t}}$$

or $\log_e 4 = \lambda t$

$$\therefore \quad t = \frac{\log_e 4}{1/324} = 1.386 \times 324 = 449 \text{ years}$$

16.166. In the chemical analysis of a rock, the ratio of two radioactive isotopes is found to be 100 : 1. The mean live of the two isotopes are 4×10^9 years and 2×10^9 years respectively. If it is assumed that at the time of formation the atoms of both the isotopes were in equal proportion, calculate the age of the rock. Ratio of atomic weight of the two isotopes is 1.02 : 1 ($\log_{10} 1.02 = 0.0086$). (Roorkee 1987)

Solution:

$$\text{Mean life} = \frac{1}{\text{decay constant}}$$

$$\lambda_1 = \frac{1}{4 \times 10^9} \text{ per year and } \lambda_2 = \frac{1}{2 \times 10^9} \text{ per year}$$

$$N_1 = N_1' e^{-t\lambda_1} \text{ and } N_2 = N_2' e^{-t\lambda_2}$$

Further $N_1' = N_2'$ (initially both were in equal proportion)

$$\frac{N_1}{N_2} = e^{-t(\lambda_1 - \lambda_2)}$$

$$\log_e \frac{N_1}{N_2} = t(\lambda_2 - \lambda_1)$$

$$\therefore \quad t = 2.303 \log_{10} \frac{N_1}{N_2} \times \frac{1}{\lambda_2 - \lambda_1}$$

Now $\dfrac{N_1}{N_2} = \dfrac{\text{ratio of masses}}{\text{ratio of atomic weights}} = \dfrac{100}{1.02}$

On substituting the values of various terms, we have

$$t = 2.303 \log_{10} \frac{100}{1.02} \times \frac{1}{\dfrac{1}{2 \times 10^9} - \dfrac{1}{4 \times 10^9}}$$

$$= 2.303\,(2 - 0.0086) \times 4 \times 10^9$$

$$= 1.834 \times 10^{10} \text{ years}$$

16.167. There is a stream of neutrons with a kinetic energy of 0.0327 *eV*. If the half-life of neutron is 700 s. What fraction of neutrons will decay before they travel a distance of 10 m? Mass of neutron = 1.675×10^{-27} kg. (IIT 1986)

Solution:

$$\text{Kinetic energy} = \frac{1}{2} mv^2 = 0.0327 \times 1.6 \times 10^{-19} \text{ J}$$

$$\therefore \quad v = \sqrt{\frac{2 \times 0.0327 \times 1.6 \times 10^{-19}}{1.675 \times 10^{-27}}} = 2{,}500 \text{ m/s}$$

Time taken in travelling a distance 10 m is

$$\Delta t = \frac{10}{2500} = 4 \times 10^{-3} \text{ s}$$

$$\text{Now } \frac{dN}{dt} = -\lambda N = \frac{-0.693}{T} N$$

$$\therefore \quad \frac{dN}{N} = \frac{0.693 \times \Delta t}{T} = \frac{0.693 \times 4 \times 10^{-3}}{700} = 3.96 \times 10^{-6}$$

16.168. Consider an element A with radioactive constant λ_1. Element A decays to B having radioactive constant λ_2. The initial amount of A is N_1^0. If the amount of A at an instant t be N_1 then find the amount of B at that instant.

Solution: The rate of decay of A and rate of growth of B is given by

$$\frac{dN_1}{dt} = -\lambda_1 N_1 \quad \text{... (1)}$$

$$\frac{dN_2}{dt} = \lambda_1 N_1 - \lambda_2 N_2 \quad \text{... (2)}$$

The solution of equation 1 is given by

$$N_1 = N_1^0 e^{-\lambda_1 t} \quad \text{... (3)}$$

Let $N_2 = C_1 e^{-\lambda_1 t} + C_2 e^{-\lambda_2 t}$ be the solution of equation 2. Here C_1 and C_2 are constants.

At $t = 0$, $N_2 = 0$ $\therefore$ $C_1 = -C_2$ ... (4)

$$\text{or} \quad N_2 = C_1 \left[e^{-\lambda_1 t} - e^{-\lambda_2 t} \right]$$

$$\frac{dN_2}{dt} = -\lambda_1 C_1 e^{-\lambda_1 t} + \lambda_2 C_1 e^{-\lambda_2 t} = \lambda_1 N_1 - \lambda_2 N_2$$

At $t = 0$, $N_1 = N_1^0$ & $N_2 = 0$

$$\therefore \quad C_1 (\lambda_2 - \lambda_1) = \lambda_1 N_1^0$$

$$\therefore \quad C_1 = \frac{\lambda_1 N_1^0}{\lambda_2 - \lambda_1} \quad \text{... (5)}$$

$$\therefore \quad N_2 = \frac{\lambda_1 N_1^0}{\lambda_2 - \lambda_1} \left[e^{-\lambda_1 t} - e^{-\lambda_2 t} \right] \quad \text{... (6)}$$

16.169. Consider two elements A & B with radioactive constants λ_1 and λ_2 respectively such that A is parent and B is daughter. If it is assumed that mean lives of the parent and daughter are of the same order of magnitude then show that the daughter substance first reaches a maximum and then decreases at the decay rate of the longer lived of the two.

Solution: With reference to Q. No. 16.168 we have

$$N_2 = C_1 \left[e^{-\lambda_1 t} - e^{-\lambda_2 t} \right]$$

The time in which the daughter reaches a maximum will be obtained by putting $\frac{dN_2}{dt} = 0$

$$\therefore \quad C_1\left[e^{-\lambda_1 t_{max}} \times -\lambda_1 + \lambda_2 e^{-\lambda_2 t_{max}}\right] = 0$$

On equating the quantity within the bracket to zero we have

$$\frac{\lambda_2}{\lambda_1} = \frac{e^{\lambda_2 t_{max}}}{e^{\lambda_1 t_{max}}}$$

$$\text{or} \quad \log_e \frac{\lambda_2}{\lambda_1} = (\lambda_2 - \lambda_1)\, t_{max}$$

$$\text{or} \quad t_{max} = \frac{\log_e \frac{\lambda_2}{\lambda_1}}{\lambda_2 - \lambda_1}$$

Ist case. If $\lambda_1 < \lambda_2$. In this case the parent has longer mean life than the daughter then the term $e^{-\lambda_2 t}$ approaches to zero at a faster rate then

$$N_2 = \frac{\lambda_1 N_1^0}{\lambda_2 - \lambda_1} e^{-\lambda_1 t} = \frac{\lambda_1 N_1}{\lambda_2 - \lambda_1}$$

$$\text{or} \quad \frac{N_2}{N_1} = \frac{\lambda_2}{\lambda_2 - \lambda_1}$$

IInd case. If $\lambda_1 > \lambda_2$. In this case the mean life of daughter is greater than that of the parent. Now N_2 will decay with its own rate of decay via equation

$$N_2 = N_1^0 \frac{\lambda_1}{\lambda_2 - \lambda_1} e^{-\lambda_2 t}$$

After a certain time the parent disappears and there remains only the daughter which decays at its own rate.

16.170. Calculate the number of α-decays for 1.5 gm of thorium 232 in two years. Radioactive constant of thorium $\lambda = 1.58 \times 10^{-18}$ /sec.
Avogadro number = 6×10^{23}/mol

Solution: Number of decays in time t is given by

$$\Delta N = N_0\left[1 - e^{-\lambda t}\right]; N_0 = \text{initial number of nuclei.}$$

$$\lambda t = 1.58 \times 10^{-18} \times 6.3072 \times 10^7 = 9.965 \times 10^{-11}$$

The calculation shows that λt is very small quantity hence

$$\Delta N = N_0\left[1 - (1 - \lambda t)\right] = N_0 \lambda t$$

On substituting the values we have

$$\Delta N = \frac{6 \times 10^{23} \times 1.5}{232} \times 9.965 \times 10^{-11}$$

$$= \frac{6 \times 1.5 \times 9.965}{232} \times 10^{12}$$
$$= 38.657 \times 10^{10}$$

16.171. Radium 226 decays of Radon 222 with a half-life of 1620 years. Radon decays to P_0^{218} with a half-life of 3.82 days. In a pure sample of radium find the number of radon half-lives elapsed when the radon reaches 80% of its equilibrium concentration.

Solution: Here Radium is parent and Radon is daughter. The half-life of parent is very large as compared to that of daughter, hence

$$N_2 = N_1 \frac{\lambda_1}{\lambda_2} \left[1 - e^{-\lambda_2 t}\right]$$

At $t \to \infty$, $N_2(\infty) = N_1 \frac{\lambda_1}{\lambda_2}$

$$\therefore \quad \frac{N_2}{N_2(\infty)} = 1 - e^{-\lambda_2 t} = 0.8$$

or $\quad e^{-\lambda_2 t} = 0.2$

$$e^{\lambda_2 t} = 5$$
$$\lambda_2 t = \log 5$$
$$t = \frac{T}{\log 2} \times \log 5; \; T = \text{half-life}$$
$$\therefore \quad \frac{t}{T} = \frac{\log 5}{\log 2} = 2.32$$

16.172. A sample consisting of radioactive Co^{58} and non-radioactive Co^{59} has specific activity equal to 2×10^{12} disintegration/sec. g. The half-life of Co^{58} is 72 days. Calculate in per cent the ratio of the mass of Co^{58} to the total mass of the sample.

$N_A = 6.02 \times 10^{23}$/mol

Solution: The specific activity is given by

$$a = \frac{dN}{(m_1 + m_2)\,dt} = \log_e 2 \frac{N_A m_1}{TA(m_1 + m_2)}$$

where N_A is Avogadro's number, T is the half-life and A the atomic weight.

$$\therefore \quad \frac{m_1}{(m_1 + m_2)} \times 100 = \frac{aTA}{N_A \log_e 2} \times 100$$
$$= \frac{2 \times 10^{12} \times 72 \times 24 \times 60 \times 60 \times 58 \times 100}{6.02 \times 10^{23} \times 0.693}$$
$$= 0.173\%$$

16.173. 20 mg of the non-radioactive isotope $_{20}Ca^{40}$ are mixed with 5 mg of the radioisotope $_{20}Ca^{45}$. How much will the specific activity of the preparation will decrease? Half-life of $_{20}Ca^{45}$ = 164 days. $N_A = 6.02 \times 10^{23}$/mol.

Solution: Before mixing the two, the activity a_1 is given by

$$a_1 = \frac{dN}{m_1 dt} = \frac{\lambda N}{m_1} = \frac{\log_e 2}{T} \times \frac{N_A m_1}{A m_1}$$

where N_A is Avogadro's number, A is the atomic weight and T is the half-life.

After mixing the two, we have the activity given by

$$a_2 = \frac{dN}{(m_1 + m_2)\,dt} = \frac{\log_e 2\, N_A\, m_1}{TA\,(m_1 + m_2)}$$

$$\text{Change in activity} = a_1 - a_2 = \frac{\log_e 2\, N_A\, m_2}{TA\,(m_1 + m_2)}$$

On putting the various values, we have

$$a_1 - a_2 = \frac{0.693 \times 6.02 \times 10^{23} \times 20 \times 10^3}{164 \times 24 \times 60 \times 60 \times 45 \times 25}$$

$$= 5.234 \times 10^{17} \text{ dis/sec. kg}$$

16.174. Determine the age of an ancient wooden item if it is known that the specific of C^{14} nuclide in them amounts to 3/5 of that in a lately felled tree. The half-life of C^{14} nuclei is 5570 years.

Solution:

The specific activity a_1 of a lately felled tree $= \dfrac{dN}{m\,dt} = \dfrac{\lambda N_0}{m}$

The specific activity a_2 of ancient wood $= \dfrac{dN}{m\,dt} = \dfrac{\lambda N}{m}$

$$a_2 = \frac{3}{5} a_1$$

$$\therefore \quad N = \frac{3}{5} N_0$$

Now $N = N_0\, e^{-\lambda t}$

$$\frac{3}{5} N_0 = N_0 e^{-\lambda t}$$

$$\frac{5}{3} = e^{\frac{0.693}{5570} t}$$

$$\therefore \quad t = \frac{2.303\,(\log_{10} 5 - \log_{10} 3)}{2.303 \times 0.3010} \times 5570$$

$$= \frac{5570 \times 2220}{3010} = 4.108 \times 10^3 \text{ years}$$

16.175. In a uranium ore the ratio of U^{238} nuclei to Pb^{206} nuclei is $\eta = 2.8$. Evaluate the age of the ore, assuming all the Pb^{206} to be a final decay product of the uranium series. The half-life of U^{238} nuclei is 4.5×10^9 years.

Solution: Let m be the mass of Pb^{206}

$$\text{Mass of } U^{238} = 2.8\,\text{m}$$

$$\text{Total mass} = 3.8\,\text{m}$$

$$\text{Now} \quad 2.8\,m = 3.8\,m\,e^{\frac{-0.693}{4.5\times 10^9}t}$$

$$t = \frac{2.303(\log_{10} 3.8 - \log_{10} 2.8)\times 4.5\times 10^9}{2.303\times 0.3010}$$

$$= 4.5\times 10^9 \times \frac{1326}{3010} = 1.982\times 10^9 \text{ years}$$

16.176. Find the decay constant and the mean life of Co^{55} radionuclide if its activity is known to decrease 4.0% per hour. The decay product is non-radioactive.

Solution:

(i) $$\text{Activity } a = \lambda N$$

$$da = \lambda\, dN$$

$$\frac{da}{a} = \frac{dN}{N} = \lambda\, dt$$

$$\frac{da}{a} = 4 = \lambda\, dt \times 100$$

$$\therefore \quad \lambda = \frac{4}{3600\times 100} = 1.1\times 10^{-5}\,\text{s}$$

(ii) $$\text{Mean life} = \frac{1}{\lambda} = \frac{360000}{4} = 90000\,\text{s}$$

16.177. It is proposed to use the nuclear reaction

$${}_1H^2 + {}_1H^2 \rightarrow {}_2He^4$$

in a nuclear reactor of 200 MW rating. If the energy from the above reaction is used with a 25% efficiency in the reactor, how many grams of deuterium fuel will be needed per day (The masses of ${}_1H^2$ and ${}_2He^4$ are 2.0141 atomic mass unit and 4.0026 atomic mass unit respectively). (IIT 1990)

Solution:

Mass defect $\Delta m = 2 \times 2.0141 - 4.0026 = 0.0256$ amu

Amount of energy released per fusion $= 0.0256 \times 931 = 23.8336$ MeV

Useful energy per fusion $= 23.8336 \times 0.25 = 5.9584$ MeV

$$= 5.9584 \times 10^6 \times 1.6 \times 10^{-19}$$

$$= 9.5334 \times 10^{-13} \text{ Joule}$$

Total energy obtained per day $= 200 \times 10^6 \times 24 \times 60 \times 60$

$= 1.728 \times 10^{13}$

Mass of deuterium per fusion $= 2 \times 2.0141 = 4.0282$

$$= \frac{4.0282}{6.02 \times 10^{23}} \text{ gm}$$

$$\therefore \quad \text{Mass of deuterium required} = \frac{4.0282}{6.02 \times 10^{23}} \times \frac{1.728 \times 10^{13}}{9.5334 \times 10^{-13}}$$

$$= 121.28 \text{ gm}$$

16.178. In an experiment on two radioactive isotopes of an element (which do not decay into each other), their mass ratio at a given instant was found to be 3. The rapidly decaying isotope has larger mass and an activity of 1.0 μ curie initially. The half-lives of the two isotopes are known to be 12 hours and 16 hours. What would be the activity of each isotope and their mass ratio after two days? (Roorkee 1990)

Solution: The initial activity A_1 and A_2 of the two masses is given by

$$A_1 = \lambda_1 N_1 \text{ and } A_2 = \lambda_2 N_2$$

After two days

$$N_1' = N_1 \left(\frac{1}{2}\right)^4 = \frac{N_1}{16}$$

and $N_2' = N_2 \left(\frac{1}{2}\right)^3 = \frac{N_2}{8}$

On dividing the two we have

$$\frac{N_1'}{N_2'} = \frac{N}{N_2} \times \frac{1}{2} = \frac{3}{2}$$

$$A_1' = \lambda_1 N_1' = \frac{\lambda_1 N_1}{16} = \frac{1 \mu c}{16} = 0.0625\, \mu c$$

$$\frac{\lambda_2}{\lambda_1} = \frac{T_1}{T_2} = \frac{12}{16} = \frac{3}{4}$$

$$A_2' = \lambda_2 \cdot N_2' = \frac{3}{4} \lambda_1 \times \frac{2}{3} N_1'$$

$$= \frac{1}{2} \lambda_1 \times \frac{N_1}{16} = \frac{\lambda_1 N_1}{32} = \frac{1}{32} = 0.03125\, \mu c$$

16.179. A nucleus X, initially at rest, undergoes alpha-decay according to the equation.

$${}_{92}X^{A} \rightarrow {}_{Z}Y^{228} + \alpha$$

(a) Find the value of A and Z in the above process.

(b) The alpha particle produced in the above process is found to move in a circular track of radius 0.11 m in a uniform magnetic field of 3 Tesla. Find the energy (in MeV) released during the process and the binding energy of the parent nucleus. Given that

$m\ (Y) = 228.03$ amu, $m\ ({}_0n^1) = 1.009$ amu, $m\ ({}_2\text{He}^4) = 4.003$ amu, $m\ ({}_1\text{H}^1) = 1.008$ amu. (IIT 1991)

Solution:

$$_{92}\text{X}^{\text{A}} \rightarrow {}_{\text{Z}}\text{Y}^{228} + \alpha$$

or $$_{92}\text{X}^{\text{A}} \rightarrow {}_{\text{Z}}\text{Y}^{228} + {}_2\text{He}^4$$

On the principles of conservation of mass number $A = 232$.

On the principles of consevation of charge number $Z = 90$.

(b) In circular path we have

$$\frac{mv^2}{r} = qvB$$

or $mv = qrB$

$\therefore$ Momentum of α-particle $= 2e \times B \times r$

$$\text{K.E of } \alpha\text{-particle} = \frac{1}{2} \cdot \frac{(qrB)^2}{m}$$

$$= \frac{1}{2} \frac{(2 \times 1.6 \times 10^{-19} \times 3 \times 0.11)^2}{4.003 \times 1.66 \times 10^{-27}}$$

$$= 0.0839 \times 10^{-11} \text{ Joule} = 5.244\ MeV$$

According to conservation of momentum the momentum of α-particle is equal and oppositc to Y.

$$\therefore \quad \text{K.E of } Y = \frac{(mv)^2}{2m_y} = \frac{1}{2} \frac{(qrB)^2}{228.03 \times 1.66 \times 10^{-27}}$$

$$= \frac{1}{2} \frac{(2 \times 1.6 \times 10^{-19} \times 3 \times 0.11)^2}{228.03 \times 1.66 \times 10^{-27}}$$

$$= 1.473 \times 10^{-14}\ \text{J} = 0.092\ MeV$$

Total energy released = K.E of α-particle + K.E of Y

$= 5.244 + 0.092$

$= 5.336\ MeV$

$$\frac{5.336}{931} = M_X - 228.03 - 4.003$$

$\therefore$ $M_X = 232.0387$ amu

Mass defect of $_{92}\text{X}^{232} = 92\ m_p + (232 - 92)\ m_n - M_X$

$= 92 \times 1.008 + 140 \times 1.009 - 232.0387$

$= 1.9573$ amu

$$\therefore \quad \text{Binding energy} = \frac{1.9573 \times 931}{232} MeV$$
$$= 7.8545\, MeV$$

16.180. Calculate the amount of energy released during the nuclear reaction ${}_3Li^7 + {}_1H^1 = 2{}_2He^4 + Q$.

Mass of ${}_3Li^7$ = 7.01823 amu, mass of ${}_1H^1$ = 1.00814 amu and mass of ${}_2He^4$ = 4.00388 amu

Solution:

$$\Delta m = \text{mass defect}$$
$$= 7.01823 + 1.00814 - 2 \times 4.00388 = 8.02637 - 8.00776$$
$$= 0.01861$$

Energy released = 0.01861 × 931 = 17.326 *MeV*.

16.181. Calculate the amount of energy released when a helium nucleus ${}_2He^4$ is formed by the fusion of two protons and two neutrons. Mass of neutron = 1.00898 amu, mass of proton = 1.00759 amu, mass of helium nucleus = 4.00277 amu.

Solution:

Expected mass of α-particle = 2 × 1.00898 + 2 × 1.00759 = 4.03314

Mass defect = Δm = 4.03314 – 4.00277 = 0.03037

Energy released = 0.03037 × 931 = 28.27 *MeV*

16.182. A free neutron undergoes a radioactive decay and forms a proton. Calculate the energy released. Mass of neutron = 1.00898 amu, mass of proton = 1.00759 amu, mass of electron = 0.00055 amu.

Solution:

$${}_0n^1 \rightarrow {}_1H^1 + {}_{-1}e^0$$

Total mass of proton and electron = 1.00759 + 0.00055 = 1.00814 amu

Mass defect = Δm = 1.00898 – 1.00814 = 0.00084 amu

Energy released = 0.00084 × 931 = 0.782 *MeV*

16.183. A star initially has 10^{40} deutrons. It produces energy via the processes

$${}_1H^2 + {}_1H^2 \rightarrow {}_1H^3 + p \text{ and } {}_1H^2 + {}_1H^3 \rightarrow {}_2He^4 + n$$

If the average power radiated by the star is 10^{16} W, the deutron supply of the star is exhausted in a time of the order of (a) 10^6 s (b) 10^8 s (c) 10^{12} s (d) 10^{16} s.

The masses of the nuclei are as follows M (${}_1H^2$) = 2.014 amu M (p) = 1.007 amu, M (n) = 1.008 amu and M (${}_2He^4$) = 4.001 amu. (IIT 1993)

Solution:

$3\,{}_1H^2 \rightarrow {}_2He^4 + {}_1H^1 + {}_0n^1$

Energy released $= [3 \times 2.014 - 4.001 - 1.007 - 1.008] \times 931.6\,MeV$

$= 0.026 \times 931.6\,MeV = 24.2\,MeV$

$$\text{Required time} = \frac{24.2 \times 10^{40} \times 10^{6} \times 1.6 \times 10^{-19}}{3 \times 10^{16}}$$

$= 1.29 \times 10^{12}$ sec

$= 10^{12}$ sec

16.184. A sample of uranium is a mixture of three isotopes ${}_{92}U^{234}$, ${}_{92}U^{235}$ and ${}_{92}U^{238}$ present in the ratio of 0.006%, 0.71% and 99.284% respectively. The half-lives of these isotopes are 2.5×10^5 years, 7.1×10^8 years and 4.5×10^9 years respectively. Calculate the contribution to activity in % of each isotope in this sample. (Roorkee 1992)

Solution:

$$\text{Activity } A = \lambda N = \frac{0.693}{T} \cdot N$$

$$N = \frac{\text{mass of sample} \times N_A}{\text{mass number}};\ N_A = \text{Avogadro number}$$

$$\text{Mass of } {}_{92}U^{234} = M_1 = \frac{0.006}{100} \times M;\ M = \text{mass of mixture}$$

$$\text{Mass of } {}_{92}U^{235} = M_2 = \frac{0.71\,M}{100}$$

$$\text{Mass of } {}_{92}U^{238} - M_3 - \frac{99.284 \times M}{100}$$

$$N_1 = \frac{0.006\,M}{100} \times \frac{6.03 \times 10^{23}}{234}$$

$$A_1 = \frac{0.693}{2.5 \times 10^5} \times \frac{0.006}{100} \times M \times \frac{6.03 \times 10^{23}}{234}$$

$$\text{Similarly } A_2 = \frac{0.693}{7.1 \times 10^8} \times \frac{0.71\,M}{100} \times \frac{6.03 \times 10^{23}}{235}$$

$$\text{and } A_3 = \frac{0.693}{4.5 \times 10^9} \times \frac{99.284\,M}{100} \times \frac{6.03 \times 10^{23}}{238}$$

$$A_1 : A_2 : A_3 = \frac{0.006}{234 \times 2.5 \times 10^5} : \frac{0.71}{235 \times 7.1 \times 10^8} : \frac{99.284}{238 \times 4.5 \times 10^9}$$

$= 0.01026 : 0.000426 : 0.00927$

Total activity $= 0.019956$

% activity of three samples are

$$\frac{0.01026}{0.019956}\times 100 = 51.41\%, \ \frac{0.000426}{0.019956}\times 100 = 2.13\%,$$

$$\frac{0.00927}{0.019956}\times 100 = 46.45\%$$

16.185. A small quantity of solution containing Na^{24} radio nuclei (half-life 15 hours) of activity 1.0 micro curie is injected into the blood of a person. A sample of the blood of volume 1 cm^3 taken after 5 hours shows an activity of 296 disintegration per min. Determine the total volume of blood in the body of the person. Assume that the radioactive solution mixes uniformly in the blood of the person (1 curie = 3.7×10^{10} disintegration per second).

(IIT 1994)

Solution:

$$\text{Half-life } T = \frac{0.693}{\lambda}$$

$$\therefore \quad \lambda = \frac{0.693}{T} = \frac{0.693}{15\times 3600}$$

$$= 1.2833 \times 10^{-5} \text{ /sec}$$

$$\text{Activity } A = \lambda N_0$$

$$3.7\times 10^4 = 1.283\times 10^{-5}\times N_0$$

$$\therefore \quad N_0 = \frac{3.7\times 10^4}{1.283\times 10^{-5}}$$

$$= 2.883\times 10^9$$

Now $\frac{dN}{dt} = \lambda N_1$; N_1 = No. of radioactive nuclei in 1 cc blood after 5 hours.

$$\text{or} \quad \frac{296}{60} = \frac{0.693}{15\times 3600} N_1$$

$$\therefore \quad N_1 = \frac{296\times 15\times 3600}{60\times 0.693}$$

$$= 3.844\times 10^5$$

Now $\frac{N_1}{N_0'} = \left(\frac{1}{2}\right)^{t/T}$; N_0' = No. of active nuclei in 1 cc of blood.

$$\therefore \quad N_0' = 2^{5/15}\times N_1$$

$$= 2^{1/3}\times 3.844\times 10^5$$

$$= 4.878\times 10^5$$

$$\therefore \quad \text{Volume of blood} = \frac{N_0}{N_0'} = \frac{2.883 \times 10^9}{4.878 \times 10^4}$$
$$= 0.591 \times 10^5 \text{ cc}$$
$$= 5.91 \text{ litres}$$

16.186. Polonium ($_{84}Po^{210}$) emits $_2He^4$ particle and is converted into lead ($_{82}Pb^{206}$). This reaction is used for producing electric power in a space mission. P_o^{210} has half-life of 138.6 days. Assuming an efficiency of 10% for the thermo electric machine, how much P_o^{210} is required to produce 1.2×10^7 J of electric energy per day at the end of 693 days. Also find the initial activity of the material (Given masses of nuclei $_{84}Po^{210}$ = 209.98264 amu, $_{82}Pb^{206}$ = 205.97440 amu, $_2He^4$ = 4.0026 amu, 1 amu = 931 *MeV* and Avogadro number = 6×10^{23} / mol. (Roorkee 1994)

Solution:

$$_{84}Po^{210} \rightarrow {}_{82}Pb^{206} + {}_2He^4$$

Mass Δm converted into energy is given by

$$\Delta m = 209.98264 - 205.97440 - 4.0026$$
$$= 0.00564 \text{ amu}$$
$$= 5.64 \times 10^{-3} \text{ amu}$$

Energy corresponding to this mass

$$= 5.64 \times 10^{-3} \times 931 \times 10^6 \times 1.6 \times 10^{-19}$$
$$= 8.4 \times 10^{-13} \text{ J}$$

Now $\lambda = \frac{0.693}{T} = \frac{0.693}{138.6} = 0.005/\text{day}$

Let M mass produces an energy of 1.2×10^7 J per day then, number of nuclei in M gm of mass is given by

$$N = \frac{M}{210} \times 6 \times 10^{23}$$

Now $-\frac{dN}{dt} = \lambda N$

$$= 0.005 \times \frac{6 \times 10^{23} M}{210}$$

$$\text{Energy produced per day} = \frac{0.005 \times 6 \times 10^{23} M}{210} \times 8.4 \times 10^{-13}$$
$$= 1.2 \times 10^7$$

$\therefore \quad M = 1$ gm

In view of efficiency of 10% required mass = 10 gm

$$\therefore \qquad -\frac{dN}{dt} = 0.005 \times \frac{6 \times 10^{23}}{210} \times 10 = \frac{10^{21}}{7} \text{ per day}$$

Let A_0 be the initial activity then after 693 days

$$A = A_0 \left(\frac{1}{2}\right)^5 = \frac{A_0}{32}$$

$$\therefore \qquad A_0 = 32A = \frac{32 \times 10^{21}}{7} = 4.57 \times 10^{21} \text{ per day.}$$

17

VALVE AND SEMICONDUCTOR

17.1. In a certain voltage range the plate current and plate potential of a diode valve is governed by the relation $I_p = AV_p + BV_P^2$ where $A = 0.15$ mA/V and $B = 0.005$ mA/V^2. Find out the current when the valve is connected to supply of emf $E = 180\ V$ through a resistance $R = 30 \times 10^3\ \Omega$.

Solution: We know that $E - I_p R = V_P$

On substituting the above value of V_p in the equation

$$I_P = AV_P + BV_P^2$$

We have $I_P = A(E - I_P R) + B(E - I_P R)^2$

or $I_P = AE - ARI_P + BE^2 - 2BI_P RE + BI_P^2 R^2$

or $BR^2 I_P^2 - I_P(2BER + AR + 1) + (BE^2 + AE) = 0$

$$\text{or } I_P = \frac{2BER + AR + 1 \pm \sqrt{(2BER + AR + 1)^2 - 4BR^2(BE^2 + AE)}}{2BR^2}$$

$$= \frac{E}{R} + \frac{(AR + 1) - \sqrt{(AR + 1)^2 + 4EBR}}{2BR^2}$$

On substituting the values of various terms, we have

$$I_P = 6 \times 10^{-3} + \left\{ \frac{5.5 - \sqrt{(5.5)^2 + 108}}{9 \times 10^3} \right\}$$

$$= (6 - 0.7) \times 10^{-3} = 5.3 \text{ mA}$$

17.2. The thermionic current density from an emitter is 0.3×10^4 A/m^2 at a temperature of 2800 K. The work function of the emitter is 4.5 *eV*. Find the value of the emission constant a for the emitter, given $e/k = 11600$.

Solution:

$$J = aT^2 e^{\frac{-eV}{kT}}$$

here V is the work function in electron volts.

$$\frac{eV}{kT \times 2.303} = \log \frac{aT^2}{J}$$

or $$\log a = \frac{eV}{kT \times 2.303} - 2\log_{10} T + \log_{10} J$$

$$= \frac{11600 \times 4.5}{2.303 \times 2800} - 2\log_{10} 2800 + \log_{10} 3000$$

$$= 8.095 - 6.8944 + 3.4771$$

$$= 4.6777$$

$$\therefore \quad a = 4.761 \times 10^4 \text{ amp/m}^2/\text{K}^2$$

17.3. A tungsten emitter is operating at 2400 K. If the emission current density is to be reduced by 25 per cent what would be the change in the work function? Express the result in electron volt. Given $e/k = 11600$.

Solution: $J = aT^2 \, e^{\frac{-eV}{kT}}$ is Richardson's equation. Let J_1 and J_2 be the two emission current densities at the work function V_1 and V_2. Then we have

$$J_1 = aT^2 \exp(-eV_1/kT)$$

and $$J_2 = aT^2 \exp(-eV_2/kT)$$

On dividing the two, we have

$$\frac{J_1}{J_2} = \frac{\exp(-eV_1/kT)}{\exp(-eV_2/kT)} = e^{\frac{e}{kT}(V_2 - V_1)}$$

$$2.303 \log_{10} \frac{J_1}{J_2} = \frac{e}{kT}(V_2 - V_1)$$

$$\therefore \quad V_2 - V_1 = \frac{kT}{e} 2.303 \log_{10} \frac{J_1}{J_2}$$

$$= \frac{2400}{11600} \times 2.303 \times \log_{10} \frac{4}{3}$$

$$= \frac{24}{116} \times 2.303 \times 0.125$$

$$= 5.955 \times 10^{-2} = 0.059 \, eV$$

17.4. Find out the I_{aV} and I_{rms} value of current in a half and full wave rectifier.

Solution: Let the alternating current be expressed as

$$I = I_0 \sin \omega t$$

For a half wave rectifier

$$I_{aV} = \frac{1}{T}\left[\int_o^{T/2} I \cdot dt\right]$$

$$= \frac{1}{T}\left[\int_0^{T/2} I_0 \sin \omega t \, dt\right]$$

$$= \frac{I_0}{T}\left[\frac{-\cos \omega t}{\omega}\right]_0^{T/2} = \frac{I_0}{\pi}$$

For a full wave rectifier

$$I_{aV} = \frac{1}{T}\left[\int_0^{T/2} I \cdot dt + \int_{T/2}^{T} I \cdot dt\right]$$

$$= \frac{1}{T}\left[\int_0^{T/2} I_0 \sin \omega t \cdot dt - \int_{T/2}^{T} I_0 \sin \omega t \cdot dt\right]$$

$$= \frac{I_0}{T\omega}\left[-\cos \omega \frac{T}{2} + \cos 0 + \cos \omega T - \cos \omega \frac{T}{2}\right]$$

$$= \frac{2I_0}{\pi}$$

For a half wave rectifier

$$(I_{rms})^2 = \frac{1}{T}\left[\int_0^{T/2} I^2 dt\right]$$

$$= \frac{1}{T}\left[\int_0^{T/2} I_0^2 \sin^2 \omega t \cdot dt\right]$$

$$= \frac{I_0^2}{4}$$

$$I_{rms} = \frac{I_0}{2}$$

For a full wave rectifier

$$(I_{rms})^2 = \frac{1}{T}\left[\int_0^{T/2} I^2 dt + \int_{T/2}^{T} I^2 dt\right]$$

$$= \frac{I_0^2}{T}\left[\int_0^{T/2} \sin^2 \omega t \cdot dt + \int_{T/2}^{T} \sin^2 \omega t \cdot dt\right]$$

$$= \frac{I_0^2}{2}$$

$$I_{rms} = \frac{I_0}{\sqrt{2}}$$

17.5. An input voltage $e = e_0 \sin \omega t$ is applied to a half wave vacuum tube rectifier with a load resistance R_L. Assuming the linear characteristics for the diode having plate resistance r_p. Find (a) average and *rms* value of the current (b) output d.c. power (c) input a.c power (d) rectification efficiency (e) ripple factor (f) total current.

Solution:

(a) $$I_{aV} = I_{d.c} = \frac{I_0}{\pi} = \frac{e_0}{\pi(r_p + R_L)}$$

$$I_{rms} = \frac{I_0}{2} = \frac{e_0}{2(r_p + R_L)}$$

(b) Output d.c power $P_{d.c}$ is given by

$$P_{d.c} = I_{d.c}^2 \cdot R_L = \frac{e_0^2 \cdot R_L}{\pi^2 (r_p + R_L)^2}$$

(c) Input a.c power $P_{a.c}$ is given by

$$P_{a.c} = I_{rms}^2 \cdot (r_p + R_L)$$

$$= \frac{e_0^2}{4(r_p + R_L)}$$

(d) Rectification efficiency η_r is defined as

$$\eta_r = \frac{\text{output d.c power}}{\text{input a.c power}} \times 100$$

$$= \frac{e_0^2 \cdot R_L}{\pi^2 (r_p + R_L)^2} \times \frac{4(r_p + R_L)}{e_0^2} \times 100$$

$$= \frac{4}{\pi^2} \cdot \frac{R_L}{(r_p + R_L)} \times 100$$

$$= \frac{40.6}{1 + r_p/R_L}$$

(e) Ripple factor γ is defined as

$$\gamma = \frac{\text{rms value of a.c component of wave}}{\text{average value of wave}}$$

$$= \sqrt{\left(\frac{I_{rms}}{I_{d.c}}\right)^2 - 1}$$

$$= \sqrt{\left(\frac{I_0/2}{I_o/\pi}\right)^2 - 1}$$

$$= \sqrt{\left[\left(\frac{\pi}{2}\right)^2 - 1\right]} = 1.21$$

(f) Total output current is found to be

$$i = \left(\frac{e_0}{r_p + R_L}\right)\left[\frac{1}{\pi} + \frac{1}{2}\sin \omega t - \frac{2}{3\pi}\cos 2\omega t - \frac{2}{15\pi}\cos 4\omega t.....\right]$$

The first term in the above equation represents the d.c current and the remaining terms are a.c harmonics. The above equation suggests that a half wave vacuum tube rectifier acts as a signal generator producing fundamental frequency and its even harmonics.

17.6. An input voltage $e = e_0 \sin \omega t$ is applied to a full wave vacuum tube rectifier. The load resistance is R_L and each diode valve has resistance r_p. Find (a) average and rms value of current (b) output d.c power (c) input a.c power (d) rectification efficiency (e) ripple factor (f) total current.

Solution:

(a) $$I_{aV} = I_{d.c} = \frac{2I_0}{\pi} = \frac{2e_0}{\pi(r_p + R_L)}$$

$$I_{rms} = \frac{I_0}{\sqrt{2}} = \frac{e_0}{\sqrt{2}(r_p + R_L)}$$

(b) Output d.c power $P_{d.c}$ is given by

$$P_{d.c} = I_{d.c}^2 \cdot R_L = \frac{4e_0^2 \cdot R_L}{\pi^2 (r_p + R_L)^2}$$

(c) Input a.c power $P_{a.c}$ is given by

$$P_{a.c} = I_{rms}^2 \cdot (r_p + R_L)$$

$$= \frac{e_0^2}{2(r_p + R_L)}$$

(d) Rectification efficiency η_r is defined as

$$\eta_r = \frac{\text{output d.c power}}{\text{input a.c power}} \times 100$$

$$= \frac{4e_0^2 \cdot R_L \times 2}{\pi^2 (r_p + R_L)^2 e_0^2} (r_p + R_L) \times 100$$

$$= \frac{8R_L}{\pi^2 (r_p + R_L)} \times 100$$

$$= \frac{81.2}{1 + {r_p}/{R_L}}$$

(e) Ripple factor γ is defined as

$$\gamma = \frac{\text{rms value of a.c component of wave}}{\text{average value of wave}}$$

$$= \sqrt{\left(\frac{I_{rms}}{I_{d.c}}\right)^2 - 1}$$

$$= \sqrt{\left(\frac{I_0/\sqrt{2}}{2I_0/\pi}\right)^2 - 1}$$

$$= \sqrt{(1.11)^2 - 1}$$

$$= 0.48$$

(f) Total output current is obtained as

$$i = \left(\frac{e_0}{r_p + R_L}\right)\left[\frac{2}{\pi} - \frac{4}{3\pi}\cos 2\omega t - \frac{4}{15\pi}\cos 4\,\omega t ......\right]$$

The above equation suggests that a full wave vacuum tube rectifier acts as a signal generator producing even harmonics.

17.7. A d.c source of voltage 300 volt is connected to two diode valves through two equal resistance of 2KΩ as shown in the Fig. 17.1. The current drawn by the two valves is 12 ma and 6 ma respectively. Find the resistance of the two valves.

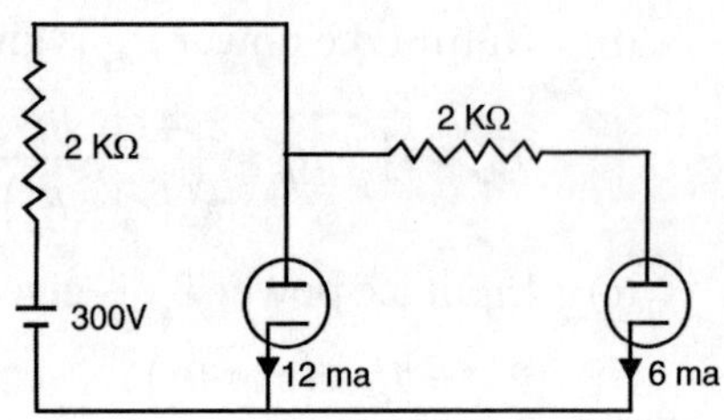

Fig. 17.1

Solution: Let R_1 and R_2 be the resistance of the two valves respectively. According to Kirchhoff's law, we have

$$300 = 2 \times 10^3 \times 18 \times 10^{-3} + R_1 \times 12 \times 10^{-3} \quad ...(1)$$

and $$300 = 2 \times 10^3 \times 18 \times 10^{-3} + 2 \times 10^3 \times 6 \times 10^{-3} + R_2 \times 6 \times 10^{-3} \quad ...(2)$$

From equation 1 we have

$$300 \times 10^3 = 36 \times 10^3 + 12\,R_1$$

or $$R_1 = \left(\frac{300 - 36}{12}\right) \times 10^3 = 22\text{K}\Omega$$

From equation 2 we have

$$300 \times 10^3 = 36 \times 10^3 + 12 \times 10^3 + 6\,R_2$$

or $$R_2 = \left(\frac{300 - 48}{6}\right) \times 10^3 = 42\text{K}\Omega$$

17.8. Find the intrinsic conductivity of Silicon at 300°K. Assume the number of electrons and holes at 300°K to be $1.5 \times 10^{16}/\text{m}^3$ and the mobilities of electrons and holes in Silicon as 1200 $\text{cm}^2/\text{v.s}$ and 500 $\text{cm}^2/\text{v.s}$ respectively.

Solution:

$$\sigma = ne\,(\mu_e + \mu_h); \quad \mu_e = \text{mobility of electrons}$$

$$\mu_h = \text{mobility of holes}$$

On substituting the values of various terms we have

$$\sigma = 1.5 \times 10^{16} \times 1.6 \times 10^{-19}\,[1200 + 500] \times 10^{-4}\ \text{mho/m}$$

$$= 1.5 \times 1.6 \times 1.7 \times 10^{-4}$$

$$= 4.08 \times 10^{-4}\ \text{mho/m}$$

17.9. Density of Silicon is 2.328 g/cm³ (a) Find the concentration of Silicon atoms (b) Phosphorus is added to it in order to make it as a *N*-type semi conductor. Assuming the conductivity to be 1 mho/cm and electron mobility as 1700 cm²/v.s. Find the concentration of conduction electrons. Atomic weight of Silicon = 28.086 gm and Avogadro number = 6.023 × 10²³/mol.

Solution:

(a) No. of silicon atoms /cm³ $= \dfrac{2.328}{28.086} \times 6.023 \times 10^{23}$

$$= 4.992 \times 10^{22}$$

(b) $\sigma = ne\,\mu_e$

$$\therefore \quad n = \frac{\sigma}{e \cdot \mu_e}$$

On substituting the values of various terms we have

$$n = \frac{1}{1.6 \times 10^{-19} \times 1700}$$

$$= 3.67 \times 10^{15}/\text{cm}^3$$

17.10. A triode valve has mutual conductance of 2.5 mA/V and plate resistance of 20,000Ω. Find the value of load resistance for a voltage gain of 30.

Solution:

$$\text{Voltage gain} = \frac{\mu R_L}{R_L + R_P} = \frac{R_P \cdot g_m \cdot R_L}{R_L + R_P}$$

$$30 = \frac{2 \times 10^4 \times 2.5 \times 10^{-3} \times R_L}{R_L \times R_P} = \frac{50\ R_L}{R_L + R_P}$$

$$30\ R_L + 30\ R_P = 50\ R_L$$

$$R_L = \frac{3}{2} R_P = \frac{3}{2} \times 2 \times 10^4 = 30000\Omega$$

17.11. The voltage amplification of a triode valve is 30 with 50 KΩ and 34 with 85 KΩ load resistance. Determine the tube constants.

Solution:

$$\text{Voltage amplification} = \frac{\mu R_L}{R_p + R_L}$$

$$30 = \frac{\mu \times 50}{R_p + 50}$$

$$34 = \frac{\mu \times 85}{R_p + 85}; \; R_p \text{ is in K}\Omega$$

On dividing the two we have

$$\frac{15}{17} = \frac{(R_p + 85)\,50}{(R_p + 50)\,85}$$

$$3 = 2\frac{(R_p + 85)}{(R_p + 50)}$$

$$3\,R_p + 150 = 2\,R_p + 170$$

$$R_p = 20 \text{ K}\Omega$$

$$\mu = \frac{30 \times 70}{50} = 42$$

$$g_m = \frac{\mu}{R_p} = \frac{42}{20 \times 10^3} = 2.1 \times 10^{-3} \text{ mA/V}$$

17.12. Find the current produced at room temperature in a pure germanium plate of area 2×10^{-4} m^2 and of thickness 1.2×10^{-3} m when a potential of 5 V is applied across the faces. Concentration of carriers in germanium at room temperature is 1.6×10^6/m^3. The mobilities of electrons and holes are 0.4 m^{2}V^{-1} S^{-1} and 0.2 m^{2}V^{-1} S^{-1} respectively. How much heat is generated in the plate in 100 sec. (Roorkee 1999)

Solution:

$$\sigma = ne\,(\mu_e + \mu_h)$$

On substituting the values we have

$$\sigma = 1.6 \times 10^6 \times 1.6 \times 10^{-19}\,[0.4 + 0.2]$$

$$= 1.6 \times 1.6 \times 0.6 \times 10^{-13}$$

Current produced in the plate

$$I = \sigma EA = \sigma\left(\frac{V}{d}\right)A$$

$$= 1.6 \times 1.6 \times 0.6 \times 10^{-13} \times \frac{5}{1.2 \times 10^{-3}} \times 2 \times 10^{-4}$$

$$= 1.28 \times 10^{-13} \text{ amp}$$

Heat developed $Q = VIt$

$$= 5 \times 1.28 \times 10^{-13} \times 100$$

$$= 6.4 \times 10^{-11} \text{ J}$$

17.13. At 25°C the potential barrier in Silicon is 0.7 V. There is a lowering of this barrier by 2.5 mv/°C. Calculate the potential barrier of Silicon at 50°C and 100°C.

Solution:

Potential barrier V at 50°C is given by

$$V = 0.7 - 0.0025 \times (50 - 25)$$
$$= 0.7 - 0.0625$$
$$= 0.6375 \text{ Volt}$$

Potential barrier V' at 100°C is given by

$$V' = 0.7 - 0.0025 \times (100 - 25)$$
$$= 0.7 - 0.1875$$
$$= 0.5125 \text{ Volt}$$

17.14. The circuit shown in the Fig. 17.2a is constructed by using four Germanium diodes. Taking the knee voltage as 0.3 volt and the resistance of the diode in the forward biasing as 1.5Ω. Calculate the current flowing in the circuit.

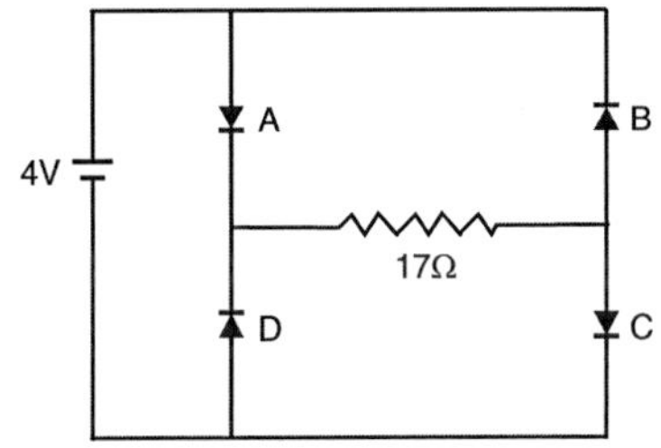

Fig. 17.2a

Solution: The two diodes B and D are in reverse biasing and therefore they will offer an infinite resistance. In other words the circuit will be open at these two diodes. The remaining two diodes A & C can be replaced by two cells of emf 0.3 volt and having internal resistance 1.5Ω. In the light of this a circuit shown in the Fig. 17.2b can be constructed.

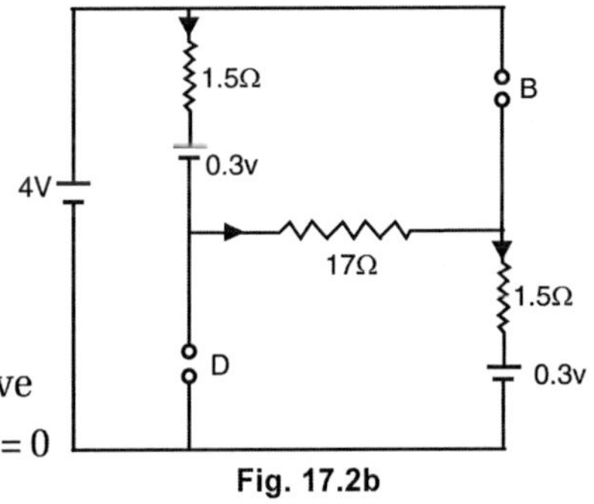

Fig. 17.2b

On applying Kirchhoff's law we have

$$4 - 1.5\,I - 0.3 - 17\,I - 1.5\,I - 0.3 = 0$$
$$20\,I = 3.4$$
$$\therefore \quad I = \frac{3.4}{20} = 0.17 \text{ amp.}$$

17.15. A Silicon diode at a forward bias of 0.96 volt carries a current of 60 ma. Determine the static and dynamical resistance of the diode at 27°C.

Solution: The static resistance r_{dc} is defined as the ratio of total applied voltage to the total current, hence

$$r_{dc} = \frac{V}{I} = \frac{0.96}{60 \times 10^{-3}} = \frac{96}{6} = 16\,\Omega$$

The dynamical resistance r_{ac} is defined as

$$r_{ac} = \frac{dV}{dI}$$

For a diode

$$I = I_s \left[e^{\frac{eV}{\eta KT}} - 1 \right]$$

where I_s is reverse saturation current.

η = constant and dependent upon the material. For Silicon $\eta = 2$

$$\therefore \quad I = I_s \left[e^{\frac{eV}{2KT}} - 1 \right]$$

$$\frac{1}{r_{ac}} = \frac{dI}{dV} = I_s \cdot e^{\frac{eV}{2KT}} \times \frac{e}{2KT}$$

$$= I_s \left(\frac{I}{I_s} + 1 \right) \frac{e}{2KT}$$

$$= (I + I_s) \frac{e}{2KT}$$

$$= \frac{Ie}{2KT}$$

or $$r_{ac} = \frac{2KT}{Ie} = \frac{2 \times 0.026}{60 \times 10^{-3}}; \quad \frac{KT}{e} \cong 0.026$$

$$= \frac{26}{30} = 0.87\,\Omega$$

17.16. In the circuit shown in the Fig. 17.3. Find out the potential difference across the terminals of the cell.

Solution: The two diodes are in the reverse biasing and therefore will offer an infinite resistance. The current flowing in the circuit is obtained by the equation

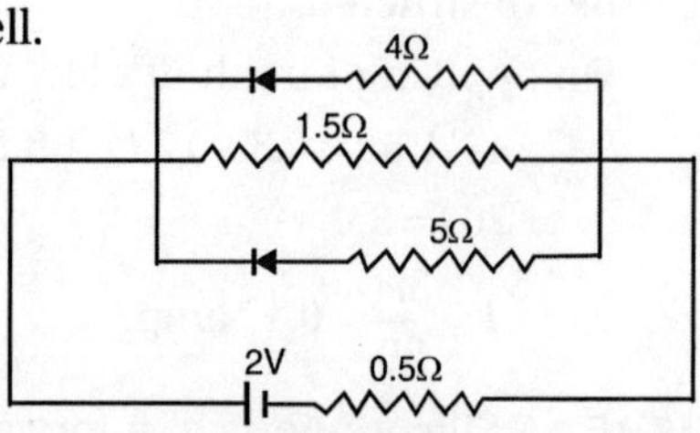

Fig. 17.3

$$2 - 1.5\,I - 0.5\,I = 0$$

or $I = 1$ amp.

Potential difference across the terminals of the cell is given by

$$V = 2 - 1 \times 0.5 = 1.5\,\text{Volt}$$

17.17. In the arrangement shown in Fig. 17.4 find the current through 3 Ω resistance. The forward resistance of the diode is 2 Ω.

Solution: Let the current flowing in 3 Ω, 2 Ω and 1 Ω resistance be I, I_1 and I_2. According to Kirchhoff's Law, we have

$$4 - 3I - 2I_1 + 2 = 0$$

or $\quad 6 = 3I + 2I_1 \qquad \text{... (1)}$

Further $4 - 3I - I_2 - 2I_2 = 0$

or $\quad 4 = 3I + 3I_2 \qquad \text{... (2)}$

From these equations we have

$$2 = 2I_1 - 3I_2 \qquad \text{... (3)}$$

or $\quad 2 = 2I_1 - 3(I - I_1)$

$$2 = 5I_1 - 3I \qquad \text{... (4)}$$

Using equations 1 & 4 we have

$$8 = 7I_1$$

or $\quad I_1 = \frac{8}{7}$ amp

Now $6 = 3I + 2 \times \frac{8}{7}$

or $\quad 3I = 6 - \frac{16}{7}$

$$3I = \frac{42 - 16}{7}$$

$$I = \frac{26}{21} \text{ amp}$$

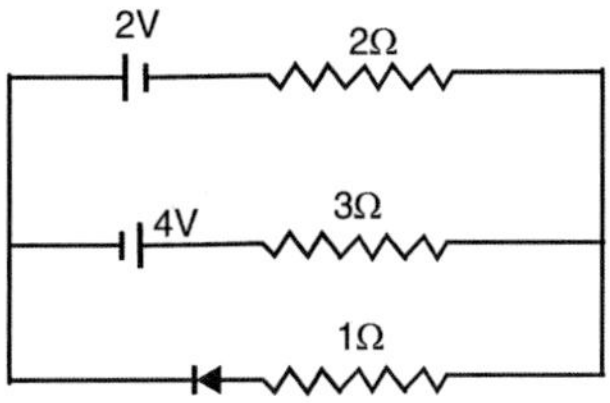

Fig. 17.4

17.18. For a triode valve the plate current i_p is given by the equation $i_p = 4 \times 10^{-6} (V_p + 30\, V_g)^{3/2}$ A. Taking $V_p = 250$ volt and $V_g = -5$ volt, determine the tube constants.

Solution: On differentiating the above equation with respect to V_g keeping V_p constant we have

$$\left(\frac{di_p}{dV_g}\right)_{V_p} = g_m = 4 \times 10^{-6} \left(V_p + 30\, V_g\right)^{1/2} \times \frac{3}{2} \times 30$$

$$= 18 \times 10^{-5} (250 - 150)^{1/2} = 18 \times 10^{-4}$$

$$\therefore \quad g_m = 1.8 \times 10^{-3} \text{ mho}$$

Similarly on differentiating the given equation with respect to V_p keeping V_g constant, we have

$$\left(\frac{di_p}{dV_p}\right)_{V_g} = \frac{1}{r_p} = 4 \times 10^{-6} \times \frac{3}{2} \left(V_p + 30\, V_g\right)^{1/2}$$

$$= 6 \times 10^{-6} \times 10 = 6 \times 10^{-5}$$

$$\therefore \quad r_p = \frac{1}{6} \times 10^5$$
$$= 16.66 \times 10^3 \text{ ohm}$$
$$\mu = r_p g_m = 1.8 \times 10^{-3} \times \frac{1}{6} \times 10^5 = 30$$

17.19. For a certain triode valve the current i_p is given by the equation

$$i_p = 87 \times 10^{-6} (\mu V_g + V_p) + 1.5 \times 10^{-7} (\mu V_g + V_p)^2$$

If $\mu = 15$ and the operating point is at $V_g = -7$ volt and $V_p = 210$ volt find i_p, r_p and g_m.

Solution:

$$i_p = 87 \times 10^{-6} (-15 \times 7 + 210) + 1.5 \times 10^{-7} (-15 \times 7 + 210)^2$$
$$= 10.8 \text{ mA}$$

On differentiating the given equation with respect to V_p keeping V_g constant, we have

$$\left(\frac{di_p}{dV_p}\right)_{V_g} = \frac{1}{r_p} = 87 \times 10^{-6} + 1.5 \times 10^{-7} \times 2 (\mu V_g + V_p)$$
$$= 87 \times 10^{-6} + 3 \times 10^{-7} (-105 + 210)$$
$$= 87 \times 10^{-6} + 10.5 \times 3 \times 10^{-6}$$
$$= 118.5 \times 10^{-6}$$
$$\therefore \quad r_p = \frac{1}{118.5} \times 10^6 = 8.439 \text{ k}\Omega$$

On differentiating the given equation with respect to V_g keeping V_p constant, we have

$$\left(\frac{di_p}{dV_g}\right)_{V_p} = g_m = \mu \times 87 \times 10^{-6} + 3 \times 10^{-7} (\mu V_g + V_p) \times \mu$$
$$= 15 \times 87 \times 10^{-6} + 45 \times 10^{-7} (-105 + 210)$$
$$= (1305 + 472.5) \times 10^{-6}$$
$$= 1777.5 \times 10^{-6} = 1.77 \times 10^{-3} \text{ mho}$$

17.20. The characteristics of a triode valve are in the form of parallel lines in the region of our interest. At grid potential of –1 volt the anode current i (in milliamperes) is represented in terms of plate voltage V (in volts) by the relation.

$$i = 0.125\,V - 7.5.$$

A grid potential of –3 V & plate potential of 300 V, the plate current is 5 mA. Find the plate resistance r_p, the mutual conductance g_m and the amplification factor μ for the triode valve.

Solution: Plate current at 300 V plate potential and –1V grid potential is given by

$$i = 300 \times 0.125 - 7.5 = 30 \text{ mA}$$

$$\Delta i = (30 - 5) \text{ mA} = 25 \text{ mA}$$

$$\Delta V_g = 2 \text{ volt}$$

$$\therefore \quad g_m = \frac{\Delta i}{\Delta V_g} = \frac{25}{2} \times 10^{-3} = 12.5 \times 10^{-3} \text{ A/V}$$

On differentiating the given equation with respect to V, we have

$$\frac{di}{dV} = 0.125 \times 10^{-3}$$

$$\therefore \quad \frac{1}{r_p} = 125 \times 10^{-6}$$

$$\text{or} \quad r_p = \frac{1}{125} \times 10^6 = 8 \times 10^3 \,\Omega$$

$$\mu = r_p g_m = 8 \times 10^3 \times 12.5 \times 10^{-3} = 100$$

17.21. We have the following observations for the linear part of the characteristic curves of a triode valve:

Grid potential V_g (in volts)	0	–2	–4	–6
Plate current i_p (in milliamperes) when $V_p = 150\text{V}$	15	12.5	10	7.5
Plate current i_p (milliamperes) when $V_p = 120\text{V}$	10	7.5	5	2.5

Find out r_p, g_m μ and voltage amplification when the load resistance is 20 kΩ.

Solution:

$$r_p = \left(\frac{\Delta V_p}{\Delta I_p}\right)_{V_g} = \frac{30}{5 \times 10^{-3}} = 6 \times 10^3 \,\Omega$$

$$g_m = \left(\frac{\Delta I_p}{\Delta V_g}\right)_{V_p} = \frac{2.5 \times 10^{-3}}{2} = 1.25 \times 10^{-3} \text{ A/V}$$

$$\mu = r_p \cdot g_m = 7.5$$

$$\text{Voltage amplification } A = \frac{\mu R}{R + r_p} = \frac{7.5 \times 20 \times 10^3}{20 \times 10^3 + 6 \times 10^3} = \frac{7.5 \times 20}{26} = 5.77$$

17.22. In a triode valve 12×10^{15} electrons are reaching on the anode per second and the power consumed in it is 384 mW. Calculate the anode current and anode voltage.

Solution:

Anode current $I_a = 12 \times 10^{15} \times 1.6 \times 10^{-19} = 1.92 \times 10^{-3}$ A

Power $= 384 \times 10^{-3} = V \times 1.92 \times 10^{-3}$

$\therefore \quad V = 200\,\text{V}$

17.23. On applying an alternating supply of 1.0 V on the grid of a triode valve a change of potential of 10 V happens across the load resistance 16000 Ω. Find the amplification factor μ assuming $r_p = 8000\,\Omega$.

Solution:

$$\text{Voltage amplification } A = \frac{\text{output voltage}}{\text{input voltage}} = \frac{10}{1} = 10$$

$$\text{Voltage amplification } A = \frac{\mu R}{R + r_p}$$

$$\therefore \quad \mu = \frac{A(R + r_p)}{R} = \frac{10(16000 + 8000)}{16000} = 15$$

17.24. In a triode valve, for grid voltage $V_g = -1.2$ V, the plate current I_p and plate voltage V_p are given by the relation $I_p = -5.0 + 0.1\ V_p$ where I_p is in ma and V_p in volts. When the grid voltage is changed to – 3.2 V and plate voltage is kept at 150 V, a plate current of 5 ma is observed. Calculate the valve constants and voltage amplification for 20 KΩ load in plate circuit.

(Roorkee 1990)

Solution:

$$I_p = -5.0 + 0.1\ V_p$$

$$\frac{dI_p}{dV_p} = 0.1\ \text{ma/volt} = 10^{-4}\ \text{amp/volt}$$

$$r_p = \frac{dV_p}{dI_p} = \frac{1}{10^{-4}} = 10^{-4}\,\Omega$$

For the value of $V_p = 150$ V, $I_p = -5 + 0.1 \times 150 = 10$ ma

$$g_m = \left(\frac{dI_p}{dV_g}\right)_{V_p} = \frac{(5-10)\times 10^{-3}}{-3.2 + 1.2} = 2.5 \times 10^{-3}\ \text{mho}$$

$$\mu = r_p \cdot g_m = 10^4 \times 2.5 \times 10^{-3} = 25$$

$$A_v = \text{voltage amplification} = \frac{\mu R}{R + r_p}$$

$$= \frac{25 \times 20 \times 10^3}{20 \times 10^3 + 10 \times 10^3} = \frac{25 \times 20}{30}$$

$$= 16.67$$

17.25. A student performs an experiment for drawing static characteristic curve of a triode valve in the laboratory. The following data were obtained from the linear position of the curves.

Grid voltage (volt)	– 2.0	– 3.5	– 2.0
Plate voltage (volt)	180	180	120
Plate current (mA)	15	7	10

Calculate the plate resistance, mutual conductance and amplification factor of triode valve. (Roorkee 1991)

Solution:

$$r_p = \frac{dV_p}{dI_p} = \frac{180-120}{(15-10)\times 10^{-3}} = 1.2\times 10^4 \text{ ohm}$$

$$g_m = \frac{dI_p}{dV_g} = \frac{(15-10)\times 10^{-3}}{-2.0+3.5} = 5.33\times 10^{-3} \text{ mho}$$

$$\mu = r_p \times g_m$$

$$= 1.2 \times 10^4 \times 5.33 \times 10^{-3} = 64$$

17.26. The plate current i_p in a triode valve is given by

$$i_p = K(V_p + \mu V_g)^{3/2}$$

where i_p is in milliampere and V_p and V_g are in volts. Calculate the value of K and the grid cut-off voltage. Given $r_p = 10^4$ ohm and $g_m = 5 \times 10^{-3}$ ohm^{-1} at $i_p = 8$ milliampere and $V_p = 300$ volt. (Roorkee 1992)

Solution:

$$r_p = 10^4 \text{ ohm}, g_m = 5\times 10^{-3} \text{ ohm}^{-1}; \mu = r_p \times g_m = 50$$

$$i_p = K(V_p + \mu V_g)^{3/2}$$

$$\left(\frac{\partial i_p}{\partial V_g}\right)_{V_p} = g_m = \frac{3}{2}K\left(V_p + \mu V_g\right)^{1/2} \times \mu$$

On substituting the values of the various terms in the above two equations we have

$$8 \times 10^{-3} = K(300 + 50\, V_g)^{3/2}$$

$$5 \times 10^{-3} = \frac{3}{2} K(300 + 50\, V_g)^{1/2} \times 50$$

On dividing the two equations we have

$$\frac{8}{5} = \frac{300 + 50 V_g}{75}$$

$$120 = 300 + 50\, V_g$$

$$V_g = -\frac{180}{50} = -3.6 \text{ volt}$$

Further $K = \dfrac{5 \times 10^{-3}}{75\,(300 + 50 \times -3.6)^{1/2}}$

$= \dfrac{10^{-3}}{15\sqrt{120}} = 6.08 \times 10^{-6}$

Cut off voltage $= -\dfrac{V_p}{\mu} = -\dfrac{300}{50} = -6\text{V}$

17.27. The plate of a triode valve of $\mu = 20$ and $g_m = 1$ mA/V is given a potential of 300 V. Find the grid voltage for which the plate current reduces to zero. If this triode is connected in parallel to another triode valve of $\mu = 30$ and $g_m = 1.5$ mA/V, find the voltage gain of the combined system when the load in the circuit is 15 KΩ. Also find the power across this load when a signal $e_g = 2\sqrt{2}\,\sin 100\,\pi t$ is applied on the grid. (Roorkee 1993)

Solution:

$$V_g = -\frac{V_p}{\mu} = -\frac{300}{20} = -15\text{ V}$$

$$r_p = \frac{\mu}{g_m} = \frac{20}{10^{-3}} = 20\text{ K}\Omega$$

Plate resistance for the second triode valve

$$r'_p = \frac{\mu'}{g'_m} = \frac{30}{1.5 \times 10^{-3}} = 20 \times 10^3\,\Omega = 20\text{ K}\Omega$$

Combined plate resistance $R_p = \dfrac{r_p \cdot r'_p}{r_p + r'_p} = 10\text{ K}\Omega$

Combined mutual conductance $G_m = g_m + g'_m$

$= 10^{-3} + 1.5 \times 10^{-3}$

$= 2.5 \times 10^{-3}$ mho

Amplification factor for the combination

$\mu_{\text{com}} = R_p \times G_m$

$= 10 \times 10^3 \times 2.5 \times 10^{-3} = 25$

$$A_v = \text{voltage gain} = \frac{\mu_{\text{com}}}{1 + R_p/R_L} = \frac{25}{1 + 10/15} = 15$$

Let e_0 be the rms value of the output voltage across the load then

$e_0 = (e_g)_{\text{rms}} \times A_v$

$= 2 \times 15 = 30\text{V}$

Power across the load $= \dfrac{e_0^2}{R_L} = \dfrac{900}{15 \times 10^3}$

$= 60 \times 10^{-3} = 60$ mW

17.28. A simple voltage amplifier uses a triode of amplification factor 10 and plate resistance 10 KΩ. If this triode is replaced by another triode valve of plate resistance 25 KΩ, the voltage amplification remains unchanged. Find the amplification factor of the second triode valve. The load resistance in each case is 20 KΩ.

Solution:

$$A_v = \text{voltage amplification} = \frac{\mu_1 R_L}{r_1 + R_L} = \frac{\mu_2 R_L}{r_2 + R_L}$$

where μ_1 and μ_2 are the amplification factor of the two valves and r_1 and r_2 are their respective plate resistance.

On substituting the values we have

$$\frac{10 \times 20 \times 10^3}{10 \times 10^3 + 20 \times 10^3} = \frac{\mu_2 \times 20 \times 10^3}{25 \times 10^3 + 20 \times 10^3}$$

$$\frac{200}{30} = \frac{20\,\mu_2}{45}$$

$$\mu_2 = 15$$

17.29. A common base *npn* transistor has emitter current $I_e = 4$ ma, reverse saturation current $I = 15$ μA and current amplification $\alpha = 0.98$. Find the values of base and collector currents.

Solution:

$$I_c = \alpha I_e + I$$

$$I_c = 0.98 \times 4 \times 10^{-3} + 15 \times 10^{-6}$$

$$= 392 \times 10^{-5} + 1.5 \times 10^{-5}$$

$$= 393.5 \times 10^{-5}$$

$$= 3.935 \text{ ma}$$

$$I_e = I_b + I_c$$

$$I_b = I_e - I_c$$

$$= 4 \times 10^{-3} - 3.935 \times 10^{-5}$$

$$= 0.065 \times 10^{-3}$$

$$= 65\ \mu\text{A}$$

17.30. A transistor having current amplification $\alpha = 0.98$ has reverse saturation current $I = 15$ μA in common base configuration. Now the same transistor is connected in common emitter configuration. Calculate the value of the collector current for the base current of 0.2 ma.

Solution:

$$I_c = \beta I_b + (1 + \beta) I$$

$$\beta = \frac{\alpha}{1 - \alpha}$$

$$I_c = \frac{\alpha}{1-\alpha}\cdot I_b + \left(1+\frac{\alpha}{1-\alpha}\right) I$$

$$= \frac{0.98}{0.02}\times 0.2\times 10^{-3} + \left(1+\frac{0.98}{0.02}\right)\times 15\times 10^{-6}$$

$$= 9.8\times 10^{-3} + \frac{1}{0.02}\times 15\times 10^{-6}$$

$$= 9.8\times 10^{-3} + 7.5\times 10^{-4}$$

$$= 10.55 \text{ ma}$$

17.31. In the circuit shown in Fig. 17.5 the base current I_B = 20 μA and the collector current I_c = 5 ma. Show that the circuit cannot be used as amplifier.

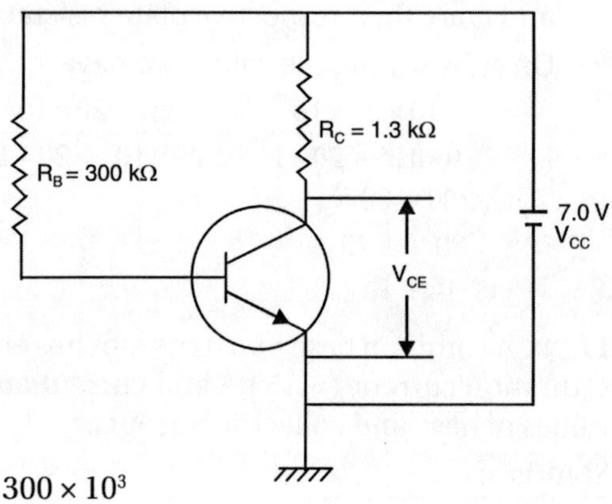

Fig. 17.5

Solution: According to Kirchhoff's Law, we have

$$V_{CC} = V_{BE} + I_B\cdot R_B$$

$$\therefore \quad V_{BE} = V_{CC} - I_B\cdot R_B$$

$$= 7.0 - 20\times 10^{-6}\times 300\times 10^3$$

$$= 1.0 \text{ volt}$$

Similarly $V_{CC} = V_{CE} + I_C\cdot R_C$

$$\therefore \quad V_{CE} = V_{CC} - I_C\cdot R_C$$

$$= 7.0 - 5\times 10^{-3}\times 1.3\times 10^3$$

$$= 7.0 - 6.5$$

$$= 0.5 \text{ volt}$$

The above calculation suggests that collector is at + 0.5 volt with respect to emitter and base is at +1.0 volt with respect to emitter. The base is at +0.5 volt with respect to the collector. Therefore emitter-base and collector-base junctions are in forward biasing. The circuit therefore cannot work as amplifier.

17.32. In the circuit of Fig. 17.6 V_{BE} and V_{CE} both are equal to zero when the input voltage at the base resistance is 6V. Calculate the values of I_B, I_C and β.

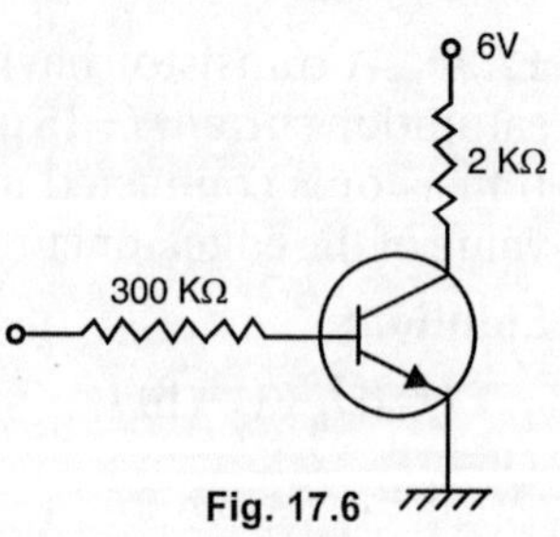

Fig. 17.6

Solution:

$$300\times 10^3\times I_B = 6$$

$$\therefore \quad I_B = \frac{6}{3\times 10^5} = 20\,\mu\text{A}$$

Similarly $2 \times 10^3 \times I_C = 6$

or $$I_C = \frac{6}{2\times10^3} = 3 \text{ ma}$$

$$\beta = \frac{I_C}{I_B} = \frac{3\times10^{-3}}{20\times10^{-6}} = 150$$

17.33. In the given circuit of Fig. 17.7 there is no input signal and the base current I_b is controlled by resistance R_1. Let us assume that there is very small potential difference between the base and emitter. Calculate the values of R_1 and R_2 using $I_c = 2$ ma, $V_{CE} = -4$ volt and $\beta = 50$.

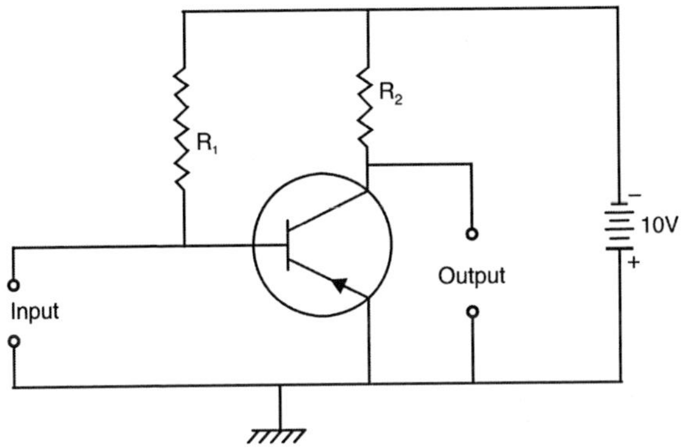

Fig. 17.7

Solution:

$$I_b = \frac{2\times10^{-3}}{50} = 40\ \mu\text{A}$$

$$R_1 \times 40 \times 10^{-6} = 10$$

$$\therefore \quad R_1 = \frac{10}{40\times10^{-6}} = 2.5\times10^5\ \Omega$$

$$R_2 = \frac{10-4}{2\times10^{-3}} = 3000\ \Omega$$

17.34. In the circuit shown in Fig. 17.8 $V_{EE}=8$V, $V_{CC}=15$V, $R_E=15$ KΩ, $R_C = 12$ KΩ, $V_{BE} = 0.5$ V and $\alpha = 0.98$. Determine the values of I_e, I_c and V_{CB}.

Solution: According to Kirchhoff's Law, we have

$$V_{EE} = V_{BE} + I_e R_E$$

$$\therefore \quad I_e = \frac{V_{EE} - V_{BE}}{R_E}$$

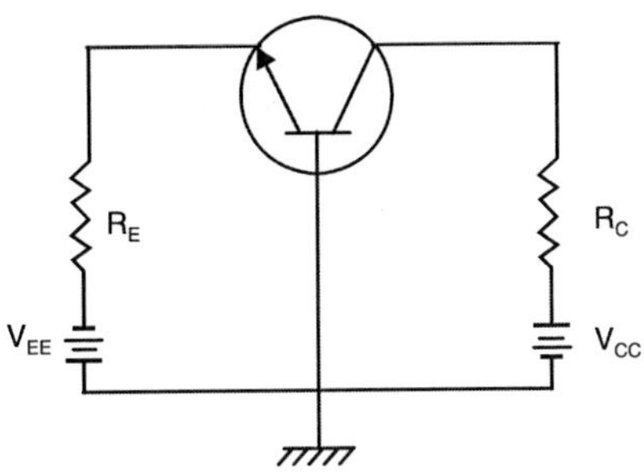

Fig. 17.8

On substituting the values of various terms we have

$$I_e = \frac{8-0.5}{15\times10^3} = 0.5 \text{ ma}$$

$$I_C = \alpha \cdot I_e = 0.98\times0.5$$

$$= 0.49 \text{ ma}$$

In the collector base circuit we have

$$I_C\cdot R_C + V_{CB} = V_{CC}$$

$$\therefore \quad V_{CB} = V_{CC} - I_C\cdot R_C$$

$$= 15 - 0.49\times10^{-3}\times12\times10^3$$

$$= 15 - 5.88$$

$$= 9.12\text{V}$$

17.35. An amplifier circuit is constructed by using a transistor in a common emitter configuration. A resistance of 1000 Ω is connected in the collector arm. Assuming β = 50, input resistance as 250 Ω and the change in the input current as 40 μA find (a) change in output voltage (b) change in input voltage (c) power gain.

Solution:

(a) $\delta V_{CE} = \delta I_C\cdot R_L$

$$= \beta\cdot\delta I_b\cdot R_L$$

$$= 50\times40\times10^{-6}\times1000 = 2 \text{ volt.}$$

(b) $\delta V_{BE} = r_i\cdot\delta I_b = 250\times40\times10^{-6} = 10 \text{ mV}$

(c) Power gain = voltage gain × current gain

$$= \frac{\beta^2 R_L}{r_i}$$

$$= \frac{50\times50\times1000}{250} = 10^4$$

17.36. In the circuit shown in Fig. 17.9, V_{CC} = 12 volt, R_B = 250 KΩ, R_C = 2 KΩ, V_{BE} = 0.5 volt and β = 50. A resistance R_E = 5 KΩ has been placed between the emitter and ground. Calculate the values of V_C, V_B and V_{CE}.

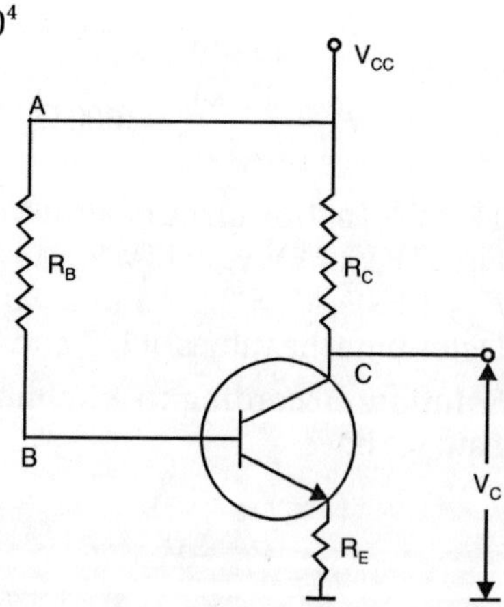

Fig. 17.9

Solution:

$$V_{CC} = I_B R_B + V_{BE} + R_E I_E$$

$$I_B = \frac{I_C}{\beta} = \frac{I_E}{\beta}$$

$$\therefore \quad I_C = \beta\cdot I_B \text{ and } I_E = \beta\cdot I_B$$

On substituting the values we have

$$12 = I_B \cdot 250 \times 10^3 + 0.5 + 5 \times 10^3 \times 50 \cdot I_B$$

$$I_B = \frac{11.5}{500} \times 10^{-3} \text{amp}$$

$$I_C = 50 \times \frac{11.5}{50} \times 10^{-3} = 1.15 \times 10^{-3} \text{ amp}$$

$$V_{CC} = V_C + I_C R_C$$

$$V_C = V_{CC} - I_C R_C$$

$$= 12 - 1.15 \times 10^{-3} \times 2 \times 10^3$$

$$= 12 - 2.3 = 9.7 \text{ volt}$$

$$V_{CC} = I_B R_B + V_B$$

$$V_B = V_{CC} - I_B R_B$$

$$= 12 - \frac{11.5}{500} \times 10^{-3} \times 250 \times 10^3$$

$$= 12 - 5.75 = 6.25 \text{ volt}$$

$$V_C = V_{CE} + I_E R_E$$

$$V_{CE} = V_C - I_E R_E$$

$$= 9.7 - 50 \times \frac{11.5}{500} \times 10^{-3} \times 5 \times 10^3$$

$$= 9.7 - 5.75$$

$$= 3.95 \text{ volt}$$

17.37. In the circuit of Fig. 17.10 we have $V_{BB} = 6$ V, $V_{CC} = 10$ V, $R_B = 200$ KΩ, $R_C = 3$ KΩ, $V_{BE} = 0.7$V, reverse saturation current I = 0.02 μA and β = 50. Calculate the three currents.

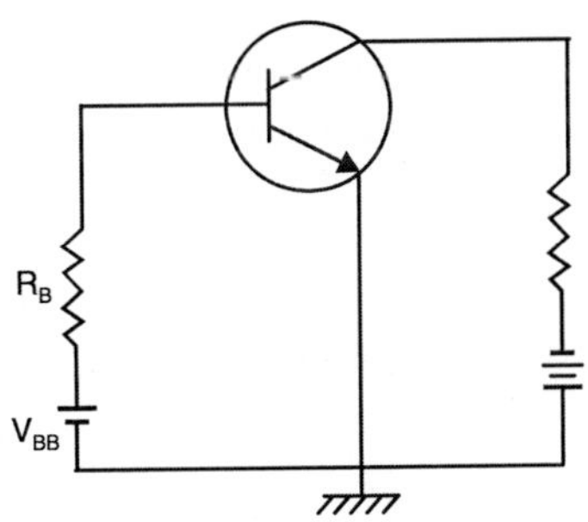

Fig. 17.10

Solution:

$$I_b \cdot R_B + V_{BE} = V_{BB}$$

$$\therefore \quad I_b = \frac{V_{BB} - V_{BE}}{R_B}$$

On substituting the values, we have

$$I_b = \frac{6 - 0.7}{200 \times 10^3} = \frac{5.3}{2} \times 10^{-5} = 2.65 \times 10^{-5} = 26.5 \mu\text{A}$$

$$I_C = \beta I_b + (1 + \beta) I$$

$$= 50 \times 26.5 \times 10^{-6} + 51 \times 0.02 \times 10^{-6}$$

$$= 1325 \times 10^{-6} + 1 \times 10^{-6}$$

$$= 1326 \times 10^{-6} = 1.326 \text{ ma}$$

$$I_e = I_b + I_c = 26.5 \times 10^{-6} + 1.326 \times 10^{-3}$$
$$= 1.352 \text{ ma}$$

17.38. A npn transistor in a common emitter mode is used as a simple voltage amplifier with a collector current of 4 mA. The terminal of a 8 V battery is connected to the collector through a load resistance R_L and to the base through a resistance R_B. The collector-emitter voltage $V_{CE} = 4$ V, base-emitter voltage $V_{BE} = 0.6$ V and base current amplification factor $\beta = 100$. Calculate the values of R_L and R_B. (Roorkee 1997)

Solution:

$$\beta = \frac{I_C}{I_b}$$

$$I_b = \frac{I_C}{\beta} = \frac{4}{100} \text{ ma}$$

$$E = R_L \cdot I_C + V_{CE}$$

$$\therefore \quad R_L = \frac{E - V_{CE}}{I_C} = \frac{8-4}{4 \times 10^{-3}} = 10^3 \,\Omega$$

Similarly

$$E = R_B \cdot I_b + V_{BE}$$

$$\therefore \quad R_B = \frac{E - V_{BE}}{I_b} = \frac{8 - 0.6}{\frac{4}{100} \times 10^{-3}} = \frac{7.4}{4} \times 100 \times 10^3$$

$$= 185 \text{ K}\Omega$$

17.39. A npn transistor is connected in common emitter configuration in which collector supply in 8V and the voltage drop across the load resistance of 800 Ω connected in the collector circuit is 0.8 volt. If the current amplification factor is $\frac{25}{26}$ determine collector-emitter voltage and base current. If the internal resistance of the transistor is 200 Ω calculate the voltage gain and power gain. (Roorkee 1994)

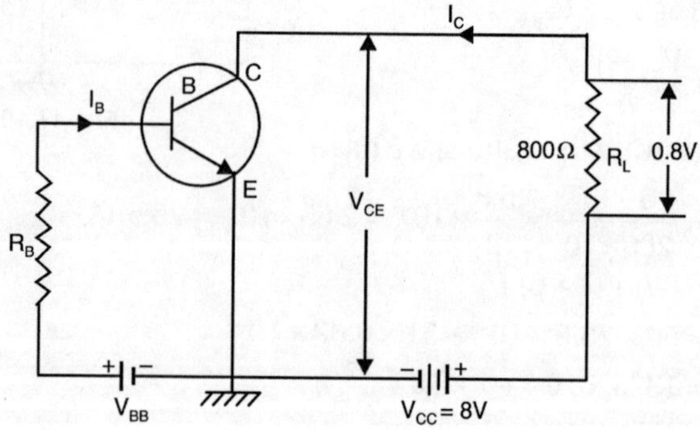

Fig. 17.11

Solution:

$$\text{Collector current } I_C = \frac{\text{voltage drop across } R_L}{R_L} = \frac{0.8}{800} = 10^{-3}\text{ amp}$$

$$V_{CE} = 8 - 0.8 = 7.2 \text{ volt}$$

$$\text{Current gain } \beta = \frac{I_C}{I_B}$$

$$\frac{25}{26} = \frac{10^{-3}}{I_B}$$

$$\therefore \quad I_B = 1.04 \times 10^{-3} \text{ amp}$$

$$\text{Voltage gain} = \beta \frac{R_L}{R_C} = \frac{25}{26} \times \frac{800}{200} = 3.846$$

$$\text{Power gain} = \beta^2 \frac{R_L}{R_C} = \left(\frac{25}{26}\right)^2 \times \frac{800}{200} = 3.698$$